D1474548

60.00
8011

American Maps and Mapmakers

DISCARD
RETIRÉ

American Maps and Mapmakers

Commercial Cartography in the Nineteenth Century

WALTER W. RISTOW

WAYNE STATE UNIVERSITY PRESS
Detroit 1985

Copyright © 1985 by Wayne State University Press,
Detroit, Michigan 48202. All rights are reserved.
No part of this book may be reproduced without
formal permission.

Library of Congress Cataloging in Publication Data

Ristow, Walter William, 1908–
American maps and mapmakers.

Includes index.
1. Cartography—United States—History. I. Title.
GA405.R57 1985 338.7′61526′9073 84-25798
ISBN 0-8143-1768-5

For Helen

Patrons

American Congress on Surveying and Mapping
W. Graham Arader III
Richard B. Arkway
Gwendolyn R. Barckley
Stephen A. Bromberg
Richard Fitch
Maurice A. Fox
Janet Green
Peter J. Guthorn, M.D.
Jonathan T. Lanman, M.D.
Laurence M. Luke
Andrew McNally III
Michigan Map Society
H. Russell Morrison, Jr.
Kenneth Nebenzahl
Old Print Shop, Inc.
The Philadelphia Print Shop
Seymour I. Schwartz, M.D.
Speculum Orbis Press
Howard E. Welsh
Eric W. Wolf

Contents

Illustrations

Chapter 10.

Chapter 11.

Chapter 12.

Chapter 13.

Chapter 14.

Chapter 15.

Chapter 16.

Chapter 17.

Chapter 18.

Chapter 19.

Chapter 20.

Chapter 21.

Chapter 26.

Chapter 27.

Chapter 28.

Acknowledgments

My personal research interest for more than a quarter of a century has been the history of American private and commercial map publishing in the century or so after the Revolution. Papers resulting from this research have been published in various professional journals. In preparing this book, it seemed appropriate to draw upon such previously published works, in whole or in part, for certain chapters. For such borrowings, I am indebted to the publications and their editors for permission to incorporate the following material.

Chapter 2 owes much to "Maps of the American Revolution," which initially appeared in the July 1971 issue of the *Quarterly Journal of the Library of Congress*, 196–215. Information included in Chapter 4 was presented at an international colloquium on "The Cartography of the Eighteenth Century and the Work of Count de Ferraris (1726–1814)" held in Spa, Belgium, September 8–11, 1976. Titled "The First Maps of the United States of America," the paper was published in 1978 by Crédit Communal de Belgique, Brussels, in *Actes Handelingen*, 179–90.

Chapter 5, detailing the cartographic career of Simeon De Witt, was included by Kirschbaum Verlag, Bad Godesberg, West Germany, in its 1968 *Kartengeschichte und Kartenbearbeitung*, 103–14, a festschrift volume honoring the late Wilhelm Bonacker on his eightieth birthday. Part of Chapter 7 contains information from "Maps of Pennsylvania and Virginia," published in the July 1966 issue of the *Quarterly Journal of the Library of Congress*, 231–42. This paper also was a source of data for Chapter 8, as was my "State Maps of the Southeast to 1833," which was published in number 6, 1966, of the *Southeastern Geographer*, 33–40. Portions of Chapter 10 are reprinted with permission from *Surveying and Mapping*, December 1962, 569–74, copyright 1962 by the American Congress on Surveying and Mapping. That chapter also draws on my "Aborted American Atlases," which appeared in the Summer 1979 issue of the *Quarterly Journal of the Library of Congress*, 320–45.

"The Ebeling-Sotzmann Atlas von Nordamerika," Chapter 11 in this volume, is derived from my paper of the same title that was published in the March 1980 number of *The Map Collector*, 2–9. Portions of Chapter 12 are taken from my article "John Melish and His Map of the United States" published in the September 1962 *Quarterly Journal of the Library of Congress*, 159–78. "Robert Mills's Atlas of South Carolina," Chapter 14 in this volume, initially appeared under that title in the January 1977 issue of the *Quarterly Journal of the Library of Congress*, 52–66. Part of Chapter 18 is derived from "Maps by Pendleton's Lithography" published in the December 1982 issue of *The Map Collector*, 26–31. "Nineteenth-Century Cadastral Maps in Ohio," which provided information for Chapter 20, was published in *Papers of the Bibliographical Society of America* 59 (1965): 306–15.

Chapter 21 is largely based on my article "The Map Publishing Career of Robert Pearsall Smith" from the July 1969 *Quarterly Journal of the Library of Congress*, 170–96. Similarly, Chapter 22 owes much to "The French-Smith Map and Gazetteer of New York State," which initially appeared in the Winter 1979 number of the *Quarterly Journal of the Library of Congress*, 68–90. Chapter 23 is based on "From Maps to Riches," which appeared in the June 1979 number of *The Map Collector*, 2–10. And, last, parts of Chapter 26 are from my paper "Alfred T. Andreas and His Minnesota Atlas" in the Fall 1966 issue of *Minnesota History*, 120–29, the official publication of the Minnesota Historical Society. All of the above noted information is excerpted or reprinted with permission, for which grateful acknowledgment is made.

My great indebtedness to the Library of Congress is evident from the number of the above-cited articles that were published in its *Quarterly Journal*. Much of the information presented in this volume, in fact, is based upon research and studies conducted during my thirty-two-year association with the library's Geography and Map Division and my responsibilities for supplementing, maintaining, and servicing its superior cartographic collections.

I have been singularly fortunate in my professional associates and colleagues in the Geography and Map Division, all of whom contributed much in knowledge, experience, and encouragement. High on the list is Clara Egli Le Gear, now in her eighty-ninth year, whose association with the Geography and Map Division extends over seven decades. During our forty years as co-workers I have profited immensely from her depth of knowledge and experience in things cartographic. Similarly, I gained much from the late Arch C. Gerlach during his sixteen years' tenure as chief of the division.

Present divisional staff members have lent assistance and support in many ways and are deserving of my warm thanks. They include John A. Wolter and Ralph Ehrenberg, currently chief and assistant chief of the division, Richard W. Stephenson, head of the Reference and Bibliography Section, and Andrew Modelski, head of the Acquisitions Unit. My thanks to them and to other former colleagues and associates.

Following my retirement in 1978, Librarian of Congress Daniel J. Boorstin made it possible for me to continue my studies in the familiar and congenial surroundings of the Geography and Map Division by naming me Honorary Consultant in the History of American Cartography. I am grateful to him for this confidence and support, which has greatly facilitated my post-retirement research.

Except in the few instances where otherwise credited, the illustrations in this volume have been drawn from the collections of the Library of Congress. Photographic negatives of original items were prepared by the library's Photoduplication Service, to which due acknowledgment is here made. The prints, however, were made by the Photographic Services department of the National Geographic Society. For this substantial contribution toward the enrichment of the book I am indebted to Carl M. Shrader, chief of Photographic Services, and to Dr. Melvin M. Payne, chairman of the board, National Geographic Society.

A book on cartographic history, needless to say, should be abundantly illustrated. This has been assured by generous contributions to the publisher from a number of friends and professional colleagues who are listed herein as Patrons. For their confidence in me and for their tangible support toward the enrichment of this volume, they have my sincere and enduring gratitude.

Herman R. Friis, former chief, Cartographic Archives, U.S. National Archives, and a longtime friend and respected colleague, graciously accepted the tedious task of critically reading the entire typescript. My heartfelt thanks go to him. Last, but far from least, I gratefully acknowledge the patient encouragement and help in all my endeavors, over more than four decades, of my wife, Helen Doerr Ristow. And as though that were not enough, as a direct contribution to this book, she accepted responsibility for the arduous task of compiling the index.

Introduction

An official of America's largest map publishing company observed in a 1976 article that "in 200 years of independence, commercial or private cartographic publishing in America has become established as a major business activity with influences and recognition far beyond its actual contribution to the gross national product. But maps are that way. Although not to the extent they should, they are likely to be found in every school, every home, every automobile and most offices; and the vast majority of these in the general community will have been produced by one of a handful of publishing companies."[1]

The mapping of America did not, of course, originate when independence was won. Maps and atlases, in great numbers, were produced during the colonial period of English, French, and Spanish possessions in North America. With some rare exceptions, such maps, plans, and charts were prepared by European surveyors, cartographers, engravers, and publishers. With the establishment of peace after the Revolution and the ratification of the Constitution, the former colonists were free to plan their own future and to forge a united nation from thirteen disparate and newly established states. Responsibility for resolving the many problems facing the new republic was assumed by the infant Congress and by the several state legislatures. The opportunities these bodies faced were balanced by a multitude of problems, many of which involved the country's expansive and varied geography.

There was, to begin with, the matter of determining accurate legal and administrative boundaries of the states. A transportation system had to be planned and developed that would strengthen internal commercial and political ties. Western lands transferred by the states or purchased from, or ceded by, Indian tribes had to be surveyed, mapped, and distributed in an equitable and, if possible, profitable manner. Turnpikes, roads, and canals were essential to encourage settlement in the trans-Appalachian territory.

These and other conditions and problems called for accurate and detailed maps and plans. Some cartographic needs were met by such federal surveying and mapping agencies as the General Land Office, Coast Survey, and the Topographical Bureau in the early decades of the Republic. Most of the maps, charts, and atlases consulted by citizens of the United States during the first one hundred years of American independence were, however, produced by private individuals and small commercial publishers. The contributions of these individuals and companies are the subject of the essays assembled in this volume. Some of these papers, published previously in various professional journals, have been revised and updated. A number of the essays, prepared expressly for this publication, describe additional facets of private and commercial mapping during the closing decades of the eighteenth century and all of the nineteenth century.

Both adversaries in the revolutionary war relied, in the early months of the conflict, on maps prepared in the prewar years by British military surveyors and cartographers. General George Washington did, in July 1777, appoint Robert Erskine geographer and surveyor general to the Continental army. When Erskine died some three years later, the post of surveyor general was filled by Simeon De Witt, one of Erskine's assistants. Several months before his death, Erskine had reported that surveys by his staff filled "upwards of two hundred and fifty sheets of paper [and] that with a proper number of hands, which I suppose to be five surveyors and two draughtsmen, such additional surveys of the roads and rivers might be taken in the course of a year, as would afford sufficient data for the forming of an accurate map of the middle States."[2] Erskine and De Witt both hoped to use the military survey data to compile maps for nonmilitary uses after the war. For various reasons these hopes were not realized. In *Mapping the Revolutionary War*, J. Brian Harley noted that "despite the farsightedness of both Erskine and De Witt, the military topographical surveys of the Revolution were largely a false dawn in the mapping of the new nation."[3] It should be reported, however, that certain of the Erskine-De Witt manuscript surveys were drawn upon by Christopher Colles in compiling his *Survey of the Roads of the United States of America*, the first parts of which were published in 1789.[4]

One of the urgent needs in Europe as well as in America following independence was for maps of the infant republic. The earliest such maps to become available were published in Europe and were modified from previously compiled works. More clearly American in origin were maps published in 1784 by two natives, Abel Buell and William McMurray. Other American cartographers who compiled early maps of their new republic were Osgood Carleton, whose map, engraved by John Norman, was published in 1791, and Abraham Bradley, Jr., who, as assistant postmaster general, published some five years later a *Map of the United States, Exhibiting the Post-Roads, the Situations, Connections, & Distances of the Post Offices*. Bradley's map, which was based on data supplied by postmasters throughout the country, constituted a distinct break from European-inspired cartography and a major advance in the development of American mapmaking. Notwithstanding Bradley's official position, his map was privately copyrighted and published.

Maps of the newly created states were also needed. The first response came from William Blodget, who published in 1789 *A Topographical Map of the State of Vermont*, which was engraved by Amos Doolittle of New Haven. Although Blodget dedicated his map to Vermont's governor, Thomas Crittendon, it was not an official undertaking. The actual surveys on which the map was allegedly based may have been the town surveys conducted under the direction of Ira Allen, Vermont's surveyor general from 1779 to 1787. Blodget also compiled and published in 1791 *A New and Correct Map of Connecticut*, which he dedicated to Samuel Huntington, governor of the state. He may have received some support and encouragement from the state for his project.

Blodget seems not to have been officially employed by either Vermont or Connecticut. Some states, however, did appoint surveyors general, even before the end of the eighteenth century. Because there were few trained American engineers or surveyors, a number of these appointees were recruited from Europe. A notable exception was De Witt, who, following his service with the Continental army, was appointed surveyor general of New York State, a post he held for half a century. In addition to planning and constructing canals, turnpikes, and other internal improvements, many of the state engineers and surveyors general conducted surveys and compiled state maps. Private individuals, however, continued to prepare and publish such maps with some official encouragement, assistance, or subvention. So effective were these combined efforts that before the Republic was fifty years old more than thirty creditable state maps had been published. The county and local source materials from which the state maps of South Carolina, New York, and Maine were compiled were also drawn upon in preparing atlases of these states. The procedures and techniques developed and utilized in preparing state maps were uniquely American and laid the foundations for an indigenous cartographic publishing industry.

Most nonofficial compilers of state maps were not professional surveyors or cartographers. For the most part, they earned their livelihood from other vocations, and the maps they produced were generally one-time efforts. Few of these individuals profited materially from their maps, and in some instances their objectives and hopes were aborted. During the last decade of the eighteenth century, commercial publishing was established in the United States by Mathew Carey, an Irish immigrant who settled in Philadelphia. Carey entered the field of cartography with his 1794 *Atlas to Guthrie's System of Geography*. Several other atlases appeared under his imprint within the next several years, most notable of which was the *American Atlas*, first issued in 1795 and republished in several revised and enlarged editions until 1809. Carey's publications stimulated competition, and atlases issued by other publishers appeared before 1800. Although Carey was a pioneer in atlas publishing, he was not exclusively concerned with cartographic works.

John Melish, a native of Scotland, was the first American publisher to concentrate his efforts wholly on producing maps, atlases, and geographical publications. From his headquarters in Philadelphia, he published an impressive number of cartographic works in the decade after 1812. Melish's maps of Pennsylvania and the United States, of which twenty-four variant states have been identified, are among his most distinguished publications. The cartographic publishing firm founded by Melish expired with his death in December 1822.

Melish's role as America's major publisher of cartographic works, though, was assumed by Henry Schenck Tanner, who had engraved a number of Melish maps. Although Tanner began his professional career as an engraver, his association with Melish aroused in him an interest in mapmaking and publishing. In 1816 Tanner joined the engraving firm established by his brother, Benjamin Tanner, and John Vallance and Francis Kearny. Very probably at the suggestion of Henry Tanner, the firm embarked on the project of compiling and publishing, in serial segments, the ambitious *New American Atlas*, the first folio of which was issued in 1818. It appears that Tanner carried the major responsibility for compiling and drafting the maps, and by 1819, when the second part of the atlas was distributed to subscribers, he had become the sole publisher of the *New American Atlas*. The fifth and last folio of maps was available late in 1823. Tanner continued to engrave maps and to retail the maps of other publishers while engaged in compiling his atlas. By the time the atlas was completed, the "List of Maps and Geographical Works Published and For Sale by H. S. Tanner" was quite impressive. He engraved the plates for the district maps in Robert Mills's *Atlas of the State of South Carolina*, as well as John Wilson's 1822 *Map of South Carolina* and several other state maps. Tanner dominated American map publishing and distribution until the early 1840s.

Although the Survey of the Coast was established in 1810, some three or four decades elapsed before any significant number of coastal charts were published by that agency and its successor, the U.S. Coast Survey. The compilation of coastal and river navigational charts, therefore, was also assumed by private citizens. Most active in publishing and selling charts and sailing guides was the firm founded by Edmund March Blunt, which was the principal source of navigational aids for more than thirty years. Guides for navigating the Ohio and Mississippi rivers were supplied by such private cartographers as Zadok Cramer and Samuel Cumings during the first half of the nineteenth century.

By 1840 there were a number of American cartographic publishers, including Fielding Lucas, Jr., of Baltimore, Anthony Finley and S. Augustus Mitchell of Philadelphia, David H. Burr of New York, Thomas G. Bradford of Boston, Lewis Robinson of South Reading, Vermont, and John Farmer of Detroit. As was true for Melish, Tanner, and most cartographers of the seventeenth, eighteenth, and early nineteenth centuries, these publishers generally printed their maps from engraved copper plates. While well-executed copper engraving produced clear and attractive maps, the technique required highly skilled engravers, was tedious and costly, and no more than one thousand or so legible copies could be printed from the plates. Maps were, therefore, quite expensive and not readily available to the average citizen.

New possibilities for reproducing maps and other graphics were opened with the invention of lithography by Alois Senefelder in 1798. This method allowed the images of an artist or cartographer to be transferred easily from paper to a lithographic stone, from which multiple copies could be printed. Lithography, much quicker and less costly than engraving, revolutionized map printing and publishing. By the time lithography was introduced in the United States in the early 1820s, the process was perfected, and American printers and imported European practitioners avidly adopted the new technique. Some of the first maps lithographically printed in the United States appeared in the late 1820s, notably those done by Pendleton's Lithography of Boston. More than two decades passed, however, before lithography loomed as a serious threat to copper-plate engraving for reproducing maps. Philadelphia, which was the principal map publishing center during the engraving period, maintained its position of leadership during the early decades of lithographic cartography. These years coincided with the opening and settlement of the West, the accelerated emigration from western Europe, canal, turnpike, and railroad construction, the discovery of gold in California, the western surveys, and the annexation of Texas. All these events stimulated interest in maps and map publishing.

In the decade or so preceding the Civil War, there was active interest in publishing county, town, and city maps. The New England, Middle Atlantic, and midwestern states were particularly well covered by county maps, many of which were offered on a subscription basis. Especially active in compiling and publishing town, county, and city maps in the prewar years were Henry Francis Walling and Robert Pearsall Smith, although a great number of individuals were involved. Some of them, like Jay Gould, later achieved success in other fields. Some county maps, as was true a half century earlier, were drawn upon as source material in compiling state maps.

Mapping during the Civil War was primarily official. Because most battles were fought in states where little detailed mapping had previously been conducted, both armies relied heavily on surveys by military engineers attached to ground forces. The Union army was better equipped in this regard because it was able to use the facilities and personnel of federal mapping agencies. Commercial map publishing was greatly curtailed in the war years, although such firms as Julius Bien, G. W. Colton, H. H. Lloyd, and others published general and localized maps of the major theaters of conflict.

When the nation was again at peace, county mapping resumed, principally in the prosperous midwestern agricultural states and in the far west. In contrast with the prewar practice in most eastern states of selling maps after they were published, midwestern county maps were sold by subscription. Because General Land Office surveyors had mapped these states by 1860, county map publishers drew heavily from the official maps. When all the more prosperous counties had been mapped, promoters and publishers devised the county atlas. This format made it possible to publish and sell atlases for counties previously covered by maps. The atlas format also permitted adding pages of descriptive text, biographical sketches, and lithographic portraits and pictures of public buildings, farmsteads, and prize livestock. Farmers and townsmen, pressured by glib subscription salesmen and artists, paid dearly to have themselves, their families, and their possessions pictured in the pages of a county atlas. Although hundreds of individuals were involved in preparing and publishing county maps and atlases, Frederick W. Beers, Louis H. Everts, Alfred T. Andreas, and D. J. Lake were among the most prolific producers of them in the postwar decades. The production of lithographic county atlases peaked between 1872 and 1877.

By the early 1870s some publishers, noting that counties were thoroughly supplied with atlases, embarked on compiling and publishing atlases of a number of eastern and midwestern states. Active in this phase of commercial cartography, which spanned the 1870s and

1880s, were Andreas, Daniel G. Beers, Frederick Beers, and Walling. The financial depression in the seventies and the development of new cartographic techniques, notably photolithography, terminated the publication of lithographically illustrated county and state atlases. General map and atlas publishing were dominated in the post-Civil War decades by the Mitchell and Colton companies. In addition to revised editions of their general atlases, both firms published railroad and immigrant guides and maps, as well as maps of states and cities. Colton was particularly active in publishing railroad maps.

The growth in the number and size of cities in the post-Civil War years spurred interest in urban plans and atlases. The needs of real estate promoters, developers, and agents were met by the publication of large-scale real estate atlases. Similarly, the interests of fire insurance and underwriting companies were served by insurance maps and atlases. Initially, a number of publishers were occupied with preparing insurance maps. Over a period of a century, however, the D. A. Sanborn Map Company, by acquiring other companies, established a monopoly in this branch of commercial cartography. The publication of panoramic maps of American cities also flourished between 1870 and 1920.

Cerography, or wax engraving, was also important to the growth of commercial cartography. Invented by Sidney E. Morse in 1834, this cartographic reproduction technique was not immediately widely used. Although Morse used the method in reproducing several of his cartographic publications, there is little evidence that he utilized electrotyping, the invention that stimulated more extensive use of the wax engraving process. The burgeoning market for cheap maps in the last three decades of the nineteenth century found in wax engraving an inexpensive and relatively simple reproductive process. Although a number of map publishers employed this technique, it achieved its greatest success after it was utilized around 1872 by Rand McNally & Company of Chicago. Founded in 1856, the company initially engaged in general printing and in preparing railroad tickets, guides, and timetables. From these it was an easy transition to compile and print railroad maps and, subsequently, maps of all types. From such modest cartographic beginnings, Rand McNally & Company has grown to be the largest commercial cartographic publisher in the United States and one of the major world producers of maps and atlases.

In the fifteen or twenty years after the Civil War, the federal government sponsored a series of surveys which were headed by Clarence King, F. V. Hayden, John Wesley Powell, and Lieutenant George M. Wheeler. Rivalry and controversies between several of the leaders of these independent surveys and other factors induced the U.S. Congress to establish the Geological Survey in 1879. One of the survey's responsibilities is preparing topographic, geologic, and other maps of the country and its various parts. For more than a century the Geological Survey has competently performed its cartographic duties. By making its excellent and detailed maps available as source material, the survey has greatly benefited and modified American map publishing. As Russell Voisin has written, "the strong, effective, commercial cartographic environment the United States enjoys today could not exist without a strong and healthy national mapping program. Without this the cartographic publisher would be deprived of his most important basic ingredient, source documentation, for it is here where it all begins."[5] This is in great contrast to commercial mapmaking in the nineteenth century, when individuals and private firms carried the major responsibility in all the steps in preparing maps and atlases for the citizens of an expanding nation.

Notes

1. Russell Voisin, "Maps for the American People," *The American Cartographer* 3 (Oct. 1976): 106.
2. Letter from Erskine to Philip Schuyler in Albert H. Heusser, *George Washington's Map Maker, A Biography of Robert Erskine* (New Brunswick, N.J., 1968), 212.
3. J. Brian Harley, Barbara Bartz Petchenik, and Lawrence W. Towner, *Mapping the American Revolutionary War* (Chicago, 1978), 35.
4. See Christopher Colles, *A Survey of the Roads of the United States of America, 1789*, ed. Walter W. Ristow (Cambridge, Mass., 1961).
5. Voisin, "Maps for the American People," 105.

1. *Our Cartographic Heritage*

Private and commercial mapmakers and publishers in the newly established United States had a rich cartographic heritage upon which to draw. From early in the sixteenth century, maps and charts of portions of North America, which eventually became part of the United States, were drawn, compiled, and published. Excellent summaries of the cartography of colonial America have been published, several of them by cartographic historian William P. Cumming.[1] Until the beginning of the eighteenth century, however, maps of the North American continent were primarily of a general type and were, for the most part, restricted to the eastern seaboard. Some of the large and decorative seventeenth-century atlases include maps of English and French America, but the few regional maps then available were based on reconnaissance surveys or were compiled from crude sketches by explorers and unverified reports of natives.

In the first half of the eighteenth century settlements in America increased in number and size, colonial administrations were strengthened and centralized, and England and France competed for control of the continent. More and better maps were needed, and British colonial cartography during these years received encouragement and support from the Lords Commissioners for Trade and Plantations. Among the more noteworthy American maps published in the mid-eighteenth century were Henry Popple's *Map of the British Empire in America with the French and Spanish Settlements adjacent thereto . . . 1733*,[2] Lewis Evans's *Map of Pensilvania, New Jersey, New York, and the Three Delaware Counties, 1749*, and *General Map of the Middle British Colonies in America, 1755* (Fig. 1–1),[3] John Mitchell's *Map of the British and French Dominions in North America . . . 1755*,[4] and Joshua Fry and Peter Jefferson's *Map of the Inhabited Part of Virginia Containing the Whole Province of Maryland . . . 1751*.[5] Because most of these maps were reprinted, issued in revised editions, or plagiarized, they continued in service throughout the revolutionary war years. Of particular distinction is the Mitchell map, which is considered by some scholars to be of primary importance in the history of the United States because it was consulted by the American and English delegates while negotiating peace in 1782 and 1783 (Fig. 1–2).

The French and Indian War ended in 1763, and under the terms of the Treaty of Paris, signed February 10, 1763, England acquired Florida from Spain and all of Canada as well as part of Louisiana east of the Mississippi River from France. For none of the former French and Spanish territories were there accurate surveys or maps. In 1764 the Board of Trade informed the British king that "we find ourselves under the greatest difficulties arising from the want of exact surveys of these counties in America, many parts of which have never been surveyed at all and others so imperfectly that the charts and maps thereof are not to be depended upon." The board recommended "in the strongest manner, that no time should be lost in obtaining accurate surveys of all Your Majesty's North American Dominions but more especially of such parts as from their natural advantages require our immediate attention."[6] The recommendation made specific reference to Atlantic Canada and east Florida. To aid in achieving its objectives, the board proposed that the British possessions in North America "be divided into a Northern and Southern District with a Surveyor General of Lands to be appointed for each."[7] The detailed proposals were prepared by Captain Samuel Holland, who had been engaged in conducting surveys of English Canada since 1758.

Born in the Netherlands in 1728, Holland joined the

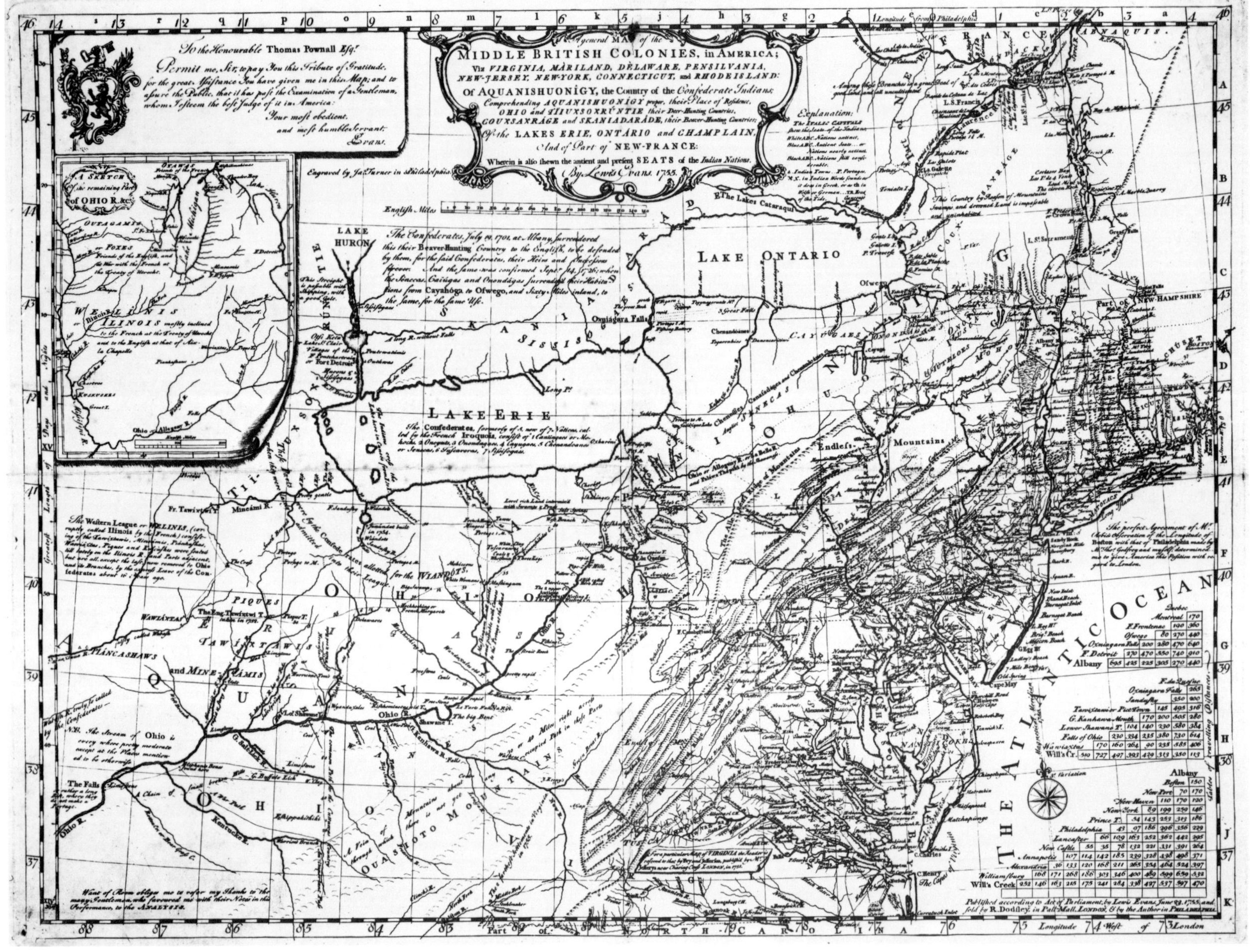

Fig. 1–1. Lewis Evans's *General Map of the Middle British Colonies in America, 1755,* was one of the most popular maps of British America. During the two decades after its publication, a number of revisions, derivatives, and plagiarisms of the map were issued.

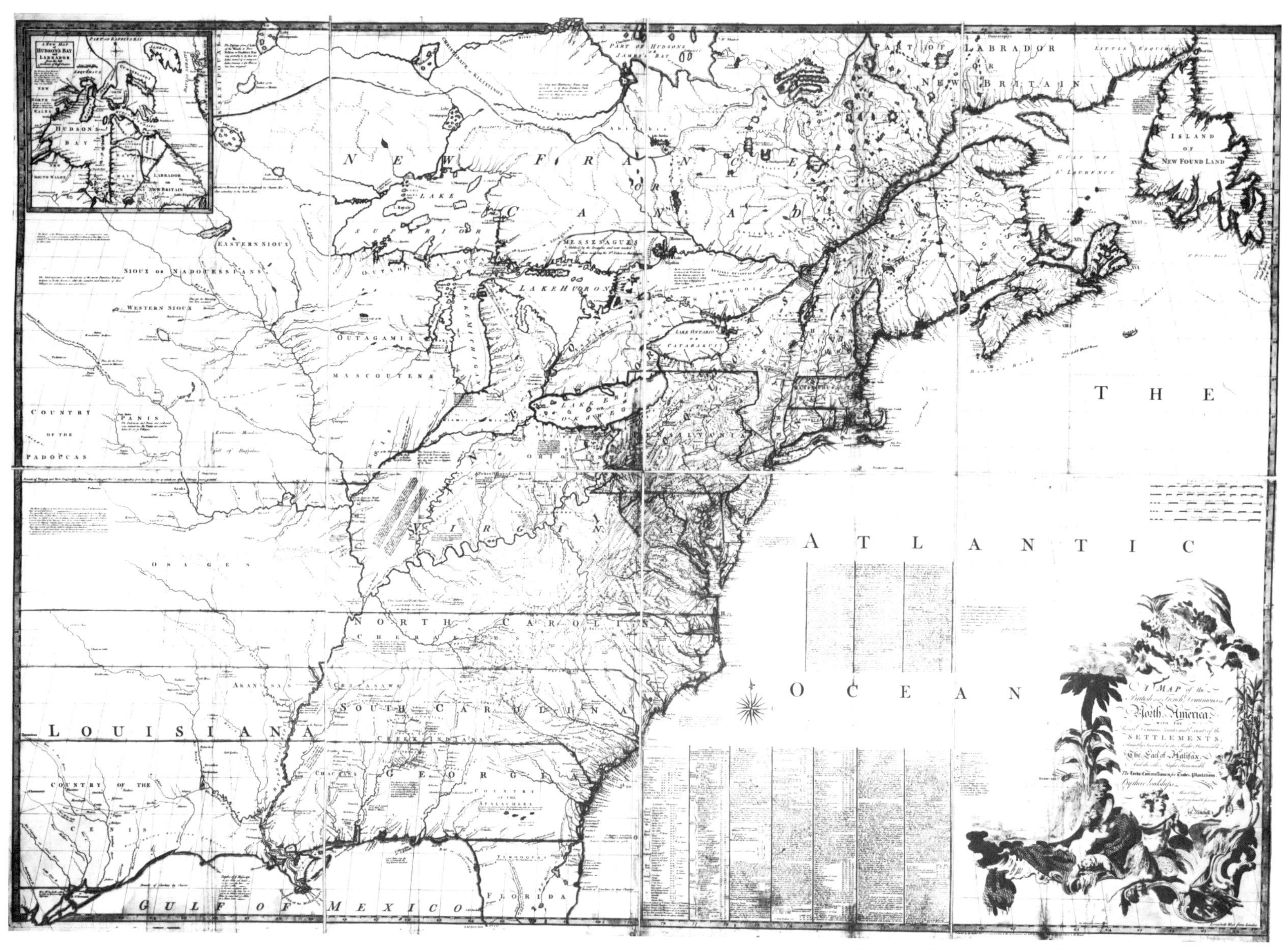

Fig. 1–2. Virginian John Mitchell's large *Map of the British and French Dominions in North America, 1755*. The map was engraved and published in England.

Dutch army at the age of seventeen and attained the rank of lieutenant in the artillery. He immigrated to England in 1754 and shortly thereafter obtained a lieutenancy in the Royal American Regiment, which was then being recruited. Holland's abilities and skills, particularly in surveying and mapping, brought him steady promotion in America. Before and after the capture of Quebec, he carried out surveys in parts of New France, New England, and the province of New York. His proposal for a scientific survey of Britain's American possessions was made to the Lords Commissioners for Trade and Plantations in 1762. It was not unexpected, therefore, that Holland was named surveyor general for the Northern District in March 1764.

In the same year William Gerard De Brahm was appointed surveyor general for the Southern District of North America. De Brahm, who was born in Germany in 1717, served in the German army and achieved the rank of captain engineer. He resigned his military commission in 1748, probably for religious reasons, for shortly thereafter he renounced the Catholic faith. With other German religious exiles, he immigrated to Georgia in 1751.

DeBrahm's engineering skill and experience were welcomed in the newly settled land, and he was soon engaged in surveying and mapping activities. He was appointed one of the two surveyors general of the Georgia Colony in 1754, and in the following year he received an interim appointment as surveyor general of South Carolina. De Brahm's four-sheet *Map of South Carolina and a Part of Georgia* was published in London in 1757 by Thomas Jefferys (Fig. 1–3). "In its accuracy for the coastal area and its thoroughness for the region covered," Cumming rates this map as "far superior to any cartographical work for the southern district that had gone before."[8]

From 1764 to 1770 De Brahm and his assistants conducted surveys and prepared descriptions of Florida, South Carolina, and Georgia. His lengthy manuscript "Report of the General Survey in the Southern District of North America,"[9] as well as a number of large manuscript maps, are preserved in British archives; a somewhat modified copy of the report is in the collections of Harvard University's Houghton Library. De Brahm's surveys were utilized by Henry Mouzon for his *An Accurate Map of North and South Carolina, With Their Indian Frontiers* published in 1775. Mouzon's map served military commanders of both the American and British armies during the revolutionary war. Few of the De Brahm or Holland land surveys were published, but a number of their manuscripts are preserved in the British Library and the British Public Record Office. Reproductions of some of the manuscripts in these two repositories are assembled in the several series of the *Crown Collection of Photographs of American Maps*.[10]

The De Brahm and Holland surveys were primarily of land areas, but some also included coastal regions because the major colonial settlements were on or near the sea. Until the end of the seventeenth century British navigators used charts prepared by Dutch or French publishers. In 1671 John Seller, a London instrument manufacturer, introduced a series of chart books under the general title *The English Pilot*. Seller's plan was carried forward by William Fisher and John Thornton and their heirs, and editions of *The English Pilot* were published for more than a century. The volume covering the coast of the Americas from Hudson Bay to the Amazon River, identified as *The English Pilot: The Fourth Book*, was published in thirty-seven separate editions between 1689 and 1794.[11] Although *The Fourth Book* continued to be used by some navigators through the revolutionary war period, it had serious limitations. Many of its charts, for example, were unaltered in new editions despite the availability of more up-to-date and accurate surveys.

Before the conflict ended British naval officers were relying upon a new and more accurate series of American coastal charts. Designated the *Atlantic Neptune*, this series was based for the most part on surveys conducted under the direction of Joseph Frederick Wallet Des

Barres. Born in Switzerland in 1721, Des Barres immigrated to England as a young man. Following training at the Royal Military Academy in Woolwich, he was commissioned in the Sixtieth Regiment, which was destined for service in North America. In the colonies he participated in the siege of Quebec and subsequently conducted surveys and prepared maps, plans, and charts of cities, fortifications, and harbors.

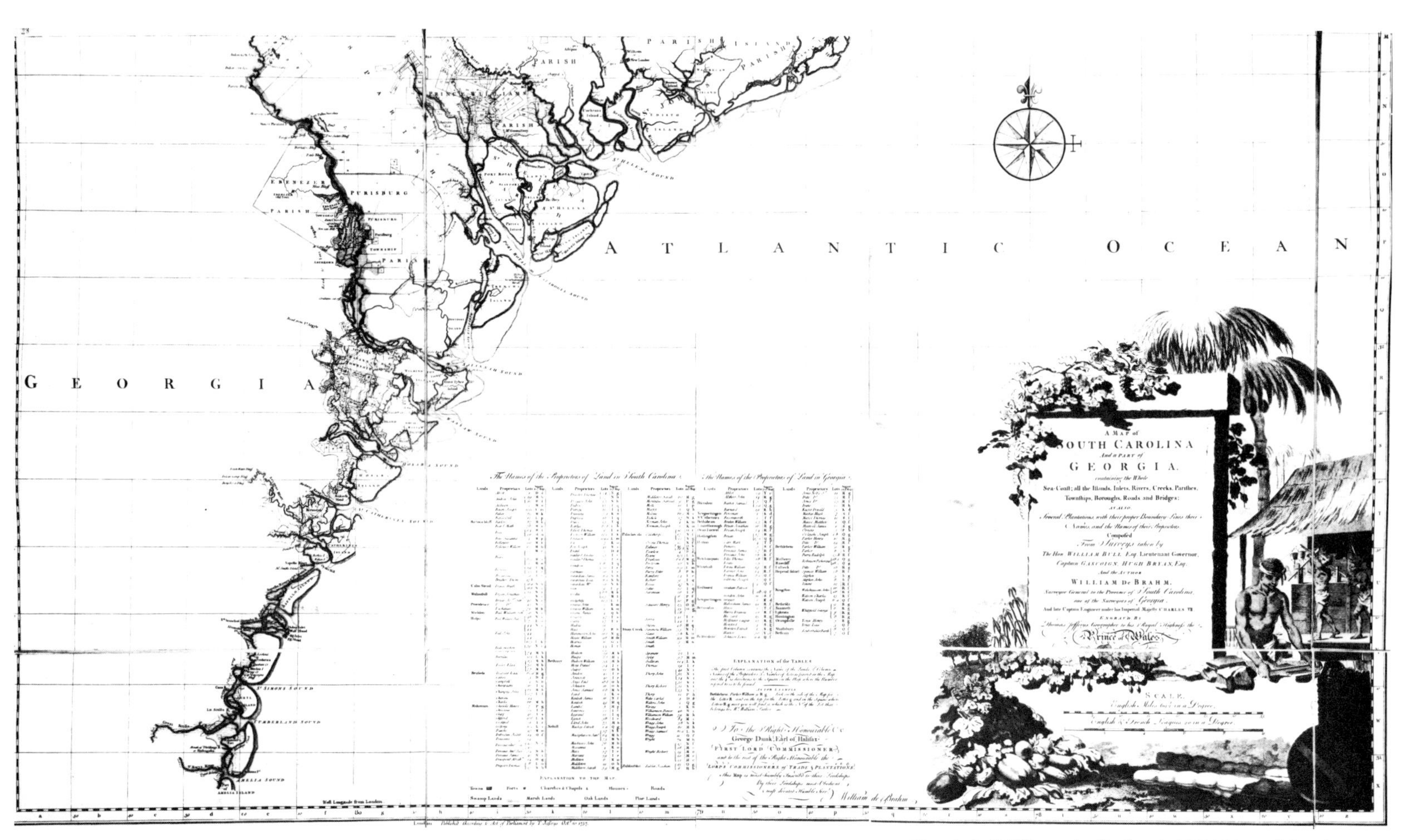

Fig. 1–3. William de Brahm's 1757 *Map of South Carolina and a Part of Georgia* is considered one of the best cartographic representations of these colonies in the pre-revolutionary war period. Only the title cartouche and a segment of the map are reproduced here.

CHARTS
of the
Coast and Harbors
of
NEW ENGLAND
from SURVEYS taken
BY
SAM.L HOLLAND Esq. Surv.r Gen.l of Lands for the Northern District of North America
AND
Geo. Sproule Cha.s Blaskowitz Jam.s Grant and Tho.s Wheeler his ASSISTANTS &c.
PURSUANT to ORDERS
from
The Right Hon.ble the Lords Commiss.rs for Trade & Plantations
Together with
Several usefull additional SURVEYS SOUNDINGS VIEWS &c taken by various Officers on the Spot
Collected composed & publish'd by Command of Government at the request of
The Right Hon.ble Vice Admiral LORD VIS.T HOWE
FOR THE USE OF THE
Royal Navy of Great Britain,
by
J. F. W. Des Barres Esq.r
SURVEYOR of the COAST and HARBOURS of NORTH AMERICA.
MDCCLXXX.

Fig. 1–4. The title page of the 1780 New England volume of the *Atlantic Neptune* illustrates the excellence of the engraving used in this important series of navigation charts.

Soon after the Treaty of Paris was signed, Rear Admiral Richard Spry, commander of the British fleet in America, called to the Admiralty's attention the lack of good charts of the Atlantic coast and waters of North America. At his recommendation, Des Barres and a staff of assistants were occupied, from 1764 to 1775, in carrying out extensive nautical surveys. During the subsequent ten years, at headquarters in London, Des Barres directed a staff of twenty or more specialists in compiling maps from the survey sketches and data and in engraving copper plates from which charts were printed. He also utilized coastal surveys prepared by other British military engineers, among them Samuel Holland.

Some 180 charts and views covering the harbors and coasts of America between Nova Scotia and Florida were ultimately included in the *Atlantic Neptune*. The charts, which served the British navy effectively during the latter months of the revolutionary war and for some years thereafter, are treasured today for their decorative beauty and cartographic excellence as well as for their historical interest.[12] *Atlantic Neptune* charts were most often assembled in bound volumes. Because charts were customarily supplied on order, no two extant volumes are identical in contents (Fig. 1–4). Many of the charts were updated periodically and there are, therefore, multiple variant editions. Sets of the *Atlantic Neptune* are preserved in a number of American and European repositories. Between 1966 and 1969, Barre Publishers, of Barre, Massachusetts, issued four series of a facsimile edition of the *Neptune*.

More limited in coverage than *The English Pilot: The Fourth Book* and the *Atlantic Neptune* is a *Collection of Charts of the Coasts of Newfoundland and Labradore &c.*, published by Thomas Jefferys, 1765–68. The nine plates in this nautical atlas were based on original surveys by James Cook and Michael Lane. Cook, who received his early training in marine surveying from Holland, subsequently gained lasting fame for his explorations and discoveries in the Pacific. There are only four recorded copies of the Cook-Lane atlas. A facsimile reproduction, from the original in the library of the University of California at Los Angeles, was published under the title *James Cook, Surveyor of Newfoundland* in San Francisco in 1965. It is enriched with a descriptive essay by the late R. A. Skelton, former head of the Map Room, British Museum (now the British Library). Captain Cook's hydrographic surveys in North America are also described in a paper written by Skelton and R. V. Tooley.[13]

In 1777 Robert Sayer and John Bennett published *The North-American Pilot* in two volumes, most of the plates of which were engraved by Jefferys. Volume one covers "Newfoundland, Labradore, and the Gulf and River St. Laurence" and is based on original surveys by Cook and Lane. The second volume includes the coasts of "New England, New York, Pensilvania, Maryland, and Virginia, also the two Carolinas and Florida." It was "drawn from original surveys, taken by Capt. John Gascoigne, Joshua Fisher, Jacob Blarney, and other officers and pilots in his Majesty's service."

By 1775 the hydrographic and topographic surveys by military and naval engineers in North America had generated a considerable body of cartographic data. Some of the surveys, particularly those of the coasts, were engraved on copper plates from which maps and charts were printed. Others were deposited in manuscript format in official British archives. Certain English publishers were privileged to draw upon the manuscripts in compiling new maps and charts.

With survey data coming regularly from engineers stationed in America and India, Britain's accelerating trade and commerce, the invention and development of new instruments and techniques for surveying, drafting, and map reproduction, London had, by the middle of the eighteenth century, become one of the world's major map publishing centers. Jefferys, publisher of De Brahm's *Map of South Carolina and a Part of Georgia* and the Cook-Lane *Collection of Charts*, among others, has been called "one of the most significant map sellers of the mid-eighteenth century."[14] Jefferys's entrée into map publishing was through engraving, and his name appears as the engraver on many maps. He published his first map in 1737, and during the next two decades he was involved in various cartographical publishing activities, with major emphasis on maps of the British Isles. Some of Jefferys's publications were independent endeavors; others were issued with co-publishers. In the early 1750s Jefferys added maps and charts of North America and the West Indies to his list of publications. Because of the success of his early American cartographical works, he earned a reputation as an authority and specialist on such maps and atlases.

About 1757 Jefferys was appointed geographer to the Prince of Wales, and when the latter became King George III in 1760, the map publisher was elevated to geographer to the king. No salary was associated with this office, nor did it insure a monopoly on publishing government maps and charts. It was, however, as J. Brian Harley has noted, "a coveted mark of status amongst mapsellers, but, in itself, did not *automatically* confer special privilege in relation to government departments, only the right to retail maps to the King."[15] Skelton believed that as geographer to the king Jefferys "enjoyed semi-official standing which gave him access to public documents and map-drafts for engraving and publication."[16] Few maps were engraved or printed in America prior to the Revolution, and colonial surveyors and mapmakers were dependent on European engravers, printers, and publishers. Thus, in the third quarter of the eighteenth century a number of American cartographers—among them Fry and Jefferson and De Brahm—had maps published by Jefferys.

After the Treaty of Paris in 1763, Jefferys again changed his area of interest to preparing surveys and maps of English counties. Harley believes that because of the greater financial outlay required to prepare original surveys, this concentration on county maps was a major cause of Jefferys's bankruptcy in 1767.[17] By forming partnerships with other publishers, he managed to continue his business until his death in 1771. A principal associate was Robert Sayer, with whom Jefferys collaborated in 1768 to publish *A General Topography of North America and the West Indies: Being a Collection of All the Maps, Charts, Plans, and Particular Surveys, That Have Been Published of that Part of the World, Either in Europe or America*. This comprehensive atlas of the British colonies in the pre-revolutionary period contains 100 maps on 109 sheets compiled from Jefferys's earlier publications and from maps and charts prepared by other European and American cartographers.

Following Jefferys's death, Sayer purchased much of his associate's cartographic stock. In a new partnership with John Bennett, Sayer published editions of Jefferys's *American Atlas: or a Geographical Description of the Whole Continent of America* in 1775, 1776, 1778, and 1783. The editions vary in size from twenty-two to thirty-four maps, a number of which were published previously in *A General Topography*. A French edition of Jefferys's *American Atlas* was published in Paris around 1792, by G. L. Le Rouge.

Notes

1. William P. Cumming, *British Maps of Colonial America* (Chicago, 1974); Cumming, *The Southeast in Early Maps, With an Annotated Check List of Printed and Manuscript Regional and Local Maps of Southeastern North America during the Colonial Period* (Chapel Hill, N.C., 1962); and Cumming, Raleigh A. Skelton, and David B. Quinn, *The Discovery of North America* (New York, 1972).
2. Henry Popple, *A Map of the British Empire in America with the French and Spanish Settlements Adjacent Thereto*, facsimile with introductory notes by William P. Cumming and Helen Wallis (Lympne Castle, Kent, England, 1972).
3. Lawrence H. Gipson, *Lewis Evans . . . to which Is Added Evans' A Brief Account of Pennsylvania, Together with Facsimiles of His Geographical, Historical, Political, Philosophical, and Mechanical Essays, Numbers I and II . . . also Facsimiles of Evans' Maps* (Philadelphia, 1939); and Walter Klinefelter, "Lewis Evans and His Maps," *Transactions of the Philosophical Society*, n.s. 61, pt. 7.
4. Edmund Berkeley and Dorothy Smith Berkeley, *Dr. John Mitchell, the Man Who Made the Map of North America* (Chapel Hill, N.C., 1973); Lawrence Martin and Walter W. Ristow, "John Mitchell's Map of the British and French Dominions in North America," in *A La Carte: Selected Papers on Maps and Atlases*, comp. Walter W. Ristow (Washington, D.C., 1972), 102–13; and *North America at the Time of the Revolution: A Collection of Eighteenth Century Maps*, with introd. notes by Louis De Vorsey, Jr., pt. 2 (Lympne Castle, Kent, England, 1974). The last book includes facsimiles of Mitchell's *Map of the British and French Dominions in North America*, Fry and Jefferson's *Map of the most Inhabited part of Virginia . . . 1775*, Claude Joseph Sauthier's *Topographical Map of the Hudson River . . . 1776*, Holland's *Provinces of New York and New Jersey . . . 1776*, and Thomas Jefferys's *Map of the Most Inhabited Part of New England . . . 1774*.
5. Joshua Fry and Peter Jefferson, *The Fry and Jefferson Map of Virginia and Maryland*, facsimile of 1st ed. (1751) in the Tracy W. McGregor Library, with introd. by Dumas Malone (Princeton, 1950), and *The Fry and Jefferson Map of Virginia and Maryland: Facsimiles of the 1754 and 1794 Printings*, 2d ed. (Charlottesville, 1966).
6. Louis De Vorsey, Jr., "William Gerard De Brahm: Eccentric Genius of Southeastern Geography," *Southeastern Geographer* 10 (1970): 21–29.
7. Don W. Thomson, *Men and Meridians, the History of Surveying and Mapping in Canada* (Ottawa, 1966) 1:100.
8. Cumming, *The Southeast in Early Maps*, 54.
9. *De Brahm's Report of the General Survey in the Southern District of North America*, ed. with introd. by Louis De Vorsey, Jr. (Columbia, S.C., 1971).
10. Archer Butler Hulbert, ed., *Crown Collection of Photographs of American Maps*. Ser. 1, 5 vols., reproductions from the British Museum (Cleveland, 1904–8), Ser. 2, 5 vols., reproductions from the British Museum (Harrow, England, 1909–12), Ser. 3, reproductions from the Colonial Office Library (London, 1914–16).
11. See Coolie Verner, *A Carto-Bibliographical Study of the English Pilot, the Fourth Book, With Special Reference to the Charts of Virginia* (Charlottesville, 1960).
12. For more detailed information on J. F. W. Des Barres and the *Atlantic Neptune* see Geraint N. D. Evans, *North American Soldier, Hydrographer, Governor: The Public Careers of*

J. F. W. Des Barres (Ann Arbor, Mich., University Microfilms, 1965); Evans, *Uncommon Obdurate: the Several Public Careers of J. F. W. Des Barres* (Toronto and Salem, 1969); and John Clarence Webster, *The Life of Joseph Frederick Wallet Des Barres* (Shediac, New Brunswick, 1933).

13. *The Marine Surveys of James Cook in North America, 1758–1768*, Map Collectors' Series, No. 37 (London, 1967).
14. J. Brian Harley, "The Bankruptcy of Thomas Jefferys: An Episode in the Economic History of Eighteenth Century Map-making," *Imago Mundi* 20 (1966): 27.
15. Ibid., 37.
16. R. A. Skelton, introd., *James Cook, Surveyor of Newfoundland* (San Francisco, 1965): 27.
17. Harley, "Bankruptcy of Thomas Jefferys," 46.

2. *The Maps of the American Revolution*

The charts, maps, and atlases described in Chapter 1 comprised the principal cartographic record of colonial North America when the revolutionary war began.[1] As with all wars, the conflict generated demands for detailed surveys of all parts of the country, with special emphasis on the principal theaters of military action. An English publisher who played a leading role in producing such pertinent charts, maps, and plans during and following the war years was William Faden. He, like Sayer, was for a brief time a partner of Jefferys when Jefferys formed several partnerships to insure himself a measure of financial solvency after his bankruptcy. Faden, the son of a successful London printer, continued to run Jefferys's business independently after Jefferys died in 1771. Shortly after this, Faden was named geographer to the king,[2] and in 1777 he published *The North American Atlas, Selected from the Most Authentic Maps, Charts, Plans, &c. Hitherto Published* (Fig. 2–1).

As geographer to the king, Faden, like Jefferys before him, had access during the war to many of the official manuscript maps and battle plans. In 1793 he assembled a selection of these maps, which he published as the *Atlas of the Battles of the American Revolution, Together With Maps Showing the Routes of the British and American Armies, Plans of Cities, Surveys of Harbors, &c. Taken During That Eventful Period by Officers Attached to the Royal Army* (Fig. 2–2). Plates for the atlas were subsequently acquired by the firm of Bartlett & Welford, which republished it in 1845.

In 1864 the Joint Library Committee of the U.S. Congress purchased from Edward Everett Hale a collection of manuscript and printed maps that had been assembled by Faden in the course of his map publishing career. The Faden Collection, now preserved in the Library of Congress Geography and Map Division, includes 101 maps (about half of them manuscripts) that relate to General Edward Braddock's expedition, the French and Indian War, colonial America, and the revolutionary war. There are brief descriptions of the individual items in the *Catalogue of a Curious and Valuable*

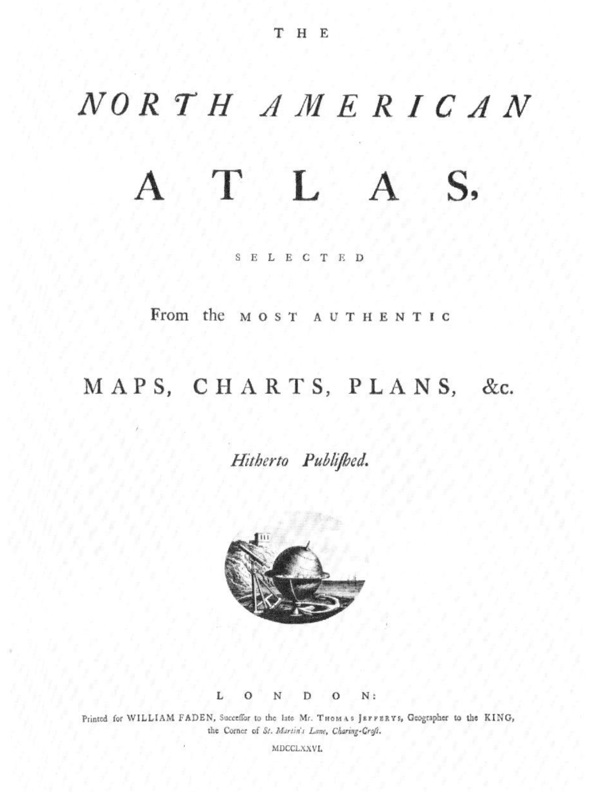

THE

NORTH AMERICAN

ATLAS,

SELECTED

From the MOST AUTHENTIC

MAPS, CHARTS, PLANS, &c.

Hitherto Publiſhed.

LONDON:

Printed for WILLIAM FADEN, Succeſſor to the late Mr. THOMAS JEFFERYS, Geographer to the KING, the Corner of *St. Martin's Lane, Charing-Croſs.*

MDCCLXXVI.

Fig. 2–1. A number of maps of colonial America prepared by British surveyors and mapmakers were assembled in William Faden's *North American Atlas.* Although the date on the atlas's title page reproduced here is 1776, the volume was not actually published until the following year.

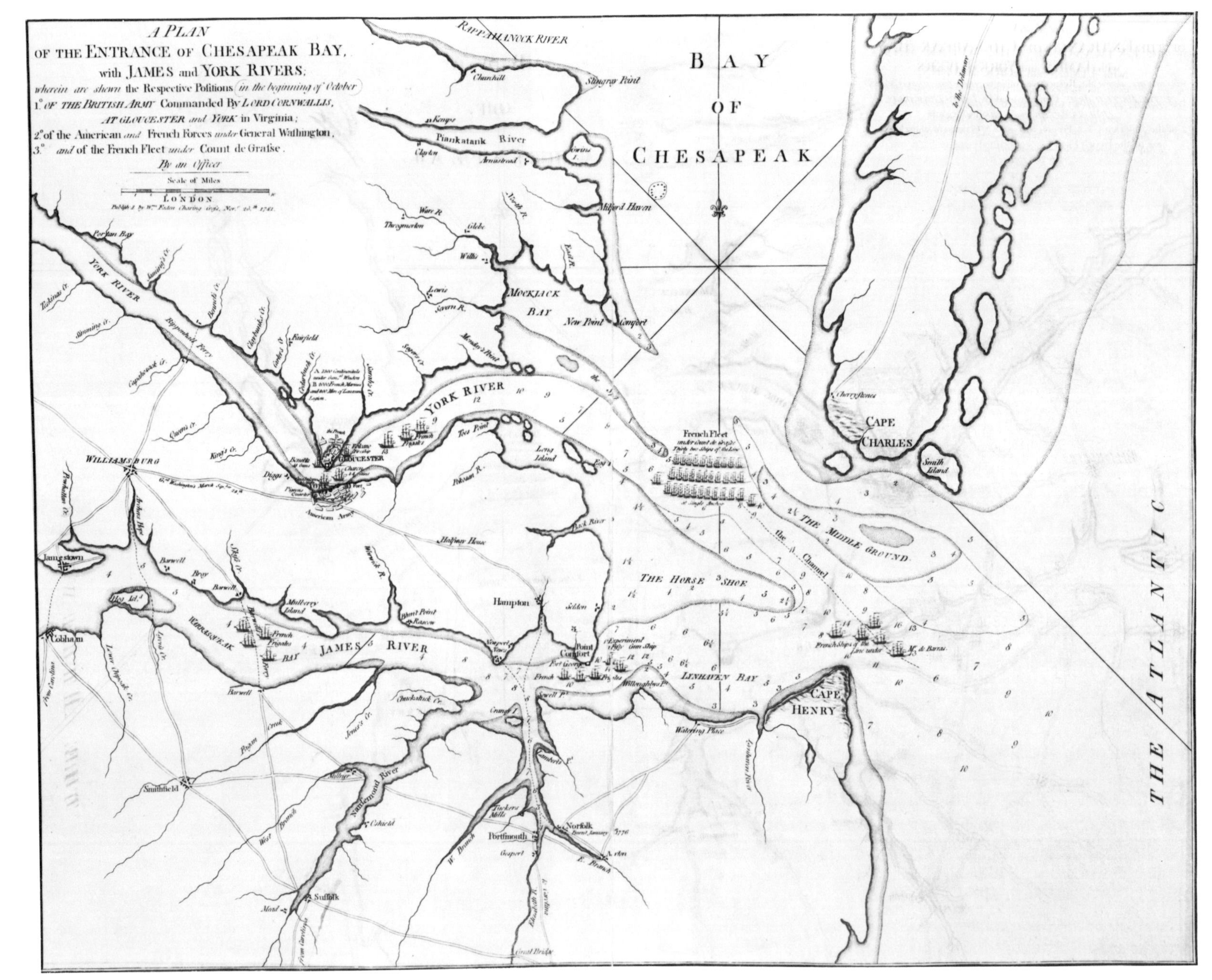

Fig. 2–2. This 1781 plan of the Chesapeake Bay area is included in Faden's *Atlas of the Battles of the American Revolution*.

Collection of Original Maps and Plans of Military Positions Held in the Old French and Revolutionary Wars, which was compiled by Hale and published in Boston in 1862.

Sayer and Bennett published editions of Jefferys's *The North-American Pilot* and the *West India Atlas* in 1775 and 1777 and in 1778 *The Western Neptune*. Of particular revolutionary war interest is Sayer and Bennett's *The American Military Pocket Atlas* (Fig. 2–3), which was published in London in 1776. According to its subtitle, it is "an approved collection of maps, both general and particular, of the British colonies, especially those which are now, or probably may be the theatre of war; taken principally from the actual surveys and judicious observations of engineers De Brahm and Romans; Cook, Jackson, and Collet, Maj. Holland, and other officers employed in His Majesty's Fleet and Armies."

Notwithstanding the availability to George Washington of many of these maps and atlases, shortly after he took command of the Continental army he wrote to the president of the Congress on January 26, 1777, that "the want of accurate maps of the Country which has hitherto been the Scene of the War, has been a great disadvantage to me. I have in vain endeavored to procure them, and have been obliged to make shift, with such Sketches, as I could trace from my own Observations and that of Gentlemen around me."[3] The general's complaint highlights a problem that was to plague the American forces throughout the war. Fighting a war of movement, Washington was particularly in need of detailed maps of the roads of the country.

Shortly after registering this complaint, Washington recommended that Robert Erskine be commissioned geographer and surveyor general of the Continental army. The Congress approved the choice, and Erskine assumed duty on July 27, 1777. Born at Dunfermline, near Edinburgh, Scotland, on September 7, 1735, he received a basic education in the local schools, followed by studies at the University of Edinburgh, where he specialized in hydraulic engineering. Erskine held various positions in Scotland and England for some fifteen years before

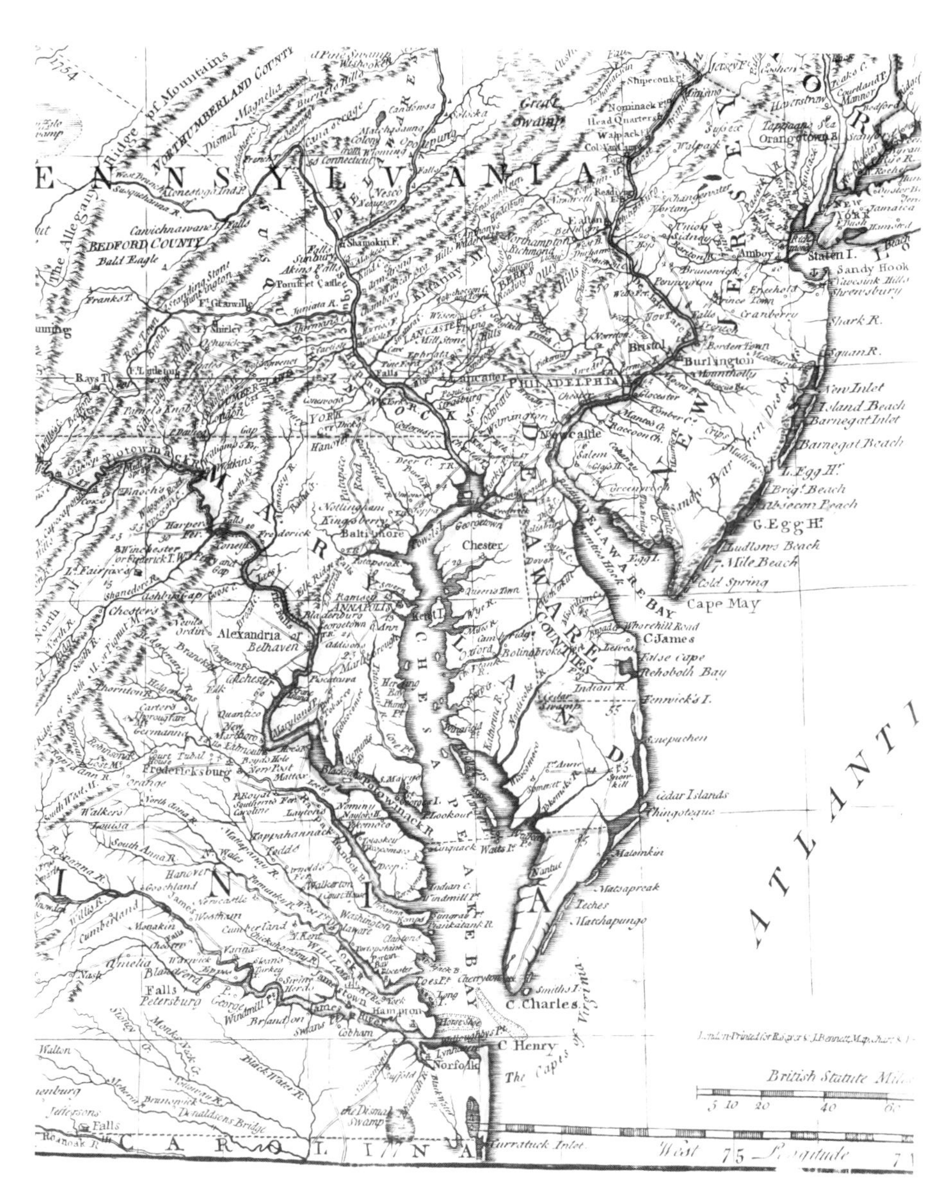

Fig. 2–3. Robert Sayer and John Bennett's *American Military Pocket Atlas* was intended for use in the field by British officers. The small volume includes six maps, among them this *General Map of the Middle British Colonies in America*, of which a portion is here reproduced.

immigrating to New Jersey in 1771, where he managed an ironmongery for a group of British investors. Several years later, when difficulties developed between England and her colonies, Erskine's sympathies were with the latter.[4]

Within a week after becoming geographer, Erskine outlined to Washington his proposed procedures. In a letter dated August 1, 1777, he wrote: "In planning a country a great part of the ground must be walked over, particularly the banks of Rivers and Roads, as much of which may be traced and laid down in three hours as could be walked over in one; or in other words a Surveyor who can walk 15 miles a day may plan 5 miles . . . six attendants to each surveyor will be proper; to wit, two chainbearers, one to carry the instrument and three to hold flag staffs. . . . Young gentlemen of Mathematical genius, who are acquainted with the principles of Geometry, and have a taste for drawing would be the most proper assistants for a Geographer."[5] One of the first young assistants recruited by Erskine was Simeon De Witt, a native of New York and a recent graduate of Queens College (now Rutgers University). He joined Erskine in June 1778. For varying periods of time during the war, some thirty or more assistant surveyors served under the geographer.

A cold and fever, probably contracted while carrying out surveys, resulted in the death of Erskine on October 2, 1780, at the age of forty-five. In a little over three years he and his assistants had prepared a hundred or more maps for Washington and his staff. In a letter written several months before his death, Erskine summarized the accomplishments of his department.

> The number of Assistant Surveyors has varied from two to six; the mean number employed for a Constancy, I suppose to be one Assistant Draughtsman, three field surveyors, and eighteen chain-bearers from the Line.
>
> From Surveys actually made, we have furnished His Excellency [Washington] with maps of both sides of the North River, extending from New Windsor and Fishkill, southerly to New York; eastward: to Hartford, Whitehaven, etc., and on the west to Eaton in Pennsylvania. Our Surveys likewise include the principal part of New Jersey, lying northward of a line drawn from Sandy Hook to Philadelphia; take in a considerable part of Pennsylvania; extend through the whole route of the Western army under Genl. Sullivan, and are carried on from New Windsor and Fishkill northward, on both sides of the river, to Albany, & from thence to Scoharie. In short, from the Surveys made, and materials collecting and already procured, I could form a pretty accurate Map of the four States of Pennsylvania, New Jersey, New York and Connecticut, and by the help of a few Magnetic and Astronomical Observations, with some additional Surveys, a very accurate one.[6]

De Witt succeeded to the office of geographer on December 4, 1780, and for the next several years he directed mapping operations for the northern Continental army. A major assignment was mapping the roads from New Jersey to Virginia in anticipation of Washington's southern campaign, which culminated with General Charles Cornwallis's surrender at Yorktown.

Thomas Hutchins was appointed geographer to the southern Continental army on May 4, 1781. Born in 1730 in Monmouth County, New Jersey, Hutchins was orphaned in his teens and shortly after joined the British army. He served in the Ohio country and apparently received training and experience in engineering, for he supervised the planning and erection of Fort Pitt. He also surveyed extensively in the Louisiana Territory and in west Florida, and fought in a number of Indian skirmishes. When the revolutionary war broke out, Hutchins was in London, where he was directing publication of his *A Topographical Description of Virginia, Maryland and North Carolina*, which was published, with an accompanying map, in 1778. Because of his sympathies with his fellow Americans, he sought to sell his commission in the British army. For alleged correspondence with Benjamin Franklin, then U.S. ambassador to France, Hutchins was imprisoned. Released after six weeks, he went to France in February 1780. Some months later he sailed for America, landing at Charleston, South Carolina, where he joined General Nathanael Greene, commander in chief of the southern Continental army.

Because of his recognized status and reputation in surveying and mapping, and at his insistence, Hutchins was designated geographer to the southern army, effective July 11, 1781, with equal rank to De Witt. At Greene's request Hutchins and his assistants began surveying Georgia early in 1782. Hutchins contracted a fever, however, and was incapacitated for most of that summer. He returned to Philadelphia in the spring of 1783, after an unproductive year in the South. No revolutionary war maps prepared by Hutchins or his assistants are known to exist.

Maps prepared under the direction of Erskine and De Witt were not published in their original state. As Harley has noted, "despite the farsightedness of both Erskine and De Witt, the military topographical surveys of the Revolution were largely a false dawn in the mapping of the new nation."[7] Christopher Colles, however, drew heavily on the Erskine-De Witt maps in compiling his 1789 *Survey of the Roads of the United States of America*.[8] The only considerable extant collection of Erskine-De Witt maps is preserved in the New-York Historical Society. They had been retained by De Witt following the war and were presented to the society in 1845 by his grandson, Richard Varick De Witt.[9]

In his *American Maps and Map Makers of the Revolution*, Peter Guthorn includes the cartographic contributions of a half dozen French military engineers who served with the Continental army. One of the most productive of these was Michel Capitaine du Chesnoy, a native of Mézières, France, who was commissioned a captain in the Corps of Engineers in April 1778 and was promoted to major later in that year. Du Chesnoy is credited with nineteen maps, most of which pertain to engagements in which General Lafayette participated. Another was François Louis Teisseidre de Fleury, who was appointed captain of engineers in Washington's army on May 22, 1777. He participated in the battles of Brandywine and Germantown and was wounded in the defense of Fort Mifflin. De Fleury also served at Valley Forge and in the battle of Monmouth before terminating his service with the American army in September 1779. Subsequently he returned to America as an officer in General Jean Baptiste Rochambeau's French army. Guthorn credits de Fleury with seven maps. Among the other French engineer-cartographers noticed by Guthorn are Jean-Baptiste de Gouvion, Gilles-Jean Kermorvan, Etienne de Rochefontaine, and Jean de Villefranche.

Guthorn does not include in his list maps by French engineers who served in the expeditionary forces commanded by the generals Jean Baptiste d'Estaing, François de Grasse, and Rochambeau. Preserved in the Library of Congress Geography and Map Division are twenty skillfully executed sepia manuscript maps and views believed to have been drawn by Pierre Ozanne. Ozanne was commissioned "sous-inginieur constructeur" ("assistant construction engineer") on d'Estaing's expedition. His drawings show the French fleet sailing out of the Mediterranean Sea and in various actions along the coast of North America and in the West Indies. Of particular interest are a view and map depicting the siege of Savannah on February 7–8, 1779. The American campaigns of General Rochambeau are well documented in *The American Campaigns of Rochambeau's Army 1780, 1781, 1782, 1783*, a two-volume work translated and edited by Howard C. Rice, Jr., and Anne S. K. Brown and published in 1972. Volume two includes 177 illustrations, most of them maps, including reproductions from original manuscript and printed maps in European and American collections. Among the latter are the Paul Mellon Library in Upperville, Virginia, the Library of Congress, Boston Athenaeum, John Carter Brown Library, Historical Society of Pennsylvania, Library Company of Philadelphia, Maryland Historical

Society, New-York Historical Society, Henry Francis du Pont Winterthur Museum, and the Smithsonian Institution.

The British army's superior cartographic resources are reflected in the more extensive extant collections of British revolutionary war maps. Throughout the conflict English commanders sent copies of campaign and battle maps to the War Office and frequently to King George III, who was an avid map collector. Some of these maps were engraved and published by private firms, such as that of William Faden, to keep the British public informed on the progress of the war. Many manuscript and printed maps are preserved in various British collections.[10]

British officers often retained in their personal possession journals, letters, reports, and maps relating to their service in America. A number of these manuscript collections have been acquired by American libraries. In 1905 the Library of Congress purchased the Admiral Lord Richard Howard Howe Collection. It includes seventy-two manuscript maps and charts of various regions along the Atlantic coast, in the West Indies, and in the Philippines. Most of the cartographic items deal with Howe's life before the Revolution, but several of them may have been consulted by the admiral in naval operations off Philadelphia and New York City during the Revolution. The largest group of British headquarters maps in the United States is in the University of Michigan's William L. Clements Library in Ann Arbor. These rich holdings include manuscript papers and maps that belonged to General Thomas Gage and General Henry Clinton. Published catalogs describe these collections, as well as other manuscript revolutionary war maps, in the Clements Library.[11] The only large body of maps prepared by the German mercenaries who fought with the British is in the Hessisches Staatsarchiv, Marburg, West Germany. Several Hessian maps in this and other archives are described by Guthorn in "A Hessian Map from the American Revolution."[12]

Many of the revolutionary war maps published in contemporary English and American periodicals are also worthy of examination. The British journals which include wartime maps are the *Gentlemen's Magazine, London Magazine, Political Magazine*, and *Universal Magazine*; the American magazines are the *Massachusetts Magazine, New York Magazine, Pennsylvania Magazine*, and *Columbian Magazine*. Many histories, biographies, and atlases published during the past two hundred years include cartographic illustrations relating to the Revolution, as do hundreds of articles published in professional journals over this period. David S. Clark has compiled a useful index to this material.[13]

More detailed than most of the cartographic works described above are maps and plans of the several theaters of the war and of individual campaigns and battles. "The extent to which battles were mapped," Harley notes, "reflects their local importance in the whole military process."[14] The significance of this aspect of revolutionary war cartography is evident in the more than two hundred printed maps described by Kenneth Nebenzahl in his *Bibliography of Printed Battle Plans of the American Revolution, 1775–1795*.[15] There are probably an equal, or larger, number of manuscript battle maps preserved in European and American libraries and official repositories, many of which are described in cartobibliographies and catalogs previously cited. With the meager engraving and printing facilities in the colonies during the war, it is no surprise that most of the contemporary printed battle maps were published in England.

The early battles and campaigns of the war in New England, including the siege of Boston and the battles of Lexington and Concord and Bunker Hill, were well covered cartographically with both printed and manuscript maps. J. De Costa's *A Plan of the Town and Harbour of Boston and the Country Adjacent with the Road from Boston to Concord*, published in London on July 29, 1775, "was the first graphic representation of the epochmaking encounters of 19 April 1775."[16] In addition to locating American and British camps and fortifications, the map depicts British regulars and colonials skirmishing and marching in the Lexington and Concord vicinity (Fig. 2–4).

The earliest revolutionary war maps printed in America were *A New and Correct Plan of the Town of Boston and Provincial Camp*, the *Exact Plan of General Gage's Lines on Boston Neck in America*, and *A Correct View of the Late Battle of Charlestown June 17, 1775* (Fig. 2–5). All three were drawn and engraved by Robert Aitken and published in 1775 in various issues of his *Pennsylvania Magazine*. Aitken, a native of Scotland, settled in Philadelphia in 1771 where he engaged in engraving, printing, and publishing. He founded the *Pennsylvania Magazine, or American Museum* in January 1775 and published it through June 1776.

The cartographic record of the battle of Bunker Hill, both manuscript and printed, is quite extensive. The collections of the Library of Congress, for example, include approximately one hundred contemporary maps which have some bearing on the battle and on the military actions in and around Boston during the early months of the revolutionary war. The earliest printed separate map of this historic engagement is *A Sketch of the Action Between the British Forces and the American Provincials on the Heights of the Peninsula of Charlestown, the 17th of June 1775*, which was published in London by Jefferys & Faden on August 1, 1775. When we remember that a sailing ship required a month or more to cross the Atlantic, engraving and publishing a map in England within a month and a half after the battle was a remarkable achievement. R. & J. Bennett, another London firm, published on November 27, 1775, *A Plan of the Battle of Bunkers Hill Fought on the 17th of June 1775, By an Officer on the Spot*. Although the officer is not identified, printed below the map is a copy of a letter, dated June 25, 1775, from General John Burgoyne to his nephew, Lord Stanley. This letter vividly describes the progress of the battle. One of the best known and informative cartographic representations of the battle is *A Plan of the Action at Bunkers Hill, on the 17th of June 1775, Between His Majesty's Troops Under the Command of Major General Howe, And the Rebel Forces. By Lieut. Page of the Engineers, Who Acted as Aide de Camp to General Howe in that Action. N. B.: The Ground Plan is From an Actual Survey by Captn. Montresor*. The map was published in London by Faden in 1777 and was included in his *North American Atlas* and the *Atlas of the American Revolution*.

Other engagements which invited considerable cartographic response were the defense of New York City and the battle of Long Island, the Philadelphia campaign, the Charleston expedition, the naval action off Newport, Rhode Island, and the Virginia campaigns which culminated with the surrender of General Cornwallis at Yorktown. The August 27, 1776, battle of Long Island is detailed in Faden's *A Plan of New York Island With Part of Long Island, Staten Island, & East New Jersey, With a Particular Description of the Engagement on the Woody Heights of Long Island . . .*, which was published October 19, 1776. The map locates the ships of the British fleet and identifies the various units of the opposing armies. Excerpts from a letter written by General William Howe to Lord Germain, printed on the map, summarizes the events of the battle.

New York City, on the eve of its capture and occupation by the British, is admirably portrayed in a *Plan of New York City in North America: Surveyed in the Years 1766 & 1767*, by B. Ratzer, London, ca. 1770. The plan shows places which were subsequently destroyed by the fire of 1776. A view at the bottom of the map sheet shows the city skyline as seen from the south. The city of Philadelphia was of particular importance because it served as the headquarters for the Continental Congress. It is shown as it then appeared on Nicholas Scull and George Heap's *A Plan of the City and Environs of Philadelphia*, which was published in London by Faden in 1777 (Fig. 2–6). Two years later, in 1779, after the British occupied the city, Faden published *A Plan of the City and Environs of Philadelphia with the Works and encampments of His Majesty's Forces under the command of . . . Sir William Howe*. The British capture of Charleston is also detailed in *A*

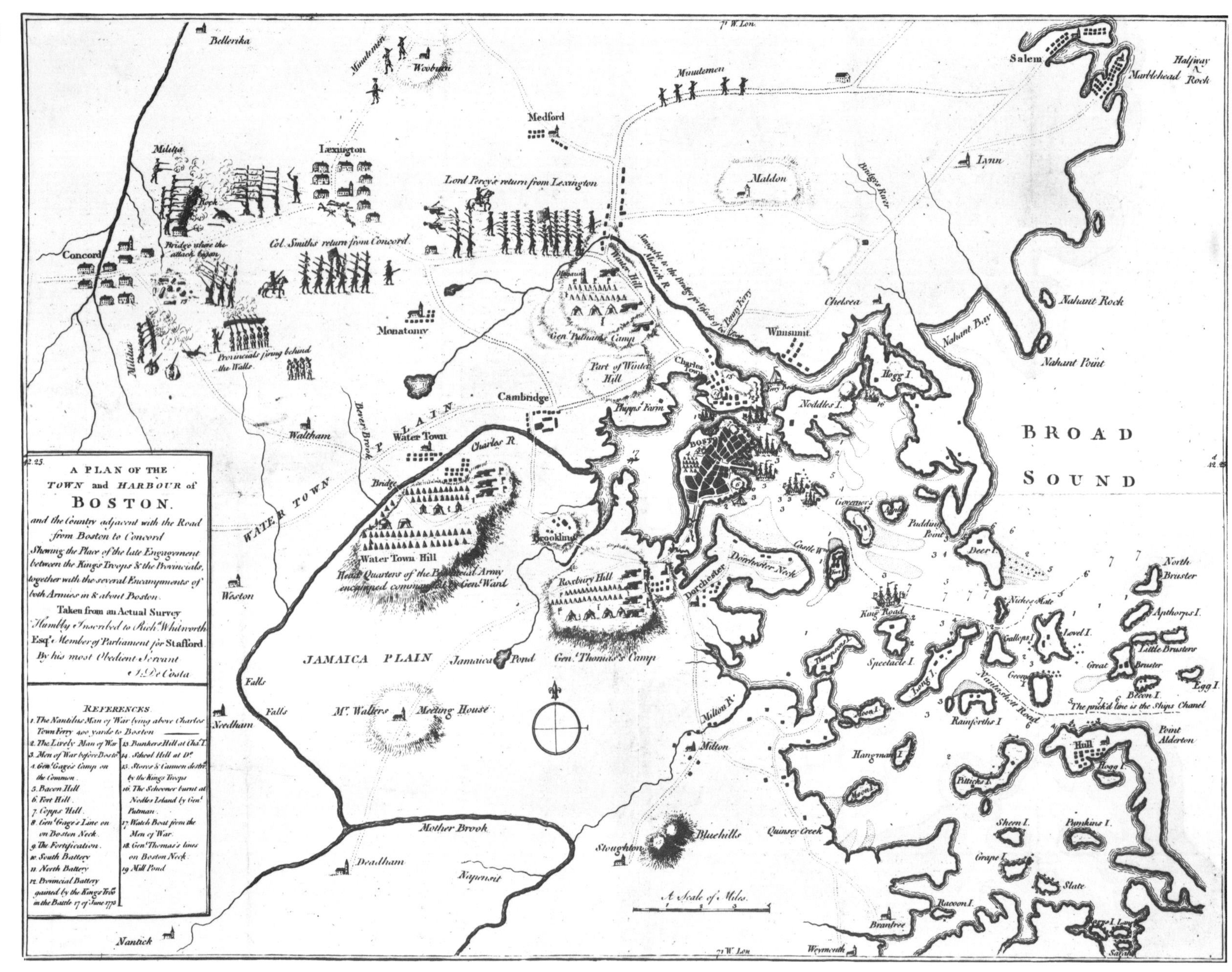

Fig. 2–4. J. De Costa's plan of Boston and its harbor was one of the earliest graphic representations of the battles of Lexington and Concord.

A CORRECT VIEW of THE LATE BATTLE AT CHARLESTOWN June 17th 1775.

Fig. 2–5. This crudely drawn perspective map by Robert Aitken was the first pictorial portrayal of the battle of Bunker Hill by an American.

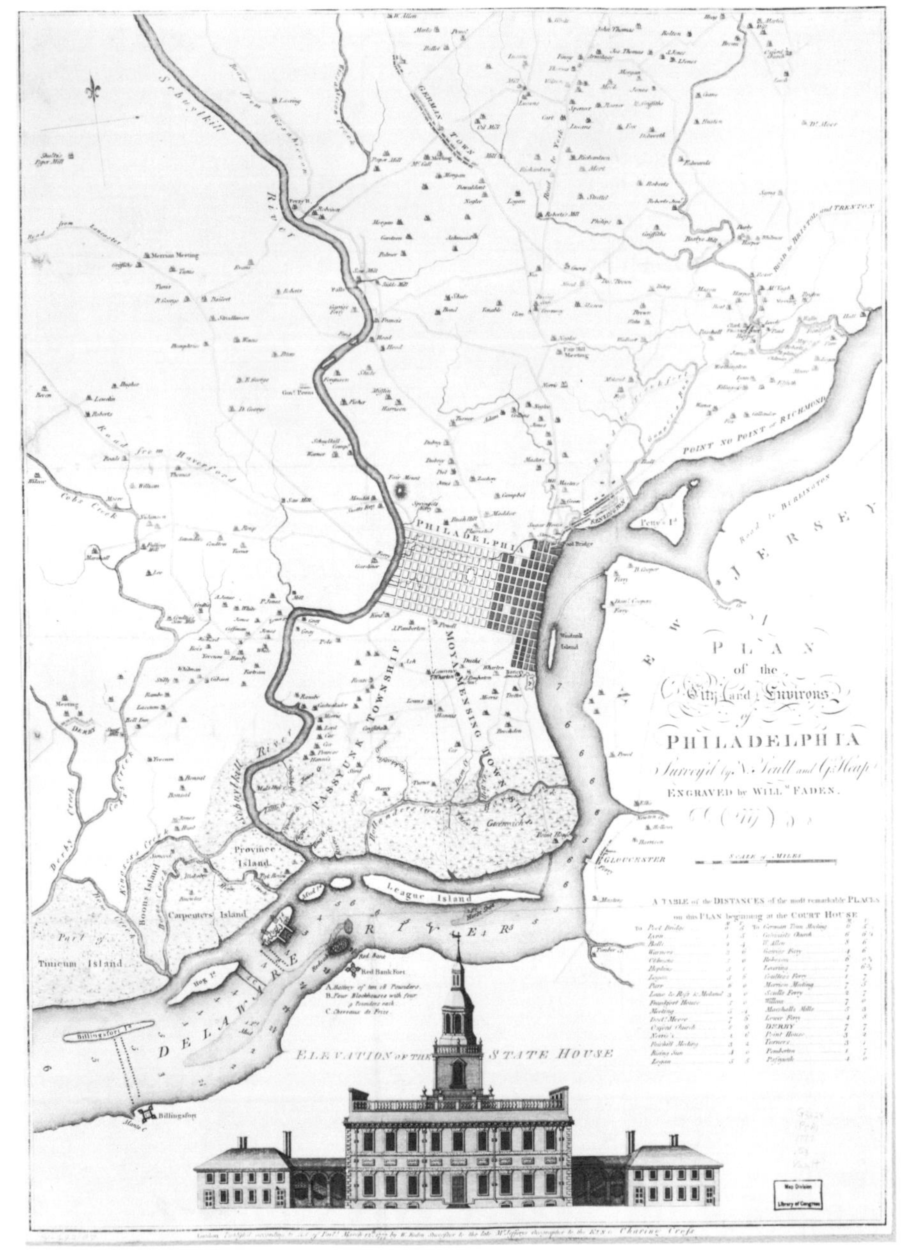

Fig. 2–6. The city and environs of Philadelphia in 1777, when it served as headquarters for the Continental Congress.

Plan of the Town, Bar, Harbour and Environs of Charlestown in South Carolina, with all the Channels, Soundings, Sailing-marks, etc., which Faden published June 1, 1780, less than three weeks after Clinton captured the city.

One of the earliest cartographic portrayals of the Yorktown surrender was published in Paris by Esnauts et Rapilly, probably in late 1781. It is titled *Carte de la partie de la Virginie ou l'armée combinée de France & des États-Unis de l'Amérique a fait prisonniere l'Armée Anglaise commandée par Lord Cornwallis le 19 Octbre. 1781 avec le plan de l'attaque d'York-town & de Glocester* (Fig. 2–7).

It is fitting, though, that one of the best maps of the British surrender at Yorktown was published in America. It is titled *To His Excellency Genl. Washington, Commander in Chief of the Armies of the United States of America, this Plan of the Investment of York and Gloucester Has Been Surveyed and Laid Down, and is Most Humbly Dedicated by His Excellencys Obedient and Very Humble Servant, Sebastn. Bauman, Major of the New York or 2nd Regiment of Artillery. This Plan was Taken Between the 22nd and 28th of October, 1781. R. Scot. Sculp.* The map was published in Philadelphia in early 1782 and was the first one to give to the American public details of the Yorktown victory. Sebastian Bauman, the maker of this map, was born in Frankfurt am Main in 1739 and was educated at Heidelberg University. Following service in the Austrian army, he immigrated to America after 1750 and fought in the French and Indian War. He joined the Continental army when the revolutionary war broke out and participated in several northern campaigns before he moved south under General Henry Knox. After the war Bauman served for a time as postmaster of New York City, where he died in 1803.

Such maps as these served both armies well during the American Revolution. They also constituted a major cartographic resource for the new republic and a foundation on which a native map publishing industry could be built. The battle and siege maps were particularly useful in the immediate postwar years, for many of them showed the street patterns of the Republic's major

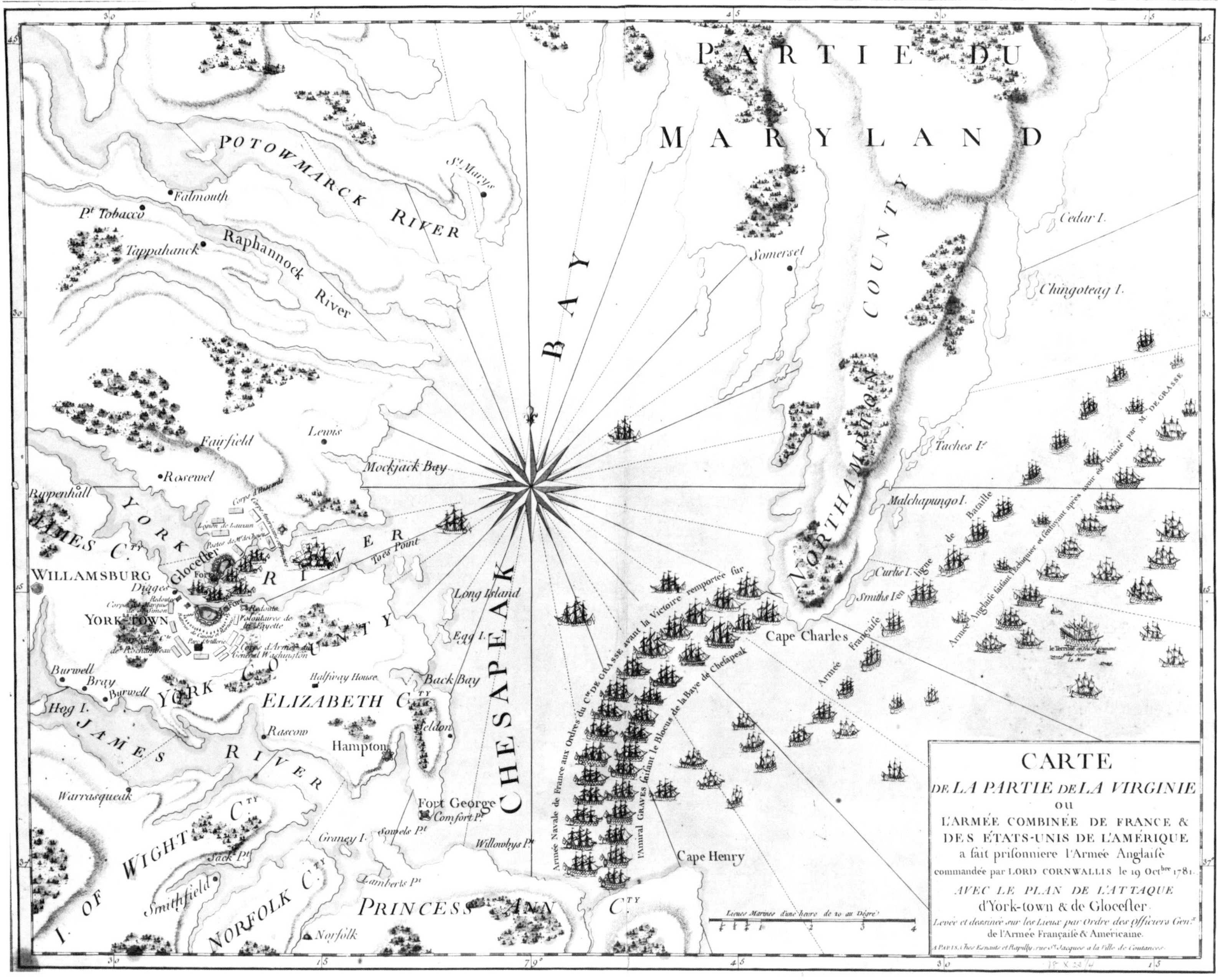

Fig. 2–7. This map of Yorktown and environs, probably published in Paris in late 1781, shows in pictorial detail the surrender of the British fleet to the French and American naval forces.

cities. It is interesting to note that contemporary revolutionary war maps still come to hand from time to time. In 1968, for example, the British Library purchased the Royal United Services Institution (R.U.S.I.) Collection. This was described as the "largest purchase of maps in [the library's] history" and the "most important archive of military maps [recently] in private hands."[17] Included in the accession was the Amherst Collection, which was presented to the institution in 1861 by the third baron Amherst, the grandnephew of Field Marshall Jeffrey Amherst, commander in chief of the British forces in North America from 1758 to 1763. There are some 130 maps and plans relating to Amherst's North American command in this collection. The Sir Augustus Frazier Collection, also part of the R.U.S.I. purchase, includes a number of maps of American Revolution interest.

In the August 1969 issue of *American Heritage*, Professor and Mrs. William P. Cumming called attention to a collection of about fifty revolutionary war maps in the private library of the duke of Northumberland.[18] These maps were in the personal collection of Hugh, Earl Percy, who was with the British army in America from 1774 to 1777, and are mainly manuscripts showing the battle terrains of several engagements of the American Revolution. Most of the maps were drawn on the scene, on the day of battle, or soon afterwards. Colonel Percy was at Lexington and accompanied General Howe when the latter evacuated Boston and sailed to Halifax. Percy again served under Howe in New York and commanded the troops that captured Fort Washington. Later he was promoted to lieutenant general and placed in charge at Newport, Rhode Island, until he returned to England in 1777.

Notes

1. For more information on the maps of the Revolution see Peter J. Guthorn, *American Maps and Map Makers of the Revolution* (Monmouth Beach, N.J., 1966); Guthorn, *British Maps of the American Revolution* (Monmouth Beach, N.J., 1972); J. Brian Harley, Barbara Bartz Petchenik, and Lawrence W. Towner, *Mapping the American Revolutionary War* (Chicago, 1978); Douglas W. Marshall and Howard H. Peckham, *Campaigns of the American Revolution: An Atlas of Manuscript Maps . . .* (Ann Arbor, Mich., 1976); University of Michigan, William L. Clements Library, *British Headquarters Maps and Sketches Used by Sir Henry Clinton While in Command of the British Forces Fighting in North America During the War for Independence, 1775–1782, a Descriptive List of the Original Manuscripts and Printed Documents . . .* by Randolph C. Adams (Ann Arbor, Mich., 1928); Kenneth Nebenzahl, *Atlas of the American Revolution . . .* (Chicago, 1974); Nebenzahl, *A Bibliography of Printed Battle Plans of the American Revolution, 1775–1795* (Chicago, 1975); Howard C. Rice, Jr., and Anne S. K. Brown, eds. and trans., *The American Campaigns of Rochambeau's Army 1780, 1781, 1782, 1783*, 2 vols. (Princeton, 1972); U.S. Library of Congress, *The American Revolution in Drawings and Prints; a Checklist of 1763–1790 Graphics in the Library of Congress*, compiled by Donald H. Cresswell, with foreword by Sinclair H. Hitchings (Washington, D.C., 1975); U.S. Library of Congress, *Maps and Charts of North America and the West Indies 1750–1789, a Guide to the Collections in the Library of Congress*, compiled by John R. Sellers and Patricia Molen Van Ee (Washington, D.C., 1981); U.S. Naval History Division, *The American Revolution, 1775–1783: An Atlas of 18th Century Maps and Charts* (Washington, D.C., 1972).
2. J. Brian Harley, "The Bankruptcy of Thomas Jefferys: An Episode in the Economic History of Eighteenth Century Map-making," *Imago Mundi* 20 (1966): 47.
3. George Washington, *Writings*, ed. John C. Fitzpatrick (Washington, D.C., 1931–44) 7:65.
4. For a detailed biography see Albert Henry Heusser's *George Washington's Map Maker: A Biography of Robert Erskine*, ed. Hubert G. Schmidt (New Brunswick, N.J., 1966).
5. George Washington, Papers, Manuscript Division, Library of Congress.
6. Heusser, *George Washington's Map Maker*, 209–10.
7. Harley, Petchenik, and Towner, *American Revolutionary War*, 35.
8. Christopher Colles, *A Survey of the Roads of the United States of America*, ed. Walter W. Ristow (Cambridge, Mass., 1961).
9. New-York Historical Society, *Proceedings* (1845): 21.
10. See British Museum, Department of Manuscripts, *Cata-*

logue of the Manuscript Maps, Charts, and Plans, and of the Topographical Drawings in the British Museum 3, pt. 2 (London, 1961); British Museum, Department of Printed Books, *Catalogue of Printed Maps, Charts, and Plans*, 12 vols. (London, 1967); and Colonial Office of Great Britain, *Catalogue of Maps, Plans, and Charts in the Library of the Colonial Office* (London, 1910).

11. University of Michigan, William L. Clements Library, *Guide to the Manuscript Maps in the William L. Clements Library*, comp. Christian Brun (Ann Arbor, Mich., 1939).
12. Peter Guthorn, "A Hessian Map from the American Revolution," *Quarterly Journal of the Library of Congress* 33 (July 1976): 219–31.
13. David Sanders Clark, *Index to Maps of the American Revolution in Books and Periodicals Illustrating the Revolutionary War and Other Events in the Period 1763–1789* (Washington, D.C., 1969).
14. Harley, Petchenik, and Towner, *American Revolutionary War*, 41.
15. Nebenzahl, *Bibliography*.
16. Ibid., vii.
17. Helen Wallis and Sarah Jeacock, "Royal United Services Map Collection," *Cartographic Journal* 7 (July 1970): 39–40.
18. William P. Cumming and Elizabeth C. Cumming, "The Treasure of Alnwick Castle," *American Heritage* 22 (Aug. 1969): 23–33.

3. *Maps of the Colonies and Provinces to 1784*

An interesting and challenging facet of post-Revolution cartography was stimulated by the pressing demands for maps of the newly established states. Some creditable maps of English colonies and provinces had, it is true, been compiled and published in the third quarter of the eighteenth century. One such map was Fry and Jefferson's *A Map of the most Inhabited part of Virginia* (Fig. 3–1), the first edition of which was published in London by Jefferys, probably late in 1753. This map had been completed by the cartographers almost two years earlier, and it carries the credit "Drawn by Joshua Fry & Peter Jefferson in 1751." There are only three known extant copies of this first edition. During the next several decades revised editions of the map were published in London, and two plagiarized French versions were issued in Paris. De Brahm, as surveyor general to the colony of South Carolina, had also completed, from his own surveys as well as those of Lieutenant Governor William Bull, Captain John Gascoigne, and Hugh Bryan, *A Map of South Carolina and a part of Georgia*. It was engraved and published in London in 1757 by Jefferys. The map is at the approximate scale of 1:310,000 and is printed on two 72-by-127-cm. sheets. De Brahm's map was reproduced in several of the large atlases published in England between 1768 and 1780.

Disputes over the English and French boundaries in northern New Hampshire inspired the preparation of a map of that province. The two collaborators were the Reverend Samuel Langdon and Colonel Joseph Blanchard. The latter commanded the New Hampshire regiment that participated in the Louisbourg expedition of 1745. Langdon was chaplain of the regiment and for his services was awarded land grants in New Hampshire. He then served as pastor of North Church in Portsmouth, New Hampshire, from 1747 to 1774. In 1756 and 1757 Blanchard conducted surveys of the province from which a map was drawn in 1757. A copy of the manuscript map is in the collections of the Library of Congress. The original draft was sent to Jefferys in London, who engraved the map and published it in 1761. It is titled *An Accurate Map of His Majesty's Province of New Hampshire in New England, taken from Actual Surveys of all the inhabited Parts, and from the best information of what is uninhabited together with the adjoining counties which Exhibits the Theatre of this War in that Part of the World, by Col. Blanchard and the Revd. Mr. Langdon* (Fig. 3–2). The imprint is "Portsmouth, N.H., Oct. 1761." Blanchard and Langdon dedicated their map "to the Right Honourable Charles Townshend, His Majesty's Secretary at War." It was reproduced in Jefferys's *A General Topography of North America and the West Indies*, which was published in London in 1768. David A. Cobb, compiler of a comprehensive catalog of New Hampshire maps, states that the Blanchard-Langdon map "is one of the more significant in the state's history."[1]

Langdon went on to become president of Harvard College in 1774, and he held this office until 1780. He was an ardent patriot and on the eve of the battle of Bunker Hill led prayers at Cambridge for the American volunteers. In 1784, Langdon, in collaboration with Abel Sawyer, Jr., issued a revised and updated version of the New Hampshire map, which they dedicated to John Hancock, then governor of Massachusetts. The copies in the Harvard College Library and the Library of Congress are among the rare survivors of this revised map.

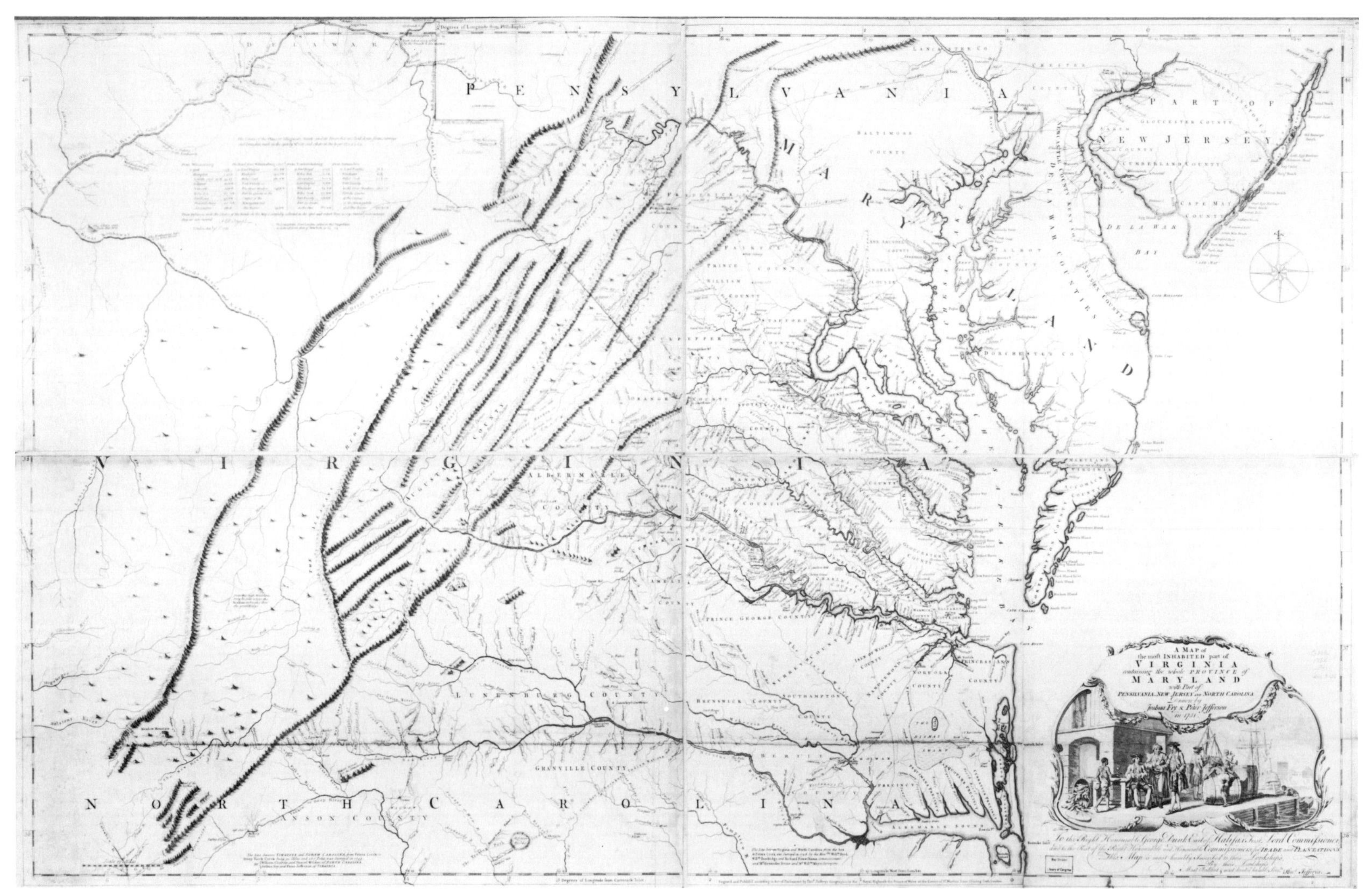

Fig. 3–1. Joshua Fry and Peter Jefferson's *A Map of the most Inhabited part of Virginia* was the most widely used map of Virginia prior to the Revolution. The first edition was probably published in 1753. The 1755 edition is reproduced here.

A map of Connecticut in the colonial period was compiled by Moses Park and published in 1766. The title dedication reads, *To the Right Honourable the Earl of Shelburne His Majesty's Principal Secretary of State for the Southern Department This Plan of the Colony of Connecticut in North America is Humbly Dedicated by his Lordship's Most Obedient Humble Servt. Moses Park*. The map is at the approximate scale of 1:275,000 and is 54 by 74 cm. in size. The place of publication and the names of the engraver or publisher are not given. Little is known about Park except that he was born in Preston, Connecticut, in 1733. Apparently two states of the map were published,

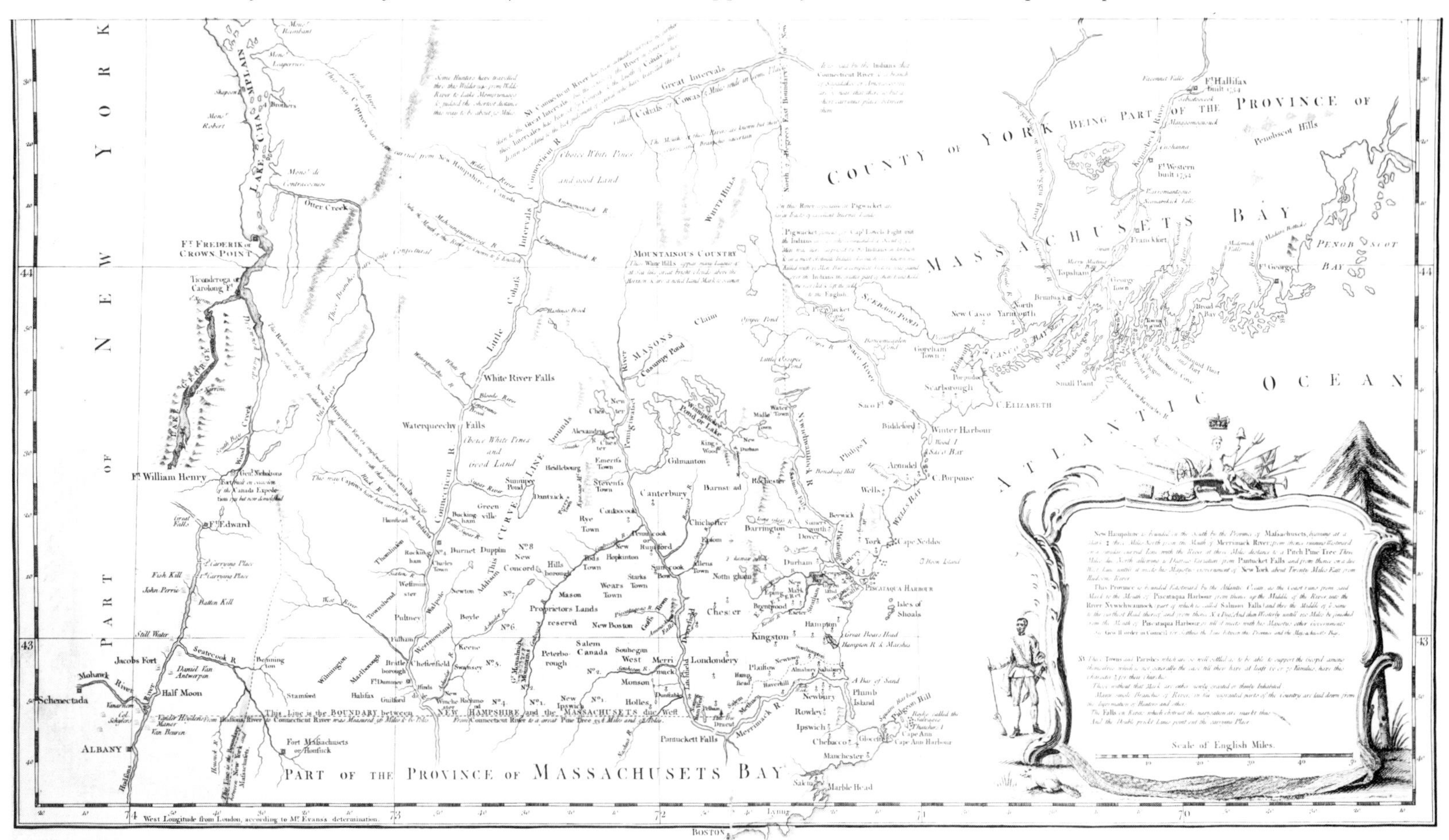

Fig. 3–2. Samuel Langdon and Joseph Blanchard's *Accurate Map of His Majesty's Province of New Hampshire in New England*. Only the southern half of the 1761 map is shown here.

the earlier of which omits latitude and longitude and does not name the Connecticut River. It also spells Shelburne without a final "e." There are also other variations of the second state.

A map of New York and New Jersey was made in 1768 by Samuel Holland at the request of a committee appointed to settle the New York-New Jersey boundary differences. This large map (135 by 52 cm.), entitled *The Provinces of New York and New Jersey with Part of Pensilvania and the Governments of Trois Rivieres and Montreal: Drawn by Capt. Holland. Engraved by Thomas Jefferys, Geographer to his Majesty. Printed for Robt. Sayer in Fleet Street and T. Jefferys in the Strand, 1768*, shows only the eastern two-thirds of the present state of New York. Revised editions of the map were published in 1775 and 1776, and modified versions were published under Governor Thomas Pownall's name in 1776 and 1777. Holland's map was also included in several editions of Jefferys's *American Atlas*.

One of the revised versions of Holland's map was prepared by Claude Joseph Sauthier and titled *A Map of the Province of New-York Reduc'd from the large drawing of that Province, Compiled from Actual Surveys by Order of His Excellency William Tryon Esqr. Captain General & Governor of the same, By Claude Joseph Sauthier; to which is added New-Jersey, from the topographical observations of C. J. Sauthier & B. Ratzer. Engraved by William Faden, (Successor to the late Mr. Thos. Jefferys) 1776*. This map was reproduced in Faden's *North American Atlas*, 1777. Sauthier, a skilled architect and landscape designer, was a native of Strasbourg. Governor William Tryon had brought him to America in 1767 to plan the North Carolina governor's palace and grounds. When Tryon became governor of New York province in 1771, Sauthier accompanied him to New York City. After the capture of New York by the British in 1776, Sauthier prepared a map of the city for General Earl Percy. When Percy returned to England, Sauthier went along as his private secretary.

The province of Pennsylvania was mapped as early as 1687 by Thomas Holme, who was appointed surveyor general by William Penn in 1682. Holme's *Map of the Province of Pennsylvania*, however, includes only a small part of the province, extending westward about thirty miles beyond Philadelphia. It is at the scale of 1:250,000 and thus provides considerable detail, showing plantations, farms, manors, liberty lands, and townships.

Three quarters of a century passed before the next provincial map of Pennsylvania was published. Its compiler was Nicholas Scull, who was born near Philadelphia in 1687. In his early years he was an official involved in negotiations with the Indians, and from 1744 to 1746 he served as sheriff of the city of Philadelphia. He filled the same office in Northampton County from 1753 to 1755. Surveying was an early interest, which was fostered by Scull's travels to the interior and his acquisition of land. Succeeding William Parsons, surveyor general of Pennsylvania province from 1741 to 1748, Scull was commissioned surveyor general June 10, 1748, and filled the post until 1761. As surveyor general, he had access to the surveys and maps prepared by county and local surveyors. Scull drew upon these data in compiling his 1759 map. Dedicated "to the Honourable Thomas Penn and Richard Penn," the map is titled *Map of the Improved Part of the Province of Pennsylvania*. It was engraved by James Turner and printed in Philadelphia by John Davis. The map, which is at the approximate scale of 1:250,000 and measures 75 by 151 cm., extends west to the Allegheny Mountains and beyond Fort Cumberland in north central Maryland. The greatest detail is given for the eastern counties, and mountains are indicated with crude hachures. A marginal note affirms that "the Author can assure the Publick that in laying down this Map, neither Care nor Pains have been wanting to place the several Parts of it as near the truth as is possible." Scull's map was reproduced in Jefferys's *General Topography*, published in 1768.

Scull died in 1762 and was survived by five sons, more than one of whom practiced the surveying profession. His grandson, William Scull, was apparently brought up by his grandfather and learned surveying from him.

In 1770 William published a *Map of the Province of Pennsylvania* (Fig. 3–3), which was also dedicated to Thomas and Richard Penn. Scull drew heavily on his grandfather's map and other manuscript and printed maps. It is at a smaller scale (approximately 1:633,690) than the 1759 map, but extends westward to include practically all of present-day Pennsylvania. The map was engraved by Henry Dawkins and printed in Philadelphia by James Nevil. In September 1776, Scull was commissioned a captain in the Eleventh Pennsylvania Regiment. Twenty-one months later he transferred to the geographer's department under Robert Erskine. Peter J. Guthorn describes five maps that were prepared by Scull during the year or so he worked with Erskine.[2] Sayer and Bennett published a revised version of Scull's *Map of the Province of Pennsylvania* in 1775, and in this form it was reproduced in the several editions of Jefferys's *American Atlas* and Faden's *North American Atlas*.

In 1770, the same year of the first publication of Scull's map of Pennsylvania, Thomas Jefferys published a map of Virginia in London. The *New and Accurate Map of Virginia Wherein most of the Counties are laid down from Actual Surveys* (Fig. 3–4) was intended to be an improvement on the Fry-Jefferson map of Virginia. Unhappily, because of certain inaccuracies on the map and the availability of revisions of the Fry-Jefferson map, the *New and Accurate Map of Virginia* was not a financial success. Its compiler, John Henry, was a native of Scotland who had been educated at King's College, Aberdeen. He immigrated to Virginia in 1727 and in the same year took out a grant of land in Hanover County. In 1731 he married the widow of John Syme, a fellow landowner in the county. During the next thirty-five years Henry acquired additional land in Virginia and fathered two sons and nine daughters. One of his sons, Patrick, became the revered Virginia patriot of the Revolution.

In an effort to increase his income, John Henry petitioned the Virginia House of Burgesses in 1766 for financial assistance for compiling and printing a new map of the colony. Although the proposal was rejected, Henry proceeded with the project. Lacking funds to make original surveys, he compiled most of the data from previously published maps. As an original feature Henry did include county boundaries on his map, as well as the names of landowners who had purchased subscriptions for it. The map was engraved and printed in London by Jefferys at Henry's own expense. It never achieved the popularity enjoyed by the Fry-Jefferson map, and judging from the small number of extant copies, subscriptions and sales of the map were limited.

As discussed in Chapter 1, the mapping of the colony of South Carolina was accomplished by De Brahm in his 1757 *Map of South Carolina and a Part of Georgia* and by Henry Mouzon in his 1775 *An Accurate Map of North and South Carolina*. In the North Carolina Colony, William Churton began collecting map data with the encouragement of Governor William Tryon. After ten years of labor Churton submitted a draft to the governor. Tryon suggested that Churton obtain more accurate surveys of the coastal regions of the colony. While carrying out such surveys, Churton died in December 1767.[3] Meanwhile, John Collet, a native of Switzerland who had served in both the French and English armies, was in 1767 appointed commander of Fort Johnson on Cape Fear, North Carolina. In December of that year Collet prepared a plan of the fort, and in the following year he was appointed aide-de-camp to Governor Tryon. The governor loaned to Collet the survey data that had been assembled by Churton. Collet then made a large manuscript draft of Churton's map, which he carried to England in 1768. This manuscript is preserved in the British Public Record Office.[4] The map was published in London in 1770 by S. Hooper under the title *A Compleat Map of North-Carolina from an actual Survey* (Fig. 3–5). Cumming observes that "not much evidence of new surveys is to be seen, however, on the Collet map." He affirms, nonetheless, that "Collet's is a handsome map, beautifully engraved by I. Bayly, and is the basis for most subsequent maps of North Carolina until many years after the Revolution."[5] Collet dedicated his map

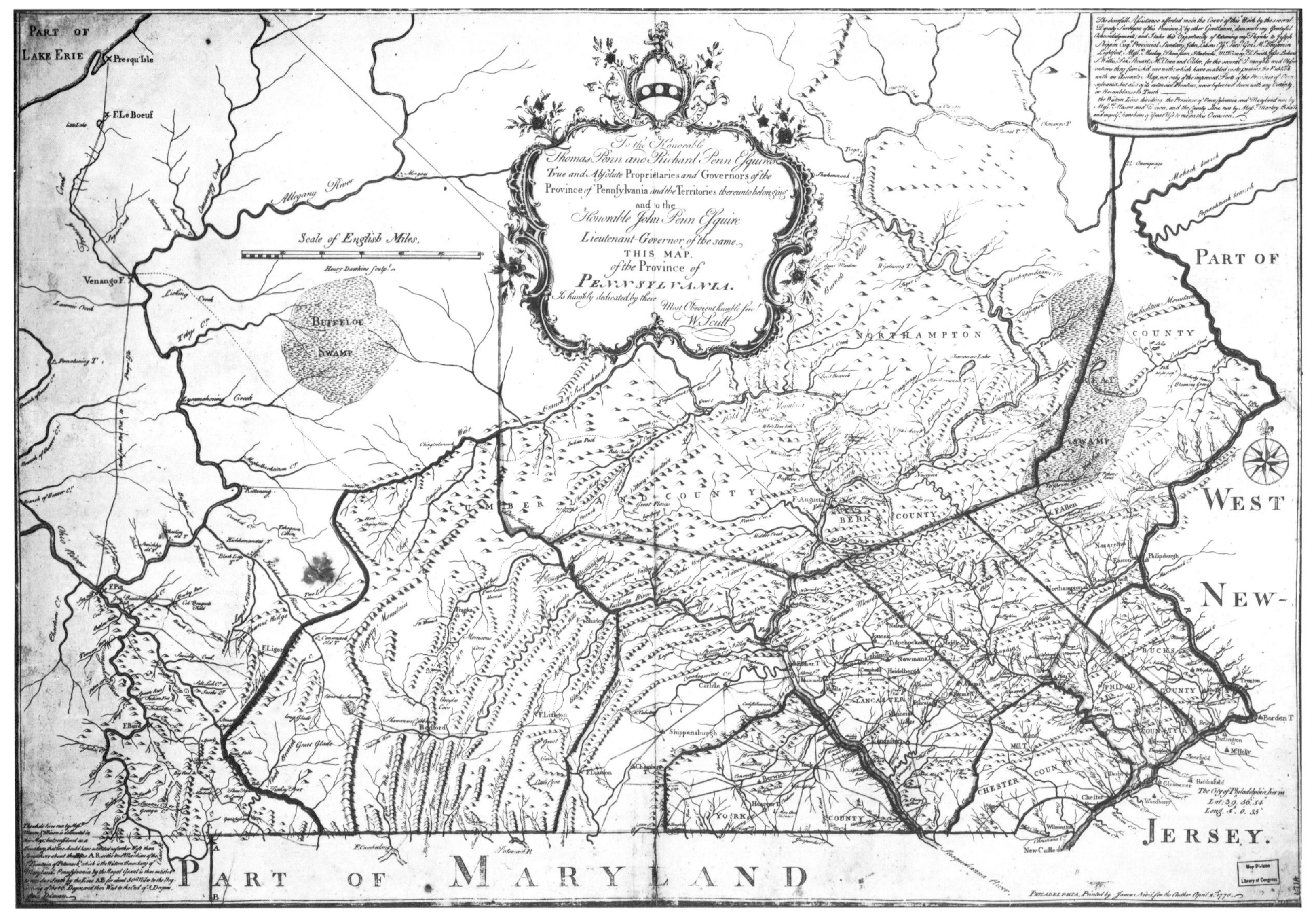

Fig. 3–3. William Scull, grandson of Nicholas Scull, compiled this 1770 *Map of the Province of Pennsylvania* by drawing heavily upon the works of his grandfather.

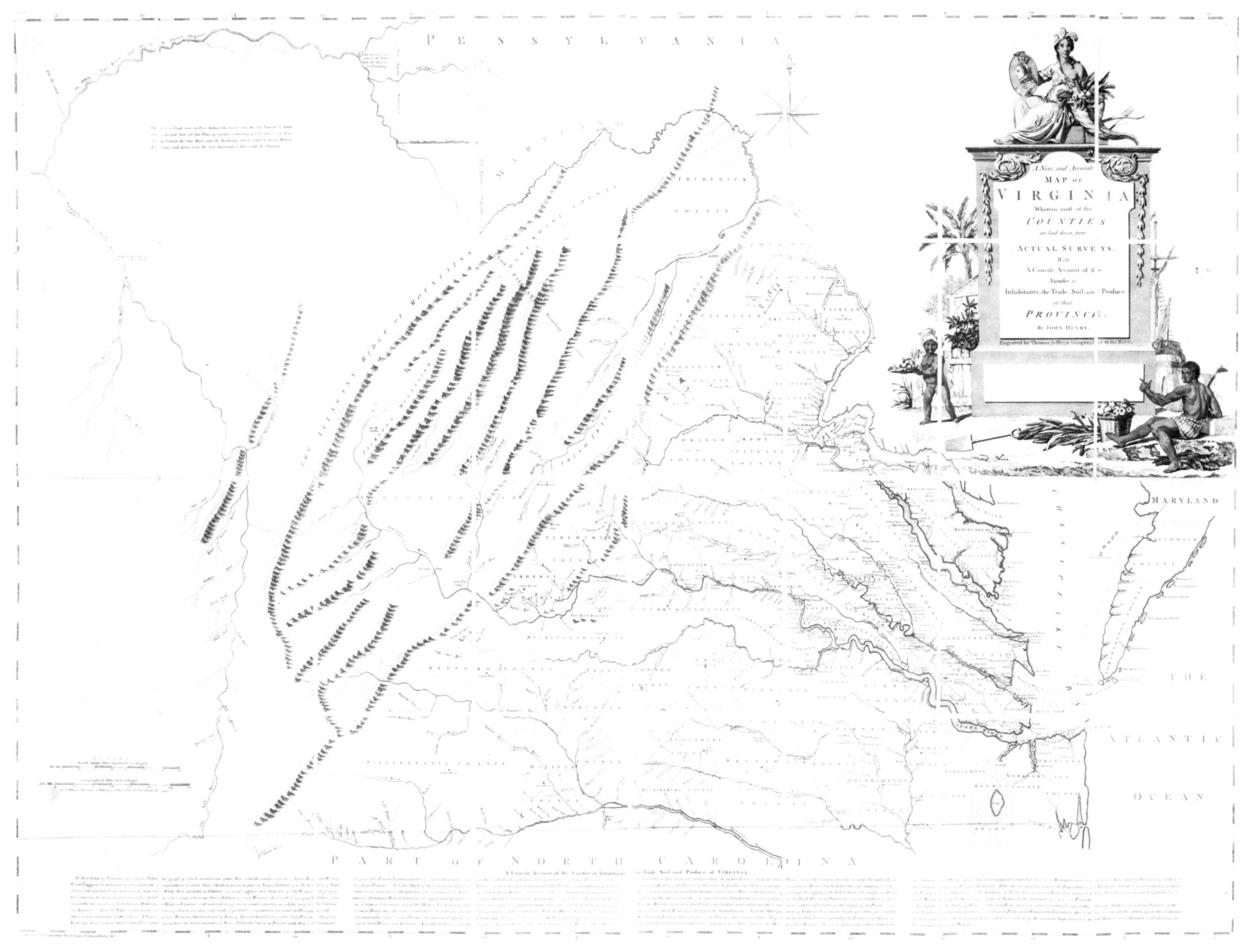

Fig. 3–4. This 1770 *New and Accurate Map of Virginia* by John Henry never achieved the popularity of the Fry-Jefferson map of Virginia.

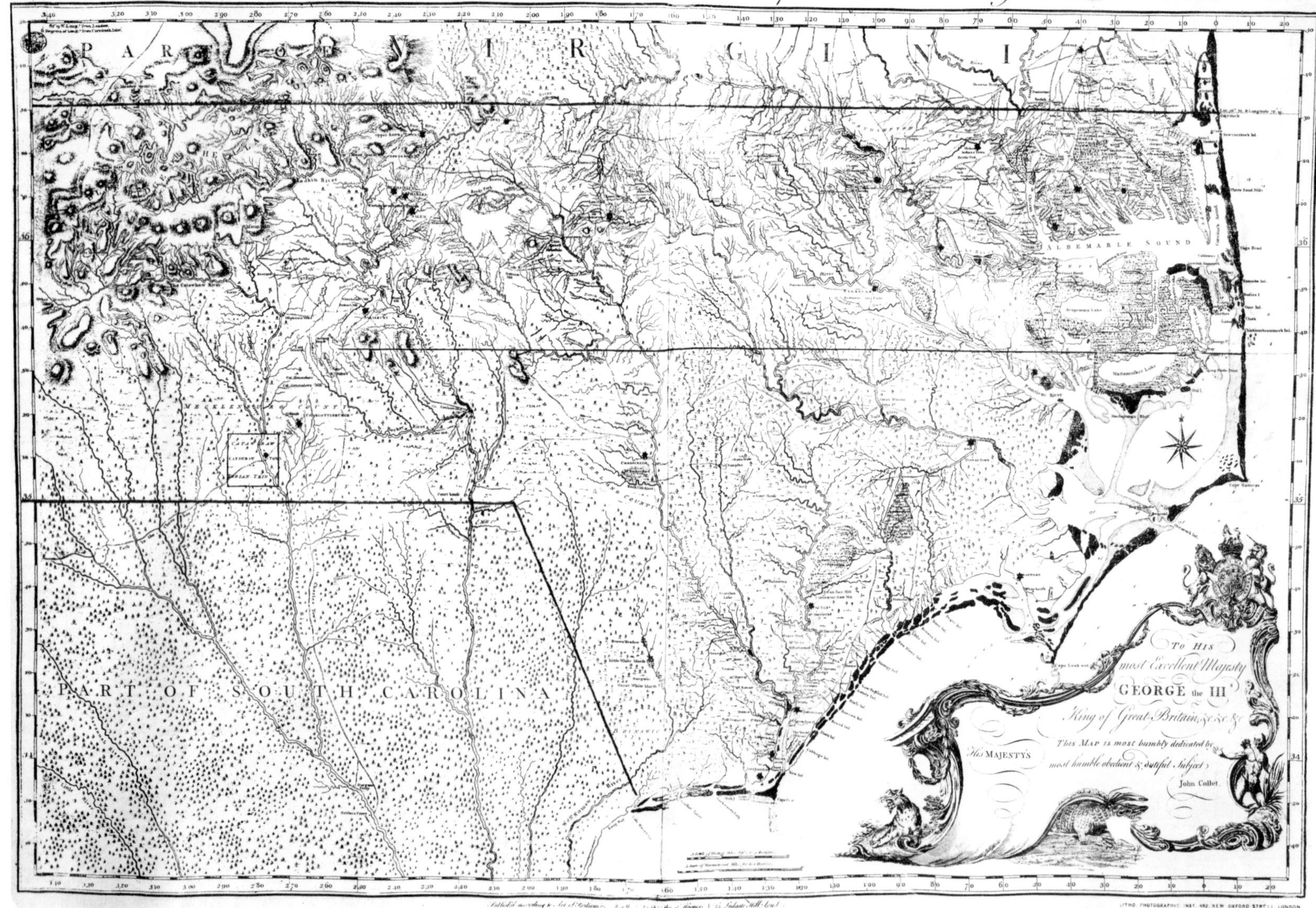

Fig. 3–5. John Collet's 1770 map of North Carolina.

"To His most Excellent Majesty George the III King of Great Britain."

For New Jersey, Faden published in 1777 a map of *The Province of New Jersey, divided into East and West, Commonly called the Jerseys*. It is 78 by 57 cm. in size and at the approximate scale of 1:420,000. Based on surveys by Bernhard Ratzer and Gerard Banker, the map includes Delaware and parts of New York, Pennsylvania, and Maryland. The map was reproduced in Jefferys's *American Atlas*, 1776, and Faden's *North American Atlas*, 1777.

Charles Blaskowitz was the author of a map of Rhode Island called *A Topographical chart of the bay of Narraganset in the province of New England, with all the isles contained therein, among which Rhode Island and Connonicut have been particularly surveyed*. It was engraved and printed in London in 1777 by Faden and inscribed by him "to the Right Honourable Hugh Earl Percy . . . with His Lordship's permission." Although the map does not show all of present-day Rhode Island, it does include Narragansett Bay and the several towns and cities on its islands and along its shores.

Bernard Romans is remembered primarily for his *Concise Natural History of East and West Florida* and its related maps, which were published in 1774. Following service with De Brahm and personal surveys in Georgia and Florida, Romans transferred his activities to New York City in order to secure subscriptions for his publications. Although he was in his late fifties when the revolutionary war broke out, he received a commission in 1776 as captain of a Pennsylvania artillery company. He retained the commission until mid-1778, but he was not happy with it. In the one expedition in which his company was involved he apparently had trouble maintaining discipline. During this stint in the military he seems to have had considerable time for intellectual pursuits, for he compiled two maps of the New England provinces, both of which are today quite rare. On neither of the maps, oddly enough, does Romans's name appear.

The *Boston Gazette and Country Journal* printed on May 19, 1777, "Proposals for Printing A New Map of the State of Connecticut with the Parts of New-York, New-Jersey, and Rhode-Island; collected from the best and latest Surveys." Later that year on October 31, the *Connecticut Gazette and the Universal Intelligencer* advertised, "Just published, and to be sold by T. Green, Price Two Dollars, Roman's [*sic*] Map of the State of Connecticut, with the Parts of New-York, New-Jersey, and Islands adjacent, that has been the late Seat of War" (Fig. 3–6). The map is at the approximate scale of 1:380,000. It is not known why Romans chose to publish it anonymously.

His second map, the *Chorographical Map of the Northern Department of North America*, was engraved, printed, and sold at New Haven in 1778. The July 6, 1778, issue of the *Boston Gazette and Country Journal* reported, "Just come to Hand, and are ready to be delivered to the Subscribers, A New Map of the Northern Department, containing the Country from Red Hook, on the North-River, (the place where the late map of Connecticut left off) to Three-Rivers, in Canada, and from the Heads of Merrimack River, in New Hampshire, to the Heads of Delaware and Susquehannah, westward, including Fort-Schuyler, and the Oneyda Lake, with part of Ontario." The map is oriented with north to the left of the sheet. The most prominent feature is the province of Vermont, which fills the top central portion of the map. One of the most interesting features of the map is its use of the name Vermont. The name was officially adopted on June 30, 1777, six months or so before Romans's map was published.

Samuel Holland produced a map of the province of New Hampshire late in the revolutionary war. The map, though, was not published until 1784, the year after the treaty of peace was signed. It is titled *A Topographical Map of the Province of New Hampshire, surveyed agreeably to the orders and instructions of the Right Honourable the Lords Commissioners for Trade and Plantations: unto Samuel Holland, Esqr., His Majesty's Surveyor General of Lands for the Northern District of North America; by the following gentlemen his deputies: Thomas Wright, George Sproule, James*

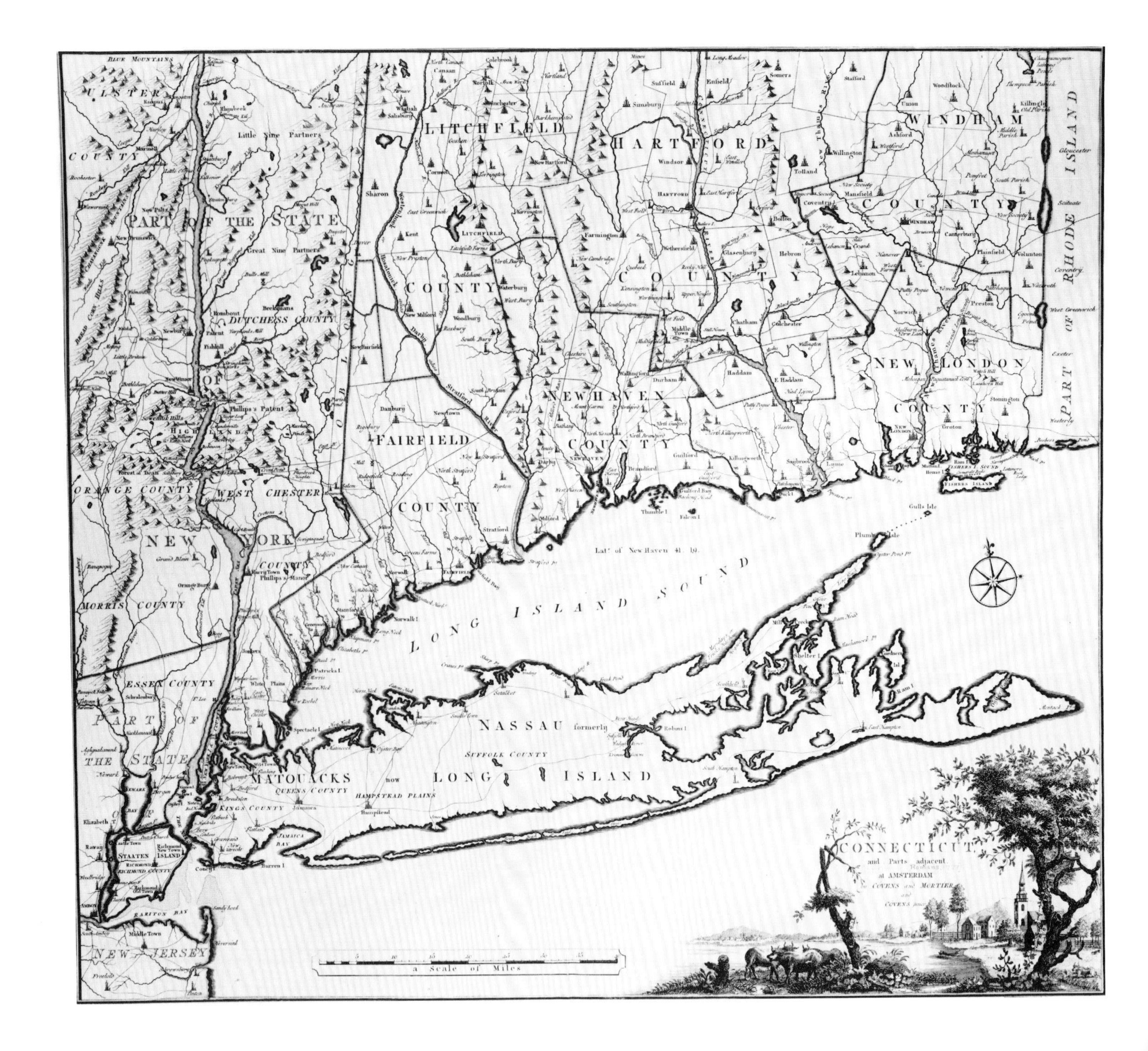
CONNECTICUT,
and Parts adjacent
at AMSTERDAM
by COVENS and MORTIER
and
COVENS junior
a Scale of Miles
LITCHFIELD COUNTY
HARTFORD COUNTY
WINDHAM COUNTY
NEWHAVEN COUNTY
NEW LONDON COUNTY
FAIRFIELD COUNTY
PART OF THE STATE OF NEW YORK
DUTCHESS COUNTY
ULSTER COUNTY
ORANGE COUNTY
WEST CHESTER COUNTY
MORRIS COUNTY
ESSEX COUNTY
PART OF THE STATE OF NEW JERSEY
STAATEN ISLAND
RICHMOND COUNTY
KINGS COUNTY
QUEENS COUNTY
MATOUACKS now LONG ISLAND
NASSAU formerly
SUFFOLK COUNTY
HAMPSTEAD PLAINS
LONG ISLAND SOUND
PART OF RHODE ISLAND
FISHERS ISLAND
Lat. of New Haven 41. 19.
BLUE MOUNTAINS
RARITON BAY
JAMAICA BAY

Grant, Thomas Wheeler, Charles Blaskowitz. London: Printed for William Faden, 1784. The map had apparently been completed some four years earlier, for there is a manuscript draft, believed to have been made in 1780, in the University of Michigan's William L. Clements Library. Holland's map, which is 118 by 80 cm. and at the approximate scale of 1:260,000 replaced the Blanchard-Langdon map of the province.

In the years immediately following the Revolution there was widespread migration from the eastern seaboard states to the lands beyond the Appalachian Mountains. Some of these people went to claim land that had been promised to veterans of the war, and some went to exchange treasury warrants for land. One migrant, Pennsylvanian John Filson (born ca. 1747), seems not to have served in the Continental army, but he had apparently acquired a supply of paper currency issued by Virginia during the war. As the young state was offering western lands to redeem its depreciated paper money, Filson moved to Kentucky in 1782 or 1783 and acquired more than twelve thousand acres of land. He settled in Lexington and occupied his time teaching school and assembling information about the territory. He presented the information he had collected in a book and a map, both of which were published in 1784. The former, entitled *The Discovery, Settlement, and Present State of Kentucke*, was published in Wilmington, Delaware, by James Adams. The map was engraved by Henry D. Pursell and printed by T. Rook in Philadelphia. Its title is *This Map of Kentucke, Drawn from actual Observations, is inscribed with the most perfect respect to the Honorable the Congress of the United States of America; and to his Excelly. George Washington late Commander in Chief of their Army. By their Humble Servant, John Filson* (Fig. 3–7). A simulated scroll at the top of the map reads: "While this Work shall live, Let this Inscription remain a Monument of the Gratitude of the Author, to Col. Danl. Boon, Levi Todd, & Jas. Harrod, Capt. Christr. Greenhoop, Ino. Cowan, & Wm. Kennedy Esqrs. of Kentucke: for the distinguish'd Assistance, with which they have honor'd him, in its Composition: & a testimony, that it

Fig. 3–6. This unsigned map of Connecticut was compiled by Bernard Romans and published in 1777 by Covens & Mortier of Amsterdam.

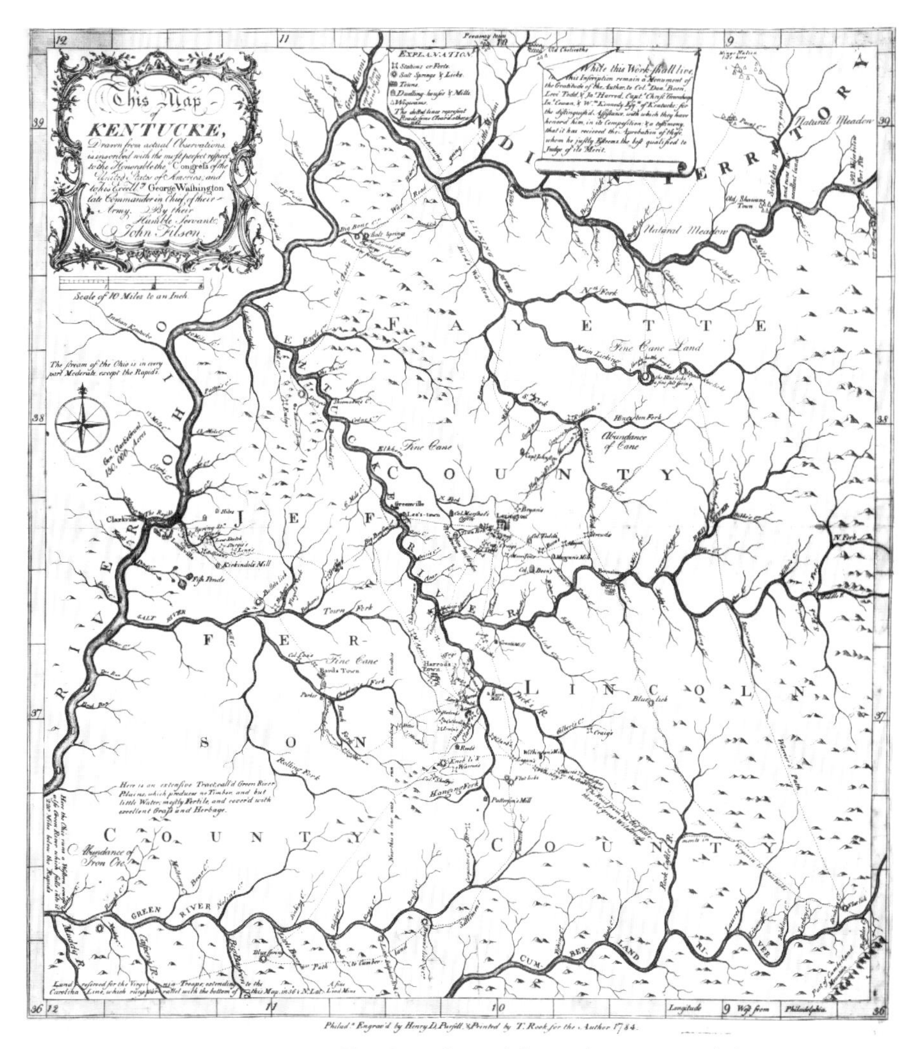

Fig. 3–7. One of the earliest maps of the trans-Appalachian country was this 1784 map of Kentucky by John Filson.

has received the Aprobation of those, whom he justly Esteems, the best qualified to Judge of its Merits."

Filson's map includes the three original counties of Kentucky: Fayette, Jefferson, and Lincoln. Because it was not based on original surveys, the map is inaccurately delineated. It is, however, minutely detailed and locates stations, salt licks, trails, forts, settlements, and other significant landmarks. The map also locates and names the major streams with many of their principal tributaries. Historian Willard Jillson noted that even "with its many errors and irregularities, this map of 1784 is by far the most usable map that had been prepared up to this time for any part of the Western interior of the United States."[6]

Filson died in October 1788 in southern Ohio while engaged in conducting surveys. It is not known whether he was killed by Indians or wild animals, for his remains were never found.

Notes

1. David A. Cobb, *New Hampshire Maps to 1900, An Annotated Checklist* (Hanover, N.H., 1981), xiv.
2. Peter J. Guthorn, *American Maps and Map Makers of the Revolution* (Monmouth Beach, N.J., 1966), 31–32.
3. William P. Cumming, *The Southeast in Early Maps* (Chapel Hill, N.C., 1962), 56.
4. Ibid., 57.
5. Ibid., 58.
6. Willard Rouse Jillson, *Pioneer Kentucky* (Frankfort, Ky., 1934), 28.

4. *The First Maps of the United States of America*

The Revolution materially affected American cartography. Political independence forced the young United States to develop its own mapping program, although for some years this remained largely dependent upon European technology, instruments, and personnel. The immediate and urgent response to the end of the war was a demand for maps of the new republic by Europeans as well as Americans. The most expeditious way to meet these demands was to revise and update maps of the former English colonies, or to use existing maps as source material in compiling new ones. The majority of such maps published during the two decades after the Declaration of Independence, therefore, offered little that was new, geographically or cartographically. The published maps did, however, acquaint an eager world with the general limits and boundaries of the new nation and its thirteen states and, in their titles, acknowledge the legitimacy of the infant American republic.

Delegates to the Second Continental Congress, assembled in Philadelphia on July 4, 1776, signed the Declaration of Independence, which united thirteen English colonies into one independent republic. The war was to continue for another seven years, and not until 1788 was the nation's Constitution ratified. The Declaration of Independence included no formal proposal for naming the new political entity, but the document was titled "The Unanimous Declaration of the Thirteen United States of America." Moreover, the Declaration concluded with the statement that "We, therefore, the Representatives of the United States of America, in General Congress, Assembled, appealing to the Supreme Judge of the world for the rectitude of our intentions, do, in the Name, and by the Authority of the good People of these Colonies, solemnly publish and declare, that these United Colonies are, and of Right ought to be Free and Independent States."

In the Articles of Confederation, drawn up late in 1777, the members of Congress agreed that "The United States of America" would be the name for the republic they had fathered a year and a half before. Two years after its original Declaration, the Continental Congress resolved on July 11, 1778, that "United States of America" would be used on its bills of exchange, thus confirming the official and legal status of the name. Less than a year later, on June 14, 1777, Congress adopted an official banner, agreeing that "the flag of the United States shall be thirteen stripes, alternate red and white, with a union of thirteen stars of white on a blue field, representing a new constellation."

In the years immediately following the signing of the Declaration of Independence, no official action was taken toward producing or adopting an official map of the American republic. The Congress was apparently deterred because of the anticipated heavy financial outlay preparing an official map might entail. De Witt did propose in June 1783 that the maps prepared by him and his staff be published, but this proposal was rejected.

Because France was the principal ally of the young United States, it was not strange that the first cartographic recognition of American independence was on a map published in Paris in 1778. Titled *Carte du Théatre de la Guerre Actuel Entre les Anglais et les Treize Colonies Unies de l'Amerique Septentrionale* (Fig. 4–1), the map is credited to J. B. Éliot, who is identified as "Ingénieurs des Etats Unis" ("United States engineers"). Another edition of the map, with the same title and date, has a

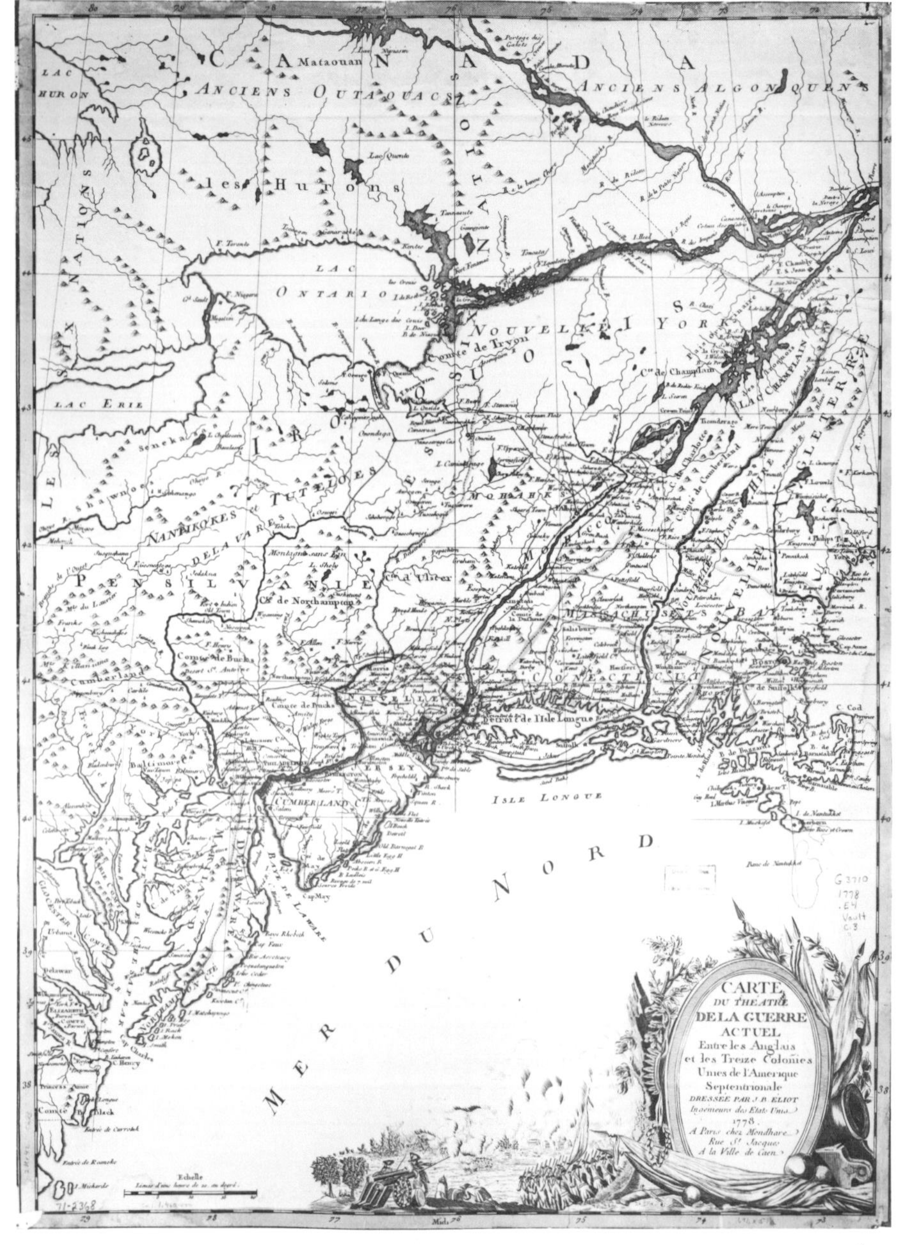

Fig. 4–1. This 1778 map by Frenchman J. B. Éliot was the earliest to include in its title "les Treize Colonies Unies de l'Amerique" ("the Thirteen United Colonies of America").

slightly different orientation, extending northward to include a larger segment of Maine and curtailing the southern extremity. This map appears to have been copied from Louis Brion de la Tour's *Carte du théâtre de la guerre entre les Anglais et les Américains*, which was published in Paris in 1777, 1778, 1779, and 1782.

A 1781 edition of Éliot's map, which duplicates the outlines of the first 1778 edition, identifies Éliot as "aide de camp du Général Washington." This map includes lists of battles and military operations up to 1781. The three versions of Éliot's map embrace only the New England and Middle Atlantic states, extending south to just beyond Chesapeake Bay. Although they do not show the entire United States of the period, the Éliot maps of 1778 and 1781 can claim priority in using the name of the new nation, in its French translation, in their title cartouches. In 1783 Mondhare of Paris published another map by Éliot which did include all of the original thirteen states as well as the lands west of the Appalachians extending beyond the Mississippi River. Its title is *Carte générale des Etats Unis de l'Amérique Septentrionale, avec les limites de chacun des dits Etats*. Although Éliot was identified as "ingénieurs des Etats Unis" and "aide de camp du Général Washington" on two of his maps, there is no evidence that he served in the Continental army. Considering that his several maps were published in France, Éliot may have served as a liaison officer from the French army in America to General Washington. The Bibliothèque Nationale Département des Cartes et Plans, however, has no proof of a French connection.

The first use of the new republic's name on an English map was by Carrington Bowles on his *New Map of North America and the West Indies, exhibiting the British Empire therein with the limits and boundaries of the United States*. It was published January 2, 1783, almost three weeks before the preliminary peace treaty was signed. The ancestry of Bowles's map may be traced to a mid-eighteenth-century map by Jean Baptiste d'Anville, via intermediate versions by Thomas Pownall and Emanuel Bowen. Bowles also published on February 4, 1783, a

New Pocket map of the United States, with the British possessions of Canada . . . according to the preliminary Articles of Peace, signed at Versailles the 20th Jany. 1783. This map, much smaller than the first, was probably an original Bowles compilation. It was later reprinted with some modifications and a decorative title cartouche. Just five days after the *New Pocket map* was issued, Sayer and Bennett published *The United States of America with the British possessions . . . according to the preliminary Articles of Peace signed at Versailles the 20th of Jany. 1783.* Recognized as the third English map of the United States, it was derived from an earlier Sayer and Bennett map that was also based on one by d'Anville. Some two months later, on April 3, 1783, John Wallis published in London a map of *The United States of America laid down From the best Authorities, Agreeable to the Peace of 1783* (Fig. 4–2). The fourth English map of the young republic, Wallis's map is of particular interest because its title cartouche includes an illustration of the U.S. flag. These maps by Wallis, Bowles (*New Pocket map*), and Sayer and Bennett are about the same size and have many common features. The final treaty of peace was signed September 3, 1783, and within the next several years a number of maps of the United States were issued by other English publishers, among them H. D. Symonds, William Darton, John Cary, Faden, Thomas Kitchin, and Thomas Bowen.

While Éliot was the first Frenchman to acknowledge American independence on his maps, Jean Lattré gets the credit for publishing the first French map of the official, newly created United States of America. Lattré, a well-known artist, cartographer, and engraver, published his attractive *Carte des Etats-Unis de l'Amerique Suivant le Traité de Paix de 1783* (Fig. 4–3) in Paris in June 1784. The cartographer dedicated the map to "His Excellence, Mr. Benjamin Franklin," the principal U.S. representative in the peace negotiations, who was well known in Parisian diplomatic and social circles. Like most contemporary map publishers, Lattré borrowed generously from previously published works, and he seems especially to have favored John Mitchell's 1755 *Map of the British and French Dominions of North America.* Lattré's *Carte des Etats-Unis* is one of the most attractively designed and executed maps of the period and reflects the talent and skill of the artist-cartographer, who was in his seventies when the map was published.

Also published at Paris in 1784 (the month and day are not designated on the map) was Brion de la Tour's *Carte des Etats-Unis D'Amérique . . . avec Les Nouvelles Limites Générales fixées par les articles preliminaires de paix . . . et confirmées par le Traité definitif du 3 7bre 1783.* Identifying Brion de la Tour as "Ingr. Géographe du Roi" ("geographic engineer to the king"), the map measures 52 by 70 cm. Like many other maps of this time, it appears to be based primarily on the Mitchell map.

After the signing of the peace treaty, most of the established French cartographers, like their British confreres, revised previously issued maps to include updated information and the official name of the infant republic. Notable in this area is the work of Rigobert Bonne and Didier Robert de Vaugondy. Few German geographers and cartographers had extensive contacts with the English colonies before the Revolution. Maps of the new nation published in Germany, therefore, were based upon earlier English or French publications. In this category is *Carte über die XIII vereinigte Staaten von Nord-Amerika, Entworfen durch F. L. Güssefeld und herausgegeben von den Homaennischen Erben, 1784.* All four copies of this map in the Library of Congress include, above the top neat line, this statement: "Les XIII Etats unis de l'Amérique Septentrionale, d'après les meilleuris & Speciales Cartes angloises qui ont parues jusqu'ici par F. L. Güssefeld" ("the thirteen United States of North America, by F. L. Güssefeld, according to the best and particular English maps published to date"). Franz Ludwig Güssefeld of Nuremberg was a prolific map producer of the late eighteenth and early nineteenth centuries. He prepared road and postal maps of a number of German provinces that were also published by the heirs of Johann Baptist Homann, an eighteenth-century Nuremberg map publisher. It is possible that the publishing house commissioned Güssefeld to compile the map of

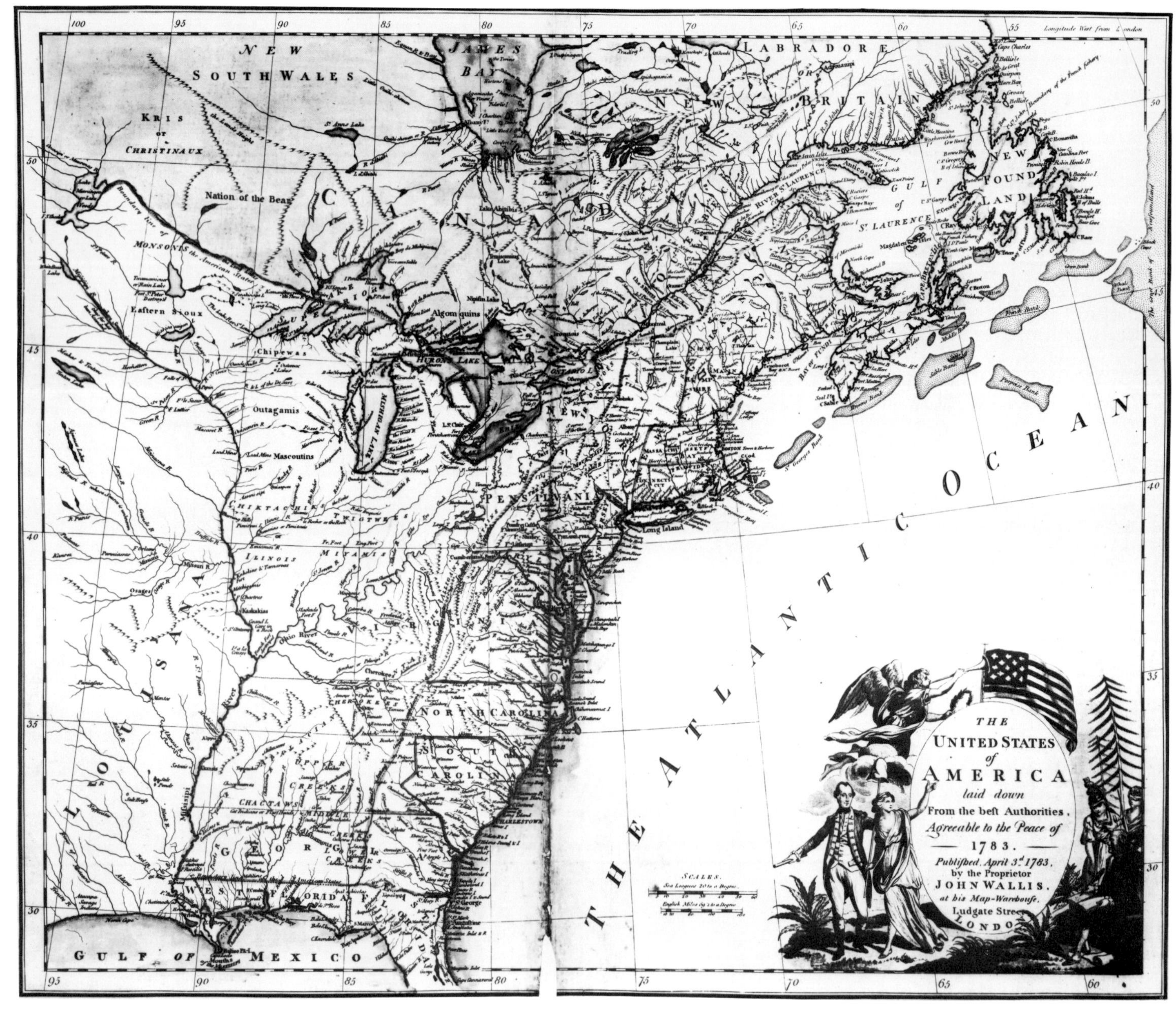

Fig. 4–2. John Wallis's map of the United States of America is notable for incorporating the flag of the new republic in its title cartouche.

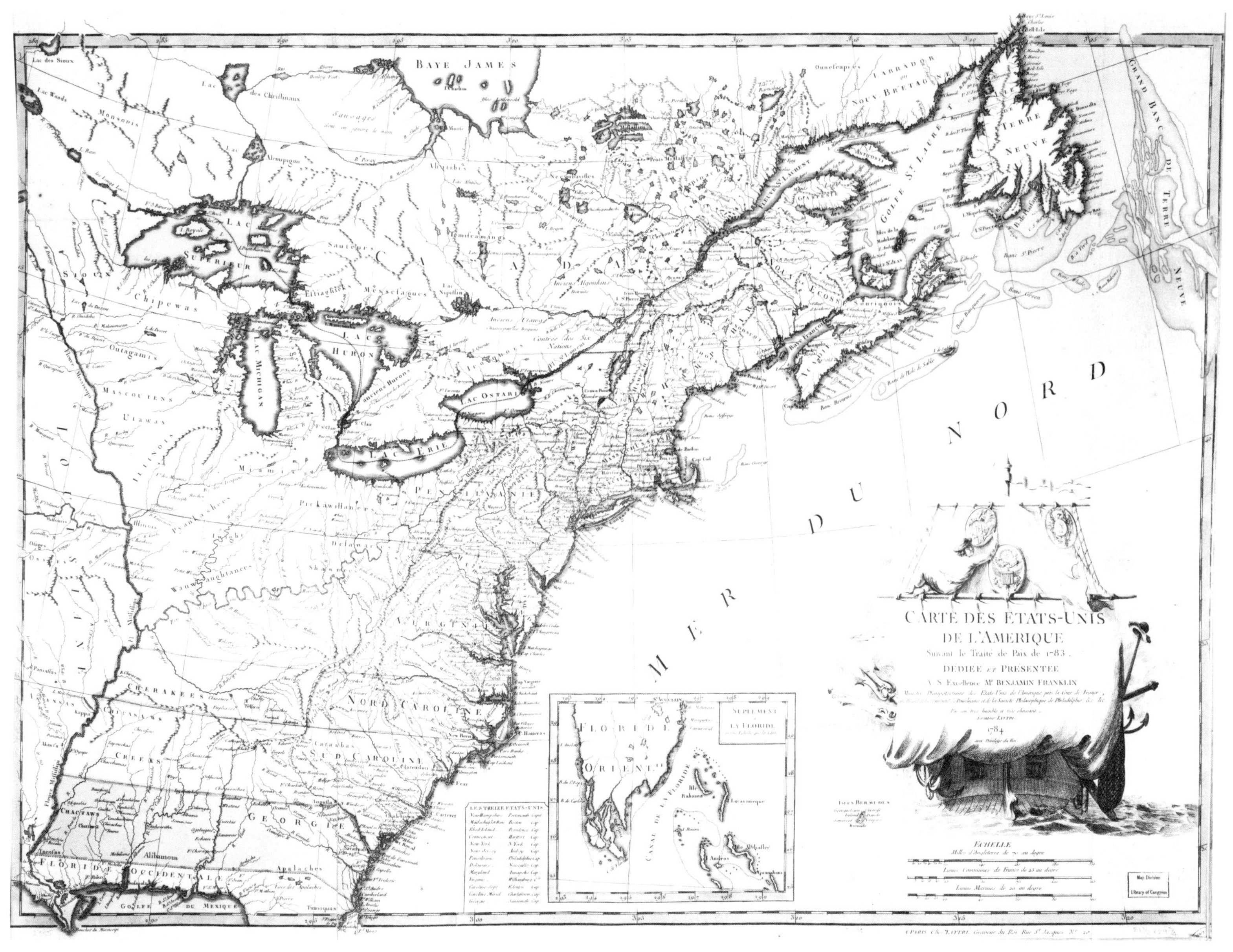

Fig. 4–3. One of the most attractive of the early maps of the United States is Jean Lattré's 1784 *Carte des Etats-Unis de l'Amerique*.

the United States, for which he drew upon existing Homann maps as well as works by French and English publishers.

Maps compiled by Americans prior to the Revolution were with few exceptions engraved and published in England. The war and the Declaration of Independence, as noted, accelerated demands for maps of the country. The first native response to these American demands was a proposal by William McMurray in the *Pennsylvania Packet* on August 9, 1783, to publish by subscription a map of the United States. More than sixteen months elapsed, however, before McMurray's map was ready and so the first American map of the young republic was published by Abel Buell. The March 31, 1784, issue of the *Connecticut Journal* noted that Buell's *New and Correct Map of the United States of America* was published and ready for subscribers. The announcement emphasized that "this Map is the effect of the compiler's long and unwearied application, diligence and industry, and as perfection has long been the great object of his labours, and it being the first ever compiled, engraved, and finished by one man, and an American, he flatters himself, that every patriotic gentleman, and lover of geographical knowledge, will not hesitate to encourage the improvement of his own country."

Buell's map is one of the largest maps discussed in this chapter. It was printed on four large sheets which, when joined, form a wall map 108 by 123 cm. in size. Because of the high attrition rate of large maps, there are fewer than half a dozen recorded extant copies of Buell's map. Two editions of the map were issued. On the first edition, the only extant copy of which is in the New York Public Library, the following imprint is lacking below Buell's name in the cartouche: "Newhaven Published According to Act of Assembly." This imprint does appear, however, on the second edition.

Buell, a native of Killingworth, Connecticut, was a jack of all trades. During his long life (1742–1822) he was occupied, with varying degrees of success, as a silversmith, jeweler and lapidary, type founder and caster, copper engraver, marble quarryman, auctioneer, privateer, and cotton miller. Early in his career he was convicted and imprisoned for counterfeiting, and shortly before the revolutionary war he was forced to flee to Florida to escape persistent creditors. Two other Buell cartographic engravings have been recorded: Moses Park's 1766 map of Connecticut and a 1774 chart of Saybrook bar at the mouth of the Connecticut River. It is also possible that during his residence in the South Buell assisted Bernard Romans in engraving charts of the Florida coast.

In compiling his large map of the United States, Buell drew freely upon the works of earlier American mapmakers, particularly Mitchell, Evans, and Hutchins. He derived information from Hutchins's *A Topographical Description of Virginia, Pennsylvania, Maryland, and North Carolina* (London, 1778) as well as from *A New Map of the Western Parts of Virginia, Maryland and North Carolina*, which accompanied Hutchins's book. Buell relied heavily upon Hutchins for data on the trans-Allegheny region. Longitude on Buell's map is reckoned from Philadelphia, then the capital city of the United States. The decorative cartouche in the lower right corner of the map is rich in symbolism. At the right Liberty is seated in an armchair beneath a tree, which provides shade from the bright rays of the sun, which is positioned above the cartouche. She holds in her right hand the scepter of liberty, on the top of which rests her cap. A small globe, with America in the forefront, is held in her left hand. At her feet is a scroll which proclaims, "Independence July IV MDCCLXXVI." In the upper left of the cartouche, a trumpeting angel salutes the flag of the new republic, beneath which the arms of the state of Connecticut are displayed by a cherub.

When McMurray's map of the United States was finally issued more than sixteen months after his initial announcement, it was entitled *The United States According to the Definitive Treaty of Peace Signed at Paris, Septr. 3d 1783* (Fig. 4–4). Following McMurray's name on the map is the designation, "late Asst. Geogr. to the U.S.,"

Fig. 4–4. William McMurray published this map of the United States in 1784. It ranks second to Abel Buell's map as being among the earliest cartographic representations of the Republic compiled by native Americans.

as McMurray had been one of Erskine and De Witt's assistants. The only extant military map credited to him is a road map of a portion of the Hudson highlands in New York State in the Erskine-De Witt series. McMurray's military service was primarily in Pennsylvania and New york, and his sole wartime cartographic contribution was made in September 1780.

Following the title on his map of the United States, McMurray gives the following data concerning the sources he drew upon: "In forming this Map, late particular Maps & Charts, were of great use; but what contributed most to its accuracy, were the unpublished Surveys, consisting of many thousand of Miles, forming a connection thro' the whole except what lies N.W. of the double dotted line which Surveys with many good Sketches, have been furnished for this usefull work, by the Geogrs. and other curious Gentlemen, who for particular purposes, had made large collections of them. This Map may therefore be said, to be composed from every Survey, and Sketch of Note, heretofore made in these Limits, conformed to the best Astronomical Observations." In his original proposal published in the *Pennsylvania Packet*, McMurray had stated that "it may not be amiss to add, that it was entirely out of the power of any person in Europe to come at the materials which compose the principal amendment in the above proposed Map, as they were only in the hands of the Geographer to the United States, and a few individuals, whose business it was, or they made it such, not only to make, but to collect all the Surveys, Plans and Observations, in their power: and who have been generous enough to give the Author of the proposed Map all their assistance, for which he returns them many thanks."

Notwithstanding these claims of using original source material, McMurray certainly incorporated data from previously published maps. Because he did utilize the official maps prepared for the Continental army, Philip Lee Phillips suggested that his map "may be called the first official map of the United States."[1] Although somewhat smaller than Buell's map, McMurray's map was also designed for wall display; it measures 67 by 98 cm. and is subject to the exposure and hazards of all wall maps. The number of extant copies of McMurray's map is, likewise, small. McMurray, like Buell, drew his prime meridian through Philadelphia. McMurray's map, too, had a relatively brief period of importance before it was superceded by other, more authoritative publications. It was, however, one of the principal sources utilized by John Fitch in compiling his map *North West Parts of the United States*, which was published in 1785. With proceeds from the sale of his map, Fitch financed experiments which resulted in his invention of the steamboat.

Osgood Carleton published, at Boston in 1791, a large map of *The United States of America laid down From the best Authorities Agreeable to the Peace of 1783* (Fig. 4–5), which was engraved by John Norman. The map is not dated, but the March 21, 1791, issue of the *Boston Gazette* carried this announcement: "Just published (and selling by J. Norman, at his Office, No. 75 Newbury-street, opposite the sign of the Lamb) A large six sheet Map of the United States of America; The Boundary Line laid down agreeable to the Peace of 1783. —Containing the Ohio Country, the Genesee Lands, Kenebec River &c. from actual surveys, with a Variety of other information." The Carleton-Norman map, also in wall-map format, is 80 by 100 cm. Longitude is reckoned from the London prime meridian.

Carleton was born in New Hampshire in 1742. Prior to 1760 he served briefly in the British army, where he acquired the fundamentals of engineering and surveying. For a short time he was official surveyor for the province of New Hampshire. He joined the Continental army in 1775, fought in the battle of Bunker Hill, and participated in various other campaigns until he was discharged, on disability, in December 1778. For a quarter of a century following the war, Carleton conducted a mathematical, cartographical, and navigational school in Boston. He compiled maps of Boston, Massachusetts, and Maine, as well as his map of the United States.

John Norman was born around 1748 in London,

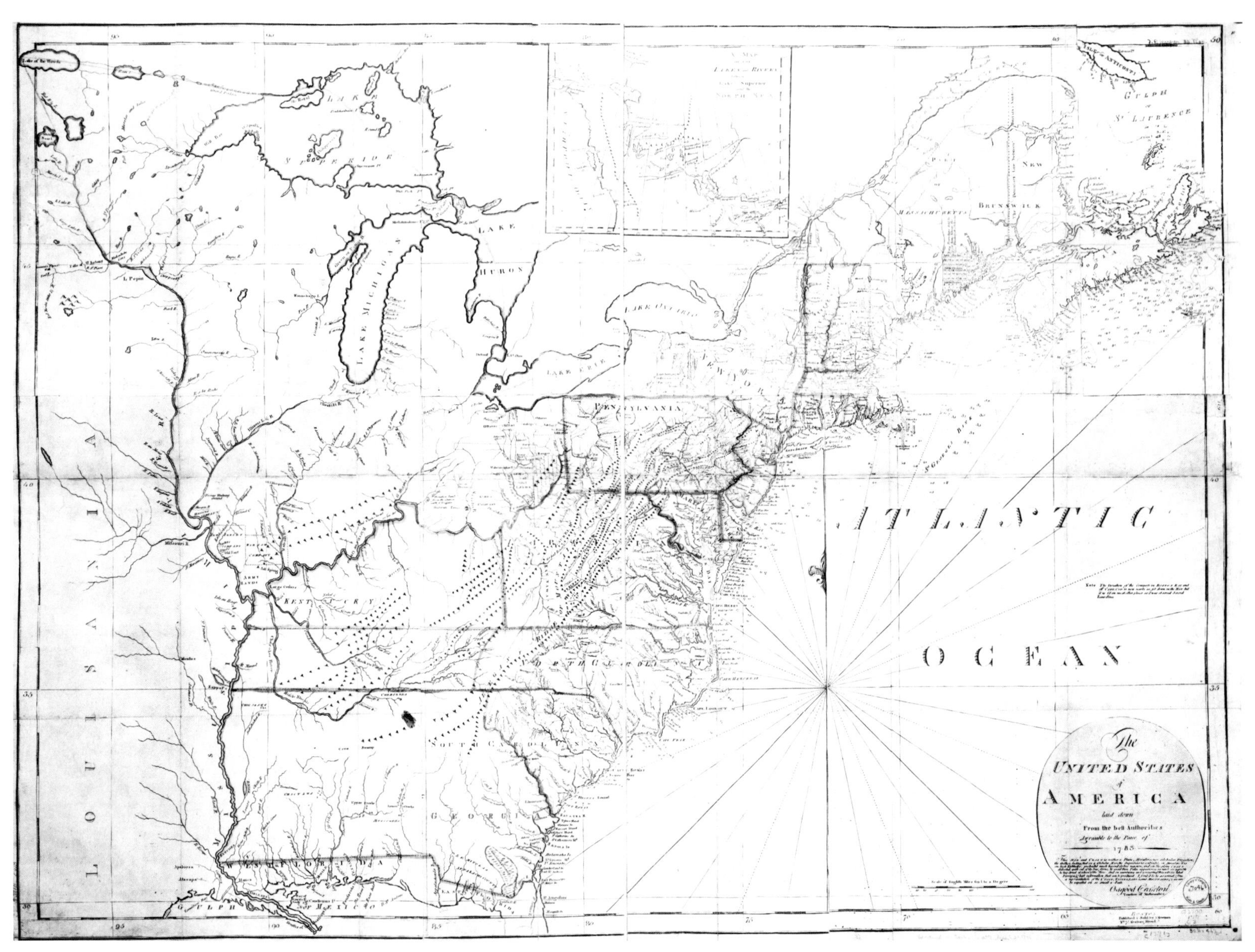

Fig. 4–5. Osgood Carleton's map of the United States was published in 1791.

where he apparently had some training as an architect, landscape engineer, and engraver. By 1774 he was in Philadelphia, advertising for work in these specializations and conducting classes in drawing. While still in Philadelphia, Norman published *A Map of the Present Seat of War, with the Harbour of New York and Perth-Amboy*. Around 1780 he moved to Boston, where he was engaged in engraving, printing, and book publishing.

The Carleton-Norman map of the United States, like others of the period, borrowed from previously published works. It includes, however, such new data as a township grid in eastern Maine, the rectangular pattern of the Gorham and Phelps lands in west central New York, and the seven ranges in southeastern Ohio, northwest of the Ohio River. Numerous soundings are given along the coast, especially off New England. Like the Buell and McMurray maps, the Carleton-Norman map extends westward beyond the Mississippi.

Fig. 4–6. Abraham Bradley, Jr., assistant postmaster general, published this semi-official map of the United States in 1796.

In 1796 Abraham Bradley, Jr., published a *Map of the United States, Exhibiting the Post-Roads, the situations, connections & distances of the Post-Offices Stage Roads, Counties, Ports of Entry and Delivery for Foreign Vessels, and the Principal Rivers* (Fig. 4–6). Because Bradley held the office of assistant postmaster general of the United States, his map had strong claim to official status. The map was largely based on new information, which Bradley no doubt obtained from postmasters in various parts of the country. It became the official map of the Post Office Department around 1825, and editions were issued as late as 1829. Bradley's map differs significantly from those published earlier in that it was not copied in whole or in part from other cartographic works. The map was printed on four sheets which, when joined, form a large wall map 86 by 95 cm. in size.

Some months prior to publication of the complete map, the northeast sheet was separately issued. Its title is *Map of the United States Exhibiting Post Roads & Distances By Abraham Bradley Junr. The first Sheet comprehending the Nine Northern States, with parts of Virginia and the Territory North of Ohio*. This sheet is not the same as the one that eventually formed part of the complete map. The registration notice on its verso reads: "No. 132 Title Page of Abm. Bradley, Junr. Map of the United States. Deposited 25th Aprl. 1796." The notice on the complete map reads: "No. 154 Title Page of Map of the United States. Deposited by Abm. Bradley junr. as Author, Sept. 26th 1796." These two Bradley maps are among the earliest maps to be registered for copyright in the United States. The northeast sheet which forms part of the complete map includes a number of changes, such as the addition of land grids in northern Maine, central New York State, and in eastern Ohio, a redrawn portion of the Ohio River, and the addition of several lakes in western Maine.

Bradley was born in Connecticut in 1767. He read law in the office of a judge, following which he settled in Pennsylvania, initially in Wilkes-Barre and later in Philadelphia. He practiced law briefly before his appointment to the Post Office Department, where he remained until 1829. His later years were spent as a secretary to a Philadelphia insurance company. Bradley died May 7, 1838, in his seventy-first year. The map he compiled and published represented the first clear cartographic break from European-dominated mapmaking and introduced a new, more distinctly American style of cartography to the United States.

Also meriting some recognition are the two maps which were folded in the 1789 edition of Jedidiah Morse's *The American Geography*: Amos Doolittle's *A Map of the Northern and Middle States Comprehending the Western Territory and the British Dominions in North America* and Joseph Purcell's *A Map of the states of Virginia, North Carolina, South Carolina, and Georgia comprehending the Spanish provinces of East and West Florida Exhibiting the boundaries as fixed by the late Treaty of Peace between the United States and the Spanish Americans*. Both maps were engraved by Doolittle, and each measures approximately 30 by 40 cm. These maps are quite undistinguished and were harshly criticized by contemporary scholars. As Ralph Brown has written in his excellent analysis of Morse's *Geography*, "critics were practically unanimous in their disapproval of the map illustrations which, unchanged for several years, became less and less useful. . . . Morse did not reach a wholly satisfactory solution to his own map problem nor can now be accredited with having made a significant contribution to American cartography."[2]

Notes

1. Philip Lee Phillips, *The Rare Map of the Northwest 1785 by John Fitch, Inventor of the Steamboat* (Washington, D.C., 1916), 27.
2. Ralph H. Brown, "The American Geographies of Jedidiah Morse," *Annals of the Association of American Geographers* 31 (Sept. 1941): 188.

5. *Simeon De Witt, Pioneer Cartographer*

A uniquely American school of cartography developed and matured in the productive half century that followed the establishment of peace. Two native-born Americans who contributed much to that development were Thomas Hutchins and Simeon De Witt. As geographers to the Continental army, they supplied maps and related information to General Washington. De Witt's output, though, was far greater than Hutchins's. Ill much of the time, Hutchins was able to carry out his duties only nominally, although he did serve on several commissions for the state of Pennsylvania, including one in which he helped survey the boundary line between Pennsylvania and Virginia. After the war and following the passage of the Land Ordinance of 1785, Hutchins was appointed geographer of the United States in charge of surveying public lands and directing the General Land Office. Serving in this capacity until his death on April 20, 1789, Hutchins was instrumental in establishing the rectangular public land survey system.[1] De Witt, on the other hand, became the surveyor general of the state of New York. In this position, he made contributions to the development of American cartography that were more extensive and comprehensive than those of his former associate.

De Witt served as surveyor general of New York State with distinction for more than fifty years (Fig. 5–1). When he was appointed to the office in 1784, much of the area west of the Hudson River was still a wilderness, and the population of the state was less than a quarter million. When he died in 1834, settlements had been established throughout the state, the Erie Canal was in busy operation, an extensive network of roads had been built, and New York embraced more than two million inhabitants. De Witt was intimately involved with most

Fig. 5–1. Portrait of Simeon De Witt in his later years by Ezra Ames from T. Romeyn Beck's *Eulogium on the Life and Services of Simeon De Witt*, 1835.

of the internal improvements that fostered such settlement and growth.

De Witt was born in Ulster County, New York, on December 25, 1756. His father, Andrew, a successful and respected physician, was descended from the early Dutch settlers. His mother was descended from Huguenots. De Witt studied under the Reverend Dirck Romeyn before enrolling in Queens College in Brunswick, New Jersey. The outbreak of the revolutionary war and the destruction of the college by British troops prevented his graduation, although he was later awarded the bachelor's degree in 1776. De Witt returned to his home after the school's close and shortly thereafter enlisted in a volunteer battalion which helped defeat the British under General John Burgoyne at Saratoga in October 1777.

Following the victory the volunteers were released, and De Witt returned to Ulster County, where he independently studied mathematics and surveying. From these studies he was called in July 1778 to serve as an assistant to Robert Erskine, surveyor general of the Continental army. De Witt then became surveyor general upon the death of Erskine in 1780. Following Cornwallis's defeat at Yorktown in 1781, there were no more major military actions, and so De Witt and his assistants spent the succeeding months at headquarters in Philadelphia preparing maps from their rough field sketches. In compliance with a congressional resolution dated October 20, 1783, the geographer was directed to deposit such maps with the secretary of war.

De Witt anticipated that these war maps might be preserved as records of the various battles and campaigns and as the most detailed and up-to-date maps then available for large segments of the country. In June 1783, therefore, he proposed to General Washington and to the Continental Congress that the military surveys be published. Jonathan Trumbull, secretary to Washington, replied on June 8, 1783: "I have mentioned to the Comm. in Chief your purpose to obtain permission for publishing a Map of the State of War in America. His Excellency directs me to inform you, that the Measure is perfectly agreeable to him, and the proposition meets his full Approbation; it being his wish to see it accomplished in an accurate Manner, and at as early a period as the nature of the work will admit."[2] De Witt's proposal, however, received a negative reaction from the Congress. As a result, the geographer again wrote to Washington, in January 1784, reporting that he

> made application last summer to Congress for permission to prepare the maps in my possession for publication. To a committee appointed on the business, I pointed out their usefulness; that though no direct advantages would result to the public, every light that could be thrown upon the transactions of the war was due, not only to the solicitude which every individual has for his country, but more particularly to the merits of our principal Commanders, whose proceedings in all these particulars have hitherto been, and will continue to be pried into, with the greatest eagerness, in every part of the globe. I represented that in every civilized nation, where circumstances prevented the progress of the liberal arts, it had been the business of government to lend every encouragement. That in the present case, the inundation of catchpenny maps, which upon the commencement of peace will be poured upon us from Europe . . . would render such an undertaking too precarious for me to venture upon at my own risque.
>
> The remarks on the opposite side were that the state of our finances was in so low a condition, that the strict economy they were obliged to observe, forbade the application of money to anything but the numerous necessities which urged their immediate relief.[3]

Concurrent with this second appeal to Washington, De Witt again petitioned Congress with a new proposal: "If the expense of bringing my maps to a further degree of perfection: by additional surveys be judged to be needless," he wrote, "I have this proposal to make. I will undertake, in the best manner I can from the materials I have, as much as can be conveniently contained in one plate, and publish it at my own risque, provided I am furnished with cash sufficient for the purpose, on account of the pay now due me from the United States. . . . From the impressions of one plate I shall be able to judge, whether it shall afterwards answer to undertake any more at present."[4] Although Washington sent a personal letter to the pertinent congressional committee, De Witt's proposal had low priority among the urgent problems that faced the young nation's leaders.

In his second letter to Washington, De Witt also asked to be relieved of his military office: "Since the army is now being disbanded, and as I suppose the public will stand no farther in need of my services . . . , I therefore ask from your excellency the last favor of a discharge, and beg acceptance of my sincerest acknowledgements for the honor I have had of serving under your excellency's particular direction. All I regret is, that the poverty of my country and myself have not permitted me to make my office more extensively useful."[5] Washington acknowledged De Witt's plea on March 3, 1784, noting that "the nature of your office being such as that Congress may possibly still have occasion for you, I cannot think myself at liberty to grant the discharge you request; but circumstanced as you are I advise that you make a final application to that Body, to know whether they are inclined to comply with your former application or whether they have any further occasion for your services."[6]

De Witt reasoned that following the war there would be a strong surge of settlers to the trans-Appalachian lands. He hoped to be involved in surveying this vast area and offered his services to the Congress. "If a new state," he wrote, "is to be laid off adjoining Pennsilvania and Virginia as has been expected, I have hopes that from the parity of the office I now hold, and that of surveyor general to such a state, congress will be inclined to transfer me to that department, especially if it be allowable to suppose them influenced by a predelection in favor of their old servants, who have done their duty with reputation under all the difficulties which the American Army had to encounter, and who have lost permanent places of employment by being engaged in a military life."[7] No action was apparently taken by Congress on this further appeal, and on May 13, 1784, De Witt resigned his army commission and accepted the appointment as surveyor general of New York State. In the following year, Hutchins was appointed geographer of the United States. De Witt's appointment to the position of surveyor general was somewhat unique in that he was one of the few experienced American-trained engineers. To fill comparable offices most state governments were obliged to import such specialists from Europe.

Soon after he arrived at his Albany office, the new surveyor general was challenged with many urgent and complex problems. One of the first to claim his attention was the correct determination of the New York-Pennsylvania boundary. At the request of Pennsylvania, the New York legislature passed in March 1785 "an Act for running out and marking the jurisdiction between this State and the Commonwealth of Pennsylvania."[8] De Witt, his uncle James Clinton, and Philip Schuyler, the former New York surveyor general, represented New York on the boundary commission. Their Pennsylvania counterparts were David Rittenhouse, the astronomer and instrument maker, and Andrew Ellicott, a member of the distinguished Maryland family of surveyors. The two parties surveyed, during the summers of 1786 and 1787, ninety miles of the common boundary extending from the Delaware River westward to the south branch of the Tioga River along the forty-second degree parallel. Because of his increasing responsibilities De Witt did not accompany the field surveyors in 1787.

Another pressing problem facing New York officials was the equitable distribution of public lands to veterans of the revolutionary war. To encourage long-term enlistments, the Continental Congress and the separate states had promised grants of land to soldiers who served for the duration of the war. As soon as they were discharged, therefore, many veterans pressed to have the promises fulfilled. To meet its commitments, New York, through a series of treaties, secured title to Indian lands in the central and western part of the state. By 1789 Indian claims were extinguished, except in designated reservations, for most of the north central part of the state. By an act passed February 28, 1789, the Board of Commissioners of the Land Office was authorized "to direct the Surveyor-General to lay out as many townships in land set apart for such purpose as will contain land sufficient to satisfy the claims of all persons who are or shall be entitled to grants of land . . . , which townships shall respectively contain 60,000 acres of land, and be laid out as nearly in squares as local circumstances will permit, and be numbered from one progressively to the last inclusive."[9] The act further directed the surveyor general to make a map of the twenty-five (later twenty-seven) townships in north central New York State that were designated as the Military Tract and to subdivide each into holdings of approximately six hundred acres.

Surveys of the townships in the Military Tract were prepared by Moses De Witt (a cousin of Simeon) and John L. Hardenburgh under the general direction of the surveyor general. In July 1791 the manuscript maps were deposited with the commissioners of the General Land Office, who were responsible for distributing the lands. The individual township maps apparently were never published. They were used by De Witt, however, in compiling a map of the north central part of the state.

This map as published includes twenty-seven townships within the Military Tract as well as lands to the south and east, embracing parts of Tioga, Herkimer, and Otsego counties. Across the top border is the title *1st Sheet of De Witt's State Map of New-York* (Fig. 5–2). The copy in the Library of Congress carries the manuscript notation "Received 17th January 1793," and was obviously the copyright deposit. This is verified in the following notice published in the January 23, 1793, issue of the *New York Daily Advertiser*: "Be it remembered that on the twenty-sixth day of October, in the seventeenth year of the Independence of the United States of America, Simeon De Witt Hath Deposited in this office, the title of a map, the right where of he claims as author." A slightly revised copy of this map in the New York State Library, Albany, shows a road crossing Fayette, Union, and Cincinnatus townships in the southeastern portion of the state. C. Tiebout is credited as engraver on both versions of the map.

In 1794 there was published *A Plan of the City of Albany, Surveyed at the request of the Mayor Aldermen and Commonality by Simeon De Witt*. The plan was engraved by J. Hutton and appears to be the only cartographic engraving in which he was involved. It is uncertain whether De Witt prepared the Albany plan as part of his official duties or on a personal contract with the city. The latter seems more likely. Illustrations of the court house and the prison decorate the left margin of the Albany plan.

Two years later, De Witt's old request to Congress that he be appointed surveyor general to the United States in order to survey and map the public lands west of the Appalachian Mountains was finally granted. In 1796, according to De Witt biographer T. Romeyn Beck, "Gen. Washington, without [De Witt's] knowledge or solicitation, nominated him to the Senate as Surveyor-General to the United States, and the appointment was cordially ratified. This event was often spoken of by Mr. De Witt to his intimate friends, as the most gratifying in his public life, as age advanced, he appeared to prize it more highly."[10] Because of his commitment to his native state, however, De Witt reluctantly declined the appointment.

In 1802 De Witt published a large (168-by-135-cm.) *Map of the State of New York*, at the scale of 1:245,000 (Fig. 5–3). Engraved on six plates, the map includes parts of

Fig. 5–2. The Military Tract in north central New York State was one of the first areas mapped by De Witt after he became surveyor general of the state. The rectangular pattern of townships is shown on this 1793 map.

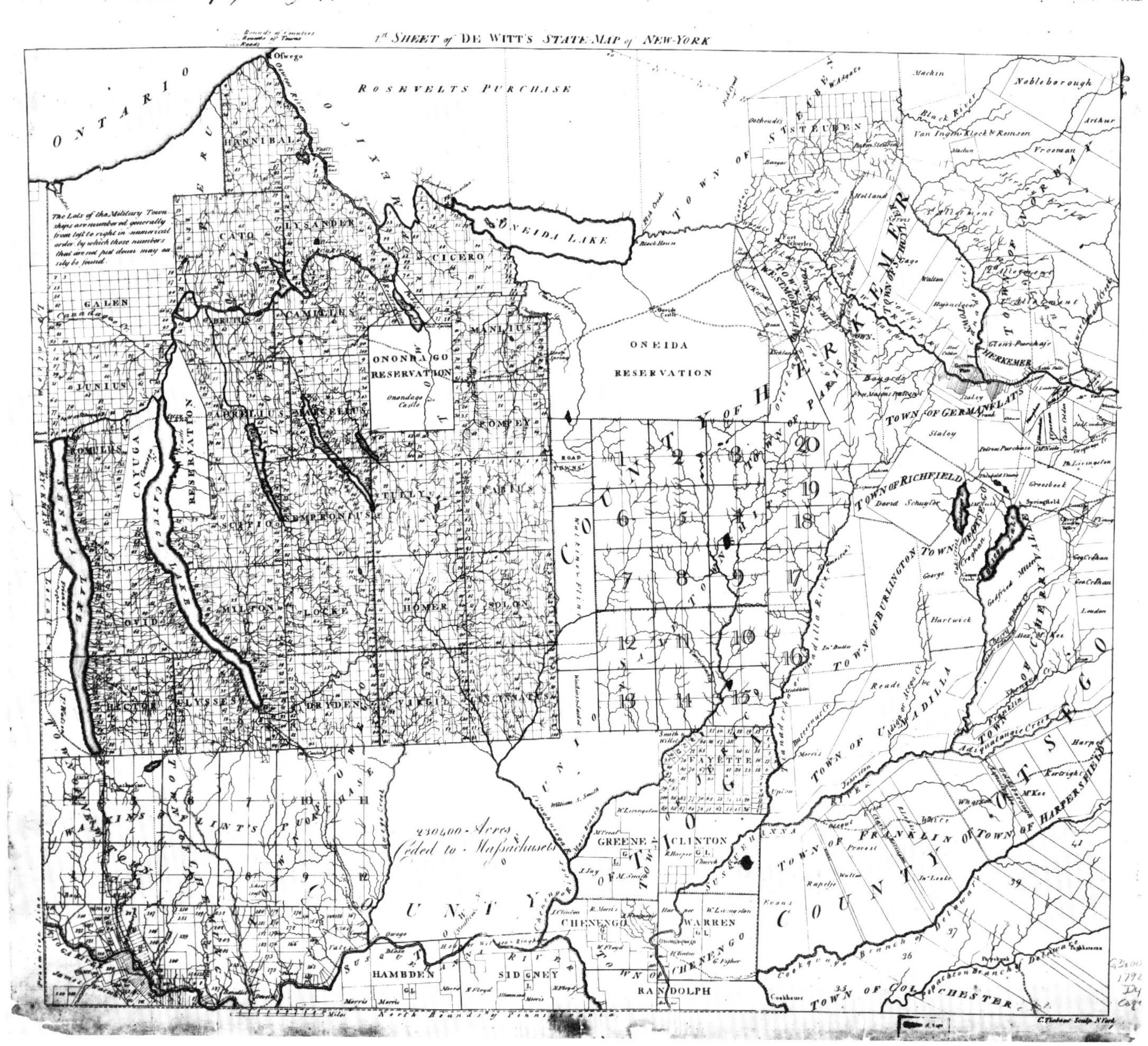
Received 17th January 1793.
1st SHEET of DE WITT'S STATE MAP of NEW-YORK
ONTARIO
ROSEVELTS PURCHASE
HANNIBAL
CATO
LYSANDER
CICERO
ONEIDA LAKE
GALEN
JUNIUS
ONONDAGO RESERVATION
MANLIUS
ONEIDA RESERVATION
AURELIUS
MARCELLUS
POMPEY
ROMULUS
CAYUGA
TULLY
FABIUS
SCIPIO
SEMPRONIUS
OVID
MILTON
LOCKE
HOMER
SOLON
HECTOR
ULYSSES
DRYDEN
VIRGIL
CINCINNATUS
The Lots of the Military Townships are numbered generally from left to right in numerical order by which those numbers that are not put down may easily be found.
230400 Acres Ceded to Massachusets
GREENE
CLINTON
CHENENGO
WARREN
HAMBDEN
SIDGNEY
RANDOLPH
FAYETTE
TOWN OF GERMANFLATS
TOWN OF RICHFIELD
TOWN OF BURLINGTON
TOWN OF FRANKLIN
TOWN OF HARPERSFIELD
TOWN OF COLCHESTER
TOWN OF STEUBEN
HERKEMER
North Bounds of Pennsylvania
C. Tiebout Sculp. N York

Vermont, New Hampshire, Massachusetts, Connecticut, and New Jersey in addition to the full expanse of New York. The part of the state shown on his 1793 *1st Sheet* was completely redrawn for the 1802 map. There are thirty-one townships in the Military Tract on the 1802 map, as compared with twenty-seven on the 1793 map. The earlier map also shows each township subdivided into some one hundred plots, each embracing approximately six hundred acres. These subdivisions are omitted in the Military Tract townships on the later map. In compiling his 1802 map, De Witt drew upon his revolutionary war maps, as well as upon the Pennsylvania-New York and other boundary surveys, the Military Tract surveys, and various manuscript and published maps. The map was one of the earliest in a series of officially sponsored, or subsidized, state maps, and New York officials were justifiably proud of it. This is evident in a resolution, passed by the New York State senate on April 1, 1803: "That his Excellency the Governor be requested to furnish the executive of each State, for the use thereof, with a map of this State lately made by the Surveyor-General."[11]

Both this map and a smaller (57-by-70-cm.) 1804 edition were engraved by Gideon Fairman. Born in Newtown, Fairfield County, Connecticut, on June 26, 1774, Fairman served an apprenticeship with a mechanic in that town. Subsequently he moved to Albany, where he learned silver-plate engraving in the shop of Isaac and George Hutton. In 1796 he opened his own engraving shop in Albany, where he was in business until 1810. In that year he moved to Philadelphia and soon began specializing in bank note engraving.

The legislative act of February 28, 1789, that authorized the survey of the Military Tract also specified that "the Commissioners of the Land Office shall likewise designate every township by such names as they shall deem proper." The designated names, all of classical Greek and Roman origin or association, were first presented to the public on De Witt's 1793 map. As a result, the surveyor general has through the years been credited or blamed for having introduced the classical names. Best evidence indicates, however, that the names did not originate from him. Maps of the townships, designated only by numbers, were submitted by De Witt to the commissioners of the Land Board on July 3, 1791. The board's minutes for that date record that "the board caused the said townships and lots thereon respectively to be numbered on the said maps agreeable to law and designated them by the following names, to wit, Township No. 1, by the name of the township Lysander and so on through the original group of twenty-five townships."[12] De Witt was not a member of the Land Board and was not present at this meeting. Historian Charles Maar asserts that "the Land Board . . . was clearly responsible for naming the military townships."[13]

De Witt resented the allegations that it was he who was responsible for the classical nomenclature. In a letter sent to a newspaper editor some years after the 1793 map was published, he stated that "the surveyor general . . . knew nothing of these obnoxious names, till they were officially communicated to him, nor had he even then any agency in suggesting them." This denial unfortunately made less of an impression upon the public than did a seven-stanza "Ode to Simeon De Witt, Esquire" that was published in the *New York Evening Post* and the *National Advertiser* in July 1819. The doggerel, published under the pseudonym "The Croakers" by Joseph Rodman Drake and Fitz-Green Halleck, two aspiring young writers, hailed the surveyor general as

> God-father of the christen'd West!
> Thy wonder working power
> Has called from their eternal rest
> The poets and the chiefs who blest
> Old Europe in their happier hour.

The closing stanza unequivocally brands De Witt as the originator of the classical names.

> Surveyor of the western plains!
> The Sapient work is thine—

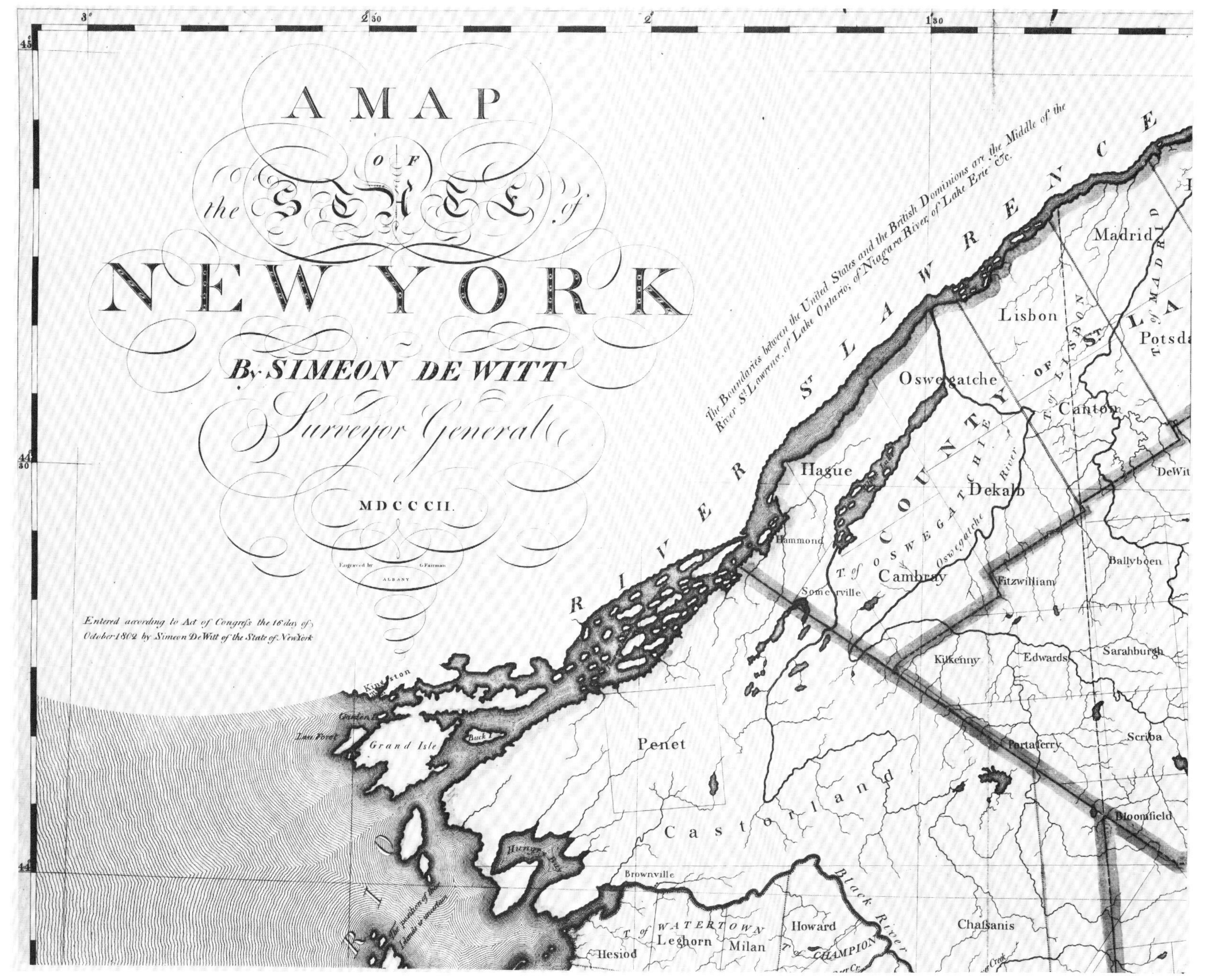

Fig. 5–3. The elaborate script title and a small segment of De Witt's 1802 map of New York.

> Full-fledged it sprang from out thy brains,
> One added touch, alone, remains
> To consummate the grand design.
> Select a town—and christen it
> With thy unrivalled name, De Witt.[14]

To his credit, it should be noted that De Witt did not affix his name to any geographical or political feature in New York State. De Witt Township, later formed of the western part of Manlius Township, was named for Moses De Witt, De Witt's cousin and one of the field surveyors of the Military Tract.

Distasteful as they may have been to some early settlers, the classical names were chosen for sound reasons. Because of the fierce and bloody conflicts between the Indians and Americans in this region during and after the Revolution, Indian names would have been even more obnoxious. Likewise, hatred of the British was still too strong to allow English names. Time tempered the distaste for the classical names, however, and the precedent set in the Military Tract was followed for other names in the western part of the state. And later when New Yorkers moved westward, they transplanted these classical names to the new settlements in the Middle West.

During the first fifteen years of his tenure as surveyor general, De Witt was personally involved in making surveys and compiling maps, but because of increasingly heavy administrative responsibilities, his cartographic contributions after 1802 were principally advisory, consultative, or supervisory. Until his death, however, there were few mapping and engineering activities within the state with which he was not associated. Thus, in 1807, he was named one of three commissioners to plan a layout of streets that would meet the future requirements of the rapidly expanding city of New York. An act passed by the state legislature on April 3, 1807, specified that Gouverneur Morris, De Witt, and John Rutherford be appointed commissioners of streets and roads in New York City "for the purpose of performing the several acts and duties hereinafter prescribed." The commissioners were empowered "to lay out streets, roads, and public squares, of such width, extent, and direction, as to them shall seem most conducive to public good" in that part of Manhattan extending from the settled portion of the city to the northern limits of the island. Specifically, they were directed "to lay out the leading streets and great avenues . . . of a width not less than sixty feet, and in general to lay out said streets, roads, and public squares of such ample width as they may deem sufficient to secure a free and abundant circulation of air among said streets . . . , and said Commissioners shall not, in any case, lay out any street of less than fifty feet in width."[15]

The commissioners engaged John Randel, Jr., to be secretary and surveyor. A competent engineer, he prepared or directed the surveys for the proposed street plan. The report of the commissioners and three copies of the street plan by Randel were presented to city officials in the spring of 1811. Randel believed that publication of his plan at that time would aid the British on the eve of the War of 1812, so it was not published. William Bridges, city surveyor, redrafted the commissioners' map, which was then neatly engraved by Peter Maverick and published in 1811 by T. & J. Swords of New York City (Fig. 5–4). Bridges's plan, which was "Entered according to the Act of Congress, Nov. 16th 1811," measures approximately 238 by 66 cm. and shows the grid of avenues and streets on Manhattan up to present-day 155th Street. A fifty-four-page booklet incorporating provisions of the act and basic parts of the commissioners report was published as a supplement to the plan. The report was accepted by state and city officials, and the rectangular or gridiron street plan which it proposed, according to writer I. N. Phelps Stokes, "became a very large factor in determining the later development of the city. Indeed, we may well say that the work of this Commission marks the division between old and modern New York."[16]

In justification of choosing a rectangular street pat-

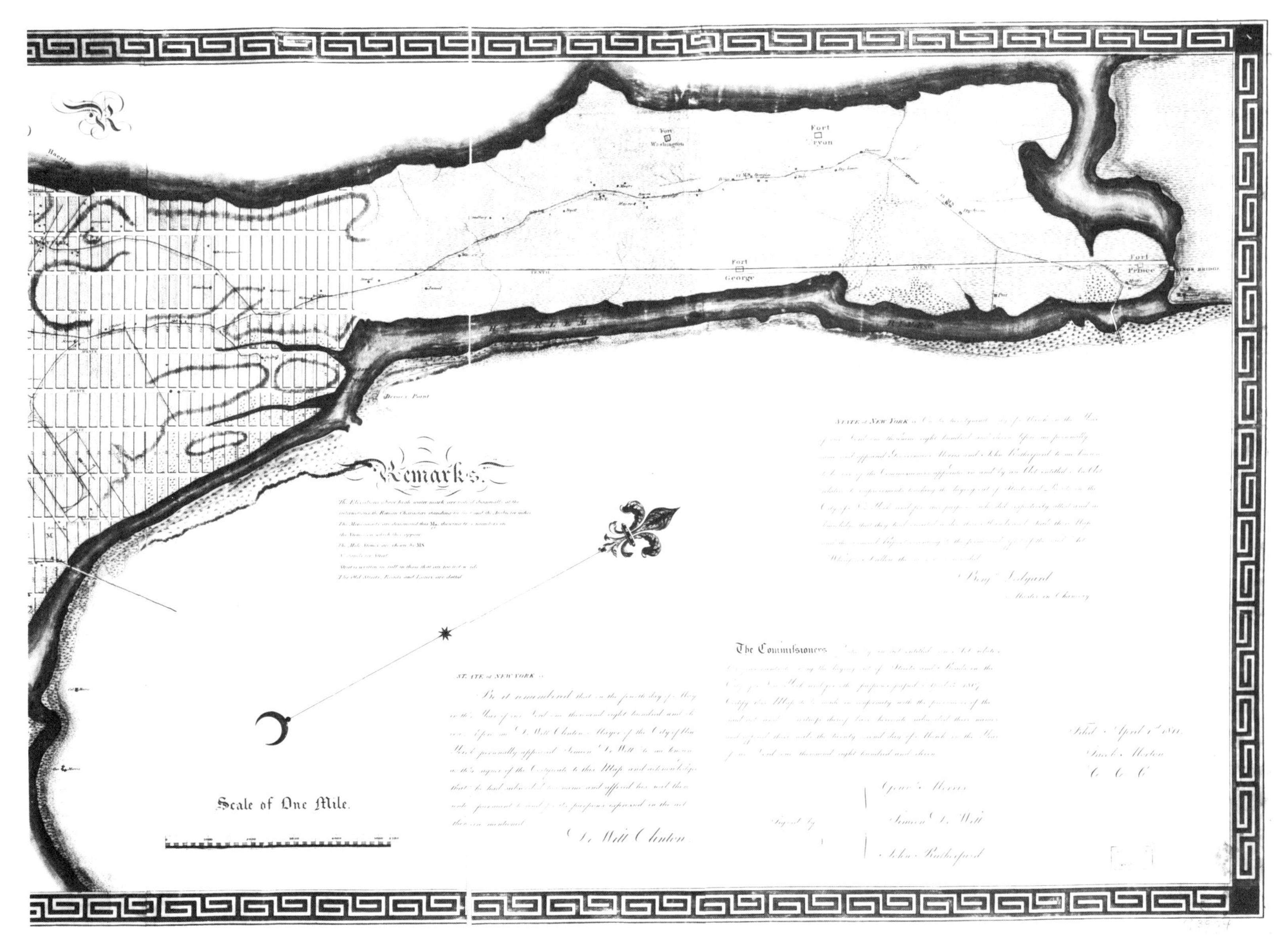

Fig. 5–4. To accommodate the expansion of New York City, a state commission, which included De Witt, planned a layout of the city's streets. The resulting plan, known as the commissioners' map, was published in 1811. Reproduced here is the right (north) half of the map.

tern, the commissioners reported that "one of the first objects which claimed their attention was the form and manner in which the business should be conducted; that is to say, whether they should confine themselves to rectilinear and rectangle streets, or whether they should adopt some of those supposed improvements, by circles, ovals and stars. . . . They could not but bear in mind that a city is to be composed principally of the habitations of men, and that strait sided, and right angled homes are the most cheap to build, and the most convenient to live in. The effect of these plain and simple reflections was decisive."[17] The commissioners also sought to explain in their report why "so few vacant spaces have been left, and those so small, for the benefit of fresh air, and the consequent preservation of health." "Certainly," they acknowledged, "if the City of New-York were destined to stand on this side of a small stream such as the Seine or the Thames, a great number of ample places might be needful; but those large arms of the sea which embrace Manhattan Island, render its situation, in regard to health and pleasure, as well as to convenience of commerce, peculiarly felicitous; when therefore, from the same causes, the price of land is so uncommonly great it seemed proper to admit the principles of economy to greater influence than might . . . have consisted with the dictates of prudence and the sense of duty."[18]

Proposals to join the Hudson River and the Great Lakes by means of a canal across New York State had been advanced before the revolutionary war ended. Pressures for such an inland waterway were intensified soon after the nineteenth century began. As surveyor general, it was inevitable that De Witt would become involved with plans for the proposed canal. The New York State legislature on February 4, 1808, appointed a committee "to take into consideration the propriety of causing an accurate survey to be made of the most eligible and direct route for a canal to open communication between the Hudson R. and L. Erie." Six weeks later the resolution was amended ordering the surveyor general "to cause an accurate survey to be made."[19] On June 11 De Witt authorized James Geddes, an engineer with extensive experience in western New York, to make the survey.

Two days later, in a letter addressed to Joseph Ellicott, an agent of the Holland Land Company, the surveyor general observed that "there are some Gentlemen . . . who are of the opinion that Nothing less ought to be contemplated than a Navigation from Erie or some Part of the Niagara River to the Waters of the Oneida Lake by a Route South of Lake Ontario, and my Object in writing to you is to obtain your Opinion about it." Ellicott replied in a letter dated July 30, 1898, that "the Tract of Country extending from the Niagara to Genessee River is remarkable for its horizontal position. . . . Nature seems to have pointed out this Route for a Canal, not only in Consequence of the little Labour comparatively speaking that would be required in digging it, but because the necessary Materials for the Construction of Locks are close at Hand . . . and Quantity of the best shaped Limestone may be produced."[20]

Excavation of the Erie Canal was started at Rome, New York, on July 4, 1817. Its planning and construction were largely directed by Americans, few of whom had specialized training in engineering. In the course of construction, however, they learned much by trial and error. The Erie Canal project, in fact, has been described as the "first American school of engineering" and a "major spark to United States progress in engineering and commerce."[21] Surveying and mapping contributed to and were beneficiaries of this progress.

The rapid settlement and growth of central and western New York State and the many internal improvements completed during the first quarter of the nineteenth century rendered obsolete De Witt's map of 1802. Because of his numerous and growing responsibilities, he was not personally able to revise the map or to compile a new one. He did, however, direct and supervise the compilation and publication in 1829 of a new map and atlas of the state prepared by David H. Burr.

Burr's project was the last significant cartographic work with which De Witt was associated. In the early

1830s the state was engaged in enlarging and improving its canal network and in preparing surveys for roads and railroads. Notwithstanding his advancing years, De Witt was intimately involved with many of these projects. When he died on December 3, 1834, he still held the office to which he had been appointed more than fifty years earlier. It had been an exciting and productive half century for the young nation and its separate states. New York State had been a leader in promoting and developing a succession of internal improvements which were essential to the expansion of settlement and economic growth. In his powerful and influential position, De Witt contributed substantially to New York's leadership in these areas. He also helped to introduce and develop a distinctively American type of mapmaking and map publishing.

De Witt's efforts were suitably recognized by New York's Governor W. L. Marcy in his message to the legislature at its sixty-eighth session in 1835:

> I deem it not inappropriate to avail myself of this occasion to pay a tribute of respect to the memory of a most faithful public servant who devoted almost the whole of a long and active life to the service of his State. In the dawn of manhood he espoused the cause of liberty, and became eminent among our revolutionary patriots. He entered the service of this State in the infancy of its government, and regarded its advancement with parental solicitude. He aided in founding and in building up most of our public institutions, and has left more, if not more enduring memorials of his useful service, than any other of our numerous public benefactors. His many private virtues shed lustre upon his public character. A life that commenced by services and sacrifices in the cause of civil liberty, and well sustained to its end by unremitting labors directed with singleness of purpose to the public good, should be held in just remembrance by those among whom it was spent, and presented, as an encouraging example to posterity.[22]

Notes

1. William D. Pattison, *Beginnings of the American Rectangular Survey System 1784–1800*, University of Chicago, Department of Geography, Research Paper no. 50 (Chicago, 1957).
2. George Washington, *Writings*, ed. John C. Fitzpatrick (Washington, D.C., 1931–44), 26:496.
3. T. Romeyn Beck, *Eulogium on the Life and Services of Simeon De Witt* (Albany, N.Y., 1835), 313–14.
4. Washington, *Writings* 27:347–48.
5. Beck, *Life and Services*, 9.
6. Washington, *Writings* 27:244–45.
7. Beck, *Life and Services*, 11–12.
8. New York, *Report of the Regents' Boundary Commission upon the New York and Pennsylvania Boundary* (Albany, N.Y., 1886).
9. Jeannette B. Sherwood, "The Military Tract," *New York History* 7 (July 1926): 173.
10. Beck, *Life and Services*, 16.
11. New York, *Senate Journal*, 26th sess., Jan. 25, 1803, 129.
12. Charles Maar, "Origin of the Classical Place Names of Central New York," *New York History* 7 (July 1926): 159.
13. Ibid., 163.
14. Ibid., 157–58.
15. William Bridges, *Map of the City of New-York and Island of Manhattan with Explanatory Remarks and References*. (New York, 1811), 5, 6, 7.
16. I. N. Phelps Stokes, *The Iconography of Manhattan Island 1498–1909* (New York, 1918) 3:478.
17. Bridges, *Map of the City*, 24.
18. Ibid., 25–26.
19. Harvey Chalmers, *The Birth of the Erie Canal* (New York, 1961), 88, 89.
20. Joseph Ellicott, "Reports as Chief of Survey (1797–1800) and Agent (1800–1821) of the Holland Land Company's Purchase in Western New York," *Buffalo Historical Society Publications* 1, no. 32 (1937): 409, 416.
21. Joseph Scothon, "New York State's Unique Canal System," *Civil Engineering* 37 (June 1967): 47.
22. New York, *Legislative Documents*, no. 2, 58th sess. (Albany, N.Y., 1835).

6. *Early State Maps: The New England States*

Maps of most of the colonies and provinces, as we have seen, had been compiled prior to the adoption of the Constitution. Many of them were compiled by British engineers or mapmakers and, with few exceptions, they were engraved and published in England. Soon after the United States was constitutionally established each of the individual states was confronted with the urgent need for an accurate and up-to-date map of its jurisdiction. The federal government was financially unable to support the compilation and publication of such maps. As historian Lloyd Brown has noted, "the principal difficulty of making an intensive topographic study of the United States before 1800 had been caused by the fact that there was so much to be done in a hurry, without precedent and before the central government was properly organized. The boundaries of the country were poorly defined and much of the interior was a dark mystery filled with unknown peril."[1] Few state governments were likewise able to fund extensive and costly land surveys or finance the engraving and publication of maps. Nonetheless, some states very early established offices for a state engineer or a surveyor general and appointed qualified American- or European-trained specialists to fill them. New York was one such state, as was seen in Chapter 5.

Few legislatures were, however, as affluent or as farsighted as New York's, and most states, therefore, relied upon private individuals to prepare maps of their official territories, granting moral encouragement and tangible assistance within the range of their capabilities. Not until the second or third decades of the nineteenth century were most of the young states able to support official mapping projects. Notwithstanding, by 1840 some thirty or more state maps had been published in one or more editions. All the former thirteen colonies were mapped at least once, and maps had also been prepared for Vermont and Maine, which respectively formed parts of New Hampshire and Massachusetts during the colonial period.

The state maps are among the earliest examples of truly American cartography. They were, for the most part, based on surveys by Americans and were compiled, drafted, engraved, printed, and published in the United States. They were, moreover, specifically designed to meet the cartographic requirements of the several states. Above all, in the methods, techniques, and procedures employed to produce state maps, American ingenuity and resourcefulness were abundantly demonstrated. Commercial map publishing, which attained a high degree of excellence and productivity during the nineteenth century, rested solidly on the foundations laid by the state maps and their makers. This indebtedness was acknowledged in 1829 by Henry S. Tanner, the foremost map engraver and publisher of his day. "Important accessions to the stock of knowledge on the geography of the United States," he affirmed, "have recently been made, by the production of excellent local and State maps."[2]

"From an actual survey," or some variant of this phrase, forms part of the title inscription on many early state maps. Crude though they may have been, the surveys represented a distinct improvement over the cartographic practices of the colonial period. Surveying was a basic and honored profession in early America, and virtually every town had an official full- or part-time surveyor who located and established the boundaries of real property. The manuscript plats or maps on which these data were recorded were retained in the county, district, town, or city archives. They were a prime source of data for the earliest compilers of state maps. State laws, in many instances, made it mandatory for county, district, or town officials to supply current maps of their jurisdictions. Such surveys, however, rarely were based on geodetic or geographic controls, and map compilers frequently had difficulty in reconciling roads and property lines along town and county borders.

Each early state map has its own distinctive and interesting story. In the production of all of them, however, there were common circumstances and conditions. All were prepared in response to the peculiar and pressing needs of the time and with the human and technical resources then available. They vary from the solitary compilations of dedicated individuals to the more sophisticated products of official engineers and surveyors. Between these extremes are examples of different levels of cooperative enterprise between private citizens and state governments.

The first state map published after ratification of the Constitution was William Blodget's *Topographical Map of the State of Vermont, from actual Survey*, which was engraved by Amos Doolittle of New Haven, Connecticut, and published at Bennington, Vermont, in January 1789. Although Blodget dedicated his map "To his Excellency Thomas Chittenden Esqr. Governor and Commander in Chief; the Honorable the Council, and the Honorable the Representatives of said State," it was a private, rather than official, undertaking. There is, moreover, nothing on the map to indicate the nature of the "actual Survey." It is possible, however, that Blodget had access to the town surveys that were deposited in the office of Ira Allen, Vermont's surveyor general from 1779 to 1787.

Blodget was born in Stonington, Connecticut, on June 8, 1754, the son of Dr. Benjamin and Mary Satterlee Blodget. His mother, two sisters, and a brother died within a year of his birth, and his father remarried in March 1755. Little is known about Blodget's early education, but it apparently included musical training on several instruments. The Blodget family seems to have moved to Providence, Rhode Island, where on December 11, 1774, William married Ann Phillis Chace. In May 1775 Blodget enlisted in Colonel Daniel Hitchcock's Second Regiment of the Rhode Island Continental Infantry.[3] Shortly thereafter, he was named ensign and in January 1776 was promoted to lieutenant while serving as military secretary to General Nathanael Greene. He became an aide to Greene, with the rank of major, in August 1776. Blodget participated in various campaigns until June 1779, when he left the army to serve as chaplain on the frigate *Deane*, a post he retained until April 1783.

Blodget seems to have had periodic financial difficulties, and he made several claims to the Congress after the war for pension payments. There is evidence that by 1786 he had established himself in Bennington, where he operated a general store and was proprietor of an iron forge. "Blodget's Forge" is identified at Bennington on Blodget's Vermont map. In the January 28, 1788, issue of Hartford's *American Mercury*, Blodget advertised "lands bought and sold, maps and drafts of any part of the State, done on the shortest notice, with accuracy."[4] Although there is early evidence of Blodget's skill in drawing, this is the first reference to his proficiency in cartography. It is likely that at this time, he was collecting data for his Vermont map and, in fact, may have already delivered the manuscript draft to Doolittle in New Haven. The *New Haven Gazette* of February 5, 1789, carried this notice:

> With pleasure we inform the friends of American improvement in arts and manufactures, that a Map of Vermont on a large scale, is just published in this city, which does honour to genius and ability of our countrymen. This map was engraved by Mr. Doolittle, for Col. Blodget, the author, and in accuracy and elegence has been seldom equalled by the productions of European artists. We are told that Col. Blodget intends to publish a map of this

> State, on a large scale, and we have no doubt that his intentions will meet with that liberal encouragement which the excellence of the above mentioned work entitles him to claim.[5]

There are two editions of this Blodget-Doolittle map of Vermont, both dated January 1, 1789. On the first edition, the letter "i" is omitted from Doolittle's name. This map is relatively rare. There is no original copy in the Library of Congress, and David A. Cobb states that "unfortunately, it is not known to be in any collection in the state of Vermont."[6]

Before the Vermont map was published, Blodget had embarked on a project to compile a similar work for the state of Connecticut. This was also noted in the *New Haven Gazette* article cited above. The mapmaker had probably by mid-1788 relocated his headquarters from Bennington to New Haven and later to Middletown, Connecticut. On January 26, 1789, Blodget sent out printed forms to a number of individuals he thought might be interested in his proposed map of Connecticut. The form reads:

> It being the will of the inhabitants of Connecticut to be furnished with an accurate Map of the State, I have undertaken the task, and presume, by advice, to request of you an answer to the following questions relative to the town of ———.
>
> Should you incline to favour my views, please to answer as many of the questions as come within your knowledge, agreeably to the characters annexed, and lay down a draft of the town with its parishional divisions, upon a scale of an inch to a mile, in as particular a manner as you can; and disclose it to the care of Mr. Josiah Meigs of New-Haven, on or before the first day of May next.

There followed a list of five questions relating to town boundaries, number and denomination of houses of worship, courses of mountains, remarkable hills, rivers, brooks, lakes, roads, etc., name and location of mills, and depth and breadth of rivers. A list of eleven code letters was given for use in locating specific features on the town map.

Blodget demonstrated ingenuity in using this method to elicit information from knowledgeable individuals in the several towns within the state. It is possible that he had previously utilized similar procedures in assembling data for his Vermont map. We do not know how many questionnaires he sent out. As in any cooperative effort of this type, there were delays in returning the completed forms. Accordingly, in a notice printed in the November 18, 1789, issue of New Haven's *Connecticut Journal*, Blodget urged that the forms be returned in order that he might get on with his compilation. He thanked those from whom he had received replies, noting that "a few only being delinquent, prevents his compilation of the whole." Apparently the notice stimulated additional responses, for on April 14, 1790, he published in the same journal his proposals for compiling the map. Blodget also noted that "the subscriber having taken great pains, for twelve months past, to collect the best materials for a compilation of this work, assisted by several of the best informed gentlemen in each town who have furnished him with accurate draughts—(and obtained from the General Assembly an exclusive right to himself of publishing the same)—hopes to meet the encouragement of the public to so arduous a task."[7] The published map was priced $2.00 for an uncolored print and $2.25 for a colored one.

Blodget's reference to an "exclusive right" granted to him by the Connecticut General Assembly relates to a resolution passed at its annual meeting in October 1789. As reported in the *Public Records*,

> upon the memorial of William Blodget . . . Showing to this Assembly that he has with Great Labour & expense nearly collected materials for a Map of this State & a Pamphlet of Explanation, and that on Account of his Residence he cannot take Ad-

> vantage of the Statute of this State for the encouragement of Literature & Genius Praying that the Benefits of said Act may be extended to him relative to said Map & Pamphlet &c as per Petition on file: Resolved by the Assembly that the Privileges and Benefits Enacted and Provided by said Statute in favour of Authors of any Book or Pamphlet Map or Chart they being Inhabitants or Residents within these United States be and the same are hereby extended to him the said William Blodget as relative to his said Map of this State & Pamphlet of Explanations.[8]

Blodget completed compiling and drafting the map in late 1790 or early 1791. It was probably published in July or August of 1791 with the title of *A New and Correct Map of Connecticut one of the United States of North America, From Actual Survey—Humbly Dedicated by Permission to his Excellency Samuel Huntington Esquire Governor and Commander in Chief of said State, By his most Humble Servant William Blodget*. The engraver is given as "Joel Allen Script et Sculpt." There is no date or place of publication on this edition. A revised and corrected second edition of the map was issued some six months later with the imprint "Middletown, Printed for the Publisher, March 1792." Both maps are at the approximate scale of 1:250,000 and measure 69 by 88 cm. Edmund Thompson in his *Maps of Connecticut* notes that "Blodget gave great attention to the delineation of rivers, showing names of small streams that had not appeared on earlier maps. But he omits within the townships many small communities that were shown by Romans."[9]

It is interesting that Blodget engaged Joel Allen rather than Doolittle to engrave the Connecticut map. Phyllis Kihn believes that "the only accountable reason for this change in engravers is that Doolittle was no doubt already at work on his own map of Connecticut."[10] Joel Allen was born in Farmington, Connecticut, in 1755. It is not known from whom he learned engraving or when he established his shop in Middletown. He died in that city in 1825.

It is unlikely that Blodget profited financially from either of his maps. There is evidence that he personally took to the road to obtain subscriptions and to make sales. Kihn relates that "the most remarkable characteristic of Blodget was his willingness to launch into a project without any prospects of remuneration. In both of these maps (Vermont and Connecticut) it was Blodget who had to bear the expense of paying the engraver, probably his largest bill. He also had to purchase the paper on which his maps were engraved, pay for the notices appearing in the newspapers, and certainly he owed someone for printing up the printed forms asking for information of the selectmen of the different towns or any other persons interested in his project."[11] There is no record that the pamphlet, mentioned in the Connecticut assembly's resolution, was ever published.

Blodget died, most probably in Hartford, Connecticut, on October 10, 1809, at the age of fifty-seven. His later years were not happy, and there are some indications that his problems were related to his intemperance. Nonetheless, his death was noticed in a number of New England newspapers, most of which called attention to his military service during the revolutionary war. Cartographically, he is remembered for being the first of the independent compilers and publishers of state maps.

Ira Allen, youngest brother of Colonel Ethan Allen, who was commander of the Green Mountain Boys in the revolutionary war, was active with other landowners in forming the convention which declared Vermont independent in 1777. Prior to that time the colonies of New York and New Hampshire both claimed the territory which later became the state of Vermont. In addition to acquiring land for himself and accepting commissions as a private surveyor, Allen served as surveyor general of Vermont, or as the territory was initially called, the New Hampshire Grants. The legislature in 1778 directed Allen to procure by advertisement in periodicals all the grants, patents, or charters to land in the territory. It is likely that Blodget drew on these materials and surveys by Allen and others to compile his *Topographical Map of the State of Vermont*.

James Whitelaw, who had served as an assistant to Allen, succeeded him as Vermont's surveyor general on October 26, 1787. He held the office until 1804, but like Allen he had engaged in private surveying during his tenure. As early as 1788, Whitelaw had begun surveys which he anticipated would result in a new map of the state. Very likely on his recommendation, the general assembly resolved at its 1790 session that

> whereas it is absolutely necessary as well for the good as for the dignity of this State that some proper method be adopted to furnish a proper map thereof,
>
> Therefore, Resolved that the selectmen of the several towns in this State make out and send to the surveyor-general before the first day of August next a proper plan of their several towns exhibiting the courses and lengths of their lines and what towns they are bounded on, also the courses of the several streams with their names, public roads and where they lead to, the situation of meeting-houses, mills and other public buildings, also the situation and names of ponds, mountains and everything necessary to make a complete map; and where towns are not organized it is hereby declared to be the duty of the proprietors' clerks or other persons who may have plans of said towns to forward them in manner aforesaid.
>
> Resolved also, that the surveyor-general be and he is hereby directed to return copies of the field-books of the several towns that have been surveyed under his direction or under the direction of the former surveyor-general to the several town clerks of the respective towns, whose duty it shall be to record same; and when the towns are not organized, to the county clerks of the several counties in which such towns lie.
>
> And it shall be the duty of said county clerks to record the same, and the secretary is hereby directed to cause the above resolution to be published immediately in both newspapers printed in this State.[12]

From the deposited town surveys Whitelaw compiled *A Correct Map of the State of Vermont From actual Survey; Exhibiting the County and Town lines, Rivers, Lakes, Ponds, Mountains, Meetinghouses, Mills, Public Roads &c. By James Whitelaw Esqr.: Surveyor General 1796. With Privilege of Copy Right* (Fig. 6–1). The map, which is 114 by 77 cm. in size and at the approximate scale of 1:240,000, was engraved in New Haven by Doolittle. A rural panorama forms the lower part of the title cartouche, and a note on the map lists the townships which were surveyed under the direction of Whitelaw, as well as those for which plans were supplied by selectmen and town clerks. "Several towns," the note reads, "neglected making returnes, and others were not very particular: which is the cause that the roads are broken off in some places." Because this map was based in part upon surveys conducted or directed by Allen and Whitelaw, it has a higher degree of accuracy than most other state maps published before 1800. Revised editions of Whitelaw's map were published in 1810, 1821, and 1824. The 1810 edition was engraved by James Wilson, the first American globemaker. Ebenezer Hutchinson of Hartford was the engraver of the 1821 and 1824 maps. The rural scene on the cartouche of the two Hutchinson editions differs from that on the original 1796 map. The Hutchinson editions also include a view of the village of Montpelier by a Mrs. S. I. Watrous.

Early in the last decade of the eighteenth century, several farsighted and public-spirited citizens of Massachusetts stimulated interest in preparing a map of that state. The first concrete proposal appears to have been made by Osgood Carleton, one of the foremost surveyors and mapmakers of early New England. In a note published in the September 26, 1791, issue of the *Boston Gazette*, Carleton suggested that there be published "by Subscription A complete Map of the State of Massachusetts, Rhode Island, and Connecticut (the District of Maine

Fig. 6–1. The title cartouche and southern part of James Whitelaw's 1796 map of the state of Vermont.

not included) to be compiled from actual surveys and the best information that can be secured from gentlemen acquainted with different parts of these states." Several years passed before Carleton's proposal received official recognition. With some additional persuasion by the Massachusetts Historical Society, the Massachusetts legislature on June 18, 1794, passed a resolution directing the "inhabitants of the several towns etc., therein, to take or cause to be taken, by their Selectmen or some other Suitable person or persons, appointed for the purpose, accurate plans of their respective towns etc., upon a Scale of 200 rods to an inch, and upon a Survey thereafter to be made, or that had actually been made, within seven years not preceding the time of passing said Resolution, and the same plans to lodge in the Secretary's Office, on or before the 1st day of June, A.D. 1795." The resolution also applied to the district of Maine which until 1820 was administratively part of Massachusetts.

Some 265 maps of towns in Massachusetts and approximately 100 of towns in the district of Maine were duly prepared and deposited with the secretary of state. At the common scale of 3300 feet to an inch the town plans show roads, churches, habitations, ponds, and rivers, as well as town boundaries. Few of the surveys were prepared by competent surveyors, and the submitted data were, therefore, inferior to those available for Whitelaw's map of Vermont.

The resolution of the legislature failed to specify how, or by whom, the town surveys were to be utilized in compiling the state map. Concerned that the data would not be processed, the Massachusetts Historical Society, at a meeting convened in January 1795, voted to "apply to the General Court for the exclusive power of compiling a map, for the benefit of the Society, from the returns which may be made of plans of towns in this State, ordered by the Legislature."[13] The request was granted subject to certain restrictions and reservations. To finance the project, the society obtained a loan from the General Court. Responsibility for compiling the map was assigned to Carleton. This assignment provided him with the opportunity to develop his aptitude in mathematics and engineering. It also gave him excellent training in surveying and mapmaking, which he utilized in running his mathematical school in Boston.

Compilation of the separate maps of Massachusetts and the district of Maine was carried out concurrently and, toward the middle of 1798, both were ready for the engraver. John Norman, a Boston craftsman who had earlier worked in Philadelphia, was engaged to prepare the plates. Norman's initial effort did not meet with official approval, and in June 1798 the Massachusetts legislature resolved "that Osgood Carleton and John Norman be allowed a further time of seven months to complete the Maps of the Commonwealth, and it is expected in the Mean time they correct all the Error in said maps, and take out the many accidental strokes in the Plate; and also that they make Margins of the Rivers, Ponds, and Sea Coasts neater, and that the whole Plate be better Polished, all of which being done they are directed to lay the same before the General Court, at their next Session, for their acceptance. Concurred by House of Representatives, June 29, 1798."[14]

The corrected plates still did not meet with the standards of the legislature and the Carleton-Norman effort was rejected. The two collaborators, therefore, privately published and sold the maps of Massachusetts and the district of Maine. The former is titled *An Accurate Map of the Commonwealth of Massachusetts Exclusive of the District of Maine Compiled pursuant to an Act of the General Court From Actual Surveys of the several Towns &c. Taken by Their Order. . . . By Osgood Carleton. Boston, Published and Sold by O. Carleton and J. Norman. Sold also by W. Norman, No. 75 Newbury St.* The map measures 89 by 119 cm. and is at the scale of 1:253,440. A note on the map explains that "as the surveys of some towns were not so full as others, the Roads and Streams of those Towns have been unavoidably discontinued." The title of the second map is *Map of the District of Maine, Drawn from the latest Surveys and other best Authorities* (Fig. 6–2). The information supplied about the compilation of this map is the same as

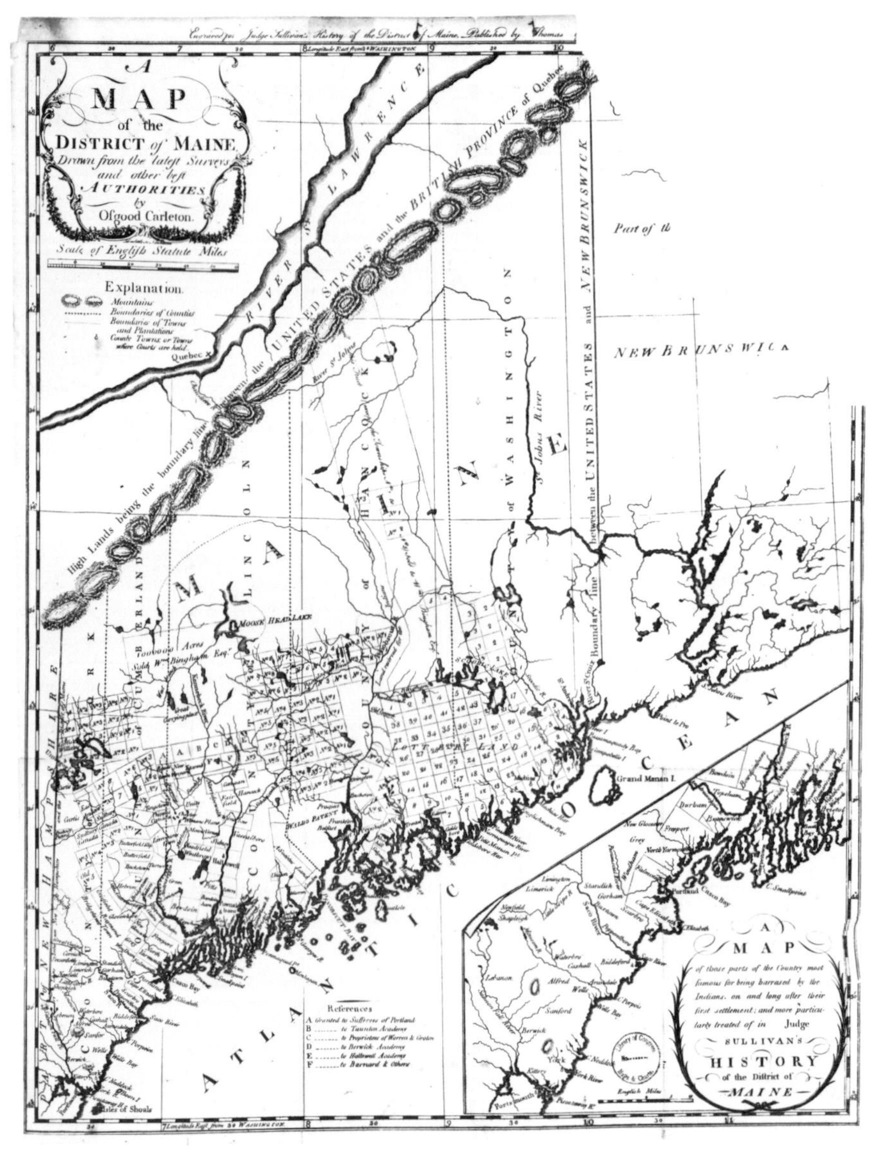

Fig. 6–2. Osgood Carleton's 1795 map of Maine, which at the time was part of the Commonwealth of Massachusetts. The map was engraved by Norman.

that on the Massachusetts map, as is the imprint of publishers and distributors. The district map also carries the disclaimer concerning roads and streams. Neither map is dated. Scholars J. Clements Wheat and Christian Brun assign to each the tentative date of 1795 because "Evans dates [them 1795] as Carleton was only selling maps in 1795."[15] In view of the rejection of the maps by the General Court in June 1798 and the grant of an extension of seven months to improve them, it is more likely that both maps were published late in 1798.

Following rejection of the Carleton-Norman plates for the Massachusetts and Maine maps, a contract for preparing new engravings was awarded to B. & J. Loring of Boston who in turn engaged Joseph Callendar and Samuel Hill to engrave the plates. Both of these engravers practiced their trade in Boston. To insure that the new engravings be of high quality, a joint committee of the Massachusetts Historical Society and the American Academy of Arts and Sciences was appointed to inspect the finished product. The inspection apparently was favorable for the *Map of Massachusetts Proper Compiled from Actual Surveys made by Order of the General Court . . . By Osgood Carleton* (Fig. 6–3) was published in 1801. Callendar and Hill are listed as engravers. The map, priced at $4.50 for an unmounted copy and $7.00 for one mounted on cloth with rollers, was published and sold by B. & J. Loring in Boston. The map of the district of Maine by the same engravers and publishers is dated 1802.

The Massachusetts senate then resolved on June 19, 1801, "that the copper plates engraved for the map of this Commonwealth be, and hereby are, granted to the Academy of Arts and Sciences and to the Massachusetts Historical Society, together with the copyright of printing, publishing, and vending the same, at the expense of said Societies."[16] The two societies apparently did not issue subsequent editions of the two maps. In 1822, however, the Massachusetts map was re-issued in Albany by Amos Lay. Except for the deletion of the original publication date of 1801 and the addition to the title

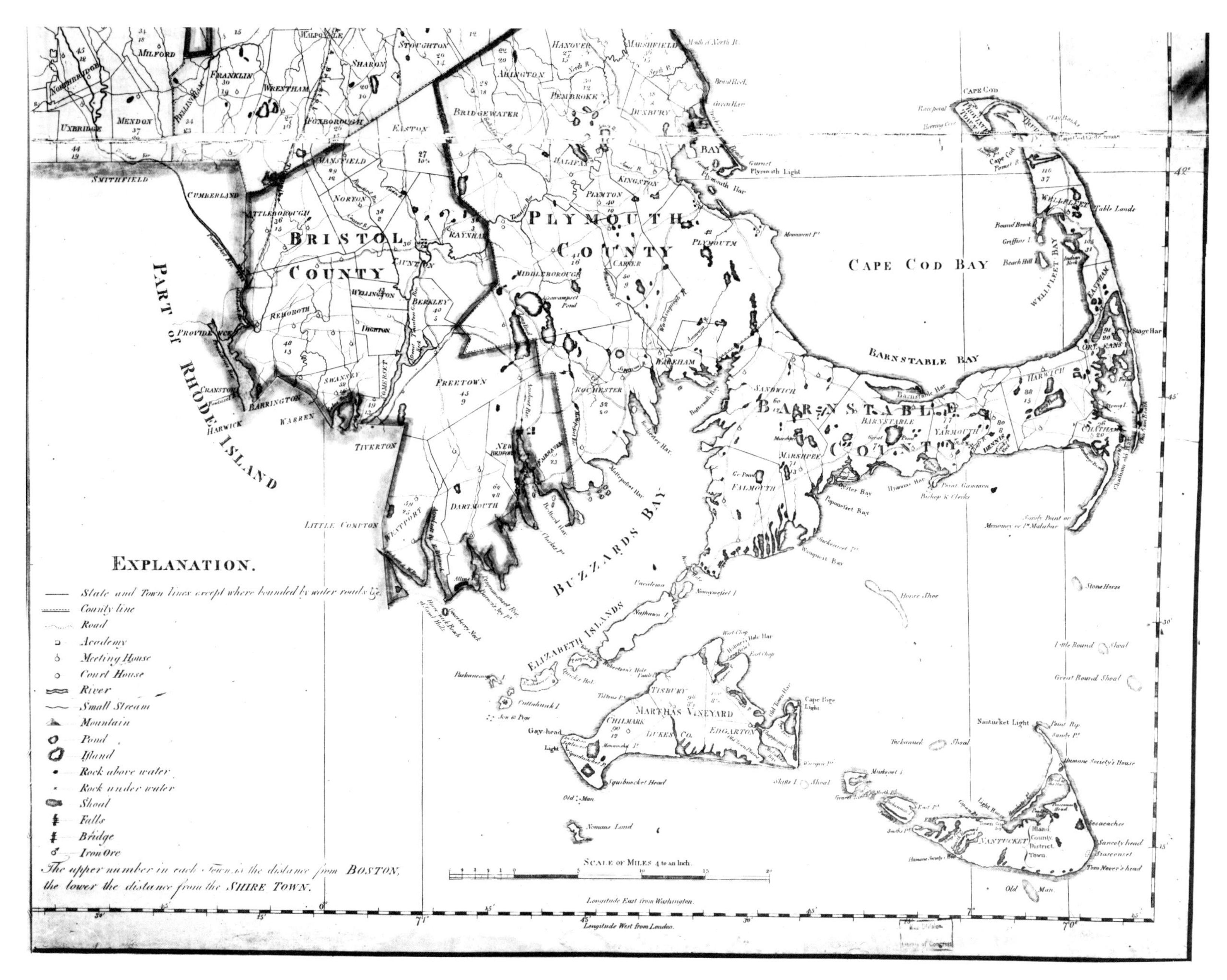

Fig. 6–3. The southeastern portion of Carleton's 1801 *Map of Massachusetts Proper Compiled from Actual Surveys made by Order of the General Court*.

cartouche of "Revised Corrected and Republished by Amos Lay, Geographer &c. Albany 1822," the principal revisions appear to be in county boundaries in the western part of the state.

The cartographical and political history of Maine were greatly influenced between the years of 1815 and 1830 by Moses Greenleaf, Jr. A historian of that state once testified that "we but do justice to a remarkable man, now almost forgotten, by saying that Mr. Greenleaf through his published writings and his accurate and beautiful maps, did more than any other man to make known to two states the value and importance of Maine while it was simply a district under the government of Massachusetts. . . . Mr. Greenleaf was the real statemaker of Maine."[17]

Greenleaf was born in Newburyport, Massachusetts, on October 17, 1777. He was the oldest child in a family of four boys and one girl. In 1790 the family moved to New Gloucester in the district of Maine where Greenleaf's father engaged in farming. In 1799 Greenleaf left the farm to open a store in New Gloucester. Shifting his place of operation first to the town of Poland, then to Kenduskeag, and finally to Bangor, he remained in business until 1806. In that year, he joined in partnership with a Mr. William Dodd, of Boston, to develop land in the area of present-day Piscataquis County. As part of his agreement with Dodd, Greenleaf was to settle with his family in the region. It was not until the summer of 1810, however, that Greenleaf moved to Williamsburg, where he spent the remainder of his life.

Handicapped in his efforts to learn more about the interior of Maine for lack of a good, detailed map of the district, Greenleaf decided to compile one. For the next five or six years he devoted every spare moment to collecting data for the proposed map. The *Map of the District of Maine from the latest and best Authorities by Moses Greenleaf 1815* was published in 1815, with a related booklet entitled *Statistical View of Maine* coming from the press the following year. Greenleaf was not able to conduct personal surveys of the entire district, so he compiled his map from earlier published works and from information received in correspondence with knowledgeable individuals. The map was dedicated "to the Honourable Legislature of the state of Massachusetts." Its engraver was W. D. Annin and its publisher Cummings & Hilliard, both of Boston.

Greenleaf recognized the limitations of his map, noting in the *Statistical View of Maine* that "in general it is correct as can be expected, until a new survey of the whole, corrected by celestial observations, under the immediate inspection of persons properly qualified for the purpose, shall furnish better materials than are now existing. And, until this is done, a perfect Map of Maine cannot be obtained." Greenleaf had received no assistance from Massachusetts in compiling his map, but after it was published that state subscribed to one thousand copies at three dollars each. This financial assistance was greatly welcomed by Greenleaf, and the purchase insured wide distribution of the map throughout Massachusetts.

When Maine was established as an independent state in 1820, Greenleaf published a revised edition of the map with the title *Map of the State of Maine* (Fig. 6–4). This revision also included a number of new towns that had been incorporated between 1815 and 1820, as well as corrections of errors discovered in the first edition. Almost as soon as this edition appeared, Greenleaf started compiling a third map to illustrate his comprehensive book, *Survey of the State of Maine in Reference to its Geographical Features*, which was later published in Portland, Maine, by Shirley & Hyde in 1829. While the map was still being compiled, the Committee on Literature and Literary Institutions of the state legislature reported "that they have examined a plan, Sketches, and Specimens of a Map and Statistical View of the State exhibited by Moses Greenleaf, and find it to be a work on which great attention and labor have been bestowed, and which promises to be executed with skill, accuracy, and judgement—and believing it to be replete with knowledge highly useful to the people and important to

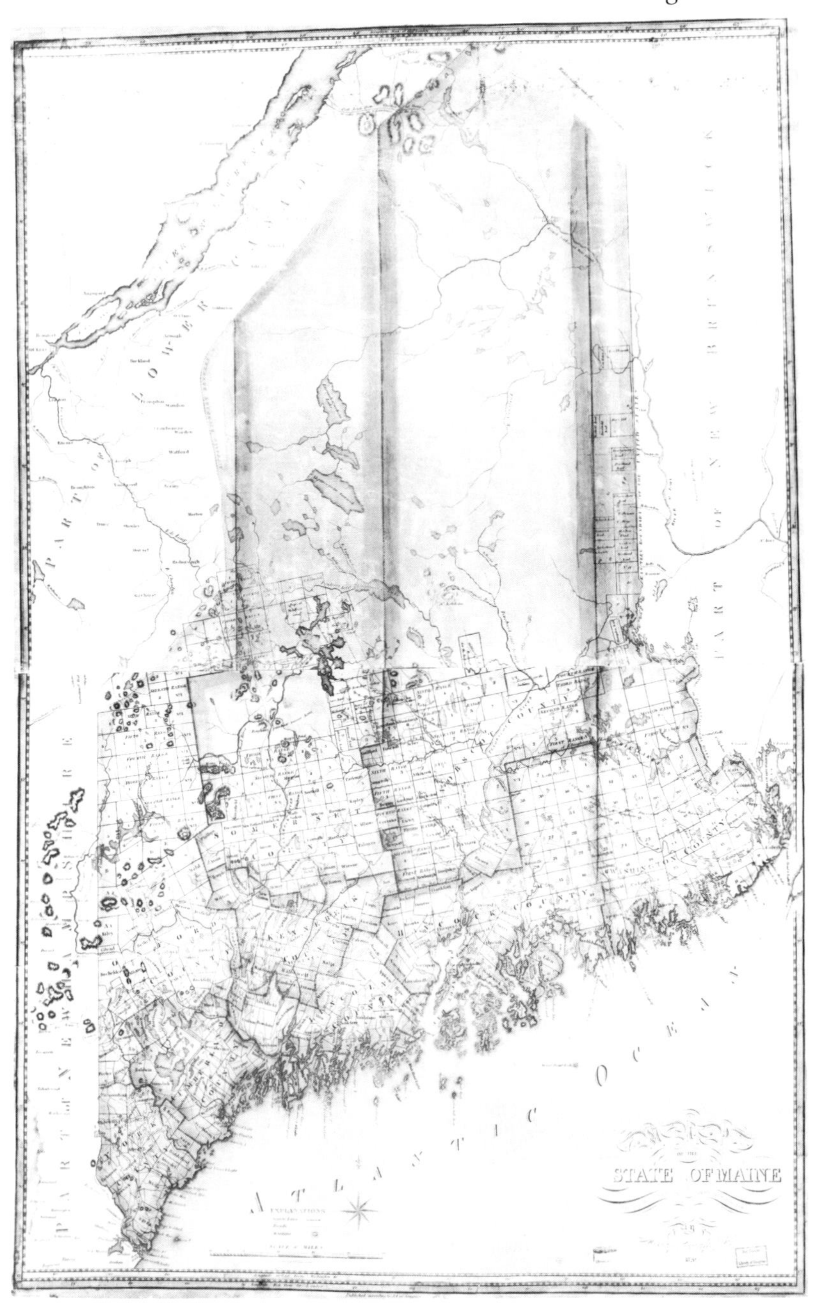

Fig. 6–4. Moses Greenleaf published his *Map of the State of Maine* soon after Maine achieved statehood in 1820.

the State, recommend it to the favorable notice and patronage of the legislature."[18] In consequence of this endorsement the legislature in 1828 granted Greenleaf $1000 "to assist him in completing and publishing his series of Maps and Statistical View of the State." The following year, when the map was published, Greenleaf received an additional $640 to cover the cost of forty copies of the map purchased for official state use.

The 1829 map, which is considerably larger than the first two editions, is titled *Map of the State of Maine With the Province of New Brunswick By Moses Greenleaf*. It was engraved by J. H. Young and F. Dankworth of Philadelphia and published in Portland, Maine, by Shirley & Hyde. Shirley & Hyde also published Greenleaf's atlas of the state of Maine in 1829. The atlas includes five maps, a series of vertical sections of the state, and some meteorological diagrams. All the maps and diagrams were engraved by William Chapin of New York. This publication, which lacks a title page, was the third atlas published of an individual state. It was preceded only by Robert Mills's *Atlas of the State of South Carolina* of 1825 and David H. Burr's *Atlas of the State of New York*, also dated 1829. Greenleaf's son and namesake published a new edition of the map of Maine in 1844. It was printed from the original copper plates cut for the 1829 map, but was updated and corrected. Three newly formed counties, Piscataquis, Franklin, and Aroostook, were added, more than eighty new towns were incorporated, and the new international boundary of the state, established by the treaty of 1842, was delimited.

Town surveys also comprised the principal compilation data for a map of New Hampshire. The state legislature in 1803 and 1804 directed that a map of the state be compiled under the direction of Secretary of State Philip Carrigain from surveys prepared by town officials and sent to the secretary's office. Actual work was begun in 1806, when a committee of the legislature reported "that the said Carrigain has completed the map of New Hampshire with great accuracy, and in a style of superior elegance."

This map, *New Hampshire By Recent Survey Made under the Supreme Authority and Published According to Law by Philip Carrigain, Counsellor at Law and late Secretary of the State*, was published at Concord in 1816 (Fig. 6–5). It is inscribed "To his Excellency John Taylor Gilman Esq. and to the Honourable the Legislature of the State of New Hampshire This Map commenced under their Auspices and matured by their Patronage is most respectfully inscribed by their Obliged Servant Philip Carrigain." The map sheet, which is 132 by 121 cm., includes inset maps of New England and the "Middle, Southern, and Western Sections of the United States," a "View of the White Mountains from Shelburne" the "Gap of the White Mountains," and an elaborately decorated title cartouche that includes the state seal. There is no engraving credit on the map.

There is no indication that Carrigain had any training in surveying. He was born in Concord, New Hampshire, on February 20, 1772, and attended Dartmouth College, from which he received a law degree in 1794. From 1805 to 1808 he held the office of secretary of state of New Hampshire and during this period was charged with compiling the state map. As in most projects of this type, many of the town surveys were made by unskilled persons and considerable ingenuity was required to reconcile the several town surveys into a coordinated map. Although the map was compiled and published under the authority of the state legislature, it carries the copyright notice, "Entered according to the Act of Congress the 14*th* day of October 1814 by Philip Carrigain of the State of New Hampshire." From 1821 to 1823 Carrigain served as clerk of the state senate. Notwithstanding his many talents, he seems to have not lived up to his capabilities. One writer notes that "Mr. Carrigain's convivial tastes and diversified pursuits did not conduce to his success in his profession."[19] He died in Concord on March 16, 1842.

No new map of Connecticut was published for two decades after Blodget's 1791 effort. In his excellent volume *Maps of Connecticut*, Thompson notes that "it was not until 1812 that an important new survey was made, that by Moses Warren and George Gillet; the work done by these two men served as a basis for much that came after them."[20] Their map is titled *Connecticut, From Actual Survey, Made in 1811; By and under the Direction of, Moses Warren and George Gillet; And by them Compiled. Published under the Authority of the General Assembly, by Hudson & Goodwin. Engraved by Abner Reed, E. Windsor* (Fig. 6–6). The map was published at Hartford in February 1812 and is inscribed, "To His Excellency Roger Griswold, Esq. And to the Honourable The Legislature of The State of Connecticut." It was copyrighted May 29, 1813, by Hudson & Goodwin.

The Warren-Gillet map has a list of some thirty-five "Explanations," including various types of mills and manufacturers, churches, academies, mines, county lines, turnpikes, and common roads. Thompson reminds us that "map makers of the late 'nineties did not record turnpikes, but continued to mark all roads as of the same class; . . . so that Warren & Gillet appear to have been the first to take notice of the new growth of turnpike companies when they published their fine map in 1812."[21]

Moses Warren, Jr., was born at Hopkinton, Rhode Island, in 1762. Two years later the Warren family moved to East Lyme, Connecticut. Warren later served in the revolutionary war, attaining the rank of sergeant. His mapping partner, George Gillet, was born in Hebron, Connecticut, in 1771 and died there on March 5, 1853. He was appointed surveyor general of Connecticut in 1813, after the map compiled by Warren and him was published. "Though the publication of this map was a private venture," Thompson wrote, "the survey had an official sanction. At the May session, 1811, of the Connecticut legislature a resolution was passed authorising 'Hudson & Goodwin at their own expense to prepare and publish from actual survey a map of this state, and for that purpose the petitioners are authorized . . . to pass over the land of individuals in the state . . . and empowered to examine the records of the state.'"[22] In

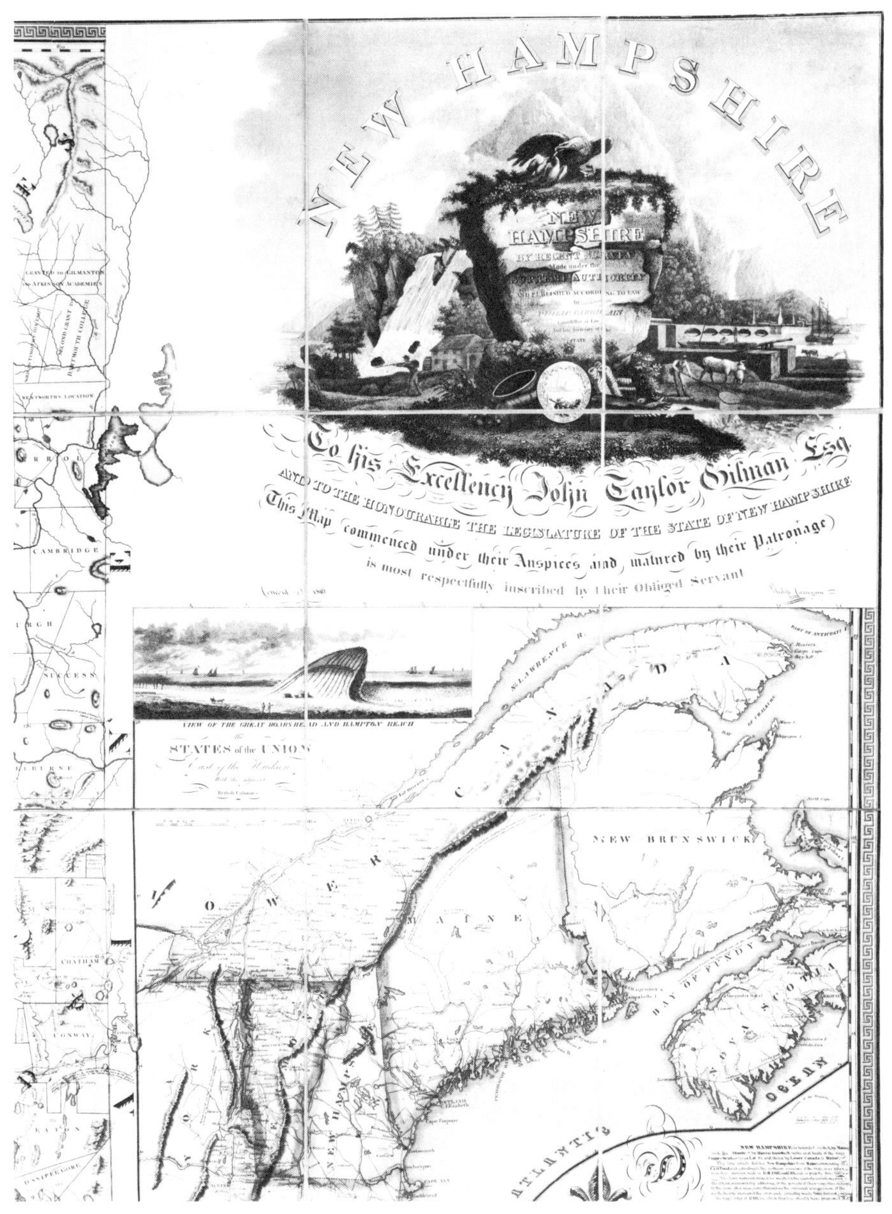

Fig. 6–5. Philip Carrigain's 1816 map of New Hampshire is considered to be the first official map of that state. J. J. Barralet drafted the map, and it was engraved by William Harrison. The title cartouche and a portion of the map are shown here.

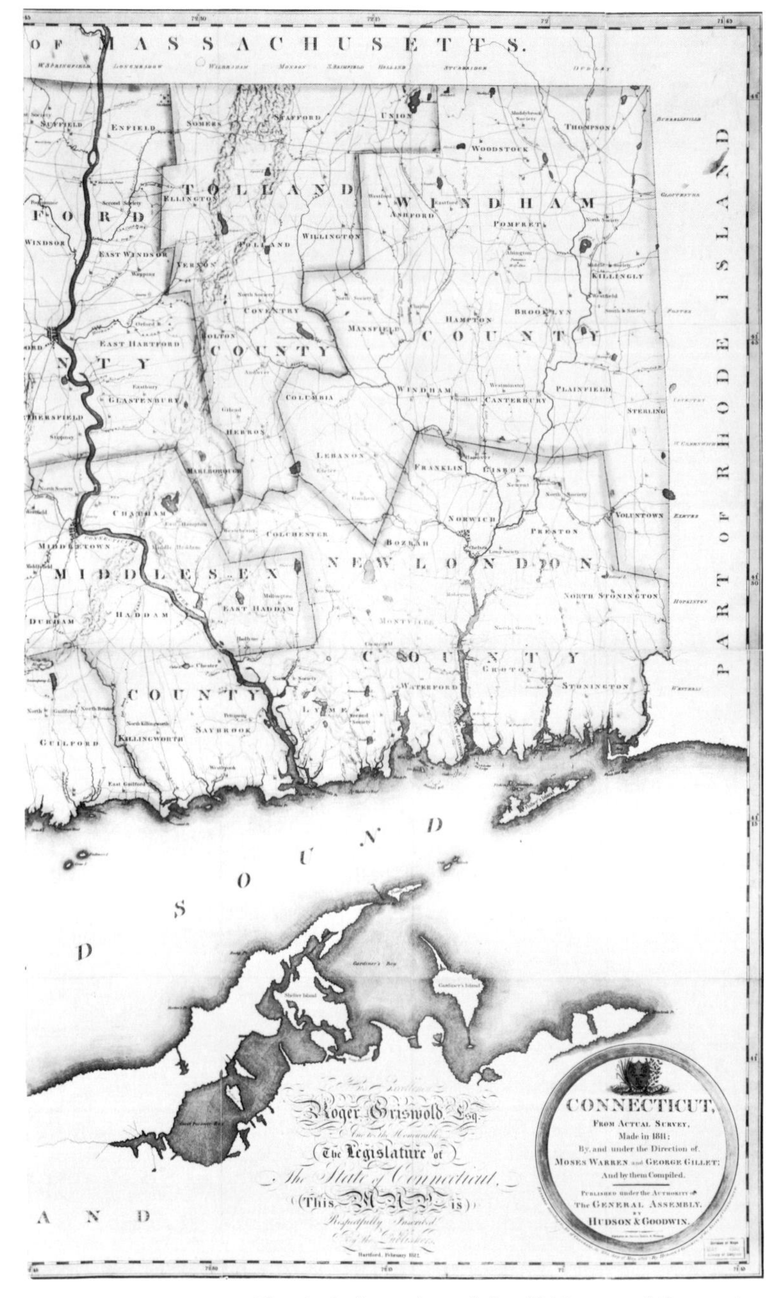

Fig. 6–6. A section of the 1812 map of *Connecticut, From Actual Survey* by Moses Warren and George Gillet.

1813 the legislature authorized the purchase of a number of copies of the Warren-Gillet map for presentation by the governor to his counterparts in other states. Thompson observes that "this purchase of copies of the map seems to have been the only state financial help given to the venture."[23]

The actual survey indicated in the map's title was very likely composed of town and county surveys, which the two collaborators had personally made in their capacities as town surveyors. Also probably included were data and manuscript maps received from other local surveyors and town and county officials. Reduced scale editions of the map were published by Surveyor General Gillet in 1820, 1829, 1833, 1842, and 1847. All of these editions were engraved and published by Asaph Willard of Hartford. Willard was associated as early as 1816 with Ralph Rawdon, who had shortly before opened an engraving shop in Albany, New York. Around 1818 he had relocated to Hartford, where he was occupied in general and bank note engraving.

By the 1830s, canal, turnpike, and railroad construction and emigration from Europe resulted in population increases and growth in the number and size of towns and cities. All these conditions created markets for maps; mapmaking, engraving, and publishing became more permanently established enterprises. This was reflected in the fact that, after 1830 or so, the publisher, rather than the surveyor or compiler, assumed responsibility for initiating maps. For example, Alfred Daggett, a New Haven engraver, published, probably in 1827, a *Map of Connecticut, From actual Survey*. To quote Thompson, "this is the earliest appearance I have found of a plate that had a long and varied career. It next appears in 1831 issued by Thrall in Hartford, with Ely's name added beside that of Daggett as engraver; in 1836 it bore the imprint of Huntington; from 1847 to 1854 that of Brown & Parsons; and finally in 1858 it carried the imprint of Brown & Gross."[24] We may speculate that by the late 1840s, engravers were beginning to feel the competition of lithographically reproduced maps and to reduce costs they modified and corrected each other's copper plates rather than issuing entirely new maps.

During the colonial period Narragansett Bay and the coastal portions of Rhode Island were mapped by Charles Blaskowitz. In the post-Revolution period, it was 1795 before Rhode Island had its own state map. It is titled *A Map of the State of Rhode Island taken mostly from Surveys by Caleb Harris* and was drawn by Harding Harris and engraved by Samuel Hill. Below the map is the imprint "Engraved for Carter & Wilkinson, Providence 1795." The map is 53 by 41 cm. in size and at the approximate scale of 1:175,000. Roads, town and county lines, and a few topographical features are shown. Apart from the information on the map, little is known about Caleb Harris, Harding Harris, or the publishing firm of Carter & Wilkinson. Hill, the only identifiable name on the map, was engraving in Boston as early as 1789. This map is the first separate map of Rhode Island and the first to show details of topography and place names in the interior of the state. A reduced version of it is found in several Mathew Carey atlases.

Initiative for the next state map of Rhode Island was taken by Amos Lay, a New York mapmaker and publisher. Lay was born around 1780, probably in New York City. Little is known about his education or training or what induced him to go into map publishing. His earliest cartographic effort appears to be the *Map of the Northern Part of the State of New York*, a joint project with Arthur J. Stansbury that was published in New York City in 1801. It includes only the northeastern part of the state and west to Cayuga County. A map with the same title but which extends west to Lake Erie and south to the Pennsylvania border was independently published by Lay in 1812. It includes details of the towns in the Military Tract, very probably derived from De Witt's 1st Sheet. An 1817 edition of Lay's map as well as an 1819 revision extend south to include New York City and parts of Pennsylvania and New Jersey. Lay also published 1822, 1823, 1824, and 1826 editions of the large New York State map. He likewise published editions of a large wall map of the United States in 1819 and 1836. On the 1819 U.S. map is the note "Title &c. designed & engraved by Thos. Starling, No. 1 Wilmington Square, London."

In 1821 Lay undertook the compilation of a map of Rhode Island. He soon found it to be an expensive task and applied to the state legislature for aid. The latter "voted that the state should purchase 12 copies of his map for $60. Two years later Lay turned the undertaking over to Ariel Van Hann of Westerly and James Stevens of Newport. Stevens did the surveying and finally assumed the entire work of publishing the map. The legislature increased the number of maps to be purchased by the state from 12 to 137. . . . The work progressed slowly, and it was not until 1831 that the map was issued, 10 years after the first recognition of it by the Legislature."[25] The title of the map is *A Topographical Map of the State of Rhode-Island and Providence Plantations; Surveyed Trigonometrically and in Detail, By James Stevens, Topographer and Civil Engineer: Newport, R.I. 1831*. It was registered for copyright in that year, but the imprint in the lower right corner reads, "Published by James Stevens, Newport, R.I., 1832." Soundings are given in Narragansett Bay, and there is a table giving 1830 census data by county for principal towns and cities. There is an extensive network of roads, and principal landowners are named.

Stevens, who was an experienced and practicing civil engineer, also found it necessary to petition the legislature for financial assistance. In the *Acts and Resolves* of the Rhode Island General Assembly for October 1829, it was reported that "whereas James Stevens of Newport has been at much expense in making actual surveys, and collecting materials for an accurate map . . . Be it enacted by the General Assembly . . . that the said Stevens be . . . authorised . . . to raise the sum of two thousand dollars . . . by lottery; . . . that he will appropriate the said sum to the completion and publication of an accurate map of the State, and will deliver . . . for the use of said State, free of all expense, one thousand copies of said Map, on or before the fifteenth day of June, A.D. 1830."

The Stevens map is the first one of a state purported to have been "surveyed trigonometrically." By 1830 the U.S. Coast Survey had completed trigonometric surveys of Narragansett Bay and the coast of Rhode Island. It is probable, therefore, that Stevens had access to these coastal surveys and extended the triangulation inland to include all of Rhode Island. His map was republished in 1846 by Isaac H. Cady of Providence "with additions & correction by S. H. Cushing & H. F. Walling."

By 1840 state governments were financially more sound, and significant improvements had been made in surveying instruments and procedures. Internal improvements, such as canals, roads, and railroads and an active interest in learning about and developing their natural resources induced some of the more affluent states to establish or strengthen their topographic and geologic departments. Triangulation surveying, as employed for Stevens's map of Rhode Island, greatly increased the accuracy of maps, and other states were not long in adopting such techniques. Until well beyond the middle of the nineteenth century, however, there was still considerable dependence upon individual surveyors, mapmakers, and publishers for the preparing of state maps.

In 1830 Massachusetts realized it needed a more accurate state map, and having the financial resources and the skills of experienced mapmakers, decided to initiate a mapmaking project. The earlier Massachusetts map, which had been sponsored by the Massachusetts Historical Society and compiled by Carleton, had numerous shortcomings and by 1830 was sadly out of date. Consequently, in March of that year the Massachusetts legislature passed a resolution requiring the city of Boston and the several towns in the state to make accurate maps of their territories on a scale of one hundred rods to an inch and deposit them with the secretary of state. That summer the governor of Massachusetts appointed Robert Treat Paine as principal engineer and Stevens as principal assistant in directing the compilation of the state map. After a short while, Stevens succeeded Paine as head of the project, with Simeon Borden as his principal assistant. Stevens then resigned in 1834, and Borden carried the project through to completion. Borden finished the fieldwork in 1838 and undertook the task of reconciling the maps prepared by local surveyors. He reported that he "found the town maps which had been returned to the Secretary so incorrectly drawn as to render it impossible, in their actual state, to make a satisfactory map from them. I was then obliged to go into the field again, with four or five assistants, and make corrections."[26]

It took from 1838 to 1842 to complete the compilation of the map from these field surveys and drafts of local surveyors, and the map was printed in 1844. Its title is *Topographical Map of Massachusetts Compiled from Astronomical, Trigonometrical, and Various Local Surveys; Made by Order of the Legislature. Simeon Borden, Superintendent. 1844. Engraved by George G. Smith, Boston.* It is 125 by 194 cm. in size and at the scale of 1:158,400. Although produced officially, the map was "published and sold by C. Hickling No. 20 Devonshire St. Boston." Some extant copies also include these imprints: "Published and sold by Eayrs & Fairbanks No. 136 Washington St., Boston" and "Published & Sold by H. F. Walling, Surveyor, 15 Joys Building, Boston." Interesting inset features of the map are a "Geological Map of Massachusetts Made by Order of the Legislature by Edward Hitchcock" and a "Plan of the Principal Triangles in the Trigonometrical Survey of Massachusetts." Borden's map was republished in 1861 by Henry F. Walling by lithographic transfer.

Borden was born in Fall River, Massachusetts, on January 29, 1798. His family moved to Tiverton, Rhode Island, when he was eight years old. He studied in the local rural schools there until he was thirteen, when his father's death terminated his formal education. Independently he studied applied mathematics and metal and woodworking. At the age of thirty he became superintendent of a machine shop, and two years later he was engaged to construct the bar to be used in measuring the baseline for the Massachusetts survey. This was his entrée to the project, which he successfully completed. In the 1840s and 1850s Borden conducted surveys for several New England railroads. He died on October 28, 1856.

Notes

1. Lloyd A. Brown, *The Story of Maps* (Boston, 1949), 276.
2. Henry S. Tanner, *Memoir on the Recent Surveys, Observations, and Internal Improvements, in the United States, with Brief Notices of the New Counties, Towns, Villages, Canals, and Railroads, Never before Delineated. Intended to Accompany His New Map of the United States*, 1st ed. (Philadelphia, 1829), 8.
3. Much of the biographical information about Blodget is derived from Phillis Kihn, "William Blodget, Map Maker 1754–1809," *Connecticut Historical Society Bulletin* 27 (Apr. 1962): 33–50.
4. Kihn, "William Blodget," 42.
5. Ibid., 43.
6. David A. Cobb, "Vermont Maps Prior to 1900, an Annotated Cartobibliography," *Vermont History* 29 (Summer and Fall 1971): viii.
7. Kihn, "William Blodget," 45.
8. Connecticut, *Public Records May 1789–Oct. 1792* (Hartford, 1948), 87.
9. Edmund Thompson, *Maps of Connecticut before the Year 1800, a Descriptive List* (Windham, Conn., 1940), 44–45.
10. Kihn, "William Blodget," 45.
11. Ibid., 43, 45.
12. Vermont, Secretary of State, *State Papers of Vermont, Volume One. Index to the Papers of the Surveyor-General* (Rutland, Vt., 1918), 2d ed. (Montpelier, Vt., 1973), 9.
13. *Proceedings of the Massachusetts Historical Society*, Vol. 1, *1791–1835* (Boston, 1899), 81.
14. *Boston Gazette*, Aug. 20, 1798.
15. James Clements Wheat, and Christian Brun, *Maps and Charts Published in America before 1800: A Bibliography* (New Haven, 1969), 36.
16. *Proceedings of the Massachusetts Historical Society*, Vol. 1, *1791–1835* (Boston, 1889), 141.
17. Samuel Lane Boardman, introd. to *Moses Greenleaf, Maine's First Map-Maker*, by Edward Crosby Smith (Bangor, Maine, 1902), xv–xvi.
18. Ibid., 68.
19. Charles Bell, *Bench and Bar of New Hampshire* (Boston, 1894).
20. Edmund Thompson, *Maps of Connecticut for the Years of the Industrial Revolution 1801–1860* (Windham, Conn., 1942), 7.
21. Ibid., 17.
22. Thompson, *Maps of Connecticut*, 39.
23. Ibid.
24. Ibid., 53
25. Howard Miller Chapin, *Cartography of Rhode Island* (Providence, R.I., 1915), 10.
26. Simeon Borden, "Account of a Trigonometrical Survey of Massachusetts, with a Comparison of Its Results with Those Obtained from Astronomical Observations, by Robert Treat Paine," *Transactions of the American Philosophical Society* 9, n.s. (1846): 34.

7. *Early State Maps: The Middle States*

New York and Pennsylvania dominated the middle states historically, physically, and economically. Both embraced extensive land areas of great diversity and richness and were strategically important during the revolutionary war. Each for a time harbored the capital of the country, and served as links between New England and the southern and western states. Both boasted major seaports and commercial centers. Both had been reasonably well mapped before the Revolution, but their dominant positions in the new republic induced them quite early to compile and publish maps of their jurisdictions. New York's surveyor general, Simeon De Witt, produced two maps of his state that were presented to the governors of all the other states. These maps served as examples and stimulated other jurisdictions to produce similar works.

Although a reduced version of De Witt's comprehensive 1802 map of New York was published in 1804, no major cartographic effort was undertaken by the surveyor general's office for some two decades. During this period several independent publishers issued maps of New York State, for which they borrowed liberally from De Witt's work. Amos Lay published maps of the state, or parts thereof, between 1801 and 1826. His maps show no place of publication and give no engraving credit. John H. Eddy, who listed himself as "Geographer" on his maps, published large ones of New York State in 1811 and 1818. Both were engraved by Tanner, Vallance, Kearny & Company of Philadelphia and printed by Samuel Maverick in New York. Apart from the information available on the maps, little biographical record has been found for either Lay or Eddy. It is known, however, that Lay died in 1851.

The Erie Canal, which opened in 1825, promised to bring growth and prosperity to the counties along its course. This prospect induced residents of the southern tier of counties to agitate for a road to insure a transportation link between the Hudson River and Lake Erie across the southern part of New York State. Yielding to these pressures, Governor De Witt Clinton recommended to the legislature that a commission be established to explore and make surveys for a road across the southern part of the state. The bill, authorizing the governor to name three commissioners to the state road commission, was passed by the legislature in April 1825.[1] Shortly thereafter three surveying parties took to the field. David H. Burr was named deputy surveyor in charge of that portion of the proposed road from Little Valley in Cattaraugus County through Jamestown to Mayville in Chautauqua County.[2]

Burr, who seems to have had no previous training or experience in surveying, was typical of the lusty, independent, and resourceful young men spawned in nineteenth century America. He was born in Bridgeport, Connecticut, in August 1803, the third of eight children of Amos and Abigail Shelton Burr. Around 1822 Burr moved to Kingsboro in Fulton County, New York, where he studied law and was admitted to the New York State bar. He did not practice law for long for in 1824 he enlisted in the New York State militia. A year later he was appointed aide-de-camp to Governor Clinton. Burr's appointment to head one of the road surveying parties was probably supported by the governor. The leader of one of the other surveying teams was Joseph Henry, who later became the first secretary of the Smithsonian Institution and a lifelong friend of Burr.

Following completion of the road surveys, Burr, perhaps again with the support of Clinton, obtained copies of the reports and maps of all three road survey parties. It is not known what prompted this action, but it is evident from subsequent events that Burr planned to utilize the survey data and maps to compile a map and atlas of New York State. He later reported that "the Legislature of the Senate in 1827, aware of the importance and public utility of an accurate map . . . upon recommendation of De Witt Clinton, then Governor of the State, after examining this Atlas, then in its incipient state, and being satisfied that the plan proposed was one which would be of great benefit to the public, and a sure means of getting an accurate State map, passed an act [on October 16, 1827], directing that whenever a set of maps was compiled according to this plan, and delivered to the Surveyor-General and Comptroller, they would revise and correct the same, and that when they were satisfied with their accuracy publish them at the expense of the State."[3] This statement suggests that Burr received official support in compiling an atlas and map of the state. It is not completely clear whether he did the compilation independently or whether he was for a time attached to the office of the surveyor general. Burr was a resident in Albany from 1826 to 1832, and on September 30, 1828, he married Susan Cottle of that city.

A communication dated February 1829 from Surveyor General De Witt to the state senate supports the possibility that Burr was working under the direction of De Witt. "For the purpose of having the maps of counties as correct as possible," the surveyor general informed the senate, "recourse was had to circular letters, one of which was addressed to the supervisor of each town, enclosing the delineation of each town as drawn by Mr. Burr, with a request to have all errors that might be preserved in it, corrected. Answers to these letters, all more or less important, have been received, and many of them have corrected very material errors, which the compiler, without such information, could not have detected. Such corrections were then communicated to the engravers and the plates altered accordingly."[4] In the same communication De Witt noted that the

> general map of the state [is] to be 50 by 60 inches, embracing, besides this state drawn according to the provisions of the act, the district of the country comprehended by Eddy's map of the state of New-York, and the adjacent parts, to be filled in the same manner as directed for the map of the state, with profiles of the canals: the title [is] to be a neat vignette with the arms of the state. The atlas [is] to consist of the following sheets: a map of each county, drawn according to the provisions of the act, on 50 sheets; a map 20 by 24 inches, embracing the same territory, and exhibiting so much of the information contained within the general map of the state, as can be shown on so small a sheet. Each of the above sheets [is] to have an appropriate title, equal in style and execution to the title of the map of Michigan, by John Farmer. The whole, [is] to be engraved equal in style and execution to the map of the New-England states in Tanner's atlas, deposited in the office of the secretary of this state, and to be completed before the first of May, 1829.[5]

In the introduction to his atlas, Burr acknowledged that "the Legislature . . . made liberal appropriations to defray the expenses, at the same time giving the Author permission to make use of all documents in any of the public offices of the State, or of the several towns and counties, which he should deem necessary in the compilation of the work."[6] Burr's *Atlas of the State of New York* and his separate map of the state did not make De Witt's deadline. Both were finally published in early 1830, although both carried the publication date of 1829. The engravings were prepared by Rawdon Clark & Company of Albany and Rawdon Wright & Company of New York City. The Albany firm was headed by Ralph Rawdon, and his brother, Freeman Rawdon, directed the New York branch. Although he apparently privately published his atlas, Burr acknowledges on the ornately embellished title page that it was compiled "under the superintendance of Simeon De Witt, Surveyor-General" (Fig. 7–1). The atlas includes a "View of the

Hudson near Fishkill" and fifty-eight plates encompassing maps of each county. It is a magnificent publication and was the second atlas of an individual state ever printed. Only Robert Mills's 1825 *Atlas of the State of South Carolina* preceded it.

In a November 25, 1829, letter to Joseph Henry, who was then on the staff of Hobart Academy, Burr complained that "the delays in the appearance of my Atlas have been greater than I anticipated but whether the delay is Chargeable to any one in particular I am uncertain but certainly not to you and I feel extremely grateful for the assistance you have so kindly afforded me. I received several copies of the Atlas on Monday & directed one to be left with Mr. Van Rensalaer (Surveyor-General's office) who promised me to show it to the Editor of the Argus. I presume he will attend to it. Mr. Crosswill has promised as soon as the work appeared to notice it."[7] The atlases delivered to Burr at the time of this letter were advance copies for on February 8, 1830, De Witt reported to the New York legislature that

> the engravings of the plates for the atlas and map of New York State have been completed, and as many copies have been printed as are required by the act for the use of the state. The colouring and mounting of the maps, and the binding of the atlases are in considerable forwardness, and progress as fast as practicable; the whole will probably be finished within six weeks. Specimens of the work are herewith presented. The style of the execution will speak for itself. No pains have been spared to have the topographical delineations as correct as possible. . . . Much credit is due to the compiler for his industry and pains in procuring materials for the work wherever they could be found, and to make it so that it may be deservedly considered as adding to the reputation of the state.[8]

In the same communication, De Witt reported the costs to the state for Burr's map and atlas: plates and

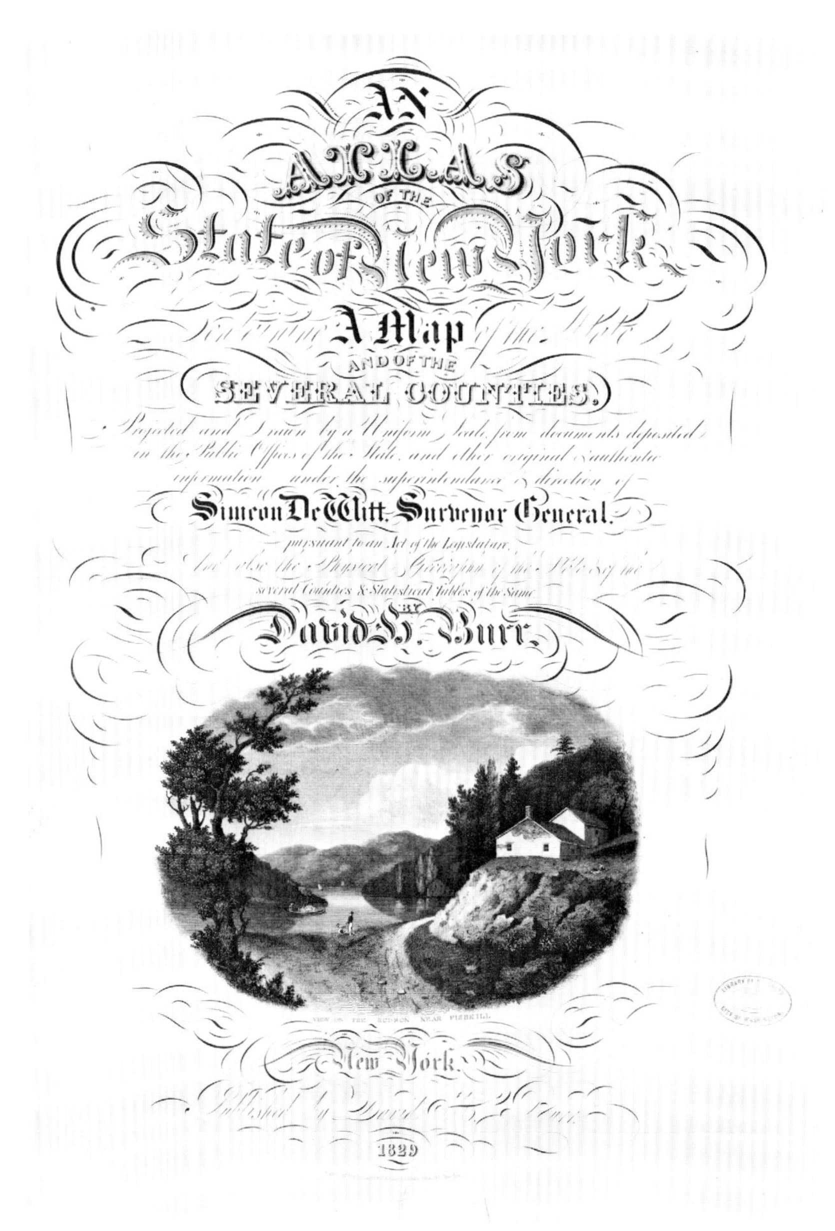

Fig. 7–1. The title page of the *Atlas of the State of New York* by David H. Burr.

engravings, $5,220.02, and paper, printing, coloring, mounting, binding, etc., $2,505.88, for a total of $7,725.90.[9] This was $274.10 below the initial authorization of $8,000. New York's legislators were apparently pleased with the products for on January 30, 1830, Assemblyman Luther Bradish "gave notice that he would, on some future day, ask leave to bring in a bill authorising the Acting Governor to procure and forward to the department of state of the United States and to each of the states in the Union, a copy of David H. Burr's map and atlas of the state."[10]

Burr's map was also not released for distribution until the end of January or early February 1830. Entitled the *Map of the State of New York And the Surrounding Country, By David H. Burr, Published by Simeon De Witt Surveyor General Pursuant to an act of the Legislature*, it is 131 by 147 cm. in size and at the scale of 1:500,000. There is a "Comparative View of Elevations" in an inset. A revised edition of this map was published by New York in 1832, with the same engravers as the original edition. Then the cartographic publishing firm of J. H. Colton & Company of New York City apparently acquired the copyright for it and published revised editions in 1833 and over the next fifteen or twenty years. S. Stiles & Company was the engraver for the Colton editions.

The success of his New York State map and atlas encouraged Burr to undertake other cartographic projects. Perhaps even before the New York atlas was published, he began to compile maps for a world atlas and by 1832 had completed and copyrighted eight maps, which were engraved by the New York firm of Thomas Illman and Edward Pillbrow. In that year Burr accepted the appointment of topographer to the U.S. Post Office Department and was unable to complete the atlas. Illman and Pillbrow assumed responsibility for its completion, and *A New Universal Atlas* was published in 1835 by D. S. Stone of New York City. The title page indicates that the atlas was "carefully compiled from the best Authorities Extant by David H. Burr."

As topographer to the Post Office Department, Burr had access to a large volume of geographical data submitted by local postmasters throughout the country. He used this information to compile his *Map of the United States of North America, With parts of the Adjacent Countries* and twelve maps of individual states or groups of states. All the maps are 124 by 92 cm., and all were copyrighted by Burr on July 10, 1839. The post office maps were sold separately or assembled with the sheets folded in quarters and bound in a large folio volume titled *The American Atlas; Exhibiting the Post Offices, Post Roads, Rail Roads, Canals, and the Physical & Political divisions of the United States of North America, Constructed from the Government Surveys & Other Official Materials, Under the Direction of the Post Master General, By David H. Burr, Geographer to the House of Representatives of the U.S.* The title page carries the same copyright registration date as the individual maps.

On most of the maps, however, Burr is described as "Late Topographer to the Post Office" and "Geographer to the House of Representatives of the U.S." The change in jobs was probably effected in 1838. The postal maps have the name "John Arrowsmith" imprinted in the lower right corners. Some Burr references indicate that he went to England in 1846 to make arrangements with the London commercial map publishing firm of John Arrowsmith to distribute the series of postal maps. The dates on the maps, however, suggest that such arrangements were probably made as early as 1839. The 1841 edition of Burr's *Map of the United States of North America* carries an Arrowsmith imprint (Fig. 7–2); the 1842 edition, though, does not.

Burr served as geographer to the House of Representatives until 1846 or 1847. In 1848 he was named U.S.

Fig. 7–2. Following the compilation and publication of his official map and atlas of New York State, Burr briefly engaged in commercial map publishing with John Arrowsmith. This 1841 map of the United States was one of Burr's commercial publishing efforts.

Map of the
UNITED STATES
OF
NORTH AMERICA
With parts of the Adjacent Countries.
BY
David H. Burr.
(Late Topographer to the Post Office)
Geographer to the House of Representatives of the U.S.
BRITISH POSSESSIONS
PACIFIC OCEAN
ATLANTIC OCEAN
GULF OF MEXICO
CARRIBEAN SEA

surveyor to the state of Florida where he directed surveys following the Seminole War. Around 1850 he accepted the position of surveyor to the state of Louisiana. He returned to Washington, D.C., within a year and some three years later became geographer to the U.S. Senate. All of these experiences made Burr well known. In consequence, in 1855 President Franklin Pierce named him the first surveyor general of Utah Territory. As the top ranking federal official in the territory, Burr was heavily involved with the legal and jurisdictional disputes between Mormon leaders and the U.S. government. Burr held the Utah appointment until 1857. He seems not to have engaged in map compilation or publishing in his later years, and little is known about his activities in the fifteen years before his death in 1875.

Pennsylvania had been mapped in the quarter century preceding the Revolution by Nicholas Scull and his grandson, William Scull. The charting of the extensive military activities and campaigns in Pennsylvania during the revolutionary war further contributed to the geographic and cartographic record of the Keystone State. In the early years of independence, Pennsylvania experienced with its neighboring states the feverish activity of establishing and extending roads and canals. Surveys conducted relative to boundary settlements also added cartographic data.

Reading Howell used this data and personal surveys conducted by himself and others as source materials to compile and publish several maps of the state of Pennsylvania. Howell was a landowner and surveyor, but little else is known about his life or career. As early as 1789, he, with several associates, surveyed and made drafts of the Delaware and Lehigh rivers. In April 1790 he was appointed one of three commissioners to explore lands near the headwaters of the Susquehanna, Lehigh, and Schuylkill rivers. On the same date, two hundred pounds were drawn from the state treasury for Howell as part of the sum of three hundred pounds granted by the assembly "to enable him to proceed in the work he has undertaken of compiling a map of this State."[11]

Howell had apparently been working on his maps as early as 1788. His first published effort was a small (47-by-67-cm.) map which is untitled but carries the inscription, "To the Legislature and the Governor of Pennsylvania This Map is respectfully Inscribed by Reading Howell." There is no date on the map, but a lengthy notice of copyright indicates "that on the thirteenth Day of June in the fifteenth year of the Independence of the United States of America, Reading Howell . . . hath deposited in this Office [District Court of Pennsylvania] the Title of a Map, the Right whereof he claims as Author. . . . A Map of Pennsylvania, & Parts connected therewith relating to the Roads and inland Navigation, especially as prepared to be improved by the late Proceedings of the Assembly (Copied from his larger map) by Reading Howell." It is not clear whether Howell's map was published in 1791, the fifteenth year of U.S. independence, or whether it was only deposited for copyright.

The "larger map" from which this map was copied is almost certainly *A Map of the State of Pennsylvania by Reading Howell, MDCCXCII* (Fig. 7–3). This 94-by-160-cm. map was respectfully inscribed by the author "To Thomas Mifflin Governor, the Senate and House of Representatives of the Commonwealth of Pennsylvania." It shows county and township lines, Indian towns and paths, roads, forges, mills, houses of worship, minerals, furnaces, "water which sinks" (sink holes), and houses. The notice of copyright indicates that the title of this map was deposited "on the eleventh day of January in the fifteenth year of the Independence of the United States." This large map was registered several months earlier than the small one.

Both Howell maps show the full extent of Pennsylvania, the first published maps to do so. The large map

Fig. 7–3. Reading Howell's *Map of the State of Pennsylvania*, 1792, was the first map of that state published after the Republic was established. Only the northeastern part of the map is here reproduced.

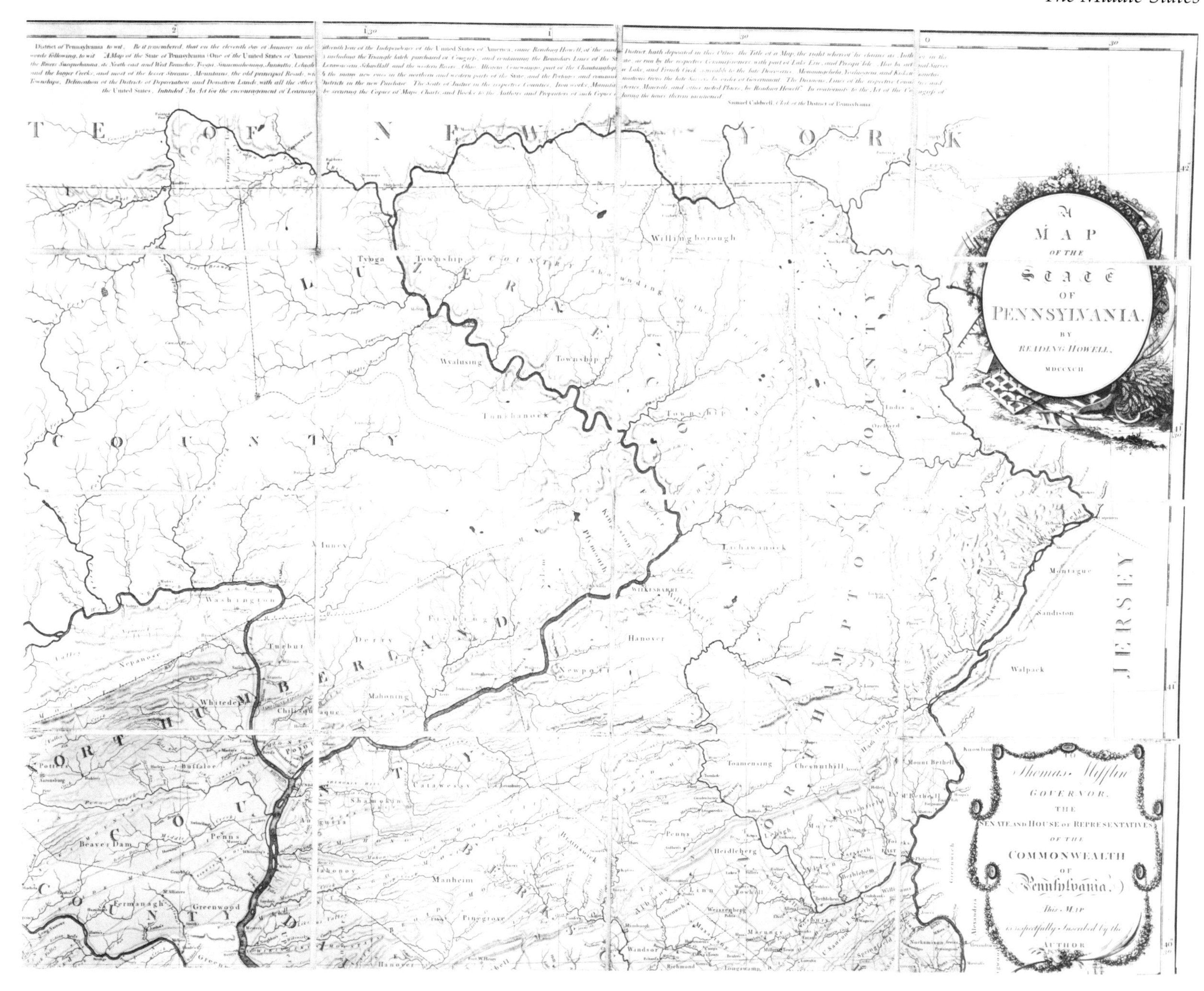
A MAP OF THE STATE OF PENNSYLVANIA, BY READING HOWELL, MDCCXCII
TO Thomas Mifflin GOVERNOR, THE SENATE AND HOUSE OF REPRESENTATIVES OF THE COMMONWEALTH OF Pennsylvania, This MAP is respectfully Inscribed by the AUTHOR
NEW YORK
JERSEY
LUZERNE COUNTY
NORTHAMPTON COUNTY
NORTHUMBERLAND COUNTY

was "Published, August 1792 for the Author; & Sold by James Phillips, George Yard, Lombard Street, London." It is not clear why Howell chose to have his map published in London rather than in Philadelphia. Later editions of the large map, issued in 1806, 1816, and 1817, were published in Philadelphia by Emmor Kimber, alone or with associates. There are 1811 and 1817 editions of the small map, the former engraved by John Vallance.

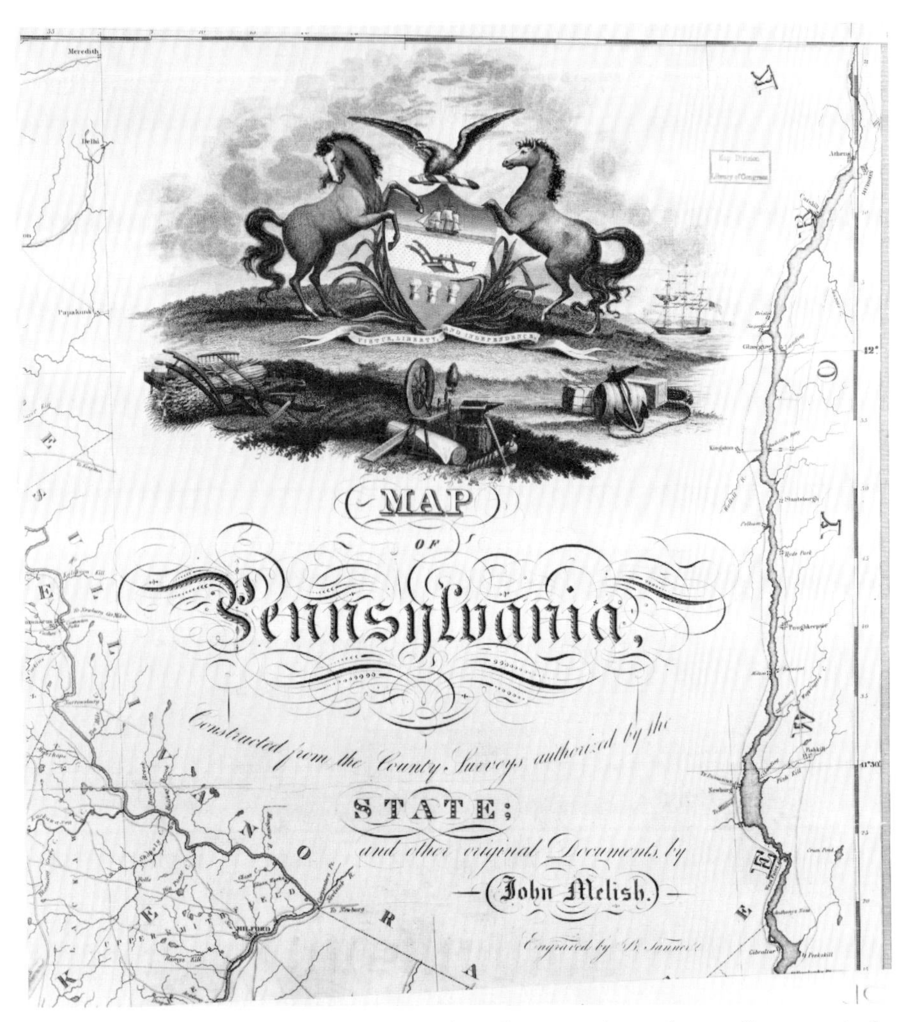

Fig. 7–4. The title cartouche and a small segment of John Melish's 1822 *Map of Pennsylvania*.

Howell's maps of Pennsylvania served the citizens of that state for about two decades. In 1822 they were superceded by the *Map of Pennsylvania, Constructed from the County Surveys authorized by the State; and other original Documents, by John Melish* (Figs. 7–4 and 7–5). This map is a model of its type for several reasons. In the planning and production stages it enjoyed a larger measure of official support and direction than did most of the other state maps. It also benefited from the experience, knowledge, and skill of its compiler, John Melish, one of the most energetic and competent commercial map publishers of his day. Finally, the Pennsylvania map was engraved by Benjamin Tanner, one of the foremost American engravers of the early nineteenth century.

Melish was thoroughly familiar with the several state maps and drew upon a number of them to compile his large *Map of the United States*, the first edition of which was published in 1816. In his *Geographical Description of the United States. . . . Intended as an accompaniment to Melish's Map*, Melish acknowledged his "recourse . . . to the . . . various State maps, from actual survey, so far as these surveys have extended."[12] He early recognized the need for official, state financed cartography. The maps produced by commercial publishers, he observed, were on small scales, of a generalized character, and compiled from original survey maps and other source data. In his *Geographical Description* he acknowledged that "the basis on which the whole of the geography of the country rests, is maps from actual survey, and its political subdivision is highly favourable to the bringing them forward in the character of State Maps." It was obvious to Melish "that every State should have its own map. It should be State property," he asserted, "subject to the control of no individual whatever." Their production, he believed, should be an official responsibility. "Individuals," he emphasized, "are not equal to the task of bringing them forward, and keeping them correct. Wherever they have embarked in the business, they have lost much time and money; and unless the states embark in it, the geography of the country cannot be brought to maturity."[13]

Melish's thoughts regarding official sponsorship and support of state maps had already been transmitted to influential members of the Pennsylvania legislature. In its session of 1814, "Mr. Isaac Weaver, of Green County, . . . moved a resolution that measures might be taken to bring forward a new map of the State. This resolution passed both houses unanimously; and during the summer, the preparatory steps were taken by the secretary of the commonwealth to ascertain the best plan of procuring the materials, and publishing the map." Because of his interest in the proposed map and his recognized competence as a geographer and map publisher, Melish's advice was solicited early in 1816 by Pennsylvania officials. The mapmaker recognized in the request for information "an excellent opportunity for the introduction of his favourite theory," and he decided, therefore, "to take a journey to the seat of the state government on the subject." He found the state officials and legislators most receptive and "it soon appeared that the disposition towards the map was favourable throughout both branches of the legislature."[14]

Together with members of a legislative committee, Melish drew up plans for the proposed act. With slight modifications, it passed both houses and with the governor's signature was enacted into law in March 1816 as "An Act Directing the Formation of a Map of Pennsylvania." Its clear, precise, and detailed provisions, specifications, and instructions could only have been formulated by a professional mapmaker.

Section one of the act directed the surveyor general "to contract with the deputy surveyors . . . for the formation of a map of each of the counties . . . ; which maps shall be on a scale of two miles and half to an inch." The physical and cultural features to be included on the map were prescribed in detail. With reference to roads, the surveyor was instructed to note "particularly

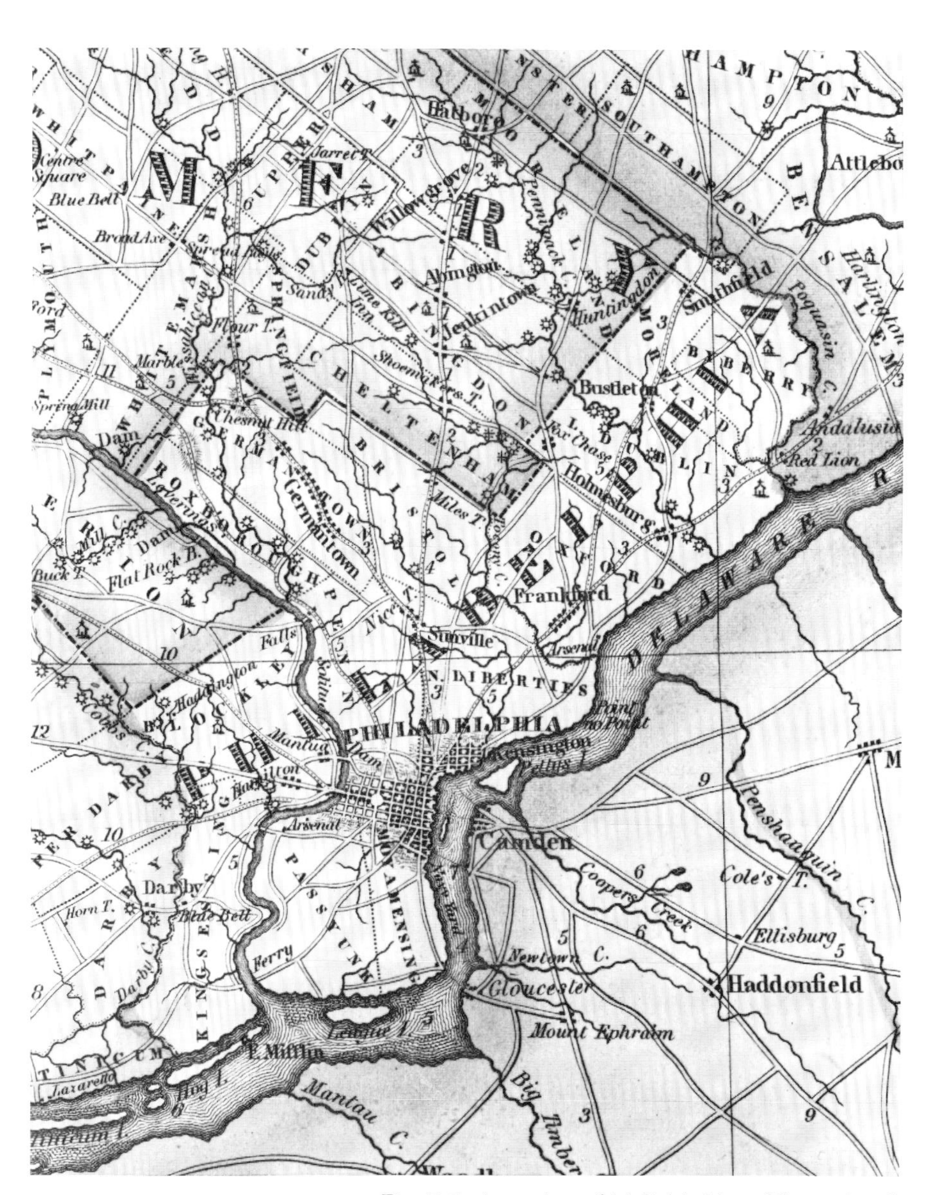

Fig. 7–5. A portion of Melish's *Map of Pennsylvania* showing the southeastern part of the state, including the city of Philadelphia and vicinity.

such as are turnpiked, and the distances in miles between the principal towns and remarkable places." The completed maps were to be "sent, as soon as convenient, to the office of the surveyor general." The expense per map was "not [to] exceed two hundred dollars for each county, unless the information . . . cannot be had in any of the public offices of the state, or of the proper county," in which case, the surveyor general was authorized to obtain the desired data "by actual survey . . . at [an] expense not exceeding in the whole six hundred dollars for each county." In section two of the act, the surveyor general following receipt of the county maps was directed to "contract for the publication of a Map of Pennsylvania, with some suitable person." There was no more "suitable person" in Pennsylvania than Melish, and the map contract was accordingly awarded to him. The surveyor general, in the third section of the act, was instructed to have hand-drawn copies of the county maps prepared for his office.

After these copies were made, the original maps were transmitted to Melish, who was instructed to use them to "make a connected map of this commonwealth, on a scale of five miles to an inch, marking thereon so many of the particulars specified in the first section of [the] act . . . and shall cause the same to be engraved on copper, in a handsome and workmanlike manner, and shall cause . . . a number to be printed . . . whereof one hundred, duly coloured, mounted, and finished, shall be delivered . . . to the surveyor-general for the use of the commonwealth."[15]

Melish and his assistants spent more than six years compiling this map of Pennsylvania. A preliminary copy was completed late in 1820 and submitted for examination. A report, dated December 15, 1820, and signed by S. D. Ingham, Pennsylvania secretary of state, and Jacob Spangler, surveyor general, notes that Melish's map "has undergone a rigid examination in all its parts . . . and we have the satisfaction to say, with much confidence, that the Map promises to be one of uncommon excellence. The whole design is in our opinion judiciously arranged and well executed, and, should the engraving be done with neatness, it cannot fail to give general satisfaction."[16]

Benjamin Tanner prepared the plates for the map, a task that required another fifteen months. Advance copies of the printed map were examined by a joint committee of the legislature and the surveyor general and were most favorably received. The committee's report dated March 23, 1822, noted that "the engraving is executed in a very strong, clear, and neatly finished manner, peculiarly suited to maps of this class, and equal, if not superior, to the style of any other map of the same class ever heretofore published."[17] Ingham and Spangler likewise testified that "they feel no hesitation in declaring that, in their view, the whole work, embracing the plan, the drawing, the engraving, and the colouring, are all evincive of the great exertion of the contractor to comply with his engagement, and that the map is worthy the expense [more than $30,000] which the State has incurred in bringing it to perfection."[18] To ensure that the published map might be as accurate and up to date as possible, proof copies were also sent to officials of all counties for criticisms, corrections, and possible additions. Melish also secured from state authorities information on the latest road surveys, and the engraved plates were corrected to incorporate these data.

In the late summer of 1822 the map was released for public sale and distribution. The *Niles Weekly Register* announced in its September 28, 1822, issue that "the long expected map of [Pennsylvania] by Mr. Mellish [*sic*] has at length appeared. It may be called a magnificent work, worthy of the great commonwealth which has so liberally furnished the means to produce it. Greater accuracy could not well have been expected than is assured in this map; made up chiefly of county surveys, taken by experienced persons, resident in the respective counties and responsible to their immediate friends and neighbors for the truth of their presentations." When assembled and mounted the six sheets, printed from an equal number of copper plates, form a map of Pennsylvania measuring 125 by 188 cm. It was available in several formats, colorings, and mountings that ranged in price from $9.00 to $12.50.

The *Map of Pennsylvania* has been described as Melish's greatest published work. This is high praise, indeed, when we note the many excellent maps and geographical works he published in the period from 1812 to 1822. The 1816 act empowered the secretary of state and the surveyor general "to contract with the publisher of the . . . state, or other persons, from time to time, for new editions of the map . . . , on such terms as they deem just and reasonable." No subsequent editions were compiled by Melish, for he died on December 30, 1822, at the peak of his publishing career. Revised editions of the Pennsylvania map were issued by the state, however, in 1824, 1826, and 1832.

As prescribed in section one of the act authorizing the map, all counties were mapped at the scale of two and a half miles to an inch. In an effort to secure maps of uniform quality and execution Melish prepared for the local surveyors "Directions for Constructing the County Maps in Terms of the Act of Assembly." The first of these instructed the surveyor to "ascertain as near as possible, the latitude of the seat of justice [i.e., the county seat], and its longitude from Washington, and run a true meridian line, and an east and west line through it, as in the specimen."[19] The various physical and cultural features to be shown on the map were specified. The surveyors were also advised to "delineate the border exactly as on the model exhibited in the specimen, and graduate the scale on the inner margin in miles of latitude and longitude." The model was a small ten-by-seven-inch "Specimen of the County Maps to be Constructed by virtue of an Act of the Legislature directing the formation of A Map of Pennsylvania." This sample map included a section of the Susquehanna River to the northwest of Harrisburg with the several adjacent mountain chains indicated with hachures. A legend picturing "Characters and Writing to be used in delineating the County Maps" occupies the upper right corner of the specimen.

Most of the surveyors seem to have followed Melish's directions, and the county maps, the manuscript originals of which are preserved in the Pennsylvania Department of Internal Affairs, are fairly uniform in appearance and format. A few of the maps do not have marginal lines; their makers obviously did not heed the carefully outlined instructions. Descriptive notes supplement the cartographic data on several county maps. Hachures or crude shading depict generalized relief for a number of counties. Township boundaries, towns and cities, churches, roads, bridges, and grist mills are among the cultural features mapped. The general format laid down by Melish served as a pattern from which evolved the large-scale county maps naming property owners and showing the extent of private land holdings that were published in great numbers in the decades immediately preceding and following the Civil War.

In addition to using them as compilation data for the *Map of Pennsylvania*, Melish hoped to publish the county maps individually. To this end, section nine of the act empowered the surveyor general "to authorise the publisher of the state map to publish the County Maps or any number of them separately, provided the same shall be done without any expense to the state." In a "Prospectus of the State Map & County Maps of Pennsylvania," an addendum to the *Geographical Description*, Melish announced that "The County Maps will be published on the large scale on which they are originally delineated, provided there be a sufficient number of subscribers to defray the expense." He suggested that they would "be exceedingly useful as pocket maps, affording at all times the means of obtaining a correct knowledge of the respective counties." The price of the county maps, printed "on fine vellum paper, or bank note paper," was not to exceed "One Dollar to One Dol-

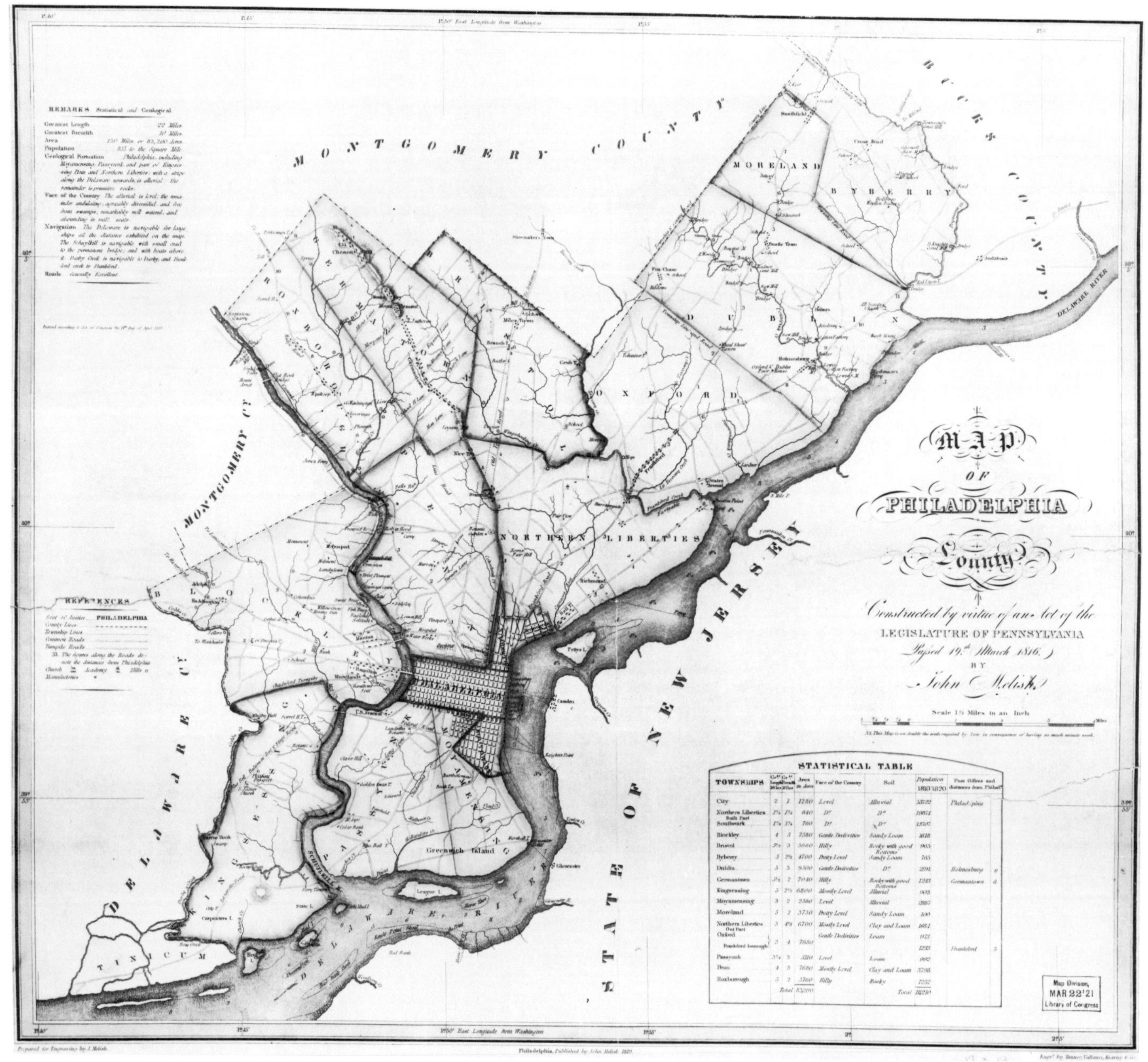

TOWNSHIPS	Gr.st Length Miles	Gr.st Breadth Miles	Area in Acres	Face of the Country	Soil	Population 1810	Population 1820	Post Offices and distances from Philad.a	
City	2	1	1280	Level	Alluvial	53722		Philadelphia	
Northern Liberties Built Part	1¼	1¼	640	D.o	D.o	19874			
Southwark	1¼	1¼	760	D.o	D.o	13707			
Blockley	4	3	7380	Gentle Declivities	Sandy Loam	1618			
Bristol	3½	3	5060	Hilly	Rocky with good Bottoms	965			
Byberry	5	2½	4700	Pretty Level	Sandy Loam	765			
Dublin	5	3	9500	Gentle Declivities	D.o	2194		Holmesburg	9
Germantown	5½	2	7040	Hilly	Rocky with good Bottoms	1243		Germantown	6
Kingsessing	5	2½	6800	Mostly Level	Alluvial	903			
Moyamensing	3	2	2560	Level	Alluvial	2887			
Moreland	5	2	3750	Pretty Level	Sandy Loam	400			
Northern Liberties Out Part	3	4½	6700	Mostly Level	Clay and Loam	1684			
Oxford	3	4	7680	Gentle Declivities	Loam	973			
Frankford borough						1233		Frankford	5
Passyunk	3½	3	5110	Level	Loam	1892			
Penn	4	3	7680	Mostly Level	Clay and Loam	3798			
Roxborough	5	2	5760	Hilly	Rocky	1252			
		Total	83200		Total	111210			

Fig. 7–6. One of the few county maps Melish actually published was this 1816 *Map of Philadelphia County,* which was based on Melish's own surveys.

lar and a half, according to the size of the counties."[20] Only a few county maps were published, however (Fig. 7–6).

Reporting on the progress of the Pennsylvania map in the third edition (1818) of the *Geographical Description*, Melish listed a number of counties for which manuscript maps had already been received. "Of these," he noted, maps of "Montgomery, Luzerne, Dauphin, Lebanon, and Huntingdon are in the hands of the engraver, and will be speedily published." Maps of Montgomery, Dauphin, Lebanon, and Huntingdon as well as of Philadelphia, Chester, Berks, Somerset, Wayne, and Pike counties were listed for sale in Melish's 1822 *Catalogue*. The Somerset County map, dated 1830 and published by H. S. Tanner, bears the credit line, "prepared for engraving by John Melish." County maps for York and Adams (1821), Lancaster (1824), Wayne (1828), and Schuylkill (1830) may have been based on surveys for the 1822 map of Pennsylvania, although none of them bear Melish's name.

Melish fervently hoped that the Pennsylvania map would serve as a model for other states to emulate. In his 1822 *Catalogue* he wrote:

> The state of Pennsylvania has now set the example, and a map of that state has been produced on a plan that has met with general approbation, and as perfect in its details as can reasonably be expected from a work of such magnitude, embracing such a vast variety of objects, and being necessarily the work of so many hands. This work is respectfully submitted to the inspection of the several state governments, as a specimen of what State Maps ought to be, and the publisher is in great hopes that before the next census is published, maps of the states will follow the example set by Pennsylvania.[21]

The entire states of New Jersey and Delaware are included on Melish's Pennsylvania map, but with infinitely less detail. Only their major cities, towns, and principal roads are shown. Because New Jersey was less settled than New York or Pennsylvania, there were fewer pressures in the post-Revolution decades to produce a New Jersey State map. The needs were sufficient, however, to induce William Watson of Gloucester County to compile and publish *A Map of the State of New Jersey* in 1812 (Fig. 7–7). The map was engraved by William Harrison of Philadelphia and dedicated "To His Excellency Joseph Bloomfield, Governor, the Council and Assembly of the State of New Jersey." It is 102 by 72 cm. and at the approximate scale of 1:300,000.

Apart from the information on the map, little is known about Watson, or the extent of the support he received from the state in compiling and publishing his map. His map certainly was not based on original surveys, and he undoubtedly used as compilation data previously published maps. The map is, however, the earliest one of New Jersey to show township boundaries. There is also a fairly extensive network of roads. Snyder notes that "Watson included an authoritative-looking six-inch engraving of the state shield, but he drew township lines and roads with little evidence of making more than off-hand assumptions in many cases. Some of the northern townships, such as Hanover and Harrington, are even shown miles away from their correct locations relative to neighbors."[22]

Well before Watson published his map, the New Jersey state legislature in 1799 had passed an act establishing a corporation called "The Company for procuring an accurate map of the state of New Jersey." The company members were to have the exclusive rights to the sale of the map for fifteen years, provided they sold all their stock and published the map by 1803.[23] Nothing came of this early official attempt to support the compilation of a New Jersey map. The legislature tried again in 1822, authorizing a loan of one thousand dollars to Thomas Gordon, "the better to enable him to procure additional surveys and defray other mapping costs, provided he repay them in two years, without interest."[24]

In 1828 Gordon published, at the approximate scale of

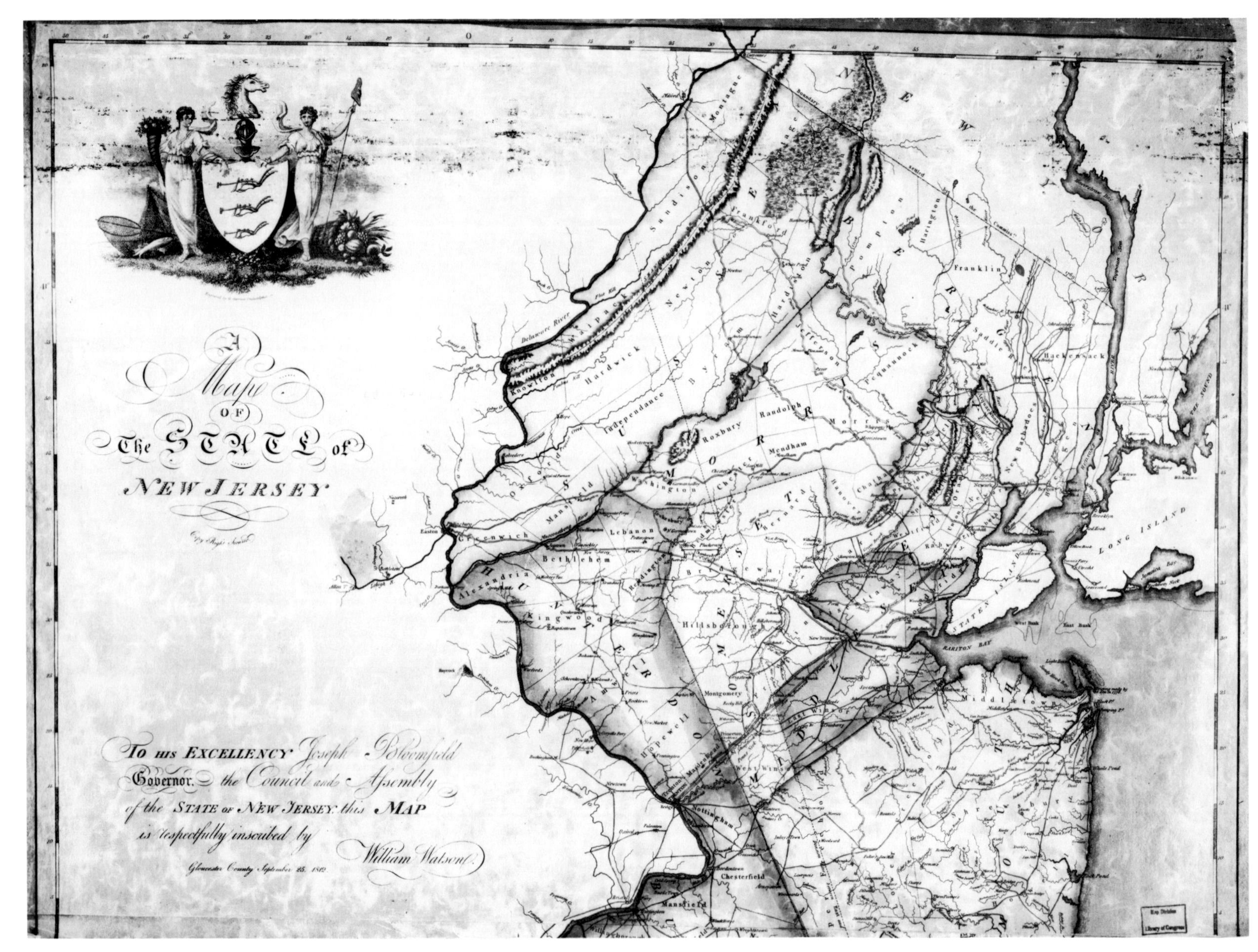

Fig. 7–7. The upper half of William Watson's 1812 *Map of the State of New Jersey.*

1:190,000, *A Map of the State of New Jersey*. It is 140 by 82 cm. in size. The map was engraved in Philadelphia by Henry S. Tanner with the assistance of E. B. Dawson and W. Allen. Townships are shown in detail, and the turnpike system is emphasized. Indicating its satisfaction with Gordon's map, the state legislature in March 1828 authorized the purchase of "125 copies for state offices and officials, for colleges and for counties."[25]

Gordon was born in Amwell, New Jersey, in 1778 but early in life relocated to Trenton. As was true of other early surveyors and mapmakers, he also practiced law and was a judge of the Court of Common Pleas. He died in 1848 and was buried in Trenton. Gordon's cartographic effort was complimented by H. S. Tanner in his *Memoir* of 1830. Tanner noted that the New Jersey part of his map of the United States

> is from the able and scientific map of Thomas Gordon. . . . This admirable map, which must have cost its author much time and money, was compiled partly from surveys made by Mr. Gordon, combined with others collected by him during the progress of his work. . . . This as well as every other good state map with which I am acquainted, has failed to reimburse the expenditure of its enterprising author; the spontaneous sales of the map, and the limited patronage bestowed on it by the Legislature of New Jersey, being, as I learn, entirely inadequate. The complete failure, and in some cases, utter ruin of those who have undertaken the construction of original state or more local maps, unaided by government patronage, should admonish novices in such matters to calculate well the cost before they attempt to bring forward large and expensive maps, which, if properly executed, cannot fail to involve them in expenses of which few, who are not experimentally acquainted with the subject, can form a just idea. State maps, or indeed local maps of any kind, whose sales must necessarily be limited, should be done by the public authorities. No individual, unless he is possessed of resources which place him above the drudgery and labour inseparable from the faithful execution of a good map, should, without mature consideration, undertake the construction of a work of this sort; he will either injure himself in the attempt to do justice to the subject, or, impelled by the want of adequate funds, send forth an imperfect work, at once injurious to himself and discreditable to his country. Fortunately for the map of New-Jersey, its author's means, (if we may judge by the evidence afforded by the map itself) were more ample than those of most others engaged in similar pursuits.[26]

A second, revised, edition of the New Jersey map was published by Gordon and Tanner in 1833. Editions "corrected and improved by Robert E. Horner" were published in 1849, 1850, 1853, and 1854. The 1850 edition received an additional one-thousand-dollar subsidy from the state.

The compiler of the first state map of Maryland was Dennis Griffith, a resident of Philadelphia. In 1795 his *Map of the State of Maryland, Laid down from an actual Survey of all the principal Waters, public Roads, and Divisions of the Counties therein* (Fig. 7–8) was engraved and published by John Vallance in Philadelphia on June 6, 1795. Vallance, a native of Scotland, was born around 1770 and settled in Philadelphia around 1791. He was engaged in engraving independently and with partners until his death in 1823. In 1792 Vallance and his partner, James Thackara, prepared the engraving for the Andrew Ellicott plan of the city of Washington, and they must have used this information on Griffith's Maryland map, for there is a large inset map of "Federal Territory" on it (Fig. 7–9). Griffith's map, which is 75 by 131 cm., is dedicated "To the Governor the Senate and House of Representatives of the State of Maryland." This may have been in acknowledgment of some financial aid or subsidy granted to Griffith by the state legislature.

Edward B. Mathews, a historian of the maps of Maryland, believed that Griffith's map was the "best compilation of existing geographical information concerning

Maryland up to the work of the Topographical Engineer of the State, J. H. Alexander, during the fourth decade of the nineteenth century." Mathews further noted that "although it is claimed by the author that 'the lines have

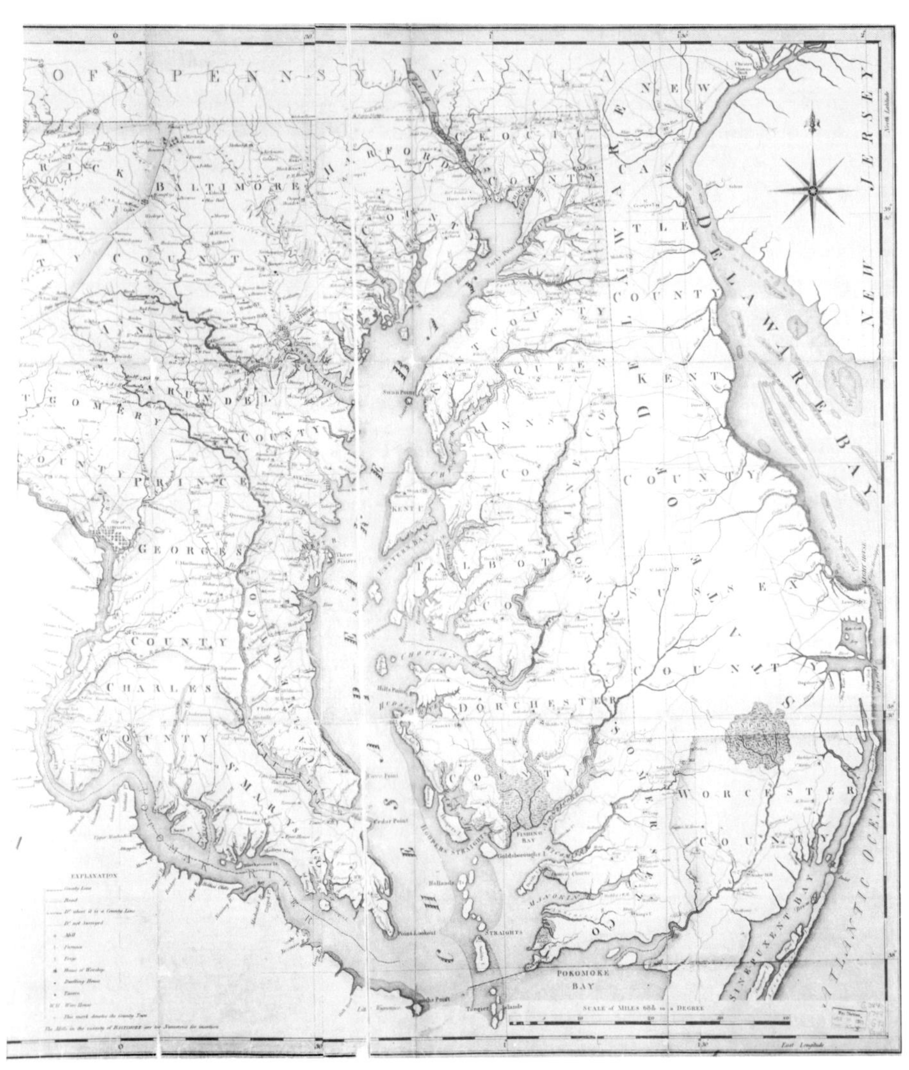

Fig. 7–8. The Chesapeake Bay portion of the 1795 *Map of the State of Maryland* by Dennis Griffith.

been laid down from an actual survey,' it has been impossible to find any reference to the man or his work beyond the date of his marriage. It seems accordingly somewhat doubtful that more than certain portions of the area were visited by him."[27] A second edition of Griffith's *Map of the State of Maryland* was published in 1813 by John Melish and was again engraved by Thackara and Vallance. There appears to have been few changes from the first edition.

Although the full extent of Delaware was included on the early maps of Pennsylvania, New Jersey, and Maryland, no separate map of the state was published before 1800. John Churchman, however, published, probably in 1787, the *Map of the Peninsula Between Delaware & Chesopeak Bays with the said Bays and Shores adjacent drawn from the most accurate Surveys*. There is no date or place of publication on the map, which is 57 by 43 cm. in size. Advertisements and notices concerning the expected publication of the map appeared in several newspapers during 1786 and 1787.

Churchman was born in 1753 in East Nottingham, Pennsylvania. A mathematician and surveyor, he conducted surveys in the Chesapeake Bay area during the revolutionary war. He apparently submitted a preliminary version of his map to the American Philosophical Society in 1779. The society committee selected to review the map reported that "we are of the opinion that he is possessed of sufficient materials, both astronomical observations and actual surveys, to enable him to construct an accurate map, and have no doubt that he has executed his design with exactness & care, but we cannot help expressing our desire of seeing the map laid down upon a much larger scale, which would render it more serviceable for promoting the Knowledge of Geography."[28] Although Churchman did not redraw his map at a larger scale, he did dedicate it to the American Philosophical Society.

In notices published in several newspapers in May and June 1789, Churchman returned "his cordial thanks to all those who so liberally furnished him during the late war with materials and other encouragement relating to his map of the peninsula between Chesopeak and Delaware bays . . . and he flatters himself (although it was performed under many disadvantages) that this small specimen of his first publication in the line of his profession will operate in the public mind, in favour of the present undertaking."[29]

This "present undertaking," was Churchman's chart and explanation on magnetic variation, initially published in 1790 and issued in a revised and enlarged edition in 1794. These studies received greater acclaim in Europe than in America, and Churchman visited the continent from 1792 to 1795 and again in 1804 and 1805 to pursue his magnetic studies and to confer with scholars at various scientific societies. He died on the return sea voyage to America following his second European sojourn on July 17, 1805.

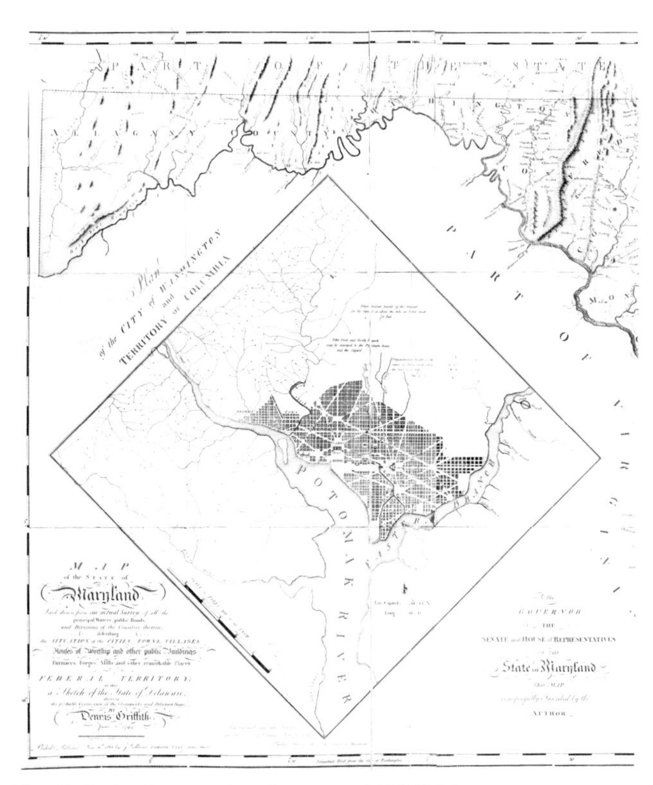

Fig. 7–9. An interesting feature of Griffith's Maryland map is this large inset map of the recently laid out "Federal Territory."

Notes

1. New York, *Senate Journal*, 48th sess. (Albany, N.Y., 1825), 616.
2. David H. Burr, "The Route from Little Valley," in New York, *Assembly Journal*, pt. G, app. K. Documents accompanying the report of the joint committee on the subject of the Hudson and Erie Road; made to the assembly February 25, 1826.
3. David H. Burr, introd., *An Atlas of the State of New York Containing a Map of the State and of the Several Counties* (New York, 1829).
4. New York, *Senate Journal*, 52nd sess., Feb. 6, 1829.
5. Ibid.
6. Burr, introd., *Atlas*.
7. U.S. National Archives, Smithsonian Institution Records, Joseph Henry Papers.
8. New York, *Legislative Documents*, 53rd sess., vol. 2, no. 189 (Albany, 1830).
9. Ibid.
10. New York, *Assembly Journal*, 53rd sess., Jan. 30, 1830, 151.
11. Hazel Shields Garrison, "Cartography of Pennsylvania be-

fore 1800," *Pennsylvania Magazine of History and Biography* 59 (1935): 281.

12. John Melish, *A Geographical Description of the United States. . . . Intended as an accompaniment to Melish's Map* (Philadelphia, 1816), 9.
13. Ibid., 172.
14. Ibid., 175.
15. Ibid., 176.
16. John Melish, *A Catalogue of Maps and Geographical Works Published and For Sale by John Melish, Geographer and Map Publisher* (Philadelphia, 1822), 5.
17. Ibid., 6.
18. Ibid., 5.
19. Melish, *Geographical Description*, 178.
20. Ibid., 182.
21. Melish, *Catalogue*, 22.
22. John P. Snyder, *The Mapping of New Jersey, the Men and the Art* (New Brunswick, 1973), 94.
23. Snyder, *Mapping*, 94.
24. Ibid.
25. Ibid., 96.
26. Henry S. Tanner, *Memoir on the Recent Surveys, Observations, and Internal Improvements, in the United States, with Brief Notices of the New Counties, Towns, Villages, Canals, and Railroads, Never before Delineated. Intended to Accompany His New Map of the United States*, 2d ed. (Philadelphia, 1830), 26–27.
27. Edward Bennett Mathews, *The Maps and Map-Makers of Maryland* (Baltimore, 1898), 398, 400.
28. *American Philosophical Society Proceedings, 1744–1838* (Philadelphia, 1884). Meeting of Aug. 20, 1779.
29. *New York Daily Advertiser*, May 21, 1789.

8. *Early State Maps: The Southern States*

Virginia, as previously noted, was mapped several times during the colonial period. The State also received favorable cartographic treatment in the several decades following adoption of the federal Constitution. Individual initiative as well as official state support contributed to this happy situation. An example of the former is the 1807 map of the state compiled under the direction of Bishop James Madison, first cousin of President James Madison. Bishop Madison (1749–1812), who was educated for the ministry, became president of William and Mary College in 1777 and remained in that post for thirty-five years. During Madison's tenure, one responsibility of the college was examining and licensing local surveyors. As a result, Madison became acquainted with many such specialists and from them learned of the need for a new map of Virginia. Unsuccessful in his efforts to secure official sponsorship, Madison personally assumed responsibility for producing the map. Madison did not know how to survey and map, and so most of the surveying, data collecting, compiling, and drafting was entrusted to associates. Original sketch maps were also received from local surveyors throughout the state.

Final compilation of the Virginia map started in 1803 or 1804 under the immediate supervision of William Prentis of Petersburg. On February 15, 1805, Madison informed Prentis that "the map of Virginia to which you have lent so considerable assistance in the collection of some of the necessary materials, is nearly completed." The map was drafted by William Davis, who was described by Madison as "a neat, correct, and intelligent draughtsman."[1] A preliminary copy of this map was submitted for inspection to the Virginia General Assembly early in 1805. It was registered for copyright at Richmond on March 4, 1807, and printed copies were distributed soon thereafter. Entitled *A Map of Virginia Formed from Actual Surveys, and the Latest as well as most accurate observations* (Fig. 8–1), the map was dedicated "To the General Assembly of Virginia . . . by their Fellow Citizens, James Madison, William Prentis, William Davis, Proprietors."

The Madison map measures 114 by 175 cm. and is at the scale of 1:440,000. It was printed from nine copper plates that were engraved by Frederick Bossler of Richmond. There is an inset map of Ohio in the upper left corner and a view of Richmond fills the upper right corner. Notwithstanding its claims for accuracy, the Madison map contained a number of errors. Plate corrections were accordingly made for reprintings after 1807. In 1818, after Madison's death, a new edition, redrafted by Davis, was published. In this form it served as Virginia's most authoritative map until 1827. With its extensive network of transportation routes, the map was, moreover, the first comprehensive road map of Virginia.

Very likely in emulation of Pennsylvania, the Virginia General Assembly passed an act on February 27, 1816, "to provide an accurate chart of each county and a general map of the Territory of this Commonwealth." The law directed the county courts "to contract with some fit person or persons for making an accurate chart of their respective counties" in accordance with provisions spelled out in precise detail. Within one year the manuscript surveys were to be deposited with the Board of Public Works. Support for a scientific survey of the state also came from Thomas Jefferson. On April 19, 1816, in a lengthy letter to Governor Wilson C. Nicholas, Jeffer-

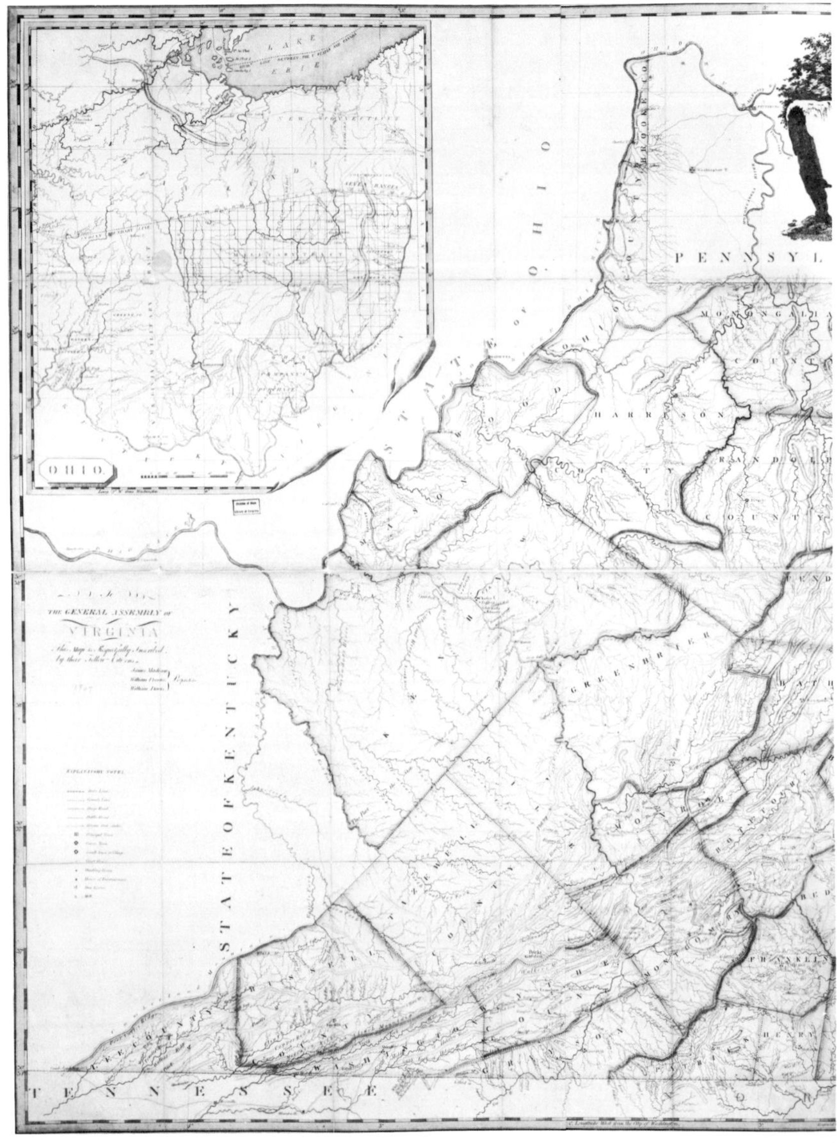

Fig. 8–1. The western portion, with an inset map of Ohio, of Bishop James Madison's 1807 *Map of Virginia.*

son outlined suggested procedures for carrying out the provisions of the act of the assembly.[2] Jefferson also was influential in engaging John Wood to supervise the county surveys and to direct compilation of the state map.

Wood, a native of Scotland, immigrated to the United States shortly after 1800. Before he was selected to supervise the county surveys, he was employed as an instructor at the Petersburg Academy. He directed work on the surveys from 1816 until his death in May 1822, by which time maps had been completed for most Virginia counties. Herman Böye, a German engineer residing in Richmond, was selected to succeed Wood, and he carried the project to completion. In February 1825 the General Assembly authorized the engraving of the map by Henry S. Tanner. Although the map bears the date 1825, it was copyrighted in April 1826, and copies were not distributed until 1827.

The elaborately embellished title, which includes views of Natural Bridge and Harpers Ferry, reads, *Map of the State of Virginia Constructed in conformity to Law from the late Surveys authorized by the Legislature and other original and authentic Documents by Herman Böye* (Fig. 8–2). In the upper left corner of the map there is an engraving of the University of Virginia campus in Charlottesville (Fig. 8–3), and a "View of Richmond from the West" fills the opposite corner. Other marginal data include a table giving the area and population of each county, a list of mountains with their altitudes and other geographical features in the state, and a lengthy essay entitled "Geological Remarks." Of particular interest is the "Memoranda to the Plan, materials and construction of this Map." From this we learn that materials consulted in compiling the map consisted "principally of the County Maps nearly all executed under the direction of the late John Wood; the surveys made at different periods for the Board of Public Works by their Engineers; the surveys in the archives of the Executive Department; . . . sundry maps and charts made for the U.S. Government, together with H. S. Tanner's maps of adjoining states, and various other documents."

The Böye-Wood map is at the approximate scale of 1:300,000 and measures 237 by 154 cm. Nine separate copper plates were required for engraving and printing it, and for this reason, it is sometimes identified as the "Nine-Sheet Map." Four hundred copies of the map were printed, of which the legislature authorized the sale of two hundred and fifty copies at twenty dollars each. The others were probably distributed to state officials and offices, colleges, and perhaps to the governors of all the other states. A reduced scale version of the map (69 by 122 cm.), published in late 1827 in an edition of eight hundred copies, sold for six dollars. By order of the governor of Virginia, the map was revised by L. von Buchholtz in 1859 and republished in both large and small editions. Writing in 1924, historian Fairfax Harrison observed that the Böye-Wood map "is altogether a credit to the Commonwealth. Not only does it show the physical features of the country, mountains, streams and roads, but it shows them correctly for the first time on any map."[3]

In his *Memoir* Henry Tanner noted

> that portion of my map [of the United States] which comprehends the entire state of Virginia . . . was taken from the large map of Virginia by Mr. H. Böye, recently published, in conformity to an act of the legislature of Virginia. . . . I deemed myself particularly fortunate in having the means of rectifying those glaring errors which the new map has exposed, and which have hitherto disfigured all our maps of this important state.
>
> Great attention has evidently been given, by the author of this excellent map, to the geographical land marks. No less than forty-seven points for latitude, and eighteen for longitude, have been fixed in the state of Virginia; and the positions of several conspicuous places have been newly rectified or verified by celestial observations, which give to the state a form and extent essentially different from the old maps.[4]

North Carolina's first state map was privately produced. It was compiled by surveyors Jonathan Price and John Strothers and published by subscription with the personal support of David Stone and Peter Browne. To Stone and Browne the *Actual Survey of the State of North Carolina* (Fig. 8–4) was "respectfully dedicated by their humble Servants Jona. Price [and] John Strothers." The map, which is at the approximate scale of 1:500,000 and measures 74 by 152 cm., was engraved by William Harrison, Jr., and printed by C. P. Harrison, both of Philadelphia. Two editions of the map were published in

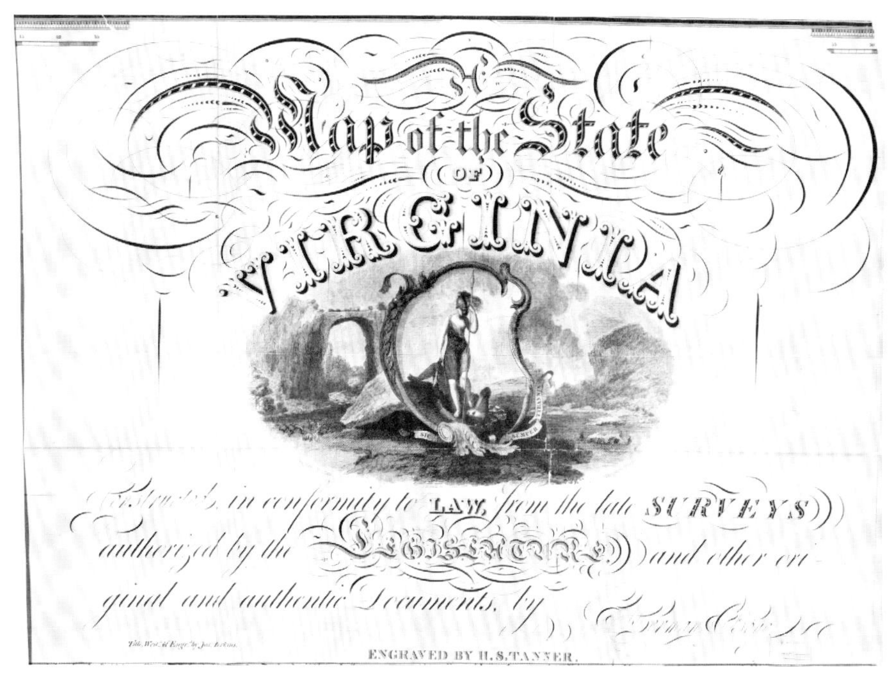

Fig. 8–2. The decorative title cartouche and a segment of John Wood and Herman Böye's 1826 map of Virginia.

1808. The date of publication is lacking on the earlier version. Both Harrisons were sons of William Harrison, a native of England who settled in Philadelphia in 1794 with his four sons, all of whom followed his professions of engraving and printing.

The Price-Strothers map was in preparation for almost two decades. Jonathan Price, an experienced surveyor, had already spent three years on surveys when he petitioned the North Carolina House of Assembly for aid in December 1792. William Christmas was employed as a draftsman on the map as early as 1792, and John Strothers, also a surveyor, became associated with the project some time before 1795. Financial difficulties were a continuing problem for these men. In response to Price's petition, the assembly granted them a loan of £500. In 1794 the compilers solicited individual subscribers in the press. They apparently received additional loans from the state, for official reports for 1802 and 1804 indicate that Price and Strothers were in arrears on three separate grants of £400, £500, and £290. State authorities were seemingly not disposed to cancel the debts or to provide funds to complete the map. As writer Mary Thornton notes, "that the map was finally published was due to the generosity of two private donors, David Stone and Peter Browne, who must have financed the greater part of the cost."[5] Stone, who at that time was governor of North Carolina, had previously been a member of the North Carolina House of Commons and of the U.S. House of Representatives and Senate. Peter Browne, a native of Scotland, was then a leader of the North Carolina bar and subsequently became a member of the House of Commons.[6]

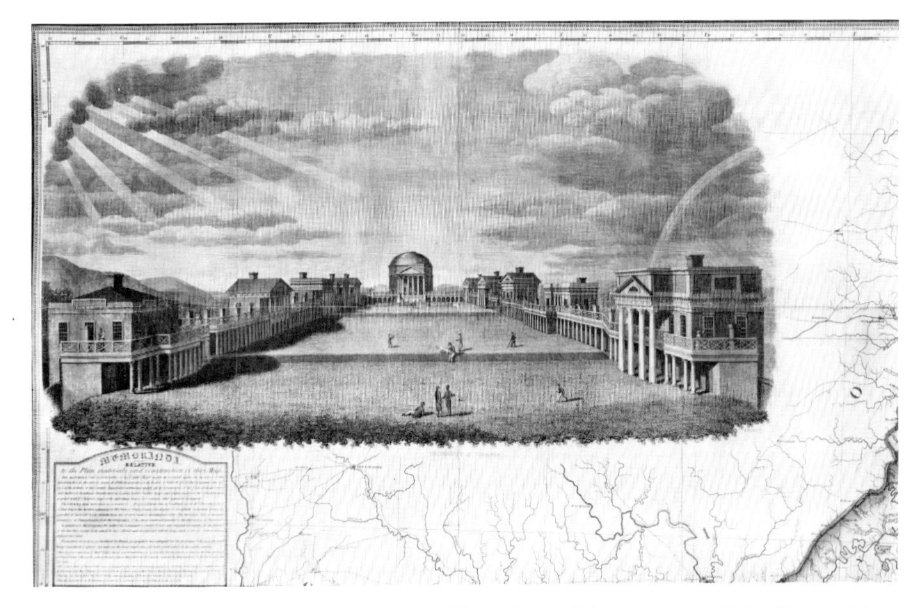

Fig. 8–3. This view of the campus of the University of Virginia in Charlottesville is a prominent feature of the Böye-Wood map of Virginia.

In 1819 Archibald D. Murphey, a foremost proponent of internal improvements in North Carolina, wrote, "it is with shame we now reflect that only a few years ago, the General Assembly refused to aid two enterprising individuals to compile a Map of the State; and that had it not been for the generous aid of two gentlemen, David Stone and Peter Browne, Esquires . . . it would not have been compiled. . . . The Map of North-Carolina, by Messrs. Price and Strothers, has an accuracy not to have been expected at the time it was compiled."[7]

With the establishment of peace following the War of 1812, the several states redirected their attention and resources to the subject of internal improvements. In North Carolina, a period of agricultural prosperity in the second decade of the nineteenth century coincided with the emergence of a group of legislators who worked out a program which they felt would improve the state's political inequalities, lessen her educational backwardness, and stimulate her trade by better transportation facilities.[8] Thus, the North Carolina legislature established a Board on Internal Improvements in 1818. Because of the small number of American-trained engineers, North Carolina, like some other states, had to seek abroad for a competent state engineer. In 1819 Hamilton Fulton and Robert H. B. Brazier, both natives of England, were respectively engaged as state engineer and state surveyor. In the same year John Couty, who had been employed in the service of Virginia's principal engineer, was engaged by North Carolina to continue the surveys that had been initiated in 1818.

Reporting on these appointments, Murphey, as chairman of the Board of Internal Improvements, explained that "it has been the object with the Board to make these Surveys auxiliary to the compiling of an accurate Map of the State. . . . It is very desirable to have such a Map." Murphey recalled that when the Price-Strothers map was published "little was known of the Geography of the Western parts of the State; and there is one-sixth part of the territory of the State, of which this Map affords a very indifferent representation."[9]

During the next decade, Fulton, Brazier, Couty, and others made surveys and prepared maps of various parts of the state. In the course of his activities, Brazier, around 1821, became associated with John MacRae, mayor of the city of Fayetteville. This association was to produce North Carolina's second state map. In 1825 MacRae presented a petition to the legislature through David L. Swain, one of its members, for a loan of three thousand dollars to make a map of the state. This request was denied, but MacRae petitioned again in 1826 and was granted five thousand dollars, of which two thousand was to be repaid within two years.[10]

Brazier apparently supplied MacRae with maps and data from the various surveys, and the latter undoubtedly received similar material from other surveyors. Then, in late 1827 or early 1828, both Fulton and Brazier received appointments from the state of Georgia, and they took with them some of the North Carolina records. In 1829 MacRae received authorization from the North Carolina legislature to compile a map of the state. To assist him, Governor John Owen on February 25, 1829, wrote to General Charles Gratiot of the U.S. Engi-

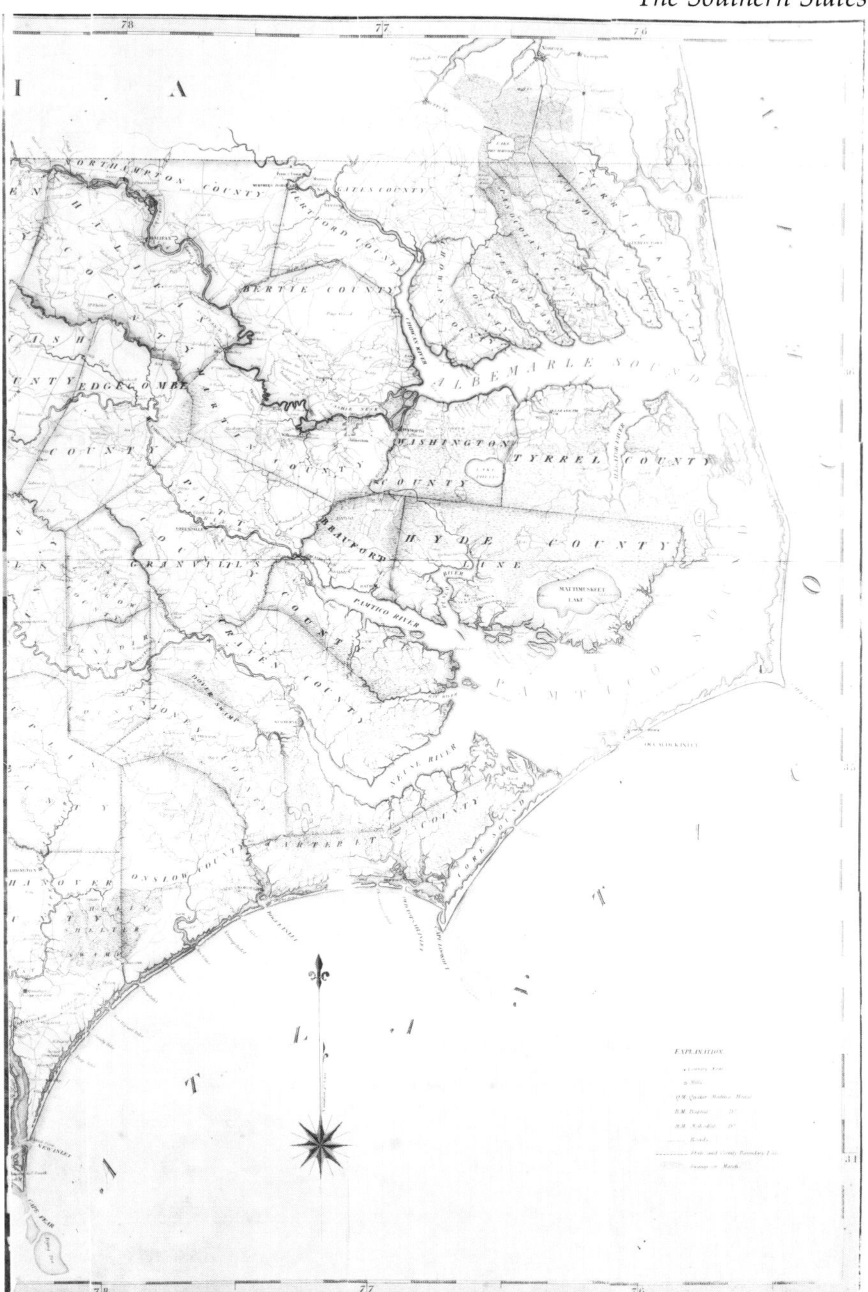

Fig. 8–4. The eastern portion of Jonathan Price and John Strothers's *Actual Survey of the State of North Carolina*, which was published in 1808.

neer Department in Washington requesting that someone be detailed to aid MacRae.[11] The following September, Lieutenant William Harford, a recent graduate of West Point Military Academy, was detailed to work with MacRae and to make any necessary surveys required to complete the state map. This occupied Harford for more than three years.

In his November 22, 1831, message to the North Carolina legislature, Governor Montfort Stokes reported that he had received "from the State of New York an elegant map of that State. . . . Several of my predecessors in office [he recalled] having recommended a suitable return for similar friendly donations, I have only to add, that a map of North Carolina, being nearly completed, an opportunity will be afforded by this State to cancel these obligations in the manner they deserve."[12] Almost two years passed, however, before there appeared *A New Map of the State of North Carolina, Constructed from Actual Surveys, authentic Public Documents and private Contributions by Robt. H. B. Brazier. Published under the Patronage of the Legislature, by John MacRae. Published by John MacRae, Fayetteville, N.C. & H. S. Tanner, Philadelphia, 1833. Engraved by H. S. Tanner, E. B. Dawson & J. Knight* (Fig. 8–5). It is 89 by 53 cm. in size and at the approximate scale of 1:400,000.

The Brazier-MacRae map was carefully scrutinized by a joint committee of the North Carolina legislature and found to be satisfactory. Early in 1833 the legislature passed a resolution authorizing the governor to purchase copies of the map for presentation to each of the states and territories.[13] Governor David L. Swain reported that this had been done in his annual message delivered to the legislature on November 18, 1833. He was moved to add that "I cannot permit myself to allude to this subject without venturing to suggest, that if a copy were procured at the public expense, and forwarded to each of the clerks of our Superior Courts, to be placed in their respective courthouses, it might have a tendency to diffuse more generally among our citizens correct knowledge of the Geography of the State, and discharge in some degree the obligation which the community is under to the enterprizing publisher."[14]

It is interesting to note that neither Fulton, Couty, or Harford received credit on the map. Tanner's engraving firm had achieved considerable success by 1830, and a number of the state maps at this time were engraved in and published by his Philadelphia plant. A corrected and updated edition of the MacRae-Brazier map was published in 1854, with the engraving done by W. Williams of Philadelphia. It shows operating railroads, railroads under construction, plank roads, and new settlements in the extreme western counties.

South Carolina's internal improvement program initially lagged behind those of other southern states. After action was finally taken by the state legislature in 1817, the program was pursued with vigor and purpose for the next decade or so. The plan adopted by the state was broad and ambitious. A primary objective was to construct an effective network of navigable canals, waterways, and roads. Between 1817 and 1828 more than two million dollars were spent on internal improvements. Surveying and mapping, essential parts of this program, were placed under the direction of John Wilson, the civil and military engineer of South Carolina. Directing the preparation of a state map was one of his major responsibilities. A native of Scotland, Wilson received his engineering education at the University of Edinburgh. He immigrated to Charleston, South Carolina, in 1800, where he married and maintained his residence for more than a decade. He received an army commission during the War of 1812 and was placed in charge of the fortifications for the city of Charleston. Following the war, Wilson served for a brief period as a major in the U.S. Corps of Topographical Engineers.

Some time in 1817 or 1818 the South Carolina General Assembly adopted resolutions aimed at producing a state map. Detailed surveys and maps of each judicial district of the state were prepared for use in compiling the map. Between 1817 and 1823 the assembly appropriated more than fifty thousand dollars to go toward its

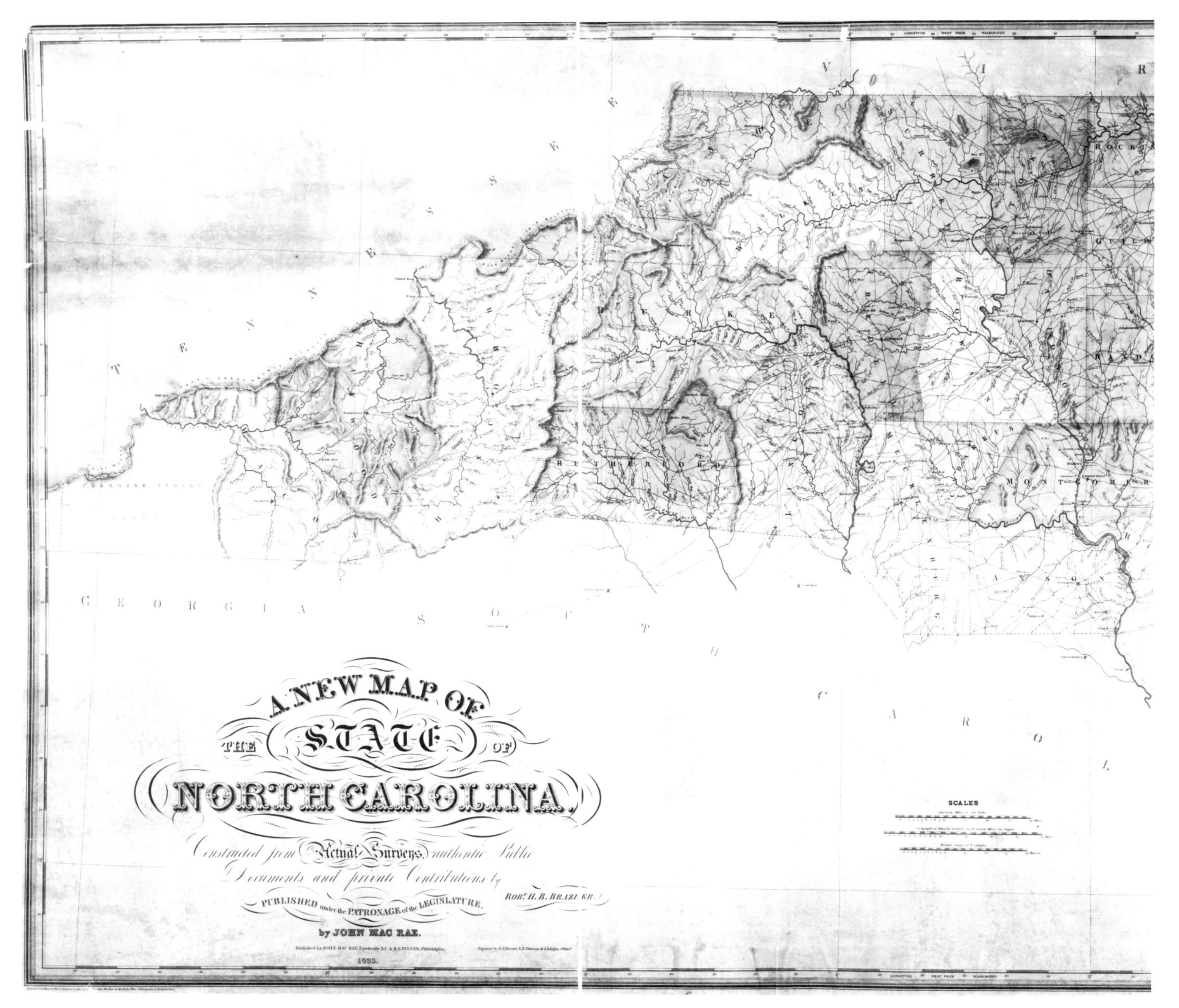

Fig. 8–5. The western half of Robert H. B. Brazier and John MacRae's 1833 map of North Carolina.

compilation and publication. In 1821 a report of the South Carolina Board of Public Works to the assembly noted that "immediately after the adjournment of the Legislature, measures were taken for completing the Map of the State. . . . The survey of Sumter district had been so inaccurately made, that an entirely new one would be required; and in several other districts some errors required correction. . . . With these additional materials and some verbal corrections, the map was placed in the hands of Major Wilson, who undertook to redraft it and to superintend the engraving in Philadelphia . . . a contract was made for engraving the plates and delivering fifty-one copies of the map, colored and mounted. These are expected to be received during the present session of the Legislature."[15]

The map was published in 1822 and entitled *A Map of South Carolina Constructed and Drawn from the District Surveys, ordered by the Legislature by John Wilson late Civil & Military Engineer of So. Caro. Engraved by H. S. Tanner, Philadelphia* (Fig. 8–6). It is 112 by 147 cm. in size and at the scale of 1:380,000. In the lower left corner there is an inset map of "Charleston Harbour." Tanner stated in his *Memoir* that "Wilson's map is decidedly one of our best and most scientific maps, and was used in correcting the adjoining parts of Georgia and North Carolina."[16] The district maps, which were the basic source material for Wilson's state map, were later revised and reprinted in Robert Mills's *Atlas of the State of South Carolina*.

Georgia was among the last of the former British colonies to have a state map. Little wonder, for when the U.S. Constitution was ratified, settlements in Georgia were limited to a narrow coastal strip and to the valley of the Savannah River. Much of the interior was still Indian territory. The need for surveys of the state was, however, early recognized. In February 1783 the Georgia legislature passed "An Act for opening the Land Office and for other purposes mentioned therein." It further provided that "there shall be a Surveyor General for the State and also a Surveyor for each County Annually chosen by the Legislature. . . . [The] County Surveyor [is] required to keep an office . . . in which . . . shall be recorded all such Platts or surveys belonging to such County . . . and the said County Surveyor shall also transmit to the Surveyor General a fair copy of the same."[17]

Daniel Sturges was named surveyor general in 1797 and held that office until 1809. He was again appointed to serve from 1817 to 1823. In 1799 Sturges designed Georgia's great seal, and in 1808 he prepared a plan of the city of Milledgeville. By this time he was already engaged in assembling the surveys and data from county surveyors, which he used in compiling his map of the state. The June 1816 issue of the *Analectic Magazine and Naval Chronicle* reported that the map of Georgia was "prepared with great labour and care by Mr. Sturges during the period of fourteen years in which he held the office of surveyor-general of Georgia." Compilation was completed in 1816, and the map was sent to the engraver in Philadelphia. John Melish reported in his *Geographical Description of the United States*, which accompanied his 1816 map of the United States, that "a new and accurate MS. of Georgia, compiled from the records of actual survey, by Daniel Sturgis [*sic*] and of which Mr. Eleazer Early, of Savannah, is the proprietor, was placed in the hands of the author of the map of the United States for publication; and he was authorized to make use of it for correcting" the U.S. map.

The Sturges-Early map was not published until late in 1818: advertisements in issues of the *Milledgeville Journal* during December of that year reported that it was ready for delivery. The *Map of the State of Georgia Prepared from actual Surveys and other Documents for Eleazer Early by Daniel Sturges* (Fig. 8–7) was engraved by Samuel Harrison, another member of the family of Philadelphia engravers. A marginal note indicates that the map was "Published and Sold by Eleazer Early Savannah Georgia and by John Melish and Samuel Harrison Philadelphia." The map measures 112 by 140 cm. and is at the scale of 1:500,000. In the lower right corner of the map there is a statistical table giving the length, breadth, size, popu-

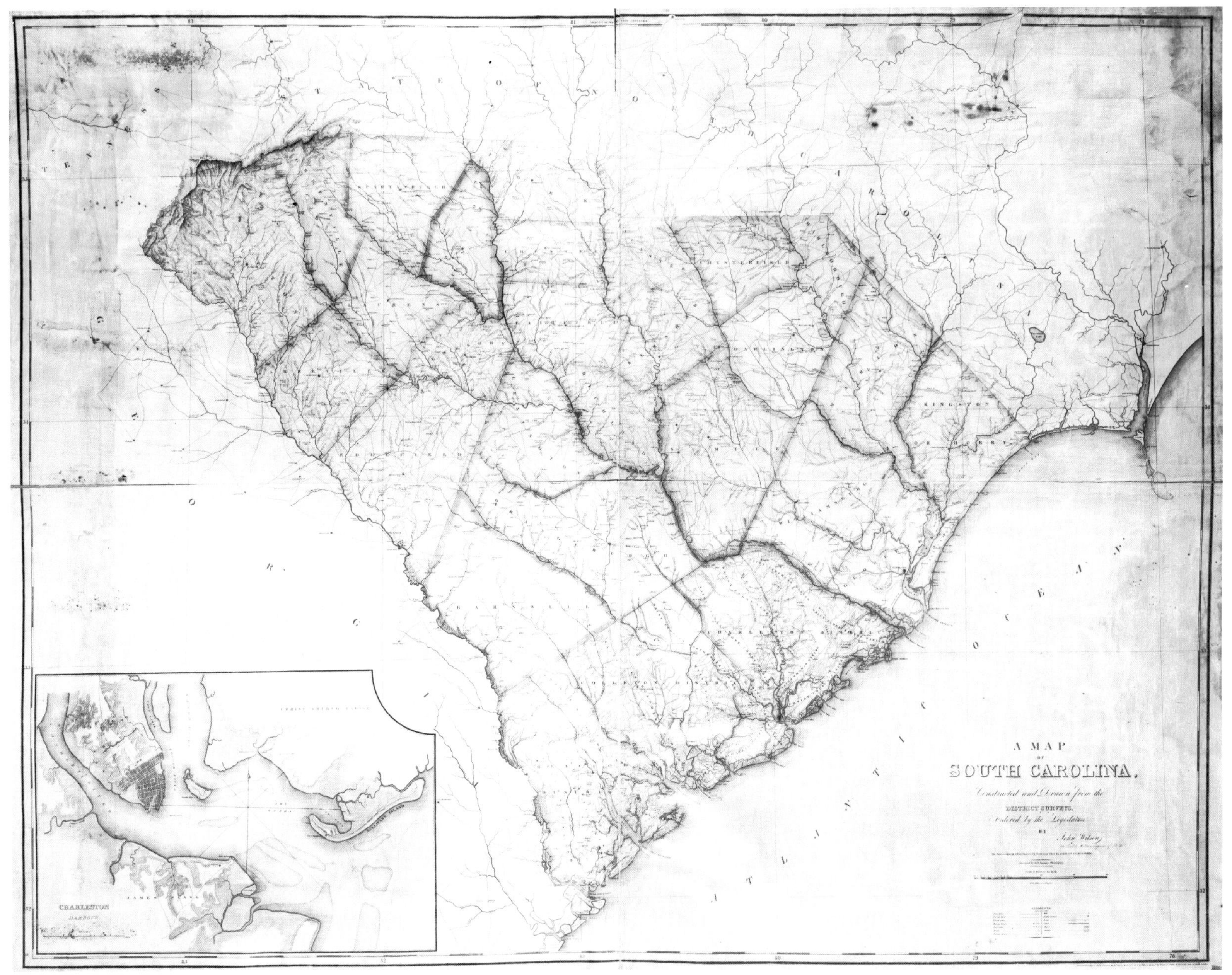

Fig. 8–6. John Wilson's map of South Carolina, 1822.

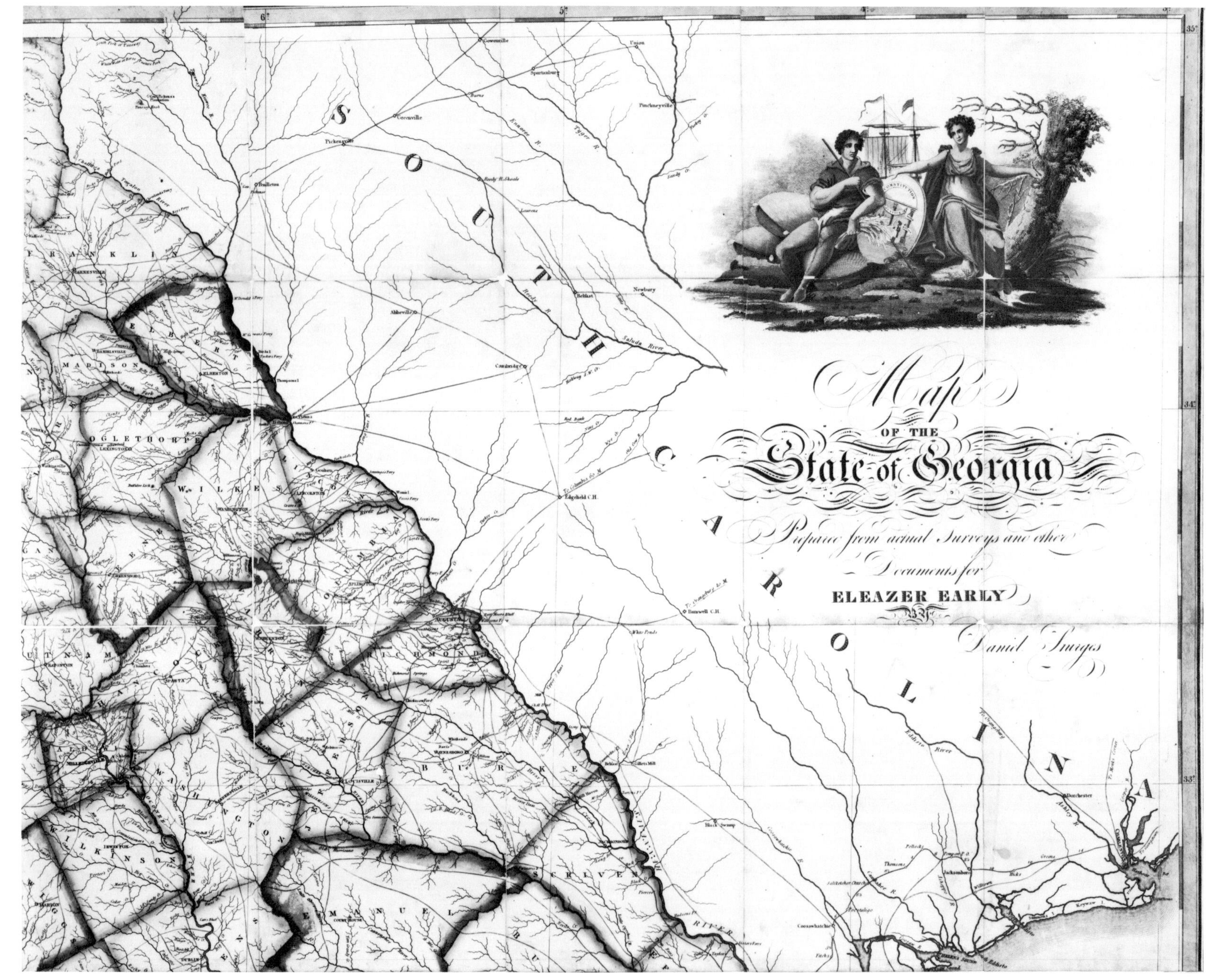

Fig. 8–7. The eastern half of Daniel Sturges and Eleazer Early's 1818 map of Georgia.

lation, and seat of justice of each county. In the left margin is a list of post offices with their distances from Washington, D.C., and Milledgeville. Two editions of the Sturges-Early map have been identified. The titles, imprints, and dates are the same in both editions. In the later version a number of changes and additions have been made in the southern Alabama portion of the map.

Eleazer Early was born in the late eighteenth century in Madison County, Virginia. Shortly before the turn of the century the Early family moved to Georgia. Peter Early, an older brother of Eleazer, served Georgia in the U.S. House of Representatives and was a judge and governor of the state. Sturges and Early were brothers-in-law, both having married daughters of James Merriwether.

In 1825, seven years after the Sturges-Early map was published, the state of Georgia established a Board of Public Works. In the following year, Hamilton Fulton was induced to leave North Carolina to accept the position of chief engineer of Georgia. But because of members' differences on whether to give priority to canal building or to railroad construction, the board was discontinued after one year of operation. Fulton, however, remained in his post until it was abolished in 1828.

By the terms of the Adams-Onis Treaty, Florida was ceded by Spain to the United States in 1819. Two more years passed, however, before the actual transfer was effected. Florida was finally organized as a territory of the United States on March 30, 1822. Within the next year the first map of Florida was published in Philadelphia. It is titled *Map of Florida, Compiled and Drawn from various Actual Surveys & Observations; by Charles Vignoles, Civil & Topographical Engineer, 1823. Engraved by H. S. Tanner & Assistants* (Fig. 8–8). It measures 70 by 60 cm. and is at the approximate scale of 1:1,300,000. Concurrent with the appearance of this map, E. Bliss and E. White of New York City published Charles Blacker Vignoles's comprehensive work *Observations upon the Floridas*.

Vignoles was born in England May 31, 1793, the descendant of a Huguenot family that had settled in Ireland around the end of the seventeenth century. Shortly after he was born, his father, a military officer, was sent with his wife and child to the British West Indies. Within the year both parents died in Guadeloupe of yellow fe-

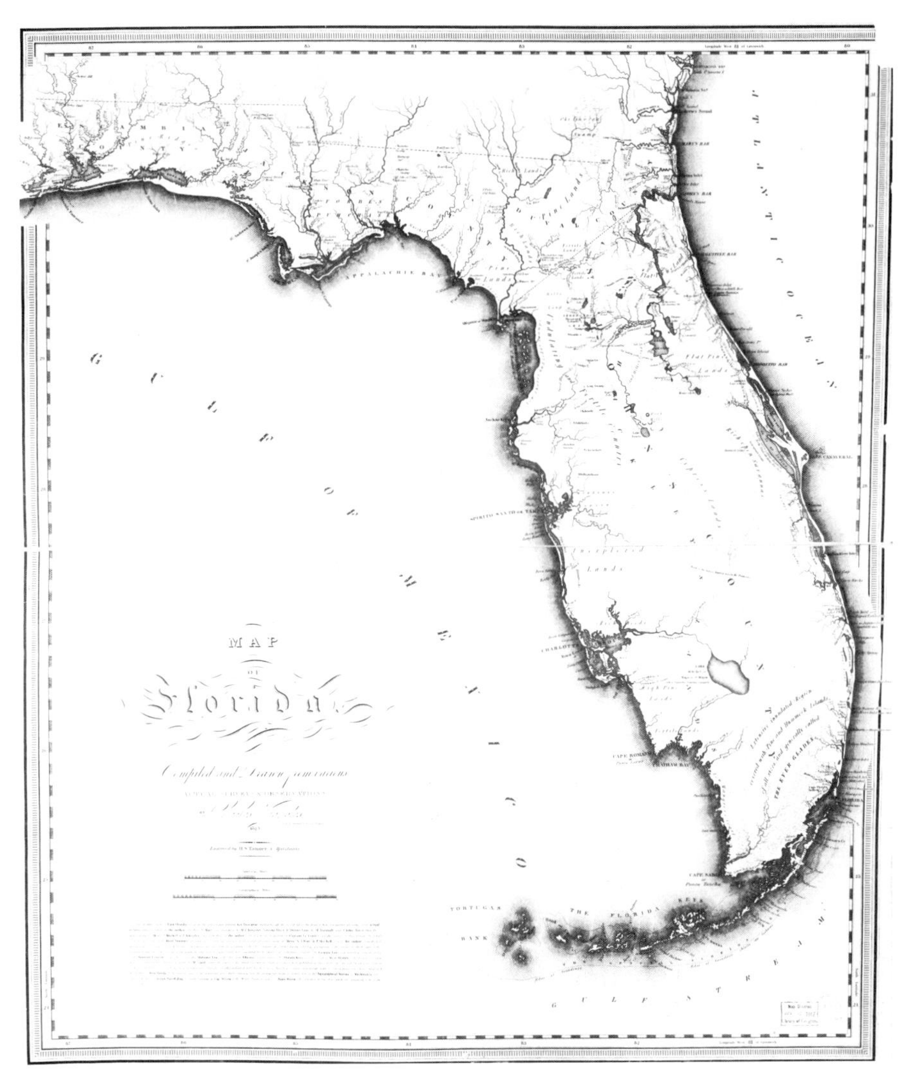

Fig. 8–8. Charles Vignoles compiled this 1823 map shortly after Florida was organized as a U.S. territory.

ver. After some months an uncle claimed the two-year-old Vignoles and placed him in the care of his maternal grandparents, Dr. and Mrs. Charles Hutton. His grandfather was an instructor at the Royal Military Academy in Woolwich. When Charles was eighteen, he and his grandfather had a serious estrangement which was never resolved. Vignoles entered Sandhurst Military Academy in 1813 but spent only a brief period there studying engineering. He received a commission as an ensign in the British army in 1814 and spent several months on active duty in the Netherlands, in the Napoleonic campaign. Vignoles's unit was ordered to Canada in mid-1814, but the ship on which he crossed the Atlantic was wrecked off Anticosti Island at the mouth of the St. Lawrence River. After participating briefly in campaigns against U.S. forces in the Great Lakes area, Vignoles's unit was reassigned to Scotland, where the young ensign was stationed for most of 1815 and 1816. The following year he was in France as aide-de-camp to General Thomas Brisbane, a post which gave him some practice in drafting and experience in preparing computations. He returned to England in 1817 and on July 13, 1817, was secretly married to Mary Griffith. Vignoles became interested in the prospect of serving as an engineering and surveying officer in South America, in the revolt of the colonists against Spanish domination. Accordingly, less than two weeks after his marriage he left his bride in England and sailed for America. After a brief stay in the Virgin Islands, he landed in Charleston, South Carolina, in October 1817.

Plans for liberating South America were abandoned when he received an appointment as assistant to South Carolina's chief engineer, Major John Wilson. Vignoles's official duties involved correcting and resurveying some of the district maps prepared by local surveyors and in compiling and drafting the state map of South Carolina. On May 31, 1818, Vignoles wrote his wife from Charleston that "during the daytime I am fully occupied in compiling and drawing out such of the surveys of the state of South Carolina as have been already completed. Some of these were done by myself, and the portion of the map I have engaged to finish is about fifty square feet."[18] With most of his responsibilities in South Carolina completed, the young engineer crossed into Florida in 1820 where he surveyed in various parts of the territory. In October 1821 he was named engineer and surveyor to the city of St. Augustine. During the next two years he conducted further surveys and explorations which resulted in his *Observations upon the Floridas* and the map of Florida. Vignoles returned to England in 1824 and had a successful career there and on the continent in planning and constructing railroad lines.

Notes

1. Earl G. Swem, "Maps Relating to Virginia in the Virginia State Library and Other Departments of the Commonwealth With the 17th and 18th Century Atlas Maps in the Library of Congress," *Virginia State Library Bulletin* 7 (1914): 85.
2. Swem, "Maps," 102, 104.
3. Fairfax Harrison, *Landmarks of Old Prince William* (1924; reprint, Berryville, Va., 1964), 640.
4. Henry S. Tanner, *Memoir on the Recent Surveys, Observations, and Internal Improvements, in the United States, with Brief Notices of the New Counties, Towns, Villages, Canals, and Railroads, Never before Delineated. Intended to Accompany His New Map of the United States*, 2d ed. (Philadelphia, 1830), 48–49.
5. Mary Lindsay Thornton, "The Price and Strothers First Actual Survey of North Carolina," *North Carolina Historical Review* 41 (1964): 481.
6. Ibid.
7. Archibald D. Murphey, *Memoir on the Internal Improvements Contemplated by the Legislature of North Carolina* (Raleigh, 1819), 47.
8. William P. Cumming, *North Carolina in Maps* (Raleigh, 1966), 25.
9. Murphey, "Memoir," 118.
10. Cumming, *North Carolina*, 25.
11. Ibid., 26.

12. Governor Montfort Stokes, *Message to the General Assembly of North Carolina* (Raleigh, 1831), 7.
13. North Carolina, *Senate Journal*, Jan. 7, 1833, 121.
14. Governor David L. Swain, *Message to the North Carolina Legislature* (1833), 11.
15. South Carolina Board of Public Works, *Report to the Legislature for the year 1821*, 99.
16. Tanner, *Memoir*, 57.
17. Georgia [Colony], "Statutes, Colonial and Revolutionary 1774 to 1805," *Colonial Records* 19, pt. 2 (1911): 203.
18. Olinthus John Vignoles, *Life of Charles Blacker Vignoles* (London, 1889), 76.

9. *Early State Maps: The Trans-Appalachian States*

Prior to the Revolution a few settlements had been made in the trans-Appalachian lands. Following the war the tide of settlers greatly increased as veterans accepted land grants in lieu of back pay for their military service. Within two decades some seventy thousand persons established new homes in the picturesque and productive Kentucky region. Some of the settlers were encouraged to cross the mountains by John Filson's 1784 map of Kentucky and his accompanying book entitled *The Discovery, Settlement, and Present State of Kentucke.*

Kentucky was granted statehood in 1792, and a year later Mathew Carey published *A Map of Kentucky from Actual Survey by Elihu Barker* in Philadelphia. The map is at the approximate scale of 1:625,000 and measures 44 by 99 cm. Bibliographers Wheat and Brun note that "the original manuscript copy of this map . . . , together with the copyright, was advertised for sale by Oliver Barker on January 8, 1793, in the (Philadelphia) *Dunlap's American Daily Advertiser* after the author's death. The printed map was announced in the (Philadelphia) *Gazette of the Unites States*, April 29, 1794, as being just published. A reduced re-engraving was published by Mathew Carey in 1795. It was republished in London in 1797, appearing in Gilbert Imlay's A Topographical Description, 3d ed."[1]

Little is known about Elihu Barker except that he died shortly after compiling his Kentucky map. Similarly, we have no record of the actual survey upon which this map was supposedly based. Willard Jillson believed, however, that Barker's map was by far the most accurate of the early maps of Kentucky. Barker, he stated, "made use of the available published sources and took the written and verbal statements of others in compiling his map."[2]

The first fine large map of Kentucky was compiled by Luke Munsell and published by him in Frankfort in December 1818. It is titled *A Map of the State of Kentucky, From actual Survey Also Part of Indiana and Illinois, Compiled principally from Returns in the Surveyor-General's Office by Luke Munsell* (Fig. 9–1). This impressive map, which is 96 by 229 cm., was inscribed "To the Legislature and People of the Commonwealth of Kentucky . . . as a testimony of their patronage and liberality in promoting its execution." The map was engraved by Hugh Anderson, who was located in Philadelphia between 1811 and 1824. David Stauffer notes that "Anderson possibly was trained abroad, and belonged to the group of Scotch engravers who came to the United States in the first decade of the last century."[3] Illustrations below the dedication and above the title were prepared by Thomas Sully, an artist who was born in Lincolnshire, England, in 1783. His parents moved to Charleston, South Carolina, with their nine children, in 1792. In 1799 Sully moved to Richmond, Virginia, where he studied and worked with his older brother, Lawrence. Sully settled permanently in Philadelphia in 1808. During the next forty years he painted portraits, a number of which are now preserved in American museums and historical societies.

Like most of the early state mapmakers, few details are known abut the life and career of Munsell. He was born in 1790, probably in or near Louisville. In December 1822 Munsell married Eliza Sneed, daughter of Achilles Sneed of Frankfort. The elder Sneed was asso-

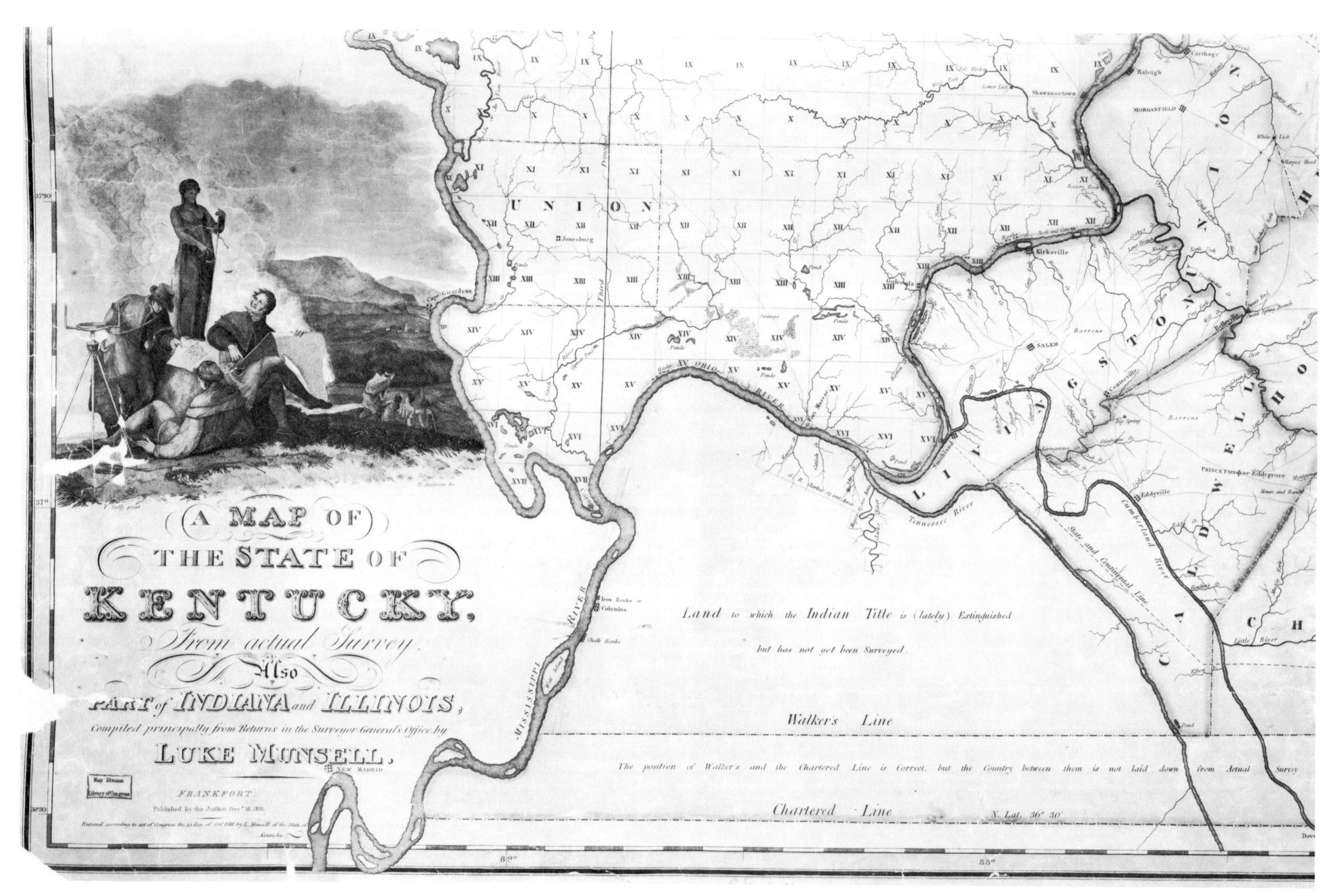

Fig. 9–1. The title and a segment of Luke Munsell's 1818 map of Kentucky. The illustration above the title depicts a group of surveyors at work.

ciated with various enterprises in the Frankfort area and served his state as a director of the Bank of Kentucky and as clerk of the court of appeals. Munsell's marriage record and other contemporary documents designate him as a doctor. Whether he was a physician or carried the title in an academic or honorary capacity is not known. From the evidence in hand it is uncertain whether Munsell was a surveyor or whether his contribution to the preparation of the map was in a supervisory capacity. Although he held several official appointments, there is no documentation that he was employed by the state of Kentucky to compile the map. Although Munsell acknowledged the "patronage and liberality" of the legislature and people of Kentucky in his map's dedication, available evidence indicates that the patronage was minimal and reluctantly dispensed.

Late in January 1819 a bill was introduced in the Kentucky General Assembly proposing "that the secretary of state be directed to purchase sufficient number of copies of Munsell's map of Kentucky to be appropriated in the following manner: Two for the representative chamber; two for the senate; one to each of the public offices in Frankfort; and one to each county court to be kept in the clerks office of the county. And that the amount of said purchase be deducted by the auditor of public accounts from the debt due by said Munsell to the state."[4] With minor changes, the proposal was approved in the house. The senate then added amendments to supply copies of the map to several state educational institutions. The amendments failed to receive house approval, and the legislative session ended without passage of the proposal. It was reconsidered by the assembly in its next session and was passed by both houses in February 1820. As approved the bill included an addendum "that an indulgence of the term of one year be granted to the said Munsell to pay the balance of the debt due by him to the commonwealth."[5] It is not clear whether Munsell's debt to the state was incurred in compiling the state map, but it seems likely that it was. In view of this, the financial aid granted the compiler seems niggardly when compared with subsidies and aid provided by other states to private and semi-official compilers and publishers of state maps.

The Kentucky-Tennessee boundary was imprecisely defined when these states were admitted to the Union, and portions of the demarcation line remained in controversy for three or four decades. In 1818–19 Kentucky's general assembly passed an act directing the governor to appoint commissioners to cooperate with similar appointees of the state of Tennessee to run and mark the boundary line between the two states west of the Tennessee River. Because the Tennessee legislature was not then in session, the appointment of commissioners for that state was delayed.

Notwithstanding, Kentucky's acting governor, Gabriel Slaughter, informed the general assembly in December 1819 that he had "deemed it [his] duty to proceed, and appointed Robert Alexander and Luke Munsell commissioners, who, with the aid of Col. William Steele and Richard Fox, surveyors, have performed the work—and the intelligence, skill, and high standing of these gentlemen furnish a pledge of its accuracy."[6] Because Steele and Fox are specifically described as surveyors, we might infer that Alexander and Munsell were not so trained or skilled. The latter, however, accompanied the surveyors in the field between June 6 and July 26, 1818. Tennessee subsequently named delegates to meet with a joint committee of the Kentucky house and senate. A compromise agreement based upon the Alexander-Munsell line was reached on February 2, 1820. In 1831 Munsell again served as a boundary commissioner for his state. With James Bright, Tennessee's commissioner, he made a survey, or resurvey, of "Walker's Line," which formed the southern boundary of Kentucky east of the Tennessee River.

Tennessee, established on land ceded to the federal government by North Carolina, joined the union as an independent state in 1796. Prior to the 1789 cession, a number of pioneers had migrated to Tennessee country. The land surveys that had been made prior to 1796 were of the inaccurate metes-and-bounds type. Tennessee, like Kentucky, was plagued with boundary controversies for a long time. Among the earliest boundary commissions and surveys with which Tennessee was involved was the one established by North Carolina and Virginia in 1779–80 to extend the boundary line between those two states. One of Virginia's commissioners was General Daniel Smith, who subsequently settled in Tennessee and became one of its leading citizens. Smith was born in Stafford County, Virginia, on October 24, 1748, and was educated at William and Mary College, where he studied engineering and surveying. He moved to Tennessee several years after serving on the 1779–80 North Carolina-Virginia boundary commission and spent the remainder of his life there.

In 1793 Mathew Carey published *A Map of the Tennassee Government formerly Part of North Carolina taken Chiefly from Surveys by Genl. D. Smith & others*. The map was engraved by Joseph T. Scott, who, like Carey, was also in Philadelphia. This small map (24 by 52 cm.) is the earliest of Tennessee based on original surveys. It was republished in 1795 in Gilbert Imlay's *Topographical Description of the Western Territory*. The Smith-Carey map is of poorer quality than most of the other state maps published shortly before the beginning of the nineteenth century. Tennessee leaders did, however, recognize the need for a good map, and some turned to Smith as the logical compiler of such a map. In April 1795 William Blount, who would later become governor of the state, wrote Smith: "You have often heard me express my anxiety for a correct map of this territory. The map already published is not correct, that which Tatham has attempted is not correct, and if it was it will probably never be published. Will you, as you have the best materials, at your leisure prior to the meeting of the General Assembly, make a map of that part of the Territory which lays west of Cumberland Mountains? . . . I know your desire for a correct map to be equal to my own and I offer no apology for making of you this request."[7] Smith did not rise to this challenge, probably because he shortly thereafter became involved in the political affairs of the new state.

More than three decades passed before a large map of Tennessee was published. It has the title *Map of the State of Tennessee, Taken From Survey By Matthew Rhea* (Fig. 9–2). It was copyrighted in 1832 and "published by the Author, Columbia, Tenn. and Sold by H. S. Tanner, Philada. Eichbaum & Norvall, Nashville, & S. D. Jacobs, Knoxville." The map is 172 by 87 cm. Below the title there is a "View of Nashville" drawn by T. V. Peticolas and engraved by W. E. Tucker. The map proper was engraved by Tanner, E. B. Dawson, and J. Knight of Philadelphia. Counties in the western part of the state have the rectangular grid pattern, indicating that this region was surveyed by General Land Office surveyors. Information about Matthew Rhea is meager. The Rhea name was, however, well known in Tennessee in this period, and there is a Rhea County in the eastern part of the state. John Rhea, possibly Matthew's father, represented Tennessee in the U.S. Congress from 1807 to 1821. Two years after the publication of Rhea's map, Eastin Morris published *The Tennessee Gazetteer or Topographical Dictionary*, which was probably intended as an adjunct to the Tennessee map. Morris was elected president of Tennessee's state bank in 1831.

The state of Louisiana, admitted to the Union in 1812, was formed of that part of the Louisiana Purchase which included the lower basin and delta of the Mississippi River. Because of its strategic location for controlling transportation on the river, this region had early invited exploration, particularly by French and Spanish expeditions. Some crude surveys were made but, as James Robertson notes, "there [was] a great lack of accurate maps of Louisiana both under the French and the Spanish. The Spaniards were fearful lest any information should leak out regarding the geographic regions. . . . Consequently what maps were made were buried in the archives. No extensive surveys were made and there were insufficient data to make an accurate map."[8] In response to a request from President Thomas Jefferson,

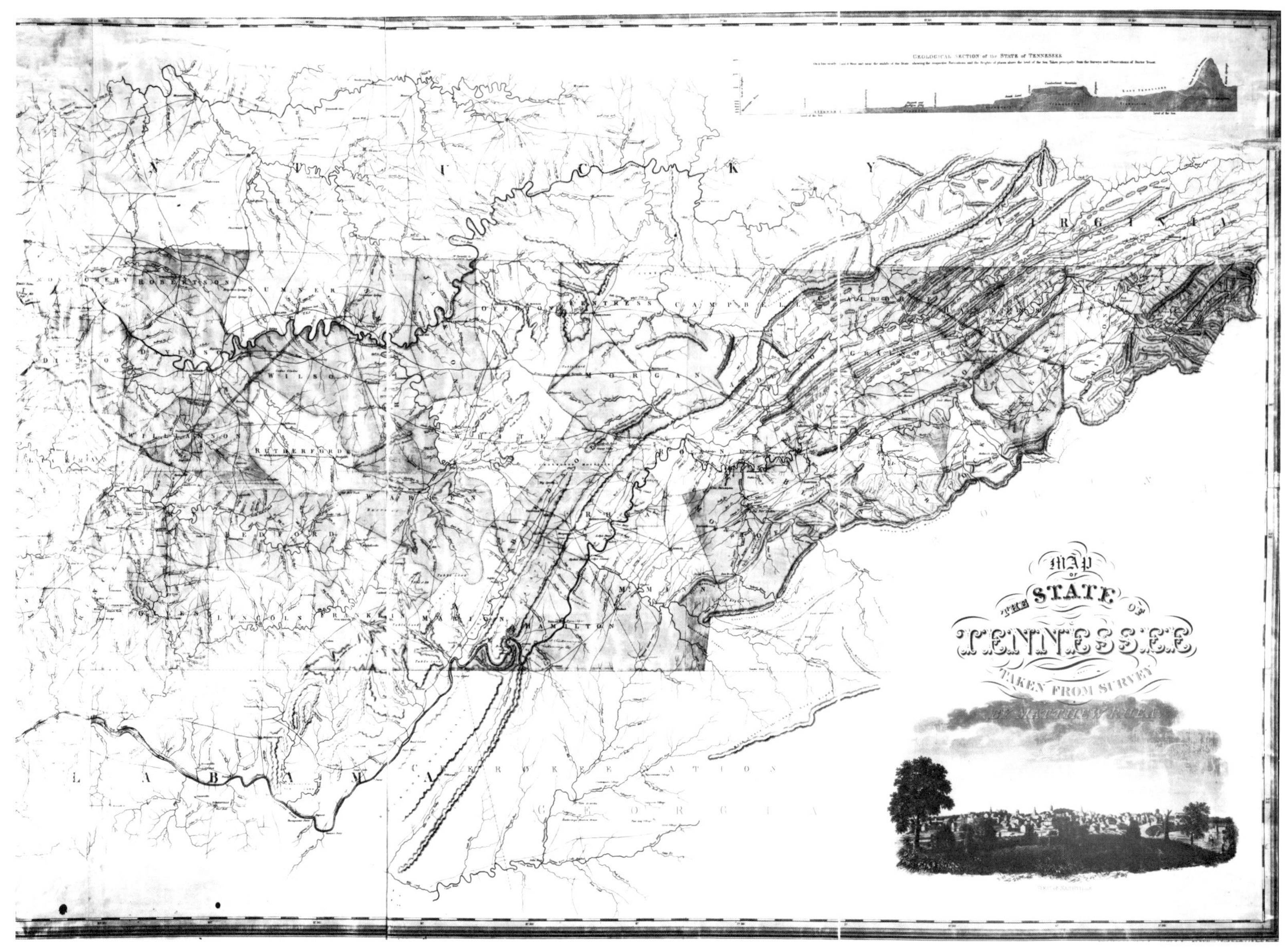

Fig. 9–2. Matthew Rhea conducted his own surveys for his map of Tennessee, which was copyrighted in 1832. Reproduced here is the title cartouche and the eastern part of the map.

Territorial Governor William Claiborne replied on August 24, 1803, that "there are I believe [no maps of Louisiana] extant that can be depended upon . . . a number of partial but accurate Geographical Sketches of that Country have been taken by different Spanish officers; but . . . it has been the policy of that Government, to prevent the publication of them."[9]

One of the earliest comprehensive maps of the present-day state of Louisiana and adjacent regions was compiled by Bartholemy Lafon and published in 1806. Lafon, who was born in France around 1765, immigrated to New Orleans in the early 1790s. He apparently had training in engineering and architecture for in 1796 he designed public baths for the city of New Orleans. The baths were never constructed, but Lafon also prepared architectural plans for several Louisiana plantation homes and assisted in laying out several new urban districts in New Orleans.[10] Soon after the United States acquired Louisiana, Lafon carried out extensive surveys of the territory in 1804 and 1805. These resulted in two maps. One of them was in seventeen sheets and comprised the Mississippi River between New Orleans and the Gulf of Mexico, showing all the passes and including a chart of the Bay of Spiritu, or Espiritu Santu. (I am not able to identify Espiritu Santu or determine its current name.) This multi-sheet map was reported in the November 22, 1805, issue of the *National Intelligencer and Washington Advertiser* to be in the hands of a Philadelphia engraver, with plans for a speedy publication. It has not been possible to locate a copy of this map. Louisiana governor Claiborne, however, was apparently referring to this map when he wrote to the U.S. postmaster general on June 17, 1805:

> The Map which I have now the Honor to enclose was made out by a Mr. Lafon from an actual survey of the country which it delineates, and ought of consequence to be very accurate. If it is so no difficulty will be experienced in the Transportation of the Mail on the Route Marked out by the red lines along the Canal of Carondelet, the Bayou Gentilli, the Chemin du Chef Menteur to the River of that name, as the Road is so far well opened; ———from hence for the present at least, the transportation must be by water through the Bayou and Lake Catherine across the Rigolets and then either up one of the Branches of the Pearl River to the residence of Mr. Fiore, or along another one to a place marked on the map Boisdore, or perhaps it would be better to terminate the Water Carriage a little to the West of this at a place marked (0) where once stood the village of Marangoin, and from whence there is the old Indian Road leading through the Pine Woods in a Northern direction. From this place Mr. Lafon tells me it is according to his Maps about 130 miles to Fort Stoddart. Should the Transportation of the Mail by water along the Bayou Catherine &c be objected to, as more tedious and expensive than a land carriage, it is the belief of Mr. Lafon that the Government may open a road from the River Chef Menteur in the direction pointed out by the Red lines through the Ile Aux Pins and the Island at the Mouth of the Marangoin to Boisdoru for $3500.[11]

Lafon's more important second map is titled *Carte Générale du Territoire d'Orléans Comprenant aussi la Floride Occidentale et une Portion du Territoire du Mississippi. Dressée d'apres les Observations les plus Reécentes Par Bmi. Lafon Ingénieur Geographe à la Nlle. Orléans* (Fig. 9–3). The place and date of publication are given in the lower right corner of the map sheet as "a la Nlle. Orleans 1806." The map extends east to include west Florida and a large part of Mississippi Territory and west to beyond the present western boundary of the state of Louisiana. The note in the above-cited issue of the *National Intelligencer* makes this reference to the *Carte Générale du Territoire d'Orléans*:

> The author has described the counties of the territory according to the new plan of division, which he has executed himself, by order of the legisla-

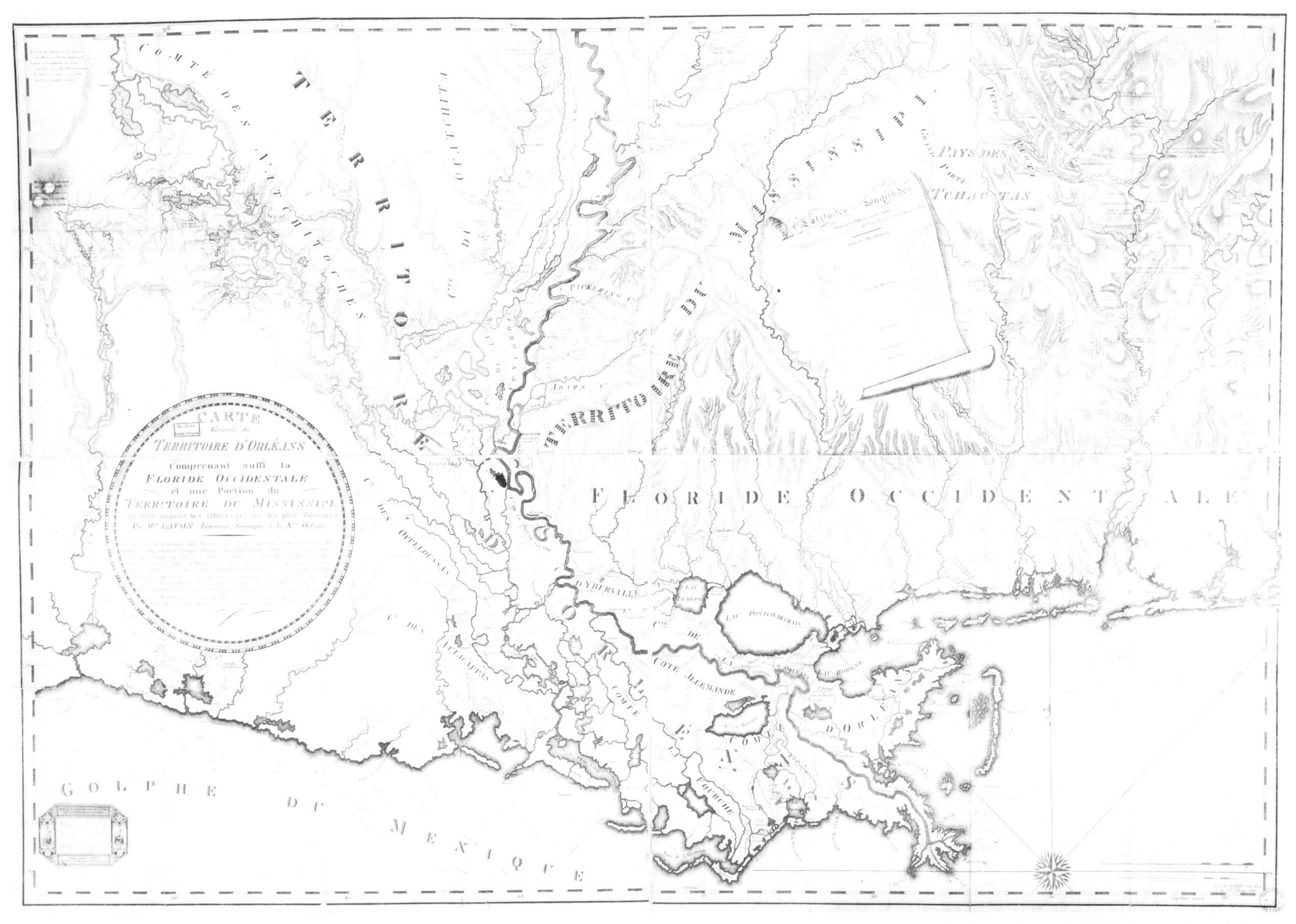

Fig. 9–3. Bartholemy Lafon's 1806 *Carte Générale du Territoire d'Orléans* is the earliest map of the territory which was to become the state of Louisiana. It also includes much of the present state of Mississippi.

> ture, he lays down the latitude and longitude of the principal points. And to render it useful to mariners, he has laid down the soundings along the coast to the Gulph of Mexico.
>
> The map is four feet wide and three feet high, its latitude extends from the mouth of the Mississippi to the limits of the territory. [The map] will be of the most expert engraving, executed by the celebrated William Harrison, of Philadelphia, and will probably be ready for delivery in the month of May, next.

William Harrison's name does not appear on the map, but he must have been the son of William Harrison, Sr. The elder Harrison could not have been the engraver, as he died in 1803. William, Jr., like his father, was born in England and came to America in 1794. Strangely, Lafon's map is also not listed in William J. Harrison's *William Harrison, Sr. and Sons Engravers, A Check List of Their Work*, which he privately published in South Yarmouth, Massachusetts, in 1978.

In January 1806, probably before his map was published, Lafon was appointed lieutenant in the Second United States Regiment. Before the end of that year he had been promoted to captain, and by mid-1807 he was serving as deputy surveyor for Orleans County. Whether he had by this date resigned from the army or held the deputy surveyor's office concurrently has not been established. Although Lafon's surveying methods were criticized in 1807 as being in the Spanish mode and inferior to official U.S. standards, he continued his cartographic pursuits.[12] In 1813, under orders of Brigadier General James Wilkinson of the U.S. Army, Lafon and an associate, Arsene Lacarriere Latour, were engaged in conducting surveys and preparing maps of southern Louisiana. News that the British were approaching New Orleans terminated this work.[13] In 1816 Latour published a *Historical Memoir of the War in West Florida and Louisiana*, which was accompanied by an atlas with nine folded maps. Historian Stanley Arthur noted that "map making with Lafon proved valuable to Latour for the experience and knowledge he thus obtained of the topography of the country to New Orleans, later enabled him to execute the splendid series of maps contained in the atlas which accompanied his Historical Memoir."[14]

During the War of 1812, Lafon prepared a number of fortress and city plans of Louisiana. Seven of these manuscript plans are preserved in the collections of the Library of Congress, and all are dated 1813 and signed "Lafon, Chief Eng." After the war Lafon was active in designing buildings and in planning and laying out new suburbs and sections in New Orleans. In the course of such actions he also bought and sold real estate properties. In 1816 his *Plan of the City and Environs of New Orleans Taken from actual Survey By B. Lafon, Geogr. & Engr.* was published. By this date Lafon was all but bankrupt.[15] He died on September 29, 1820.

The next significant contribution to Louisiana's cartographic development was the publication of William Darby's *Map of the State of Louisiana with part of the Mississippi Territory, from Actual Survey* (Fig. 9–4). It was published in Philadelphia on May 1, 1816, by John Melish. This map, which is 80 by 112 cm., outlines parishes and shows roads, rivers, lakes, crude hachured relief, swamps, soundings along the coast, and settlements. It includes not only all of the state of Louisiana, but also much of Mississippi Territory and segments of west Florida and eastern Texas. Published concurrently with the map was Darby's booklet entitled *A Geographical Description of the State of Louisiana*. In it Darby acknowledged that "Lafon's Map, published in 1805 [actually 1806], considering the then state of geographical knowledge respecting Louisiana, possesses much real merit."[16]

William Darby was born in Lancaster County, Pennsylvania, on August 14, 1775, of Irish parents who settled in that region shortly before the American Revolution. In 1786 the Darby family migrated to Ohio, where Darby's boyhood was spent. Although he had a minimum of formal education, Darby read widely, and at the age of eighteen he was employed as a school-

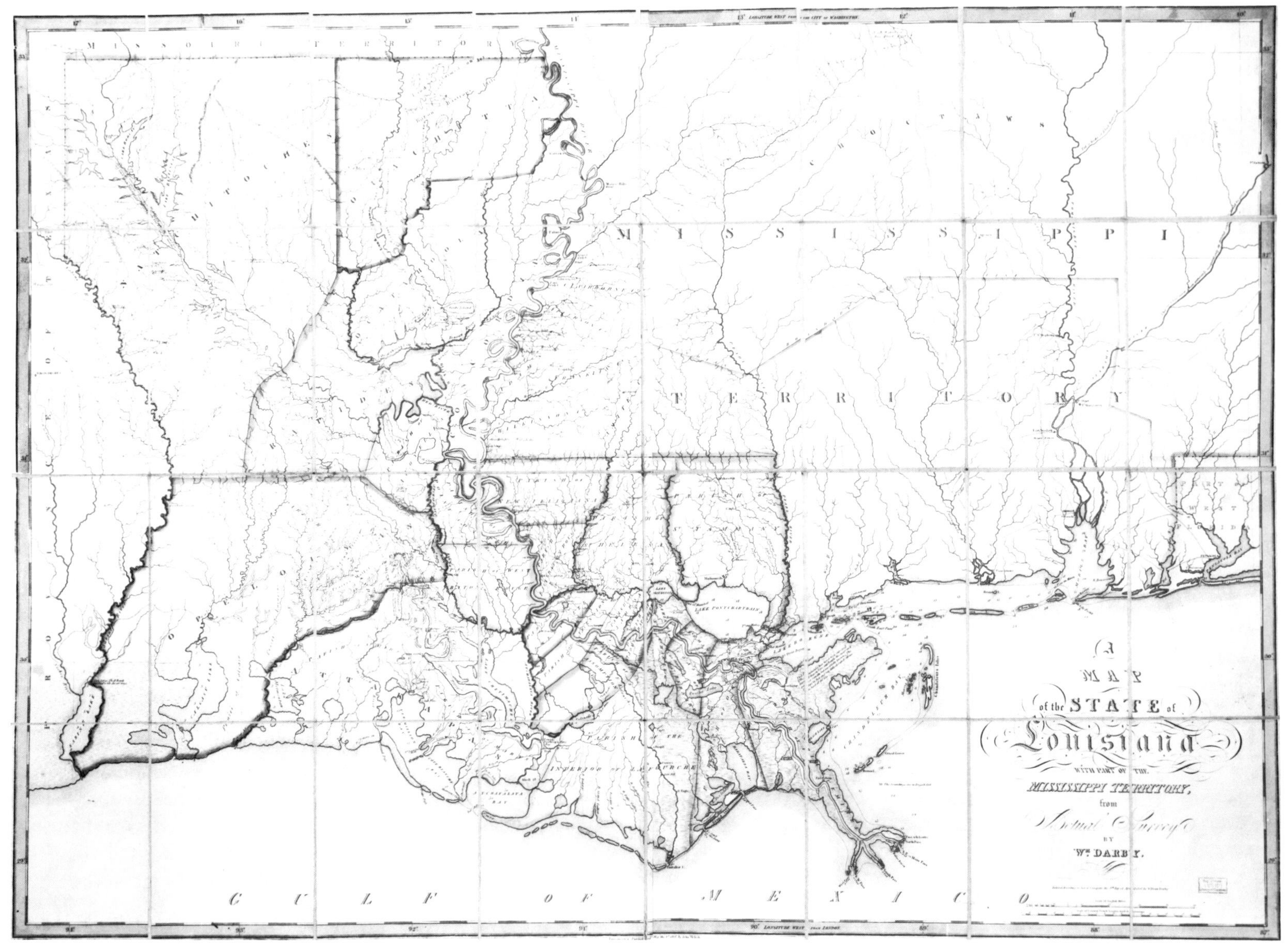

Fig. 9–4. William Darby's map of Louisiana was published by John Melish, who used much of its information for his 1816 U.S. map.

teacher. Following his father's death in 1799, Darby moved to Natchez in the Mississippi Territory, where he became established as a cotton planter. Heavy losses caused by a fire in 1804 led him to accept employment for several years as a deputy surveyor for the U.S. General Land Office. According to Darby's own testimony, it "was in the first part of the year 1800 that I first formed the design to make a map of, and write a statistical account of the region of country, including the State of Louisiana and parts adjacent. From the time mentioned to the month of August 1811, I kept the plan in view, though only incidentally collecting material; but thence forward, until late in 1814, my attention was turned and my time devoted almost exclusively to the project." Because he found existing maps and surveys unsatisfactory, Darby "in the latter of 1811, . . . made an extensive tour over the northern part of what is now the State of Louisiana . . .[and] made in 1812 and 1813, a regular survey, which was commenced at the flag-staff of Fort Claiborne, at Natchitoches. . . . The element obtained by this survey, incorporated with all other requisite data, which I had been enabled to obtain, constituted the element for my map and statistical account of Louisiana and part of then Mississippi Territory."[17]

The War of 1812 interrupted his work on the map of Louisiana and, during the campaign of 1814–15 Darby served as a topographer on the staff of General Andrew Jackson. Unsuccessful in securing official support for publishing his map and statistical report, Darby went to Philadelphia in 1815 in search of a publisher. Both map and booklet were published by Melish in 1816, with the understanding that Melish could use data from the Louisiana map in compiling his general map of the United States. This proved to be an unhappy concession for Darby, for Melish's map became the official map in boundary negotiations between Spain and the United States in 1819. Melish accordingly received recognition and credit, while Darby, whose survey and map were the sources for the geographical data for the region under negotiation, remained unmentioned. In his later years, while serving in Washington as a low-salaried government clerk, Darby unsuccessfully petitioned Congress for some compensation for his contribution.

Louisiana, which joined the Union as a public land state in 1812, provided the first opportunity to test the U.S. land policy in a region that was not included in the original public domain. Because of complex topographic, historic, legal, and legislative problems, the policy was not easily applied to the region. As Harry Coles noted, "prior to admission of Louisiana in 1812 and for at least a half-century afterwards, efforts to fit the Federal land legislation to the peculiar conditions of the region brought many administrative entanglements which required remedial legislation and resort to the courts for settlement of numerous disputes."[18] It is clear, therefore, why no land offices were opened in Louisiana until after the War of 1812. Although some public land surveys were undertaken in the territory as early as 1807, they are quite inaccurate, and the "representation of the country on township maps was delayed for many years."[19] For this reason no township boundaries are shown on Darby's map of Louisiana.

The second edition of Darby's map of Louisiana was also published by Melish, probably toward the end of 1816. The third and fourth editions, undated but probably issued in 1818 and 1819, were published jointly by James Olmstead of New York City and Benjamin Long & Company of New Orleans. In 1818 Kirk & Mercein of New York City published another of Darby's works, the *Emigrant's Guide to the Western and Southwestern States and Territories*, which includes two maps. Sales of this volume brought Darby modest prosperity and enhanced his reputation as a geographer. In 1821 T. H. Palmer of Philadelphia published Darby's *Memoir of Florida*. Between 1829 and 1838, under the pseudonym Mark Bancroft, he wrote and published a number of magazine articles about the West.[20]

Darby's second wife, Elizabeth Tanner Darby (a sister of Benjamin and Henry S. Tanner), died in July 1847. In the following year on November 14, 1848, his brother-in-law Benjamin Tanner died in Baltimore. Shortly thereafter, on a visit to that city Darby renewed his acquaintance with Benjamin's daughter, Mary, whom he married April 10, 1849. They spent most of the following years in Washington, D.C., where Darby was employed in the General Land Office. He died in that city on October 9, 1854.[21]

The first map of Louisiana to incorporate data from rectangular surveys is *A Map of the State of Louisiana with Part of the State of Mississippi and Alabama Territory By Maxfield Ludlow Chief Clerk Surveyor Genls. Office South of Tennessee Engraved by Charles and J. G. Warnicke.* The map is undated, but it was registered for copyright on July 28, 1817, in the eastern district of Pennsylvania. However, because "Part of the State of Mississippi" is included in the title, we may assume that the map was actually published after December 10, 1817, when Mississippi became a state. Maxfield Ludlow's map is 101 by 190 cm., is slightly larger than Darby's, and extends farther east than Darby's to include a larger part of Florida Territory. It also shows the rectangular survey pattern for much of north and central Louisiana and for the central part of Mississippi. In contrast with Darby's map, Ludlow's does not show parish boundaries. In the lower left corner of the Ludlow map is a diagram illustrating the "mode of numbering the Townships & Ranges on the West side of the Mississippi River." A note on the map records that "the Townships have been surveyed By Authority of the United States, under the direction of the following gentlemen[:] Isaac Briggs, Seth Pease & Thomas Freeman Surveyor Generals South of the State of Tennessee."

Little is known about Ludlow beyond the information contained on the map. It is known that he was employed by the federal government as a surveyor, and as early as 1806 he conducted surveys in the Fire-Lands in northern Ohio. These were lands granted to Connecticut citizens whose property had been burned during the revolutionary war. It is possible that he was responsible for compiling and drafting his Louisiana map and may have participated in conducting the surveys on which it was based. Likewise, there is little information concerning the map's engravers, Charles and John G. Warnicke. It is only known that John died in Philadelphia on December 29, 1818.

The mapping of the Ohio River and its valley has had a rich and varied history. "The story of the Ohio River and its valley," wrote Lloyd Brown in his book *Early Maps of the Ohio Valley*, "has no real beginning, and there are no indications that it will ever have an ending. . . . No man or men can claim its discovery, for along its banks are traces of prehistoric life and a long-lost civilization which left artifacts, mounds, and skeletons indicating that long before the advent of the European white man the Ohio River was a good place to live by, surrounded with all the bounties that nature could bestow upon primitive man."[22] The Ohio Valley, accessible via the St. Lawrence River and the Great Lakes, up the courses of the Mississippi and Ohio rivers, and via trails and mountain gaps from the eastern seaboard, was early a crossroads and a prize for explorers from several European countries. To support their respective claims to this promised land, maps were made by the early pioneers and voyageurs. Quoting Brown again, "the mapping of the interior of North America and the early exploration of the Great Lakes and the Ohio River were accomplished by an assorted group of people. . . . In view of the assortment of people from all walks of life who penetrated the interior of the continent it seems incredible that any accurate maps were produced in the seventeenth and eighteenth centuries. Yet such was the case, and there is no great mystery about it."[23]

When the Land Ordinance of 1785 established the General Land Office and specified that public lands be surveyed prior to settlement, it also provided that these public lands be divided into rectangular units one mile square, which were then to be subdivided into sections and quarter sections. Ohio was selected as the test ground for initial surveys under this rectangular system. In July 1786 surveys were begun in the region to the west of the Ohio River in the southeast section of present-day Ohio. Seven ranges were laid out by the surveys, and from these bases lines were extended dividing the land into townships approximately one mile square (Figs. 9–5 and 9–6).[24]

From this inauspicious beginning the public land surveys were progressively extended west and south, and by 1830 they had reached the Mississippi and beyond. The first director of the public land surveys, U.S. geographer Thomas Hutchins, died in 1789, only a few years after the rectangular survey was begun. For various reasons, no successor to Hutchins was named until 1799, when General Rufus Putnam was appointed first surveyor general of the United States. Putnam was succeeded in 1803 by Jared Mansfield, who instituted improvements and refinements in the rectangular survey system, most notably the introduction of principal meridians and parallels of latitude.

Mansfield's appointment coincided with Ohio's organization as a state. The new state was settled rapidly and, apparently, the settlers constituted a profitable market for maps. To meet the demand some four or five separate maps of the state, as well as a number of atlases, were published within the first quarter century of statehood. The first of the maps was the *Map of the State of Ohio taken from the returns in the office of the Surveyor General by Jared F. Mansfield*, which was deposited for copyright in October 1806. Mansfield's map, which is 89 by 64 cm. in size, shows the rectangular survey pattern extending over most of the eastern and southwestern portions of the state.

As surveyor general of the United States, Mansfield had access to the original sketches and notes prepared by government surveyors. The map, however, was apparently privately prepared and published by Mansfield and engraved by William Harrison. During these years when the federal government was not equipped to publish maps, it was not uncommon for officials of the U.S. or state governments to privately publish maps based on official surveys. This was true of the maps of Abraham Bradley, Jr., and David H. Burr.

B. Hough and Alexander Bourne acquired the copyright for Mansfield's map around 1814 and in 1815 published an enlarged version titled *Map of the State of Ohio from Actual Survey by B. Hough and A. Bourne*. It was engraved by Henry S. Tanner and "Published 1st May 1815 by B. Hough & A. Bourne [Chillicothe] and J. Melish, Philadelphia." The Hough-Bourne map, which measures 124 by 113 cm., was the first map to show all the areas actually surveyed within the state. Little is known about Hough or Bourne beyond the information printed on their map. It is probable, however, that they were General Land Office surveyors.

In 1820 Bourne published a small *Map of the State of Ohio including the Indian Reservations, Purchased and laid out into Counties and Townships in 1820. Drawn by J. Kilbourne, Engraved by A. Reed, E. Windsor, Conn. July 1820.* John Kilbourne was born in Connecticut in 1787 and graduated from the University of Vermont. In 1804 he migrated to Worthington, Ohio, a town laid out by his uncle Colonel James Kilbourne.[25] John became principal of the Worthington Academy and, during the next decade he wrote and published several geography textbooks. Around 1815 he abandoned teaching and opened a bookstore in Columbus. There he published in 1816 the first edition of the *Ohio Gazetteer*, a topographical dictionary, which went through a number of editions. The 1818 edition was the first to include a map of the state. Kilbourne died in 1831, but the *Gazetteer* continued to be published until 1841. A separate map of the state, 63 by 61 cm. in size, was compiled and published by Kilbourne in 1822.

In 1828 Horton Howard published what scholar Thomas Smith calls "a very important map in the history of the mapping of the state of Ohio in the nineteenth century."[26] The map, which is 72 by 87 cm., is titled *Topographical Map of the State of Ohio, Published by Horton Howard, Columbus, Ohio 1828*. It was engraved by

William Woodruff of Cincinnati and compiled by Alfred Kelley, a central figure on the Ohio Canal Commission, from surveys of the state authorized by the commission between 1822 and 1824. Woodruff, a skilled engraver, worked in Philadelphia until 1824, when he relocated to Cincinnati. In addition to engraving portraits and maps, he published and sold some of the best examples of midwestern cartography during the early 1830s. In 1831 Woodruff engraved a corrected and revised edition of the Hough-Bourne map for N. & G. Guildford of Cincinnati. Rural and river views decorate its title cartouche.

Compilation and publication of state maps by private individuals declined notably after 1830. There were several reasons for this. By this date the public land surveys had been extended through most of the states beyond the Appalachians and east of the Mississippi River. With excellent survey maps and data available there was no longer a need for individuals to compile and publish maps. As we have seen, both Melish and Henry S. Tanner had recommended against such private endeavors, noting that few individuals had profited from such projects, and many had suffered financially because of such attempts. Another deterrent to individual cartographic projects was the establishment and expansion of commercial map publishing houses after 1812. With the source material provided by the public land surveys and other data, firms such as those founded by Melish, Tanner, and others could adequately supply the cartographic needs of the public. We will examine more closely such commercial cartographic publishing houses in subsequent chapters.

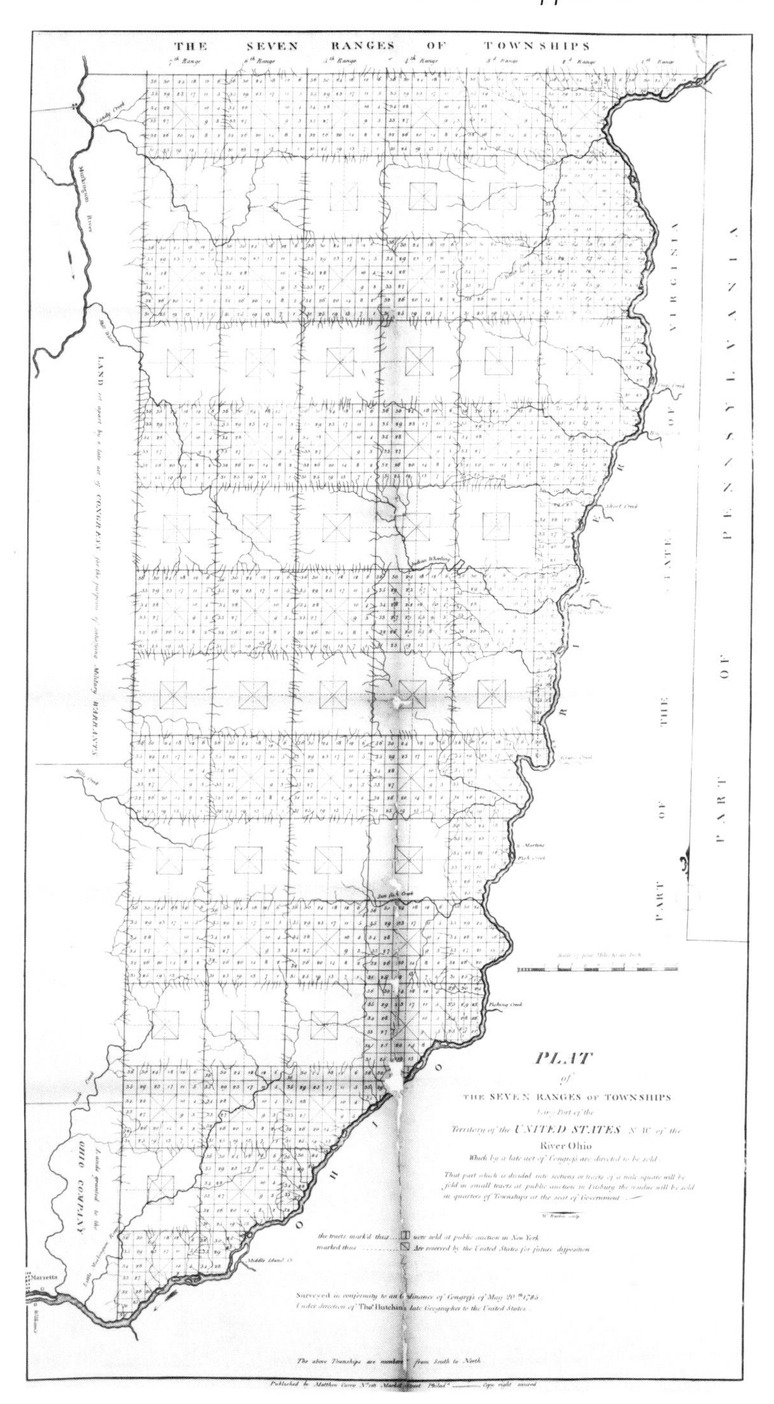

Fig. 9–5. This *Plat of the Seven Ranges of Townships* in southeastern Ohio was based on the first 1786 surveys, on the rectangular pattern, carried out under the direction of Thomas Hutchins. The plat was first published in the 1812 edition of Mathew Carey's *American Atlas*. It was engraved by William Barker.

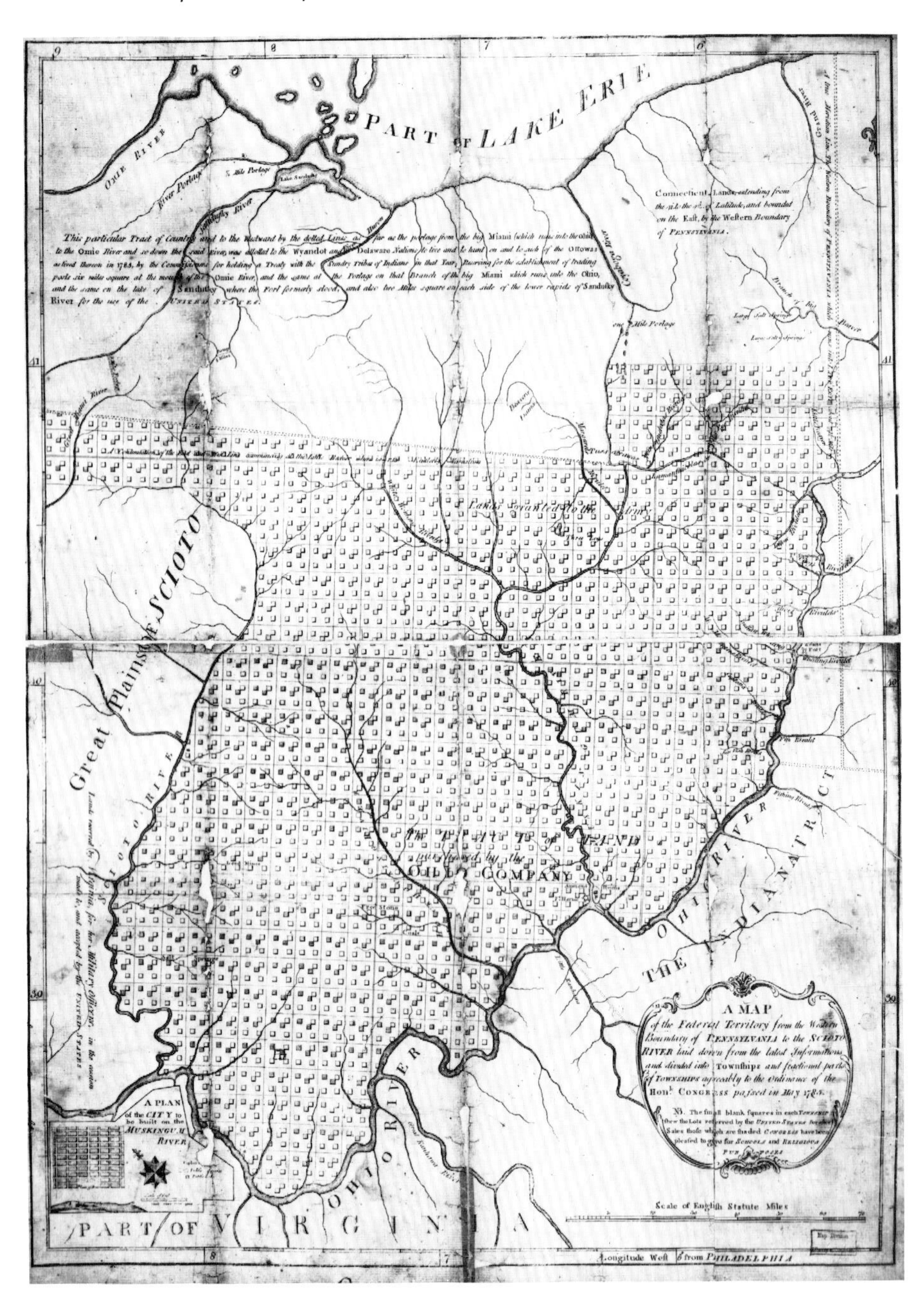

Fig. 9–6. Manasseh Cutler's *Map of the Federal Territory from the Western Boundary of Pennsylvania to the Scioto River*, 1788, was derived, in part, from the as yet unpublished *Plat of the Seven Ranges of Townships*.

Notes

1. James Clements Wheat and Christian F. Brun, *Maps and Charts Published in America before 1800: A Bibliography* (New Haven, 1969), 139–40.
2. Willard Rouse Jillson, "Elihu Barker's Map of Kentucky," *Kentucky Historical Society Register* 21 (Sept. 1923): 322–23.
3. David McNeely Stauffer, *American Engravers upon Copper and Steel* (New York, 1907) 1:9.
4. Kentucky, *House Journal*, 1818–19, Jan. 29, 1819, 238.
5. Ibid., 1819–20, Feb. 12, 1820, 410.
6. Ibid., 23–27.
7. General Daniel Smith, "Papers," *American Historical Magazine* 6 (1901): 213.
8. James A. Robertson, *Louisiana under the Rule of Spain, France, and the United States 1785–1807* (Cleveland, 1911) 2:140.
9. Clarence E. Carter, ed., *Territory of Orleans, 1803–1812*, vol. 9 of *Territorial Papers of the United States* (Washington, D.C., 1940), 16.
10. Mary Cable, *Lost New Orleans* (Boston, 1980), 16, 71–72, 76.
11. William Charles Cole Claiborne, *Official Letter Books 1801–1816*, ed. Dunbar Rowland (Jackson, Miss., 1917) 3:97–98.
12. Carter, *Territorial Papers*, 746–47.
13. Stanley C. Arthur, *Jean Laffite, Gentleman Rover* (New Orleans, 1952), 99.
14. Ibid., 100.
15. Stanley Faye, "Privateersmen of the Gulf and Their Prizes," *Louisiana Historical Quarterly* 22 (Oct. 1939): 1068.
16. William Darby, *A Geographical Description of the State of Louisiana* (Philadelphia, 1816), iv.
17. Henry O'Rielly, "Pioneer Geographical Researches, Explorations and Surveys in the Louisiana Purchase," *Historical Magazine*, ser. 2 (Oct. 1867), 223.
18. Harry L. Coles, Jr., "Applicability of the Public Land System to Louisiana," *Mississippi Valley Historical Review* 43 (June 1956): 40.

19. Coles, "Applicability," 44.
20. J. Gerald Kennedy, *The Astonished Traveler: William Darby, Frontier Geographer and Man of Letters* (Baton Rouge, 1981), 62, 136–40.
21. Ibid., 124–25.
22. Lloyd A. Brown, *Early Maps of the Ohio Valley, a Selection of Maps, Plans, and Views Made by Indians and Colonials from 1673 to 1783* (Pittsburgh, 1959), 1.
23. Ibid., 8.
24. William D. Pattison, *Beginnings of the American Rectangular Land Survey System, 1784–1800*, Research Paper No. 50 (Chicago, 1957), 37–67.
25. This information is mainly derived from Thomas H. Smith, *The Mapping of Ohio* (Kent, Ohio, 1977).
26. Ibid., 171.

10. *Early American Atlases*

Concurrent with the production of the earliest state maps in the last decade of the eighteenth century, there was a flurry of activity in the publication of atlases of the new republic. The atlas format is a popular one because it assembles between two covers a number of related maps. It also provides a most convenient arrangement for retaining and preserving maps in a home library.

English colonial activity in North America during the eighteenth century was reflected in the many maps and atlases that issued from British cartographical publishing houses, particularly in the third and fourth quarters of the century. Thomas Jefferys's atlas, *A General Topography of North America* (London, 1768), was not a systematic effort, but it does contain one hundred maps of the English colonies by various cartographers and engravers. About 40 percent of the maps are of regions presently within the United States. After Jefferys died in 1771, his successors continued to publish his atlas under the title *American Atlas*. Issues were printed in 1775, 1776, 1778, and 1782. They variously include twenty-two to thirty-four plates.

Following the Revolution, there was considerable activity in the United States by American mapmakers and publishers. One of them, Mathew Carey, was a pioneer in producing cartographic works. A native of Dublin, Carey immigrated to Philadelphia in 1784 and in the following year established a print shop and publishing house. His initial interests were in journals and serials, but by 1792 he began to publish monographs. His earliest cartographic publication, issued in 1794, was *A General Atlas for the Present War*, which reprinted seven maps from the English work *Atlas to Guthrie's System of Geography*. These maps show regions that were involved in the Napoleonic campaigns. They were reengraved by William Barker and Joseph T. Scott, both of Philadelphia. They were among a number of engravers who worked on contract for Carey.

In 1795 Carey published *The General Atlas for Carey's Edition of Guthrie's Geography Improved*. William Guthrie's popular textbook was originally issued in London in 1770. In addition to Carey's editions of it, the textbook had been put out under the imprint of English and French publishers. The sixteen maps of American states included in Carey's *General Atlas for the Present War* were reprinted with five others in his *American Atlas*, which was also published in 1795. The *American Atlas* is the earliest atlas of the United States. The engravers of the maps in both atlases were Barker, Scott, James Thackara, and John Vallance, all of Philadelphia, Samuel Hill of Boston, Amos Doolittle of New Haven, and Benjamin Tanner of New York. Samuel Lewis, who compiled and drafted a number of the maps, had a reputation as a geographer, draftsman, mapmaker, and penman (Fig. 10–1). Subsequent to his work for Carey, he engaged in map compilation and publishing independently and with the English publisher Aaron Arrowsmith.

A second edition of Carey's *American Atlas*, expanded to twenty-six maps, was published in 1809, and revisions of the *American Pocket Atlas*, initially published in 1796, appeared in 1801, 1805, 1813, and 1814. Carey's *General Atlas* was likewise published in a number of revisions until 1818. Carey retired in 1822 and the publishing business was continued by his son, Henry C. Carey, and his son-in-law, Isaac Lea. Cartographic publishing, however, was of decreasing importance under the Carey & Lea imprint.

The elder Carey had made a number of innovations in cartographic publishing during his nearly forty-year-long career. Commenting on them, writer J. Brian Harley notes that the documentary records for the early American map trade are generally sparse, but the Carey accounts make it clear, in analysing the rise of cartography after 1783, that we must rid our minds of the model of the larger London cartographer in this period. . . . Carey, . . . rather than employing draftsmen, map engravers, map colorers, and mounters directly in his own workshop, operated an urban cottage industry, organizing on an out-worker basis the various processes of atlas-making among the publishing trades clustered in Philadelphia. In terms of business structure this represents

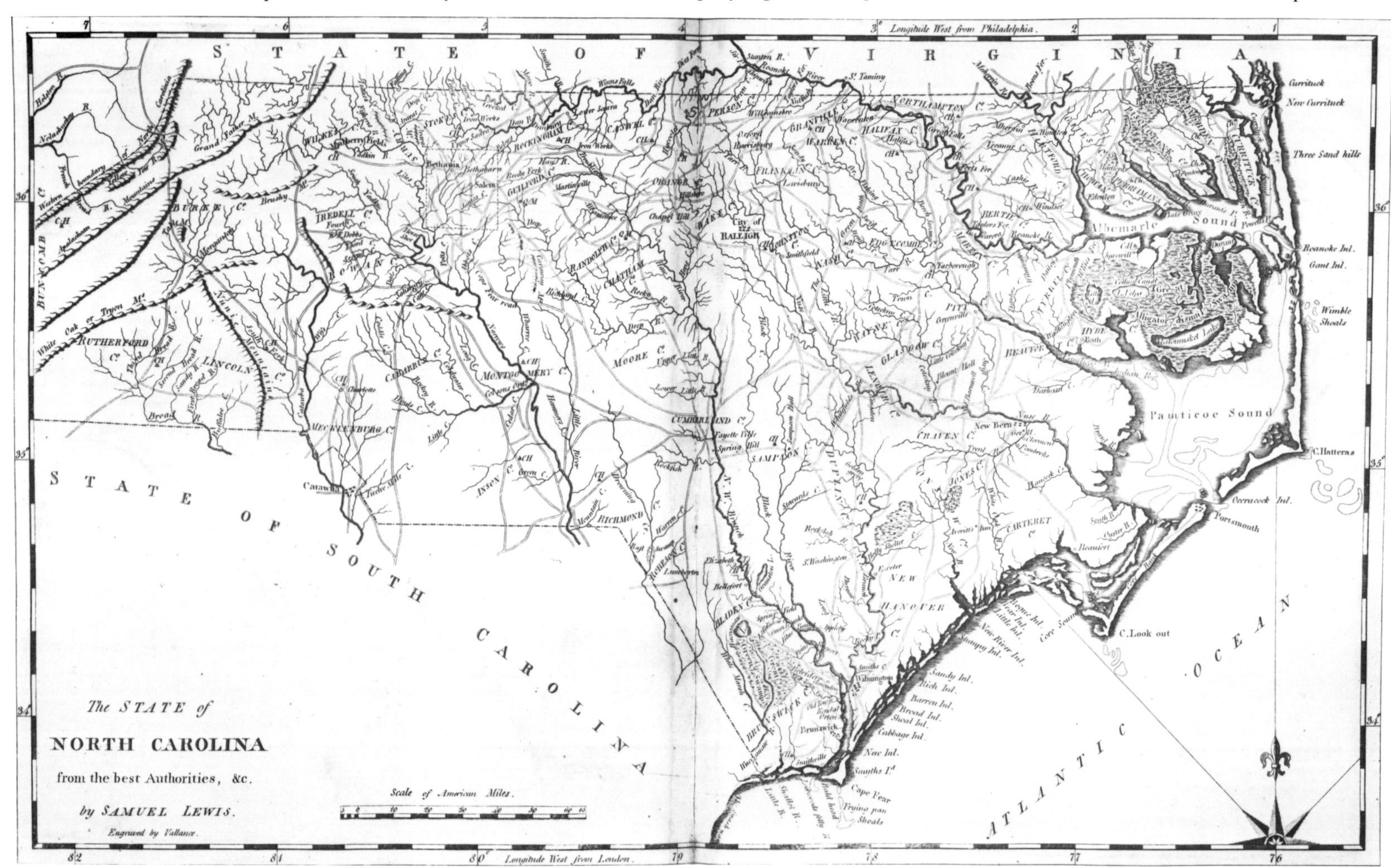

Fig. 10–1. The map of North Carolina by Samuel Lewis from Carey's 1795 *American Atlas*. Lewis compiled at least ten of the maps in the atlas.

> an embryonic state in the development of American commercial cartography, characterised by complex linkages between various craftsmen both in Philadelphia and, to a less extent, reaching out to more distant specialists in Boston and New York, the other two main centres of publishing in this period. Carey's most tangible achievement was his organization of these self-employed artisans into an effective production line for his atlases. Four main stages were involved: first there was a geographer who compiled the map and prepared a fair drawing; second, there was the engraver who transferred this image to the copperplate; third there was a printer trained to use a roller-press for map work; and fourth, there were 'map finishers,' diverse small craftsmen, who were responsible for preparing the maps according to individual customer requirements.[1]

This practice was also followed by later American cartographic publishers such as John Melish and Henry S. Tanner, who like Carey were operating on extremely limited budgets.

Another Carey practice also emulated by subsequent publishers was to use the same maps in several different atlas publications. Harley discloses that

> Carey became a master in the art of packaging the same maps for reissue in several of his publications. A set of atlas maps could thus be bound throughout the text of [Guthrie's] *Geography* or, if the customer preferred it, be brought together in a separately bound atlas; the same maps also appeared in Carey's other atlases which, moreover, ran into more than one edition. A map of Virginia (1794), compiled by Samuel Lewis and engraved by James Smither, was published not only in *Carey's American Atlas*, *Carey's General Atlas*, and *The General Atlas for Carey's Edition of Guthrie's Geography Improved*, but also in the second edition of Thomas Jefferson's popular *Notes on the State of Virginia*.[2]

In further comments on Carey, Harley asserts that

> a few general observations can be made about the rise of American map-making in the 1790s. In Boston, New York and Philadelphia, many of the preconditions for a flourishing map trade, in terms of appropriate craft skills capable of adaptation, were already in existence and in this sense Carey was the catalyst. Moreover, the policy of using local craftsmen made good commercial sense and for a publisher so undercapitalised as Carey it is doubtful if there was a real alternative. Local engravers could be employed more cheaply than their counterparts in London or Paris, and there was a further advantage of immediate surveillance of the work and speed of production.[3]

The *American Atlas* concept introduced by Carey was shortly adopted by other publishers in Europe as well as in the United States. *An American Atlas Containing eight maps and a Plan of the City of Washington, Engraved by J. Russell. London: Sold by H. D. Symonds, Paternoster Row, and J. Ridgeway, York-street, Saint James' Square* was published in London in 1795. The eight maps are of North America, South America, the West Indies, the United States, the New England states, the southern states, and Kentucky. A double-page plan of the city of Washington is also included. Although the city had been laid out by 1795, the federal government had not yet been transferred from Philadelphia.

Similar in format and size to the Carey atlas and its English copy is W. Winterbotham's *The American Atlas*, published at Philadelphia in 1796 by John Reid. It was designed as an accompaniment to Winterbotham's *An Historical, Geographical, Commercial and Philosophical View of the United States*. The atlas includes twenty maps of which nine were engraved by Benjamin Tanner, five by David Martin, two by D. Anderson, one by John Scoles, and one by John Roberts. There is no engraving credit on two of the maps. At this date Tanner was still working in New York City, as were apparently all the other engravers who prepared plates for the atlas.

Also published in 1796 was *An Atlas of the United States Containing a Large Sheet Map of the Union . . . , by Joseph Scott*. It was published by Frank and Robert Bailey of Philadelphia. This small, pocket-size atlas includes a fold-out map of the United States and eighteen individual state maps. All nineteen maps were originally included in Scott's *United States Gazetteer*, which was published in 1795. The U.S. map also comprises the frontispiece of Scott's *Geographical Dictionary of the United States of North America*, which was published in Philadelphia in 1805. None of the maps in Scott's publications have drafting or engraving credits.

John Melish published *A Military and Topographical Atlas of the United States* in Philadelphia in 1813. Six of its eight maps were engraved by Henry S. Tanner. An enlarged edition of Melish's *Military and Topographical Atlas*, containing twelve maps, was published in 1815. Tanner added map publishing to his engraving activities around 1820. His first major venture into the atlas field was the *New American Atlas*, published in 1823, which still ranks as one of the most magnificent atlases ever published of the United States. It was also one of the last American atlases published during the period in which maps were still engraved.

Also published in 1823 was Sidney E. Morse's *An Atlas of the United States on an Improved Plan; Consisting of ten maps*. Its engravers and publishers were Nathaniel and Simeon S. Jocelyn of New Haven. The preface to this small atlas gives sources consulted in compiling the several maps. There is a separate index for each map. Sidney Edwards Morse (1794–1871) was the son of Jedidiah Morse (1761–1826), a Congregational minister and the compiler and publisher of a very successful series of geography books between 1784 and 1825. Sidney Morse's atlas may have been planned as a supplement to his father's *Geography*.

Sidney Morse began a career in journalism in 1812, and in 1815, at the age of eighteen, he became editor of the *Boston Recorder*, the first religious newspaper in the country. With his younger brother, Richard Cary Morse, Sidney founded the *New-York Observer*, a weekly religious journal, which he edited and published until 1858. Morse had an inventive mind and, in 1834, with Henry A. Munson, he invented cerography, a new method for reproducing and printing maps. Cerography, or wax engraving, was a much more expeditious and less expensive process than copper-plate engraving, but the resulting maps were far inferior in quality and appearance.[4]

To publicize his invention and its products, Morse, in collaboration with Samuel Breese, compiled a series of thirty-two maps which were issued as supplements to the *New-York Observer*. The first cerographic map, which was of the state of Connecticut, appeared in the June 29, 1839, issue. Also in that issue, Morse announced plans to print additional maps that could be assembled to form an atlas. Thus, between 1842 and 1845 the *Observer* printed map supplements that were then assembled to form the *Cerographic Atlas of the United States* (Fig. 10–2). The maps in the atlas are uncolored and undistinguished in appearance. Moreover, they are printed on poor quality soft paper which has deteriorated with time. Few copies of the atlas have therefore survived. Other Morse applications of the wax engraving technique produced the *Cerographic Bible Atlas*, published in 1844, and the *Cerographic Missionary Atlas*, dated 1848. Morse also issued, through Harper & Brothers of New York City, a *North American Atlas* and several editions of a *School Geography*, which also contained cerographic maps.

Aborted American Atlases

The atlases previously described were more or less successful. During the early years of the Republic other

Fig. 10–2. The map of Connecticut from Sidney E. Morse and Samuel Breese's *Cerographic Atlas of the United States*.

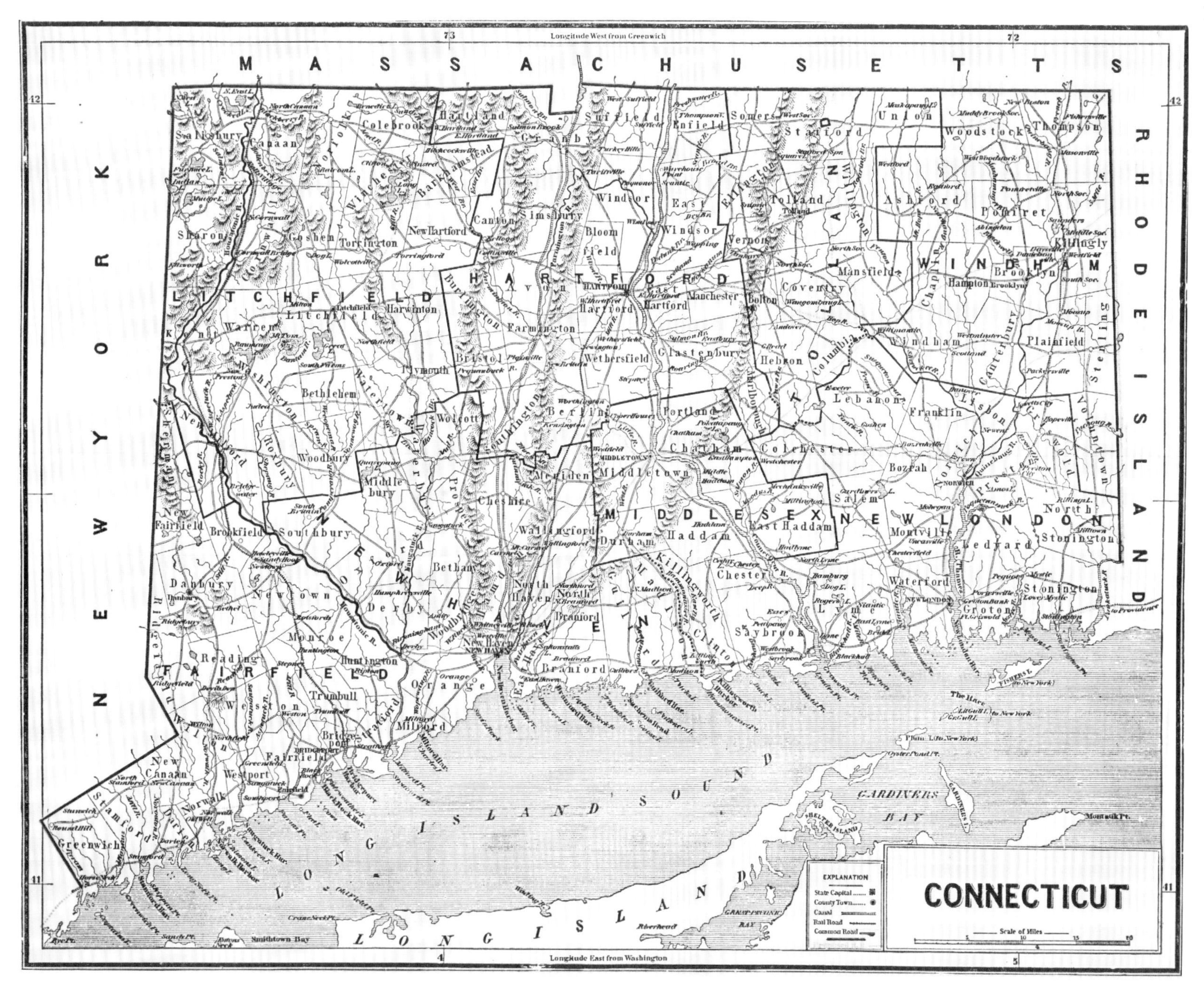
Longitude West from Greenwich
73
72
42
41
MASSACHUSETTS
NEW YORK
RHODE ISLAND
LITCHFIELD
HARTFORD
TOLLAND
WINDHAM
FAIRFIELD
NEW HAVEN
MIDDLESEX
NEW LONDON
LONG ISLAND SOUND
LONG ISLAND
GARDINERS BAY
EXPLANATION
State Capital
County Town
Canal
Rail Road
Common Road
Scale of Miles
CONNECTICUT
Longitude East from Washington
4
5

individuals also sought to compile and publish atlases for which they thought there would be a demand. Unhappily, the demand was not always there, and the evolutionary course of American cartographic history is littered with the remnants of a number of aborted atlases. These unsuccessful attempts are, nonetheless, essential chapters in the history of American map publishing.

After the Revolution, land communication was of major importance in uniting the infant states. It had not been as important before independence, because each colony communicated mainly with the mother country across the Atlantic. As a result, highway mileage in the early days of the Republic was quite limited, probably not exceeding five or six thousand miles of readily usable roads. With few exceptions, the roads were unsurfaced, deeply rutted, muddy, and impassable following heavy rains. Highway maps were nonexistent and travelers had little guidance in planning or embarking on journeys beyond their immediate environs. It is not surprising, therefore, that road guides were among the earliest projected American atlases.

Walker and Abernethie's "Travelling Map of South Carolina"

The Charleston, South Carolina, *Columbian Herald* of September 27, 1787, contained

> proposals for publishing by subscription, Walker and Abernethie's Travelling Map of the Public established high Roads throughout the state of South Carolina, beginning at the city of Charleston, and extending to the boundary lines of said State, from actual survey. . . . This Map will be laid down by a scale of one inch to one mile, and on which will be delineated all roads leading out of, or crossing, the high roads; the true course these roads go, and to what particular place; names of gentlemen's plantations that are known or that lay contiguous to the said roads; every place of public worship, all rivers, creeks, rivulets, causeways, swamps, branches or gullies; every stage or inn, all mills or millseats, the cleared and wood lands; the districts, country and parochial lines, touching or crossing the said roads; with every other distinguishing mark or observation that will render this Map a true directory, as well as an amusement to the traveller.
>
> Also on this Map will be pointed out and distinguished by a particular mark & figure, where every mile post should be erected, which will enable the parochial commissioners to fix mile posts throughout the whole state.

It was planned to publish the "Travelling Map" in "volumes, each volume to contain 2,000 miles, which will be 40 copper plates, 19 inches long and 9 wide, to be bound in sheep skin, which will make them portable for travellers, as they can be easily rolled and carried in the pocket,—An index and direction how to use the map will be prefixed." The price to subscribers was set at six dollars for the first volume. It was assumed that "all the roads in the state may be comprised in 2, or at most in 3 volumes." The original subscription only bound subscribers to take one volume. It was "left to their opinion to complete the plan or not as they please, by extending their subscriptions to the subsequent volumes, after the publication of the first."

There is a suggestion that the "Travelling Map" had official support, for subscriptions were "taken in at the secretary of state's office, common pleas office, surveyor general's office, tax office," as well as by various shopkeepers and "by gentlemen properly authorized." The notice closed with the admonishment that "those gentlemen who may be inclined to encourage the above undertaking, are requested to subscribe as soon as possible, as the publishers intend to begin the survey, whenever a sufficient number of names appear on their list."

Probably at the same time as the proposals notice was published in the *Columbian Herald*, there was distributed to prospective subscribers a *Specimen, of an Intended trav-*

elling Map, which was identified as "Plate X, Page 19, Walker & Abernethie, Roads of South Carolina, 1787" (Fig. 10–3). The specimen has three strip maps arranged side by side that trace the "Road to Watboo Bridge from Charleston by Goose Creek & Strawberry Ferry." The maps on the specimen appear to be the only part of the "Travelling Map" that was ever published. Two extant copies of the specimen are known. One is preserved in the South Carolina Historical Society in Charleston and the other is in the Library of Congress.

Little is known of the prospective publishers of the "Travelling Map." Both were natives of Scotland who settled in Charleston around 1783. Thomas Walker was from Edinburgh and appears to have been trained as an architect and tombstone cutter. Tombstones signed by him are found in Charleston graveyards. He once advertised an evening school course for teaching the rules of architecture, although there is no record of any architectural works he may have designed. Thomas Abernethie opened an engraving and copper-plate printing shop in Charleston shortly after he arrived there. The credit "Abernethie Sculpt. Charleston" appears on three maps that illustrate David Ramsay's two-volume *The History of the Revolution in South Carolina*, which was published in Trenton in 1785. In an advertisement published December 22, 1786, Abernethie noted that he had moved "to 227 Meeting-street, where he solicits a continuance of the favors of his friends; he takes the liberty of acquainting them that he continues the business of a Land Surveyor, and from the experience he has acquired in an extensive line of business in Europe, has no doubt of giving entire satisfaction to those gentlemen who are desirous of having their estates accurately surveyed and neatly planned."[5] His services also included "maps and Plans copied, diminished or enlarged, and neatly mounted on linen, either with rollers or for the pocket."

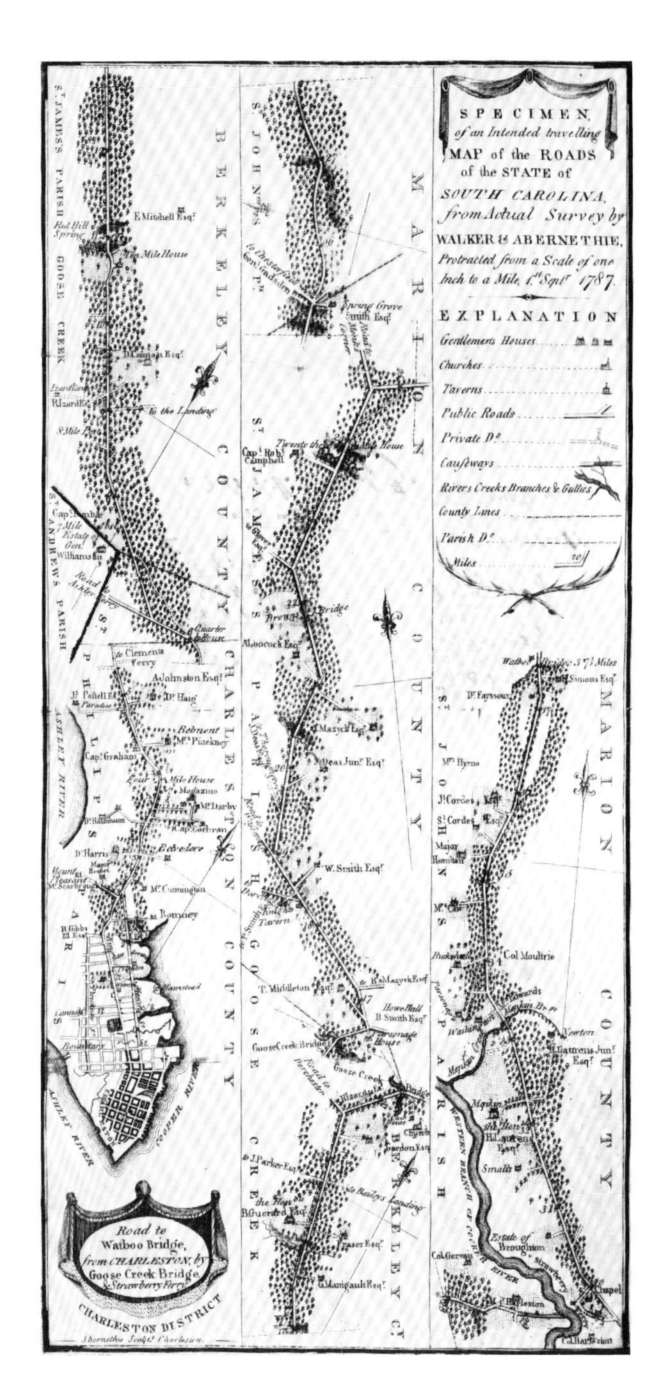

Fig. 10–3. Thomas Walker and Thomas Abernethie's *Specimen, of an Intended travelling Map of the Roads of the State of South Carolina, from Actual Survey*, 1787.

Almost a decade later on February 3, 1795, Abernethie was still advertising the same services from a different location, at 42 Queen Street. He thanked "a generous public" for past favors "and hopes that he will receive a continuation of the same, having for the present made it his determination to make his residence in town."[6] Abernethie died early in November 1795; on the thirteenth of that month the goods and chattels belonging to his estate were appraised. Included were seven slaves and a "lot of Surveying Instruments & Engraving Tools."[7] On September 5, 1796, there were offered for sale "a Copper-Plate Printing Press, Belong to the estate of the late Mr. Thomas Abernethie. By order of the executrix, Colcock & Patterson."[8]

Walker and Abernethie appear to have patterned their projected "Travelling Map" after the road books of England, Scotland, and Ireland that were published in the 1770s. Maps in their specimen resemble the cartographic style and technique used in *Maps of the Roads of Ireland* by fellow Scotsmen George Taylor and Andrew Skinner. This road book was published in 1778 from surveys made during the previous year. Taylor and Skinner had earlier published road books of Scotland and northern England. Following completion of their Irish road survey, Taylor and Skinner joined the British army, or were impressed into service by the military engineers, and were sent to America. Preserved in the British Library is the manuscript *Map of New York & Staten Islds. And Part of Long Island Surveyed by Order of His Excellency General Sir Henry Clinton K. B. Commander in Chief of His Majesty's Forces. 1781*. This detailed map was "Surveyed & Drawn, by George Taylor & W. Skinner. Surveyors to His Excellency the Commander in Chief."

No copies of the "Travelling Map of South Carolina" are known and we must, therefore, assume that the ambitious project never advanced beyond the proposal stage and the preparation of the specimen with its three strip maps. The Constitution of the new republic was not ratified until September 1788, and the nation was in the midst of a severe postwar economic depression. There were obviously few individuals who were willing or able to subscribe six dollars for a road book that was then only in the preliminary planning stage. Moreover, it is questionable whether Walker and Abernethie could have made a profit on their projected atlas. The first volume was to contain forty sheets, each nineteen by nine inches, which were to be printed from engraved copper plates. Engraving was a highly skilled and tedious technique, and the cost of producing one plate at that time was probably about one hundred dollars. The engraving bill alone for one volume, therefore, would have amounted to about four thousand dollars. Wages for several surveyors and the cost of ink, paper, sheepskin, printing, and binding might have required an investment of as much as eight thousand dollars by Walker and Abernethie. Between fifteen hundred and two thousand subscriptions would have been required to cover costs and ensure a modest profit for the two partners. This number was, obviously, not achieved, and the "Travelling Map" became the first of several aborted atlases in the first thirty years after the United States was established.

Colles's Survey of the Roads of the United States of America

Two years after Walker and Abernethie published proposals for their "Travelling Map," Christopher Colles published a broadside which set forth "Proposals for Publishing a Survey of the Roads of the United States of America." The *Survey of the Roads of the United States of America* was to be published on a subscription basis, with each purchaser paying "one quarter of a dollar at the time of subscribing (to defray several incidental charges necessary for the work) and one eighth of a dollar upon the delivery of every six pages of the work"[9] The first three plates of the *Survey* and its title page dated 1789 were distributed with this broadside. It is very likely that plates 4, 5, 6, and 7 were also issued in the same year. Like Walker and Abernethie's "Travelling Map," Colles's *Survey* consists of strip maps arranged side by side, two or three to a plate. Collectively, the strip maps on each plate cover approximately twelve miles. The cartographic execution and engraving are inferior to the workmanship on the three strip maps in the Walker and Abernethie specimen.

Although Colles apparently received few subscriptions for the *Survey*, he continued work on the project and completed eighty-three plates by about 1792. This fell a bit short of the one hundred pages which had been promised in his proposals. The eighty-three plates, with two or three strip maps on each, cover approximately one thousand miles of roads extending from Albany, New York, on the north and Stratford, Connecticut, on the northeast to Yorktown and Williamsburg, Virginia, on the south. The map scale is 1:110,000 as compared with the 1:62,500 scale on the Walker and Abernethie specimen.

Colles was English by ancestry but was born in Dublin, Ireland, on May 9, 1739 (Fig. 10–4). In addition to receiving a good general education, he had specialized training in mathematics, geography, engineering, and science with Dr. Richard Pococke, a distinguished cleric. Under the tutelage and sponsorship of his uncle, William Colles, Christopher worked on several canal and engineering projects in Ireland. He was restless, however, and changed jobs and residences frequently, even after his marriage to Anne Keough of Kilkenny, Ireland, on January 14, 1764, and the birth of several children.[10]

After the deaths of Pococke in 1765 and his uncle in 1770, Colles's principal ties to Ireland were severed, and he decided to immigrate to America, where he hoped to find ample opportunities for his scientific and engineering training and experience. With his wife and four children, Colles sailed from Cork in May 1771. They landed at Philadelphia on August 10, 1771, and remained there for about three years. In Philadelphia Colles advertised for work in his various specialities—designing hydraulic engines and buildings, land surveying, canal building, and teaching. He apparently received few commissions and was obliged to give public lectures to support his growing family.

Early in 1774 Colles moved his family to New York, where he again engaged in lecturing. Shortly thereafter, however, he made a proposal to the city council to prepare a water-supply system. The outbreak of the revolutionary war put an end to the project, but the city did compensate Colles with £150 in 1788. Colles left New York City during the war, probably because of his close association with John Lamb and other members of the Sons of Liberty. He spent the war years in upper New

Fig. 10–4. Portrait of Christopher Colles by John Wesley Jarvis, ca. 1809. Courtesy of the New-York Historical Society.

York State, northern New Jersey, and perhaps eastern Pennsylvania and western Connecticut. There is no evidence that he ever served in the Continental army, but he may have instructed artillery officers in mathematics at the invitation of his friends General Henry Knox and General John Lamb. During these years, Colles also designed and constructed a perambulator which consisted of a series of cogwheels that recorded the number of revolutions of the wheel to which they were attached. The number of revolutions multiplied by the perimeter of the wheel gave a fair measure of the distance traversed.

While still in exile from New York City, Colles probably spent some of his time surveying the roads in New York State. Surveys for the maps on the first three plates of the *Survey* may have been completed in 1783, when Colles reestablished residence in New York City. Soon after his return, he was deeply involved in various enterprises and activities. Of particular note were his several proposals for inland waterway projects, one of which was concerned with removing obstructions on the Mohawk River. Colles also spent some time surveying roads, primarily in New York State. When he distributed his proposals for the *Survey* in 1789, he may have already completed surveys for most of the roads between New York City and Albany.

Few individuals subscribed to the *Survey*, and by early 1790 Colles had to turn to printers and booksellers to solicit subscriptions. He also presented to the New York State legislature on March 13, 1790, a petition "praying aid of the Legislature to enable him to proceed in an intended survey of roads, by means of a perambulator." The petition "was read and referred to Mr. J. Brown, Mr. Havens, and Mr. Clowes." The legislature's next response was to refer it to the "general government for support of his business."[11] Colles hastened to follow up on this recommendation and on March 30, 1790, a second petition "was presented to the House and read, [in which he was] praying to be employed by Congress in a survey and publication of the Roads of the United States."[12] The Congress referred this petition to the Post Office Department which, understandably, had a deep interest in transportation matters. After testing the accuracy of Colles's perambulator, Postmaster General Samuel Osgood reported back to the Congress on April 27, 1790, that

> the assistance requested of the public, at the rate of one eighth of a dollar per mile, will amount to about three hundred and seventy-five dollars; for the extent to be surveyed cannot vary much from three thousand miles. The Postmaster General, upon a due consideration of the benefits that will result from the execution of the work undertaken by the memorialist, is of the opinion that the public interest will be promoted by granting him the aid prayed for, in proportion to the distance surveyed, and the publication, after being qualified to its having been done with the proper attention.[13]

It is regrettable that because of the more pressing demands upon the Congress and the country's resources, no action was taken on the favorable report of the postmaster general, and Colles had to seek other assistance in striving to complete his road book.

It is likely that Colles personally conducted the surveys represented on plates 1–33 of the road book. They map the roads from New York City to Stratford, Connecticut, from Stratford to Poughkeepsie, New York, from New York City to Poughkeepsie, from Poughkeepsie to Albany, and from Albany to Newborough. The west side of the Hudson River from Newborough to Paulus Hook is unmapped and plates 34–39, apparently reserved for this stretch of road, are not in the road book.

The remaining plates, 40–86, with the additions of 45*, 46*, and 47*, map the roads extending south from New York City through eastern Pennsylvania and western New Jersey via Wilmington, Delaware, Baltimore, Maryland, and Alexandria, Virginia, to Williamsburg and York, Virginia (Fig. 10–5). There is no evidence that

Colles ever traveled south of Philadelphia, and so plates 40–86 are probably not based on his personal surveys. Studies have shown that the information on the last fifty plates closely parallels that on maps prepared for General Washington under the direction of Simeon De Witt.[14]

It seems evident that Colles was somehow able to borrow the official military maps, from which he prepared the last fifty plates of his road book. The most plausible explanation is that Colles obtained the maps with the aid of his friends Lamb and Knox. The latter, who directed the artillery in the Continental army, was appointed secretary of war by the Continental Congress in 1785, and the appointment was reaffirmed following Washington's inauguration as president in 1789. The military maps were quite likely in the custody of the War Department by this time, and we may assume that Knox granted Colles permission to consult them.

Colles practiced mapmaking in Ireland and so he quite certainly drafted the strip maps for the *Survey*. Since the title page carries the credit "C. Tiebout, Sculp.," some bibliographers have assumed that Tiebout also engraved the maps. He may have prepared some of the early plates, but variations in style and symbolism suggest that more than one engraver may have been involved. Moreover, Tiebout went to London in 1793 for further study, and some of the plates are believed to have been engraved after this date. In his *History of New York City*, Benson Lossing states that "Colles constructed and published a series of sectional road maps which were engraved by his daughter."[15] This may be true, for two plates of the *Geographical Ledger*, a later Colles project, are credited to "Eliza Colles, Sculp."

The *Survey* is generally regarded as the earliest road map of the United States. There may be some objection, therefore, to including it among aborted atlases. Although approximately 80 percent of the projected plates

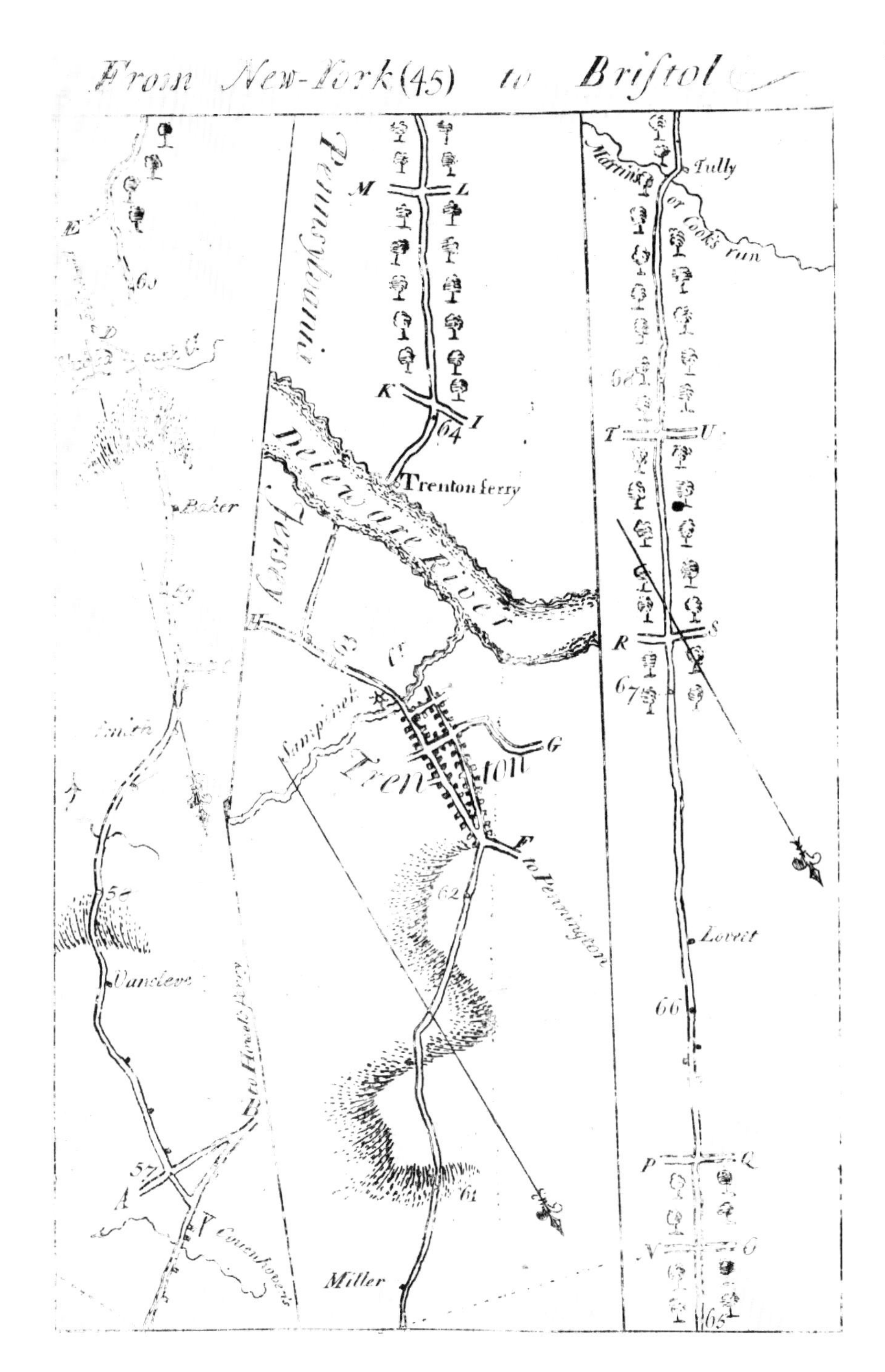

Fig. 10–5. Plate 45 from Colles's *Survey of the Roads of the United States of America*, 1789.

were completed, large sections of the United States were not mapped, such as New England, the west bank of the Hudson River, and the states south of Virginia. Although in the proposals subscribers were "considered as engaging to take 100 pages," only eighty-three plates were published. There are fewer than twenty extant copies of the *Survey* in its most complete eighty-three-plate form and only an additional ten or fifteen less complete sets—further evidence that few subscriptions were received.

Colles's Geographical Ledger

The strongest evidence that the *Survey* was aborted comes from Colles himself, who initiated another cartographical project, the *Geographical Ledger*, which was to incorporate the *Survey*. In 1792 Colles was involved in planning canals on the upper Connecticut River, and his acceptance of this commission may have been a factor in terminating his work on the *Survey*. Work on the South Hadley Canal was not completed until the autumn of 1794, but Colles probably returned to New York City late in the summer of 1792. It is possible, though, that the *Geographical Ledger* idea had been spawned in Colles's brain before he participated in the Connecticut River canal project. Certainly the complicated indexing system used in the ledger must have been well thought through before he returned to New York. By 1794 five large maps, as well as some sixty-five pages of text, indexes, and references were published under the imprint of John Buel, who was located at 24 Little Q Street in New York. This material was registered for copyright by Colles on June 7, 1794. This collection was entitled *The Geographical Ledger and Systemized Atlas; Being an United Collection of Topographical Maps, Projected by One Universal Principle, and Laid Down by One Scale, Proposed to Be Extended to Different Countries as Materials Can Be Procured*. The title page (Fig. 10–6) also includes an outline of the three projected sections of the *Geographical Ledger*:

I. Alphabetical references for pointing out the situation of lakes, islands, shoals, mills, mines, churches, iron-works, forts, bridges, fords, ferries, country seats, extensive tracts of land, and other remarkable objects.
II. An alphabetical index, refering to the different parts of the map, whereby any city, town, river, creek, island, lake, &c. can be speedily found by inspecting a very small space, without the pains of searching over the whole map.
III. An actual survey of a number of roads, specifying the true situation of every river, creek, church, mill, bridge, ford, ferry and tavern thereon, and their distances in miles, exactly engraved upon copper.

Section III is evidence that Colles hoped to incorporate the plates of the *Survey* within the *Geographical Ledger*. In fact, extant portions of the ledger include plates 40–47 of the *Survey*, which map the roads between New York City and Philadelphia.

The subtitle of the ledger indicates that the plates of the atlas were to be laid down on one scale, and that it was proposed that the atlas coverage be extended to different countries as material can be produced. The ledger, therefore, was planned as an open-end or loose-leaf atlas which might eventually embrace the entire world. Individuals could subscribe for all the plates or "take such particular sheets as they may require," as stated in its introduction.

The map sheets are laid out on a conic projection, the construction of which is explained in the introduction and illustrated with two grids, one of which carries the credit "Eliza Colles, Sculp." The maps are drawn at the scale of 1:625,000, and each sheet covers two degrees of latitude and four degrees of longitude. Longitude is reckoned from London, Paris, and Philadelphia. The sheets have no margins, and the cartographic data extend to the edges of the sheets. Thus, adjacent sheets can be readily conjoined with no interruption of information. To facilitate such assembly, there are columns of three numbers in the upper left corner of each sheet. The middle number identifies the actual sheet on which the numbers appear, the upper number is that of the north adjoining sheet, and the lower number indicates the sheet immediately to the south.

The original plan envisioned a list of references and an index for each sheet. "Letters ranged in alphabetical order," the introduction notes, "are set on the top and left hand side of each sheet, by reference to these from an index, any place may be conveniently and expeditiously found. In order to accumulate a large proportion of information upon a small extent of map, single letters are placed upon the surface of each sheet; by reference to these letters, the position of different objects can be accurately and quickly determined, without crowding the map with a multitude of words." Colles stated in the introduction that because of this elaborate indexing system, "I have given this work the name of the *Geographical Ledger*, as the situation of places can be found (by means of the index and reference) as speedily as a merchant can find any particular account in his ledger." Directions for using the index, which precede the index to sheet 1548, note that "for greater facility in finding the names of places, the two leading letters of each work is set in the margin of each column. [Further,] a number of different single letters are placed on the surface of the maps and referred to by a table of references and by the index."

There are extant today only five map sheets of the *Geographical Ledger* and this is perhaps all that were published. They are numbered 1369, 1458, 1459, 1548, and 1549 and collectively map all of New England, excluding northern Maine, most of New York State extending west beyond the Genessee River, eastern Pennsylvania, northern New Jersey, and a small section of Canada which adjoins Vermont, New Hampshire, and Maine. All five sheets, very likely the copies registered for copy-

THE

GEOGRAPHICAL LEDGER

AND

SYSTEMIZED ATLAS;

BEING

AN UNITED COLLECTION OF TOPOGRAPAICAL MAPS, PROJECTED BY ONE UNIVERSAL PRINCIPLE, AND LAID DOWN BY ONE SCALE, PROPOSED TO BE EXTENDED TO DIFFERENT COUNTRIES AS MATERIALS CAN BE PROCURED.

BY CHRISTOPHER COLLES, OF NEW-YORK.

CONTAINING

I. Alphabetical references for pointing out the situation of lakes, islands, shoals, mills, mines, churches, iron-works, forts, bridges, fords, ferries, country seats, extensive tracts of land, and other remarkable objects.

II. An alphabetical index, refering to the different parts of the map, whereby any city, town, river, creek, island, lake, &c. can be speedily found by inspecting a very small space, without the pains of searching over the whole map.

III. An actual survey of a number of roads, specifying the true situation of every river, creek, church, mill, bridge, ford, ferry and tavern thereon, and their distances in miles, exactly engraved upon copper.

NEW-YORK—*Printed by* JOHN BUEL, *No. 24, Little Q. Street.*

—1794—

Fig. 10–6. The title page of Colles's *Geographical Ledger*.

right, are in the Library of Congress. The New York Public Library has a copy of sheet 1459, and a copy of 1549 is in the collections of the New-York Historical Society. There is a small segment (approximately one-eighth) of sheet 1549 in the Connecticut Historical Society. The New York Public Library also holds sixty-five text and index pages, two grids which explain the conic projection employed for the *Geographical Ledger* maps, and plates 40–47 of the *Survey*. Included are the title page, the contents outline, a general introduction, and references and indexes for plates 1458, 1459, 1548, and 1549. This is the most complete extant holding of printed matter for the *Geographical Ledger*. Several years ago the Library of Congress acquired twenty-one printed pages, including the title page, the introduction, two grids for the conic projection, the reference and index for plate 1549, and *Survey* plates 40–47. The American Philosophical Society Library also has a segment of the printed pages of the *Geographical Ledger* as well as the same *Survey* plates.

Colles noted in the introduction that in compiling the *Ledger* maps he utilized "Samuel Holland's elegant map of New Hampshire, and that of Claude Joseph Sauthier's of the state of New York." For plate 1458, which maps north central New York State, Colles appears to have also consulted Simeon De Witt's *1st Sheet of De Witt's State Map of New-York*, which was deposited for copyright in January 1793. Plate 1458 of the *Ledger* embraces much of the data, including the controversial classical place names, that were first presented on De Witt's map. The index to plate 1458 lists all the classical names under "Military lands of New York, agreeable to a law of 25, 1792." There are some minor differences between De Witt's map and Colles's map, and it is possible that Colles consulted an early manuscript version of De Witt's map in the surveyor general's office. Colles almost certainly personally compiled and drafted the five *Geographical Ledger* maps. Inasmuch as he does not seem to have utilized William Blodget's 1789 map of Vermont, compilation for the northern part of New England was probably completed no later than 1790. It is interesting to note that the ledger's maps are primarily of regions not covered by Colles's *Survey*.

In discussing the *Survey*, it was suggested that Colles's daughter may have engraved at least some of the strip maps for the road book. There is confirmation in the *Geographical Ledger* and its five maps that she was an engraver. Her credit line, "Eliza Colles Sculp.," appears on the projection grid in the text portion of the ledger as well as on map plates 1369 and 1549 (Fig. 10–7). Also, the engraving style on plates 1458 and 1459 is similar to that on the plates bearing Eliza Colles's name. It is quite likely, therefore, that she engraved these plates also. Plate 1548, which maps south central New York State, northeastern Pennsylvania, and northwestern New Jersey, shows differences in technique from the other four maps. On the plates credited to Eliza Colles, for example, hills are shaded on the eastern slopes, whereas on plate 1548 the hill shading is on the western or left slopes. Plate 1548 also includes more names and fewer index letters than the other plates do. It is probable that this was the last of the five sheets to be completed, and that public displeasure with the complicated indexing system had induced Colles to place more conventional data on plate 1548.

On the plates signed by Eliza Colles, an italic style of lettering is used for many of the place names. A number of plates in the *Survey* also show a strong preference for italics. It is possible, therefore, as Benson Lossing stated, that Colles's daughter did engrave at least some of the plates for the *Survey*. The *Survey*'s title page carries the credit line "C. Tiebout, Sculp.," and it is believed that he also engraved some of the map plates. We may speculate that Colles's daughter may have served as an apprentice to Cornelius Tiebout and then later assumed the responsibility for engraving some of her father's maps. Following his work on the *Survey*, which is among his earliest engravings, Tiebout prepared a map of New York City which was published in the 1789 city directory. Between 1789 and 1793 he also engraved land-

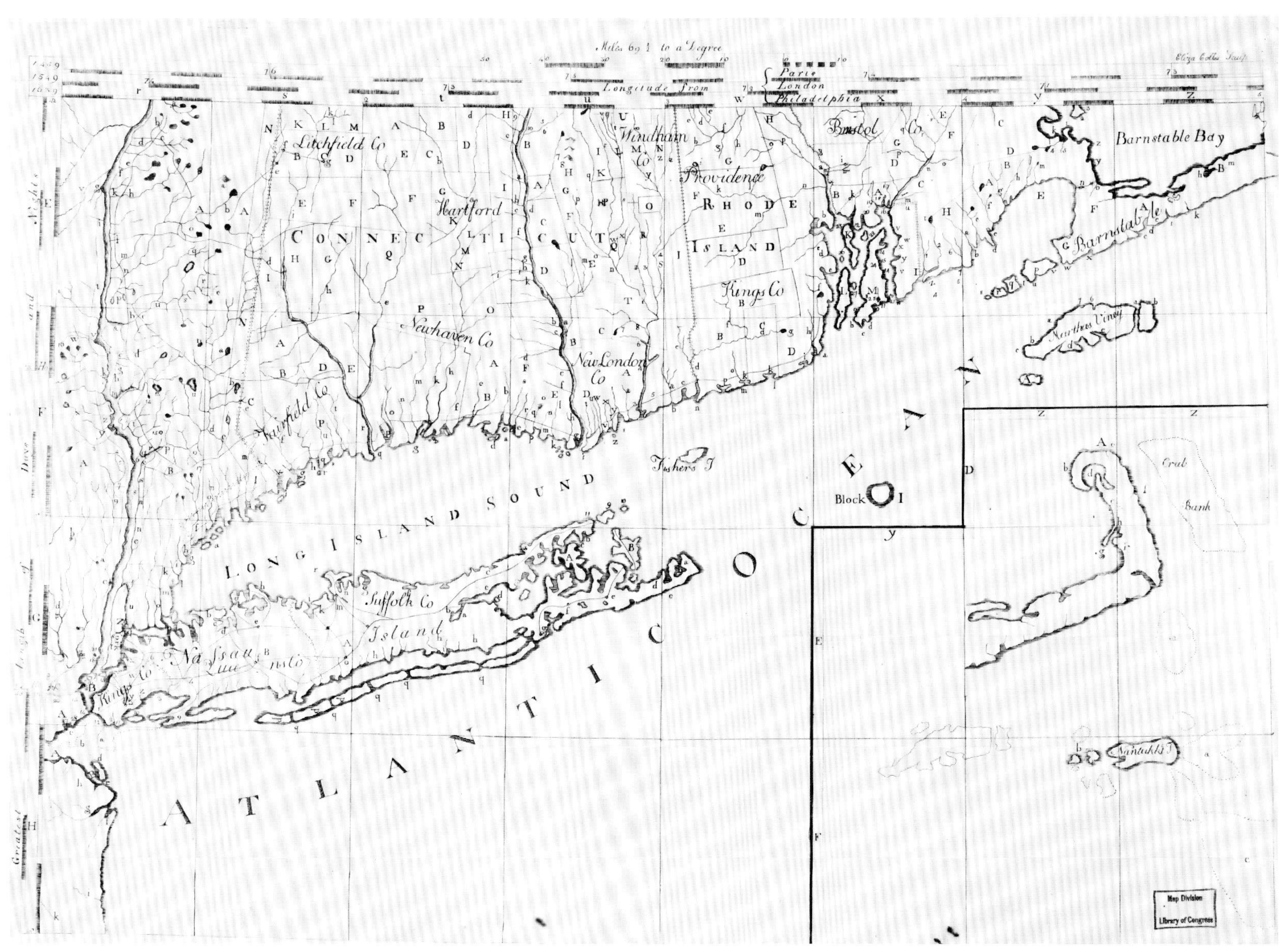

Fig. 10–7. Plate 1549 of the *Geographical Ledger*. Eliza Colles's name appears in the upper right corner.

scape scenes and several maps for New York magazines. He also engraved the *1st Sheet of De Witt's State Map of New-York* in 1793. Tiebout may have been the source from which Colles obtained a preliminary copy of this map. Tiebout also engraved the *Plan of the City of Washington* that was published in the June 1792 issue of the *New York Magazine and Literary Repository*. In 1793 Tiebout went to London, where he studied under James Heath.

Little is known about Eliza Colles (who may be America's first woman engraver) apart from her contributions to the *Survey* and the *Geographical Ledger*. Colles and his wife are reported to have had eleven children, several of whom died in infancy. Although Eliza is not listed in the family's unpublished genealogical table prepared around 1948 by Richard Colles Johnson, a descendant in Chicago, other sources record that Elizabeth, or Eliza, as she was called, was born to the Colleses in 1776. This was before their wartime exile from New York City. Thus, Eliza was only thirteen years old in 1789 when she learned engraving from Tiebout and eighteen when she engraved the map plates for the ledger. We know Eliza was alive in 1793 and 1794, when she engraved the *Geographical Ledger* plates. One source indicates that she died in 1799 at the age of twenty-three. It is probable that she was one of the many victims of the yellow fever epidemics which New York City experienced in the middle and late nineties.

The *Geographical Ledger* was Colles's final cartographic effort. Both it and the *Survey* were more ambitious undertakings than the times and the economy of the country could support. They were also beyond the financial resources of the visionary, but impoverished, engineer. The citizens of the infant United States were not yet ready for Colles's progressive ideas.

Melish's State Sheet Maps

John Melish is another individual whose plans for an American atlas were unrealized. Melish envisioned publishing an integrated series of maps that would meet most geographical needs. His plans were set forth in several of his publications, which were issued as supplements to his major maps. In the first edition of *A Geographical Description of the United States*, Melish wrote, with reference to his *Map of the United States*, that

> in truth it is absolutely impossible to make a general map of such an extended country as this to answer a particular purpose. The object of this map was to serve as a subject for general reference, and as a key map to the local maps of the several states and territories from actual survey. It is believed that it will be found well calculated to answer this purpose, because its geographical accuracy can be depended upon. . . . Besides the maps from actual survey . . . another kind of map will be found very useful, particularly to travellers. These are *sheet maps* of the several states and territories; and as they will answer remarkably as accompaniments to the present general map [of the United States], a series of them will be brought forward as quickly as good material can be collected.

Later in this same volume Melish reemphasized that

> it is proposed to publish as soon as possible, as accompaniments to the Map of the United States, a series of sheet maps, to embrace each state and territory in the Union, showing the counties, post towns, post roads, and multitude of minute particulars, which could not be introduced into a general map.
>
> Also a Series of Sheet maps of other countries, and A Series of Sheet Charts of the most interesting waters.
>
> These sheet maps and charts will be all uniform in plan and size, so that the possessors may bind them, or any number of them, into an Atlas.[16]

This plan was also outlined by Melish in the second and third editions of the *Geographical Description of the United States* and in his *A Geographical Description of the*

World, which was published in 1818. In the latter volume, he seems to have limited the atlas plan to the United States. After discussing his progress in producing state maps from actual survey, he wrote,

> the second branch [i.e., the preparation of sheet maps of the states] depends partly on the progress of the first [the actual surveys], and partly on judicious compilation. The subject has been embraced, as far as practical, in every point of view; and pretty good sheet maps have been brought forward of Indiana, Tennessee, Ohio, and Kentucky. It is in contemplation to bring forward two sheet maps with description, one of the southern and another of the northern section of the western country; and it is expected that these will speedily be followed by *a sheet map of each state and territory, on a uniform plan*, SO AS TO FORM AN ATLAS OF THE UNITED STATES.[17]

This statement confirms that Melish quite definitely planned to issue an open-end or loose-leaf atlas made up of his projected series of state maps. Publishing state maps "on a uniform plan," that is, at the same scale, was a new and revolutionary idea that facilitated size comparisons. A scale of 1:940,000 was adopted. The cartographer began to compile and issue sheet maps of selected states. The first of these, the *Map of Indiana*, was based on surveys by Burr Bradley and was published in 1817. Little is known about Bradley, who was probably a U.S. General Land Office surveyor. Indiana had just been admitted to the Union in December 1816, and the principal settlements were in the south between the Wabash and Ohio rivers. Rapid settlement in the new state encouraged the publication of an "improved" edition of this Indiana map in 1820.

Melish's Indiana map carries the imprint, "Philadel-

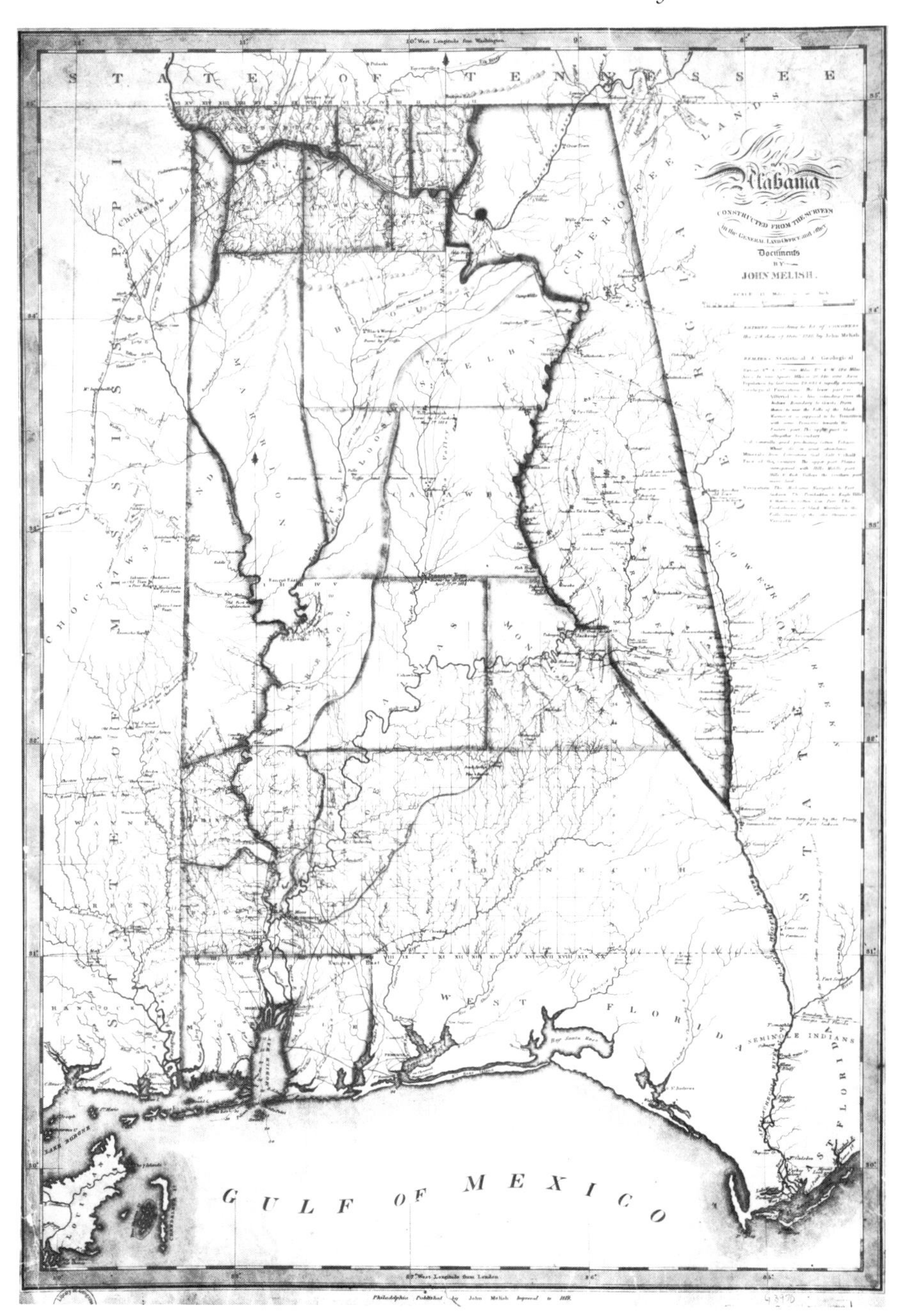

Fig. 10–8. John Melish's 1819 map of Alabama, one of a series of sheet maps that he hoped to assemble and publish as an atlas of the United States.

phia, Published by John Melish & Sam'l. Harrison." In 1816 Melish had formed a partnership with Harrison, who was the third son of master engraver, William Harrison. It is possible that Harrison worked on some of the other state sheet maps in addition to that of Indiana. The partnership, however, was terminated by the death of Harrison on July 18, 1818, at the age of twenty-nine. Harrison's death was undoubtedly a major factor in slowing down Melish's output of the state sheet maps.

However, a sheet map of Illinois, "constructed from the Surveys in the General Land Office and other documents," was published by Melish in 1818, with an "improved" edition issued in 1819. The anticipation of Alabama's entering the Union was the inspiration for Melish's sheet map of that state in 1818 (Fig. 10–8). It was followed by a revised edition in 1819, when statehood was granted, and a third edition in 1820. For the Alabama map, Melish drew heavily upon William Darby's *Map of the State of Louisiana With Part of the State of Mississippi and Territory of Alabama, from Actual Survey,* which was published by Melish in 1816. Darby's map was also the major source used by Melish for his sheet maps of Mississippi and Louisiana published in 1819 and 1820, respectively. General Land Office surveys were also consulted in compiling these maps. Some Melish catalogs also list sheet maps of Ohio, Kentucky, and Tennessee. The colored state sheet maps each sold for $1.50. Because of Melish's death in 1822, his dream of an atlas of the United States, with state maps on a uniform scale, was not realized.

Notes

1. J. Brian Harley, "Atlas Maker for Independent America," *Geographical Magazine* (London) 49 (Sept. 1977): 767.
2. Ibid., 769.
3. Ibid., 771.
4. For comprehensive descriptions of cerography and wax engraving see David Woodward, *The All-American Map, Wax Engraving and Its Influence on Cartography* (Chicago, 1977).
5. *Charleston Morning Post*, Dec. 20, 1786.
6. *Charleston City Gazette*, Feb. 3, 1795.
7. *Charleston County Inventories C, 1793–1800* (Columbia, 1969, microfilm), 153.
8. *South Carolina Gazette*, Sept. 5, 1796.
9. Christopher Colles, *Proposals for Publishing a Survey of the Roads of the United States of America by Christopher Colles of New York* (New York, 1789). This is a broadside.
10. For additional information about Christopher Colles see the facsimile edition, Colles, *Survey of the Roads of the United States of America*, ed. Walter W. Ristow (Cambridge, Mass., 1961).
11. New York, *Assembly Journal*, Mar. 13, 1790.
12. *House Journal*, 1st and 2d Congs., Mar. 30, 1790, 185.
13. *American State Papers: Class 7, Post Office Department, Documents Legislative and Executive of the Congress of the United States 1789 to 1833*, ed. Walter Lowrie and Walter S. Franklin (Washington, D.C., 1834), 3.
14. See Colles, *Survey*.
15. Benson Lossing, *History of New York City* (New York, 1884), 75.
16. John Melish, *A Geographical Description of the United States* (Philadelphia, 1816), 13–14, 172.
17. John Melish, *A Geographical Description of the World* (Philadelphia, 1818), 5.

11. *The Ebeling-Sotzmann* Atlas von Nordamerika

The independence of the United States created a desire for information about the new republic in Europe as well as in America. In the several decades following the ratification of the Constitution, therefore, a number of Europeans traveled extensively in the United States. After returning to their native lands, some published their accounts and impressions of the country and its citizens. One of the most comprehensive, detailed, and sympathetic geographic descriptions of America, however, was compiled and published by a European who never set foot on the American continent. This ambitious and dedicated individual was Christoph Daniel Ebeling, a professor of history and classical languages at the Hamburg Gymnasium (Fig. 11–1).

Ebeling was born November 20, 1741, in Garmissen near the city of Hildesheim in western Germany. He enrolled in the University of Göttingen in 1763 expecting to study for the ministry. Theology did not, however, stimulate him intellectually; he found history, literature, and the English language more appealing studies. After graduating from Göttingen, Ebeling tutored in Leipzig for a year or two, then accepted a teaching position in a commercial academy in Hamburg in 1769. In the following year he was named director of the academy, an office he held for more than two decades. In 1784 Ebeling was appointed to the Hamburg Gymnasium, where he taught for the remainder of his life. During this busy career, Ebeling also found time to gather data for the colossal task he had undertaken of summarizing the geography and history of the United States for his German-speaking compatriots.

The dedicated scholar early developed an interest in the evolution of free states, and this concern, coupled

Fig. 11–1. Portrait of Christoph D. Ebeling by Peter Suhr. Courtesy of Hamburg Staatsbibliothek.

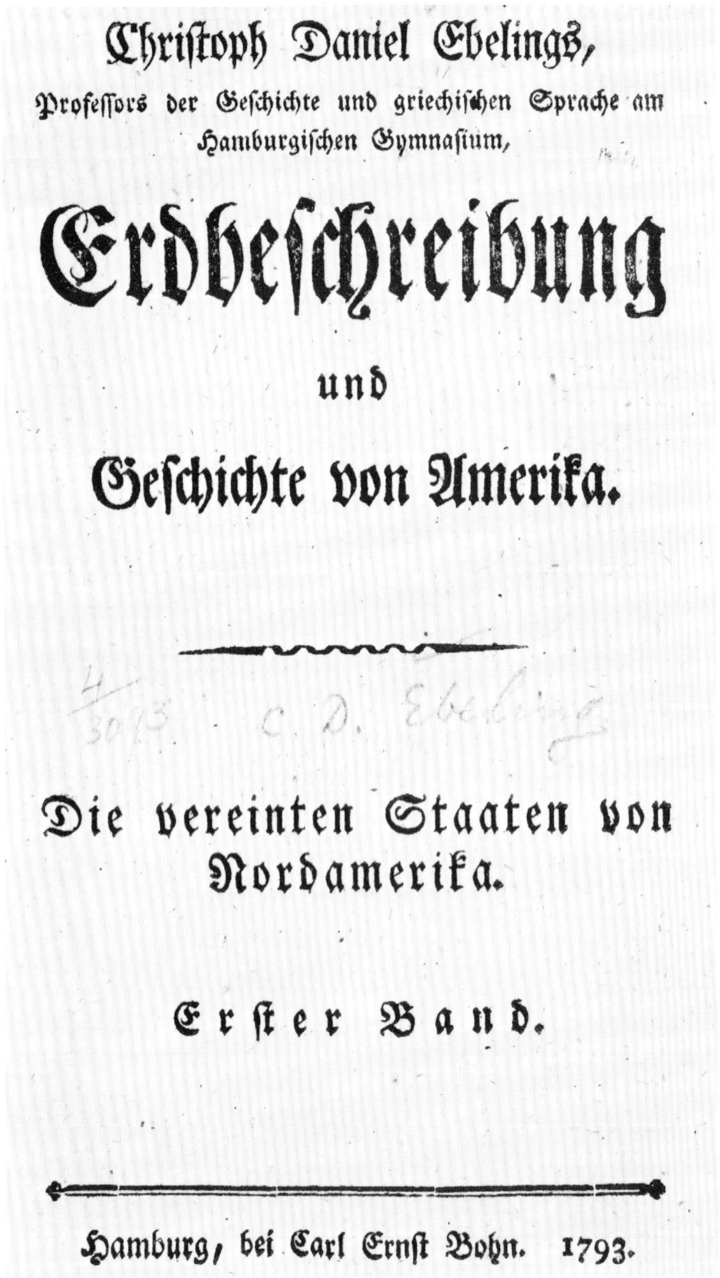

Christoph Daniel Ebelings,
Professors der Geschichte und griechischen Sprache am
Hamburgischen Gymnasium,

Erdbeschreibung

und

Geschichte von Amerika.

Die vereinten Staaten von
Nordamerika.

Erster Band.

Hamburg, bei Carl Ernst Bohn. 1793.

Fig. 11–2. The title page of volume one of Ebeling's *Erdbeschreibung und Geschichte von Amerika* published in 1793. This volume describes the geography of New Hampshire and Massachusetts.

with his knowledge of the English language, directed his attention to America. As early as 1777–78, Ebeling published a book entitled *Amerikanische Bibliothek*, which included German translations of selected political, descriptive, and statistical tracts in English. This suggests that he had, even before the Revolution, started to gather information about the country across the Atlantic. It was a pleasurable, if demanding, task, which consumed a considerable portion of his time and effort for almost half a century. He also assembled one of the most comprehensive Americana libraries of his time. Ebeling accomplished this by carrying on a voluminous correspondence and exchange with American colleagues, few of whom it was his good fortune to meet.[1]

Ebeling avidly read, thoroughly absorbed, and critically analyzed data on America and drew upon it in compiling his multivolume *Erdbeschreibung und Geschichte von Amerika, die vereinten Staaten von Amerika*. The first volume of this geographical and historical study of the United States was published in Hamburg by Carl Ernst Bohn in 1793 (Fig. 11–2). It summarized the geography and history of the states of New Hampshire and Massachusetts. Over the next quarter century, Ebeling published six additional volumes of the *Erdbeschreibung*, the last of which appeared in 1816. Rhode Island, Connecticut, Vermont, and New York were the subjects of volume two, published in 1794. Volume three, which was issued two years later, added supplementary information on New York State and also covered New Jersey. Pennsylvania was featured in the fourth volume, published in 1797, and Delaware and Maryland were dealt with in volume five, which came out two years later. Pennsylvania was further described in volume six, published in 1803, because of the large number of Germans who had settled there. Many of them had described the state's features for relatives and friends in their native land in letters and pamphlets. After an interval of thirteen years, volume seven, which described the state of Virginia, was finally published. The delay was caused in part by restrictions on shipping and mailing resulting from the Napoleonic Wars in Europe and the War of 1812 and the British blockade in America. Ebeling had also planned to compile additional volumes of the *Erdbeschreibung* dealing with the southern and trans-Appalachian regions. These objectives were never realized for want of reliable information on those regions and because the hard-working professor died on June 20, 1817.

The intended scope of the *Erdbeschreibung* was outlined by Ebeling in a letter dated June 26, 1794, and addressed to Ezra Stiles, president of Yale College. Ebeling wrote that he planned to cover each of the fifteen states of the Union and the western territories in their own volumes, the United States as a whole in a separate volume, Spanish America in three volumes, and possibly the other European colonies in three or four more volumes. "It is," Ebeling informed Stiles, "an arduous task that I undertook, but I was incited to persevere by the animating beauty of the subject, the many imperfect and false accounts Europe has of your country, and the possible good effect which a faithful picture of a truly free republic founded upon the most solid foundations, could produce in the most part of Europe, so very remote from such happiness as you enjoy."[2]

The seven published volumes of the *Erdbeschreibung* included no maps, which greatly limited their utility. Ebeling recognized this deficiency and in 1795, two years after the first volume was published, he announced plans for compiling the *Atlas von Nordamerika*.[3] As planned, the atlas was to include eighteen plates, sixteen of them state maps at fairly large scales. To compile and draft the maps, Ebeling engaged Daniel Friedrich Sotzmann. In a letter dated April 29, 1796, Ebeling informed his American correspondent Dr. William Bentley that "there is publishing now under my inspection a collection of Maps of America by the Geographer of the Academy at Berlin Mr. Sotzmann an able mapmaker." In the same letter he reported that "three [maps] are now in the Hand of the Engraver viz. (1.) *New Hampshire* according to Holland and Mr. Belknap . . . (2.) *Vermont* after Whitelaw and Blodget (3.) *Connecticut* after Blodget, Carey and two other maps published in Holland. (4.) *Pennsylvania* according to Howell is to follow these, for I intend not to go on, before there exist in your country, such maps as there are approved of as extant."[4]

The Ebeling-Sotzmann maps of Vermont, New Hampshire, and Connecticut (Fig. 11–3) are all dated 1796. The map of Massachusetts, which is undated, was probably also published in that year. Roman numerals in the upper margins identify the maps as follows: Vermont, No. XVI, New Hampshire, No. II, Massachusetts, No. III, and Connecticut, No. VI. The significance of the Roman numeral XVI for Vermont is not clear. This state was one of the first mapped, and it should logically have had a lower number. Writing to Joel Barlow on September 15, 1795, Ebeling reported, "Mr. Sotzmann and my map of Connecticut is ready and by a proof sheet I find it is very well done as to the most possible exactitudes (as far as we can judge) as to design and engraving. I shall send it to you soon."[5]

Four maps were then published in 1797, Rhode Island, No. V, New Jersey, No. VIII, Pennsylvania, No. IX, and Maryland and Delaware (on the same sheet), No. X (Fig. 11–4). Maine, numbered IV, was the only map published in 1798 (Fig. 11–5), and the final one to appear was New York, which is dated 1799 (Fig. 11–6). The latter is not numbered, but should probably have been labeled VII in the series. Ebeling informed Bentley in a letter dated September 16, 1798, that the maps "of Maine and New York are engraving."[6] In a subsequent letter written September 7, 1799, the geographer told Bentley that he was forwarding to him maps of Maine and New York. "That of New York," the German professor complained, "coasts [*sic*] me a great deal of trouble and many expenses to the Editor as the erection of new

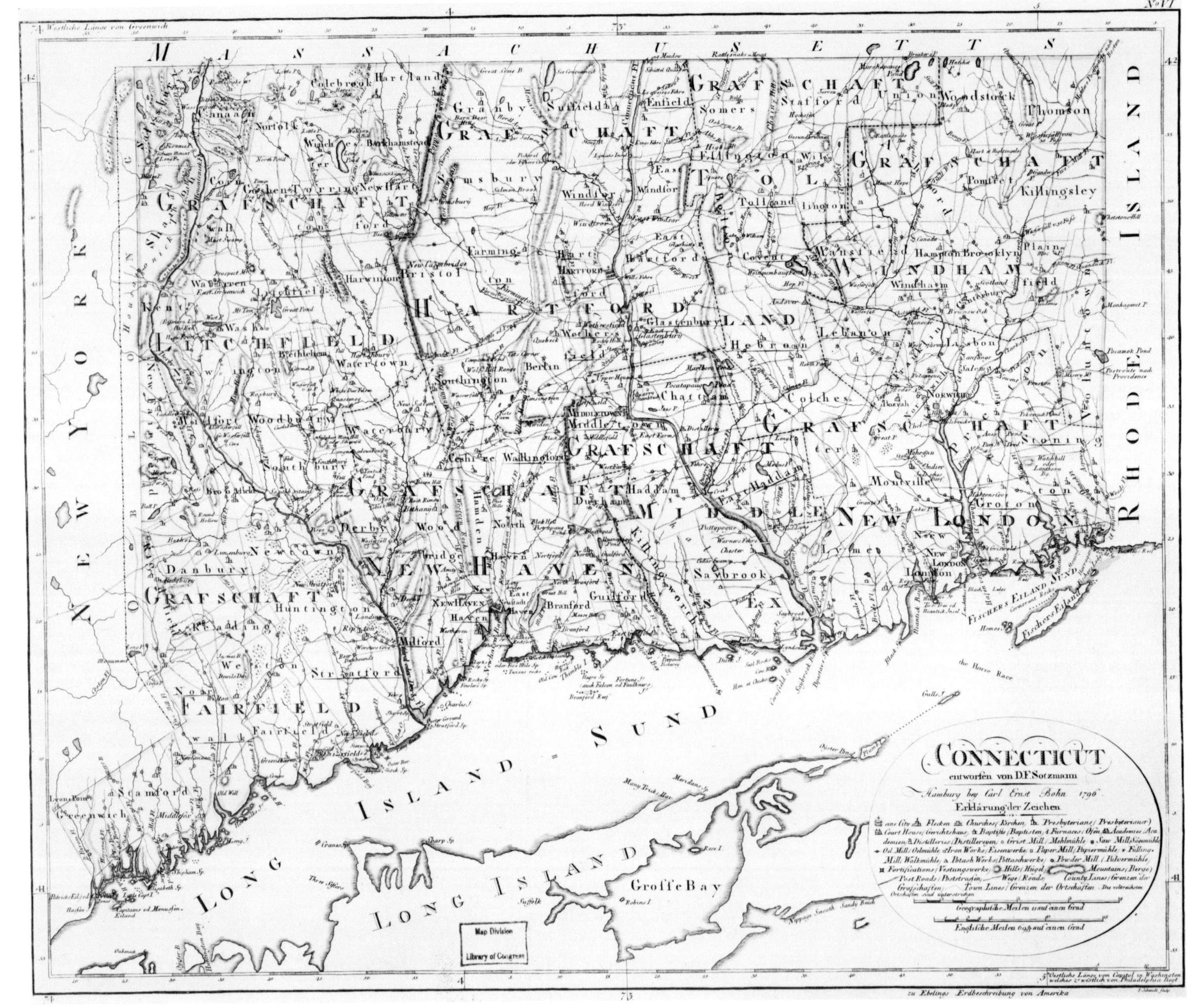

Fig. 11–3. This 1796 map of Connecticut was one of the earliest published for the projected *Atlas von Nordamerika*.

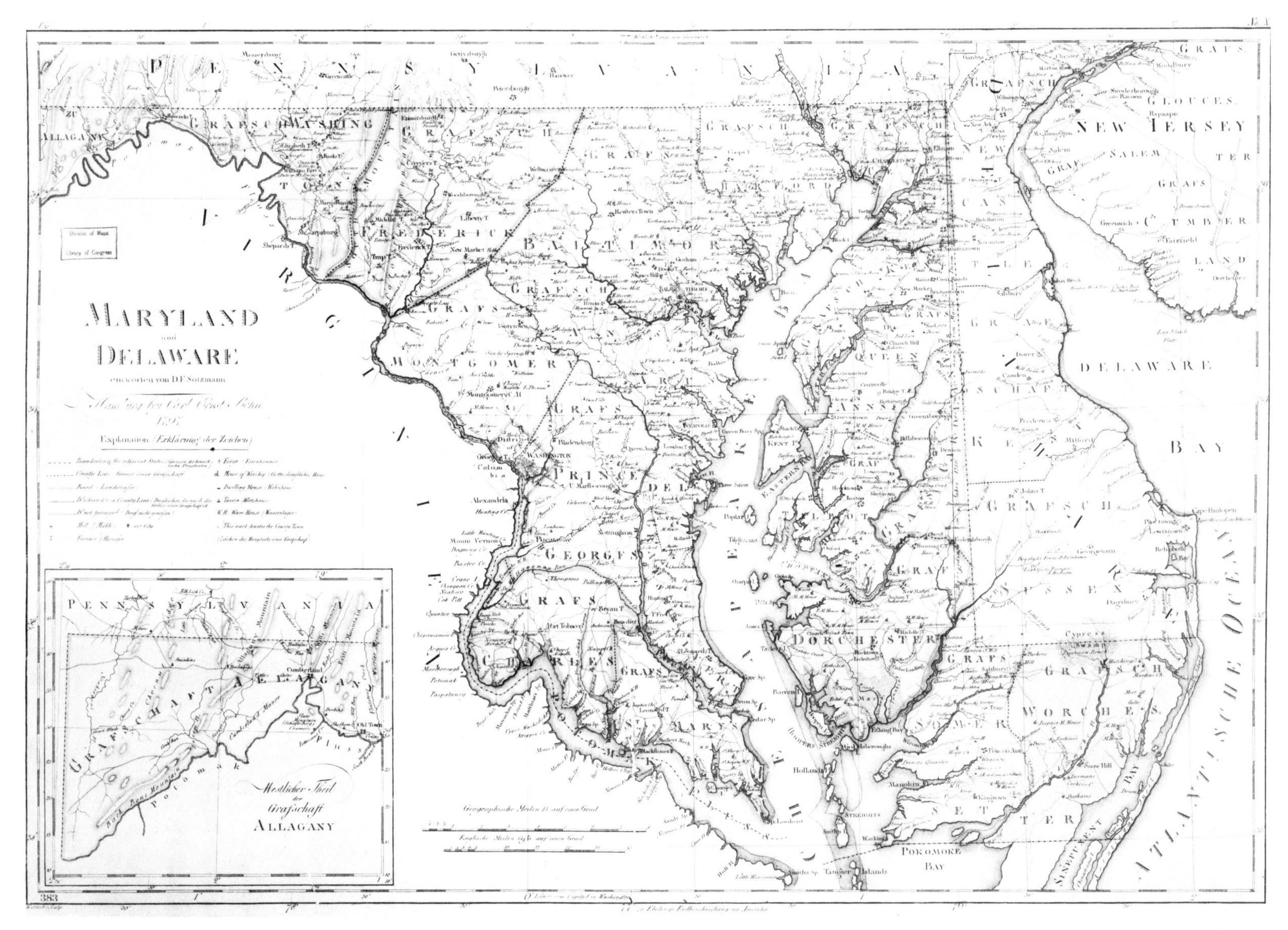

Fig. 11–4. The 1797 Ebeling-Sotzmann *Maryland and Delaware* map.

Counties obliged us to engrave a considerable part anew."[7]

Most useful to Ebeling and Sotzmann in compiling their series of state maps were the maps of the separate states from actual surveys that had been published in America. Several such maps were available by 1799, primarily for the New England and Middle Atlantic states. In addition to the sources Ebeling himself acknowledged using, Ebeling and Sotzmann certainly consulted the maps of Maine by Osgood Carleton (1795), Mary-

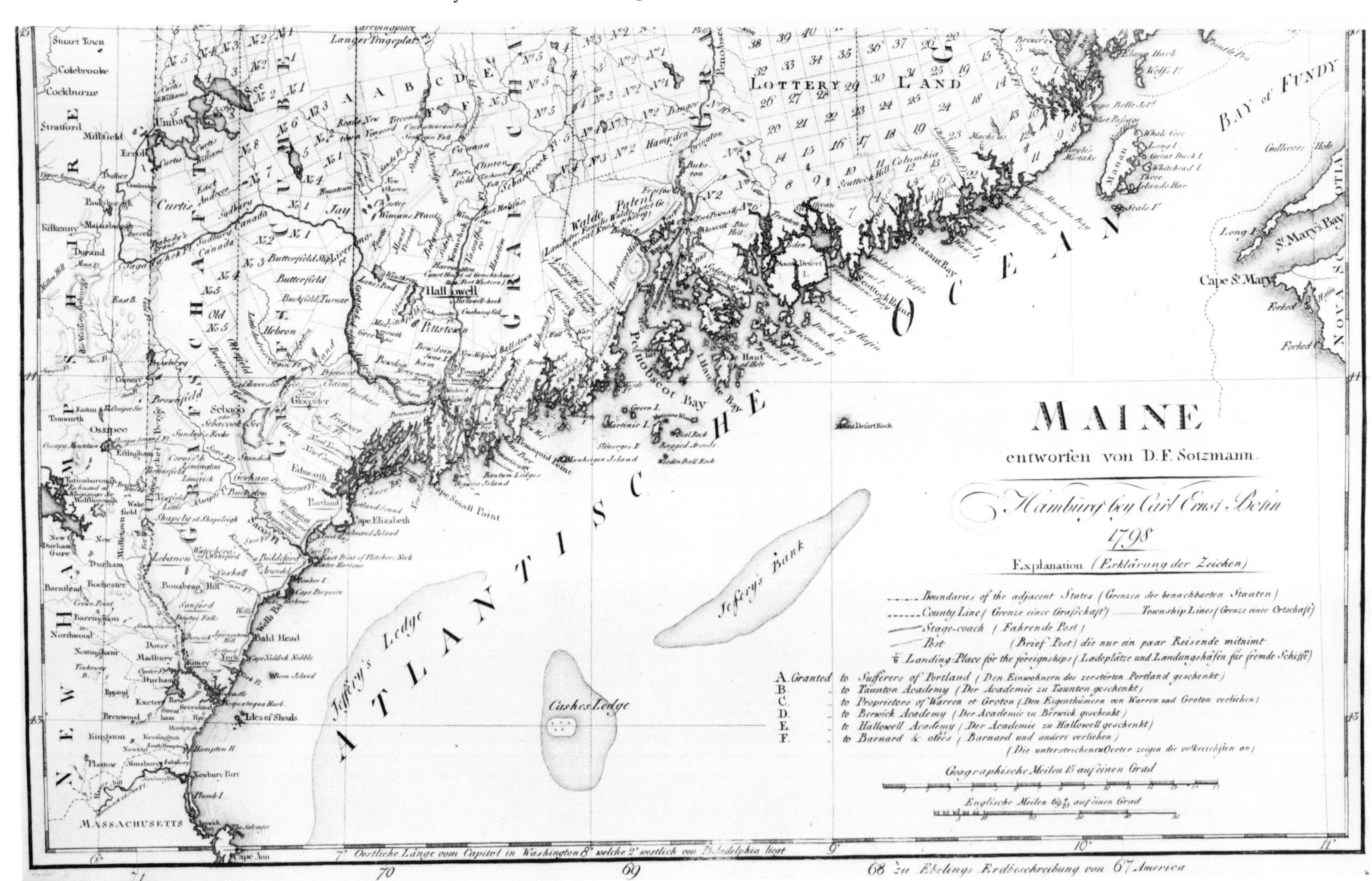

Fig. 11–5. The southern half of the Ebeling-Sotzmann map of Maine. Note the lengthy list of explanations.

Fig. 11–6. The Ebeling-Sotzmann map of New York.

land and Delaware by Dennis Griffith (1794), and the first sheet of Simeon De Witt's map of New York State (1793.). Ebeling and Sotzmann's cartographic project was, unfortunately, too early to benefit from the many excellent state maps that were published in the first two decades of the nineteenth century. In his letters Ebeling indicated an awareness of several that were in process, notably Bishop James Madison's map of Virginia, which was not published until 1807, and Jonathan Price and John Strothers's map of North Carolina published in 1808. Writing in 1940, the late Ralph H. Brown concluded that "the Ebeling-Sotzmann maps of the states are a fusion of a number of others, to which many data were newly added. Hence they form a special series, whose lineage cannot be traced from outward appearances. . . . The derivatives are likely to be more useful than the principal bases, however, because of the valuable details added from other sources. The fusion was so successfully accomplished that it seems impossible to trace with certainty the derivation of the many map details."[8]

The maps compiled by Ebeling and Sotzmann vary in size from 36 by 44 cm. (Connecticut) to 41 by 70 cm. (Pennsylvania). There are likewise differences in scale from 1:1,200,000 (New York and Maine) to 1:200,000 (Rhode Island). The maps of Vermont, New Hampshire, and Connecticut are at the scale of 1:380,000, Massachusetts, New Jersey, and Maryland and Delaware are at 1:500,000, and Pennsylvania is at 1:735,000. At the tops of the maps, longitude readings are from the Greenwich prime meridian, and at the bottoms they are from the Washington, D.C., prime meridian. All the maps have legends. County and township lines are shown, and there are numerous towns and cities. For some states, names of the most densely populated settlements are underlined. All of the ten state maps carry the credit "von D. F. Sotzmann" and all have the reference "zu Ebelings Erdbeschreibung von Amerika." The maps of Connecticut, Massachusetts, New Hampshire, and Vermont were engraved by Paulus Schmidt; Maine, Maryland and Delaware, New York, and Pennsylvania were engraved by W. Sander; and Rhode Island was engraved by H. Kliewer. There is no engraving credit on the New Jersey map.

The Ebeling-Sotzmann maps supplement the descriptions of the states covered in the first six volumes of the *Erdbeschreibung*. No map was published for Virginia, the subject of volume seven. Although Madison's map of Virginia was published in 1807, Ebeling wrote to Barlow as late as April 17, 1812, that "what is most necessary to me, is *Bishop Madison's Map of Virginia* and the Newest Census *1810*." He added, "my description of Virginia is finished, but what figure shall it make when the Virginians see that I don't even know exactly where their counties of Cabell, Giles, Mason, Monroe, Nelson, Page, Brooke, etc. are situated."[9] Writing to Bentley on May 22, 1815, Ebeling reported that "I have got Madison's map of Virginia. But I find only 96 counties therein. I know that a new one *Cabell* is created lately, but cannot find out its situation. You will greatly oblige me, by informing me thereof."[10] Professor Ebeling was in poor health for the last several years of his life, and, if the desired information was received, he was apparently not able to use it to complete his compilation of the Virginia map before his death.

The *Erdbeschreibung* and the eighteen-plate atlas which was to supplement it both fell short of the goals set by Ebeling. Only a small number of American collections, among them the Library of Congress and Harvard University, have copies of all ten published maps. This series, thus, is among the rarest of cartographic Americana for the closing decade of the eighteenth century. Ebeling wrote to the Reverend Jeremy Belknap on October 1, 1796, and promised that "as soon as six maps are completed, I shall join a memoir giving an account of the materials and authors made use of in the construction of each map."[11] That this memoir was, apparently, never completed is a great loss to the cartographic history of America for the closing decade of the eighteenth century. We are fortunate, however, that Ebeling's comprehensive library of books, newspapers, and maps was purchased shortly after the professor's death by Israel Thorndike, a Boston merchant who presented the Ebeling Collection to the Harvard University Library.

In his article on early U.S. maps, Ralph Brown wrote that "concerning Sotzmann, little can be learned."[12] Thanks to Wolfgang Scharfe's summary of the cartography of Brandenburg, however, we do have authoritative information about Sotzmann and his cartographical career.[13] Sotzmann was born in the Berlin suburb of Spandau in 1754 and received his early education there under Engineer Captain Materne. He received training in engraving as well as drafting, and his earliest cartographic contributions were, in fact, as an engraver. From 1773 to 1778 Sotzmann was a building superintendent in Potsdam, and in 1778 he was appointed secretary and architect of the General-Tabaksadministration. Sotzmann was named geographer of the Akademie der Wissenschaft in Berlin in 1786. He became secretary and controller of the academy's Engineering Department, later the Military Department, in 1787 and was pensioned in 1826. He died in 1840 at the age of seventy-two.

Sotzmann's most productive years as a cartographer were from 1783 to 1806, when he produced a number of medium-scale district and province maps, as well as city plans. For much of this period he was associated in his mapping endeavors with Carl Ludwig Oesfeld (1741–1804). Scharfe rates Oesfeld and Sotzmann as two of the most distinguished cartographers in the German-speaking countries in the early years of the nineteenth century. Their works, he believes, provided the impulse and laid the foundations for the cartographic style of many official and private mapmakers until well into that century. Although most of Sotzmann's maps were of German regions, shortly after 1783 he prepared (probably as a book illustration) a small map entitled *Die Vereinigten Staaten von Nord-America nach der von Wm. Faden 1783 herausgegebenen Charte*, which was published in Berlin by Haude und Spener. Also probably prepared to illustrate a book was the *Karte des nördlichsten America nach der zweiten Ausgabe von Arrowsmith's grosser Mercators-Karte in acht Blatt gezeichnet von D. F. Sotzmann 1791 . . . gestochen von Carl Jäck Berlin*. These maps may have brought Sotzmann to Ebeling's attention and persuaded the latter to enlist the cartographer to prepare maps for the *Atlas von Nordamerika*. The arrangements between the two men were probably completed in 1795.

Brown strongly believed that Ebeling "personally directed the preparation of the ten sheets known to have been issued—a responsibility that has not been generally appreciated. The traditional reference to this series as the 'Sotzmann maps' tacitly attributes them to the lesser contributor, partly because Sotzmann's name appears prominently on each of the maps but also because Ebeling's letters to American correspondents have not been inspected with a view to determining the role he actually played in their preparation."[14] Unquestionably, Ebeling exercised strong direction concerning the information to be presented on the state maps. It is unlikely, however, that a cartographer of the stature of Sotzmann would have accepted only the role of a map draftsman. He was located in Berlin while Ebeling was in Hamburg, and it may be assumed, therefore, that Sotzmann was given a fairly free hand in compiling the maps following Ebeling's directions and wishes. This assumption need not, in any way, detract from the great contribution made by Ebeling in assembling and recording geographical information about the United States in the early decades after the nation was established.

Notes

1. A number of Ebeling's letters are printed, or paraphrased, in William Coolidge Lane, ed., "Letters of Christoph Daniel Ebeling to Rev. Dr. William Bentley of Salem, Mass. and to Other American Correspondents," *American Antiquarian Society Proceedings*, Oct. 21, 1925, 272–451.
2. Ibid., 281.
3. *Nieuwe Algemeene Konst- en Letter-Bode* 4 (Haarlem, 1793),

53, and *Intelligenzblatt der Algemeinen Literatur-Zeitung,* (Jena, 1795), 756–60.

4. Lane, "Letters," 289.
5. Ibid., 286.
6. Ibid., 306.
7. Ibid., 342.
8. Ralph H. Brown, "Early Maps of the United States: The Ebeling-Sotzmann Maps of the Northern Seaboard States," *Geographical Review* 30 (July 1940): 471–79.
9. Lane, "Letters," 414–15.
10. Ibid., 419.
11. Ibid., 289n.
12. Brown, "Early Maps," 478.
13. Wolfgang Scharfe, *Abriss der Kartographie Brandenburgs 1771–1821* (Berlin, 1972).
14. Brown, "Early Maps," 472.

12. *John Melish*

Although Mathew Carey was the first American publisher to issue cartographical works, maps and atlases were not his primary publishing interests. His output in this specialty was rather limited. It remained for John Melish to devote exclusive attention to publishing geographical and cartographical products. Melish played a foremost role in bringing together from many and varied sources the geographical and cartographical knowledge of the period, and presenting it systematically and graphically for the edification and enlightenment of the citizens of the young republic.

Melish was born in Scotland in 1771 and orphaned at an early age. He was apprenticed to a leading textile manufacturer in his native city of Glasgow. When not working he pursued studies at the University of Glasgow, and, in due course, his enterprise and abilities were rewarded with his admission to a partnership in the textile company. Official business took Melish to the West Indies in 1798, and eight years later he visited the United States. Georgia was his primary goal, but personal and business interests also took him through most of the other seaboard states before he returned to Scotland in 1807.

During the next several years the British government, through various short-sighted policies, impeded the commercial relations between Great Britain and the United States. With the aim of salvaging some of his investments and commercial outlets in America, Melish, accompanied by his son, returned to the United States in 1809. Convinced that political and economic conditions were not likely to improve, Melish decided in 1811 to sever his ties with Scotland and to settle in the United States. Toward the close of the year his wife joined him in New York, and together they reviewed future economic options.

Melish's first inclination was to engage in farming, and with this objective in mind he toured most of the settled areas of the country, as well as some lands beyond the Appalachians, to evaluate their agricultural potential. The trip covered more than 2,400 miles and extended as far west as Ohio, north to upper New York State, and east to Massachusetts. On this tour and during his earlier American travels, Melish compiled extensive and perceptive notes. This detailed information, Melish concluded, might also be of value to others. The resulting fact-filled two-volume work entitled *Travels in the United States of America in the Years 1806 & 1807, and 1809, 1810, & 1811* was published in Philadelphia (where Melish finally settled) in 1812. It contains excellent descriptions of the physical and cultural environment, as well as many shrewd and incisive personal observations.

In the preface to volume one of the *Travels*, Melish wrote, "as I have always considered books of travels to be very defective when unaccompanied by maps, I have spared no labour, nor expence, to have a good set of maps to illustrate this work. They have been drawn with great care from the best materials to which I could get access, aided by much local information; and the engraving has been executed by the first artists in Philadelphia."[1] The eight maps that illustrate the two volumes, all drawn by Melish, apparently aroused in him a deep interest in cartography and directed him to a career as a geographer and map publisher (Fig. 12–1).

The "first artists in Philadelphia" were the engravers John Vallance and Henry S. Tanner. Seven of the *Travels* maps, one in volume one and six in volume two, were engraved by Vallance. Tanner engraved the eighth map, that of the United States, which served as the frontispiece for volume one. Tanner, the younger brother of

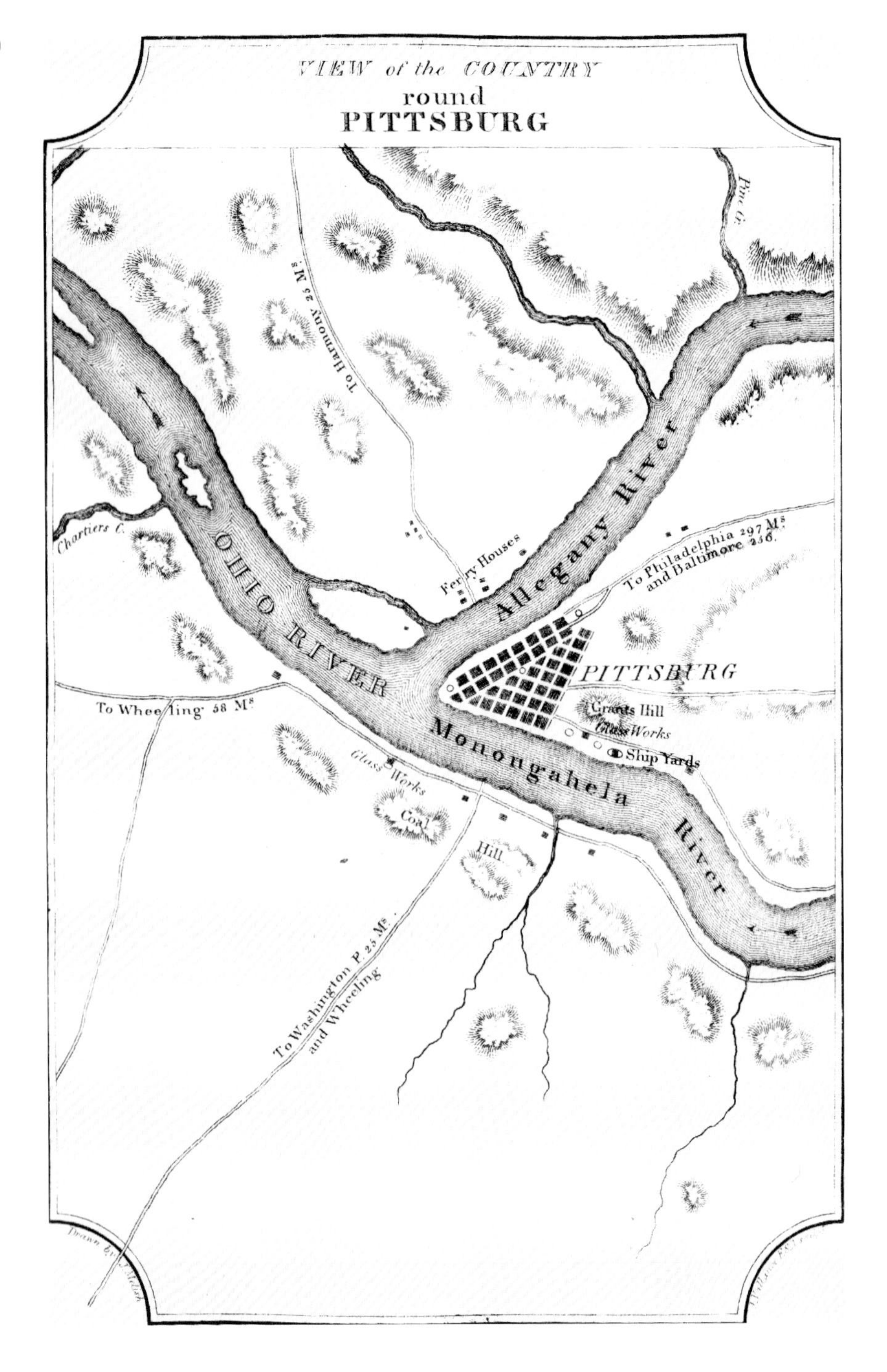

Fig. 12–1. The map of Pittsburgh and vicinity from John Melish's 1812 *Travels in the United States of America*.

Benjamin Tanner, was born in New York City in 1786. He had moved to Philadelphia in 1810 to join his older brother, who was an established engraver trained by Peter Maverick, one of New York City's earliest engravers. Tanner served as Benjamin's apprentice. Benjamin Tanner and Vallance were for a time in partnership. The map of the United States is believed to be Henry Tanner's first professional attempt at map engraving. It was also his first association with Melish, an experience which apparently impelled him to follow Melish's profession of map and atlas publisher. The careers of Melish and Tanner were closely intertwined, and they played major roles in laying the foundations of American commercial map publishing. One of Melish's first imprints after embarking on his publishing career was a portable version of Tanner's U.S. map. It was sold folded within 13.5-by-9-cm. covers, accompanied by a twenty-four-page *Statistical Account of the United States, with Topographical Tables of the Counties, Towns, Population, &c. From the Census of 1810*.

"While engaged in drawing" the maps for *Travels*, Melish informs us, "a much valued friend suggested the propriety of drawing a general map of the seat of war [the War of 1812], and proferred the use of a very ample set of maps in his possession."[2] The resulting map, titled *Map of the Seat of War in North America*, was also engraved by Tanner. It was initially published in 1813 in a folded format and accompanied by a twenty-page *Description of the Seat of War in North America*. The map includes only that part of the country located north of the city of Washington, D.C., and east of Wisconsin Territory. The *Map of the Seat of War* differs significantly from the U.S. map in *Travels*. The former extends west only to Lake Michigan and south to northern Virginia. The map in *Travels* has its southern limit in northern Florida and its western limit just beyond the Mississippi River. It was a popular item, and, as Melish reported, "it sold so rapidly, that the first plate was soon worn out, and a new one has since been brought forward, enlarged and much improved."[3] This success encouraged Melish to compile a *Map of the Southern Section of the United States* (Fig. 12–2) and a *Map of the American Coast, from Lynhaven Bay to Narraganset Bay*, both of which were engraved by Tanner and published in folded formats. An eighteen-page *Description of the Seat of War in the Southern Section* accompanied the *Map of the Southern Section*.

Before the end of 1813, Melish bound together the *Map of the Seat of War*, the *Map of the Southern Section*, and the *Map of the American Coast*, with their respective texts, the *Map of Detroit River and Adjacent Country*, the *Plan of Quebec and adjacent country*, and the small, page-size maps *Country round the Falls of Niagara*, *East End of Lake Ontario*, and *Plan of Montreal*. He published this collection as *A Military and Topographical Atlas of the United States; Including the British Possessions & Florida*. Tanner prepared the engraved plates for all but one of the maps. The exception, the *View of the Country round the Falls of Niagara*, was engraved by Vallance and had previously appeared in volume two of *Travels*. The *Map of the Seat of War* included in the atlas was considerably revised.

The *Military and Topographical Atlas* also proved to be a popular sales item. After peace was reestablished between England and the United States in 1815, Melish published an enlarged and revised edition of the atlas, enhanced by several new maps (Fig. 12–3). Added are the maps *East End of Lake Ontario and River St. Lawrence*, *River St. Lawrence From Williamsburg to Montreal*, *Map of the Seat of War Among the Creek Indians*, and *Map of New Orleans and adjacent country*. The first three carry Tanner's engraving credits; the engraver of the New Orleans map is not identified.

The Philadelphia city directory for 1814 has an entry for the engraving firm of Tanner, Vallance, Kearny & Company. The senior member of the firm was Benjamin Tanner. In his early commercial map publishing years, Melish appears to have had close relations with the several members of this firm, as well as with other Philadelphia engravers. In this respect he was following the cottage-industry pattern established by Carey.

In 1813–14 Melish collaborated with Vallance and Henry Tanner to compile and publish *A New Juvenile Atlas and Familiar Introduction to the Use of Maps: With a Comprehensive View of the Present State of the Earth*, which was printed in Philadelphia by G. Palmer. The Advertisement on the page opposite the title page explains that

> this work was originally published in London by Laurie and Whittle, celebrated map publishers. The leading object in the compilation of it was to communicate in the clearest manner, A GENERAL IDEA OF THE USE OF MAPS, and of the RELATIVE SITUATION, with the COMPARATIVE IMPORTANCE, of all the different STATES and NATIONS of the EARTH. . . . In the present edition the maps and geography of the latest London edition have been carefully revised and improved; and the American part has been much altered, in consequence of the revolutions that have taken place in that quarter. A NEW MAP AND DESCRIPTION of the UNITED STATES, including part of Canada and Florida is added.

The atlas includes ten maps: five were engraved by Henry Tanner, four by Vallance, and one by Thackara. The atlas has descriptive text as well as a table of "Post Towns in the United States, with distances from Washington, D.C." In publishing an American edition of a British atlas, Melish was again emulating Carey. The atlas also indicates that Melish was by 1813 fully committed to map and atlas publishing. With the Advertisement on the page opposite the title page, there is a list of his recently published works. Thirteen titles are listed, including *Travels*, individual maps, and a *Statistical Account of the United States*. The *New Juvenile Atlas* was published January 1, 1814, too early for the list to have included the revised edition of the *Military and Topographical Atlas*.

Although he contracted with Vallance and Tanner to

Fig. 12–2. Melish's 1813 *Map of the Southern Section of the United States* that was subsequently included in his *Military and Topographical Atlas of the United States*.

engrave a number of his early maps, Melish also engaged other Philadelphia engravers. Thus, William Darby's map of Louisiana, which Melish published in 1816, was engraved by Samuel Harrison. Conversely, Tanner and his associates did not work exclusively for Melish. Among Tanner's early engravings, for example, are two maps prepared for an atlas to accompany *Mayo's Ancient Geography and History*, which was published in 1813 and 1814 editions by John P. Watson of Philadelphia.

Perhaps because his own travels disclosed a need for a road guide of the country, Melish compiled and published *A Description of the Roads in the United States* in 1814. This fifty-four-page booklet includes no maps, but lists in tabular form the major roads leading from Washington, D.C., to various parts of the country, the most important lateral or crossroads, and the principal state roads. Mileages between cities are given.

With the War of 1812 still in progress, Melish observed in the Advertisement following the title page of *A Description of the Roads* that

> the events of war, though often distressing in their nature, produce at least one good effect; they excite curiosity, and become subservient in a high degree to the dissemination of useful information. No country is more favourable for the promulgation of geographical knowledge than the United States of America. The great extent of territory—the vast quantity of unsettled lands—the rapid progress of society,—and the circumstances of the inhabitants being equally interested in the prosperity of the country,—all combine to produce a stimulus of a very powerful nature; and this has been so much excited during the present contest with the British government, that the desire for extensive information is general.

During the remaining years of his life, Melish accepted full responsibility for assembling and disseminating

A

MILITARY AND TOPOGRAPHICAL

ATLAS

OF THE

UNITED STATES;

INCLUDING

THE BRITISH POSSESSIONS & FLORIDA:

COMPREHENDING

1. Map of the Seat of War in North America.
2. Description of do.
3. Map of the Southern Section of the United States, including Florida, &c.
4. Description of do.
5. Map of the Coast from Norfolk to Newport.
6. Map of Detroit River.
7. Plan of Quebec and adjacent country.
8. Official Documents relative to the operations of the British army in reducing the Canadas.
9. Chart of the St. Lawrence river from Sacket's Harbour to French Mills.
10. Map of the River St. Lawrence, and adjacent country, from French Mills to Montreal.
11. Map of the Seat of War among the Creek Indians.
12. Map of New Orleans and adjacent country.

BY JOHN MELISH.

PHILADELPHIA,

PUBLISHED BY JOHN MELISH.

G. PALMER PRINTER.

1815.

Copy 2

Fig. 12–3. The title page of the second edition of Melish's *Military and Topographical Atlas of the United States*.

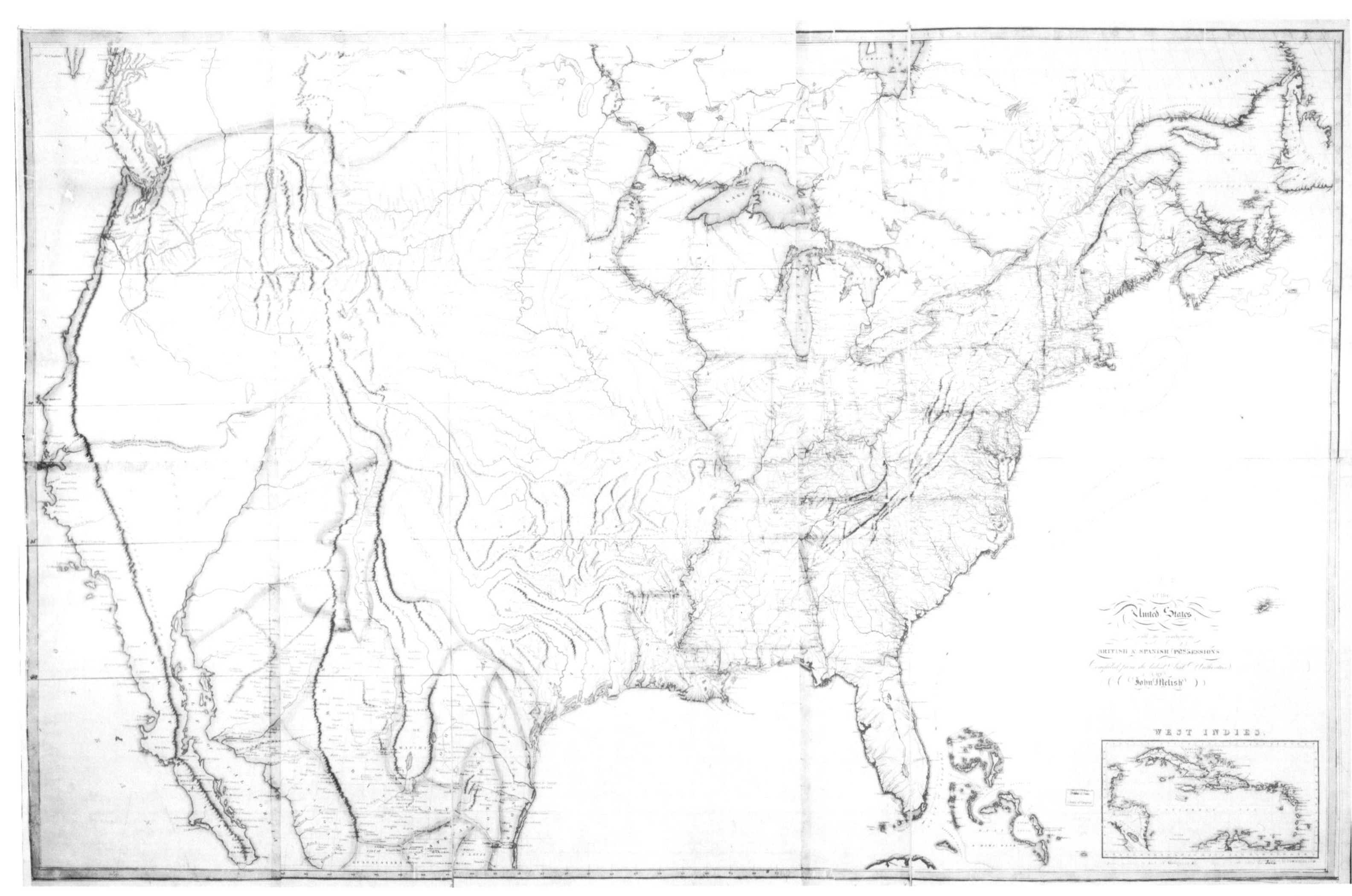

Fig. 12–4. Melish's *Map of the United States with the contiguous British & Spanish Possessions*. This is the fifth state of the 1816 edition of the map. Melish's U.S. map was one of the first to show the full extent of the country from the Atlantic to the Pacific.

geographical and cartographical information. Appended to the *Description of the Roads* booklet is a list of "Works Lately Published by John Melish, Geographer and Map Seller," which had been expanded to sixteen titles, plus three titles in the "Hands of the Engraver."

Revised and enlarged editions of *A Description of the Roads* were published by Melish in 1816 and 1822 as *The Traveller's Directory through the United States*. Both of these editions have a fold-out map of the United States. The one in the 1816 edition was engraved by Tanner, and the one in the later edition was engraved by Harrison. A. T. Goodrich of New York City published an 1826 edition of Melish's *Traveller's Directory*.

Except for revisions of previously published items, Melish issued no major cartographical or geographical works in 1815. In that year he embarked on the very ambitious project of compiling and publishing a large wall map of the entire country. The plan was outlined in the *Prospectus of a Six Sheet Map of the United States and Contiguous British and Spanish Possessions*, which was distributed to prospective purchasers early in the year. In the prospectus Melish declared that "in the course of his geographical studies, he was frequently led to regret, that there was no map in existence presenting an entire view of the United States territory; and having occasion to consult a great variety of documents in constructing his maps relative to the late war, he formed an opinion that a *Map of the United States in connection with the British and Spanish Possessions*, constructed with special reference to the events of war, would be a great desideratum in geographical science." Melish promised that the map would be compiled from "the best and latest State maps, and various local and MS. Maps from actual survey, Bradley's and Arrowsmith's general maps; materials in various Travels through the country, particularly Pike's and Lewis' and Clark's, and various materials in the public offices at Washington."[4]

In his booklet *Geographical Intelligence*, published in 1818, Melish gave credit to a friend for the idea of compiling a map of the United States. "During the progress of the war," he wrote, "a very respectable *Friend* in Philadelphia, when talking of the Map of the Seat of War, said 'I wish friend John, thee would make a Map of the Seat of Peace.' The hint was not lost. The author had seen the good effects of maps, particularly when accompanied by descriptions, and he resolved to condense into one grand view the *whole of the United States territory*, including the *British Possessions* and *Spanish Possessions* contiguous to it, to be ready as soon as possible after these regions became the 'Seat of Peace.'"[5]

Initially, Melish envisioned a map in four sheets, each 64 by 51 cm., and limited to that portion of the United States east of the Pacific watershed. This plan was later revised, as the cartographer explained in his supplement to the U.S. map, the *Geographical Description of the United States*. "It was intended," he explained, "to carry the map no farther west than the ridge dividing the waters falling into the Gulf of Mexico, from those falling into the Pacific Ocean. A subsequent view of the subject pointed out the propriety of adding the two western sheets so as to carry it to the Pacific Ocean."

As published in 1816, the six-sheet map measures 92 by 145 cm., and it is at the scale of 1:3,800,000. Six of the seven states of the map carry the notice "Entered According to Act of Congress the *6th* day of June 1816" (Fig. 12–4). However, it is unlikely that any copies of the map were distributed this early. The June 15, 1816, issue of *Niles Weekly Register* reported that "The indefatigable Mr. Mellish [*sic*] is about to furnish us with a new and very interesting map of the United States and their territories, with the adjacent British and Spanish possessions. . . . A proof impression was shewn to the editor a few days ago." From this we may infer that the first state of the map was not published until some time after June. Some authorities believe that the first copies of the map were not offered for sale until November 1816.

Melish's map of the United States apparently found a ready market, and early printings were quickly sold out. At least three states of the map were issued before the end of 1816. No extant copies of the map are dated 1817, but the last four states dated 1816 were probably not distributed until the succeeding year. The *Map of the*

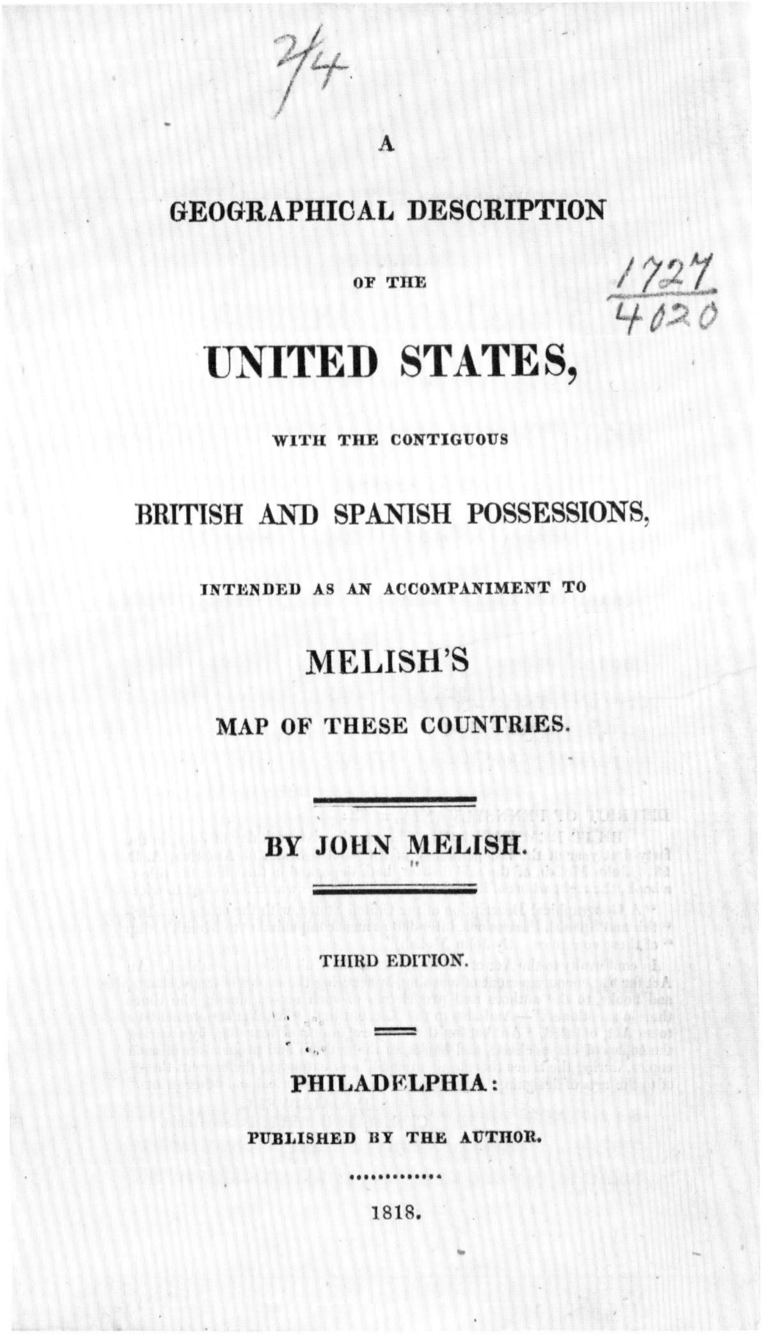

2/4

A

GEOGRAPHICAL DESCRIPTION

OF THE

1727
4020

UNITED STATES,

WITH THE CONTIGUOUS

BRITISH AND SPANISH POSSESSIONS,

INTENDED AS AN ACCOMPANIMENT TO

MELISH'S

MAP OF THESE COUNTRIES.

BY JOHN MELISH.

THIRD EDITION.

PHILADELPHIA:

PUBLISHED BY THE AUTHOR.

....

1818.

Fig. 12–5. The title page of the third edition of Melish's *Geographical Description*.

United States is one of Melish's major publications, and it is a significant milestone in the history of American commercial cartography. The early editions were printed from six copper plates engraved by Vallance and Tanner. The map's supplemental booklet, *A Geographical Description*, contains a particularly useful and valuable summary of the source materials consulted in compiling it. "The various state maps from actual survey" supplied data for the more settled areas of the country while "information regarding the territories was principally procured from the land office at Washington." The travel reports of Meriwether Lewis and William Clark, Zebulon Pike, and other explorers and surveyors furnished data about the western lands. Summing up, the *Geographical Description* affirms that "the author has been most generously supplied with information from every quarter; and he has used every exertion to avail himself of it, so as to produce a view of the country which he hopes will be as valuable to his fellow citizens as it is gratifying to himself."

The *Geographical Description* proved to be as popular as the map, and a second 1816 edition was published. The second edition carries the imprint, "T. H. Palmer, printer." A third edition of the *Geographical Description* was published in 1818 (Fig. 12–5). In his "Observations on the Third Edition," Melish recalled that "it is now about fifteen months since the first edition of this work issued from the press. Two whole editions have been disposed of, and the demand continues unabated. . . . In pursuance of the author's general plan, editions of a moderate size, only, are prepared, so as to afford frequent opportunities of bringing forward new matter." The cartographer noted that frequent changes were being made in the geography of the country, particularly in the western regions. To keep abreast of these changes he explains that "by the plan adopted, the subject is always new. The map is printed off 100 at a time, and before a new hundred be printed, the plates are carefully revised, and if there be room for corrections or improvements, these are made accordingly." Melish also promised that "a new edition of the description will be made annually, and every thing new that occurs in the course of the year will be inserted in the new edition." He was not, however, able to keep this promise, and only one subsequent edition of the *Geographical Description* was published. It was in 1826, several years after Melish's death.

He did continue to issue a number of editions and printings of the *Map of the United States*. Thus, five variants of an 1818 edition have been identified, each of which carries the notation "Improved to the 1*st* of January 1818." It was the 1818 edition that was consulted by the official arbiters in delimiting the boundary between the United States and Spanish territory in 1819. Article three of the Adams-Onis Treaty, which was signed at Washington on February 22, 1819, specified that

> the boundary line between the two countries, west of the Mississippi, shall begin on the Gulf of Mexico, at the mouth of the River Sabine, in the sea; continuing along the western bank of the river to the 32d degree of latitude; thence by a line due north to the degree of latitude where it strikes the Rio Roxo of Natchitoches or Red River; thence, following the course of the Rio Roxo westward to the degree of longitude 100 west from London and 23 from Washington, thence crossing the said Red River, running thence by a line due north to the River Arkansas; thence following the course of the southern bank of the Arkansas to its source, in latitude 42° north; and thence in that parallel to the South Sea. The whole being as laid down in Melish's Map of the United States, published at Philadelphia, improved to the first of January, 1818.

It will be recalled that the data for this portion of the map were derived by Melish from William Darby's *Map of the State of Louisiana* published in 1816.

In 1819 two states of the *Map of the United States* were published, both designated, "Improved to the 1st of April, 1819." The earlier edition is the first to show, by means of a dash-dot engraved line, the United States-Mexico boundary as fixed by the Adams-Onis Treaty. Eight variants of the 1820 edition of the United States map have been identified. The first two carry the notice "Entered according to Act of Congress the 6*th* day of June 1820." On the other six the date is given as the "16*th* of June." No copies of the first two were actually distributed, as the publisher apparently decided to make major modifications after the proof sheet was deposited for copyright. The reasons for the changes are noted in the 1826 edition of the *Geographical Description*, which was prepared by Melish but not published until after his death.

> When the late treaty was negociated with Spain which had reference to the map in fixing the southwest boundary, it was determined to bring forward *an entire new edition of the Map*, exhibiting Florida as a part of the United States, and making all alterations that had taken place in the country, up to the time of publication; and from a conviction that Mexico would soon become independent, and would eventually be of great importance to the United States, it was determined to add another sheet exhibiting a complete view of that very interesting country, with all the most important West India Islands. This was accordingly executed, and the supplement was so enlarged as to exhibit a view of the whole West Indies, with Guatimala, the Isthmus of Panama, and the northern provinces of South America, now forming part of the Republic of Colombia.[6]

Adding the West Indies and all of Mexico increased the overall size of the second state of the 1820 edition of the map to 109 by 145 cm. This was the first state to be published in the enlarged format. Subsequent variants dated 1820 are distinguished primarily by additions and corrections in the Latin American portions of the map.

There is no extant 1821 edition of the United States map. This was very likely because Melish and his associates were engaged in compiling the state map of Pennsylvania at that time. When the Pennsylvania map was in the hands of the engraver, the cartographer again turned to revising the *Map of the United States*. A new edition, of which two states are identified, was published in 1822. Both are annotated, "Improved to 1822," below the title cartouche. The principal modifications on the map are in the upper Mississippi Valley, in the far West, where corrections were made in the Columbia River drainage system, and along the United States-Mexico border. An 1823 edition of the U.S. map was "Published by James Finlayson, Agent Philada. Successor to John Melish."[7]

Although Melish devoted considerable effort between 1815 and 1822 to the *Map of the United States*, he also compiled and published a number of other maps during these years. It was during this time that he began publishing his integrated series of state maps which he hoped would result in an atlas of maps on a uniform scale. In 1817, a year after the first edition of the United States map appeared, Melish published a large map of the world on the Mercator projection, which was engraved by Samuel Harrison. Apparently to add an experienced engraver and a younger associate to his firm, Melish had formed a partnership with Harrison in 1816. A second edition of the world map was published in 1818 (Fig. 12–6) and was accompanied by the booklet *A Geographical Description of the World*. Both the map and booklet have the imprint "Published by John Melish and Samuel Harrison." In the preface of the booklet Melish explains that

> with a view of doing more ample justice to the business, the author some time ago associated himself with Samuel Harrison, map engraver in this city, and a specimen of their joint labour is now presented to the public, in the map of the World. From this the utility of the establishment may be inferred, and having the power within themselves of executing every part of the map business with the utmost celerity and accuracy, the public may rest assured that every effort will hereafter be used to render the business in every respect worthy of that high degree of public patronage which it has received.

A Geographical Description of the World has a small folded world map as its frontispiece. Two variants of this map are found in copies of the work. The earlier, engraved by Tanner, appeared previously in *The New Juvenile Atlas*. It bears a June 1, 1816, publication date. A map of the world on a double-hemisphere projection is the frontispiece in other copies of the *Geographical Description of the World*. It has the imprint "Published 1st Sepr. by J. Melish, Philadelphia. Improved to 1818."

Lists of Melish's published works also include *A General Atlas and Geography*, which contains "a general description of all the countries in the world, with numerous statistical tables, and a series of coloured maps." Another publication, the *Universal School Atlas*, includes eight maps and was published in 1818. It was designed to accompany the cartographer's *Universal School Geography*, which was published in the same year. In *Geographical Intelligence*, Melish wrote,

> having now completed [the U.S. and world maps] the next object was to form a plan by which they might have full practical effect, particularly among the youth of the United States. The mode adopted for this purpose was to prepare a School Atlas and Geography. This work has just issued from the press, and as every effort has been used to render it *clear* and *perspicuous*, it is confidently believed it will be the means, particularly when connected with the *Map and Description of the World* and *Map and Description of the United States*, of giving more facility to the study of geography than ever has been done heretofore.

Melish spent most of 1822 expanding his *Geographical Description of the United States*, and with some minor updating, the five-hundred-page volume was published in 1826 by A. T. Goodrich of New York. In the preface, written by Melish in 1822, he noted that "the *Description* having answered a valuable purpose, it was determined to bring forward a new and improved edition as soon as possible after access could be had to the United States census of 1820. This, it was presumed, could be comprised in a work of 250 pages; but, on arranging the necessary details, it has swelled to more than 500 pages. . . . To this has been added 12 local maps, so as to illustrate some of the most important positions in the country." The closing paragraph of the preface could serve as an epitaph for Melish: "Having had access to the best geographical materials, and having used his utmost endeavors to put them into a form calculated to instruct his fellow citizens, the author respectfully consigns this work to their care, believing that his labour will not have been in vain."

Melish died while he was still vigorously engaged in "promulgating geographical information." The publisher Goodrich notes in his preface to the 1826 *Geographical Description* that "a memoir of the late John Melish was intended to have been inserted in this edition, but unforeseen circumstances have prevented it, and confines this brief note to the single remark that he closed his active and valuable life in the city of Philadelphia on the 30*th* of December, 1822." Within a week after Melish's death, newspapers announced an auction at which was offered "all the entire stock in trade of the late John Melish, comprising a valuable collection of Engraved Copper Plates, with the copy rights and impressions from the said plates." A fortnight or so later, "all the neat household furniture of John Melish, deceased, comprising sideboard, breakfast and dining tables, chairs, carpets, bedsteads, beds and bedding, with a variety of other articles not enumerated as well as kitchen furniture" was auctioned.

Thus the Melish map firm came to an abrupt end. The

Fig. 12–6. The title cartouche of Melish's 1818 map of *The World*.

plates and stock of his publications were purchased by A. T. Goodrich. Except for the 1826 edition of the *Geographical Description*, though, there is no indication that Goodrich reissued any of Melish's publications. A small U.S. map carries the note "improved to 1824 by Jno. G. Melish and published by A. T. Goodrich & Company, New York." This John G. Melish was probably Melish's son, and this small map appears to be his sole cartographic contribution.

An obituary in a Philadelphia paper attested that Melish's "works in the sciences of geography and political economy are universally known, and their importance has been acknowledged by the highest characters of our country."[8] Five U.S. presidents are known to have possessed copies of Melish's large map of the country,[9] and Melish was personally acquainted with several of these gentlemen. The variety and number of Melish's publications are truly remarkable, particularly so when we recall that he accomplished all of them within scarcely more than a decade. For his significant cartographic contributions Melish merits recognition as one of the founders of American commercial map publishing.

Notes

1. John Melish, *Travels in the United States of America in the Years 1806 & 1807, and 1809, 1810, & 1811* (Philadelphia, 1812) 1:x.
2. John Melish, *A Military and Topographical Atlas of the United States* (Philadelphia, 1813), 3.
3. Ibid.
4. John Melish, *Prospectus of a Six Sheet Map of the United States and Contiguous British and Spanish Possessions* (Philadelphia, 1815).
5. John Melish, *Geographical Intelligence* (Philadelphia, 1818), 4.
6. John Melish, *A Geographical Description of the United States. . . . A New Edition, Greatly Improved* (New York, 1826), iii–iv.
7. For a table identifying the variant states of the United States map see "John Melish and His Map of the United States," in *A La Carte*, comp. Walter W. Ristow (Washington, D.C., 1972), 162–82.
8. *Democratic Press*, Jan. 1, 1823.
9. John Adams, John Quincy Adams, Thomas Jefferson, James Madison, and James Monroe.

13. *Henry S. Tanner*

The two decades between 1820 and 1840 have been called the "Golden Age of American Cartography." During these years commercial map publishing, based upon copper-plate engraving, reached its zenith. A principal contributor to the golden age and one of the most productive and successful cartographic publishers of the period was Henry Schenck Tanner. Based in Philadelphia for forty years, he engaged in general and specialized engraving, independently and with partners. He started his career with his brother, Benjamin, and in 1811 the two were in partnership as general engravers. David Stauffer states that about this time Tanner "engraved outline illustrations for some of the magazines of that city, though he was chiefly engaged upon map and chart work."[1]

Tanner's early association with Melish profoundly influenced him to devote his career exclusively to map publishing. While Tanner worked on several Melish projects, he did not limit himself in doing work only for Melish. In addition to engraving the two maps for the atlas accompanying *Mayo's Ancient Geography and History*, he also engraved *A Map of Wayne & Pike Counties, Pennsylvania*, which was registered for copyright by Jason Torey, and B. Hough and A. Bourne's *Map of the State of Ohio From Actual Survey*, which was copyrighted June 27, 1814, but probably not published until early 1815. In 1815 and 1816 Tanner prepared twenty-seven engraved plates for *A New and Elegant General Atlas Containing Maps of Each of the United States*, which was published by Fielding Lucas, Jr., of Baltimore. There is no publication date on the title page, but the atlas's map of Virginia is dated 1816, and this is usually accepted as the publication date of the whole work. Tanner's contributions include a double hemisphere world map, maps of Europe, Asia, Africa, North America, South America, the West Indies, and Canada, as well as maps of most of the states. Many of the maps were compiled and drafted by Samuel Lewis, who had earlier prepared maps for several Carey atlases. Twenty-two of the maps in the *New and Elegant General Atlas* were engraved by Samuel Harrison, Melish's one-time partner.

Around 1816, the Tanner brothers, with Vallance and Francis Kearny, formed an engraving partnership under the name Tanner, Vallance, Kearny & Company. Individually and collectively the partners continued to prepare engravings for Melish, among them a plan of the city of Philadelphia, the *Map of the United States*, and the 1822 official state map of Pennsylvania. Francis Kearny, a member of the engraving partnership, was born in Perth Amboy, New Jersey, around 1780. At the age of eighteen he was apprenticed to the engraver Peter Maverick in New York City. Kearny then moved to Philadelphia in 1810. Inasmuch as Benjamin Tanner had earlier worked with Maverick, it may have been through him that Kearny and Tanner became acquainted. According to Stauffer, Kearny achieved his greatest success in line, stippled, and aquatint engraving for magazines, annuals, and books.

Although all the partners of Tanner, Vallance, Kearny & Company had experience in engraving maps, Henry Tanner appears to have been the one most interested in cartographic engraving. While he was engaged in engraving Melish's *Map of the United States*, it seems probable that he conceived the idea of compiling and publishing an American atlas. Lacking financial resources to undertake the task independently, Tanner persuaded his partners to sponsor the publication of the *New American Atlas*. Very likely Tanner assumed major responsibility for planning the atlas, compiling the maps, and engraving some of them. It was soon apparent to the other partners that compiling and publishing the *New American Atlas* would be an extremely costly undertaking. The decision was made, therefore, to publish it serially, which would insure some financial returns before the entire volume was completed.

Accordingly, the firm issued the first series of maps in July 1819. Included were maps of the world and Europe, each on one sheet, and a two-sheet map of South America. The world map has the imprint "Published by H. S. Tanner, Philada." and the credit "Engraved by Tanner, Vallance, Kearny & Co." The imprint on the map of Europe is "Engraved & Published by Tanner, Vallance, Kearny & Co. Philadelphia." The map of South America was registered for copyright by Tanner, Vallance, Kearny & Company, but was "Engraved & Published by H. S. Tanner, Philadelphia." This initial group of maps was enclosed in a gray paper folio, on the cover of which was printed the title, *The New American Atlas No. 1. Containing Maps of the World, Europe, and South America in Two Sheets, Arranged from the most Authentic Documents. Philadelphia: Published By Tanner, Vallance, Kearny & Co. No. 10, Library Street, 1819.*

This folio also included a large broadside addressed "To the Public," which stated the objectives of the atlas, outlined the projected arrangement of the maps, and described the plan for distributing the folios and publishing the atlas. The editor noted in the broadside that "our geography is so rapidly progressive, that no European publication can keep pace with our improvement and the extension of our settlements. The subject must be brought to maturity in our own country, and, such is now the respectable state of the Arts here, that we can assert with confidence, that we possess the materials and skill sufficient to exhibit a topographical representation of the United States, infinitely superior, as it regards correctness and detail, and every way equal in style, to any European publication of the kind." Subscribers were informed that "the publishers of this work have been collecting materials, preparatory to the execution of it, for several years, in which they have been assisted by some of the ablest geographers in this country and Europe." The broadside further explained that "it was originally intended by the proprietors of this work to publish each state and territory on a separate sheet; it has since, however, been considered most proper, as being best calculated for a work of general and convenient reference, to supercede that plan by the one now adopted [i.e., to have two or more states on one map], by which the Atlas can be furnished at 30 dollars, instead of 65 dollars, the price first proposed."

The broadside also prescribed the following "Terms of Publication":

I. The size of each sheet will be about 25 by 23 inches, and drawn from the latest and most authentic documents. They will be engraved in the first style of map engraving, and shall in every branch of its execution be purely American.

II. The Maps will be printed on the first quality vellum paper, and coloured in an elegant and appropriate manner.

III. The Atlas will be completed in five numbers; each to contain four sheets, except the last, which will contain five, including an engraved title sheet. They will be delivered to Subscribers, folded on guards, at Six Dollars each number, payable on delivery.

IV. Persons collecting subscribers for six copies, and becoming responsible for the payment, shall be entitled to a seventh, gratis. The maps may be had separately at the rate of two dollars each sheet.

The wish was also expressed that "the proprietors . . . desirous of rendering [the atlas] as correct as possible, embrace this method of respectfully soliciting the aid of gentlemen residing in the interior, who may be in possession of any geographical information, not before published, by communicating the same to the publishers."

The second folio of maps distributed late in 1819 included maps of Asia, America, New York, Ohio, and

Indiana. All are on double sheets. The credit line on the New York map reads, "Engraved and Published by H. S. Tanner, Philadelphia." The other four maps were engraved and published by Tanner, Vallance, Kearny & Company. "By H. S. Tanner" was added to the cover of the folio, suggesting that Tanner was assuming the major responsibility for the atlas. The third folio, issued in 1821, confirms this, for the imprint reads, "Published by Henry S. Tanner, No. 29, South Tenth Street." In a broadside note, "To the Public," inserted in this folio, Tanner informed subscribers that "it will be remembered that the publication of the work was commenced by Messrs. Tanner, Vallance, Kearny and Co. in whose names the two first numbers were issued. In consequence of a dissolution of that connection, I became the sole proprietor of the Atlas. The arrangements necessary in this case contributed in part to suspend, for a time, the operations necessary to its advance." Maps of New England, Virginia, Maryland and Delaware, Louisiana and Mississippi, and Africa in this folio all carry the imprint, "Engraved and Published by H. S. Tanner, Philadelphia." Concluding his statement, Tanner reminded subscribers that

> the first prospectus was issued in 1818, and the publication of the first number shortly followed. It soon, however, became obvious that an increase of patronage was indispensable to the support of an enterprise attended with so much expense, and particularly where in an incipient state of the work so much capital must be employed in the collection of necessary information, drawings, engravings, &c. before an adequate return of funds could be expected. Some idea of the amount already vested, and that necessary to insure the final completion of the work, may be conceived, when it is mentioned, that the expenses attending the first sheet in the present number [i.e., the map of New England] exceed two thousand two hundred dollars; this sum may perhaps stagger the belief of those who are unacquainted with the difficulties and expenses inseparable from an original undertaking of this nature.

Accelerating production costs may have induced Tanner's partners to withdraw from the atlas project. We may wonder, however, how the geographer was able independently to compile the remaining maps, to engrave plates for them, and to publish them. Tanner had, of course, receipts from the first three folios of the atlas. In addition, he sold some of the atlas maps as separates, along with his other cartographic publications. Printed on the back cover of the third folio is an extensive "List of Maps and Geographical Works Lately Published and for Sale by H. S. Tanner, No. 8 Sanson-street, Philadelphia." This address differs from that printed on the cover of the folio.

The fourth folio, distributed in the latter half of 1822, enclosed a large four-sheet map of North America, which was deposited for copyright by Tanner on May 27, 1822. Below the title on the southwest sheet there is a large engraving which illustrates the Natural Bridge of Virginia and Niagara Falls. The North America map was printed by William Duffee and "Engraved & Published by H. S. Tanner, Philadelphia, 1822" (Fig. 13–1).

All six maps in the fifth folio were registered for copyright on August 20, 1823, and were probably not distributed to subscribers until late in that year. Included are maps of several states in the following groupings: Pennsylvania and New Jersey, Kentucky and Tennessee, North and South Carolina, Georgia and Alabama, Illinois and Missouri, and Florida. All were "Engraved by H. S. Tanner & Assistants" and "Published by H. S. Tanner, Philadelphia." Also included within this folio is an elaborately engraved title page, an index of the twenty-two map plates (actually eighteen maps), and a lengthy and comprehensive "Geographical Memoir" that extended over seventeen large folio pages. The memoir was signed by Tanner on September 10, 1823.

The title of the atlas is *A New American Atlas Containing Maps of the Several States of the North American Union, Pro-*

jected and drawn on a Uniform Scale from Documents found in the public Offices of the United States and State Governments and other Original and Authentic Information. An engraving depicting the "First Landing of Columbus in the New World" decorates the title page (Fig. 13–2). Below the illustration is this quotation from Antonio Herrera's *Life of Columbus*: "Columbus and his followers landed and kneeling they all kissed the ground they had so long desired to see; they next erected a Crucifix and prostrating themselves before it, returned thanks to God for conducting their voyage to such an happy issue." Joseph Perkins is credited with having drawn and engraved the lettering on the title page, and the illustration is by "Barralet del. Humphrys sc."

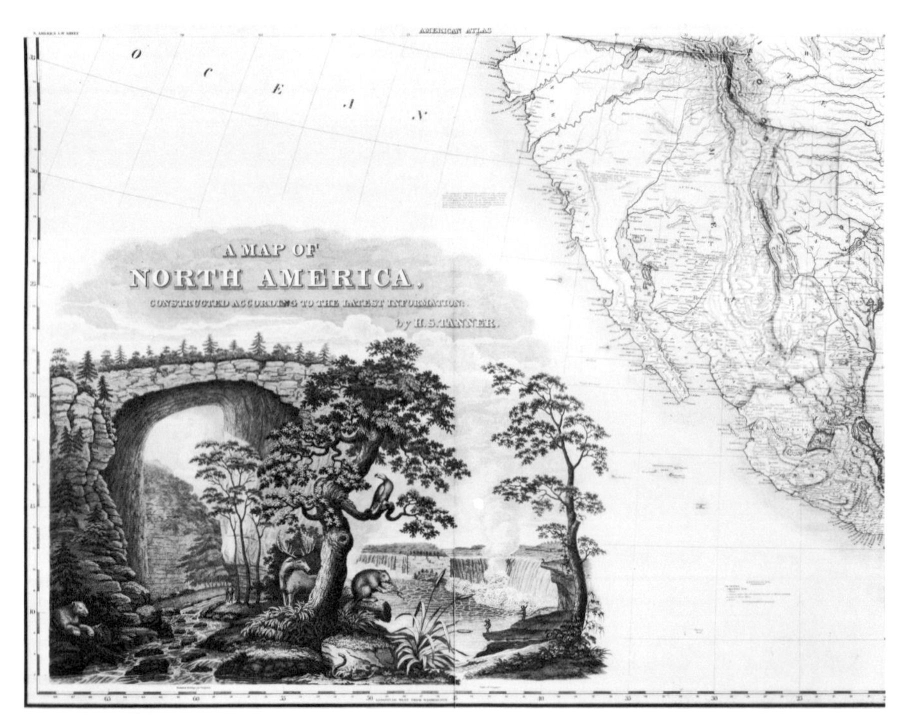

Fig. 13–1. The title cartouche of Tanner's 1822 map of North America. The map was also published in Tanner's *New American Atlas* in 1823.

Joseph Perkins was born in Unity, New Hampshire, on August 19, 1788, and graduated from Williams College in Massachusetts. Shortly thereafter he settled in Philadelphia, where he was occupied with bank note and script engraving. Perkins moved to New York City around 1826 and for several years, from about 1828 to 1831, was in partnership with Asher B. Durand engraving bank notes. Perkins died in New York City on April 27, 1842. John James Barralet was born in Dublin, Ireland, of French parents around 1747. In 1795 he settled in Philadelphia, where he engaged in painting portraits and landscapes and in preparing illustrations for books. Barralet died in Philadelphia on January 16, 1815. William Humphrys was also a native of Dublin, who was born in 1794. He immigrated to Philadelphia, where he learned engraving from George Murray. In 1827, Humphrys went to England, where he had some success engraving stamps and bank notes. He died in Genoa, Italy, in 1868, where he had gone because of poor health. Because Barralet died in 1815, it is obvious that the view of the landing of Columbus was prepared some years before it was selected to illustrate the title page of Tanner's *New American Atlas*. It is not known when the view was first published, but it continued to illustrate atlas title pages as late as 1859.

The "Geographical Memoir" encompasses a wealth of information, and it is an exceedingly valuable feature of the atlas. In detailing the source materials consulted in compiling the atlas maps, Tanner provided subscribers with a comprehensive inventory of the cartographic resources then available in the United States. He devoted a full page of the memoir to justify his choice of using the prime meridian of the city of Washington on the large four-sheet map of North America. His primary reason was the lack of agreement on an international prime meridian, and he, therefore, deemed it appropriate to employ the one which ran through the capital city of the United States. It will be recalled that Melish had also used the Washington, D.C., meridian. However, for all the foreign maps, as well as for the state maps, he used the Greenwich prime meridian. Tanner also explained in the memoir that "the end proposed to be effected by the publication of the American Atlas was, to exhibit to the citizens of the United States a complete geographical view of their own country, disencumbered of that minute detail on the geography of the eastern hemisphere, which is usually introduced in our Atlasses, to the exclusion of matter more immediately interesting to those for whom they are intended."

The major contribution of the *New American Atlas* was its state maps. Tanner had initially hoped to have individual maps, on a uniform scale, for each state. Because of financial limitations only two states, Florida and New York, appear alone, while the others are presented in groups. The state maps are at the scale of 1:940,000. They show such physical features as rivers, lakes, swamps, and mountains. The cultural features portrayed include administrative boundaries, cities, towns, roads, and canals. The maps were manually colored to differentiate administrative or judicial divisions. The *New American Atlas* was bound in half leather and priced at thirty dollars a copy. Its maps were also sold separately, mounted flat, or folded in slip cases, and in various groupings (Fig. 13–3).

Tanner's atlas received commendatory reviews in American and foreign newspapers and journals. The *United States Gazette* (Philadelphia) reported in its September 1823 issue that "Mr. Tanner of this city has completed the last number of the New American Atlas. . . . It is decidedly one of the most splendid works of the kind ever executed in this country, and we sincerely hope that public spirit will reward this expensive undertaking. . . . This Atlas is a work, which has been desired a long time by the public, and from the manner in which the work is executed, it is richly deserving of extensive

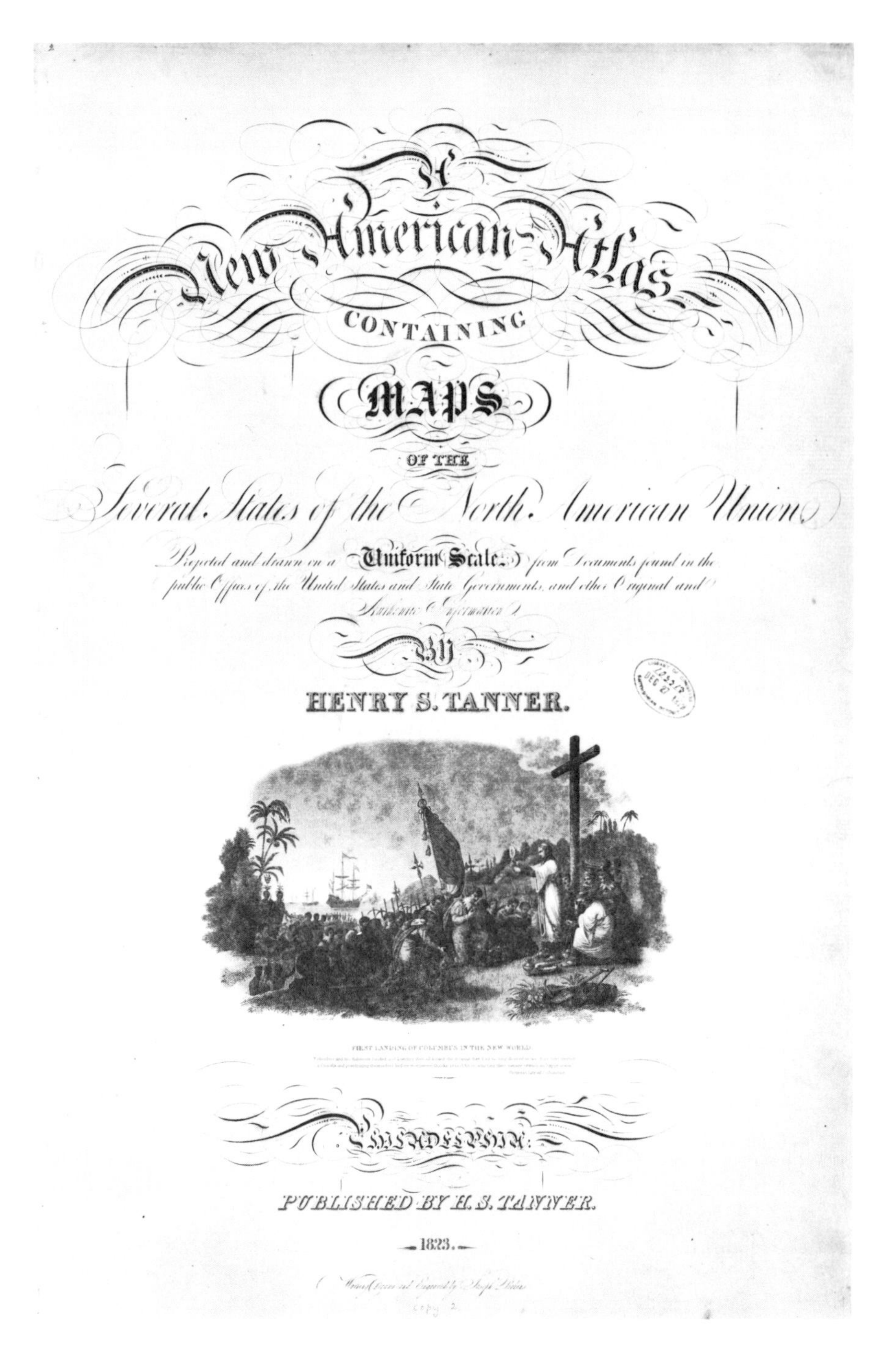

Fig. 13–2. The title page of Tanner's *New American Atlas*. This was the most distinguished atlas published in the United States during the engraving period.

Fig. 13–3. This 1830 map of Virginia originally appeared in the *New American Atlas* and, like the other maps in the atlas, was issued as a separate.

patronage."[2] New York City's *National Advocate* expressed the belief in its August 25, 1824, edition that "never certainly has either America or Europe, produced a geographical description of the several States of the Union, so honorable to the Arts, and so creditable to the nation as *Tanner's American Atlas*."[3]

The distinguished scholar Jared Sparks wrote in the April 1824 issue of the *North American Review* that "on the whole as an *American Atlas*, we believe Mr. Tanner's work to hold a rank far above any other, which has been published. The authentic documents to which he had access, the abundance of his materials, the apparent fidelity, with which they are compiled, the accurate construction of his maps, and the elegance with which they are executed, all these afford ample proofs of the high character of the work, of its usefulness as a means of extending the geographical knowledge of our own country, and of its claims to public patronage."[4]

Alexandre Barbie du Bocage, a reputable French geographer of the period, reviewed the *New American Atlas* in the bulletin of France's Société de Géographie. He took exception to Tanner's choice of the Washington prime meridian for the map of North America, and he thought that the preliminary maps, that is, those of non-American countries, were not prepared with the same care as were the American maps. The reviewer detected a number of errors in the maps of foreign countries, and he concluded that Tanner appeared to have sought to escape responsibility for them in his memoir. Barbie du Bocage did concede that Tanner's treatment of the U.S. maps was an effort "for which we are able to do nothing but applaud." He concluded that "although unequal in its presentation, this Atlas merits your examination with the interest you place on all that is good and useful. It is particularly designed for Americans, it is this section therefore to which the author has given major attention and thus deserves our praise."[5]

The success of the *New American Atlas* greatly enhanced Tanner's reputation as a publisher of cartographic works. The death of Melish on December 30, 1822, some eight or nine months before the last plates of the atlas were distributed, removed Tanner's principal competitor. Tanner quickly filled the vacuum created by Melish's death, and in the next two decades, he firmly established the foundations of American commercial map publishing.

As new geographical data became available, Tanner updated and revised the maps comprising the *New American Atlas*. An article in the *New York Mercantile Advertiser* for August 26, 1824, explained Tanner's plan for keeping the maps current.

> It has frequently been regretted that to keep pace with the continually progressive detail of Topographical information in the purchase of Atlases, costs a great deal of money every two or three years, and now, since the publication of "Tanner's American Atlas," they are likely so much more to abound, as to create new difficulties in the selection; but the more so, as by lively coloring, and bulk of matter, we are too apt to think we buy *cheap*, while the *variety of scale*, on which the State Maps are generally drawn, is too often overlooked, to the *serious of injury* of youth, in their instruction, and all the inconveniences which attend topographical comparison of every description.
>
> To put an end in a great measure to these perplexities, we find it is the determination of Mr. Tanner (who besides being a scientific Geographer, is an eminent Engraver, and continually receiving the new geographical information from all quarters as it occurs) to alter his Plates as changes may require, and to sell the altered sheets separately, and colored to correspond with his Atlas, at two dollars fifty per sheet; the effect of which will be, that persons possessing the Atlas, may every two, three, or four years, take out an old sheet or two as changes may render necessary, to be replaced by the new, and thus, at a very trifling expense, *always preserve their Atlasses up to the present time*, for perhaps twenty-five years to come, or so long at least, as the Plates will admit of alteration. This we conceive, is an advantage so *uncommon*, that we deem it sufficiently worthy to present

> to the public, as an additional recommendation of "Tanner's American Atlas," which is drawn on a uniform scale for the States, and is in every respect the *Master Piece of the Day*.[6]

Although the entire atlas as a whole was apparently revised only once, in 1825, some editions of it have maps dated as late as 1839. Tanner also published the non-American maps under the title *A New General Atlas Consisting of Maps of the Several Grand Divisions of the Known World* in 1828. The atlas maps were also sold individually, mounted on rollers for use in schools, and were assembled and bound in various atlas combinations. In 1828 he published, in a small-format volume, *A New Pocket Atlas of the United States with the Roads and Distances Designed for the Use of Travellers*. The individual state maps appear to have been adapted from the plates in the *New American Atlas*. It is interesting to note that a "List of Maps for Sale by H. S. Tanner," printed in the atlas, includes more than eighty titles.

Concurrently with his work on the *New American Atlas* and its several revisions, Tanner undertook the ambitious task of compiling a large map, the *United States of America*, which he copyrighted on June 10, 1829. It is at the scale of 1:2,000,000, somewhat larger than Melish's 1816 map of the United States. Melish's map, however, extends westward to the Pacific Ocean, whereas Tanner's map terminates just west of the upper course of the Missouri River. Tanner's *United States of America* has thirteen inset maps, principally of cities and their environs, a statistical table, and profiles of a number of canals and railroad rights of way. The map was engraved by Tanner, E. B. Dawson, and W. Allen. The title cartouche, which pictures a country landscape with two deer in the foreground, was engraved by J. W. Steel (Fig. 13–4). Tanner again used the Washington, D.C., prime meridian for his U.S. map.

Tanner followed Melish's example by supplementing his map with a comprehensive *Memoir on the Recent Surveys, Observations, and Internal Improvements in the United States*. Some of the information was borrowed from the "Geographical Memoir" in the *New American Atlas*. Unlike Melish, who printed only one hundred copies of his U.S. map at a time and made essential plate corrections and additions between printings, Tanner published regular editions of his *United States of America*. We do not know how many copies were printed for each edition, but there were probably no less than five hundred. On July 30, 1830, Tanner copyrighted the second edition of his U.S. map. The principal additions appear to be in the northern parts of the states of Indiana and Illinois and in the southern parts of the territories of Michigan and Wisconsin. A second edition of the *Memoir on the Recent Surveys* was also published in 1830. The variations from the first edition are minimal.

Subsequent editions of the *United States of America* were published in 1832, 1834, 1836, 1841, and 1844. They successively show new towns, railroads, canals, and states in the upper Mississippi River valley and the western Great Lakes region. The shape of Lake Michigan underwent several modifications in the various editions. Data from the 1840 census is a feature of the 1841 edition. On the 1844 edition—the last one printed—the imprint has been changed to "New York, Published by H. S. Tanner."

The *United States of America* was reviewed by A. H. Brué in the bulletin of the Société de Géographie. Brué, like the French reviewer of the *New American Atlas*, took exception to Tanner's use of the Washington prime meridian. He, however, commended Tanner for his excellent choice of source material and reminded readers that Tanner was made a corresponding member of the prestigious Société de Géographie for his work on the *New American Atlas*. Since that honor, Brué added, "Tanner has not ceased to add to his noteworthy geographical and cartographical contributions."[7]

Shortly after the first edition of his U.S. map appeared, Tanner published, in 1831, *A New and Authentic Map of the World, Embracing all the Recent Discoveries and exhibiting particularly the Nautical Researches of the most*

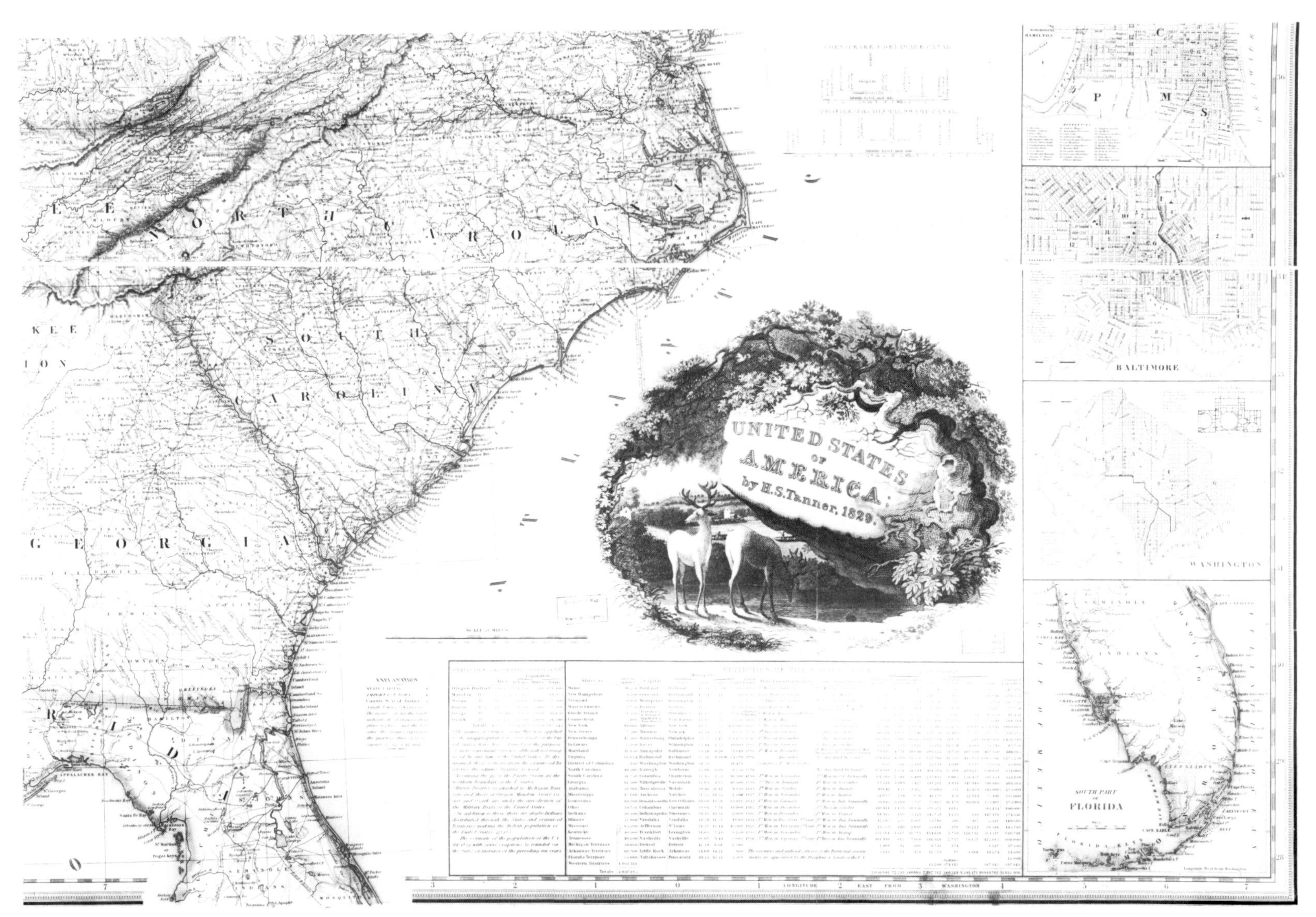

Fig. 13–4. The title cartouche and a segment of Tanner's 1829 map of the *United States of America*.

Distinguished Circumnavigators . . . by H. S. Tanner. Engraved by E. B. Dawson, Philadelphia. In contrast with Melish's large map of the world, which was on the Mercator projection, Tanner employed the globular, or double-hemisphere, projection. A second edition of the world map was registered for copyright on January 1, 1834. No other versions have been identified.

As Tanner's publishing activities expanded, he devoted progressively less time to engraving. In 1825, the district maps for Robert Mills's *Atlas of the State of South Carolina* were already being engraved by "H. S. Tanner & Associates." Similarly, Tanner was assisted in engraving the U.S. map by E. B. Dawson, W. Allen, and J. Knight. By the time his map of the world was published in 1831, Tanner's transition from engraver to geographer and publisher had been fully effected. In addition to compiling, engraving, and publishing Tanner's maps and atlases, Tanner's firm also engraved and published maps prepared by others. Among these were John Wilson's *Map of South Carolina*, 1822, Charles Vignoles's *Map of Florida*, 1823, John Wood and Herman Böye's *Map of the State of Virginia*, 1826, Thomas Gordon's *Map of the State of New Jersey*, 1828, and Robert Brazier and John MacRae's *New Map of the State of North Carolina*, 1833. Tanner and his associates also published city plans.

The *New American Atlas* continued to be published and updated until 1839, but because of its relatively high price of thirty dollars, sales were apparently below Tanner's expectations. Tanner, therefore, planned a smaller atlas, and in 1832 he distributed copies of a proposal for publishing *A New and Elegant Universal Atlas* by subscription. Its scope was much more comprehensive than that of the *New American Atlas*, with foreign countries presented on individual maps or in groups, and with separate maps of each of the continents. Individual maps were planned for each state. The page size was reduced to 36 by 28 cm. (compared with 53 by 38 cm. for the *New American Atlas*), which enabled the publisher to sell the *New and Elegant Universal Atlas* for fifteen dollars.

The first of a planned fourteen folios was distributed in January 1833 and included five maps, three of which were of states and two of foreign countries. In the proposal that accompanied it, Tanner stated that he had "been collecting materials, preparatory to the execution of [the atlas] for several years, in which he has been assisted by some of the ablest geographers in this country and in Europe. In addition he will avail himself of all the recent and important discoveries in both hemispheres, to enable him to execute the proposed Atlas in a manner every way satisfactory to the public." Tanner hoped that "the work will be completed as soon as circumstances will permit, consistently with accuracy and elegance of execution." Nonetheless, three years passed before the final folio was delivered to subscribers. Together, the folios included a total of sixty-five plates, on a number of which there are insets, making a total of 117 maps.

Eight different engravers prepared the plates for the atlas, and eighteen of the maps carry no engraving credit. J. & W. W. Warr contributed eighteen plates, Knight prepared nine, Dawson and W. Brose each engraved seven, and lesser contributions were made by F. Dankworth, E. Gillingham, D. Haines, and W. Haviland. It is not known whether all these artists were regularly employed by Tanner, or whether they were engaged on a contract basis. The latter is more likely. The maps in the atlas are hand colored to show administrative subdivisions. The job of coloring was probably also given to an independent contractor, and it is likely that stencils were used to insure greater speed and neatness. With the completion of the last folio in 1836, the atlas was issued in bound format with the new and lengthy title *A New Universal Atlas Containing Maps of the various Empires, Kingdoms, States and Republics of the World, With a special map of each of the United States, plans of cities &c. Comprehended in seventy sheets and forming a series of One Hundred and Seventeen Maps, Plans and Sections by H. S. Tanner. Philadelphia, Published by the Author, 1836.* This volume includes several maps not distributed in the folios, such as the maps of Oceania, Palestine, and the cities of New York, Philadelphia, and Washington. The title page has the same illustration, the "First Landing of Columbus in the New World," that decorated the *New American Atlas*.

A revised edition of the *New Universal Atlas* was published in 1839. Among its new features was a one-page index, which follows the author's notice in the front of the volume. In the notice, Tanner states that "at length we have the satisfaction to announce the completion of our laborious work. . . . Maps nearly double the size of those originally proposed were issued; and in place of four maps, each number contained five, six, and, in one case not less than eight maps." Modifications and additions were made on some maps, especially for some of the newer states in the upper Mississippi River valley. For such maps, the copyright registration was updated to 1839. The 1843 edition of the *New Universal Atlas* has a new publisher's imprint, "Published by Carey and Hart 1842," which replaces Tanner's name as publisher on the title page. Tanner is still noted as the author. Following the title page there is a "Publishers Notice," dated September 25, 1843, which reads:

> The Plates comprising this highly valuable work have been recently subjected to a complete examination and revision, the present Edition is in consequence greatly improved beyond any that has preceded it. A great amount of interesting matter has been added to nearly every part of it, particularly in the geography of our own country, thus rendering it, by the accuracy contained in the *Universal Atlas*, the distinctness of the Engraving and the splendour of the Colouring, the most elegant and complete work of the land that has yet appeared in the United States, and greatly superior in its representation of American Geography, to any similar production elsewhere. The reputation of the author is so well known and established throughout the Union, that it is deemed superfluous to dilate on that topic, the Publishers, however, cannot refrain from stating their belief that in this his latest effort, Mr. Tanner has produced a work which reflects on him the highest credit, and proves that his fame as a Geographer is richly merited.

Tanner had quite likely completed most of the revisions for this edition of the *New Universal Atlas* before its publication was assumed by Carey & Hart. This is evident from the copyright registrations, which remain in Tanner's name on most of the maps. On certain maps the imprint has been changed to "Published by Carey & Hart, Philadelphia." The imprints on the other maps in the atlas read, "Published by H. S. Tanner, Philadelphia, Tanner & Disturnell, 124 Broadway, New York."

The 1844 edition of the *New Universal Atlas* was also published by Carey & Hart. The copyright notice on this edition reveals that it was "Entered according to the Act of Congress in the year 1844 by Carey & Hart in the Clerk's Office of the District Court for the Eastern District of Pennsylvania." The Carey & Hart imprint appears on many of the maps in this edition, although most of them still carry the Tanner imprint. There has been a slight rearrangement of maps in the 1844 edition: city plans follow the maps of their respective states, instead of being placed at the end of the atlas, as in earlier editions. This edition includes plans of the cities of New York, Philadelphia, and Washington, D.C., and new maps of Wisconsin, Indiana, Illinois, Iowa, Missouri, and Texas. New place names have also been inserted on maps of states and territories in the South and Southwest. The firm of Carey & Hart was a successor of the printing and publishing house founded by Mathew Carey in Philadelphia in 1784. It was not concerned primarily with cartographic publishing, although Carey had issued several atlases in the 1790s. There is no record of other cartographic publications by Carey & Hart, and they may have taken over the *New Universal Atlas* as a personal favor to Tanner.

Another milestone in the development of Tanner's *New Universal Atlas* was its two 1846 editions. In the first edition, the publisher's imprint on the title page reads, "Philadelphia, Published by S. Augustus Mitchell, N.E. Corner of Market & 7th Streets, 1846." The atlas, however, still carries the 1844 Carey & Hart copyright notice. The Carey & Hart imprint was, however, deleted on all maps, with the Mitchell identification substituted on several. Tanner's copyright notice also remains on a number of the maps, although it is completely deleted on the second 1846 edition. Changes were made on several maps, and there are color variations from the previous edition. A map of Texas has also been added. It was copyrighted by C. S. Williams in 1845 but carries the Mitchell imprint.

The most significant change, however, is in the type of reproduction. In earlier editions of the *New Universal Atlas*, the maps were printed from engraved copper plates and the plate marks are clearly visible. In the Mitchell edition, plate marks are lacking, and decorative green borders frame all the maps. The maps appear to have been reproduced by transferring the engraved images to lithographic stones, from which the maps were then printed. The lithographic transfer process had only recently been introduced to the United States, and its use for the *New Universal Atlas* is an early application of it in American commercial map publishing. It is interesting to note that another edition of the *New Universal Atlas* was published by Mitchell in 1846. In this edition Tanner's name has been deleted entirely. Editions of the *New Universal Atlas* were published until 1859. Some of the later ones were issued by other publishers.

Individual maps from the *New Universal Atlas*, particularly of the states, were sold separately by Tanner and frequently folded to pocket size and enclosed between board covers. In 1835 Tanner assembled twenty-eight state maps in a bound volume and published them under the title *Atlas of the United States Containing Separate Maps of Each State and Territory of the North American Union*. This appears to be the only edition of this atlas. The extensive publishing activities of Tanner are evident in the "List of Charts and Geographical Works," which supplements the atlas. The list includes 123 titles. Brief mention should also be made of Tanner's publication of the *Atlas Classica, Being a Collection of Maps of the Countries Mentioned by the Ancient Authors Both Sacred & Profane*. There is no date on the title page, but it was probably published around 1840.

During this golden age of cartography, the United States witnessed a great deal of activity related to the improvement of communication and transportation. Turnpike, canal, and (in the latter part of the period) railroad construction, carried out both by private enterprise and the federal and state governments, was generally grouped under the subject of internal improvements. Maps showing the ever-expanding U.S. transportation network were in great demand, and Tanner published a number of such cartographic items for the country as a whole and for individual states (Fig. 13–5). One such item is his *Map of the Canals & Railroads of the United States, reduced from the large map of the United States. Entered according to Act of Congress, the 16th day of June 1830. Engraved by J. Knight*. The state maps, which were issued mounted for wall display or folded within covers as pocket maps, were printed from the plates originally prepared for the *New Universal Atlas* (Fig. 13–6).

Tanner also published a whole series of travelers guides. In 1834 he issued the *Travellers Guide or Map of the Roads, Canals, and Rail Roads of the United States, With the distances from place to place* (Fig. 13–7) and the first edition of *The American Traveller; or Guide Through the United States Containing Brief Notices of the Several States, Cities, Principal Towns, Canals and Rail Roads, &c. With Tables of Distances by Stage, Canal, and Steam Boat Routes*. Subsequent editions of the *American Traveller* with Philadelphia imprints were published until 1839. The eighth edition, dated 1842, has a double imprint: "Philadelphia, H. Tanner, Jr.; New York, T. R. Tanner." These two Tanners were very likely the sons of Henry Tanner. The

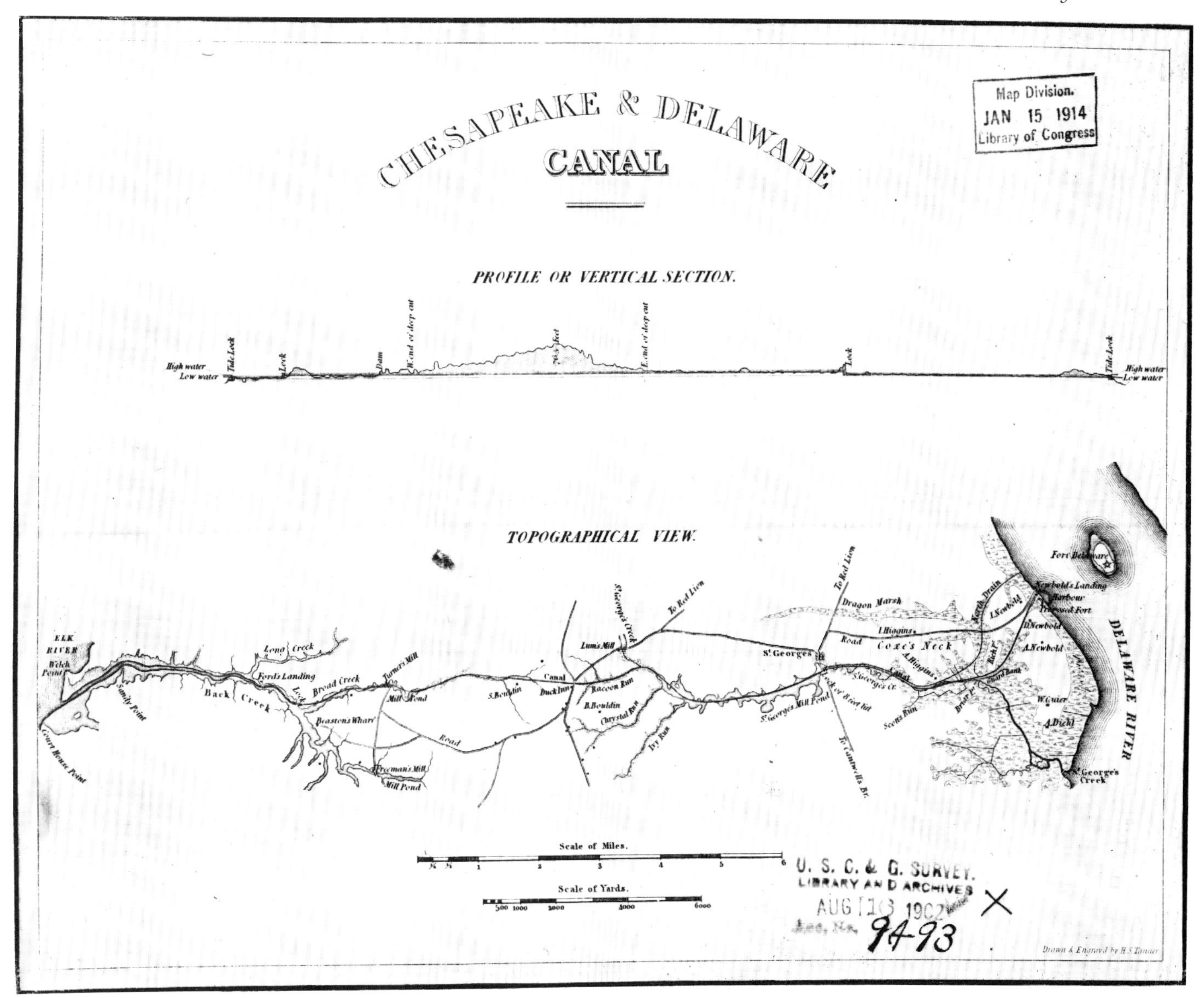

Fig. 13–5. One of Tanner's transportation maps is this undated one showing the Chesapeake and Delaware Canal.

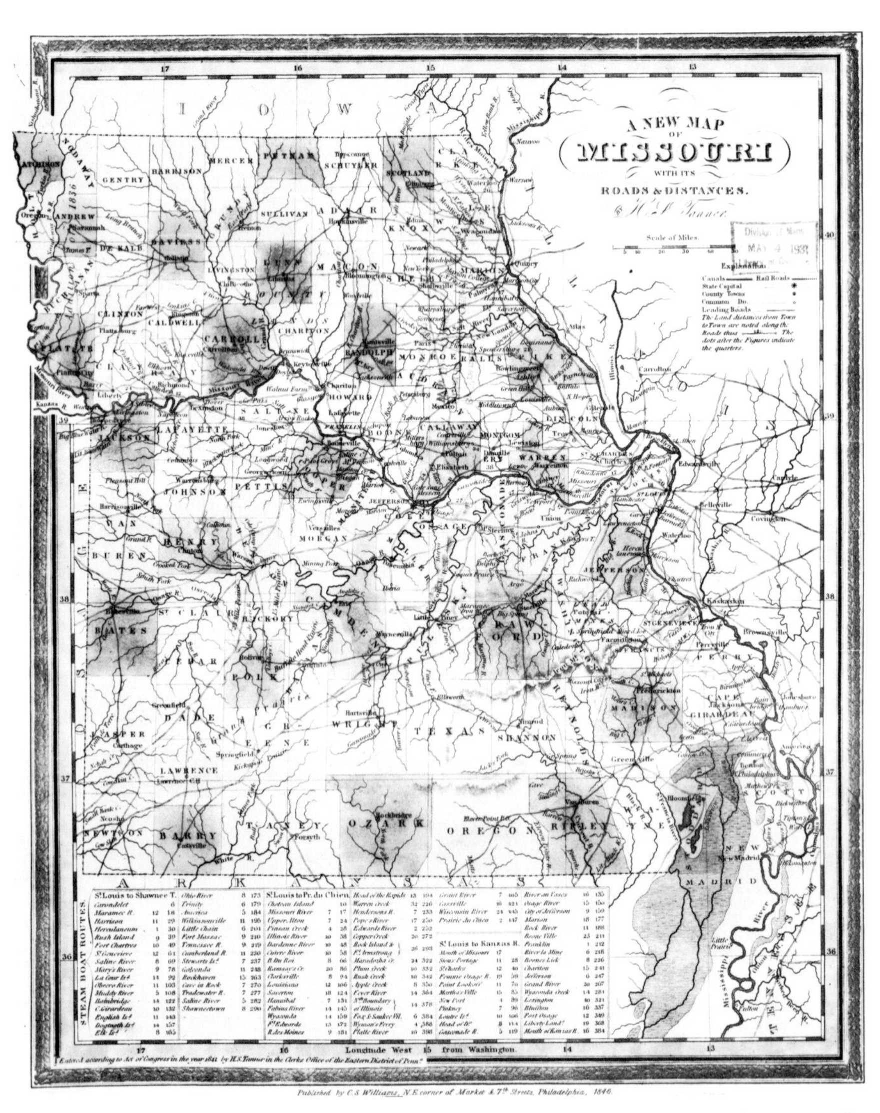

Fig. 13–6. Issued as a separate in 1846, this map of Missouri was a revision of a state map that first appeared in the *New Universal Atlas*.

imprint on the tenth edition of the *American Traveller*, however, is "By H. S. Tanner, New York, Pub. at the Map Establishment, 1846."

Stauffer notes that "in 1843, Henry S. Tanner removed to New York and there engaged in the engraving and publishing of Maps, charts, etc. He contributed geographical and statistical articles to various periodicals, and published guide-books for a half-dozen sections of the United States."[8] That Tanner was winding down his activities in Philadelphia to make this move is evidenced by Carey & Hart's publication of the 1842 edition of his *New Universal Atlas*. Although Stauffer states that Tanner engaged in the engraving and publishing of maps and charts in New York, there are very few Tanner cartographic publications with a New York City imprint. One work that does have a New York imprint is the 1844 edition of his large U.S. map. Its imprint reads, "New York, Published by Henry S. Tanner." The map's copyright statement notes that it was "Entered according to Act of Congress in the year 1843, by H. S. Tanner in the Clerk's Office of the Southern District of New York." Earlier editions of the map were entered in the clerk's office of the eastern district of Pennsylvania.

Editions of Tanner's *Map of the United States of Mexico* were published in 1846 and 1847 and were also registered for copyright in the clerk's office of the southern district of New York. This map, first published in Philadelphia in 1825, was an enlargement of the southwest portion of his large map of North America, which first appeared in the *New American Atlas*. A modification of Tanner's map of Mexico was published in twenty-four editions between 1846 and 1858 by John Disturnell.[9] In view of the apparent close association between Tanner and Disturnell, we may infer that Disturnell published this map with Tanner's permission.

It is unknown why Tanner gave up a seemingly prosperous cartographic publishing business in Philadelphia and relocated in New York City. The most likely explanation may be that he was reluctant to shift his reproduction facilities from engraving to lithography. The lat-

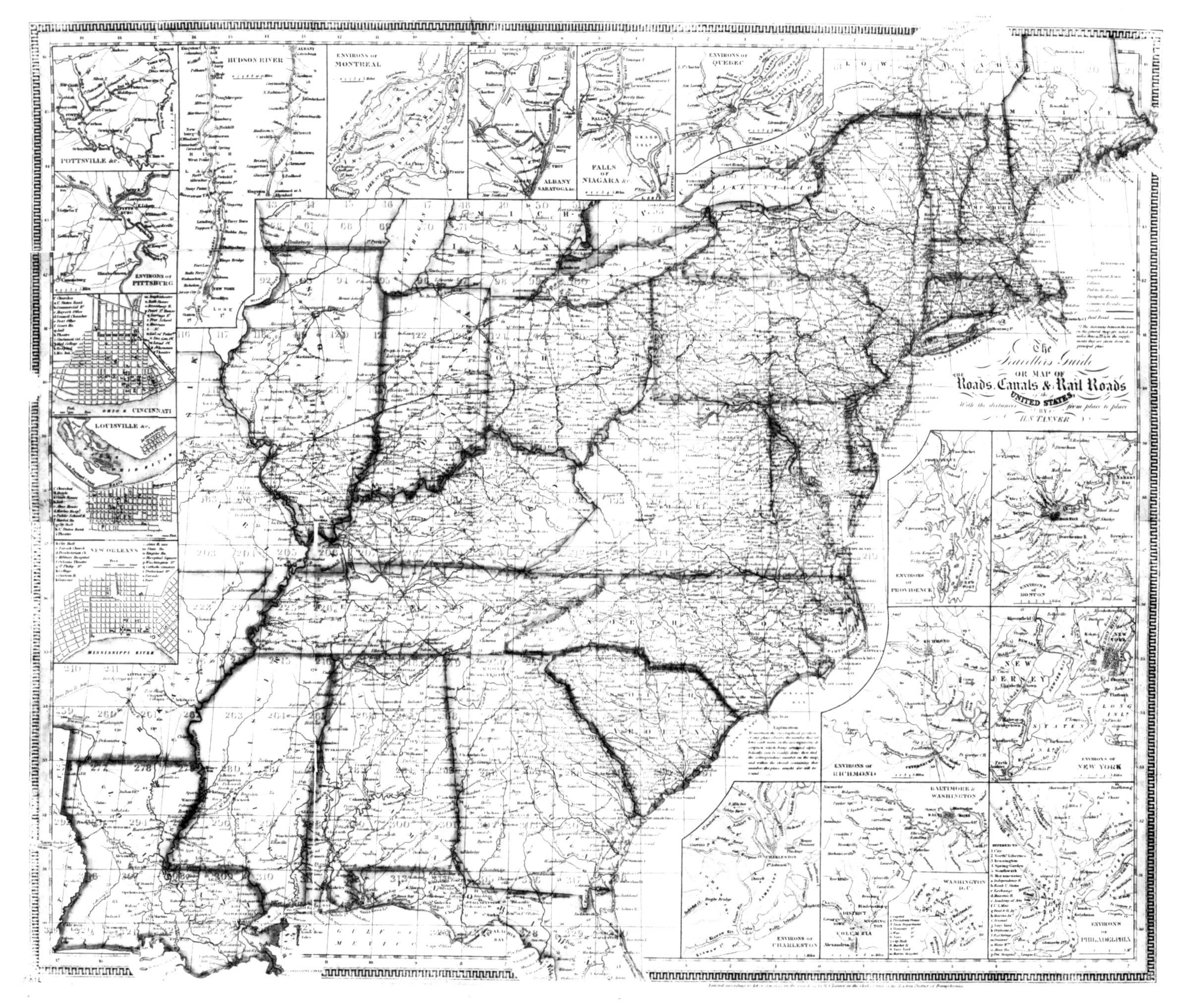

Fig. 13–7. An early example of the inclusion of railroads on a U.S. map. This map was published by Tanner in 1834.

ter technique, invented by Alois Senefelder in Munich in 1798, was introduced in the United States around 1819 or 1820. It was first employed for printing maps in 1822, but it was another five years before lithographic cartography established a limited foothold in the United States. During the next two decades, engraving maintained its supremacy, largely because lithography still relied principally on the use of heavy stones. Images were drawn on these stones, usually in reverse, with greasy ink or crayons. It was not until around 1845, when the transfer process was adopted and zinc printing plates replaced the stones, that lithography's supremacy over engraving in map reproduction was assured.

When Tanner relocated in 1843, he was fifty-seven years old and apparently disinclined to reorganize his map publishing company. After 1835, in fact, many of his maps and atlases were revised editions of works issued previously. He seems to have focused on descriptive texts after 1834, when the first edition of his *American Traveller* was published. He may have continued to prepare guidebooks, which were probably published by Disturnell or some other established New York City firm. Whether he continued this activity until his death in Brooklyn on May 18, 1858, is not known.

During his career, Tanner did much to strengthen the foundations of U.S. commercial cartography and contributed much to America's golden age of map publishing. In 1935 W. L. G. Joerg observed that although Tanner was "a commercial map maker, he was truly a scientific geographer. He produced, for his time, the outstanding map representations of the territory of the United States based on a critical study of the source material."[10]

Notes

1. David McNeely Stauffer, *American Engravers upon Copper and Steel* (New York, 1907) 1:265.
2. *United States Gazette* (Philadelphia), Sept. 1823.
3. *National Advocate* (New York City), Aug. 25, 1824.
4. Jared Sparks, "A New American Atlas, Containing Maps of the Several States of the North American Union," *North American Review* 18, n.s. 9 (1824): 387–88.
5. Alexandre Barbie du Bocage, "Review of *New American Atlas*, by Henry S. Tanner," *Société de Géographie Bulletin* 7 (1827): 223–36.
6. New York Mercantile Advertiser, Aug. 26, 1824.
7. A. H. Brué, "Rapport fait a la Société de Géographie, dans sa séance, du 18 fevrier, 1831, sur une carte des Etats Unis, par M. Tanner," *Société de Géographie Bulletin* 15 (Mar. 1831): 93–116.
8. Stauffer, *American Engravers*, 265.
9. For a detailed analysis of the relationship between Tanner's and Disturnell's maps see Lawrence Martin and Walter W. Ristow, "John Disturnell's Map of the United Mexican States," in *A La Carte* (Washington, D.C., 1972), 204–21.
10. W. L. G. Joerg, "Henry S. Tanner of Philadelphia: His Place in American Geography, 1815–1850," *Association of American Geographers Annals* 25 (March 1935): 46.

14. *Robert Mills's Atlas of South Carolina*

A significant first in American cartography was the 1825 publication of Robert Mills's *Atlas of the State of South Carolina,* the earliest atlas of an individual state. A slightly revised edition of this atlas was published by Mills in 1838, and facsimile editions were issued in 1938 and 1965. No other atlas of South Carolina has been published since this atlas's issue more than a century and a half ago. Atlases of Maine and New York were published in 1829, and more than thirty-five years elapsed before any other state atlases were published.

Although Mills's atlas was in this sense unique, it was intimately associated with the internal improvement trend that flourished in the United States during the first several decades of the nineteenth century. There was a great westward migration during this period, which increased the pressures on individual states to establish transportation and communication routes between the frontier and the seaboard cities. State governments responded by initiating programs to improve the navigability of rivers and to construct canals and roads. A number of states appointed official engineers or surveyors and boards of public works to plan and direct internal improvement projects.

Basic to all such programs were accurate and up-to-date maps and, beginning in 1789 and continuing through the next four or five decades, maps of each state were compiled, some in several editions. Because few budgets could support coordinated surveying and mapping programs, most of the state maps, as we have seen, were compiled by private initiative, usually with some form of legislative encouragement or subsidy. A frequently used procedure was to enact a law requiring such second-level administrative jurisdictions as counties, towns, or districts to prepare maps of their confines and to deliver copies within a specified time period to the secretary of state or to the state engineer or surveyor. From the manuscript county maps the state surveyor, commercial publisher, or private individual would compile the state map.

Only four states—New York, Pennsylvania, South Carolina, and Virginia—produced state maps as official projects during the early decades of the nineteenth century. In 1830, in his evaluation of contemporary cartographic materials, Henry Tanner noted that "the Legislature of South Carolina, in imitation of the laudible examples [of Pennsylvania and Virginia] has produced a map of the state, that must confer lasting honour on the promoters of the work as well as the state at large."[1] Because preparation of Mills's atlas was intimately related to John Wilson's 1822 map of South Carolina, it is pertinent to review its inception and development.

Maps of twelve states had been published before a survey of South Carolina was considered. The initial proposal, contained in a memorial prepared by George Blackburn, a mathematics professor at South Carolina College, was presented to both houses of the legislature in December 1815. The memorial was favorably received by the select committee to which it was referred. The committee's report acknowledged

> that the advantages to be derived from a correct map of the State, are so numerous, and universally admitted as to render unnecessary a particular detail. To every portion of our citizens a correct knowledge of the relative situation and extent of our Districts, rivers, swamps, mountains, roads,

> and Towns must be of great utility. To military men a *minute* knowledge of the surface of the ground is highly important. . . .
>
> To those who are engaged in Agriculture and commerce, this knowledge is of scarce less importance. But above all it is important to the members of the Legislature, to whom is committed the power of forming congressional and Judicial Districts—of making roads and canals—of opening rivers & swamps—in a word of superintending and controlling all the arrangements of the State, both civil & military.
>
> These obvious advantages when added to those general considerations, which have influenced all enlightened nations in their efforts to extend the bounds of knowledge and science, cannot fail, it is confidently believed, to induce the Legislature to adopt some measure for procuring a correct map of the State.[2]

Both houses of the legislature approved the proposal, with the recommendation that the map project be under the personal supervision of the governor. An annual appropriation of five thousand dollars was authorized, and Blackburn was engaged to carry out astronomical, geodetic, and topographic surveys.

A report on Blackburn's first-year activities was considered by a joint committee of the state legislature in December 1816. The committee agreed "that Mr. Blackburn has collected much information which may be found important in forming a map of the state." The members concluded, however, that

> your committee are under an impression that an actual survey of every part of this State is essential to the formation of such a map as the Legislature appears to have desired, they therefore recommend that there be taken an actual survey of the Judicial Districts. . . . That the Governor be authorised to appoint a surveyor to each District . . . and further that when the above materials shall be ready the Governor be authorised to employ a draftsman to form a map of the whole.[3]

Blackburn's survey was unique, and few other state maps compiled during the early decades of the nineteenth century were laid out on such a precise mathematical base.

This project and related internal improvement programs had the support of a succession of energetic and farsighted governors. David R. Williams was in the executive office when Blackburn conducted his survey. Governor Andrew Pickens, Jr., who succeeded Williams, carried out the directive of the 1816 legislature by appointing a number of district surveyors. Pickens's report to the house and senate asserted: "[I] spared no exertion . . . and visited every district, believing that, by this means I would be able, the more readily, to employ the persons best qualified" to conduct the district surveys.[4]

Nineteen surveyors were ultimately engaged to prepare maps of the twenty-eight districts in the state. Edgefield was the only district completed in 1817. Its surveyor, William Anderson, mapped Barnwell, the adjoining district, in 1818; Chester, surveyed by Charles Boyd, and Marion, surveyed by Thomas Harlee, were also completed that year. Although Chesterfield District, by John Lowry, was the only survey finished in 1819, that year witnessed great activity, with seventeen surveyors in the field; the twenty-one maps which they prepared were completed in 1820. The final map, of Sumter District, is dated 1821. This large district in the east central part of the state within the poorly drained coastal plain "defeated two surveyors before it was successfully mastered by Stephen H. Boykin."[5]

All the surveyors were employed on contract, and their payments ranged from seven thousand dollars for small districts to eighteen thousand dollars for large ones.[6] Stephen Boykin, Marmaduke Coate, and Thomas Harlee each surveyed three districts; two surveys each were conducted by Thomas Anderson and the team of Charles Vignoles and Henry Ravenel; and the balance of the surveyors each accounted for one map, probably of their home districts. The surveyors of Darlington and Marlborough districts were not identified.

Several years ago Charles E. Lee, director of the South Carolina Archives Department, discovered in the archives sixteen of the original district survey maps which had remained unidentified for almost a century and a half. The maps are substantially diverse in cartographic and drafting techniques. All show considerable detail in river and stream systems, cities and towns, and rural land ownership. Some include marginal notes, legends, and titles. Regrettably, none of the field books submitted by the surveyors with their maps have survived. Also preserved in the archives is "a little manuscript volume entitled 'James M. Elford's Astronomical Observations of S.C. Made by Order of His Excellency Governor John Geddes 1820. . . .' Research reveals it as ground work (determination of latitudes and longitudes for key points) for the map of the state."[7] Along with Blackburn's calculations, this volume provided the control data for the map of South Carolina. The data were also utilized in the extensive internal improvement program launched by the state in 1819. The responsibility for compiling and drafting the map of South Carolina from this material and the twenty-eight manuscript district maps was assigned to John Wilson, the state's civil and military engineer.

The 1821 report of the South Carolina Board of Public Works, which includes calculations for a number of places in the state, notes that

> immediately after the adjournment of the last Session of the Legislature, measures were taken for completing the Map of the State. . . . The survey of Sumter district had been so inaccurately made, that an entirely new one would be required; and in several other districts some errors required correction. . . . With these additional materials, and some verbal corrections, the map was placed in the hands of Major Wilson, who undertook to redraft it and to superintend the engraving in Philadelphia, and the correction of the plates, as the artists proceeded in the work. With Henry S. Tanner . . . a contract was made for engraving the plates and delivering fifty-one copies of the map, colored and mounted. These are expected to be received during the present session of the Legislature.
>
> To render the work as correct as possible, it was desirable, that the longitude of the capital of the State, should be correctly ascertained. For this purpose, the instruments in the possession of the Board were placed in the hands of the very able Professor of Mathematics and Astronomy, in the South-Carolina College. The college Observatory was found so defectively constructed and so unsteady, that no reliance could be placed on observations made in it. A small but very useful observatory was therefore erected at the College, to receive the astronomical circle and other instruments. With these advantages, Professor Wallace made his very accurate observation on the solar eclipse of the 27th August last.[8]

The map was finally published as the *Map of South Carolina, Constructed and Drawn from the District Surveys, Ordered by the Legislature, by John Wilson, late Civil and Military Engineer of So. Cara. Engraved by H. S. Tanner, Philadelphia. . . . 10th Day of April, 1822.* "The Astronomical Observations by Geo. Blackburn & I. M. Elford" are acknowledged, but no credit is given to Professor Wallace.

The map was an expensive project. In addition to the initial five thousand dollars the legislature appropriated in 1816 to begin the district surveys, another five thousand was expended in 1817, with an increase of nine thousand for each of the subsequent two years. The cost of completing the surveys in 1820 approximated twenty-five thousand dollars. With additional costs for engraving and printing, the total expenditure for the state map was upwards of ninety thousand dollars.[9]

At least twenty-five hundred copies of Wilson's map were printed. The journal of the South Carolina house for December 19, 1822, noted that twenty-two hundred copies of the map were in the possession of the superintendent of public works. The house voted to request the governor to submit one copy to each of the other state and territorial governors and authorized a reduction in the map's price to five dollars.[10]

In 1820, to direct and coordinate the extensive internal improvement program with which the state map was associated, the South Carolina legislature established a Board of Public Works. Elected members of the board included Joel R. Poinsett as president, Abram Blanding, John Lyde Wilson, and Robert Mills. Born in Charleston, South Carolina, on August 12, 1781, Mills received his early education at Charleston College and more advanced instruction in Charleston and Columbia under the tutelage of the Irish-born architect, James Hoban. In 1798 Hoban moved to Washington, where he was engaged in constructing official buildings. Mills joined him in 1800, and there became acquainted with Thomas Jefferson, spending much of his time at Monticello drafting plans and reading in Jefferson's extensive library. In 1803 Mills became an assistant to Benjamin Latrobe, the newly appointed architect of national buildings. During the five years he was associated with Latrobe, he received further valuable architectural and engineering training.

Mills married Elizabeth Barnwell Smith in 1808 and shortly thereafter set up his own architectural practice in Philadelphia. He designed a number of public and private buildings in that city before moving to Baltimore in 1812, where he remained until 1820. His achievements in Baltimore include the Washington monument as well as a number of residences, churches, and public buildings. While based in Baltimore, he also designed homes and office buildings in Richmond, Virginia.

After moving back to his home state in 1820, Mills was elected to the South Carolina Board of Public Works. During the next decade he was occupied with various engineering and architectural projects, including constructing roads and canals and designing public buildings. After the board's functions were transferred in 1822 to the superintendent of public works, Mills continued working on state and private projects on a contract basis. Public buildings designed by Mills are still in use in Charleston and Columbia.

As a practicing engineer and architect and a member of the Board of Public Works, Mills was deeply interested in and strongly supportive of the map project. He also displayed an early concern for making the district maps more widely available. It is likely that the idea for a state atlas incorporated in the 1821 report of the board, may have been proposed by Mills. In discussing the state map, the report notes that because of possible errors and anticipated changes in the landscape, it would be essential to revise the map periodically. For this reason the board opposed transferring the copyright of the map to a private publisher:

> Should the State keep the copy right, the means of correcting and improving the map to correspond with the improvements of the country, appear to the Board to be both cheap and easy. Let all the district maps be engraved; place in the hands of the commissioners of the roads of each district an atlas containing all the district maps, and several copies of their respective districts. On these, as new settlements arise, new roads are laid out, or other objects present themselves, they can be delineated, and the errors of former surveys can be corrected.[11]

In its report for 1822 the Board of Public Works returned to the matter of the district surveys. "After the compilation of the general map was completed," it was noted, "these surveys were carefully put up, and are subject to the order of the legislature. It was suggested in the last annual report, that these surveys might be worthy of preservation and multiplication by having them engraved."[12] The board went so far as to determine that the cost of printing five hundred copies of each district map would be $1,780, and that an atlas containing the twenty-eight maps could be sold for $10. Because the legislature did not appropriate funds for this proposed atlas and because of the discontinuation of the Board of Public Works, Mills apparently decided to undertake the atlas project himself. In this he had the cooperation of his former board associate Abram Blanding, who had been appointed to the newly created post of superintendent of public works. In considering Blanding's report for 1823, a committee of the house observed:

> By the report of the Superintendent it appears that he had entered in to a provisional contract with Mr. Robert Mills for publishing the District Surveys, by which the State will obtain at least twelve Atlasses without any expence. Your Committee beg leave to recommend that this contract be sanctioned and that the Superintendent be directed to subscribe for Fifty Copies of the Atlas, which would be the means of enableing the State for the trifleing sum of Six hundred dollars to furnish an Atlas to each Board of Commissioners of Roads.[13]

The recommendation was approved by both houses of the legislature, and copies of the manuscript district surveys were turned over to Mills.

During the next eighteen to twenty months, Mills redrafted the manuscript maps. This was no simple task because, as noted earlier, the originals were prepared by nineteen different surveyors with varying degrees of skills and aptitudes. Some maps drawn at larger scales had to be converted to the 1:125,000 standard adopted for the atlas. The uniformity of cartographic style and lettering suggests that all the district maps in the atlas were personally drawn by Mills. With one or two exceptions, the original surveyors are given credit. All the maps, however, carry the inscription "Improved for Mills' Atlas" and the date 1825.

It is uncertain how much Mills improved on the manuscript surveys. Certainly he was thoroughly familiar with the major physical features of the state and with the internal improvements completed after the original surveys were made between 1817 and 1821. He may possibly have conducted personal field checks in some districts and had other maps checked by local surveyors. Comparing the district maps in the atlas with their corresponding surviving manuscript maps reveals some minor differences in cartography and orthography.

As Tanner engraved Wilson's map of South Carolina, he also received the contract to engrave the district maps. Mills probably forwarded the redrafted maps to him as they were completed. The earliest maps may have been sent to Tanner around the beginning of 1825, with the final ones not delivered until a year or so later. Although all the district maps are dated 1825, the atlas was not published until early the following year. On December 15, 1825, the journal of the South Carolina senate records that it was "resolved that the Superintendent of Public Works be authorized to pay to Robert Mills fourteen dollars for each of the fifty copies of his Atlas instead of twelve dollars heretofore authorized and subscribed for in behalf of the State."[14] Because the legislature only met for three or four weeks at the end of each year, Mills was not able to present his atlas to the senate until November 30, 1826. The presentation letter, recorded in the senate journal, reads:

> To the honorable the President of the Senate of South Carolina
>
> Sir,
>
> I have the honor to present for the acceptance of the hon. the senate of South-Carolina, this copy of my Atlas of this State, which accom-

panies this letter; and to request that you will be pleased to inform the senate of my high consideration and respect. Very respectfully, sir,

I have the honor to salute you,

Robt. Mills,

Columbia, Sept. 29, 1826[15]

In the early fall of 1826, eighty copies of the atlas had been delivered to the state, for which Mills received twelve hundred dollars. Mills also presented to the senate on December 1 a petition "offering to provide for the distribution of the district maps among the people, at such prices as may enable every citizen to possess the map of his own district."[16] The petition was referred to a special committee, and we may assume that Mills's request was granted. Also during the 1826 session, Mills presented to the legislature "his large Map of South Carolina and . . . Statistics of South Carolina."[17] The *Statistics of South Carolina, 1826* volume, which includes more than eight hundred pages of descriptive, historical, and statistical information about the state, was compiled by Mills as a supplement to his atlas.[18] Since there appears to be no extant copies of Mills's large map of South Carolina, it is possible that the presentation copy was in manuscript. Whether manuscript or printed, the map as well as the copies of the atlas presented to the house and senate were probably destroyed by General William T. Sherman when he burned the South Carolina State capitol during the Civil War.

Mills's atlas includes twenty-eight district maps, all at the uniform scale of 1:125,000 (Figs. 14–1 and 14–2). Among the physical features mapped are bays, coastal features, rivers, creeks, branches, falls, cypress ponds, fishponds, bluffs, gullies, lakes, islands, mountains, swamps, mineral springs, and woods. Names of property owners are given in all districts, making the atlas particularly useful today for genealogical and historical research. The maps are also rich in names of cities, towns, and villages and in such cultural features as battlefields, blacksmith shops, bridges, cabinet shops, canals, causeways, churches, schools and colleges, cotton factories, court houses, doctors' offices, ferries, fords, grist and saw mills, mines and quarries, dams, potteries, public houses and inns, tan yards, and rice fields. Maps in some extant copies of the atlas have a pink wash around district borders. The plates are printed in black ink and are arranged alphabetically by district names, although this pattern is not adhered to rigidly. The map of Charleston District, for example, is presented first in some extant volumes (Fig. 14–3). Because the district maps are of varying sizes, nine of the twenty-eight plates are folded, while the other nineteen appear on double pages. All plates are fastened to tabs and bound within covers measuring 56 by 38 cm. From the few surviving volumes in their original binding, it appears that the most common format consisted of board covers with red leather spines and corners.

On the title page of the atlas there is a small map of the state bordered on three sides by descriptive information and chronological and statistical tables (Fig. 14–4). Above the map is printed the title *Atlas of the State of South Carolina, Made under the Authority of the Legislature; Prefaced with a Geographical, Statistical and Historical Map of the State.* Mills is described as an engineer and architect, and the atlas is dedicated "To the Honorable the Senate and House of Representatives of South Carolina." The map measures 24 by 29 cm. and is at the approximate scale of 1:1,650,000. The imprint under it reads, "Published by F. Lucas, Jr. Baltimore for Mills' Atlas. B. T. Welch & Co. Sc." The map was printed by John D. Toy, also of Baltimore.

It is curious that Mills chose Fielding Lucas to provide the map for the Atlas's frontispiece and title page instead of Tanner, who had done all the other maps. From his eight-year residence in Baltimore, Mills undoubtedly

Fig. 14–1. The map of Richland District from Mills's atlas. Based on surveys by Marmaduke Coate, it includes South Carolina's capital city, Columbia.

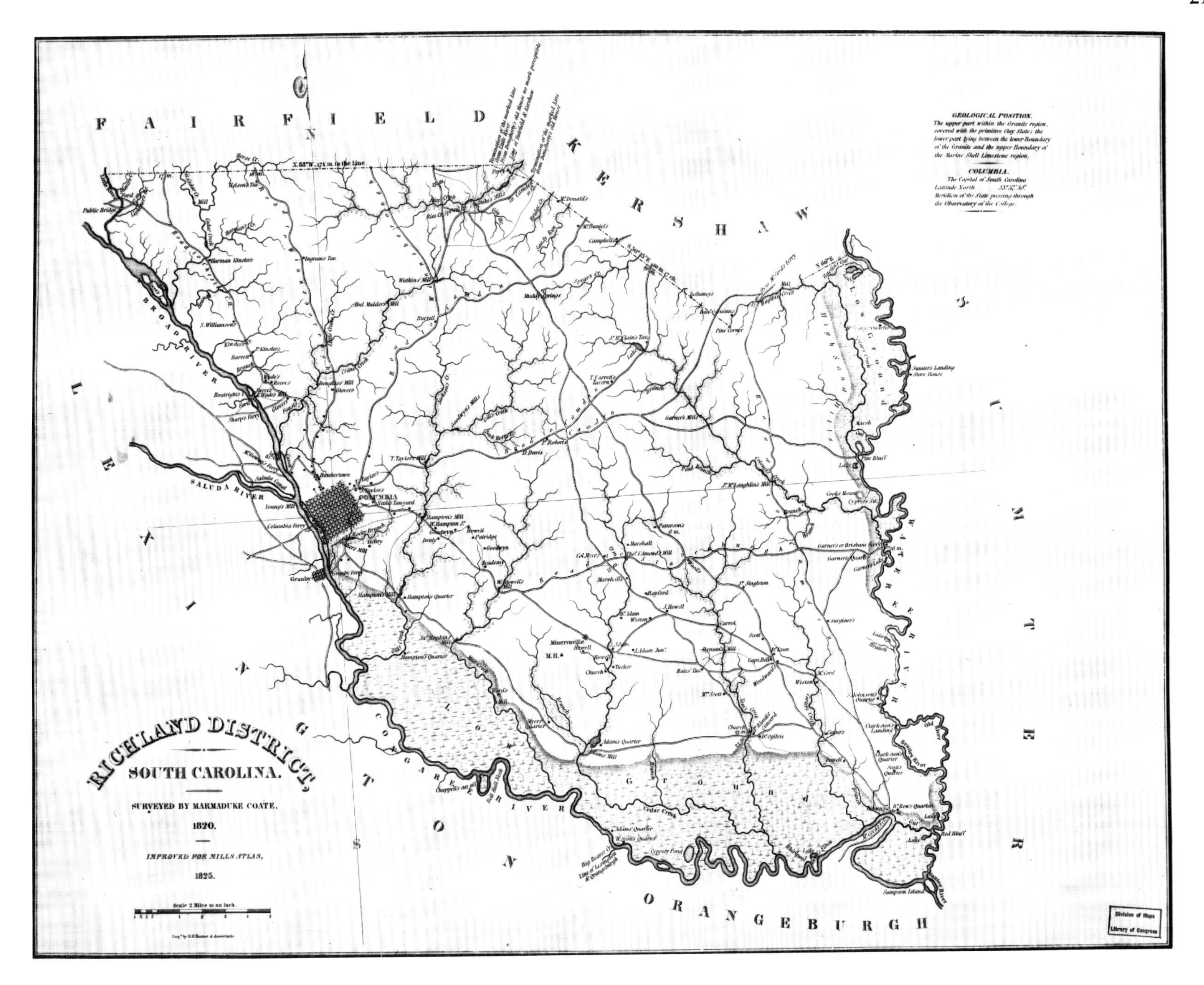
RICHLAND DISTRICT,
SOUTH CAROLINA.
SURVEYED BY MARMADUKE COATE,
1820.
IMPROVED FOR MILLS ATLAS,
1825.
Scale 2 Miles to an Inch.
GEOLOGICAL POSITION.
The upper part within the Granite region, covered with the primitive Clay Slate; the lower part lying between the lower Boundary of the Granite and the upper Boundary of the Marine Shell Limestone region.
COLUMBIA.
The Capital of South Carolina
Latitude North 33°.57'.30".
Meridian of the State passing through the Observatory of the College.
FAIRFIELD
KERSHAW
SUMTER
ORANGEBURGH
LEXINGTON
BROAD RIVER
SALUDA RIVER
CONGAREE RIVER
WATEREE RIVER
COLUMBIA
Granby
Division of Maps
Library of Congress

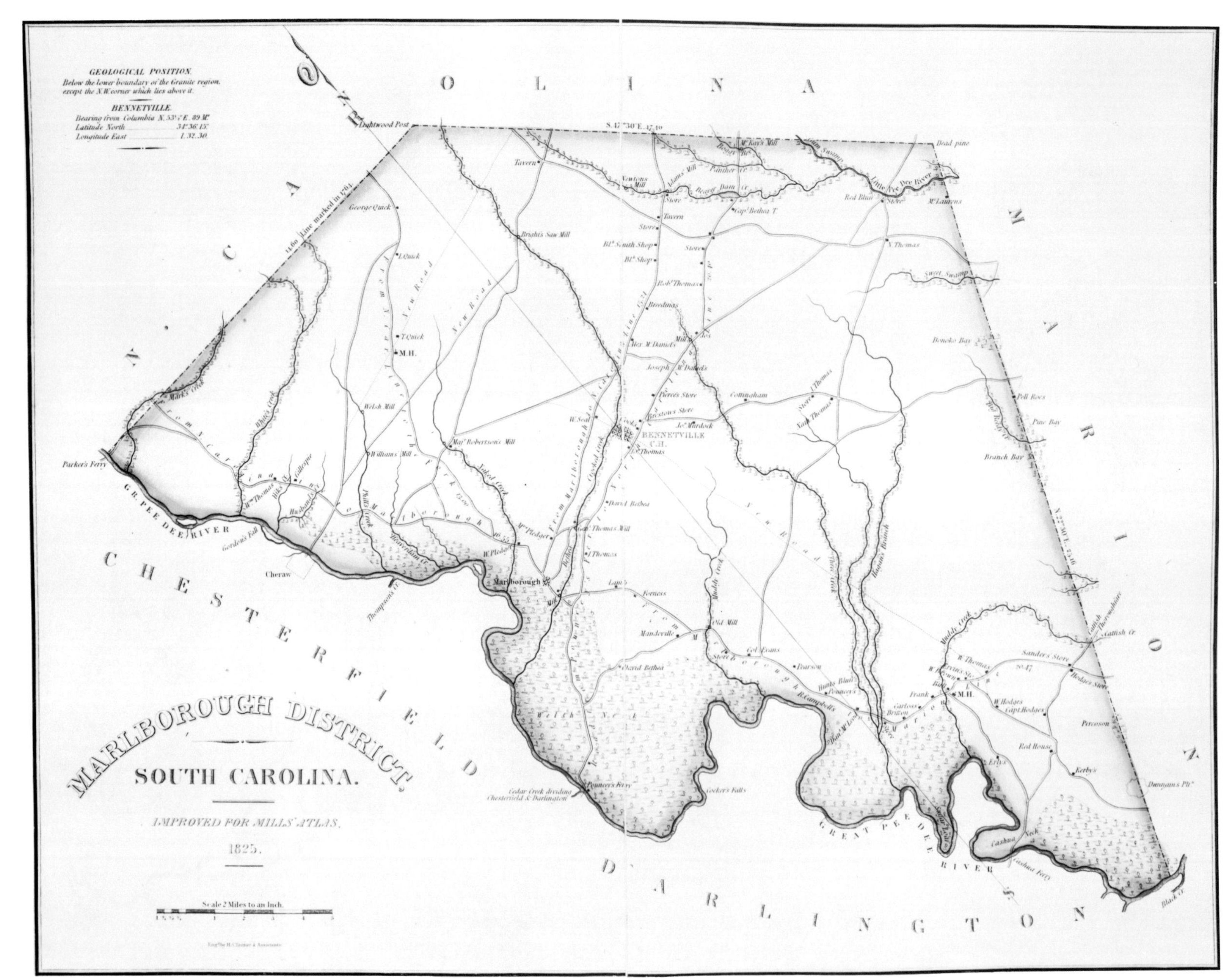

Fig. 14–2. The map of Marlborough District in Mills's atlas. It includes no surveyor's name.

Fig. 14–3. A portion of the Charleston District map showing the city of Charleston and its vicinity.

ATLAS OF THE STATE OF SOUTH CAROLINA,

MADE UNDER THE AUTHORITY OF THE LEGISLATURE;

PREFACED WITH A GEOGRAPHICAL, STATISTICAL AND HISTORICAL MAP OF THE STATE.

BY ROBERT MILLS, OF SOUTH CAROLINA, P. A.

ENGINEER AND ARCHITECT.

TO THE HONORABLE THE SENATE AND HOUSE OF REPRESENTATIVES OF SOUTH CAROLINA, THIS WORK IS RESPECTFULLY INSCRIBED BY THE AUTHOR.

Situation, Boundaries and Extent.

SOUTH CAROLINA lies between 32° 4′ 30″ and 35° 12′ North latitude, and 1° 30′ and 6° 54′ West longitude, from the Capitol at Washington, (or 78° 25′ and 83° 49′ West longitude from Greenwich.) It is bounded on the north and north-east by North Carolina, south-east by the Atlantic Ocean, and south-west by Georgia, from which it is separated by the Savannah, Tugaloo, and Chatuga rivers. It is 189 miles long, and 160 broad; containing 30,213 square miles, or 19,455,680 acres.

Face of the Country, Soil, Health, &c.

In relation to the above particulars, the state may be considered under seven divisions. 1st. The Sea Islands. These extend along the coast, from Winyaw Bay to the Savannah River, and are of comparatively recent formation. They are generally composed of sand, which in many parts is mixed with a rich vegetable loam of great fertility, producing abundantly black seed, or Sea Island cotton, corn, &c. The sweet and sour orange grow here with but occasional injury from the frost of very cold winters. These islands are surrounded by salt water, and are therefore generally healthy. They form delightful residences amidst abundant wealth.

2d. The tide flows up the rivers of this state, from thirty to forty miles from the ocean, and about one third of this distance only the water is salt. The immense marshes bordering on these rivers, are therefore capable of being irrigated every flood tide with fresh water. These constitute the valuable rice lands called *tide swamp*, which [illegible]

Rivers and Canals.

[illegible]

Chief Towns.

[illegible]

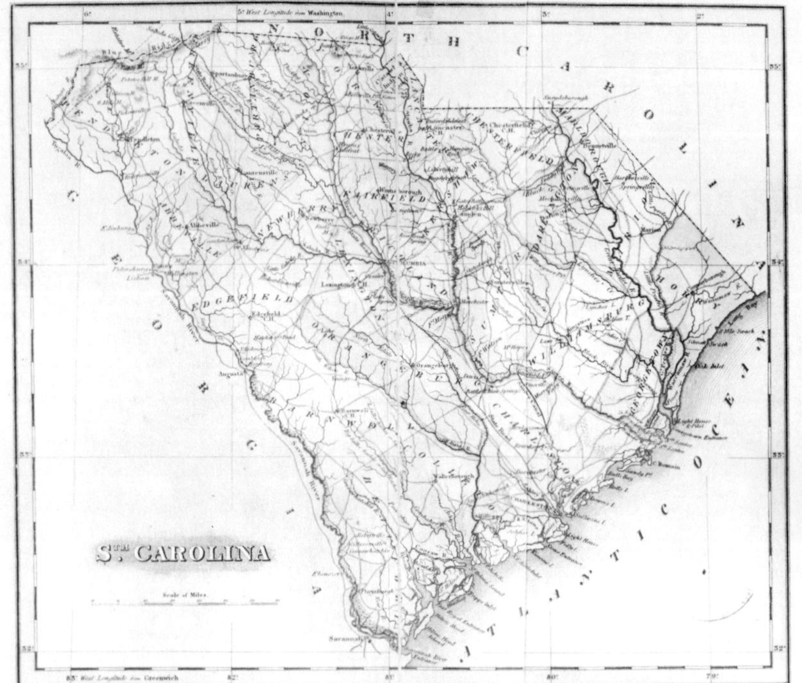

Published by F. Lucas Jr. Baltimore for Mills Atlas

Civil Division and Population.

There are 28 law districts, 22 equity districts, 44 fiscal and election districts, and nine congressional districts in the State.

DISTRICTS.	TOWNS.	WHITES.	SLAVES.	FREE BLACKS.	TOTAL.
Charleston	Charleston	19,396	57,221	3,595	80,212
Abbeville	Abbeville, Vienna	13,488	9,615	64	23,167
Barnwell	Barnwell	8,162	6,336	232	14,730
Beaufort	Coosawatchee, Beaufort	4,679	27,339	181	32,199
Chester	Chesterville	9,611	4,542	56	14,189
Chesterfield	Chesterfield, Cheraw	4,412	2,062	171	6,645
Colleton	Walterborough, Jacksonborough	4,341	21,770	293	26,404
Darlington	Darlington, Society Hill	6,407	4,473	69	10,949
Edgefield	Edgefield, Hamburg	12,864	12,198	57	25,119
Fairfield	Winsborough	9,378	7,748	48	17,174
Greenville	Greenville	11,017	3,423	90	14,530
Georgetown	Georgetown	1,830	15,546	227	17,603
Horry	Conwayborough	3,568	1,434	23	5,025
Kershaw	Camden	5,628	6,692	112	12,432
Lancaster	Lancasterville	5,848	2,798	70	8,716
Lexington	Lexington, Granby	5,267	2,801	15	8,083
Laurens	Laurensville	12,735	4,878	49	17,682
Marion	Marion	6,662	3,463	86	10,201
Marlborough	Bennetville	3,230	3,053	142	6,425
Newberry	Newberry	10,177	5,749	178	16,104
Orangeburg	Orangeburg	6,760	8,829	64	15,653
Pendleton	Pendleton	22,140	4,715	167	27,022
Richland	Columbia	4,499	7,627	195	12,321
Spartanburg	Spartanburg	13,655	3,308	26	16,989
Sumter	Sumterville, Statesburg	8,844	16,143	382	25,369
Union	Unionville	9,786	4,278	62	14,126
Williamsburg	Kingstree	2,795	5,864	57	8,716
York	Yorkville	10,251	4,590	95	14,936
In 1820, by census		237,460	258,475	6,806	502,741

List of Governors.

UNDER THE PROPRIETARY GOVERNMENT.

1670. William Sayle.
1671. Joseph West.
1671. John Yeamans.
1674. Joseph West.
1682. Joseph Morton.
1684. Joseph West.
1684. Richard Kirk.
1684. Robert Quarry.
1685. Joseph Morton.
1686. James Colleton.
1690. Seth Sothell.
1692. Philip Ludwell.
1693. Thomas Smith.
1694. Joseph Blake.
1695. John Archdale.
1696. Joseph Blake.
1700. James Moore.
1703. Nathaniel Johnson.
1709. Edward Tyne.
1710. Robert Gibbs.
1712. Charles Craven.
1716. Robert Daniel.
1719. Robert Johnson.
1719. James Moore.

UNDER THE TEMPORARY REPUBLICAN GOVERNMENT.

1719. Arthur Middleton.

UNDER THE ROYAL GOVERNMENT.

1721. Francis Nicholson.
1725. Arthur Middleton.
1730. Robert Johnson.
1735. Thomas Broughton.
1737. William Bull.
1743. James Glen.
1756. William H. Littleton.
1760. William Bull.
1762. Thomas Boone.
1765. William Bull.
1766. Charles G. Montague.
1769. William Bull.
1775. William Campbell.

UNDER THE REPUBLICAN GOVERNMENT.

1775. John Rutledge.
1778. Rawlins Lowndes.
1779. John Rutledge.
1782. John Mathews.
1783. Benjamin Guerard.
1785. William Moultrie.
1787. Thomas Pinckney.
1789. Charles Pinckney.
1790. Charles Pinckney.
1792. Arnoldus Vanderhorst.
1794. William Moultrie.
1796. Charles Pinckney.
1798. Edward Rutledge.
1800. John Drayton.
1802. James B. Richardson.
1804. Paul Hamilton.
1806. Charles Pinckney.
1808. John Drayton.
1810. Henry Middleton.
1812. Joseph Alston.
1814. David R. Williams.
1816. Andrew Pickens.
1818. John Geddes.
1820. Thomas Bennet.
1822. John L. Wilson.
1824. Richard J. Manning.

Military System.

The militia of the state are organized according to the acts of Congress on this subject. The Governor is the commander-in-chief. There are 5 major-generals, commanding as many divisions; 10 brigadier-generals, commanding as many brigades; and 165 colonels, lieutenant-colonels, and majors, having command of the 43 regiments. There are also attached to each brigade a regiment, battalion or squadron of artillery and cavalry, commanded by their own colonels or majors. The staff consists of one adjutant and inspector-general, one quarter-master-general, 5 deputy adjutant, 5 deputy inspector, and 5 deputy quarter-master-generals; besides the usual assistant deputies, &c. The whole effective force of the state is as follows:

Of Infantry and Riflemen	34,000
Cavalry	4,200
Artillery	2,100
Total	40,300

Government.

[illegible]

Education.

[illegible]

Chief Towns continued.

[illegible]

Commerce and Manufactures.

[illegible] States, it amounted to 67,914 tons; in foreign vessels, 18,211. The average annual nett amount of revenue derived from this source by the United States, in the years 1815, '16, '17, and '18, exceeded $1,237,000. In the last year, the duties collected in Charleston amounted to only $736,000, owing to the vast quantity of cotton and rice shipped to the Northern ports, which pay no duties, and which go to swell the amount of exports at those ports. Very little attention is paid to manufactures in the state, except in the domestic way.

Chronological History of South Carolina.

1663. Charles II. grants to the Earl of Clarendon, and seven others, all lands lying between the 31st and 36th degrees of North latitude, and extending West to the South or Pacific Ocean.
1669. The celebrated John Locke frames a constitution and body of laws for Carolina; never accepted by the people.
1670. First settlement in South Carolina takes place, under William Sayle at Port Royal.
1671. Old Charleston located, west side of Ashley River, still called Old Town.
1673. The first invasion of the province by the Spaniards, at Port Royal.
1680. Present city of Charleston founded. In this year also, the first Indian war took place, which nearly ruined the colony.
1682. Second settlement of Port Royal, by a colony from Scotland, under Lord Cardross.
1682. The Presbyterians form the first Church; the Episcopalians the second, and the Baptists the third.
1686. The Spaniards dislodge the Scots settlements at Port Royal.
1690. The first public act of the people passed against the oppression of the Proprietary Government. They petition parliament to be taken under the royal protection.
1692. The first Episcopal and Congregational Churches built.
1694. Rice first introduced by Thomas Smith.
1700. First public library established, begun by the liberality of Dr. Bray.
1700. First appearance of the small-pox and yellow fever.
1701. The French Protestant Church founded.
1702. The first military expedition undertaken against the Spaniards, by Governor Moore.
1703. The war with the Apalachian Indians takes place under Governor Moore.
1706. The French and Spaniards make a combined attack on Charleston, but defeated. Col. Palmer, with 300 men, invades the Spanish settlements in Florida, with success.
1710 & 1712. Laws for founding and erecting a free school in Charleston, passed.
1712. A dangerous conspiracy of the Indians in North Carolina, defeated by Col. Barnwell.
1715. The Yemassee war takes place, which threatens the destruction of the province; but the Indians were defeated by Governor Craven at the Salt Catchers.
1719. The proprietors relinquish their right and interest in the province to the crown. At the same time the province is divided into North and South Carolina.
1719. About this time the pirates that infested the trade of the province to the West Indies, crushed and destroyed.
1723. Service first performed in the present St. Philip's Church.
1730. Newspapers began to be printed in the province.
1731. The first Presbyterian Church built.
1733. The celebrated George Whitfield visits the province.
1741. Indigo first planted by Miss Eliza Lucas.
1759. The German Protestant church built. Same year the Jews also built a synagogue.
1761. Service first performed in St. Michael's Church.
1763. The two Floridas ceded by Spain to Great Britain. Same year the boundaries of the back settlements were fixed.
1765. Stamp act passed in the British parliament.
1774. First act passed in the province to oppose with force the royal authority, 6th July.
1775. Prohibition to import British goods, February 1st.
1775. The people enter into an association to defend their liberties.
May. The first military force raised for the defence of the province; consisting of two regiments of foot, and a regiment of rangers.
Sept. First revolutionary military enterprise, the taking possession of Fort Johnson, by Colonel Motte.
Nov. 12. Commencement of hostilities in South Carolina, by the Tamar and Cherokee vessels of war.
Dec. The royal vessels of war driven out of the harbour by Col. Moultrie.
1776. An army marched into the back settlements against the royalists.
March. A temporary independent constitution formed by the act of the provincial Congress. The first in the Union.
June 28. The first blood spilt in defence of liberty in South Carolina, at Fort Moultrie, during the British naval attack on that place. The enemy repelled by Col. Moultrie, with severe loss to their navy, (then the terror of the world,) which results in giving undisturbed peace to South Carolina for three years.
June 30. The Cherokee Indians, excited by the British, begin their massacres on the frontiers. Plan of a simultaneous attack of Indians and tories, from the sea coast to the mountains, discovered, and defeated.
July 15. The Indians and tories defeated by Major Downes, in the back settlements.
1777. Scale of depreciation of the paper currency begun; [illegible]
1778. This and the last year the province enjoyed a lucrative commerce and profound peace. Lafayette arrives in this state.
Feb. 6. Treaty of amity, commerce and alliance, signed between France and the United States.
1779. General Lincoln takes command of all the southern forces, and establishes his first post at Purysburg.
1779. The British attacked and defeated at Port Royal by General Moultrie.
Colonel Pickens defeats the British and tories near Kettle Creek.
Feb. 28. General Ash surprised and defeated at Briar Creek (Savannah river,) by General Prevost.
May 11. Charleston invested by General Prevost. Count Pulaski defeats the British cavalry.
[illegible]
1790 June 3. The present constitution of the State adopted at Columbia, and went into operation on the first Monday of Jan. 1791.
1791. Feb. Right of primogeniture abolished, and an equal distribution of intestates' estates granted.
1791. The Roman Catholics organized into a church.
1791. Gen. Washington visits Charleston.
1800. County courts abolished, and district courts established, in the several counties or districts.
1801. Santee Canal finished, and goes into operation. The work began in 1793.
1802. Vaccination introduced by Dr. Ramsay.
1802. South Carolina College founded at Columbia. Carried into effect in 1804.
1807. Right of suffrage extended to all citizens, without requiring a property qualification.
1811. General free-school system established, for educating poor children, at the public expense of 37,000 dollars annually.
1812. Bank of the State established.
1812. War with England declared, June 18.
1814. Treaty of Peace signed at Ghent, December 24.
1817. System of internal improvement established.
1825. Mar. 8. General Lafayette visits this State after an interval of forty-seven years.

Religion.

The Methodists are the most numerous religious denomination in this State. Next to these are the Presbyterians; then the Baptists; after these the Episcopalians and German Lutherans. The Jews form a respectable society in Charleston and Columbia.

was acquainted with Lucas, who was active and prominent in civic affairs. His relations with Tanner were probably purely professional. In his informative paper "Baltimore Mapmakers," Frank N. Jones states, "Lucas had lost the bid for providing the South Carolina legislature with an atlas to his good friend, Robert Mills, but Mills engaged Lucas to provide the maps, which were engraved for him by B. T. Welch and printed by John D. Toy."[19] Jones errs, of course, in suggesting that Lucas provided the district maps for the atlas.

Mills's selection of the Lucas map of South Carolina was very likely based on convenience and economy. Cost was a primary concern to Mills, who personally assumed most of the financial risk of publishing the atlas. Although Tanner had engraved Wilson's large map of the state, to reduce it to the size desired by Mills and to prepare an engraving would have been quite costly. A reduced map of South Carolina, derived wholly from Wilson's map, was joined with a map of North Carolina on one plate in Tanner's *New American Atlas*, but to issue the South Carolina map separately would also have necessitated reengraving. Moreover, this map of the two Carolinas was at the scale of 1:1,300,000. Much larger than the Lucas map, it would have limited the amount of descriptive data that could have been included on the two pages of the atlas. In addition, Lucas had used the map in his 1823 *General Atlas Containing Distinctive Maps of All the Known Countries in the World*. With no need for further compilation and engraving, Lucas no doubt was able to supply Mills with copies of the map at a modest cost.

Actually, two variants of this map were used in Mills's atlas. The earlier version, which is quite rare, is identical to the map in Lucas's 1823 atlas. It is reproduced in the Lucy Hampton Bostick and Fant H. Thornley facsimile edition of Mills's atlas published in 1938. The state road which extends northwest from Charleston through Columbia to the state boundary in Greenville County is not labeled. The imprint below the map reads, "Drawn & Published by F. Lucas Jr. Baltimore, B. T. Welch & Co. Sc." On all other copies of the map, the state road is identified in five places and "for Mills' Atlas" has been added to the imprint following "Baltimore." The map has been corrected throughout to reflect such internal improvements as the extension of the highway system and the construction of canals. Such physical features as mountains, tributary streams, and springs have also been added. A number of historic sites, like revolutionary war battlefields and new place names have been inserted in the northern districts.

Fig. 14–4. The title page of Mills's atlas.

As a former member of the Board of Public Works and an experienced architect, Mills was thoroughly familiar with South Carolina and the internal improvements that had been effected during the years before the publication of his atlas. We may presume, therefore, that he found the original version of Lucas's map outdated and unsatisfactory. A small number of atlases containing this version of the map apparently were distributed before Mills noted its inaccuracies and arranged for it to be replaced. Fortunately, most extant copies of the atlas include the updated map. A third variant of Lucas's small map was inserted in copies of Mills's *Statistics of South Carolina, 1826*. It is identical to the corrected map in the atlas except for the imprint, which reads, "Published by F. Lucas, Jr., Baltimore, for Mills' Atlas, & Statistics."

Because the cost of publishing the atlas was apparently greater than he had anticipated and since sales were disappointing, Mills sought aid from the legislature in 1827. In his petition he noted that "the petitioner has at great labor and expense prepared and published all the Maps of the Districts of the State in the most finished style of Engraving but in consequence of the depression in the price of the staple article of the State [i.e., cotton], and the pecuniary difficulties resulting therefrom the sales of the maps have been much retarded, and your petitioner deprived of the means of meeting his Engagements with his Engraver."[20] The house, regrettably, denied Mills's request for a loan of fifteen hundred dollars.[21]

This petition also suggested the possibility of additional map purchases by the state in lieu of a loan. The senate acted favorably on this suggestion, and a special committee recommended that "if any of the boards of commissioners of roads in this state have not been as yet, supplied with the maps of their own & of the adjacent districts, that the Superintendent be instructed to procure the requisite number, & that he be authorized to obtain from the Comptroller-general the requisite warrant for the purpose on the treasurer of the upper or lower divisions as may be desirable."[22] This action provided some relief for Mills's financial plight, but it was obvious to him by this time that further professional opportunities in South Carolina were limited. In as early as October 1826 he had dispatched letters to contacts in Washington offering his talents and services to the federal government. An 1826 letter to the secretary of war notes: "My engagements in South Carolina will close with the year." And in a March 31, 1827, letter to Charles Nourse, son of the register of the federal treasury, Mills requested assistance in securing an appointment with the Engineer Corps: "I have been engaged for the last six or seven years in the work of Internal Improvement in this state—This work is now brought nearly to a close, and my professional engagements with it—I have felt desirous to enter into the service of the general government even at a moderate salary rather than run the risk of getting a large salary from the state governments, as I believe there would be more stability in office under the former."[23]

Nothing apparently came of these requests, and in 1828 Mills again petitioned the South Carolina legislature to purchase more copies of the atlas and the district maps. In an August 15, 1829, letter to President Andrew Jackson, Mills stated that "the opportunity now offered by your residence at Washington enables me to gratify my wishes . . . and I shall regard it as a favor, your acceptance of the *Atlas of the State of South Carolina*, with a *Statistical* history of the same, both of which I have directed to be forwarded to you from Philadelphia."[24] In a postscript, Mills informed the president that he was applying to the secretary of war for possible employment. This effort brought favorable action, for in 1830 Jackson appointed Mills federal architect and engineer. Mills spent the remainder of his life in Washington where he designed a number of federal buildings as well as the Washington Monument.

He did not, however, abandon his interest in his atlas. In 1837 he petitioned the South Carolina legislature for two thousand dollars to redeem the copper plates of the district maps, which were still held by the engraver. The legislature agreed, and with the ownership of the copper plates assured, Mills prepared a revised edition of the *Atlas of the State of South Carolina*, which was published in 1838. The map on the frontispiece and title page in this edition is credited to a new printer, "J. & W. Kite printers, Philadelphia"; the imprint reads, "Published by A. Finley, Philada."; and the descriptive and statistical data have been updated to 1838. "South" is spelled out in the name of the state in the lower left corner of the map, whereas it is abbreviated as "Sth" in earlier editions. The South Carolina railroad running from Charleston to Augusta has been added, and in the northwest two new districts, Pickens and Anderson, have been formed from Pendleton District. A major change in the district maps is the alteration of Pendleton District, which is divided and retitled, "Pickens & Anderson, formerly Pendleton District." The original 1825 publication date has been removed from all the atlas's district maps. The 1838 edition of the atlas apparently had a limited sale. There is, for example, no copy of it in the Library of Congress, and only two extant copies are identified in the second volume of Clara E. Le Gear's *United States Atlases* (1953), which lists holdings in more than 130 United States libraries.

Although he failed in 1840 to sell the engraved plates of the district maps to the South Carolina General Assembly, Mills still hoped to salvage something of his investment in time and money. In 1847 he petitioned the legislature for aid in preparing new and updated editions of the district maps. The tabling of this request by the legislature seems to have terminated his association with the *Atlas of the State of South Carolina*.

A slightly reduced, limited-edition facsimile of the atlas (350 copies) was published, as noted earlier, in 1938 by Lucy Hampton Bostick and Fant H. Thornley. Francis Marion Hutson of the Historical Commission of South Carolina prepared the introduction. This edition, which reproduces the early version of the Lucas map, contains an index with names listed alphabetically by district. In 1965 Robert Pearce Wilkins and John Keels published another facsimile edition in Columbia, South Carolina, with an introduction by Charles E. Lee. It appears in an enlarged format that permits a number of the maps to be bound without folds. The pink wash borders are omitted on the district maps in this edition and, unlike the 1938 facsimile, it has no index.

Mills's *Atlas of the State of South Carolina* is a cartographic milestone because of its use of astronomical and scientific surveys for its district maps, its position as the first atlas of an individual state, and the impetus it gave to local and regional cartography in the early decades of the nineteenth century. It is also a tribute to the dedicated and self-sacrificing personal effort of Mills in carrying through to completion its compilation and publication. Monuments to Mills's architectural genius still survive in several American cities, but his South Carolina atlas is the sole record of his contribution to the history of cartography in the United States.

Notes

1. Henry S. Tanner, *Memoir on the Recent Surveys, Observations, and Internal Improvements, in the United States, with Brief Notices of the New Counties, Towns, Villages, Canals, and Railroads, Never before Delineated. Intended to Accompany His New Map of the United States*, 1st ed. (Philadelphia, 1829), 31.
2. South Carolina, *House Journal*, Dec. 13, 1815, 128–29.
3. Ibid., Dec. 16, 1816, 170.
4. Robert Mills, *Mills' Atlas of South Carolina: An Atlas of the Districts of South Carolina in 1825*, introd. Charles E. Lee (Columbia, 1965).
5. Ibid.
6. Ibid.
7. Ibid.
8. David Kohn and Bess Glenn, eds., *Internal Improvements in South Carolina, 1817–1828* (Washington, D.C., 1938), 99–100.
9. Tanner, *Memoir*, 31.
10. South Carolina, *House Journal*, Dec. 19, 1823, 201.
11. Kohn and Glenn, *Internal Improvements*, 104.
12. Ibid., 146.
13. South Carolina, *House Journal*, Dec. 19, 1823, 202.
14. South Carolina, *Senate Journal*, Dec. 15, 1825, 178.
15. Ibid., Nov. 30, 1826, 26.
16. Ibid., Dec. 1, 1826, 31.
17. South Carolina, *House Journal*, Dec. 13, 1826, 170.
18. Robert Mills, *Statistics of South Carolina, Including a View of Its Natural, Civil, and Military History, General and Particular* (Charleston, 1826).
19. Frank N. Jones, "Baltimore Mapmakers," *Surveying and Mapping* 21 (Dec. 1961): 489.
20. Kohn and Glenn, *Internal Improvements*, 118–19.
21. South Carolina, *House Journal*, Dec. 10, 1827, 171.
22. South Carolina, *Senate Journal*, Dec. 14, 1827, 186.
23. Robert Mills, *Some Letters of Robert Mills, Engineer and Architect* (Columbia, 1938), 8–9.
24. Ibid., 14.

15. *Charts and Guides for Navigating Coasts and Rivers*

Of great importance to the English colonists in America were the navigation charts and pilot books used for guiding the ships that carried them to their new homes. The earliest ship captains who sailed to North America and explored its coasts and harbors had no nautical aids. Slowly the collective experience and records of many navigators were assembled into descriptive guides or were translated into sailing charts. Because there were no official charting agencies prior to 1796, when the British Admiralty established a hydrographic office, the initial responsibility for compiling and publishing navigation books and charts was assumed by private individuals and firms. It was not until late in the seventeenth century that systematic guides and charts were available for sailing across the North Atlantic Ocean and along the coast of North America.

In 1671 John Seller introduced the first part of *The English Pilot: The Fourth Book*, which "was the first great atlas of wholly English origin to deal exclusively with American waters; . . . its production involved some of the most noted map makers and publishers of the time, and . . . through successive editions its maps illustrated the unfolding geographical knowledge of the American coast within a century of exploration and settlement."[1] Seller had planned to make *The English Pilot* a comprehensive series of charts, but the project proved to be beyond his financial and physical capacities. He was obliged to obtain the assistance and collaboration of other compilers and publishers, among them John Thornton, William Fisher, and John Wingfield. Fisher and Thornton ultimately carried out Seller's plan for a series of chart books, and to them belongs the credit for establishing *The English Pilot* as one of the principal navigation atlases in the closing decades of the seventeenth century and the beginning of the eighteenth.

The first edition of *The English Pilot: The Fourth Book* was published in 1689 by Fisher and Thornton. Most of the charts it contains were compiled by the latter. Thornton was a noted cartographer and publisher of charts for the Hudson's Bay Company.[2] Between 1689 and 1794, thirty-seven separate editions were published of *The Fourth Book*, and it was the primary navigation aid during those years for British pilots sailing the North American coasts. Detailed sailing directions were also supplied with its charts. Only minor corrections, however, were made to the atlas's charts during the more than one hundred years it was used, to the misfortune of many navigators.

A contribution to coastwise navigation during the colonial period was made by Cyprian Southack, a navigator who spent much of his career in American waters. His work, *The New England Coasting Pilot*, was published between 1729 and 1734.[3] Southack was born in London on March 25, 1662, and migrated to Boston in November 1685. He engaged in various maritime activities, including that of privateer in 1689 when England and France were in conflict over their American holdings. He returned to privateering in 1703, when he supported British forces in various encounters along the New England and Nova Scotia coasts. In as early as 1694, Southack had prepared a chart of Boston harbor. It was the first of twenty or more coastal and harbor charts he made based on his own surveys. Several of his charts were included in later editions of *The English Pilot: The Fourth Book* and in other navigation atlases.

Southack's most ambitious work was *The New England*

THE

NEW ENGLAND

COASTING PILOT

FROM

SANDY POINT of NEW YORK,

UNTO

Cape CANSO in Nova Scotia,

And ~~Part of~~ all the *Iſland* of *BRETON*. The harbour of Louisbourg

WITH THE

COURSES and DISTANCES from Place to Place, and TOWNS on the Sea-Board; HARBOURS, BAYS, ISLANDS, ROADS, ROCKS, SANDS: The Setting and Flowing of TIDES and CURRENTS; with ſeveral other DIRECTIONS of great Advantage to this Part of Navigation in NORTH-AMERICA.

By Capt. *CYPRIAN SOUTHACK*,

Who has been Cruizing in the Service of the Crown of GREAT BRITAIN *Twenty-two Years.*

DIRECTIONS *to Sail from the Light-houſe at* Boſton *in* New England.

	Miles.
FRom the Light-houſe to Point ALDERTON, South by Eaſt	1 [illegible]
From the Light-houſe to the South-Eaſt Point of EGG Rock, Eaſt half North	0 [illegible]
From the Light-houſe to Little BRUSTERS, South-Eaſt, Point North Eaſt, [illegible] Eaſt	1 [illegible]
From the Light-houſe to GRAVE's Rocks, North-Eaſt, [illegible] Eaſt	2 [illegible]
From the Light-houſe to the Weſtermoſt Part Spit of Sand, Weſt half North	1 [illegible]
From the Light-houſe to GEORGE's ISLAND, South Part, Weſt half South	2
From the Light-houſe to RAMFORD's ISLAND, South-Eaſt Part, Weſt by South	3 [illegible]
From the Light houſe to PETICT ISLAND, North-Eaſt Part, South-Weſt by Weſt	2 [illegible]
From the Light-houſe to CHAMBERLAIN's Point of Rocks, South-Weſt half Weſt	1 [illegible]
From PETICT ISLAND, Eaſt Part, to NANTASKET Road, North-Weſt and by Weſt	1 [illegible]
From the Light-houſe to HARDIAN's Rocks, South-Eaſt, the Eaſt Part	2 [illegible]
From the Light-houſe to CONIHASSET Rocks, Eaſt Part, is South-Eaſt and by Eaſt	9
From the Light-houſe to St. SURREN's Rocks, Weſt-South-Weſt, [illegible] Weſt	1 [illegible]
From the Light-houſe to NAHANT Rock, North by Eaſt	7 [illegible]

If you come out of the Sea in the Winter Seaſon, and there ſhould be a great deal of Ice, my Advice is to hale a-ſhore at the South Part of RAMSFORD ISLAND, on the ſoft Ground juſt by NANTASKET Road; and there is no Ebb nor Flood will do you Damage: There ſtay till you can get up to BOSTON.

Fig. 15–1. The title page of Cyprian Southack's *New England Coasting Pilot*.

Coasting Pilot, which includes eight charts (Fig. 15–1). The atlas has no text pages, but there are one hundred descriptive notes on the faces of the charts. The coastal regions covered extend from Sandy Point, New York, to Canso in Nova Scotia and include part of Cape Breton Isle (Fig. 15–2). Southack died in Boston on March 27, 1745, at the age of eighty-three. In around 1758, William Herbert and Robert Sayer published, in London, on one large sheet, the eight charts that comprised *The New England Coasting Pilot*. The composite chart was reissued by Mount, Page & Mount in 1775.

The English-French struggle for control of North America during the middle of the eighteenth century stimulated cartographic surveys both on land and along the coast. British military and naval engineers were particularly active, notably so following the French and Indian Wars and England's acquisition of Canada. From around 1756 until the outbreak of the American War of Independence, surveys of the coasts and harbors of Newfoundland, Nova Scotia, the St. Lawrence River, and the coasts of New England were made by various military engineers, among them Samuel Holland, James Cook, William Gerard De Brahm, and Joseph F. W. Des Barres.

Des Barres began surveying after the end of the French and Indian Wars. He and a staff of assistants were commissioned to conduct hydrographic surveys and compile charts. In 1774 Des Barres returned to England with the manuscript charts and other data he had accumulated during eight years of surveying. Pursuant to the advice of Admiral Lord Richard Howe, commander of the British fleet in America, the king ordered that the charts be published. For the next decade, under the direction of Lord North, first lord of treasury and the Lords Commissioners for Trade and Plantations, Des Barres and a staff of assistants, which variously in-

Fig. 15–2. Chart of Cape Cod from *The New England Coasting Pilot*. Southack's charts were the earliest prepared by an American navigator.

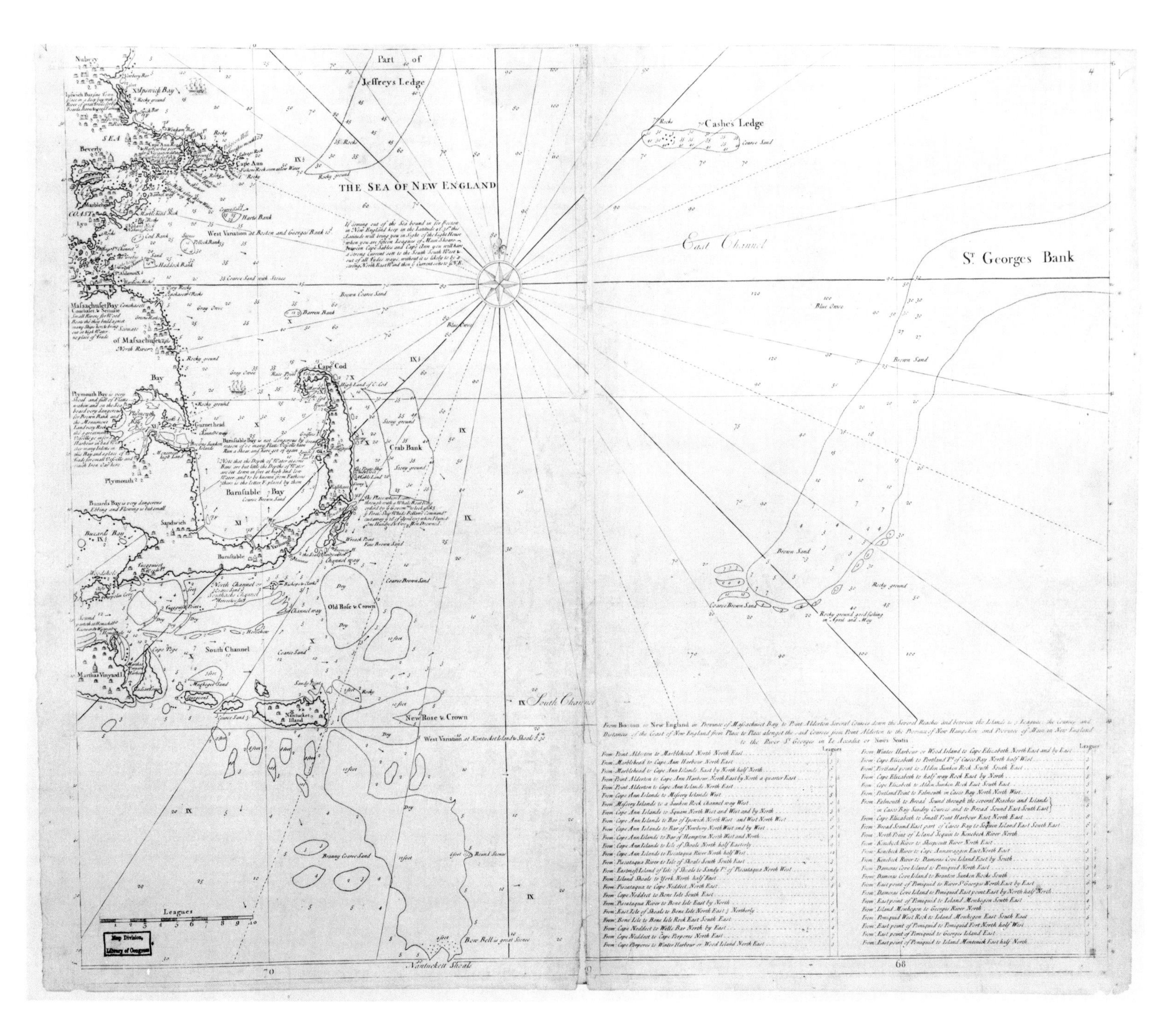
Part of
Jeffreys Ledge
Cashes Ledge
THE SEA OF NEW ENGLAND
St. Georges Bank
East Channel
Cape Ann
Cape Cod
Crab Bank
Barnstable Bay
Plymouth
Sandwich
Barnstable
South Channel
Old Rose & Crown
New Rose & Crown
Marthas Vinyard
Nantucket Island
Brown Sand
Rocky ground
West Variation at Boston and Georges Bank
Leagues
Nantuckett Shoals
Map Division, Library of Congress

cluded from five to twenty-three individuals, were occupied in compiling, drafting, engraving, printing, and publishing the North American charts. They utilized not only the surveys prepared by Des Barres, but also the land and hydrographic surveys completed by Holland, Cook, and others. By the end of 1776, charts for Book I, which covered the coasts and harbors of Nova Scotia, were ready for printing. Book II, charting the coasts and harbors of New England, was ready by the end of 1777. Known collectively as the *Atlantic Neptune*, the charts proved of inestimable value to the British navy during the revolutionary war.

Des Barres and his staff continued with the project through 1784, publishing successively Book III (the gulf of the St. Lawrence River and the estuaries of Cape Breton and St. John), Book IV (the coast of North America south of New York), and Book V (various views of the North American coast). The *Atlantic Neptune* series includes 180 different charts, as well as a number of views. It proved to be the standard guide for navigating the North American coasts until well into the nineteenth century. Corrections were made on individual charts, and new editions were issued periodically. Some charts have as many as twelve variants. Of extant volumes of the *Atlantic Neptune*, no two are identical. This is apparently because series of charts were assembled and bound to order for a particular ship or voyage.

The *Atlantic Neptune* has been described as "a magnificent contribution to hydrography and a classic of the minor arts."[4] Extant volumes and individual charts and views are treasured for their historical interest as well as for their technical execution and skilled draftsmanship. Des Barres biographer John Webster wrote that "the work must be regarded as one of the most remarkable products of human industry which has been given to the world through the arts of printing and engraving. . . . Apart from the practical value of the Atlantic Neptune, the artistic excellence of the views alone would give it high rank. This feature is solely due to Des Barres. He drew with great sensitiveness and had an exquisite sense of color. Many of his aquatints, whether in monotone or color, are of the highest quality."[5] Although it is unlikely that many American navigators in the years following the Revolution had *Atlantic Neptune* volumes or its separate charts, some copies of these publications did find their way into American hands. They proved to be a major compilation source for the pilot books and nautical atlases printed and published in the United States until the early years of the nineteenth century.

The first post-Revolution American nautical atlas was published in Boston by Matthew Clark in 1790. It is untitled but has a dedication to "His Majesty John Hancock esquire Governor and Commander in Chief of the Commonwealth of Massachusetts." The nine charts of the volume cover the coast of America from the Gulf of Florida (now called the Straits of Florida) north to New England. The *Atlantic Neptune* was probably the primary compilation source. All the charts bear some sort of certification of accuracy by Osgood Carleton. The *Chart of the Coast of America from New York to Rhode Island* bears Carleton's assurance that "I have examined this Chart and find the Head Landings & Angles confined to their true Latitudes & Longitudes & the Data Mathematically true & I approve of it as a true and accurate Chart" (Fig. 15–3). Several of the charts credit "J. Norman" or "Josh. Seymour" as engravers. The certifications on the charts in Clark's atlas attest to the high regard in which Osgood Carleton was held by his contemporaries.

Their experience with Clark's atlas apparently induced Carleton and John Norman to collaborate in compiling, engraving, and publishing their own marine atlas. The first edition of the resulting *The American Pilot* was published by Norman in Boston in 1792 (Fig. 15–4). There is a testimonial by Carleton on the title page as well as on several of the eleven charts. It is believed that he compiled at least some of the charts, utilizing as source material the *Atlantic Neptune* and other earlier English charts.

Norman engraved the charts in addition to publishing

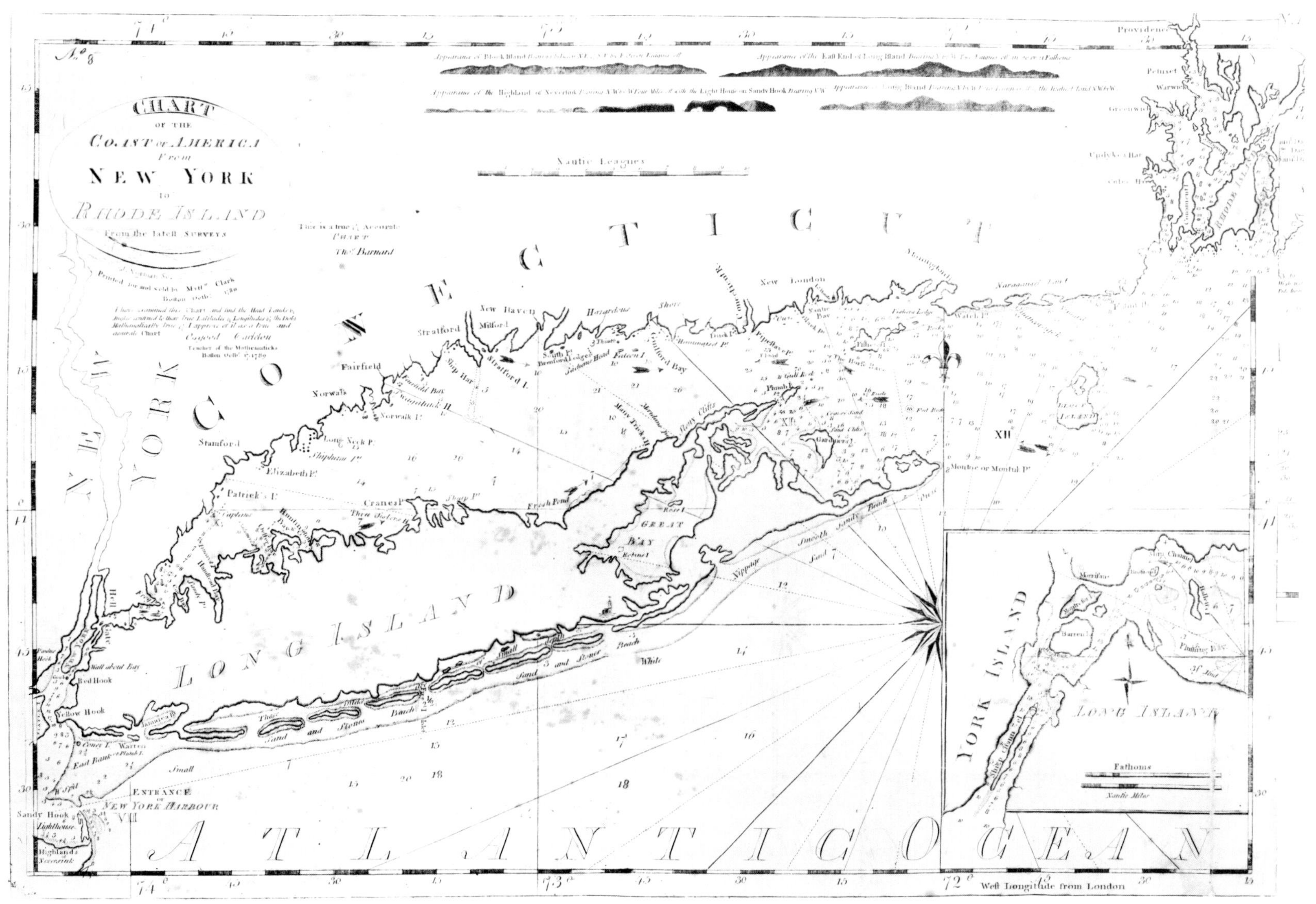

Fig. 15–3. Chart of the coast from New York to Rhode Island from Matthew Clark's nautical atlas. Numerous soundings are shown in Long Island Sound.

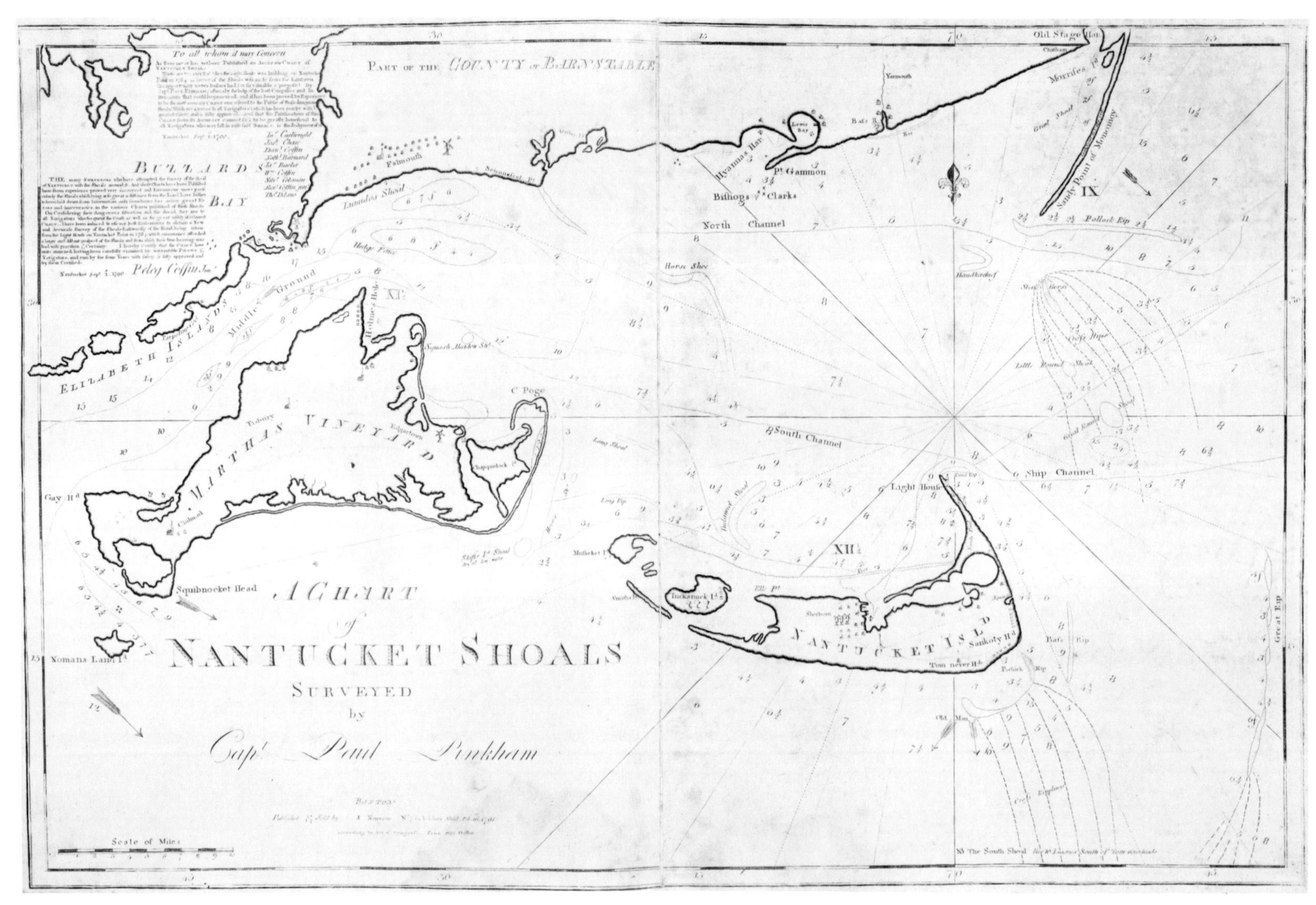

Fig. 15–4. Captain Paul Pinkham's chart of the Nantucket shoals from Norman's *The American Pilot*.

the atlas. He operated an engraving and printing shop and had previously published and sold *A Map of the Present Seat of War* in 1776. He also engraved a series of portraits for *An Impartial History of the War*, which was published in 1782. With a Mr. White he founded the *Boston Magazine*, the first number of which was issued in November 1783. During the next several years Norman engraved a number of illustrations for the magazine. He also printed and sold the first directory of the city of Boston in 1789. His name does not appear in this edition, but it is in the second edition published in 1796.

The American Pilot was reissued in 1794 by Norman, with essentially no changes. Like the first edition, it includes the two pages of "Particular Directions for Sailing to and entering the Principal Harbours, Rivers &c." by Carleton. Although Norman lived until June 8, 1817, this edition was the last to carry his imprint. Two subsequent editions, issued in 1796 and 1803, were published by William Norman, probably Norman's son. The 1796 edition includes two new charts: one of the Chesapeake and Delaware bays taken "chiefly from Anthony Smith, Pilot of St. Marys" and one of the coast of New England and Georges Bank "from Holland's actual surveys." Several charts in the earlier editions of *The American Pilot* were dropped. No new charts were added to the 1803 edition, but the sailing directions were expanded to four pages.

The navigation books and atlases heretofore described still owed much to their English predecessors. In the closing years of the eighteenth century, however, Edmund March Blunt embarked upon the publication of truly American navigation books and charts. For seven decades, pilot books and charts published by Blunt and his heirs guided American and foreign navigators. Blunt, who was born in Portsmouth, New Hampshire, on June 20, 1770, was one of twelve children of William and Elizabeth March Blunt. Shortly after his birth the family moved a short distance down the coast to Newburyport, Massachusetts, where he spent his boyhood. Little is known about Blunt's education and training, but we may infer that from his mother, the daughter of a minister, he received a good grounding in religion. The maritime activities of Newburyport also certainly interested him and helped guide his future career.

Blunt's initial business venture occurred in 1793 when he opened a bookstore which he named the Sign of the Bible. In the same year, in partnership with Howard S. Robinson, Blunt founded a newspaper, the *Imperial Herald*, which was printed in Blunt's shop. He subsequently bought out his partner and with his cousin, Angier March, expanded the newspaper. In 1795 Blunt, whose flourishing book publishing and rental library enterprises were taking up more and more of his time, sold his newspaper interests to March. Also contributing to Blunt's retirement from journalism was his accelerating involvement with navigational publications. From his own observations as well as from conversations with ship captains, Blunt was shocked by the inaccuracies in most of the existing navigation guides. In 1796, therefore, he published the first edition of the *American Coast Pilot*, which on the title page gave credit to Captain Lawrence Furlong as the compiler. Whether Furlong was responsible for the work or whether Blunt used his name to give nautical authority to the volume, we do not know. At any rate, in the eighth (1815) and subsequent editions of the work, Furlong's name was deleted.

The contents in the first edition of the *American Coast Pilot* were described by Blunt in the *Newburyport Herald*:

> The American Coast Pilot containing the courses and distance from Boston to all the principal harbours, capes and headlands included between Passamaquoddy and the capes of Virginia with directions for sailing into and out of all the principal ports and harbours with the sounding on the coast; also a Tide Table shewing the time of the high water at full and change of the moon in all the above places together with the courses and distances from Cape Cod and Cape Ann to the Shoal of Georges and from said capes out to the south of

> East channel and the setting of the current on the eastward and westward; also the latitude and longitude of the principal harbours and headlands, &c. by Capt Lawrence Furlong. Also courses, directions, distances etc. from the capes of Virginia to the River Mississippi from the latest survey, and observations. (Approved by experienced Pilots and Coasters.) The first edition printed at Newburyport by Blunt and March. Sold by them and the principal book stalls in the U.S. 1796.[6]

Although Blunt's *American Coast Pilot* undoubtedly contained information derived from American seamen, it also drew heavily upon such earlier works as Norman's *American Pilot*, the *Atlantic Neptune*, and, to a lesser degree, even on *The English Pilot: The Fourth Book*. The first four editions of the *American Coast Pilot* contained only text, but small charts of individual harbors were included in the fifth edition, published in 1806, and in subsequent issues. In all, twenty-one editions of the *American Coast Pilot* were published up to 1861, those after 1827 carrying the imprint of E. & G. W. Blunt, sons of Blunt.

In 1799 Blunt plagiarized and published John Hamilton Moore's *The New Practical Navigator*, an English publication dating back to 1772. In 1800 Blunt published a second revised edition of *The New Practical Navigator*. In the preface to this edition he acknowledged that "the American editor has not presumed to revise and enlarge a work of such high authority, without duly consulting several Gentlemen of the first mathematical and nautical talents in our country." Among his advisors and consultants were Nicolas Pike, Osgood Carleton, and Nathaniel Bowditch. Bowditch's contribution to the volume was a chapter on the methods of finding the longitude at sea. This was the beginning of his long association with Blunt.

Before preparing a third edition of *The New Practical Navigator*, Blunt again engaged Bowditch to correct Moore's book. The revisions were so extensive that the work was renamed *The New American Practical Navigator*, and Bowditch was listed as editor on the title page. It was published in 1802. A second edition of this work was published in 1807, and a third in 1811. Most copies of the third edition were burned in the great fire of Newburyport on May 31, 1811, which destroyed the Blunt building. Blunt had actually already sold the building and had relocated his company to New York City earlier in the year, but his stock was still destroyed. His shop in New York was called the Sign of the Quadrant and was situated at 202 Water Street. From this new location Blunt printed the third edition of the *New American Practical Navigator*. Subsequent editions (which total thirty-five) were printed in New York. Bowditch continued as editor through the ninth edition dated 1837. Following his death on March 16, 1838, his son, Jonathan Ingersoll Bowditch, assumed the post of editor.

Believed to be the earliest separate chart published by Blunt is *A Chart of George's Bank, Including Cape Cod, Nantucket and the Shoals lying on their Coast, with Directions for Sailing over the same &c. Surveyed by Capt. Paul Pinkham* (Fig. 15–5). It was engraved by Amos Doolittle in New Haven and published in 1797. In the lower right corner is the inscription, "Engraved & Printed for Edmund M. Blunt Proprietor of American Coast Pilot 1797." The accuracy of the chart is certified by six sea captains: Henry Bates, Joseph Higgins, Batch Swain, William Tabent, Peter Swain, and John Hussey. A comprehensive note to the public is printed in the right margin of the sheet and was signed at Nantucket in 1796 by Paul Pinkham.

Blunt's trade at his Newburyport store included navigation charts as well as nautical instruments and pilot books. Some charts offered for sale were issued by other publishers, some of which Blunt reprinted. Many of Blunt's charts were based on surveys made by the U.S. Navy prior to the War of 1812. They were, however, withheld from sale until the war had terminated. By 1822 Blunt's sales catalog listed eighteen charts.

In 1816 Blunt's youngest son, Edmund, prepared an original survey of New York harbor. At the time he was only seventeen years old. Then he and his brother,

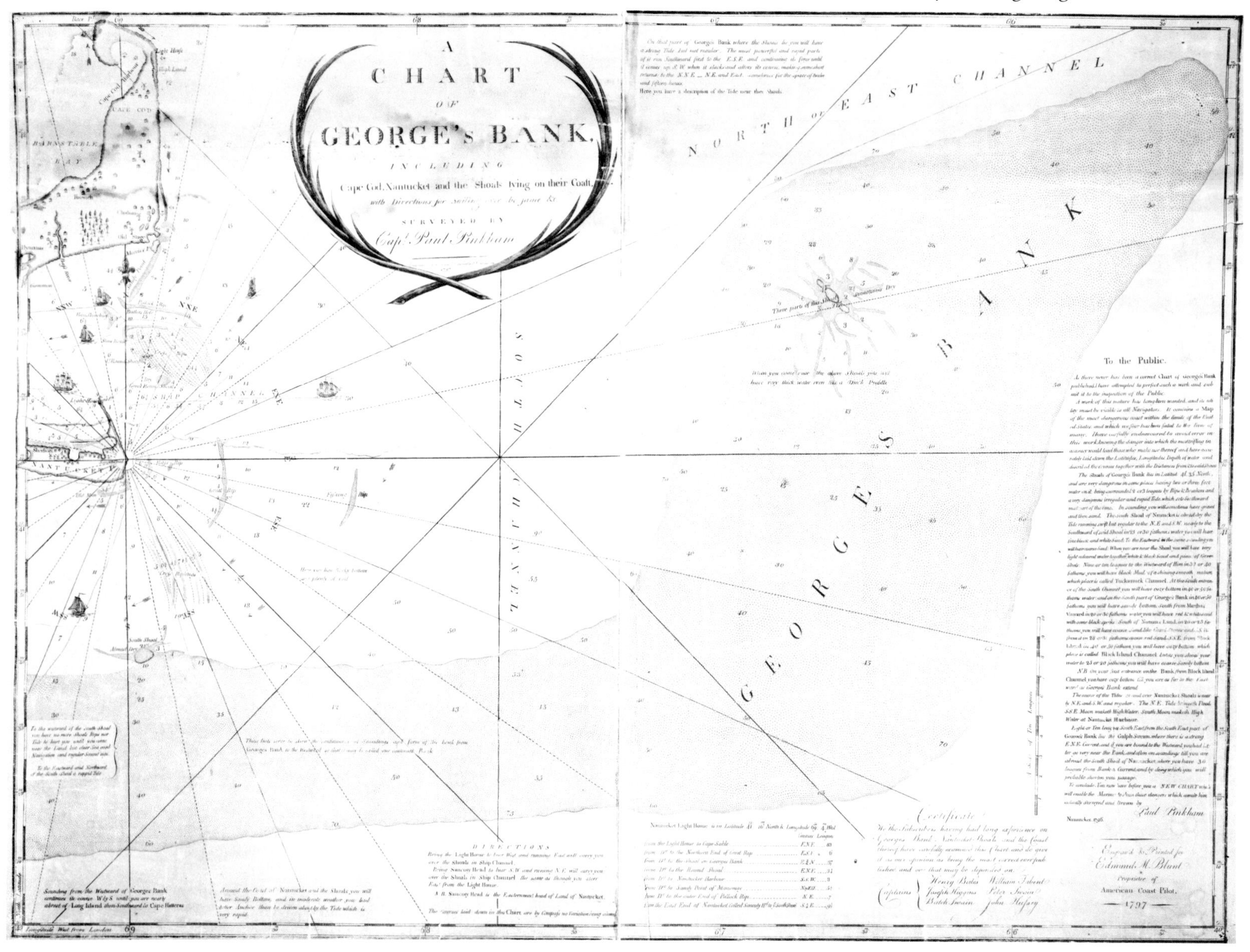

Fig. 15–5. Pinkham's *Chart of George's Bank* published by Blunt in 1797. Like others of the period it includes considerable descriptive data on its face.

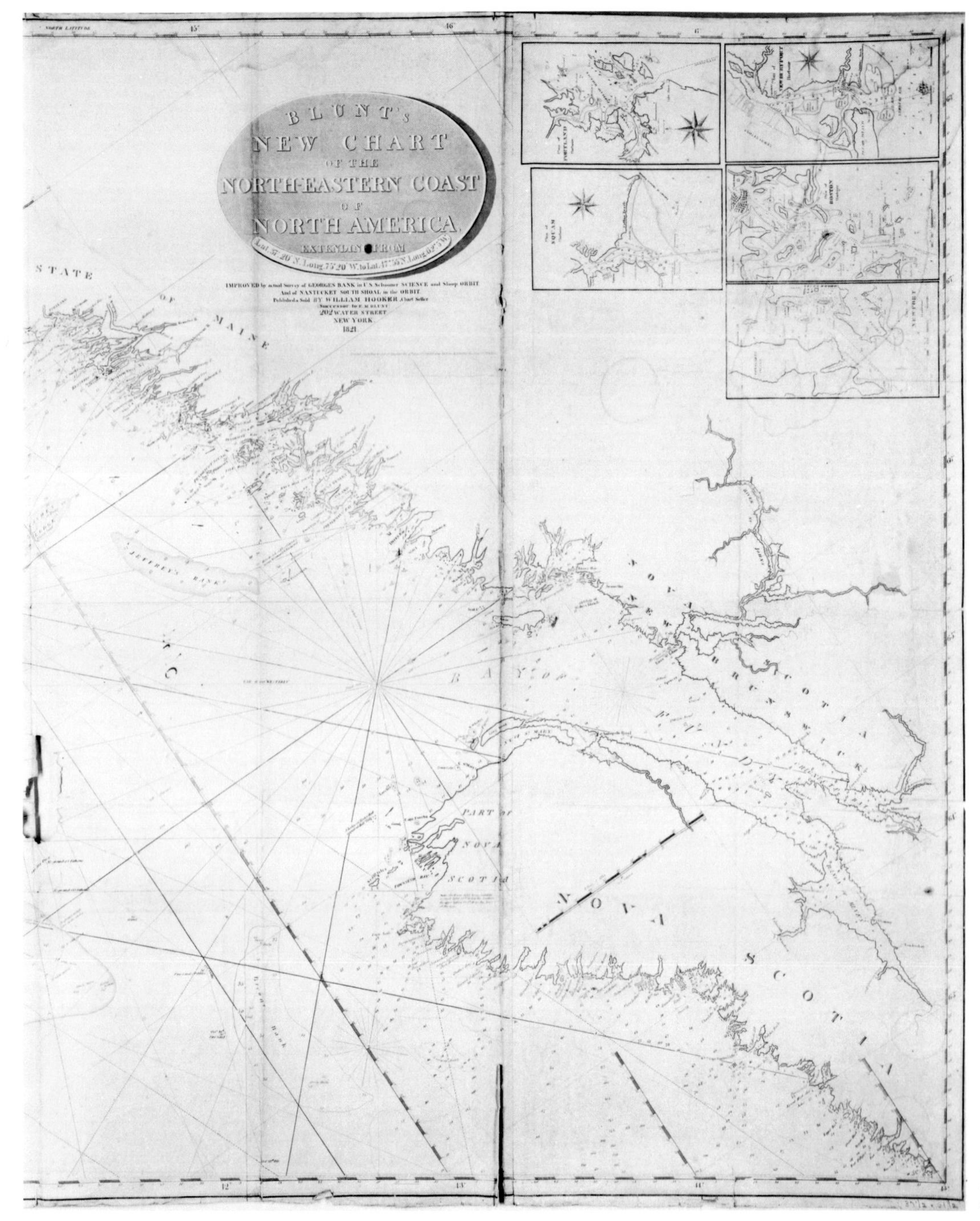

Fig. 15–6. Blunt's 1821 chart of the northeastern coast of North America.

George William, assisted several U.S. Navy officers in surveying the Bahama Bank in 1819 and 1820. In 1821 Edmund and Blunt surveyed Nantucket and Georges shoals. Edmund also made surveys of Long Island Sound in 1830. In that year the Blunt firm published an atlas entitled *Blunt's Charts of the North and South Atlantic Oceans, the Coast of North and South America and the West Indies*. It includes fourteen plates. In the collections of the Library of Congress there are fifty different Blunt charts, some of which were issued in five or more revisions (Fig. 15–6).

Blunt retired in 1826 or 1827, and the business was carried on by Edmund and George William under the name E. & G. W. Blunt. Blunt lived in retirement at his home in Ossining, New York, where he died on January 4, 1862.

Edmund was appointed first assistant in the U.S. Coast and Geodetic Survey in 1833 and spent the next thirty-three years with that federal agency. His official duties apparently did not prevent him from assisting his brother George in managing the family business. There was, inevitably, a close relationship between the Blunt company and the U.S. Coast and Geodetic Survey. George also maintained close ties with the U.S. Navy and served it as a civilian adviser during the Civil War.

In 1830 the U.S. Navy had established the Depot of Charts and Instruments, which in 1854 became the U.S. Naval Observatory and Hydrographic Office. Hydrographic surveys had begun in 1837 with a survey of Georges Bank. In June 1866 the office was reorganized as the U.S. Navy Hydrographic Office, drawing upon experience gained during the Civil War. This hastened the demise of E. & G. W. Blunt, which could not compete with the federally funded office. Other contributing factors were the death of Blunt in 1862 and of Edmund in 1867. As a result, George sold the copyright for the *American Coast Pilot* to the U.S. Treasury Department and that for *The New American Practical Navigator* to the U.S. Navy Hydrographic Office. In the tenth edition of the *American Coast Pilot*, published in 1822, the elder Blunt had written that he "had undertaken a duty, the performance of which belongs rather to a nation than to an individual." Harold L. Burstyn concurred, observing that "the United States was slow to recognize its responsibility, giving a glorious seventy year career to the first family of American hydrography."[7]

In 1837 Captain Seward Porter copyrighted and published a series of nine charts of the coast of Maine. In contrast with the extensive sequence of charts and navigation books published by the Blunts, this appears to have been a one-time effort for Porter. The charts, which are numbered one to nine, are all titled *Chart of the Coast of Maine*. They show the following coastal regions: (1) Passamaquoddy, (2) Machias Bay, (3) Gouldsboro to Addison and Moore Peck's Reach (Fig. 15–7), (4) Mount Desert Island, (5) Penobscot Bay, (6) Medomac River, (7) Sheepscot River, (8) Harpswell, Orr's Island, and New Meadows River to Kennebec, (9) Casco Bay. The Library of Congress collection also includes a reproduction of a tenth manuscript chart, which had not yet been published when Porter died. It charts the Saco River to Cape Elizabeth and Portland harbor.

Five of the charts are at the scale of 1:48,000. The scale of chart 5 is 1:130,000 and chart 8 is 1:100,000. There are two scales on chart 10, 1:52,000 for the coastal region and 1:26,500 for Portland Harbor. No scale is indicated on chart 7, which also lacks soundings. Charts 5 and 10 also are without depth indications, which appear near the coast on all the other charts. Porter's handwritten notice of copyright registration appears on chart 1 in the Library of Congress set. The actual printed registration is at the bottom of charts 8 and 9.

Porter was born in Freeport, Maine, in 1784. He is reported to have had eleven brothers, all of whom became masters of vessels. After several years at sea, Porter and his brother Samuel became ship owners and merchants in 1812, with offices at Union Wharf in Portland. During the War of 1812, the brothers engaged in privateering. Their brig, *Dash*, was one of the most successful privateers. On its third cruise, however, the *Dash*

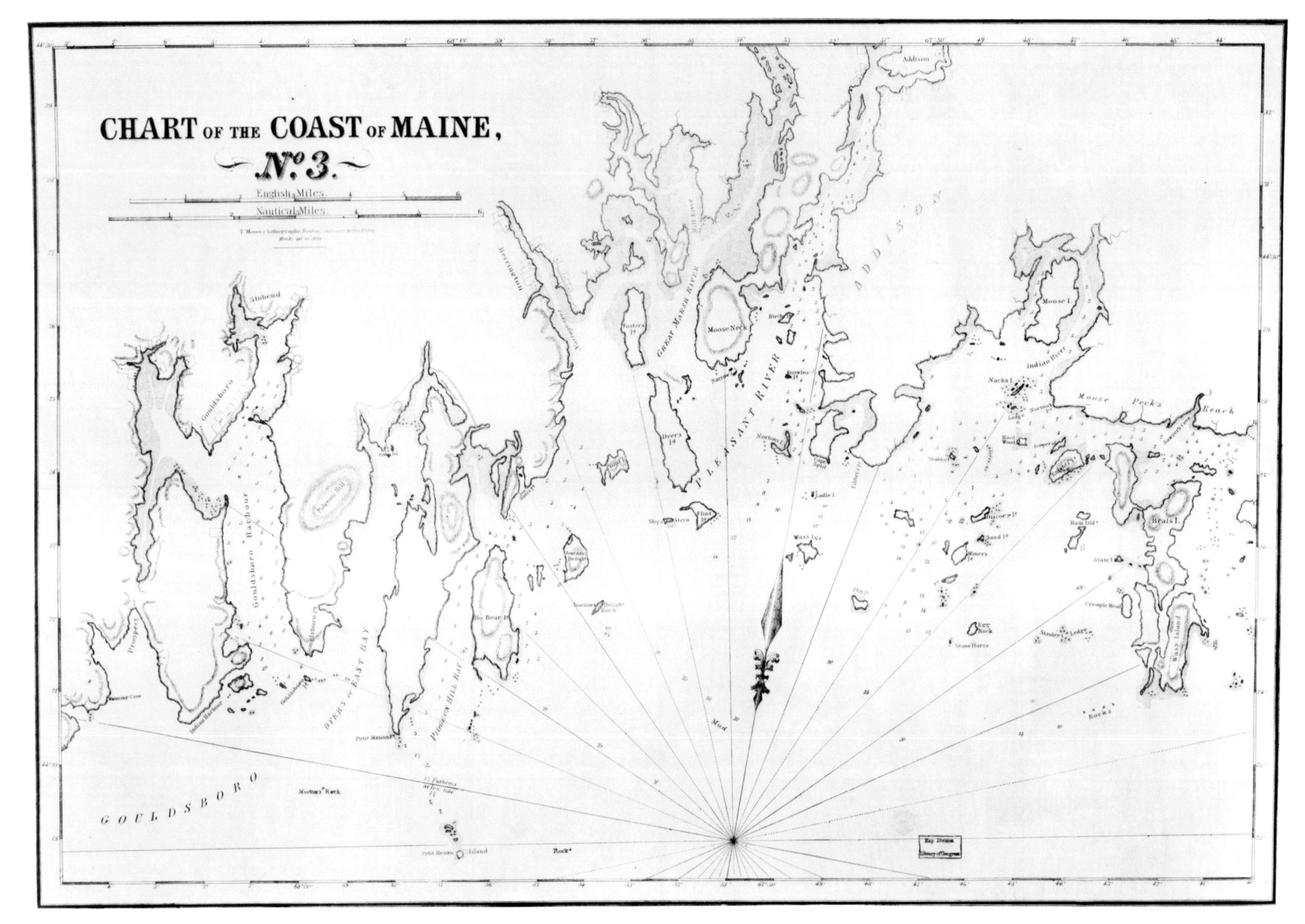

Fig. 15–7. Chart 3 of Captain Seward Porter's chart of the coast of Maine, 1837.

was lost, along with three Porter brothers, Ebenezer, John, and Jeremiah.

Porter moved to Bath around 1822 where he built the first steam-operated mill. He also built the first steamboat that served the Maine coast. With Samuel, Porter operated the steamer *Connecticut* on the New York, Boston, and Portland route. Some accounts indicate that he acquired an unfinished survey of the Maine coast done by a British naval officer, and that this was the source of the data used for his chart series. It seems likely, however, that he also utilized the *Atlantic Neptune* charts. Apart from their significance as early American coastal charts, the Porter charts are of interest in having been reproduced by lithography. All the charts, except 7 and 10, have the imprint, "T. Moore's Lithography, Boston, successor to Pendleton." Charts 3 and 8 have the additional credit of "Moody del. on stone." Porter died in 1838, a year after the publication of his charts.

Pendleton's Lithography, founded in Boston in 1825 by John and William Pendleton, was the first successful and long-lived lithographic printing shop in the United States. It continued in operation until 1836, when William Pendleton sold the business to Thomas Moore, who had been his bookkeeper. Moore carried on until 1840. Pendleton's Lithography and Moore's Lithography produced a variety of lithographic works, including maps and charts. A number of lithographic artists and printers received their training at the Pendleton-Moore establishment, among them, apparently, Moody. Moody's interest in lithography appears to have been ephemeral, for he is not listed in David M. Stauffer's *American Engravers upon Copper and Steel* (1907). George C. Groce and David H. Wallace have an entry for a John E. Moody in their *Dictionary of Artists in America*, but he was listed in the 1840–42 Boston directory as an engineer.[8]

A number of other individuals compiled and published charts of portions of the coast of the United States in the early decades of the nineteenth century. In 1812 *A New Chart of Massachusetts Bay, Drawn from the Latest Authorities by S. Lambert, Salem* was published by Cushing & Appleton, also of Salem. The 60-by-79-cm. chart includes soundings and outlines the major banks. It is oriented with north on the right, and has two decorative compass roses and several unadorned ones. Captain Samuel Lambert also compiled a *Chart of Nantucket Shoals and George's Bank*, which was published in 1813. Pasted on the verso of one copy of the Massachusetts Bay chart there is a broadside certifying the accuracy of both of these Lambert charts. The broadside is signed by members of a "Committee chosed by the Salem East India Marine Company," which included Samuel G. Derby, Moses Townsend, Joseph Ropes, Nathaniel Silsbee, Nathaniel Bowditch, Robert Emery, James Devereux, Israel Williams, John Collins, and William Lander. The broadside notes that the Lambert "charts are not mere copies from old ones, on a different scale, but contain much information which later observations and an increased navigation have afforded. The coast is principally drawn from Holland's Surveys and Bowditch's Navigator; the depth of water, and quality of the bottom, are almost entirely derived from the examination of the log books and journals, and the immediate communications of experienced and observing navigators."

An 1822 edition of the chart has the title *A New Chart of Massachusetts Bay and Part of the Coast of Maine, Drawn from the Latest Authorities by Samuel Lambert, Hydrographer, Salem*. There appears to be no difference from the earlier chart except the title, which reflects Maine's admission into the Union in March 1820. Lambert also compiled *A New Chart of the Coast of Connecticut, New York, New Jersey and Delaware, Drawn from the Latest Authorities*. It was engraved by Thomas Wightman and published in 1815 by Cushing & Appleton. Wightman, a native of England, was associated with the Boston engraver Abel Bowen at about the time this chart was engraved. Like the Massachusetts Bay chart, it is oriented with north on the right of the sheet. There are soundings and two large decorative compass roses. In the lower left corner of the sheet there is an inset chart of the Delaware River and Delaware Bay. Apart from the information on his several charts, we have no information about Lambert.

Private surveyors and publishers also produced a few charts of the South Atlantic and gulf coasts in the first quarter of the nineteenth century. In 1818 Robert Blackford undertook the compilation of a series of charts of the U.S. coast "extending from Florida to Nova Scotia." His chart of the coast of Nova Scotia and New England, which was engraved by S. Stockley, was dedicated to President James Monroe. It is the only one of the series which I have examined; and it may be the only one that was actually completed and published. In 1823 Richard Patten, an instrument and chart maker in New York, published *A Particular Chart of the Coasts of West Florida Louisiana Mississippi & Alabama Including New Orleans, Mobile & Pensacola From the English Admiralty Surveys*. He also compiled and published other charts, including one of the New England coast in 1827. On this chart, Patten plagiarized information concerning the relocation of the Nantucket South Shoal from the 1821 surveys by Blunt and Cheever Felch. Blunt brought suit against Patten in U.S. Circuit Court in 1828 and was awarded damages. Silvio Bedini notes that "there was much unfairness in the trial . . . it was well known that Blunt had habitually 'borrowed' from English charts without acknowledgement during most of his career."[9]

A Map of Mobile Bay in the State of Alabama Comprising the Rivers & Creeks was compiled and published in 1820 by Curtis Lewis of Alabama. It gives soundings, locates shoals, and traces several "ships tracks" through the bay. In the lower right corner of the chart there is a large compass rose and, in the upper right segment, there are "Directions for coming into Mobile Bay, Mobile, and Blakely." A second edition of the chart published around 1840 and credited to "Curtis Lewis & Capt. Welsh," was lithographed by Miller & Co., 15 Broad Street, New York.

The Great Lakes for various reasons did not invite the interest of American private or commercial map and chart makers. Prior to the French and Indian Wars, the lakes and their tributaries and land areas were under French control. Before the American Revolution British military engineers had rebuilt some of the old French forts at various sites along the shores of the lakes, but no hydrographic surveys of the waters and harbors had been undertaken. During the War of 1812, Lakes Huron and Erie, in particular, witnessed several naval encounters between English and American ships. There is no indication, however, that charts or maps played a major role in the outcome of these engagements. Undoubtedly, officers of vessels on both sides had some familiarity with the lakes, as well as with the accumulated experience of earlier navigators of these waters. The data, though, were most likely in the form of descriptive guides, for there is no record of available navigation charts in this period.

The situation had not improved much a decade or so after the war. In the summer of 1826, Thomas McKenney, an agent of the U.S. Office of Indian Affairs, took a trip to the lakes for the purpose of negotiating a treaty with the Chippewas at Fond du Lac in Wisconsin Territory. During a stop at Detroit he wrote a letter home describing his traverse of Lake Erie:

> I knew its length, its breadth and depth, and yet I must confess that I had no more correct conception of the lake as it appeared to me than if I had never had the slightest acquaintance with its dimensions. All my previous conception of a lake fell so far short of its actual vastness and ocean-like appearance, as to be wholly absorbed in the view of it. I could but wonder what my opinion of lakes will be, after I shall have seen and navigated Huron and Superior. Lake Erie, though considerably smaller than either, is a vast sea, and often more stormy, and even dangerous, than the ocean itself.[10]

The size of the lakes and the cost in time and money to survey and chart them undoubtedly deterred nonofficial cartographic compilers and publishers. The lakes, moreover, did not invite individual travel or transport, and people going west had to travel by established boat lines. Lake travel accelerated greatly after the completion of the Erie Canal in 1825 and the establishment of steamboat lines on the Great Lakes several years earlier.

The year 1825 was a milestone in the history of charting the Greak Lakes. Commodore William F. Owen and Captain Barrie of the British Royal Navy completed in that year the charting of several principal harbors on the Canadian shore. Begun in 1815, their surveys were the first scientific governmental hydrographic surveys based on triangulation made of the Great Lakes. Earlier maps and charts of these waters, generally on small scales, were made by French and British military engineers. In 1828 Admiral Henry W. Bayfield of the Royal Navy initiated hydrographic surveying and charting on Lakes Ontario and Erie. Bayfield's work continued until 1835, and the resulting charts were particularly useful as they marked several important channels not surveyed by Owen and Barrie. Henry Adrienne of Oswego, New York, published a *Chart of Lake Ontario from Actual Survey by Augustus Ford, U.S.N.* in 1836. The chart was drawn on stone by Haring and printed at the lithographic plant of Sarony & Major, 117 Fulton Street, New York City. We may wonder how an officer of the U.S. Navy was able to prepare a chart for issue by a private publisher. In any event, it appears to be one of the earliest charts published by an American commercial company of any of the Great Lakes.

Charting the Great Lakes took another major stride forward in 1841, when the U.S. Congress appropriated fifteen thousand dollars for the "Hydrographical Survey of Northern and Northwestern Lakes." This led to the establishment of the Lake Survey under the U.S. Army Corps of Engineers. Headquarters, initially in Buffalo, New York, were moved to Detroit in 1845. Since this date surveys have been conducted and charts published of the coasts, harbors, and shipping channels of all the Great Lakes. By 1889 the initial charting program had been completed. Canadian and American surveys have produced some 250 charts, which are periodically corrected and updated. In 1970 the Lake Survey was transferred to the National Ocean Survey, a unit of the National Oceanic and Atmospheric Administration of the U.S. Department of Commerce.

Notwithstanding the establishment of an official agency for surveying and charting the Great Lakes, several charts were privately published after 1841. Robert Hugunin's *Chart of Lake Erie*, dated 1843, was deposited for copyright on January 3, 1844. It was lithographically printed by Hall & Mooney of Buffalo. No information has been uncovered about Hugunin. *A New Chart of Lakes Michigan, Huron and St. Clair, Compiled from the Latest and Most Reliable Surveys by Capt. M. Caldwell* was published at Buffalo in 1855. Although the printer is not indicated, the chart appears to have been lithographically reproduced. There are no soundings, but there is a large compass rose and shipping lanes between ports are shown.

Almost from the time of the first American settlements, the frontier beckoned enterprising Americans, who followed westward trails, traces, crude roads, and rivers. After the Revolution, and particularly following the Louisiana Purchase and the Lewis and Clark Expedition, the westward movement gained considerable momentum. The surge beyond the Appalachians was greatly accelerated in the boom years that followed the War of 1812. The Ohio River was the preeminent east-west artery, and this was the principal transportation route followed by the pioneers. Whatever their place of origin, they took one of two land routes, the turnpike to Pittsburgh or the national road to Wheeling, both of which terminated at the river. Little wonder that these two cities became important commercial centers and river ports.

By the closing years of the eighteenth century navigation on the Ohio River was booming, and there was an urgent need for reliable navigation guides. The first such recorded publication was the *Ohio Navigator*, which was published in Frankfort, Kentucky, by Hunter & Beaumont in March 1798. A second edition of this guide was published in August 1798. Both editions were priced at twenty-five cents. A more enduring work was Zadok Cramer's *Ohio Navigator*, the first two editions of which were published in Pittsburgh in 1801. Indicative of the popularity of Cramer's guide is the lack of extant copies of the 1801 editions. There are only a few surviving examples of the third edition, titled *The Ohio and Mississippi Navigator* and dated 1802. The fifth and all subsequent editions are titled *The Navigator, or, the Trader's Useful Guide*.

Cramer was born in New Jersey, of Quaker parents, in 1773. He spent most of his early years in Washington County, Pennsylvania, where he learned the bookbinding trade. In the spring of 1800, he moved to Pittsburgh were he opened a bindery. Shortly thereafter he purchased a book store, and not long after that he was operating a circulating library and a printing shop. Cramer's earliest publications were almanacs, but the one that firmly established his reputation was the *Ohio Navigator*. Charles W. Dahlinger noted that the young printer and publisher "had been in Pittsburgh but a short time when he realized the necessity for a publication giving detailed information for navigating the western rivers. He daily saw swarms of immigrants pass through the place, bound west and south, who lingered there attempting to learn, not only about navigating the rivers, but of the country to which they were bound. He proposed to furnish the information and set about collecting data for the purpose."[11]

The result was the *Ohio Navigator*. In its third edition, Cramer stated that the information used in the guide had been derived "from the journals of gentlemen of observation, and now minutely corrected by several persons who have navigated those rivers for fifteen and twenty years." One of the primary sources drawn upon by Cramer appears to have been Thomas Hutchins's *A Topographical Description of Virginia, Maryland and North Carolina*, which was published in London in 1778. Cramer also noted that he had "the assistance of several of the most eminent pilots and navigators and the use of late manuscript journals of gentlemen of observation."

The 1806 edition was the first to include maps, or rather, charts. They are cartographically crude renderings, which were reproduced from wood blocks. The same charts appear in all subsequent editions of the *Navigator*. On the charts the Ohio River is shown in heavy black, and islands are patches of white. To clarify his directions, Cramer seems to have initiated the practice of identifying islands with numbers. Ninety-eight islands were numbered in the Ohio River and 125 in the Mississippi south of Cairo, Illinois. The *Navigator* has twenty-six charts, half of which map the Ohio River and the other half the course of the Mississippi River below where the Ohio flows into it. The guidebook also includes detailed navigating directions, as well as a variety of general, historical, and geographical information. Each edition grew in size, with the 1814 edition reaching a maximum of 360 pages. Subsequent editions decreased in size, with the twelfth, and last, edition in 1824 having 275 pages.

John Spear became a printing and publishing partner of Cramer in 1808, and William Eichman joined the firm in 1811. Cramer was apparently induced to take partners because of his declining health. Afflicted with tuberculosis, he moved in 1811 to Natchez, where he hoped the milder climate might effect a cure. While in that city, he issued editions of the *Louisiana and Mississippi Almanack* in 1811 and 1813. He also established a warehouse and commission store for storing and selling various manufactured goods. Cramer died on August 1, 1813, in Pensacola, Florida, where he had again apparently moved because of his illness. His business interests, including the publication of the *Navigator*, were carried on for another decade or so by his wife and his partners.

Contemporary and later writers praised Cramer's *Navigator* for its utility. In his *Keelboat Age on Western Waters*, Leland D. Baldwin wrote that "the *Navigator* was complete in its advice to the immigrant and trader, even giving them directions as to their purchase of a boat."[12] Solon Buck, writing in 1914, believed that "the *Navigator* of Zadok Cramer in its numerous editions was one of the most useful guide-books ever published and the detailed descriptions, revised from time to time, make it especially useful to the historical student today."[13]

It was inevitable that a guidebook as useful and valuable as the *Navigator* would have its imitators. The most successful of such publications was Samuel Cumings's *The Western Navigator: Charts of the Ohio River in its Whole extent, and of the Mississippi River from the Mouth of the Missouri to the Gulf of Mexico, Accompanied by Directions for the Navigation of the Ohio and Mississippi*. The first edition was published in two volumes in Philadelphia by E. Littell in 1822. Volume one contains fifteen maps of the Ohio River and eighteen of the Mississippi, some of which are printed two to a double-page plate. The maps, larger and of much better quality than those in Cramer's *Navigator*, were compiled and engraved by Henry S. Tanner. Volume two contains navigation directions and is patterned after Cramer's guidebook. Cumings's two-volume *Western Navigator* was priced at ten dollars, in contrast to Cramer's one-volume *Navigator* at one dollar. The high price of Cumings's work may have limited its sales, for a second edition was published in Cincinnati in 1825 in a reduced one-volume format. The title of this edition was changed to *The Western Pilot*, which was the title of all subsequent editions through 1854.

Notes

1. Coolie Verner, *A Carto-Bibliographical Study of the English Pilot, the Fourth Book, with Special Reference to the Charts of Virginia* (Charlottesville, 1960), vii.
2. Eva G. R. Taylor, *The Mathematical Practitioners of Tudor and Stuart England* (Cambridge, 1954), 257.
3. For a detailed description see Clara Egli Le Gear, "The New England Coasting Pilot of Cyprian Southack," *Imago Mundi* 11 (1955): 137–44.
4. Geraint N. D. Evans, *Uncommon Obdurate: The Several Public Careers of J. F. W. Des Barres* (Toronto and Salem, Mass., 1969), vii.
5. John Clarence Webster, *The Life of Joseph Frederick Wallet Des Barres* (Shediac, New Brunswick, 1933), 27.
6. Leigh Jackson Russell, "Edmund March Blunt," *Essex Institute Historical Collections* 79 (Apr. 1943): 107–8.
7. Harold L. Burstyn, *At the Sign of the Quadrant: An Account of the Contributions to American Hydrography Made by Edmund March Blunt and His Sons* (Mystic, Conn., 1957), 21.
8. George C. Groce and David H. Wallace, eds., *The New-York Historical Society's Dictionary of Artists in America 1564–1860* (New Haven, Conn., 1957), 451.
9. Silvio Bedini, *Thinkers and Tinkers, Early American Men of Science* (New York, 1975), 361–62.
10. William Ratigan, *Great Lakes Shipwrecks and Survivals* (Grand Rapids, Mich., 1960), 135.
11. Charles W. Dahlinger, *Pittsburgh, a Sketch of Its Early Social Life* (New York, 1916), 174–75.
12. Leland D. Baldwin, *Keelboat Age on Western Waters* (Pittsburgh, 1941), 58.
13. Solon Justus Buck, *Travel and Description 1765–1865. Together with a List of County Histories, Atlases, and Bibliographical Collections and a List of Territorial and State Laws* (Springfield, Ill., 1914), 9.

16. *Urban Plans and Atlases*

The birth and growth of American cities is an interesting subject for study. The origins of most Old World cities are lost in the vague uncertainties of prehistory. By the time North America was settled, European cities had benefited from many centuries of planning and evolution. This experience was drawn upon by the promoters and colonists of the settlements in the New World. John Reps, a distinguished authority on the history of town planning and development, notes that "from the beginning of American settlement the planning of towns played an important role in the development of colonial empires by the European powers contending for the prize of the New World. Whether as market centers, bases for the exploration and exploitation of natural resources, military camps for the subjugation of a region, ports for fishing and trade, or havens from the religious persecutions of Europe, the founding of towns occupied a key position in colonial policy." Reps also reminds us that "the first settlers brought with them concepts of towns and cities derived from European experience. These ideas of the proper patterns of streets, building sites, and open spaces, and the institutional arrangements of land tenure were transplanted to an environment that differed sharply from Europe."[1]

Town planning and settlement were among the first objectives of the English colonists who immigrated to North America. Maps or plans were undoubtedly made of the early towns and villages but with few exceptions these existed only in manuscript copies, in limited numbers that were retained in official files. It was not until the late seventeenth century that a map of an Anglo-American city was printed. This plan is entitled *A Portraiture of the City of Philadelphia in the Province of Pennsylvania, 1683*. It was prepared by Thomas Holme and "sold by Andrew Sowle in Shoreditch London." Holme was born in Waterford, Ireland, in 1624 and is believed to have served under Admiral William Penn, father of Pennsylvania's William Penn, in the Hispaniola campaign of 1654. Some five or six years later he was active in advancing the Society of Friends in Ireland. Holme also appears to have engaged in surveying while in Ireland. On April 18, 1682, he was appointed surveyor general of Pennsylvania by William Penn. Within four days of receiving news of this assignment, Holme sailed for Pennsylvania and arrived there in June. Following Penn's directions, Holme surveyed and laid out a plan for the new city of Philadelphia with a gridiron pattern of streets. The manuscript plan was sent to London where it was engraved, printed, and published in 1683 in a book entitled *A Letter From William Penn . . . To the Committee of the Free Society of Traders of Pennsylvania residing in London . . . To Which is Added, An Account of the City of Philadelphia Newly Laid Out.*

More than half a century elapsed before another significant plan of Philadelphia was prepared. Published in 1752 it is entitled *A Map of Philadelphia and Parts Adjacent. With A Perspective of the State-House.* The plan extends north to beyond Germantown, south to Tinicum Island, and west to beyond Derby and Merrion Meeting. Produced by Nicholas Scull and George Heap, it was the first plan of Philadelphia to be printed in America. Seven years after this plan was published, Scull completed his *Map of the Improved Part of the Province of Pennsylvania*. In 1755 Heap prepared *A Prospect of the City of Philadelphia* under Scull's direction. It is a four-sheet panorama of the city, as viewed from the Jersey shore. Because of the large size of the print (224 cm. long when the sheets are joined) it was engraved in England.

During the early decades of the eighteenth century several plans of Boston were published. The first and most noteworthy of these was John Bonner's *Map of the Town of Boston in New England*, which was engraved by Francis Dewing and published in 1722. This edition is described by Reps as "the first printed plan of Boston to survive as well as the earliest extant town plan published within the present boundaries of the United States."[2] Bonner's plan, 41 by 59 cm. in size, proved to be exceedingly popular, and eight states of it were issued until 1769 (Fig. 16–1). All except the first were published by William Price. Many changes and additions were made on the plan's successive issues.

In 1728 William Burgis prepared and published a *Plan of Boston in New England* (Fig. 16–2), which he dedicated "To His Excellency William Burnet, Esq.," the governor of the Massachusetts colony in that year. The plan was engraved by Thomas Johnston, whose name is incorrectly inscribed on it as "Johnson." Reps notes that "Burgis obviously copied most of the details of Bonner's earlier work, plagiarism being traditional among cartographers then as now. Even the letters used in the legend to designate the churches were the same, the only difference being the addition of Christ Church, built in 1723."[3] The Burgis plan apparently was not able to compete successfully with the various states of Bonner's rendering, and only one edition of it is known.

During the colonial period, printed plans were available for only the larger cities. Some plans were engraved and printed in England but, by the second and third decades of the eighteenth century, American surveyors, engravers, and printers were preparing creditable city plans. The first plan of New York engraved and printed in that city is *A Plan of the City of New York from an Actual Survey Made by James Lyne*. The survey was probably made in 1730, and the plan drawn in the same year. It was very likely printed in the following year by William Bradford, who was at that time the provincial printer. Although the plate from which the plan was printed is one of the first examples of copper engraving in New York City, the engraver of the Lyne-Bradford plan is not known. Bradford dedicated the plan "To His Excellency John Montgomerie, Esq. Capt. Genl. & Govr. in Chief of his Majesti's Provinces of New York & New Jersey." The plan is known in but one state, and the original exists in only a small number of copies. Several facsimilies have, however, been reproduced over the past two and a half centuries.

Even rarer than the Lyne-Bradford plan is *A Plan of the City of New York From an Actual Survey Anno Domini M,DCC,LV*. The survey was made by F. Maerschalck, the city surveyor, and the plan was published by G. Duyckinck in 1755. I. N. Phelps Stokes states that "a microscopic comparison of this plan with the Bradford plan proves beyond the shadow of a doubt that the two were printed in part from the same plate."[4] The Maerschalck plan is also known in only one state, of which there are only three extant copies. A number of facsimiles of it have been published.

Cartographic output was particularly prolific in the years before and during the American Revolution. This was stimulated largely by the many military and naval surveyors and mapmakers who accompanied British forces participating in the French and Indian and revolutionary wars. Because the larger colonial cities were strategic objectives, particularly in the latter conflict, military cartographers focused considerable attention on preparing plans of urban centers. American surveyors and mapmakers also engaged in urban mapping during the years preceding the War of Independence. One such American effort was the 1774 plan *To the honorable House of Representatives of the Freemen of Pennsylvania . . . of the City and Liberties of Philadelphia, with the Catalogue of Purchases Is Humbly Dedicated by their most Obedient Humble Servant John Reed*. This plan, which is 80 by 155 cm. in size, was engraved by James Smithers and printed in Philadelphia by Thomas Man. Also of the pre-revolutionary war period is *A Plan of the City of Philadelphia the Capital of Pennsylvania, from an Actual Survey by Benjamin Easburn, Surveyor General, 1776* (Fig. 16–3). Engraved by P. Andre, the plan was published in London on November 4, 1776, by Andrew Dury. It is un-

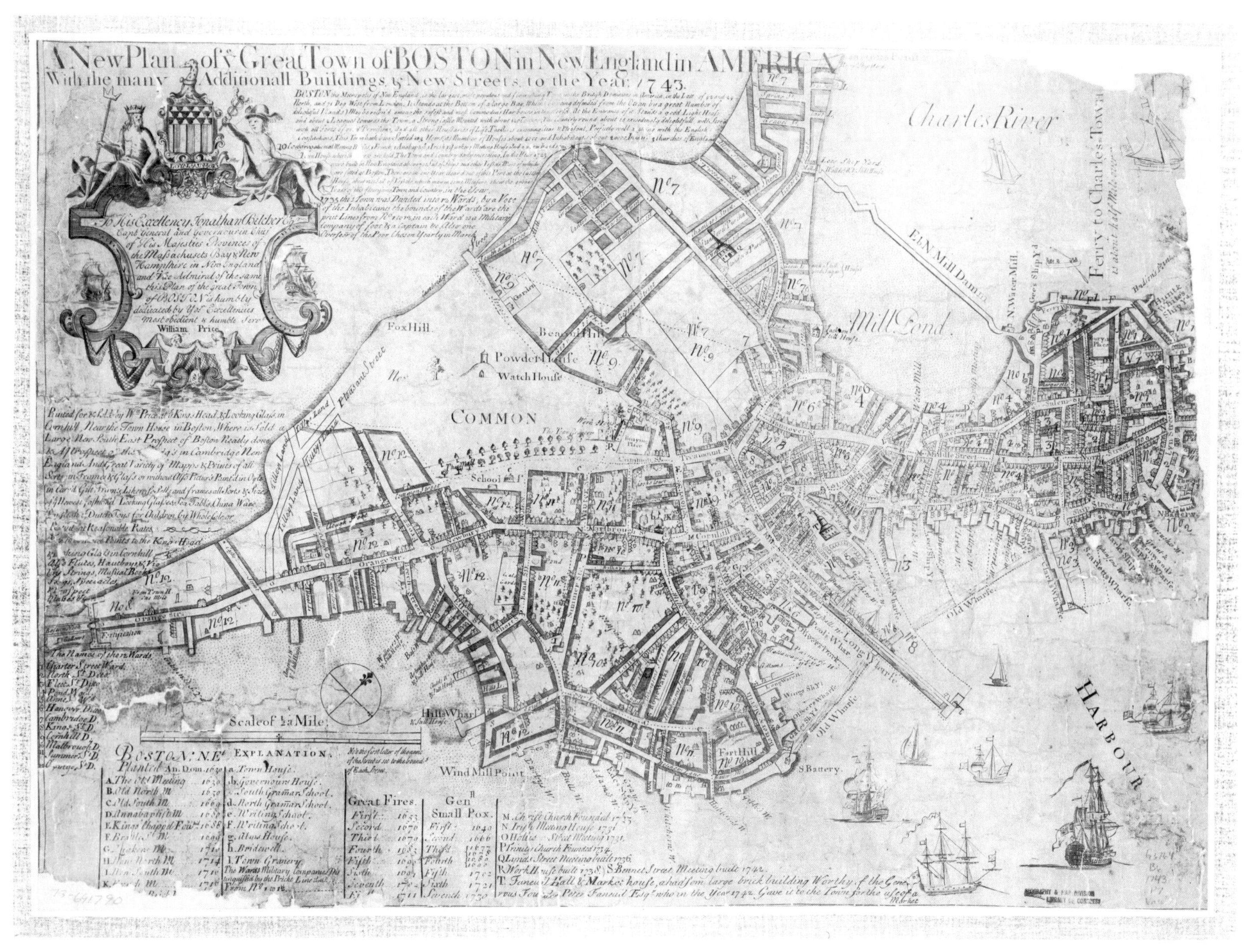

Fig 16–1. The 1743 edition of John Bonner's map of Boston published by William Price.

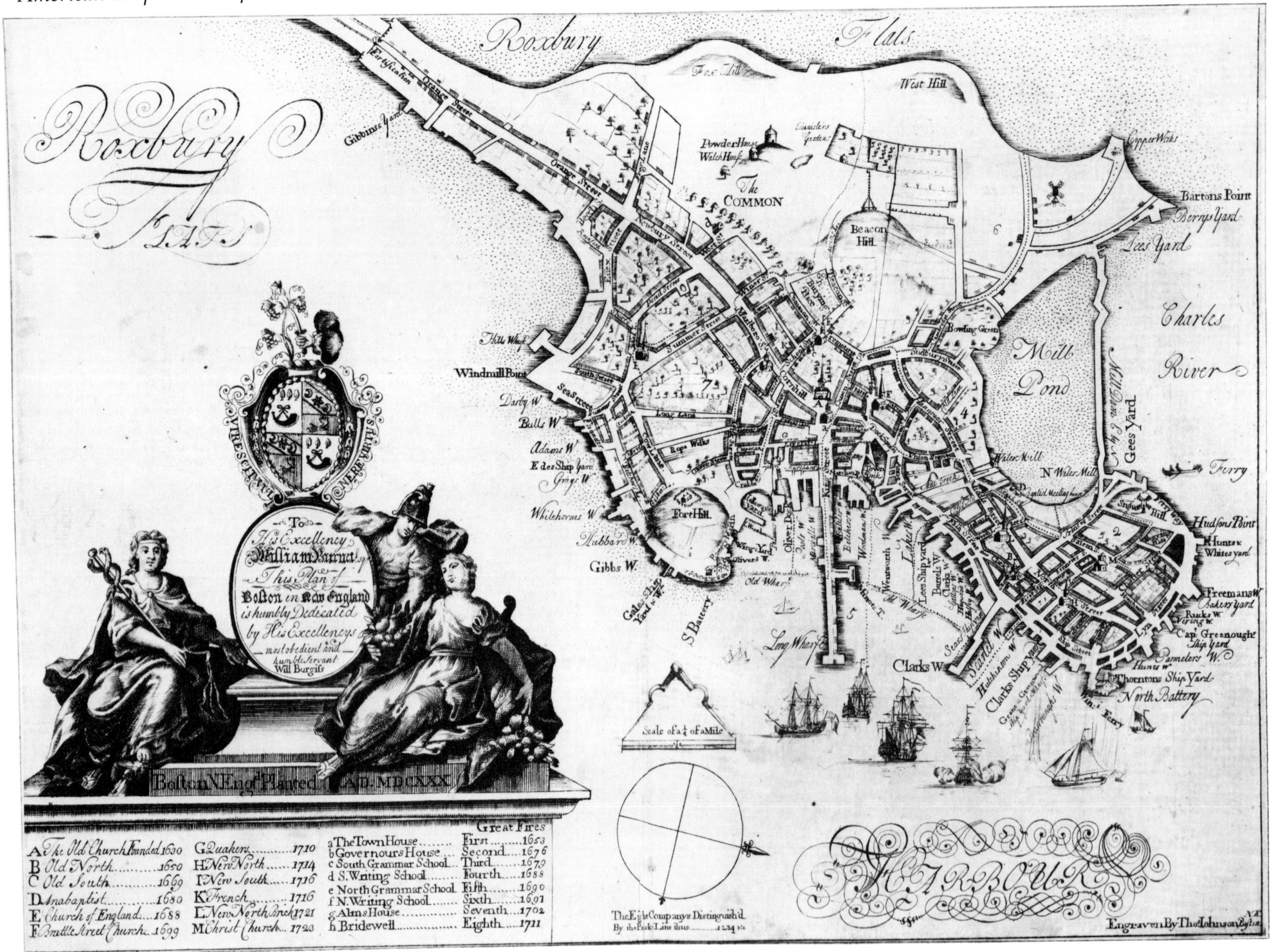

Fig. 16–2. William Burgis's 1728 *Plan of Boston in New England* was largely derived from Bonner's map of Boston.

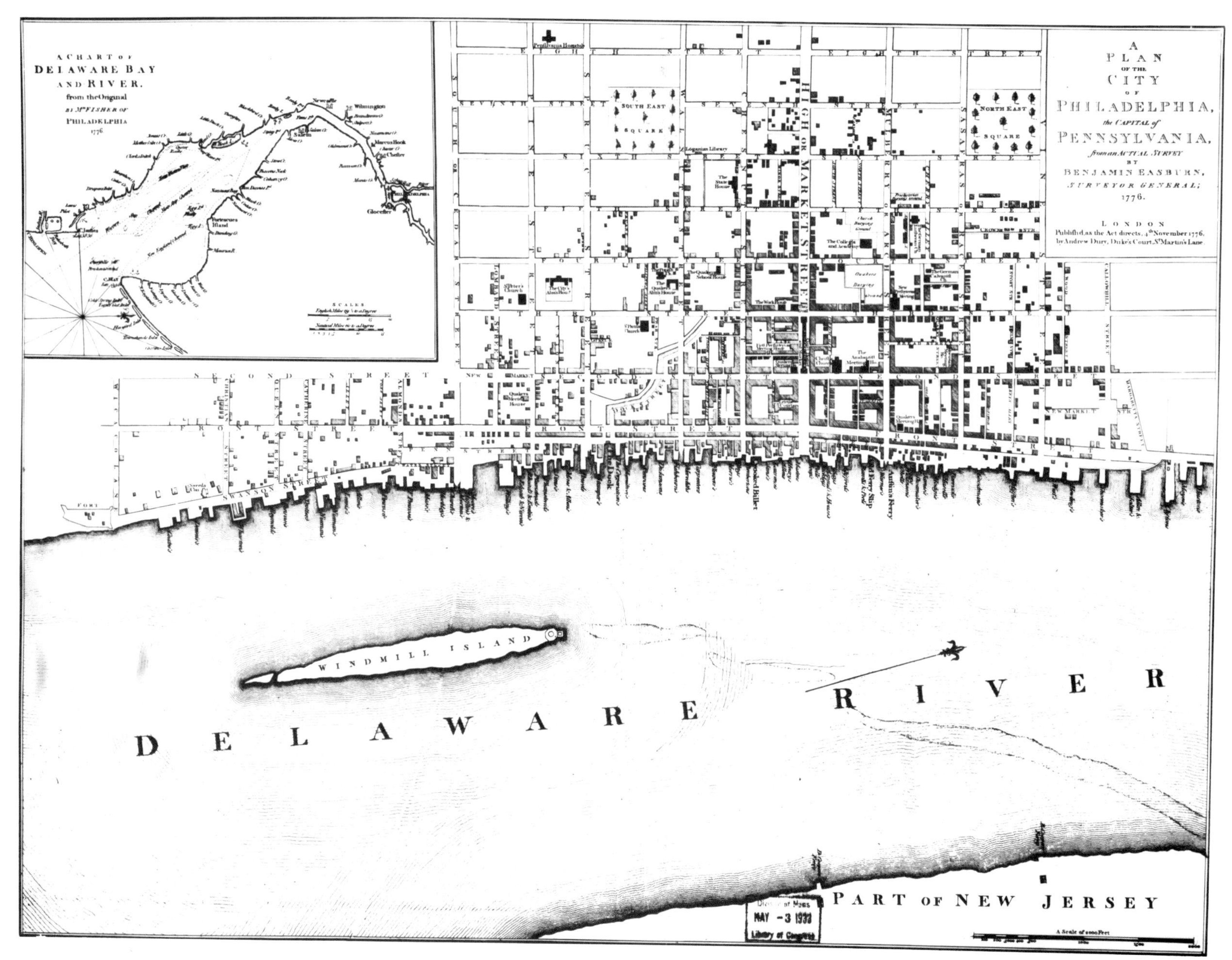

Fig. 16–3. Benjamin Easburn, surveyor general of Pennsylvania, prepared this plan of Philadelphia in 1776.

likely, therefore, that many copies were available to residents of Philadelphia before the end of the war.

Boston was one of the earliest theaters of the War of Independence, and British military cartographers produced several plans of the city. Published in London by Dury in 1775 was *A Plan of Boston and Its Environs shewing the true Situation of His Majesty's Army*. The map also notes that it was "Drawn by an Engineer at Boston, Oct., 1775." The engraver was John Lodge. Another wartime product is *A Plan of Boston in New England with Its Environs*, which was published in London in 1777. The plan was drawn by Henry Pelham in 1775 and 1776 and engraved by Francis Jukes in aquatint on two sheets. When the two sheets are joined, the plan measures 72 by 108 cm. Pelham, a colonial Loyalist, had at his disposal various military maps and sketches. Reps regards his plan as "one of the finest maps of Boston and vicinity ever produced."[5]

We are also indebted to British military engineers for several excellent plans of New York City in the revolutionary war years. Surveys for the earliest of these were carried out early in 1766 by Lieutenant John Montresor, chief engineer under General Thomas Gage, commander in chief of British forces in America. Montresor completed a manuscript draft of his survey in February 1766, which he presented to General Gage. At the end of October in that year Montresor returned to England on a six-month leave and took with him the drafts of several plans he had prepared, including that of New York City. These were submitted to the map engraving establishment that had been founded by John Rocque. Although Rocque had died in 1762, his widow, Mary Ann, carried on the business assisted by several engravers. One of them, P. Andrews, prepared the plate for Montresor's plan of New York in 1767. It is titled *A Plan of the City of New York & Its Environs*. In 1775 Andrew Dury acquired a number of Rocque plates, among them that of Montresor's plan. He published the plan, with no changes except in the imprint and date, in 1775.

Lieutenant Bernhard Ratzer of the Royal American Regiment infantry unit, supplemented Montresor's surveys while Montresor was in England. Ratzer sent his completed draft to Thomas Kitchin in London who engraved and published the map in 1769 (Fig. 16–4). On it the engineer's name was mispelled "Ratzen," and the plan is, therefore, generally identified as the Ratzen Plan. Ratzer continued his survey in the environs of New York City which, with the original area shown on the Ratzen Plan, he assembled into a large three-sheet map, which was also engraved by Kitchin and published by him in 1770. In addition to the lower portion of Manhattan Island, the plan includes the western extremities of Long Island and a segment of New Jersey. At the bottom of the map sheet there is "A South West View of the City of New York Taken from the Governour's Island." Stokes described this Ratzer map as "one of the most beautiful, important, and accurate early plans of New York."[6] A second state of this map was published in January 1776 by Jefferys & Faden of London.

The most pressing cartographic demand after peace was reestablished was for maps of the new republic and of its several states. Urban mapping accordingly had low priority, and the needs of city dwellers, therefore, continued to be met for a decade or so afterwards by the plans already published. The expansion of existing towns and cities and the establishment of new settlements beyond the Appalachians induced a number of private surveyors and cartographers to prepare maps of various American cities during the last decade of the eighteenth century.

One of the earliest post-Revolution city plans of Charleston, South Carolina, was published in London in 1790 for the London-based Phoenix Assurance Company, Ltd. Soon after the Treaty of Paris was signed, the Phoenix Assurance Company extended its fire insurance coverage to the United States and West Indies. The Charleston plan, which is believed to be the earliest fire insurance map published, was prepared from a survey made for Phoenix by Edmund Petrie in 1788. It is titled

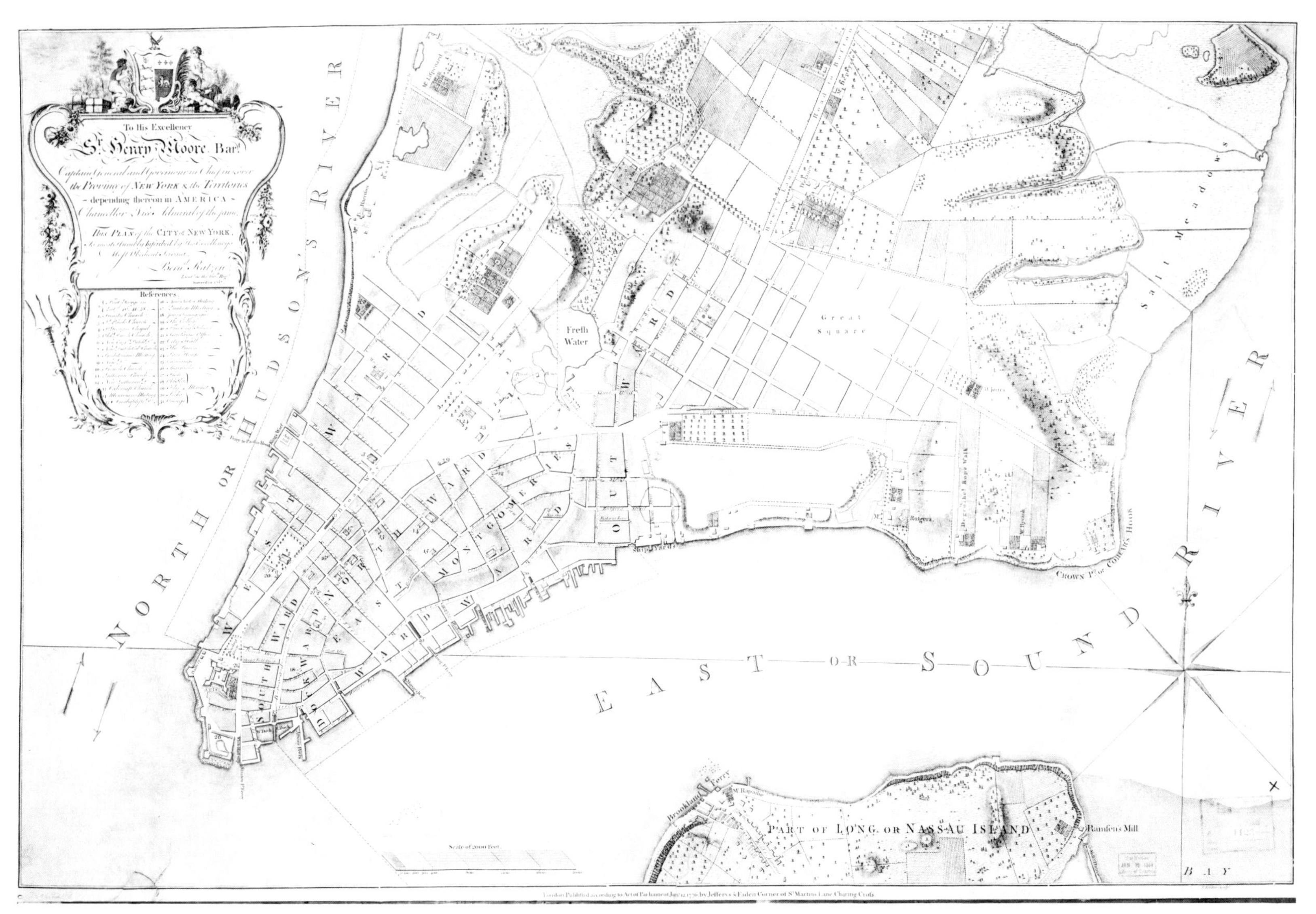

Fig. 16–4. This 1769 plan of New York City is often identified as the "Ratzen" plan because the name of the compiler, Bernhard Ratzer, is misspelled on it.

Ichonography of Charleston, South Carolina, and was compiled "At the Request of Adam Sunno, Esq. for the use of the Phoenix Fire-Company of London,Taken from Actual Survey, 2d August 1788 by Edmund Petrie. Published 1st Jany. 1790 by E. Petrie No. 13 America Square" (Fig. 16–5). The map lists and locates public buildings and wharves, gives the width of streets and lanes, identifies ninety-nine private and commercial properties, and locates the fire station and public wells. Prepared as it was for the specialized use of a London insurance company, the plan very likely had little distribution in the United States. Only two or three extant copies are known today. Phoenix is known to have prepared surveys of half a dozen other American cities, but no copies have been located, and they may have been retained only in manuscript drawings in company files.

On July 16, 1790, President George Washington signed into law "An Act for establishing the temporary and permanent seat of the Government of the United States." Early in 1791 the president selected Pierre Charles L'Enfant as the chief engineer to plan the new capital city and Andrew Ellicott as the surveyor. On June 28, 1791, Washington, accompanied by L'Enfant and Ellicott, surveyed the site selected for the city and examined L'Enfant's preliminary plan. Differences developed between L'Enfant and government officials, however. As a result, the completed draft of the city was prepared by Ellicott and was then turned over to the engravers. Two engravings were actually prepared of Ellicott's draft, one by the Philadelphia firm of Thackara & Vallance (Fig. 16–6) and the other by Samuel Hill of Boston. The Hill engraving was completed first and copies of the printed plan were exhibited at the second sale of public lots on October 8, 1792. "The Boston and Philadelphia engravings were distributed widely in the United States and abroad, some five thousand copies being in circulation by the end of 1792."[7]

One of the most active American-born cartographers in the post-revolutionary war decades was Osgood Carleton. Among his published works are maps of Massachusetts, Maine, and New Hampshire, charts of the New England coast, and plans of the city of Boston. His earliest plan of the city was prepared in 1795 and was first published in John West's *Boston Directory* of 1796. Although a scroll beneath the map's title carries the credit "S. Hill, sc.," some authorities believe that the map was engraved by Joseph Callendar. Carleton's plan was also published in the 1798 and 1800 editions of the directory and was issued separately at a larger scale in 1797 and 1800.

Several maps and plans of Boston were prepared between 1814 and 1830 by John G. Hales, who was born in England in 1785. He published in 1814 a *Map of Boston in the State of Massachusetts, Surveyed by J. G. Hales Geogr. & Surveyor*. The plan, which shows the lines of estates and individual houses and specifies the construction materials used in building them, was engraved by T. Wightman, Jr. There are few extant copies of this plan. In 1821 Hales published the book *Survey of Boston and its Vicinity*, which included his *Map of Boston and Its Vicinity From Actual Survey by John G. Hales*. The map was engraved by Edwin Gillingham; it was also issued separately. An 1819 edition of the map in the collections of the Library of Congress has the imprint "Boston Published by John G. Hales, Proprietor, & by J. Melish Philadelphia 1819." Melish's name is omitted on the 1820 and 1829 editions. Hales died on May 20, 1832, at the age of forty-five.

One of the first maps of New York City published after the Revolution is a *Plan of the City of New York*, which was compiled by James McComb, Jr., for the 1789 edition of *The New York Directory*. The map, which was engraved by Cornelius Tiebout, was also issued separately. Tiebout, it will be recalled, also engraved the title page and at least some of the plates of Christopher Colles's *Survey of the Roads of the United States*, the first parts of which were also published in 1789. A native of New York City, he was one of the earliest American-born engravers. A later map, the *New & Accurate Plan of the City of New York in the State of New York in North America*, was published in 1797 and is described by Stokes as "one of the most

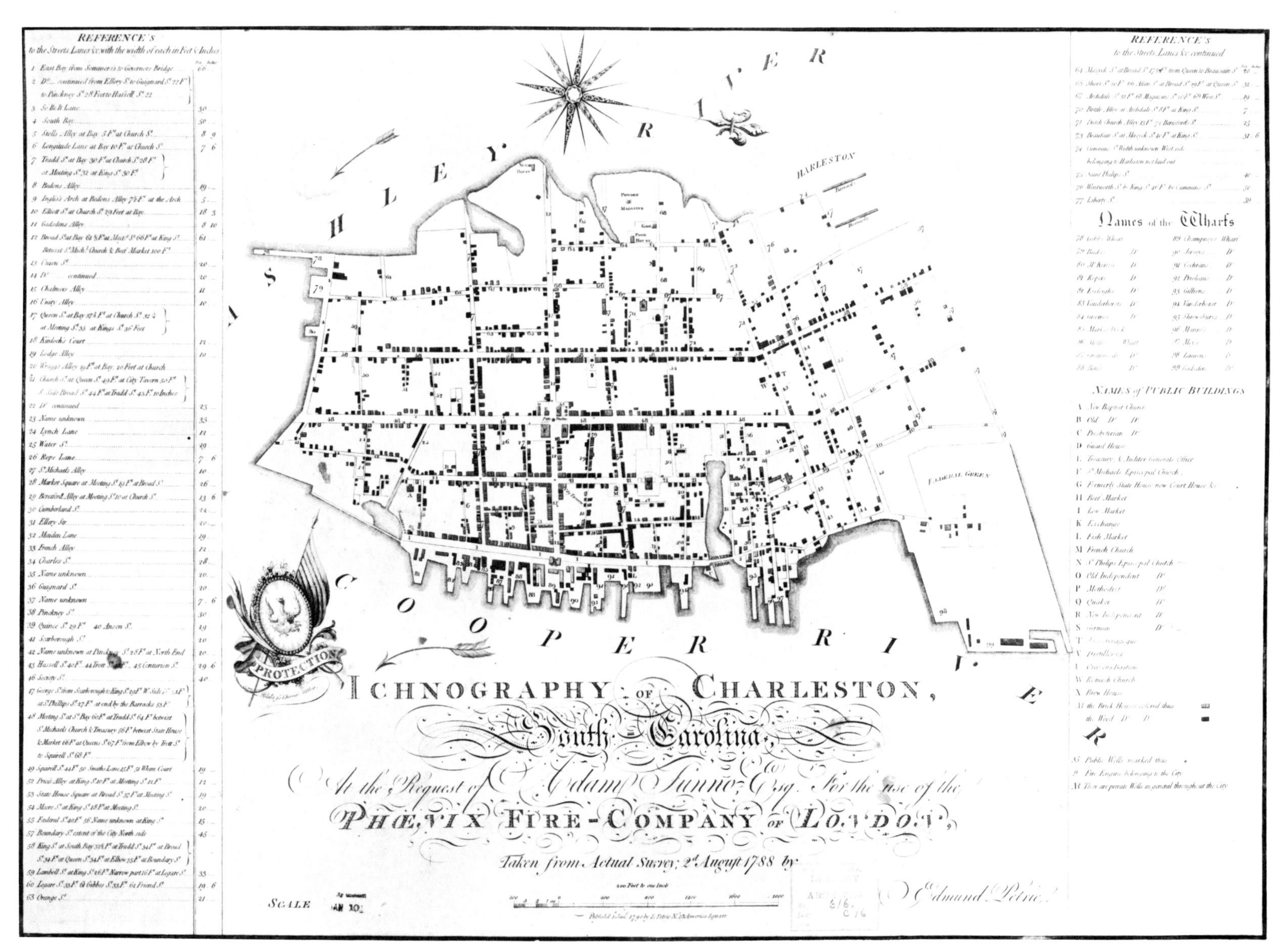

Fig. 16–5. Edmund Petrie's 1790 *Ichonography of Charleston, South Carolina* is considered to be the earliest American fire insurance map.

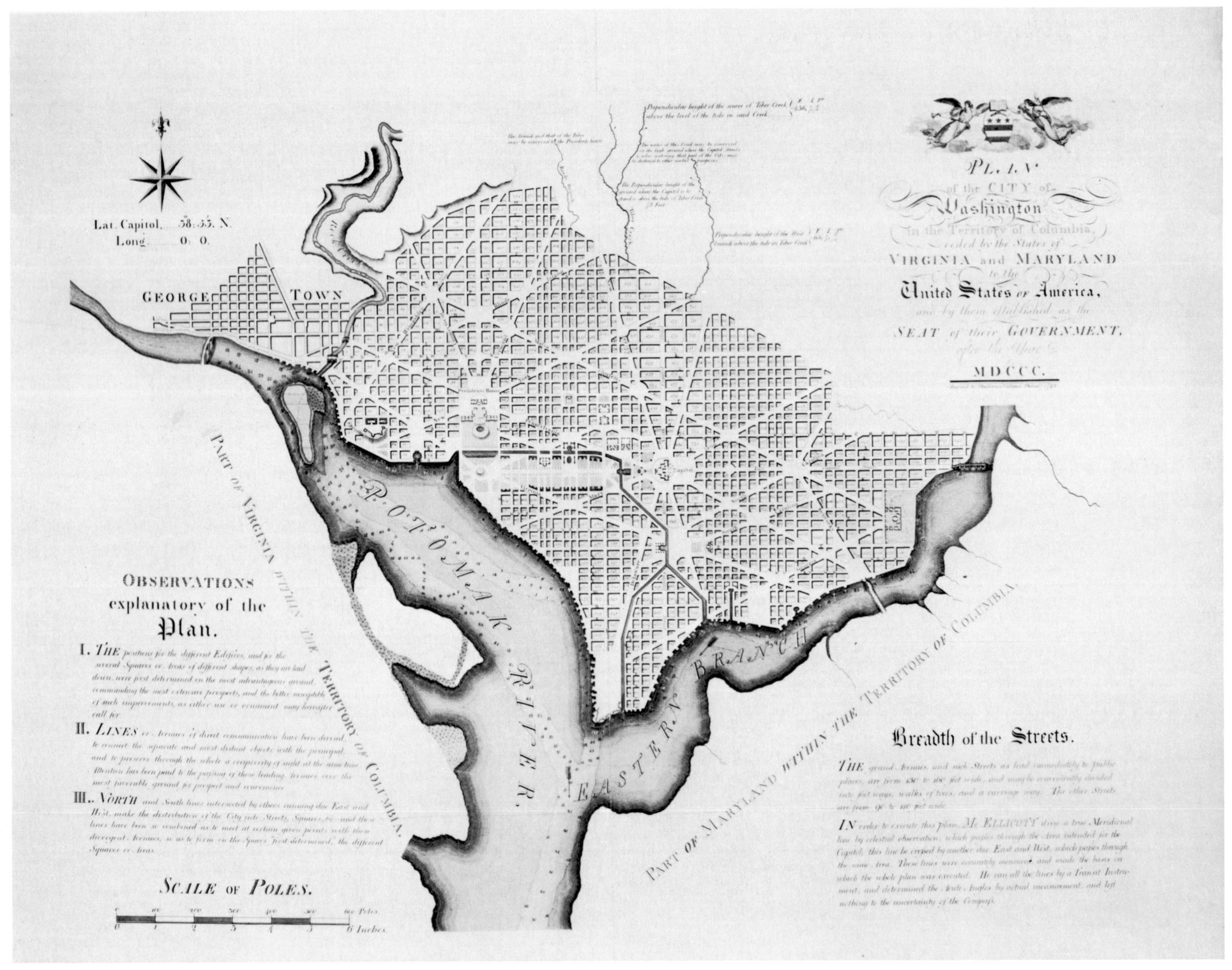

Fig. 16–6. Andrew Ellicott's plan of Washington, D.C., was the earliest published plan of the city. This 1792 plan was engraved by Thackara & Vallance and published in Philadelphia.

accurate and beautiful engraved plans of the city, and particularly interesting on account of its tiny bird's-eye views of some of the most important buildings, as well as for the clear idea which it gives of the country lying along the East River in the vicinity of Crown Point."[8] The map, which was compiled by B. Taylor and engraved by John Roberts, is sometimes referred to as the Taylor-Roberts Plan. Stokes identifies only four extant copies of the plan. A reproduction of it was published in David T. Valentine's *Valentine's Manual* of 1853. In December 1797, New York City officials engaged Casimir T. Goerck and Joseph F. Mangin to survey and map the city. The resulting plan was engraved by Peter Maverick and published in November 1803 as *A Plan and Regulation of the City of New York*. Original copies of the Goerck-Mangin plan are extremely rare, but a reproduction of it was made for *Valentine's Manual* of 1856.

The plans of New York City heretofore described portray only the lower portion of Manhattan Island. By the first decade of the nineteenth century, however, population pressures induced the legislature and governor of New York State to pass an act providing for the laying out of streets and roads throughout the full extent of Manhattan Island. To carry out the act, a commission was created and three commissioners were named: Gouverneur Morris, Simeon De Witt, and John Rutherford. In late 1811 John Randel, Jr., secretary and chief surveyor of the commission, completed three manuscript drafts of a *Map of the City of New York and Island of Manhattan as Laid Out by the Commissioners Appointed by the Legislature April 3d 1807*. Randel intended to have the map engraved and published but, because of the War of 1812, he decided to defer publication so copies of the map would not provide strategic information to the enemy. However, William Bridges, the architect and surveyor of New York City, received permission to prepare the commissioners' map for publication. The map, beautifully engraved by Peter Maverick on six sheets, was issued in separate sheets and as a mounted wall map (Fig. 16–7). In the latter format it measures 235 by 63 cm. Needless to say, Randel was deeply incensed by Bridges's action and criticized the accuracy of the map. He then canceled his own publication plans. Bridges also published in 1811 a small booklet entitled *Map of the City of New York and Island of Manhattan with Explanatory Remarks and References*, which was printed by T. & J. Swords, of 160 Pearl Street. The small volume includes the legislative acts of April 3, 1807, and March 24, 1809, remarks of the commissioners, references explaining the map, and a list of subscribers.

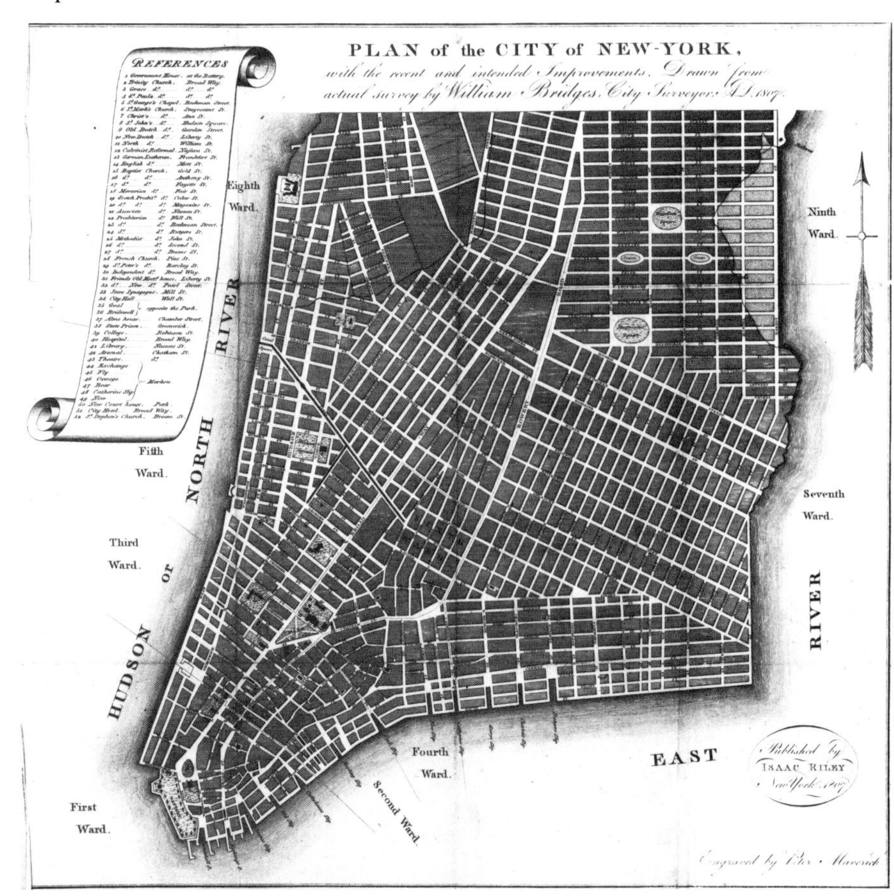

Fig. 16–7. William Bridges's 1807 plan of New York City.

Maps and plans were also prepared for a few towns and cities in upper New York State in the post-revolutionary war decades. Simeon De Witt published *A Plan of the City of Albany* in 1794 that was "surveyed at the request of the Mayor Aldermen and Commonality" of the city. The plan, which includes drawings of the courthouse and the prison, was engraved by Isaac Hutton.

During the last decade of the eighteenth century when Philadelphia served as the capital of the United States, several interesting and attractive maps and plans were published of the city. Two of the surveyors and cartographers who created these maps were French exiles and one was a British military engineer who had returned to the United States after the revolutionary war. The French Revolution forced some engineers and surveyors to flee France and settle in such French possessions as Saint-Domingue (now Haiti) in the West Indies. The uprising in 1791 forced an exodus of whites from Saint-Domingue to the United States. Among these exiles were the two French engineers A. P. Folie and Charles P. Varlé. They undoubtedly were acquainted, but in America they pursued separate and competitive careers.

In February and March 1794, a proposal was printed in several journals to publish "by subscription, a grand plan of the city of Philadelphia and its environs; taken from actual survey by A. P. Folie, geographer from St. Domingo." The proposal promised that "nothing will be omitted to render this useful and desirable work acceptable to an enlightened public, that is in the author's power." The original draft of the plan was on view at B. Davies's bookstore, where subscriptions were also received. Because of this reference, the resulting map is sometimes referred to as "Davies's Map," although his name does not appear on the published plan. The cartouche of the plan reads, "To Thomas Mifflin Governor and Commander in Chief of the State of Pennsylvania This Plan of the City and Suburbs of Philadelphia is respectfully inscribed by The Editor 1794." Folie is credited as the draftsman in the lower left corner, and the engravers "R. Scot & S. Allardice Sculpsit" are credited in the opposite corner.

The Englishman John Hills produced a number of maps and plans for the British army during the revolutionary war.[9] Discharged from military service in 1784, he chose to remain in the United States. He settled first in Princeton, New Jersey, but relocated to New York City in 1785. In the following year, he moved to Philadelphia, where he remained for the next several years. In 1796 Hills published a map of Philadelphia, 94 by 67 cm. in size, for which he had offered subscriptions as early as 1790. Its title is *This Plan of the City of Philadelphia and Its Environs, shewing the improved Parts* (Fig. 16–8). The map "is Dedicated to the Mayor, Aldermen and Citizens thereof, By their most obedient Servant, John Hills, Surveyor and Draughtsman, May 30th 1796." In the lower left corner of the plan is the imprint "Philadelphia, Published and Sold by John Hills, Surveyor & Draughtsman, 1797." The plan was "engraved by John Cooke of Hendon, Middlesex, near London."

In 1808 Hills published a circular map of Philadelphia within a ten-mile radius, its center being the hub of the city. It is titled *A Plan of the City of Philadelphia and Environs Surveyed by John Hills in the summers of 1801, 2, 3, 4, 5, 6, & 7*. The map was engraved on nine plates by William Kneass, who was born in Lancaster, Pennsylvania, on September 25, 1781. He was occupied as an engraver in Philadelphia for about three decades and, in January 1824 he was appointed engraver and diesinker at the U.S. Mint. From 1818 to 1820 Kneass was in partnership with James H. Young, another early nineteenth-century engraver. The Philadelphia map was "dedicated with Gratitude and Esteem To His numerous & Respectable Subscribers By Their Obliged Friend John Hills." Peter J. Guthorn notes that "the map has been encountered in several variations, some of which are inconsistent in detail."[10]

Charles P. Varlé was born in southern France around 1770. He received training in engineering in Toulouse

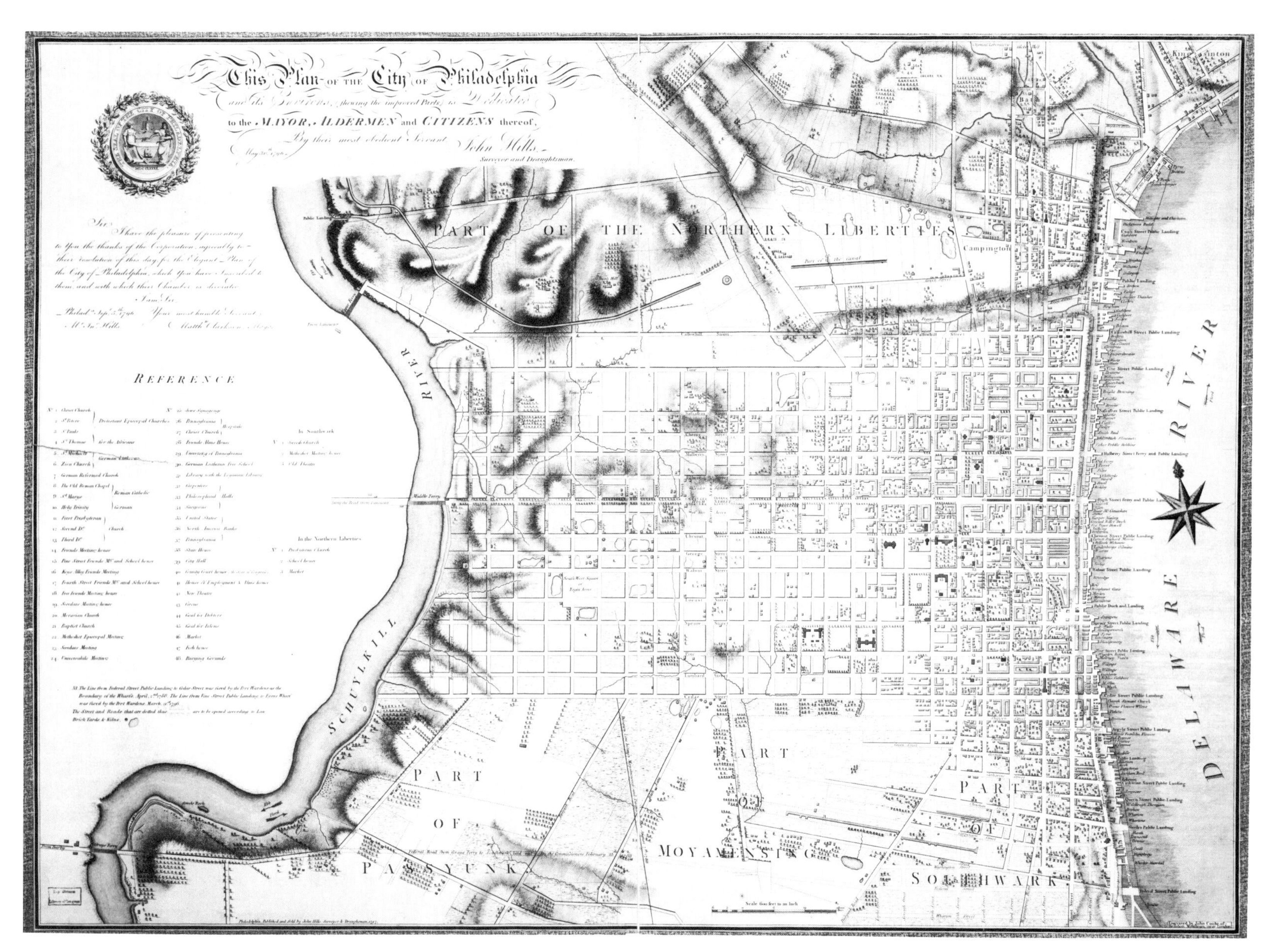

Fig. 16–8. John Hills's 1796 *Plan of the City of Philadelphia*.

and worked on several canal, turnpike, and bridge projects before the French Revolution induced him, like Folie, to immigrate to Saint-Domingue. In Saint-Domingue he worked for M. de Vincents, chief engineer of the island until the slave revolt of 1791 forced both Varlé and de Vincents to flee to the United States. Varlé settled in Philadelphia where, under the name Peter C. Varlé, he served as a private in the Fourth Company of the Third Regiment of the Philadelphia Militia.[11] Varlé appears to have used the name Peter, or the initials P. C., for about ten years; after 1807 all references to him are under the name Charles Varlé. Varlé then obtained an appointment in the War Department through de Vincents. De Vincents had served in the American Revolution and was an acquaintance of George Washington and General Henry Knox, the secretary of war. Varlé remained in government service only a short time, for he left to repair a canal on property owned by Knox in Maine.

Upon returning to Philadelphia in 1796, Varlé conducted surveys for a map of that city. The map, which is undated, is thought by some authorities to have been published in 1796, with a second edition in 1802. Most probably there was but one edition, published in 1802. The map inscription reads, "To the citizens of Philadelphia this new plan of the city and its environs is respectfully dedicated by the editor." In the lower left corner is the credit "P. C. Varlé, geographer & engineer, del." The map was engraved by Robert Scot of Philadelphia.

Folie and Varlé also prepared plans of the city of Baltimore, Maryland. In 1792 Folie published the *Plan of the Town of Baltimore and Its Environs Dedicated to the Citizens of Baltimore Taken upon the spot by their most humble Servant A. P. Folie French Geographer* (Fig. 16–9). It was engraved by James Poupard, a Philadelphia jeweler, goldsmith, and engraver. Some accounts indicate that he also was a French citizen and had immigrated to the United States from the island of Martinique. Varlé, who had moved to Maryland in 1798, published his map of Baltimore in 1833. Titled *Plan of the City of Baltimore including the South Baltimore Cos. Grounds*, it was accompanied by his pocket guide *A Complete View of Baltimore with a Statistical Sketch*. This was Varlé's last cartographic work, and in his pocket guide he said goodbye to the citizens of Baltimore:

> Gentlemen—After having spent among you about thirty years of my life, in the enjoyment of as much happiness as my fondest wishes could anticipate, I have at length resolved to visit France, my native country, which has always been the steadfast friend and ally of this republic. In bidding adieu to your hospitable city, I leave behind me the present work, as a feeble testimony of the interest I have always taken in its welfare and happiness. That you may receive from heaven a continuance of prosperity, and that your present flourishing conditions may not experience the slightest diminution, will ever be the fervent prayer of your obedient servant, and fellow citizen, Charles Varlé.[12]

Varlé probably died in France, for there is no evidence of subsequent activities in the United States.

In 1801 the Baltimore firm of Warner & Hanna published an attractive colored plan of the city. It was based on the Folie plan but, unlike the Folie map, it includes only the city proper. The plan was engraved by Francis Shallus of Philadelphia, who worked as an engraver between 1797 and 1821. Maps based on the Warner & Hanna plan illustrated several early Baltimore city directories. In 1947 the Peabody Institute of Baltimore published an attractive facsimile of this plan, which was reproduced by collotype by the Meriden Gravure Company of Meriden, Connecticut.

The July 27, 1811, issue of the *Federal Gazette and Baltimore Daily Advertiser* printed a proposal to publish by subscription an entirely new survey of Baltimore by T. H. Poppleton, described as a "Practical Land Surveyor and Draughtsman." Surveys no doubt began in 1811 or 1812, but it was not until February 1818 that Maryland's general assembly commissioned the cartog-

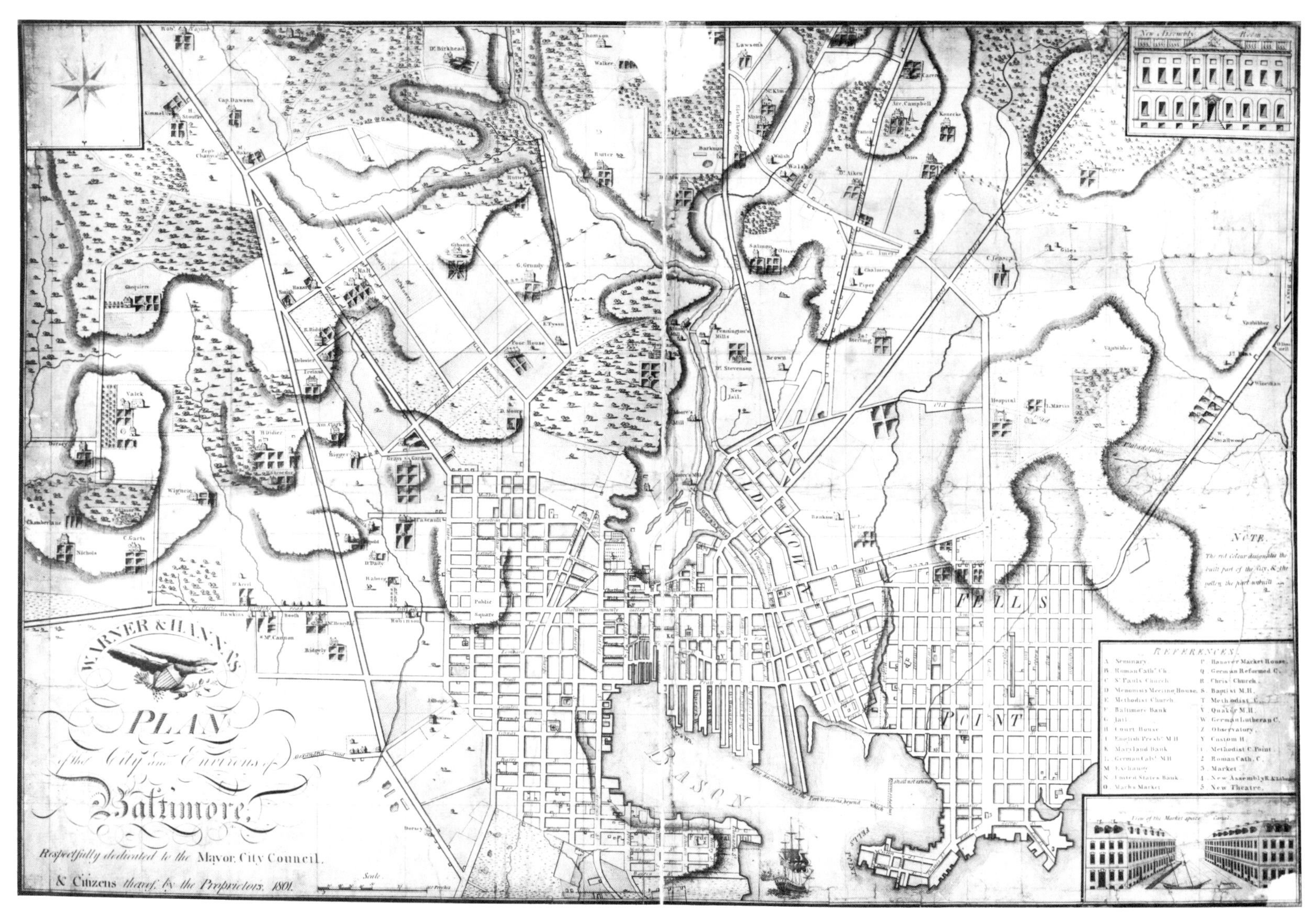

Fig. 16–9. This attractive *Plan of the City and Environs of Baltimore* was published in 1801 by the Warner & Hanna firm. It was based in part on A. P. Folie's 1792 plan and was engraved by Francis Shallus of Philadelphia.

rapher to print the map. The large plan was engraved on two sheets, which when joined measure 136 by 110 cm. Its title inscription is *This Plan of the City of Baltimore as enlarged & laid out under the direction of the Commissioners Appointed by the General Assembly of Maryland in Feby. 1816 Is Respectfully Dedicated to the Citizens thereof By Their Obt. Servt. T. H. Poppleton Surveyor to the Board C. P. Harrison, Script. Sculpt. New York 1823*. The map is framed by thirty-seven marginal illustrations of public buildings, churches, monuments, and fountains drawn by Joseph Cone. There is also an inset of a 1729 plan of the city with perspective views dated 1752 and 1822. The engraver of the plan, Charles P. Harrison, was born in England in 1783. He was a member of the Harrison family of engravers. Little is known about the surveyor Poppleton apart from the several maps he produced. After his 1811 proposal did not yield an adequate number of subscriptions to justify publishing the Baltimore plan, he apparently went to New York City. In 1817 the firm Prior & Dunning published his *Plan of the city of New-York the greater part from actual survey made expressly for the purpose (the rest from authentic documents) by Thos. H. Poppleton, city surveyor*. It was engraved by Peter Maverick and printed by Samuel Maverick. The plan only shows lower Manhattan below Thirtieth Street. Editions of the plan, "revised and corrected by Wm. Hooker," were published in 1826 and 1829, the former by Prior & Dunning, the latter by Prior & Brown.

All of the city plans heretofore described were prepared, with few exceptions, by individual surveyors, cartographers, and engravers. In some instances they received encouragement and support from legislatures or city councils. In general, however, they were moved to prepare the maps and plans to fill the obvious needs and demands for such essential tools in an awakening democracy. They, no doubt, hoped to derive financial gain from their endeavors, but these hopes were rarely realized. By the second and third decades of the nineteenth century, though, these private entrepreneurs were gradually replaced by several commercial map publishing companies operating in the United States. These establishments included those founded by John Melish, Henry S. Tanner, Fielding Lucas, Jr., David H. Burr, John Farmer, Edmund Blunt, and Blunt's son-in-law William Hooker.

While some publishers, such as Melish and Tanner, were primarily engaged in general map or atlas publishing, they did produce a small number of city plans as well. Farmer concentrated his cartographic efforts on the midwestern states and on plans of Detroit. In his later years Lucas devoted much of his publishing efforts to plans of Baltimore. Following his move from Newburyport, Massachusetts, to New York City in 1811, Blunt expanded his activities to publish guides and maps of New York City as well as coastal charts. Hooker engraved plans of the city for early editions of Blunt's *Strangers' Guide to New York*, but by 1824 he was publishing his own *New Pocket Plan of the City of New York*, which was issued in a number of editions until the mid-1840s. Hooker also engraved city plans for other publishers, among them Humphrey Phelps, Peabody & Company, and A. W. Wilgus of Buffalo, New York.

The population of the country had increased appreciably by the 1840s, and the new settlements established beyond the Appalachians were developing into important urban centers. Few trans-Appalachian cities, however, had engraving, printing, or publishing facilities before the middle of the nineteenth century, and plans of these cities are rare. This was not the case with the older, more established cities of the eastern seaboard. City directories and guidebooks illustrated with maps and plans were popular publications. John Disturnell entered the guidebook publishing field around 1833 and produced a number of such works in numerous editions over the next several decades. J. Calvin Smith and T. R. Tanner (son of Henry Tanner) were at times associated with Disturnell in preparing some of them. Disturnell also issued city plans separately, folded within simulated leather covers. In addition, he published several plans of New York City by Burr in the late 1830s. As noted earlier, Tanner shifted from map and atlas publishing to guidebook publishing in the early 1840s, with his headquarters in New York City. New York appears to have been the most active center for urban map publishing until the middle of the nineteenth century. The earliest publications of J. H. Colton were city plans. Some of these were included in Colton atlases, and some were issued as separates. Plans of New York, Brooklyn, Jersey City, and other urban centers were published by the Colton firm as late as 1886.

The introduction of lithography in the United States revolutionized urban plan publishing, as it did the other branches of commercial cartography. The first lithographically printed map published in the United States illustrated an article in the 1822 volume of the *American Journal of Science and Arts*. Although a number of small lithographic shops were established in Boston, Philadelphia, and New York, engraving persisted as the major technique for reproducing maps and other graphics until the 1840s.

The first lithographic shop to survive for a decade or more was the one established in Boston in 1826 by William and John Pendleton. Among the early cartographic items lithographed by this firm were the plans of Massachusetts towns that had been prepared in accordance with an 1830 act of the state legislature by local surveyors. These plans served as Simeon Borden's source material in compiling his 1844 *Topographical Map of Massachusetts*, which was engraved and not done by Pendleton's. Between 1830 and 1836, a number of the town plans were published, and they were reproduced and printed by lithographic techniques in most instances. Pendleton's Lithography and its successor, Moore's Lithography, printed most of these plans. The Pendleton firm also lithographed plans of other cities, including the 1835 *Plan of the Town of Halifax, Nova Scotia*. John Pendleton did not stay in business with his brother long and operated a lithographic shop in New York City for a brief period. A cartographic item bearing a New York Pendleton imprint is the plan *The City of New York drawn from actual surveys as furnished by the several city surveyors 1834. Pendleton, lithographer No. 192 B.Way*.

Except for the larger metropolitan centers, the market for city and town plans was not great enough to offset the cost of engraving. Lithography, being a simpler and less costly technique, was particularly suited to printing city maps and plans. Because the cartographer's or draftsman's drawing could be mechanically transferred to a lithographic stone or zinc plate, the relationship between the cartographer and the lithographer could be less intimate than that between the cartographer and the engraver. The lithographer reproduced various other graphics in addition to maps and only infrequently engaged in publishing. His task in the work flow pattern was limited to transferring the graphic to the lithographic stone, from which multiple copies could be printed.

An early lithographer was Nathaniel Currier. After an apprenticeship with Pendleton's Lithography and M. E. D. Brown, he established his own lithographic shop in New York City. Among his earliest lithographic works was the *Map of Cleveland and Its Environs*, which was surveyed and published by Ahaz Merchant in October 1835. The imprint credit is "N. Currier's Lithy. New York." Other Currier city plans include a map of Port Kalamazoo, Michigan, ca. 1837; the 1837 map of the *City of Detroit Michigan from Late and Accurate Surveys*; and the map of Centreville, Michigan, ca. 1840.

Prosper Desobry, another New York City lithographer, printed the *Map of Brooklyn Kings County, Long Island, From an Entire New Survey by Alexr. Martin 1834* and John V. Suydam's *Plat of Madison the Capital of Wisconsin, 1836*, among other maps. P. A. Messier, whose lithographic shop was at 28 Wall Street in New York City, has a credit line on the map *Chicago with the Several Additions Compiled from the recorded plats in the Clerk's Office Cook County, Illinois*, which is believed to have been published in 1836.

Philadelphia, which had been the center of commercial map publishing during the later engraving decades, made an easy transition to lithography. One of the earliest engravers to make the switch was Cephas Childs. He had served as an engraving apprentice to Gideon Fairman and then opened his own shop in Philadelphia in 1818. In 1829, in association with John Pendleton and Francis Kearny, he established a lithographic shop. Childs went to Europe in 1831 to study the technique. After his return, he invited to Philadelphia Peter S. Duval, a young lithographer he had met in France. G. H. Lehman was also employed by Childs, and in 1835 he and Duval formed their own lithographic firm. An early product of their shop is the *Plan of the City of St. Louis, Surveyed & designed by R. Paul, 1823, revised and corrected in June 1835.* The copyright was secured in November 1835 "by Rene Paul of Missouri."

A series of maps and plans was published between 1830 and 1843 by Britain's Society for the Diffusion of Useful Knowledge. They were reproduced from steel plates prepared by drypoint steel engraving. Among the cartographic publications of the society were plans of the cities of New York (1840), Philadelphia (1840), and Boston (1842). Prepared primarily for a limited English membership, these and other plans published by the society probably had a limited distribution in the United States.

During the 1840s plans of cities engraved on steel were also published in the United States. Among these is the *New Map of the City of New York With Part of Brooklyn & Williamsburgh*, which was compiled by J. Calvin Smith and first copyrighted in 1839 by Disturnell. Later editions published until 1848 were copyrighted and published by Henry Tanner at his map store located at 237 Broadway in New York City. The map was "engraved on steel by Stiles, Sherman & Smith."

By the late 1840s, there were a number of lithographic printing establishments in Philadelphia, New York, Boston, and other American cities. Engraving was by this time being rapidly superceded by lithography as the preferred technique for reproducing maps, plans, and other graphics. One lucrative field for lithography was printing maps and plans for real estate developments in, and adjacent to, the rapidly expanding towns and cities. Citizens of such urban regions, particularly business and professional leaders, also constituted a rich market for city plans. Many of the plans were prepared by local surveyors or draftsmen as one-time efforts. In a few instances, however, gifted and motivated individuals developed profitable careers in compiling and drafting city plans.

In this category was J. C. Sidney, a native of England who immigrated to Philadelphia around 1845. Initially, he did not find employment as an engineer, which he was by training, and worked as a part-time assistant in the Library Company of Philadelphia. The librarian, John Jay Smith, and his son, Robert Pearsall Smith, were instrumental in introducing the anastatic process, a variation of lithography, to the United States in 1846. Their anastatic printing shop, operated by the younger Smith, had only limited success. The experience, however, enabled the Smiths to establish contacts with a number of lithographers and printers in Philadelphia. Utilizing the drafting skills of Sidney, they embarked on a program of map publishing. The first product of this association was the *Map of the Circuit of Ten Miles Around the City of Philadelphia . . . From Original Surveys by J. C. Sidney, C. E.*, which was copyrighted and published in 1847 by Robert Pearsall Smith. The map was lithographed by N. Friend and printed at Peter Duval's establishment. The circular map included, in the two lower corners of the sheet, illustrations of Girard College and Laurel Hill Cemetery. This map was described by the elder Smith as "the most successful of our maps."[13] In addition to colored and uncolored editions printed on paper, the map was printed in silk, linen, and muslin souvenir editions. No extant copies of the cloth editions have been identified, perhaps because they were printed with nonpermanent ink.

Under Smith sponsorship, Sidney prepared over the next decade plans of Philadelphia, Boston, New York City, Trenton, Baltimore, Albany, and Troy, as well as maps of eight or ten counties in the states of New York, New Jersey, Pennsylvania, and Maryland. No maps by Sidney are dated after 1851, although several by Sidney and James Neff were published as late as 1855. The Sidney maps and plans are of wall map dimensions and are attractively and neatly composed, lithographed, and printed. With few exceptions they were copyrighted and published by Robert Pearsall Smith.

Between 1841 and 1870 the common council of the city of New York published a series of annual manuals, which included a record of the year's administrative actions. Because the manuals were compiled and edited from 1842 to 1866 by David T. Valentine, it is known as *Valentine's Manual*. In addition to official records and chronology, the manuals included historical essays and were illustrated with contemporary and historical maps, views, and portraits. Most of the maps which illustrated the manuals were lithographed by George Hayward, a lithographer in New York City from 1834 to 1872. Contemporary administrative and topographic maps as well as facsimiles of historical maps were lithographed by Hayward.

During the middle decades of the nineteenth century a number of maps and plans of New York City were published, some as separates and others as illustrations for, or supplements to, directories and guidebooks. Humphrey Phelps, Charles Magnus, J. H. Colton, Disturnell, J. Calvin Smith, Ensign, Bridgman & Fanning, Horace Thayer & Company, Julius Bien, Matthew Dripps, and H. H. Lloyd are among the individuals and firms who compiled, lithographed, or published plans of the city in the two decades prior to the Civil War.

Plans of New England towns were compiled and published by Henry Francis Walling during the late 1840s and early 1850s. Among his works were plans of Cambridge, Massachusetts, and Providence and Pawtucket, Rhode Island. During and following the Civil War, he also published from his New York City headquarters several guidebooks, among them an 1867 guide with a map of the city of New York.

Most city maps and plans published before the Civil War were reproduced in lithographic printing plants in Boston, New York, and Philadelphia. There were a few lithographic shops, however, that had been established in other cities. One of them published the *Map of Chicago and Vicinity Compiled by Rees & Rucker, Land Agents*. It was drawn by William Clogher and was published in 1849 by "Julius Hutawa, Lithr., Map Publishing Office. N. Second Street 45, St. Louis, Missouri." The large *Map of the City of Albany . . . Published by Sprague & Co. . . . Albany and M. Dripps . . . New York* (1857) was surveyed and drawn by E. Jacob and was "Engd. on stone by Hoffman, Pease & Tolley, Albany, New York." The Cincinnati lithographers Middleton, Wallace & Co. reproduced the *Map of the City of Cleveland* in 1855. It was published in that city by Spear, Dennison & Company. E. Weber & Company of Baltimore, established by Edward Weber in 1835, printed maps for federal documents from 1845 to 1848. Following Weber's death in 1848, his nephew, August Hoen, assumed leadership of the company and, in 1853, the name was changed to A. Hoen & Company. Hoen also printed local maps, among them the 1853 map of the *Revised Location of the Boundary Avenues authorized by the ordinance of the Mayor & City Council of Baltimore*. The map, which was drawn by "Aug. Paul, Civil Engr.," was approved on March 4, 1853, by the city commissioners, one of whom was Fielding Lucas, Jr.

Mapmakers and lithographers also followed the prospectors of the California gold rush. In 1850 the firm of William B. Cooke and Josiah J. Le Count was established in San Francisco. A map of that city published around 1853 has the credit, "Lith. of Josiah J. Le Count San Francisco." Britton & Rey, another early San Francisco lithographic firm, published the *Map of the City of San Francisco* in 1856.

Although William Pendleton operated the first successful American lithographic shop in Boston in the 1830s, that city was eclipsed as a lithographic center by Philadelphia and New York in the 1840s and 1850s. Thus, the large wall *Map of the City of Boston, Massts.*

1852, Surveyed and Drawn by I. Slatter & B. Callan, Civil Engineers was "Engraved and Printed at Ferd. Mayer's Lithography, No. 93 William St. New York." The map was jointly published by Matthew Dripps, 103 Fulton Street, New York, and L. N. N. Ide, 138½ Washington Street, Boston. The Mayer firm also lithographed the *Plan of the City of Knoxville, Tennessee*, which was drawn by R. W. Patterson and published in 1855.

As noted earlier, one of the earliest fire insurance maps was published in 1790 for the Phoenix Assurance Company, Ltd., of London. Maps for American insurance firms came much later. Although a fire insurance company was founded in Philadelphia as early as 1752, the number of American companies was limited before the War of 1812. After peace was established, entrepreneurs were encouraged by the increase of population and the growth of cities. Companies were organized in New York, Philadelphia, Hartford, and a few other cities. Until the 1840s, most of the insurance companies were small and locally owned. Properties could be personally inspected by agents, and there was little need for insurance maps. Nonetheless, a few such maps were published, among them *The Firemen's Guide, a Map of the City of New-York Showing the Fire Districts, Fire Limits, Hydrants, Public Cisterns, Stations of Engines, Hooks & Ladders, Hose Carts, Etc.*, which was published in 1834 by Prosper Desobry under the direction of U. Wenman.

A disastrous fire in New York City in 1835, which caused losses of more than twenty million dollars, bankrupted most of the smaller fire insurance companies. New state and city laws requiring substantial reserve funds encouraged the formation of larger companies whose coverage areas extended over several cities. This made the personal inspection of risk properties difficult or impossible and created an urgent demand for fire insurance maps. Around 1849 or 1850, George T. Hope, an officer of the Jefferson Insurance Company of New York City, began compiling a large-scale map of several New York City wards. He interested other insurance companies and set up a committee to supervise the project. William Perris, a British-trained engineer, was engaged to make the surveys and draft the maps. In 1852 maps of the seventh, tenth, and thirteenth wards, at the scale of 1:600, were published by Perris and Augustus Kurth and lithographically reproduced by Korff Brothers of 30 Cedar Street. During the next decade a number of bound volumes were issued by the committee under the general title *Maps of the City of New York Surveyed under the directions of Insurance Companies of Said City*. Insurance maps and atlases were also published before the Civil War of Newark, New Jersey, Philadelphia, and St. Louis, among others. The Civil War limited the production of insurance maps, but activity accelerated in this specialization after 1865.

The Aetna Insurance Company, in 1866, engaged D. A. Sanborn, a young surveyor from Somerville, Massachusetts, to prepare insurance maps of several Tennessee cities. Prior to this assignment, Sanborn had conducted surveys and compiled an atlas of Boston that was later published in 1867 under the title *Insurance Map of Boston*. The success of this atlas and his work for Aetna induced the young surveyor to establish the D. A. Sanborn National Insurance Diagram Bureau in New York City in 1867. Over the next four or five decades, the Sanborn Company (later the Sanborn Map Company) expanded its insurance mapping activities to towns and cities throughout the United States (Fig. 16–10). It also absorbed or secured control of all the other insurance mapping companies and by 1915 had established a monopoly in this cartographic specialization.[14]

The decade after the Civil War was a boom period, particularly in the states and territories west of the Appalachian Mountains. This boom was accelerated by favorable land laws such as the Homestead Act, increased immigration, railroad construction, and the Industrial Revolution. A major consequence of this period was the growth in size and number of urban centers and the increased importance of towns and cities in the social and political life of the nation. Immediately preceding and during the war years, engraving had been almost completely replaced by lithography as the preferred method for reproducing maps and other graphics. With the establishment of lithographic printing plants in such midwestern cities as Cincinnati, St. Louis, Detroit, Chicago, and Milwaukee, there was more activity in commercial map and atlas publishing. Many towns and cities had by the mid-1860s engaged official surveyors to prepare surveys and plans of their jurisdictions among other tasks. Such plans were frequently utilized by cartographic and directory publishers in compiling their commercial maps and plans. The quality of most of these urban maps and plans, unhappily, declined appreciably during the remaining decades of the nineteenth century.

It is worth noting that some of the best maps and plans of American towns and cities in the decade or so before the Civil War appear as insets on county maps. The New England, Middle Atlantic, midwestern, and Pacific coast states were particularly well covered by county maps. Following the war, county and state atlases were also published for these areas and included city and town maps. Some county atlas publishers produced atlases of cities. Frederick W. Beers, who was a particularly prolific producer of county atlases, published an atlas of New York City and its vicinity in 1867 and 1868 editions. His other city atlases included those of Worcester, Massachusetts, in 1870, Long Island in 1873, Newton, Massachusetts, in 1874, Saratoga and Ballston, New York, in 1876, and Richmond, Virginia, in 1877. For J. B. Beers & Company he compiled and drafted *New York City from Official Records and Surveys*, which was published in five volumes from 1876 to 1885.

From his Philadelphia headquarters, Griffith Morgan Hopkins produced several atlases of counties in Mary-

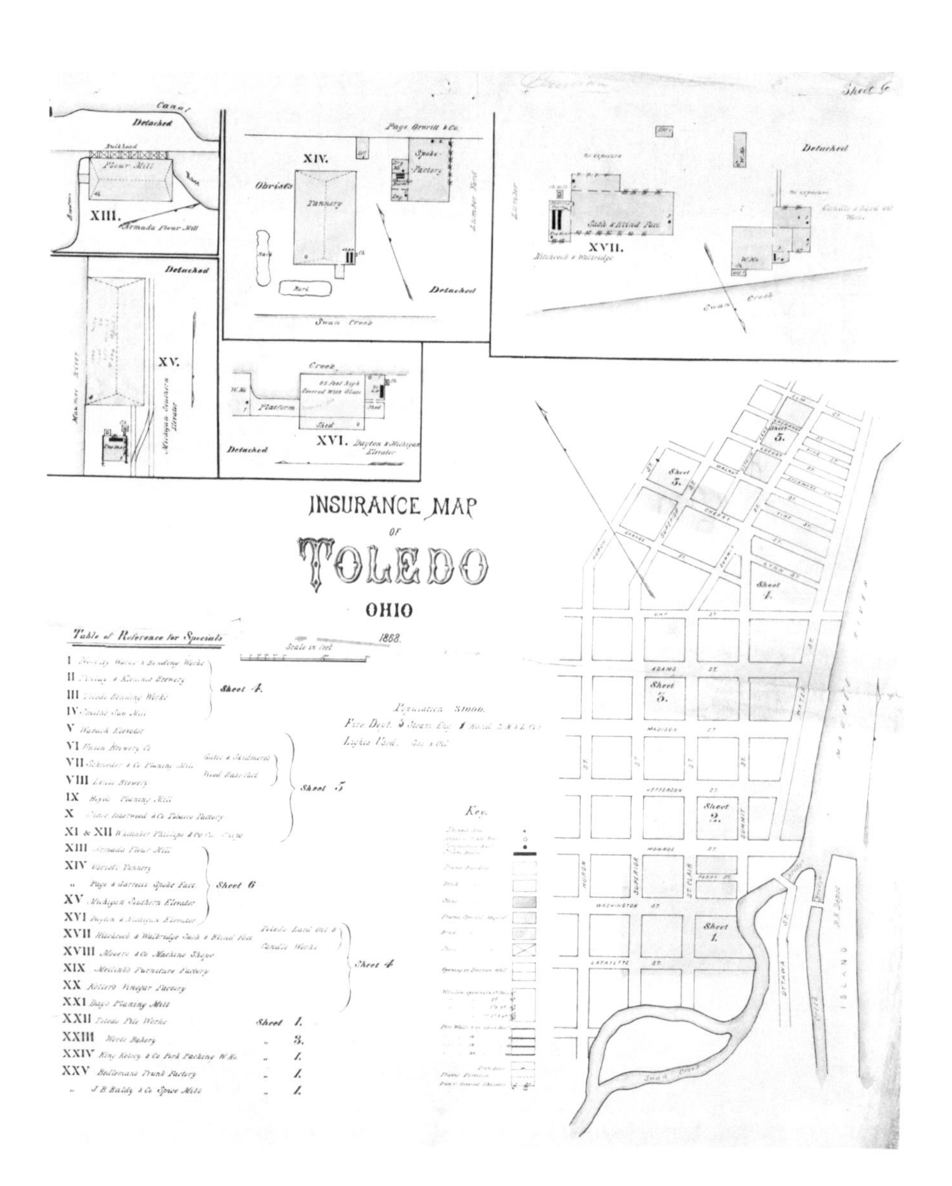

Fig. 16–10. This 1868 Sanborn map of Toledo, Ohio, was based, in part, on surveys by Daniel Carter Beard, who later became one of the founders of the Boy Scouts of America.

land, Pennsylvania, and Virginia, as well as a number of city atlases in the 1870s. In the 1880s and 1890s he seems to have devoted his major efforts to publishing atlases of cities in New England, New York, Pennsylvania, New Jersey, and in several southern and midwestern states. He retired in 1900, but his firm continued producing atlases and plat books under the management of his brother, Henry W. Hopkins. Upon Henry Hopkins's retirement in 1907, George B. C. Thomas assumed leadership. The company fortunes declined during the depression years and, in 1943, the G. M. Hopkins Company was purchased by the Franklin Survey Company of Philadelphia. The Hopkins name was, however, continued on some atlases beyond this date. Hopkins atlases are of the real estate type, with the large-scale plates showing lot and block numbers and dimensions, street widths, and names of property owners.

After the Civil War, the interest in and the need for insurance maps and atlases revived. In the 1870s a number of individuals and small companies entered the field. William A. Miller concentrated on New Jersey cities, publishing atlases of Elizabeth, Paterson, Plainfield, Rahway, Union City, and West Hoboken between 1872 and 1874. New Jersey cities also invited the attention of Spielmann & Brush, which produced insurance atlases of Jersey City, Hoboken, West Hoboken, and Union City in 1873 and 1874. Scarlett & Scarlett, active between 1889 and 1891, published insurance atlases of Newark, Harrison, and Kearny, Essex and Mercer counties, and the New Jersey coast resorts. The most active Midwest publisher of insurance atlases was the Charles Rascher Insurance Map Publishing Company of Chicago. Between 1885 and 1893 Rascher published insurance atlases of Chicago, Kansas City, Kansas, and Kansas City, Missouri, St. Paul and Duluth, Minnesota, and several cities in Michigan. St. Louis, Missouri, and several communities in southern Illinois were mapped for insurance clients between 1874 and 1898 by the Alphonso Whipple Company.

Except for the Sanborn Company, the most prolific and successful publisher of insurance maps and atlases was the firm founded by William Locher and Ernest Hexamer in Philadelphia in 1857. Until 1915, this company (which operated for most of this period under the name Ernest Hexamer & Son), published detailed insurance atlases of the city of Philadelphia. Hexamer remained independent longer than any other insurance atlas publisher, but in 1915 it too was absorbed by the Sanborn Map Company. For more than a half century afterwards, Sanborn went unchallenged in publishing insurance maps and atlases. Sanborn surveyors mapped more than twelve thousand U.S. towns and cities, and the company published some seven hundred thousand frequently revised maps, some bound in atlases and others issued as separate sheets. The company achieved its peak production in the 1930s but continued to supply its maps and atlases to fire insurance companies for several more decades. By 1960, however, it was evident that the industry was undergoing major changes and the detailed maps prepared by the Sanborn Map Company were no longer required. In 1962 the National Bureau of Fire Underwriters's map committee reported that "there is a general (not unanimous) view that residential mapping is not considered essential by the companies or bureaus, nor is it considered essential to have town maps for those communities which are predominantly residential but that business and industrial areas for all other towns and cities warrant map service."[15] The report was actually more optimistic than the demand for insurance maps warranted; within five years Sanborn had virtually ceased issuing new maps.

Closely related in format and content to fire insurance atlases are real estate atlases. In contrast with the extensive coverage of the former, real estate atlases are limited primarily to large urban centers. Real estate atlases were first introduced around 1880. Principal publishers of such works were the firms of Elisha Robinson and Roger H. Pidgeon in New York City and George Washington and Walter Scott Bromley in Philadelphia and later in

New York City. Five or six volumes were required to cover the larger cities such as New York and Philadelphia, but there are also condensed one-volume real estate atlases of these cities. Like insurance atlases, real estate atlases reached their maximum development in the period between the two world wars. The most active producer of real estate atlases in recent years has been the firm of George William Baist in Philadelphia.

A particularly interesting phase of city mapping in the United States was the drawing, printing, and publishing of panoramic maps and bird's-eye views (Fig. 16–11). The production of panoramic maps peaked around 1880, although they were published as early as the 1830s and were popular until World War I. Various factors nourished this distinctive chapter in the history of American commercial cartography. Reps noted that "the rise of cities in the American West coincided with the introduction and development of lithography as a quick and inexpensive method of producing printed images. The hundreds of new communities that sprang into existence as the West was settled provided attractive subjects for artists and publishers who began to specialize in the business of providing city views. To them we owe much of our knowledge of the nineteenth century urban scene."[16] Elizabeth Singer Maule described the peculiar nature of these bird's-eye views.

> A bird's eye view is a combination map and mock aerial photograph of a community. Seemingly, its creators sketched from about 2,000 feet aloft in a balloon, using what amounts to an isometric perspective and always drawing streets at an angle to the borders. The results in almost each case are surprisingly two-dimensional, akin to the engineer's drawing rather than a fine artist's more three-dimensional work. In the foreground, closer to the viewer, perspectives appear flatter than in the middle distance, while in the far distance the horizons seem to rise to the sky. The foregrounds seem oddly out of place and almost unnecessary. The towns are always neat, busy, and progressive. Their streets and terrains are nearly flat, their unimportant buildings similar, their trees all lollipops. They are peopled with stick-figure citizens who crack whips over teams of Noah's ark horses. Locomotives belch coloring-book smoke, and ships crowd together in rivers and harbors, endowing the towns with vitality and bustle. The towns intentionally appear more active than they were, an attribute wholly in harmony with the boastful atmosphere of the middle and late nineteenth century. It is atmosphere so naive that it merely adds to the charm of the views and to their popular appeal.[17]

In the two decades before the Civil War, panoramic maps were primarily reproduced in lithographic printing plants in Philadelphia and New York. The accelerated immigration to the Middle West after the war brought a number of skilled lithographers from Germany who settled and set up lithographic plants in Chicago, Cincinnati, St. Louis, Madison, and Milwaukee. During the last third of the nineteenth century, Madison and Milwaukee achieved some importance in lithography and in publishing panoramic maps. It is estimated that some five or six thousand panoramic maps were published in the United States in the eight decades between 1830 and 1910. The amazing aspect of this production is that most of them were prepared by a relatively small number of view artists. The most prolific artists were Albert Ruger, Thaddeus M. Fowler, Lucien R. Burleigh, Henry Wellge, Oakley H. Bailey, Joseph J. Stoner, Adam Beck, and Clemons J. Pauli.

Panoramic maps were printed in fairly limited editions ranging probably from five hundred to several thousand copies. They included small towns and villages as well as large metropolitan centers. In the late 1870s the highly successful Currier & Ives lithographic firm published a number of panoramic maps of the larger cities. Prices for bird's-eye views varied from fifty cents to six or seven dollars each. Some were in monochrome while others were multicolored. Initially, the

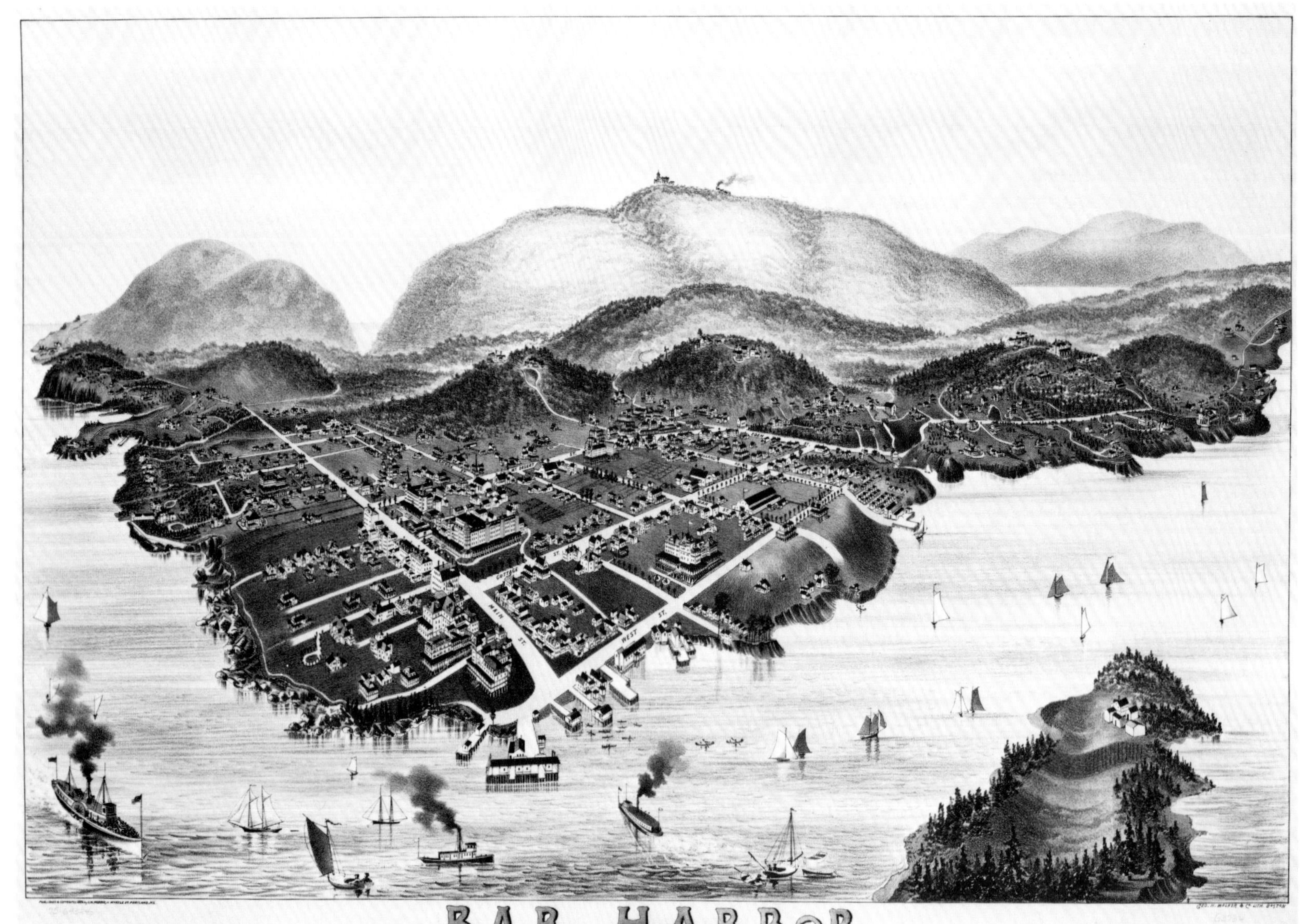

MAIN ST.
WEST ST.
BAR HARBOR
MT. DESERT ISLAND
MAINE

coloring was applied manually, possibly with stencils, but later views were done by chromolithography.

The California gold rush also stimulated interest in bird's-eye views and panoramic maps. The earliest ones of western cities were printed in established lithographic plants in New York, Philadelphia, and Boston. By the mid-1850s, however, several lithographic presses and publishers had been opened in San Francisco by Britton & Rey, Kuchel & Driesel, and A. L. Bancroft & Company, among others.

Panoramic plans of cities have become popular items with collectors in recent decades, and their prices in dealers' catalogs today range from three hundred to eight hundred dollars. They have also been featured in a number of exhibits, and catalogs and lists of holdings in various libraries have been compiled and published.[18] Thomas Beckman observed that today bird's-eye views "are becoming increasingly valuable and are actually sought by art museums, historical societies, libraries, and private collectors. These independently published views, with their amalgamation of precise topographical draftsmanship and skilled printing, constitute a unique and captivating form of American popular art."[19]

Fig. 16–11. This 1886 panoramic map of Bar Harbor and Mount Desert, Maine, was lithographed and printed by George H. Walker & Company of Boston and published by G. W. Morris of Portland, Maine.

Notes

1. John W. Reps, *Town Planning in Frontier America* (Princeton, N.J., 1960), 3.
2. John W. Reps, "Boston by Bostonians: The Printed Plans and Views of the Colonial City by Its Artists, Cartographers, Engravers and Publishers," in *Boston Prints and Printmakers 1670–1775*, ed. Colonial Society of Massachusetts (Richmond, Va., 1973), 8.
3. Ibid., 15.
4. I. N. Phelps Stokes, *The Iconography of Manhattan Island 1498–1909* (New York, 1915) 1:274.
5. Reps, "Boston by Bostonians," 52.
6. Stokes, *Iconography*, 1:346.
7. Coolie Verner, "Surveying and Mapping the New Federal City: The First Printed Maps of Washington, D.C.," *Imago Mundi* 23 (1969): 67–68. For more information see Richard W. Stephenson, "The Delineation of a Grand Plan," J. L. Sibley Jennings, Jr., "Artistry as Design, L'Enfant's Extraordinary City," and Ralph E. Ehrenberg, "Mapping the Nation's Capital, The Surveyor's Office, 1791–1818," in *Quarterly Journal of the Library of Congress* 36 (1979): 207–24, 279–319.
8. Stokes, *Iconography*, 1:442.
9. See Peter J. Guthorn, *British Maps of the American Revolution* (Monmouth Beach, N.J., 1972), 24–27.
10. Peter J. Guthorn, *John Hills, Assistant Engineer* (Brielle, N.J., 1976), 42.
11. Pennsylvania Archives, *Muster and Pay Rolls, Pennsylvania Militia, 1790–1800*, sixth ser., vol. 5 (Harrisburg, 1907), 514.
12. Richard W. Stephenson, "Charles Varlé, Nineteenth Century Cartographer," paper presented at the annual convention of the American Congress on Surveying and Mapping, 1972.
13. John Jay Smith, *Recollections*, ed. Elizabeth Pearsall Smith (Philadelphia, 1892), 224.
14. For a comprehensive summary of the Sanborn Map Company see Walter W. Ristow, "United States Fire Insurance and Underwriters Maps, 1852–1968," *Quarterly Journal of the Library of Congress* 25 (1968): 194–218; reprinted in *Surveying and Mapping* 30 (1970): 19–41. Also see U.S. Library of Congress, *Fire Insurance Plans in the Library of Congress, Plans of North American Cities and Towns Produced by the Sanborn Company, A Checklist* (Washington, D.C., 1981).

15. "Report of the Committee on Maps," *Proceedings of the National Board of Fire Underwriters*, ninety-fifth annual meeting (New York, 1962), 139.
16. John W. Reps, *Cities on Stone, Nineteenth Century Lithograph Images of the Urban West* (Fort Worth, Tex., 1976), 2.
17. Elizabeth Singer Maule, *Bird's Eye Views of Wisconsin Communities* ([Madison], Wis., 1977), 1.
18. John Cumming, *A Preliminary Checklist of 19th Century Lithographs of Michigan Cities and Towns* (Mount Pleasant, Mich., 1969); Thomas Beckman, *Milwaukee Illustrated: Panoramic and Bird's-Eye Views of a Midwestern Metropolis, 1844–1908* (Milwaukee, Wis., 1978); John R. Hebert, *Panoramic Maps of Anglo-American Cities: A Checklist of Maps in the Collection of the Library of Congress, Geography and Map Division* (Washington, D.C., 1974); John Reps, *North American Views and Viewmakers: A Union Catalogue of Lithographic Prints of Cities and Towns, 1834–1926*, Columbia, Mo., forthcoming.
19. Beckman, *Milwaukee*, [8].

17. *Other Map Publishers of the Engraving Period*

John Melish and Henry Tanner, as we have seen, were the foremost commercial map publishers in the first half of the nineteenth century. A number of other surveyors, compilers, engravers, and publishers, however, made significant contributions to American cartography during the period when printing from engraved copper plates was still the dominant technique for reproducing maps. Companies founded by several of these individuals successfully made the transition from engraving to other types of reproduction and remained in operation until late in the nineteenth century. References have been made to several of these mapmakers in earlier chapters, but they warrant more specific recognition here.

One of the earliest of these mapmakers was Samuel Lewis. He was born around 1754, but his place of birth is unknown. Lewis taught drawing and writing, and he was also known as a geographer. His first cartographic works were the maps he compiled for Mathew Carey's 1795 American edition of Guthrie's *Geography*. For Carey's *American Atlas*, published in 1795, Lewis prepared more than half the maps.

Shortly after the turn of the century, Lewis formed a partnership with Aaron Arrowsmith, an English publisher of maps and atlases. The first product of their association was *A New and Elegant General Atlas, Comprising All the New Discoveries, to the Present Time; Containing Sixty-three Maps. Drawn by Arrowsmith and Lewis*. The atlas was published in 1804 by several branches of the same company: John Conrad & Company in Philadelphia; M. & J. Conrad & Company in Baltimore; Rapin, Conrad & Company in Washington, D.C.; Somervell & Conrad in Petersburg, Virginia; and Bonsal, Conrad & Company in Norfolk, Virginia. All the American maps were drawn by Lewis. They were engraved by a number of individuals, among them, Benjamin Tanner, William Harrison, Jr., and David Fairman (brother of Gideon Fairman). The maps of foreign areas had quite likely appeared previously in Arrowsmith atlases.

Jedidiah Morse, who initiated his *American Geography* in 1789, had included in its earlier editions only a few small maps. Criticism of the meager cartographic illustrations led him to arrange with Lewis and Arrowsmith to issue an edition of their atlas to accompany the 1805 edition of his *Geography*. Thus, the title page of the 1805 *New and Elegant General Atlas* notes that it is "Intended to Accompany the New Improved Edition of Morse's Geography, But equally well calculated to be used with his Gazetteer, or any other Geographical work." The atlas was published in Boston by Thomas & Andrews. This publisher also issued the 1812 and 1819 editions of the atlas, which also were intended to accompany Morse's *Geography*. Each of the four editions of the *New and Elegant General Atlas* has sixty-three maps, and the maps appear to be unchanged in the several issues. Morse compiled his geographies from the writings of others. He had no cartographic competence and was content to accept the unrevised atlas prepared by Lewis and Arrowsmith.

Morse had, several years prior to 1805, approached Lewis about the possibility of preparing maps for inclusion in his *Geography*. Ralph Brown described Lewis as "perhaps the most enterprising commercial map-maker of the period . . . [who] offered to prepare [for Morse] 'eighteen small maps at twenty-five dollars each in my commercial style and thirty-one dollars each, in my most Elegant, so as to make them pictures.'"[1] Nothing came of these negotiations, and Morse subsequently settled for the *New and Elegant General Atlas*.

Lewis also compiled several large maps of the United States. The earliest, dated 1815, is *A New and Accurate Map of the United States of North America, Exhibiting the Countries, Towns, Roads &c. in each State. Carefully Compiled from Surveys and the most Authentic Documents. By Samuel Lewis. Philadelphia, Published by Emmor Kimber, 1815*. The Library of Congress has only the southeast quarter of the map, which includes the title cartouche. This part of the map was deposited for copyright and may have been the only portion completed in 1815. The map in its complete form is dated 1816 and measures 173 by 185 cm. It extends west only just beyond the state of Louisiana and the central part of Minnesota. An 1819 edition of the map was "published by Kimber and Sharpless for Emmor Kimber."

In 1819 Lewis also issued his *Travellers Guide. A New and Correct Map of the United States including portions of Missouri Territory, Upper & Lower Canada, Nova Scotia, New Brunswick, the Floridas, Spanish Provinces &c. Collected and compiled from the most Undoubted Sources by Samuel Lewis, Geographer and Draftsman, 1819*. It was published by Henry Charles of Philadelphia. The map's dimensions are 74 by 105 cm.

Lewis expanded his area of coverage to include the West Indies and the northern tip of South America in his 1820 *Correct Map of the United States With the West Indies, from the best Authorities*. The map, which is 97 by 95 cm., was published in Philadelphia by William Cammeyer, Jr. An 1821 edition of the map was "Published, Printed, and Coloured by Henry Charles, Philadelphia." This was apparently Lewis's last cartographic contribution, for he died in October 1822, while visiting his son-in-law, James Johnston, in Lancaster, Pennsylvania.

During the first half century after the Revolution there was considerable interest in atlas publication. The contributions of Henry S. Tanner in this medium were of major significance, but several other individuals produced noteworthy atlases. Among these was Fielding Lucas, Jr., who was born in Fredericksburg, Virginia, on September 3, 1781. Little is known about his early education or training, but, in 1798, at the age of seventeen, he went to Philadelphia where, judging from his subsequent activities, he may have worked in bookstores or publishing houses. In 1804 he moved to Baltimore and appears to have been involved in publishing and selling books, maps, and atlases. He became, in 1807, the junior member of the book-selling firm of Conrad, Lucas & Company, which operated for less than three years before terminating in 1810. In the following year Lucas established his own business at 138 Market Street, where he carried on successfully until his death in 1854.

In 1813 Lucas enlisted in the army and served briefly in the Twenty-seventh Maryland Regiment under Captain John Kane. Even before enlisting Lucas appears to have been cognizant of the growing interest in cartographic publications, and apparently he had already started compiling an atlas. Titled *A New and Elegant General Atlas Containing Maps of Each of the United States*, the volume was probably published in 1816. There is no date on the title page, but the map of Virginia in the volume is dated 1816. The atlas has fifty-four plates and, despite its limiting title, it has maps of the world and foreign countries as well as maps of the United States. The foreign maps, very likely derived from European sources, were engraved by Samuel Harrison, a member of the Harrison family of engravers. Most of the U.S. maps were drawn by Lewis and engraved by Henry S. Tanner. For the maps in the atlas Lucas seems to have drawn on the skill and experience of draftsmen and engravers with whom he had become acquainted during his residence in Philadelphia. A second edition of *A New and Elegant General Atlas*, published in 1817, has two maps that were prepared by Lucas.

Perhaps also a result of his contacts in Philadelphia, Lucas prepared some twenty maps which comprised 40 percent of the total number of plates for *A Complete His-*

torical Chronological and Geographical Atlas. This volume was published by Carey & Lea in Philadelphia in 1822. Subsequent editions, dated 1823 and 1827, also include Lucas's maps. Concurrent with this project Lucas compiled a major atlas of his own which was published in 1823 under the title *General Atlas Containing Distinct Maps of All the Known Countries in the World, Constructed from the Latest Authority*. It contains a number of maps "Drawn & Published by F. Lucas Jr." (Fig. 17–1). Some of the maps had previously been published in his *New and*

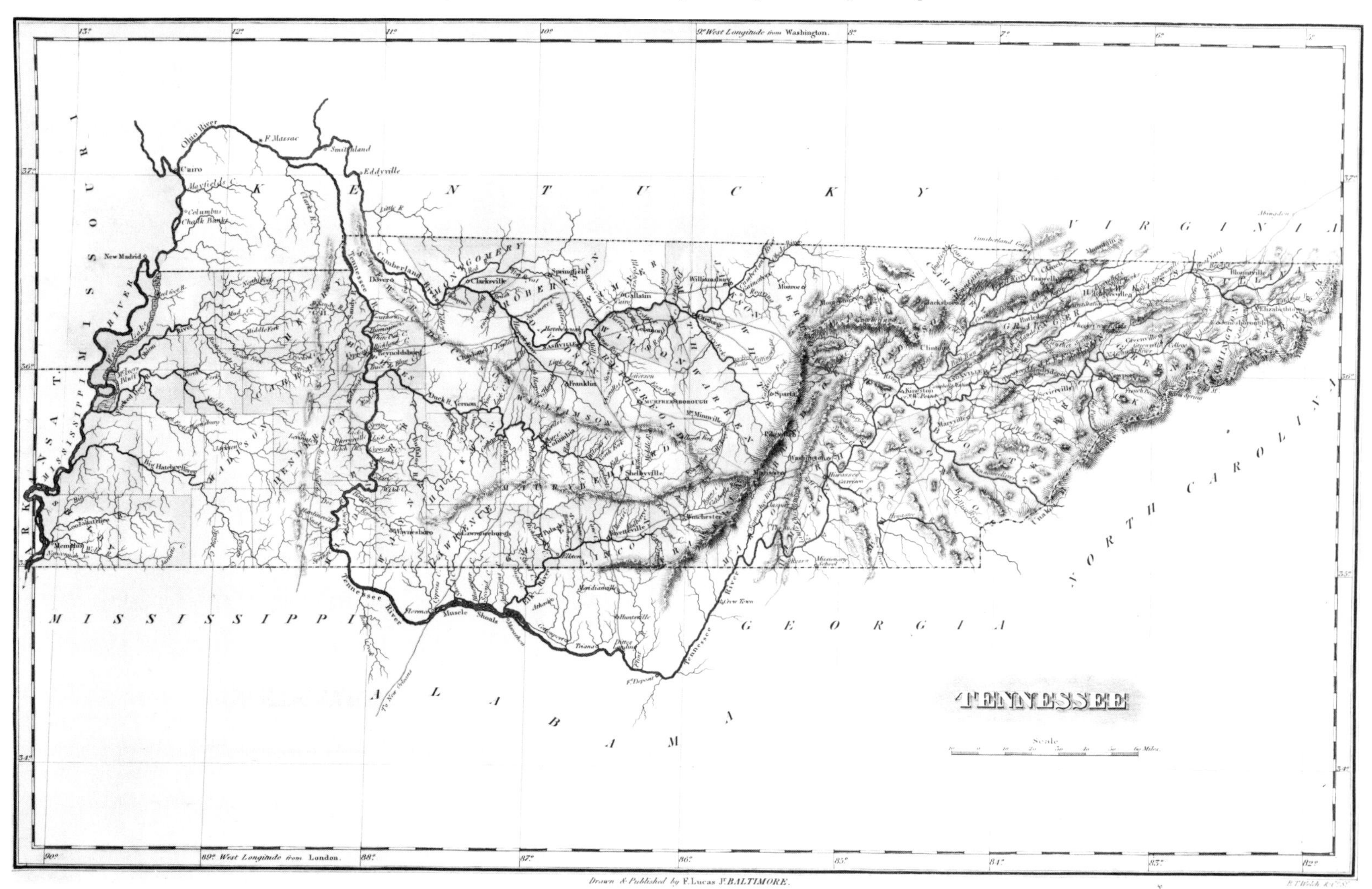

Fig. 17–1. Map of Tennessee from Lucas's *General Atlas*.

Elegant General Atlas, but the engraving credits of Harrison and Tanner were removed. There were good reasons for these deletions. Harrison died in 1818, and Tanner formed his own map publishing firm in 1819. Tanner had also published his *New American Atlas* in 1823, the same year in which Lucas's *General Atlas* was issued. The two works were jointly reviewed by Jared Sparks, who considered both "honorable to the country, and among the strong marks of our literary and scientific advancement; they are trophies of American enterprise, which it becomes a discerning public to regard with favor, and reward with personal patronage; and we hope the authors will be encouraged to pursue their labors, which they have thus far presented with sound credit to themselves, and so much benefit to the community."[2]

Lucas's *General Atlas* has 104 maps as compared with the 54 in the *New and Elegant General Atlas*. The new maps in the *General Atlas* include those of the newly formed states and territories in the West and maps of the West Indies, Mexico, and parts of South America. The engraving credits on the new maps include B. T. Welch & Company of Baltimore and Young & Delleker of Philadelphia. Lucas then assembled the 21 maps of the West Indies from the *General Atlas* into a separate volume entitled *A New General Atlas of the West India Islands*, which was published in 1824. In the following year Lucas supplied Robert Mills with the map of South Carolina that served as the title page and frontispiece for his *Atlas of the State of South Carolina*. The map, a revision of the South Carolina map in the *General Atlas*, was engraved by B. T. Welch & Company.

Lucas seems to have abandoned atlas publishing around 1830 and thereafter concentrated his efforts on compiling and publishing local maps. His maps of Maryland and Baltimore were issued in numerous editions to as late as 1852 (Fig. 17–2). In addition to his own maps and atlases, Lucas offered for sale at his Baltimore bookstore works of other publishers. He also sold state maps folded into pocket-size covers of board or imitation morocco. They were particularly popular around the middle of the nineteenth century and sold for fifty or seventy-five cents each.

The increasing importance of lithography in map printing and reproduction was no doubt a factor in Lucas's decreasing interest in map and atlas publishing. Nonetheless, maps bearing his imprint are dated as late as 1852. He died on March 12, 1854, in his seventy-third year. During a half century of residence in Baltimore he attained a degree of eminence above and beyond his success as a publisher and bookseller. He served as a director of the Baltimore and Ohio Railroad from 1835 to 1854, as the president of the Board of School Commissioners from 1837 to 1838, and as president of the Second Branch of the City Council from 1838 to 1841.

Another American cartographic publisher of this period was Anthony Finley of Philadelphia. Little is known about his background, but he was probably born around 1790. Judging from contributors to his atlases, he apparently moved in the same Philadelphia circles of engravers and compilers as other contemporary publishers. Finley also borrowed freely from European sources in compiling his atlas.

His first publication, issued in 1824, is titled *A New General Atlas, Comprising a Complete Set of Maps, representing the Grand Divisions of the Globe, Together with the several Empires and States in the World. Compiled from the Best Authorities and corrected by the Most Recent Discoveries, Philadelphia*. Its ornate title page was designed and engraved by Joseph Perkins, who also prepared the title page illustration for Tanner's *New American Atlas*. Perkins also engraved, in a beautiful cursive hand, the table of contents for Finley's atlas. All of the atlas's sixty plates were engraved by Young & Delleker (Fig. 17–3).

The *New General Atlas* was favorably reviewed in the July 1824 issue of the *North American Review*. The reviewer observed that

> the number of elegant maps and atlases which have come from the press within a short time in the United States, is a most flattering proof of the

increased attention of the community to the important study of geography, and the liberal enterprise and zeal of our publishers and artists. The present work is very much on the plan of Mr. Kneas's [*sic*] splendid Cabinet Atlas. . . . It contains sixty maps, about half of which are devoted to the American continent, and the remainder to other parts of the world, chiefly to Europe. . . . The engraving is done almost uniformly with remarkable distinctness and the face of the maps is frequently beautiful, not overloaded with a confusion of useless names.[3]

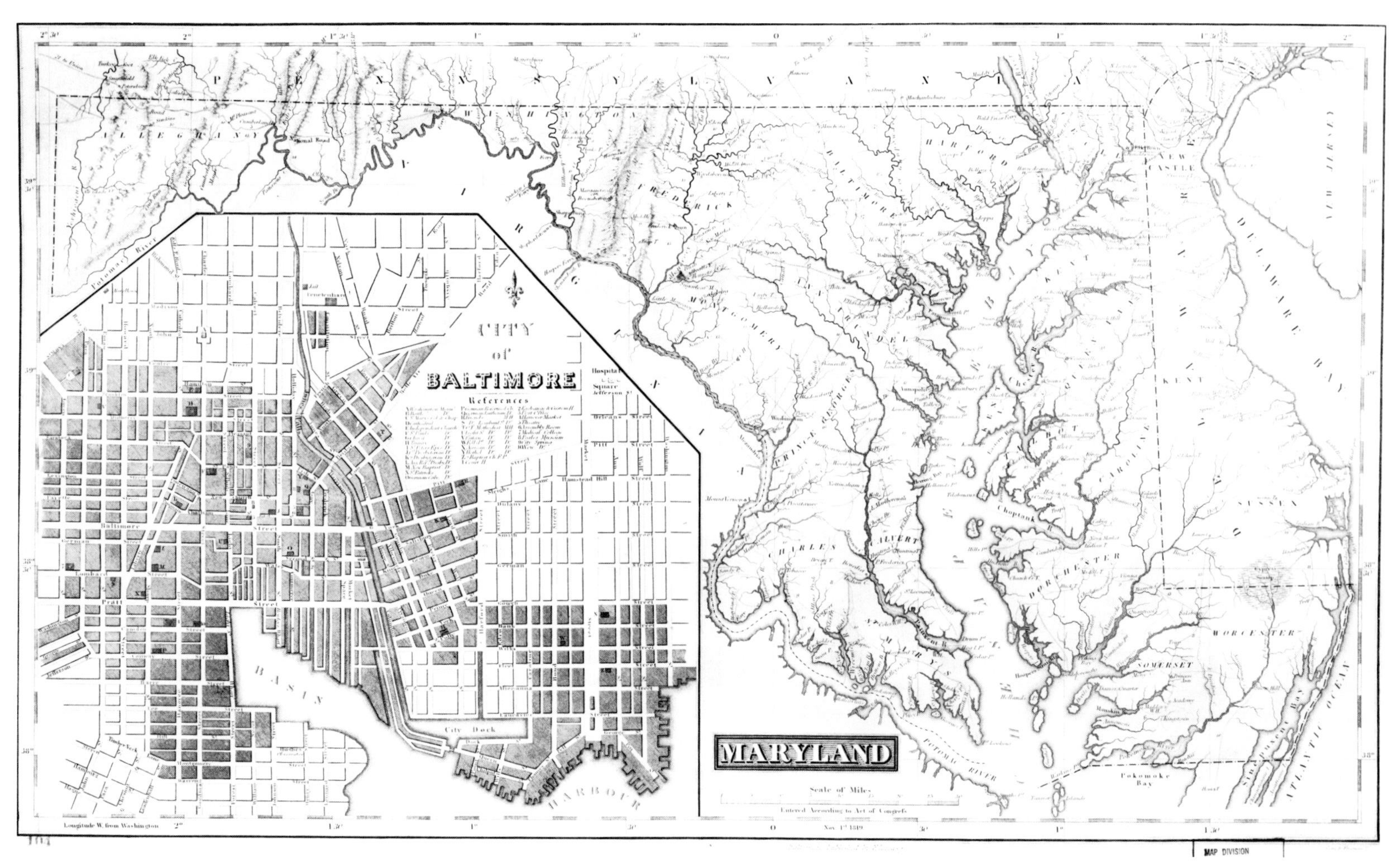

Fig. 17–2. One of Lucas's maps of Maryland and Baltimore. It was published in 1826.

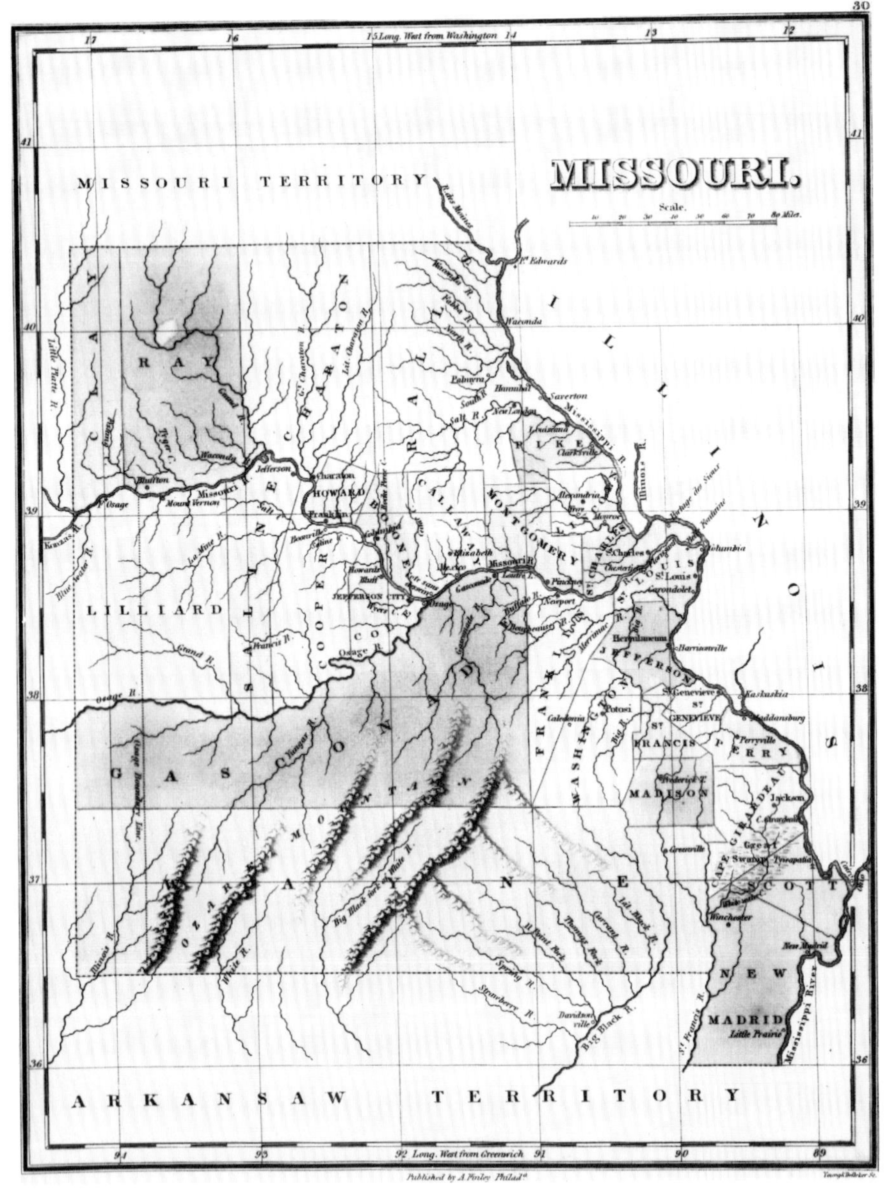

Fig. 17–3. This map of Missouri, published by Finley in 1824, was engraved by Young & Delleker.

The 1826 edition of the *New General Atlas* also has sixty plates, but new data have been added on the maps of the states in the upper Mississippi valley. In the 1829 edition, the page size has been enlarged slightly and charts showing the comparative heights of mountains and the lengths of rivers have been added. The Library of Congress owns a copy of this 1829 edition, which is bound with Finley's *Atlas Classica; or Select Maps of Ancient Geography, Both Sacred and Profane*. Two of the *Atlas Classica's* ten maps have explanatory text derived from the atlas of Le Sage, a French publisher. The maps are probably from the same source, although they have been reengraved by Thackara, Vallance, Tanner, and P. E. Hamm.

The *Atlas Classica* does not appear in the 1830, 1831, or 1833 editions of the *New General Atlas*, but it does appear again in the 1834 edition. Beginning in the 1831 edition of the *New General Atlas* a map of Liberia is included. The 1834 edition of the *New General Atlas* was the last. Finley's 1826 *A New American Atlas Designed Principally to Illustrate the Geography of the United States of North America* appears to have been a one-time effort. Most of the atlas maps carry the credit "Drawn by D. H. Vance," and all were engraved by J. H. Young. The same plates, with dates and publisher's name changed, were used by S. Augustus Mitchell in 1831 for an atlas published under the same title as Finley's 1826 volume.

Thomas Gamaliel Bradford (1802–87) of Boston served as an assistant editor of the *America Encyclopedia* before entering the field of atlas publishing. His first effort, published in 1835 and titled *Atlas Designed to Illustrate the Abridgement of Universal Geography, Modern & Ancient*, is a condensed version of a French work by Adrian Balbi. Bradford's atlas is a small and rather undistinguished volume with thirty maps and four pages of comparative tables and statistical charts. None of the maps carries an engraver's credit. Half are of North America and the United States, and the remainder show the world and various foreign countries. The atlas was published concurrently by William D. Ticknor of Boston,

Freeman Hunt & Company of New York, and De Silver Thomas & Company of Philadelphia.

In 1838 Bradford produced a much more elaborate atlas in an enlarged format (50 by 40 cm.) which was published by Weeks, Jordan & Company of Boston. The title, on an ornately engraved and hand-colored title page, is *An Illustrated Atlas Geographical, Statistical and Historical of the United States and Adjacent Countries*. The title page was engraved by James Archer of Boston, who had previously prepared most of the plates for John H. Hinton's *History and Topography of the United States*, which was published in 1834. Bradford's *Illustrated Atlas* includes maps of North America, the United States, and the West Indies; individual maps of the states and the provinces of Canada; and plans of the cities of Washington, D.C., New Orleans, Cincinnati, Philadelphia, New York, Boston, and Louisville and Jefferson, Kentucky. Most of the maps were engraved by G. W. Boynton, who operated an engraving company in Boston during the 1830s and 1840s. The maps are attractive and legible and are printed on strong paper.

Two other editions of the atlas, with extensive descriptive text added, were published in 1838, one by Weeks, Jordan & Company and the other by E. S. Grant & Company of Philadelphia. In addition to the hand-colored engraved title page, both of these atlases have a second black-and-white title page, which was "stereotyped and printed by Folsom, Wells, and Thurston, printers to Harvard University, Cambridge." There are forty maps in each edition. Bradford's *Illustrated Atlas* was one of the first American general atlases to supplement the maps with lengthy geographical descriptions.

In 1842 the *Universal Illustrated Atlas* was published in Boston by Charles D. Strong. It was edited by Bradford and S. G. Goodrich. Like the second and third 1838 editions of the *Illustrated Atlas*, it includes two title pages. An interesting innovation in the *Universal Illustrated Atlas* is that it was lithographed. The lithography was done by B. W. Thayer of Boston. This edition is printed on a weak pulp paper. A note by Goodrich on the verso of the second title page states that "to render the work *Universal*, several maps have been added . . . and to make it conform to the present state of facts, and to exhibit the results of the recent Census of the United States, a Supplement is added."

Although publishers of atlases achieved the greatest prominence, several private and commercial cartographers had a measure of success preparing maps of local areas in the early decades of the nineteenth century. One such individual was Lewis Robinson. He was born in South Reading, Windsor County, Vermont, on August 19, 1793. As a boy, Robinson worked on his father's farm for nine months of the year and attended school in the three winter months. His education also included one term at Duttonsville Academy and one year at a high school in Granville, New York. Robinson learned engraving and printing as an apprentice to Isaac Eddy, whose shop was in Weathersfield, in the southeastern part of Windsor County.

Robinson's son Calvin Robinson related that "soon after his father came of age . . . he engaged in the business of book publishing, establishing a printing office at Greenbush. He published a number of works there, mostly educational, which were well up to the times in merit, and style of finish."[4] In around 1823 or 1824, Robinson established a copper-plate printing shop in South Reading, where he produced scriptural illustrations and maps. The first Robinson cartographic product on record is the small *Improved Map of the United States Corrected and Published by Lewis Robinson 1825*. It is a rather crude rendition of an 1813 map published by Shelton & Kensett of Cheshire, Connecticut, and engraved by Amos Doolittle. Robinson had revised the map to show the new states of Alabama, Mississippi, Louisiana, and Missouri.

Robinson's first original cartographic contribution was the *Map of Vermont & New Hampshire*, which he published in 1828. This was apparently a popular item, and some twelve revisions were issued to 1859 (Fig. 17–4). He also published separate maps of Vermont and New

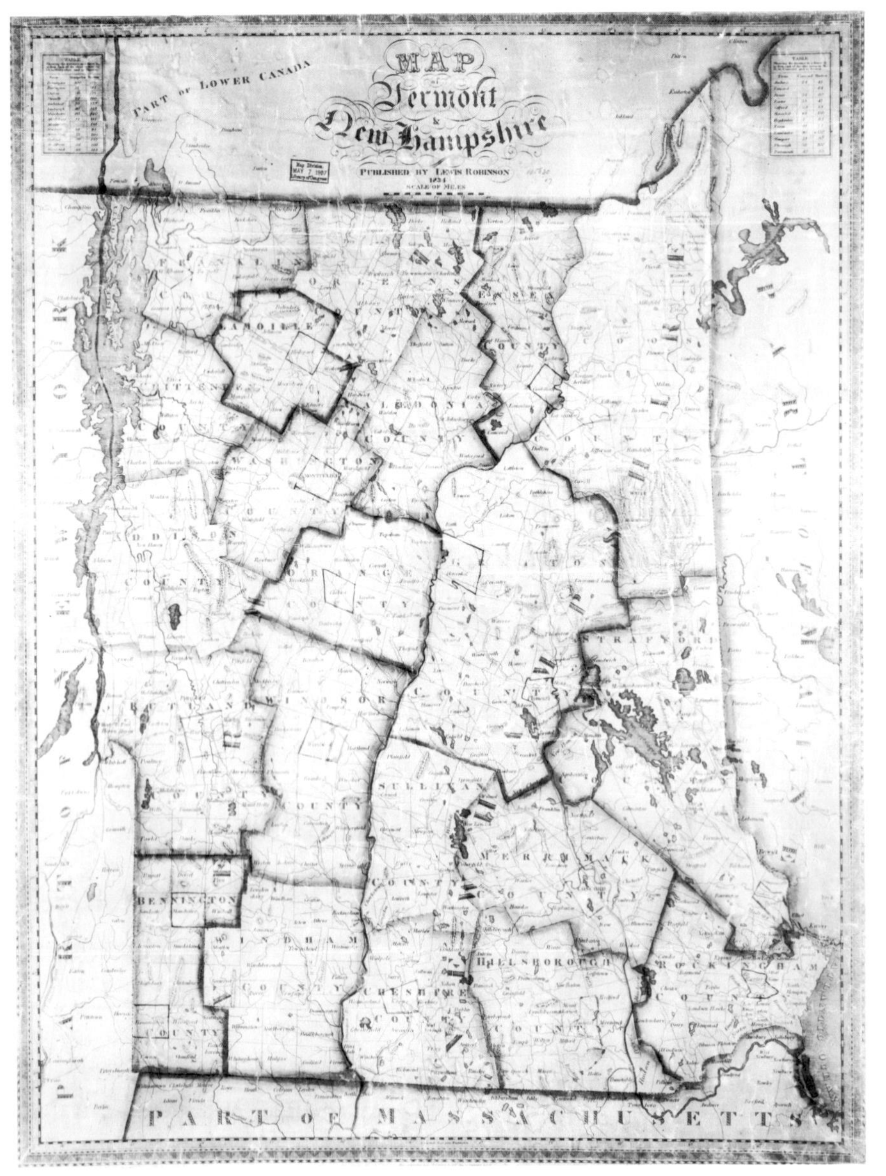

Fig. 17–4. The 1834 edition of Lewis Robinson's *Map of Vermont & New Hampshire.*

Hampshire, the first editions of which are dated 1834. *An Improved Map of Maine* was published in 1835, apparently the only edition. Robinson published a large (120-by-101-cm.) *Map of the United States* in 1833, which was engraved by J. G. Darby of Burlington, Vermont. Revised editions, which include inset maps, are dated 1835, 1836, and 1846.

In 1836, with two of his brothers-in-law, Levi and Samuel Manning, Robinson established a large map publishing business in Akron, Ohio. Robinson's company issued maps of Indiana (1830), Akron (ca. 1836), Ohio (1837), and the Connecticut Reserve (1838) and sold them throughout the western states.[5] A comprehensive list of maps published by Robinson, including multiple editions, is included in George Dalphin and Marcus McCorison's article entitled "Lewis Robinson—Entrepreneur." Many of Robinson's maps were sold by itinerant peddlers at prices ranging from fifty cents to three dollars (for the large 1833 U.S. map) each. In New England, as might be expected, his maps of New Hampshire, Vermont, and Vermont and New Hampshire were particularly good sellers. Dalphin and McCorison report that during the years for which records exist Robinson sold 6,774 maps to peddlers, of which 5,995 carried his imprint. In addition he sold 10,761 prints, of which 533 were printed by him.[6]

Except for his large map of the United States, none of Robinson's maps has an engraving credit. Robinson probably engraved them himself. The title cartouches are ornately engraved, but the maps are not particularly attractive. All those examined have state and county boundaries outlined in rather garish colors. Many of the maps appear to have been mounted on cloth and attached to rods at the top and bottom margins to facilitate hanging on a wall. Some were sold unmounted and in a folded format.

Robinson phased out his map business in 1861, probably because of the competition of lithography in map reproduction. In limiting his production primarily to local maps, which were sold by horse and wagon peddlers, Robinson was able to survive longer than most publishers of engraved maps. Concurrently with his map publishing, Robinson engaged in several other activities, among them keeping a store, operating a starch mill, manufacturing wooden articles, and selling prints. He undoubtedly continued some of these after his map business expired. He died in South Reading on November 16, 1876.

Robinson apparently did not copyright his maps, and many of them, therefore, are not in the collections of the Library of Congress. Because they were sold to farmers and town dwellers, in many instances as wall hangings, the attrition rate was very great. As a result, the number of extant copies of Robinson maps is shamefully small. Dalphin and McCorison note that "of nearly 6,000 maps which he sold between 1840 and 1855 only 65 have been located in 1962. Fifty libraries and collections have been solicited for holdings of Robinson maps. No doubt other copies and perhaps unrecorded editions exist in smaller collections, but little remains to bear evidence of the extraordinary energy of Lewis Robinson, entrepreneur, of the small Vermont hill town of Reading."[7]

Another map publisher of this period was Charles Varlé. Discussed in Chapter 16 in relation to his city plans, Varlé also merits mention here for publishing engraved county, state, and U.S. maps. After moving to Maryland in 1798, he served as the superintendent of the Susquehanna Canal Company. For the next decade he was involved with canal activities in Maryland and Delaware. In connection with proposals for a canal joining Chesapeake Bay with Delaware Bay, Varlé is believed to have prepared an unsigned map entitled *A Map of the State of Delaware and Eastern Shore of Maryland With the Soundings of the Bay of Delaware from Actual Survey & Soundings Made in 1799, 1800, & 1801 by the Author*. The map accompanied a report entitled *Candid Considerations Respecting the Canal Between the Chesapeake and Delaware Bays*, which was published in Baltimore in 1827.[8]

In 1808 Varlé published by subscription *A Map of Frederick and Washington Counties, State of Maryland*. The map, which was engraved by Francis Shallus of Philadelphia, was not actually delivered until 1809. It sold for three dollars. For an additional two dollars subscribers could have "their plantations, mills &c. marked on the map." Dated 1809 is Varlé's *Map of Frederick, Berkeley, & Jefferson Counties in the State of Virginia*, which was engraved in Philadelphia by Benjamin Jones (Fig. 17–5). It was accompanied by a thirty-four page *Topographical Description . . . in which the Author has Described the Natural Curiosities of Those Counties*, which was printed in Winchester, Virginia, in 1810 by William Heiskell.[9] These two Varlé county maps were forerunners and helped set the pattern for the numerous county land ownership maps that were published between 1840 and 1880.

Varlé published in Baltimore in 1817 a large (111-by-146-cm.) *Map of the United States Partly From New Surveys*. The map, which is at the scale of 1:1,900,000, was engraved in Philadelphia by J. H. Young and is one of the earliest examples of his engraving. There is an inset map of North America and two vertical profiles from the Pacific to the Atlantic on the map sheet. The U.S. map extends west only to beyond the state of Louisiana and the upper course of the Mississippi. A revised edition was published in 1819 that incorporated a number of corrections and additions, although the imprint date was unchanged. The plates for Varlé's map were subsequently acquired by R. Stebbins & Company of New York, which published an edition of the map "corrected to the year 1835."

John Farmer and his sons and heirs made noteworthy contributions to American commercial cartography from 1825 to 1915. Farmer, like Lewis Robinson, maintained his map company and survived the competition from lithography in the declining years of copper-plate engraving. He did this by limiting his maps to a few states, by publishing them in numerous editions, and by selling many of them through itinerant salesmen. Nonetheless, in its later years, the Farmer company did have some of its maps reproduced by lithography. During nine decades, three generations of Farmers pub-

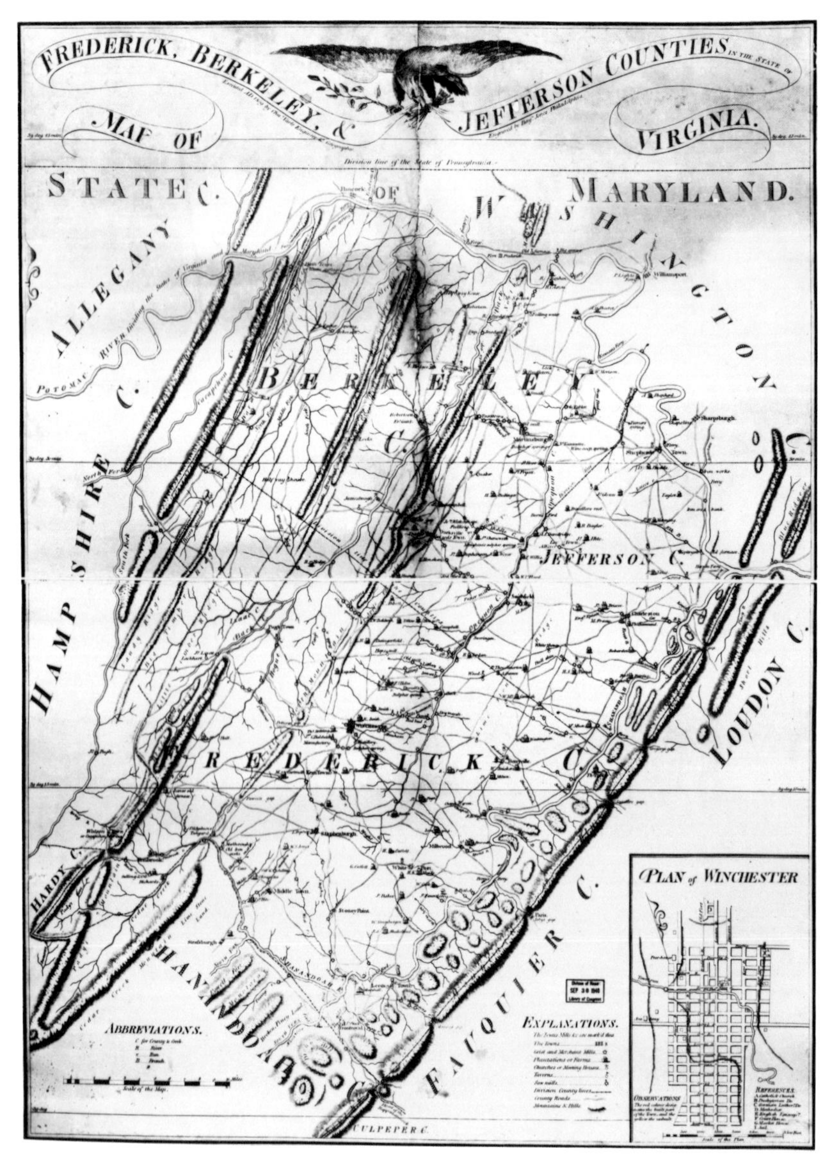

Fig. 17–5. Charles Varlé's 1809 map of the Virginia counties of Frederick, Berkeley, and Jefferson.

lished some eighty different maps, primarily of Michigan, Wisconsin, and neighboring states in the upper Mississippi River and Great Lakes regions.

John Farmer was born in Half Moon, Saratoga County, New York, on February 9, 1795.[10] He was educated in and near Albany, and surveying was among the subjects he pursued. His first job was a teaching position in Albany, but he resigned in December 1821 to accept the appointment of principal in a Detroit school. Farmer began his duties as principal of the Lancastrian School on December 10, 1821, at an annual salary of five hundred dollars. He resigned this position January 26, 1824, and moved to Ohio, where he taught classes in surveying. His engagement was a brief one, and late in 1824 he was back in Detroit, where he was engaged by Orange Risdon to draft a *Map of the Surveyed Parts of the Territory of Michigan*. The map, copyrighted in January 1825 and engraved in Albany by Rawdon Clark & Company, was not ready for distribution until a year or so later. No credit was given to Farmer on the published map.

This cartographic experience and his belief that the opening of the Erie Canal in 1825 would stimulate western migration and trade convinced Farmer that there would be a market for maps. He accordingly established his own map company in 1825 in Detroit. Apparently confident that the business would prove successful, Farmer returned briefly to his birthplace, where, on April 5, 1826, he married Roxanna Hamilton.

Prior to his marriage Farmer had completed a map of Michigan, which was sent to engravers Balch & Stiles of Utica, New York, in June 1825. The map was copyrighted on August 29 and published in September. This was before the distribution of the Risdon map and gave support to the Farmer company's claim that it published the first map of Michigan. The map was reissued in 1830 and in many subsequent editions (Fig. 17–6). The 1830 map was issued separately and as an accompaniment to Farmer's *Emigrants Guide, or Pocket Gazetteer of the Surveyed Part of Michigan*, which was published in Albany. A second edition of the guide was published in 1831.

Another 1830 publication was a *Map of the Territories of Michigan and Ouisconsin*, which was engraved by Rawdon Clark & Company. In 1831 Farmer then prepared, for a congressional report, a *Plat of the City of Detroit as laid out by the Govr. and Judges*, which was lithographically reproduced by Bowen & Company of Philadelphia. Although an 1835 edition of the Michigan map, with some corrections and additions, has the credit "En-

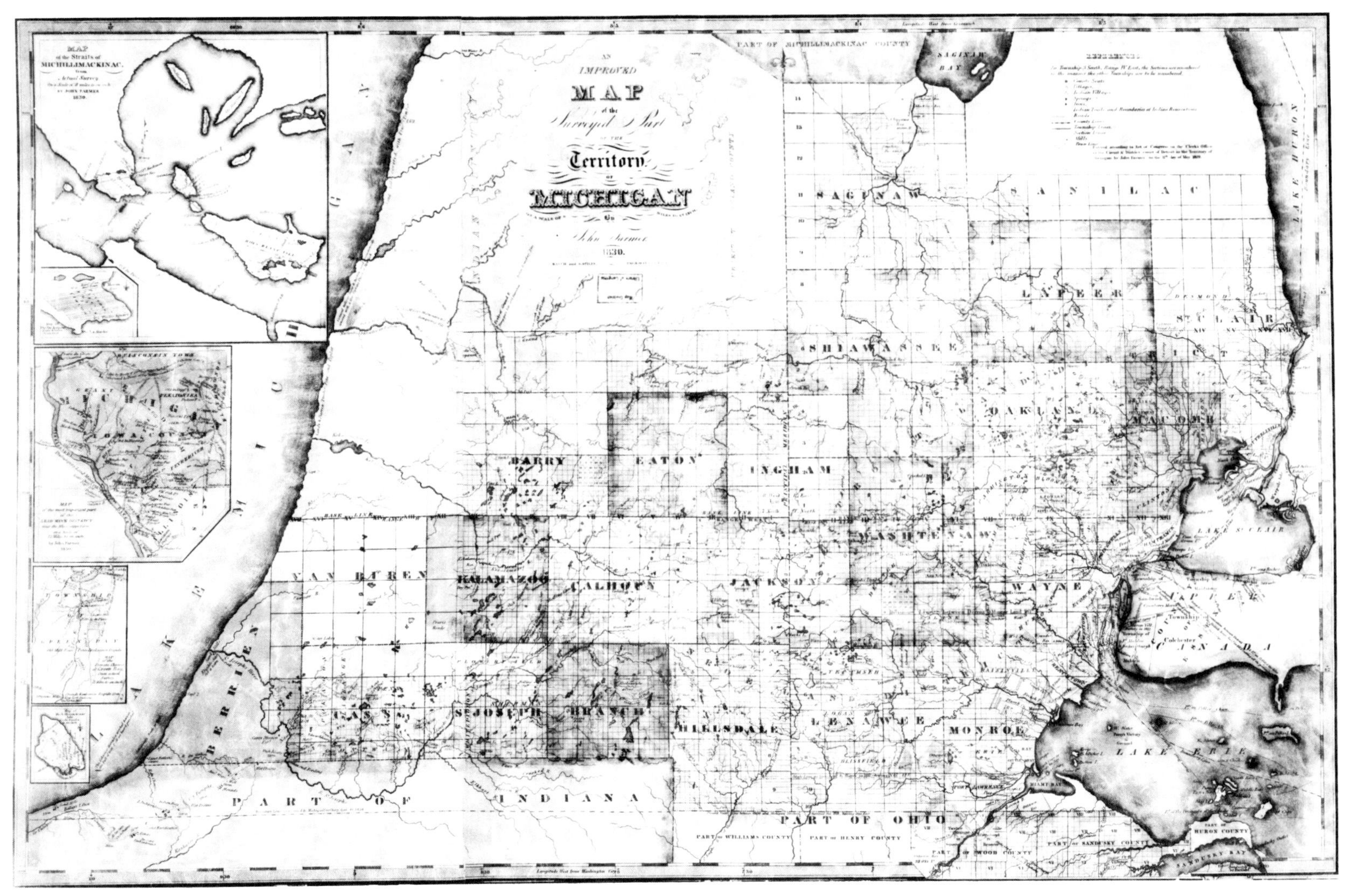

Fig. 17–6. The 1830 edition of John Farmer's Michigan map. It was engraved by Balch & Stiles.

graved by J. Farmer," the 1835 edition of the *Map of the Territories of Michigan and Ouisconsin* was still engraved by Rawdon Clark & Company.

In 1889 Farmer's son Silas Farmer wrote that "these maps and gazetteers of 1830 and 1836 circulated extensively in the East, and had a more marked effect in stimulating the unprecedented emigration of those days than any and all other private enterprises. . . . No other Territory or State, in its infancy, was so accurately represented or so thoroughly advertised by means of reliable maps as Michigan. The maps and gazetteers of Mr. Farmer contributed largely to this work, and his publications, though issued by private enterprise, were none the less a great public advantage."[11]

Despite such cartographic success, Farmer sold the copyrights for his maps in 1835 to the large map publishing firm of J. H. Colton in New York City. For the next two decades Colton published and sold Farmer maps. Farmer's decision to sell his copyrights was prompted by the inconvenience and expense of having to send his maps for engraving to such distant cities as Albany and Utica. Between 1836 and 1844 Farmer abandoned commercial map publishing and devoted his time and talents to public employment. From 1837 to 1841 he was the surveyor of Wayne County, and in April 1838 he was elected school inspector and chairman of the Detroit School Board. He served in these educational positions until 1842. He also variously served from 1839 to 1842 as the sealer of weights and measures, the street commissioner, and the district and city surveyor. During these years Farmer also perfected his engraving technique, which he had learned in 1835, and collected data for use in compiling more maps.

Map compilation, engraving, and publishing were still his first loves, and in 1844 he reestablished his map publishing company. In that year he published a *Map of the State of Michigan and the Surrounding Country*, which was a great improvement upon his earlier map of the state. This map proved to be a very popular item, and some twenty-three editions of it were issued to 1873. All of Farmer's maps published after the 1844 Michigan map were on the same scale and were combined, in some editions, to include several states. Thus, an 1846–47 edition of the Michigan map includes the upper part of Wisconsin and Lakes Superior and Michigan. Similarly, Farmer published an 1848 *Map of the States of Michigan and Wisconsin embracing a great part of Iowa and Illinois and the whole Mineral Region with a Chart of the Lakes*. The several sections of this map were also sold separately.

Most of the maps published by Farmer after 1844 include his engraving credit. It is likely, however, that as the business expanded other employees prepared corrections and additions to the plates. It is not clear whether Farmer printed his maps directly from copper plates, or whether the images were transferred to lithographic stones or to stereotype or electrotype plates. It is unlikely that copper plates would have survived the numerous impressions required by his volume of sales.

Farmer's intensive work habits and his long hours of labor spent in preparing and engraving maps caused him to have a nervous breakdown early in 1859. He was admitted to a local hospital, but, after a brief period of confinement, he leaped over the railing of a second-story porch or balustrade and was badly injured in a twenty-foot fall. The injuries proved to be fatal, and he died on March 26, 1859. His map publishing business was carried on by his widow, Roxanna, their two sons, John H. and Silas, and their daughter, Esther. The death of Farmer and the start of the Civil War, however, limited publishing operations for several years.

In 1862 the company published *Farmer's Rail Road & Township Map of Michigan and Chart of the Lakes, From U.S. Surveys & Other Authentic Sources. Drawn and Engraved by John Farmer, C.E. Completed and Published by R. Farmer & Co. (Successors to the late John Farmer.) 85 Monroe Avenue, Detroit*. In appearance and type of reproduction the map differs from earlier Farmer maps. Although Farmer is credited as the engraver, the printing appears not to have been effected from a copper plate. This is one of the few maps issued under the R. Farmer & Company imprint. Near the upper left corner of the map is an advertisement for "J. H. Farmer, M.D., Dentist, Corner of Monroe Avenue and Farmer Street, Detroit."

This map was also published in an 1864 edition, which has the imprint "S. Farmer & Co., (Successors to the late John Farmer)." Silas had purchased the interests of his mother, sister, and brother in 1863. The 1863 *Guide Map of the City of Detroit* has the imprint "Published by S. Farmer & Co. Successor to John Farmer & R. Farmer & Co." and was "entered [for copyright] according to Act of Congress in the Clerk's Office of the United States Court for the District of Michigan, this 12th day of August, 1863, by S. Farmer & Co., as Publishers and Proprietors." The Farmer maps *Lake Superior and the Mining Regions*, the *Railroad and Township Map of Wisconsin*, and the *Railroad and Township Map of Michigan and Chart of the Lakes* were copyrighted in 1867. The last map was reissued in 1871, 1872, and 1873 editions. Below the title cartouches on each of these three editions is the statement:

> Silas Farmer & Co. also publish the ONLY Sectional Maps of Michigan and Wisconsin that claim to or do show the small Lakes, Streams, Marshes, Swamps, etc. The universal testimony of all who have seen or used our Sectional Maps is that they are more minute, reliable, and accurate than the maps of any other publisher, and are of great utility, and eminently serviceable to every Pioneer, Lumberman, and in fact to every person dealing in Real Estate, or who desires Truthful Information concerning the States named. The First Map Ever Made of the region of country now comprised in the States of Michigan and Wisconsin was made by the predecessor of this firm, Mr. John Farmer, in 1826.

Silas Farmer also published in 1873 *The New Official State Map of Michigan, Compiled from United States, State & County Surveys & Records Published by Silas Farmer & Co. 31 Monroe Ave. Cor. of Farmer Str., Detroit, Michigan, 1873*. Beneath the cartouche is the credit "Engraved by the Calvert Lith. Co. Detroit." A box in the margin includes the information that the map was published in three formats: (*a*) on rollers, with cloth backing, at $3.00; (*b*) on rollers, with cloth on the edges, at $2.00; and (*c*) in elegant cloth covers as a pocket map, at $1.50. The map also carries a notice that agents are wanted. In 1871 Calvert & Company published a map of Michigan and Wisconsin that S. Farmer & Company alleged had been plagiarized from Farmer maps. After a year of litigation the two firms agreed to a settlement, and thereafter the map, lithographed by Calvert & Company, was published under the S. Farmer imprint.

In addition to map publishing, Silas Farmer had a strong interest in history. He was appointed to the honorary position of historiographer of the city of Detroit in around 1878. He conscientiously sought to fulfill his duties in this post, conducting extensive historical research that resulted in the 1884 publication of his two-volume *History of Detroit and Michigan*. Two revised editions were later issued, and in 1890 Farmer's *History of Detroit and Wayne County* was published. Farmer also published *The Michigan Book* in 1901. This was his last effort, for he died suddenly on December 28, 1902. His mother had died on May 11, 1890, at the age of ninety.

Farmer's son, Arthur John (born in 1876), took over the business. He seems to have lacked the cartographical interest and business ability of his father and grandfather. He continued to reissue previously published Farmer maps, and between 1905 and 1915 he copyrighted several automobile road maps, atlases of southern Michigan, and a Michigan railroad atlas, which went through several editions. The maps copyrighted by Arthur John Farmer, however, lack the cartographic excellence and legibility of earlier Farmer maps. They may have been reproduced by the wax engraving and electrotyping process, which was widely used for printing maps in the United States in the early decades of the twentieth century. Farmer quit the map publishing business around 1915, and the plates for Farmer maps passed to C. M. Burton.

Fig. 17–7. A Wilson terrestrial globe, ca. 1822.

James Wilson is important to the history of cartography for having manufactured and sold the first American globes. He was born in Londonderry, New Hampshire, on March 15, 1763. Wilson received a rudimentary education in that town and worked on his father's farm for several years. When he was about twenty-one years old he purchased one hundred acres of uncleared land near Francistown, New Hampshire, built a cabin, and began farming. In 1796 he moved to Bradford, in Orange County, New Hampshire.

Several years later, on a visit to Dartmouth College in Hanover, Wilson saw a pair of English terrestrial and celestial globes. They so intrigued him that he resolved to duplicate them. This occupied his time for the remainder of his long life. To learn about world geography he purchased an eighteen-volume set of the *Encyclopaedia Britannica*. Wilson then walked to New Haven, Connecticut, to learn the basics of engraving from Amos Doolittle. He made revisions on the plate for James Whitelaw's 1810 map of Vermont, which was engraved by Doolittle. In 1813 he engraved *Whitelaw's Map of the Northern Part of the United States and the Southern Part of the Canadas* and in the same year, with the help of Isaac Eddy, he engraved the plate for a large chart entitled *Chronology Delineated to Illustrate the History of Monarchial Revolutions*.[12] With some background training as a blacksmith he was able to cast the hardware for the globes and to construct his own lathes, tools, and presses. He also made his own ink, glue, and varnish, and designed his own maps and gore patterns.

All this labor and effort produced in 1810 his first pair of globes, which he sold. Henceforth the globe factory seems to have gone into full production with the assistance of Wilson's three oldest sons, Samuel, John, and David. In around 1815, the sons opened a branch of the factory in Albany, New York, although the head office remained in Bradford, Vermont, under Wilson's supervision. There is no record of how many Wilson globes were manufactured and sold between 1810 and 1835 (Fig. 17–7). Ena Yonge in her *Catalogue of Early Globes* describes seventy-six Wilson globes preserved in the United States.[13] By far the largest number of these globes are 33 cm. in diameter. There are, however, some Wilson globes with diameters as small as 8 cm. and as large as 51 cm.

All three of Wilson's sons associated with him in the globe business predeceased him, and so control of the Albany plant passed to Cyrus Lancaster, who had been with the firm for a number of years. In 1835, Lancaster married Samuel Wilson's widow, Rebecca. It is not known how long Lancaster continued to operate the business, but he lived until 1862. Wilson died in 1855 at the age of ninety-two.

Notes

1. Ralph Brown, "The American Geographies of Jedidiah Morse," *Annals of the Association of American Geographers* 31 (1941): 187–88.
2. Jared Sparks's review can be found in the *North American Review* 18, n.s., 9 (1824): 390.
3. *North American Review* 19 (July 1824): 261–62.
4. Calvin Robinson, "Lewis Robinson," in Gilbert A. Davis, *Centennial Celebration Together with an Historical Sketch of Reading, Windsor County, Vermont* (Bellows Falls, 1874), 144–46.
5. Ibid., 145.
6. George R. Dalphin and Marcus A. McCorison, "Lewis Robinson—Entrepreneur," *Vermont History* 30 (Oct. 1962): 297–313.
7. Ibid., 310.
8. Richard W. Stephenson, "Charles Varlé, Nineteenth Century Cartographer," paper presented at the annual convention of the American Congress on Surveying and Mapping, 1972.
9. Ibid., 2.
10. Much of the data on the Farmers are derived from William L. Jenks, "A Michigan Family of Map-makers," in Louis C. Karpinski, *Bibliography of the Printed Maps of Michigan 1804–1880* (Lansing, 1931), 16–22.
11. Silas Farmer, *The History of Detroit and Michigan* (Detroit, 1889) 1:698.
12. Dalphin and McCorison, "Lewis Robinson," 298.
13. American Geographical Society, *A Catalogue of Early Globes Made Prior to 1850 and Conserved in the United States*, ed. Ena L. Yonge (New York, 1968).

18. *The Lithographic Revolution*

Most of the maps and atlases heretofore described were reproduced from engraved copper plates. This technique for reproducing illustrations was introduced around the middle of the fifteenth century. It was first applied to cartographic printing in 1477 with reproduction of twenty-six maps for the Bologna edition of Ptolemy's *Geographia*. Engraving is an intaglio process in which the image is inscribed with a sharp tool, called a burin, into a soft metal, usually copper. The plate, on which the image is inscribed in reverse, is inked and then is wiped clean, leaving ink only in the depressions. A damp sheet of paper is laid on the plate and pressure is applied in a press to insure that the ink is properly transferred to the paper.

Engraving is a highly skilled art, and considerable time is required to prepare a map plate. Also, the soft copper wears quickly and does not survive more than a thousand or so impressions. The process did, however, produce clear and attractive maps. Moreover, changes and corrections could readily be made on the plates. John Melish, for instance, printed no more than one hundred copies of his U.S. map before updating the plates with new geographical information.

Engraving had, nonetheless, certain limitations. It was a tedious and costly process and incapable of supplying the volume of maps demanded by the growing population by the mid-1800s. Various individuals had, accordingly, sought to develop more expeditious and less costly processes for duplicating maps and other graphics. Alois Senefelder of Munich was the first to achieve promising results with his invention of lithography in 1796.

In contrast with woodcut reproduction, a relief technique, and copper-plate printing, lithography is a planographic process. The desired image is drawn in reverse on a highly polished fine-grained stone with greasy crayon or ink. From the stone (or a properly prepared metal plate) the image is transferred to paper in a press. The transfer is effected by chemical rather than physical action. Senefelder, in fact, called his invention "chemical printing," but when the technique was introduced in France in 1804, it was called "lithographie," which became its permanent name.

In his manual, *A Complete Course of Lithography*, published in 1819, Senefelder explained how lithography differed from earlier printing methods. He noted that

> the chemical process of printing is totally different from both. Here [i.e., in lithography] it does not matter whether the lines be engraved or elevated; but the lines and points to be printed ought to be covered with a liquid, to which the ink, consisting of a homogeneous substance, must adhere, according to its chemical affinity and the laws of attraction, while, at the same time, all those places that are to remain blank, must possess the quality of repelling the colour. These two conditions, of a purely chemical nature, are perfectly attained by the chemical process of printing; for common experience shews that all greasy substances . . . or such as are easily soluble in oil . . . do not unite with any watery liquid, without the intervention of a connecting medium; but that, on the contrary, they are inimical to water, and seem to repel it. . . . Upon this experience rests the whole foundation of the new method of printing . . . because the reason why the ink, prepared of a sebaceous matter, adheres only to the lines drawn on the plate, and is repelled from the rest of the wetted surface, depends entirely on the mutual chemical affinity, and not on mechanical contact alone.[1]

During the first two decades following Senefelder's invention, lithography was subjected to much experimentation by various individuals principally in Germany, France, and Austria-Hungary. There was little agreement on what constituted the lithographic process. As Michael Twyman observed, "what we now know as lithography began as a number of imitative processes whose only common factor was the lithographic stone. . . . Sometimes the stone was etched or incised to imitate a wood engraving, and sometimes it was etched or engraved as on copper." He believed, however, that "by 1823 the planographic method seems to have been generally accepted as the most natural method of printing from stone, except for specific types of work like lettering, mechanical drawing, maps, and architectural plans, where greater precision was required."[2]

The introduction of lithography to the United States was delayed until the process had outgrown its experimental period and was firmly established as a viable procedure for reproducing graphics. Generally accepted as the earliest example of American lithography is a small drawing by Bass Otis published in the July 1819 issue of the *Analectic Magazine*.[3] Although Otis is credited with having prepared several other small lithographs, his interest in the new process was apparently of brief duration. By 1820 lithography had become quite standardized in Europe, which meant, observes Peter Marzio, that "the entire craft was imported to the United States. Americans received it enthusiastically, honoring it as a fixed science with a complete set of laws. Rather than attempting to improve the process, their initial reaction was to look for stones that worked as well as the magical specimens from Solenhofen [the Bavarian town where the first lithographic stones were quarried]."[4]

In 1822 the *American Journal of Science and Arts* published a "Notice of the Lithographic Art, or the art of multiplying designs, by substituting Stone for Copper Plate, with introductory remarks by the Editor." This notice was accompanied by a number of illustrations printed from lithographic stones prepared by the New York City firm of Barnet & Doolittle. William Barnet and Isaac Doolittle, the two young partners, had acquired a basic knowledge of the lithographic process in Paris, where Barnet's father, Isaac Cox Barnet, was then serving as the American consul. Barnet and Doolittle opened their shop in 1821 at 23 Lumber Street in New York City. It is recognized as the first American lithographic printing company. Although the firm was short-lived, it is significant because among its lithographs for the *American Journal of Science and Arts* was a page-size map titled "Barton on the Catskills." Illustrating D. W. Barton's article "Notice of the Geology of the Catskills," the map is the earliest example of lithographic cartography in the United States.

Two years after this publication, Anthony Imbert, a French artist who had immigrated to New York City, prepared four lithographic maps as illustrations for Cadwallader D. Colden's *Memoir*, which celebrated the completion of the Erie Canal. Two of the maps were facsimiles of earlier works. In an appendix to the *Memoir*, Colden noted that

> a considerable number of the printed plates of this work are in lithography . . . it was necessary to have at least two Maps; one of the State of New York, exhibiting the course of the Canal, and its relation to the neighboring country, but especially to the navigable waters of it; the other to show its connexion, not only with the water courses of the United States, but those of the whole Northern Continent. For this purpose, the artist . . . was charged with preparing Drafts of Maps . . . and calculated to be executed in Lithography by Mr. Imbert; . . . and the result is seen in those respectable specimens of Lithographic mapping; they are, in many respects, superior to the general style of copperplate maps, for utility and effect, and rivalling it in neatness.[5]

A number of small lithographic printing shops were established between 1825 and 1835 in Boston, New York, Philadelphia, and Washington. Some were operated as adjuncts to established copper-plate engraving firms while others were entirely new enterprises. Maps were printed by several of these firms, most of which did not survive long. The most enduring was Pendleton's Lithography, which operated successfully for fifteen years. Rollo Silver, an authority on early printing in the United States, affirms that the lithographic "art became permanently established in this country with the founding . . . in 1825 of the firm of William S. and John Pendleton."[6]

Information about the origins of the Pendletons and of their lithographic establishment is uncertain and conflicting. One version holds that William and John were sons of a Liverpool-born sea captain who had settled in New York City with his English wife around 1789. Both boys were born in that city, William in 1795 and John in 1798. According to Christopher C. Baldwin, who claimed to have received the information from William, Captain Pendleton was lost at sea shortly after the birth of the second son.[7] Little is known about their early lives in New York City. From their later experiences, we may infer that William's education included music and that John's included drawing. Later William, and perhaps John, was apprenticed to engravers to learn that trade. George Groce and David Wallace noted that in 1816 both boys were employed to install gas lighting systems in the Peale Museums in Philadelphia and Baltimore. In 1820 they toured a number of American cities exhibiting Rembrandt Peale's large painting "The Court of Death."[8] As a result of these contacts, some biographers credit Peale with having assisted the Pendletons in establishing their lithographic shop in Boston.[9] Baldwin, disagreeing with this account, reported that William went to Washington in 1819 where he worked as an engraver for about a year. He was joined there by his brother, and the two then traveled in the West.[10] This trip occurs at the same time as the tour taken to exhibit Peale's painting.

Following the tour, John returned to Philadelphia while William went to Pittsburgh. Not successful in finding employment there as an engraver, William taught flute and piano lessons to support himself. William next appears in Boston. There is a lack of agreement as to whether he went there directly from Pittsburgh or whether he took brief sojourns in New York City or Canada. Citing John W. A. Scott, a one-time employee of the Pendletons, Charles Taylor noted that "William S. Pendleton came to Boston from Canada with Alexander Mackenzie, a copper plate engraver, who had failed in business in Montreal. This was in about 1819 or 1820. Mackenzie went into partnership with Abel Bowen in 1821, and presumably William Pendleton worked for them, and the next we hear of Pendleton was when he was in partnership with Bowen on Harvard Place in 1825."[11] In view of William's earlier activities, his arrival time in Boston was very likely no earlier than 1821.

John, on the other hand, spent some time in Europe, probably in 1824 or 1825, where among other pursuits he learned the principles of lithography. The November 5, 1825, issue of the *Boston News Letter* announced that "this beautiful and highly useful art which has lately made great advancement in Europe, we are happy to announce is in successful operation in this city, being introduced by Mr. J. Pendleton, who has made it his study in Europe." A month later, the *Boston Monthly Magazine* noted that

> specimens of this art have, from time to time, reached us, and excited considerable attention among our artists, as well as curiosity amongst our lovers of the arts; but still nothing was done to bring lithography into this country until within a few months, when Mr. John Pendleton commenced an establishment for lithography in this city. . . . Mr. Pendleton is a young gentleman of taste and talents, from the State of New York, who on a visit to Paris, on business of an entirely different nature, and becoming pleased with lithography, put himself immediately under the first artists of France, and acquired, as we believe, a thorough

> knowledge of the art and the principles on which it is founded. With this stock of information, and with a great love of the profession, and, in addition, a good supply of the proper stone and other materials for the pursuit of the art, he came to Boston and engaged with his brother, a copper-plate printer of established celebrity.

Apparently John was for a time occupied with lithographic printing independently, perhaps utilizing space provided by his brother and Bowen. Not long afterwards, though, notices signed by William and Bowen in the February 4 and 11 issues of *Bowen's Boston News Letter and City Record* announced the dissolution of the firm of Pendleton & Bowen, effective January 31, 1826. The periodicals also reported that William would continue the engraving business with his brother, "who will add to the Establishment the advantages of Lithographic Printing. . . . To those whose occasions require Fac Similes, Maps, Circulars, &c., to which this art is peculiarly adapted, Specimens will be exhibited."[12]

The Franklin Institute of Philadelphia early evinced an interest in lithography and offered encouragement to practitioners of the new art. The committee in charge of the institute's third exhibition in October 1826 awarded, "for the best specimen of lithography executed in the United States, . . . to W. & J. Pendleton, Boston, the silver medal." The subject of the lithograph, unfortunately, was not indicated. In the following year, the Franklin Institute awarded "to Rembrandt Peale, of Boston for . . . his beautiful portrait of Washington, executed by him, and printed at the Press of Messrs. Pendleton of Boston, esteemed the best specimen of American lithography ever seen by the committee of fine arts—silver medal."[13]

Some of the earliest products of Pendleton's Lithography were portraits, a number of which were published in various issues of the *Boston Monthly Magazine*. In 1828 the firm printed portraits of Presidents Washington, Jefferson, Adams, Madison, and Monroe from stones prepared by Nicolas E. Maurin, a skilled Paris lithographic artist. The lithographs were reproductions of paintings by Gilbert Stuart. This was the first series of lithographs to portray a sequence of American presidents, and it was referred to as the "American Kings."[14] During its years of operation, the Pendleton firm prepared a wide range of lithographic reproductions, including maps. Perhaps the earliest Pendleton cartographic effort was a facsimile of John Foster's *Wine Hills Map* of New England, which was originally published in 1677. The facsimile was prepared by Moses Swett, a lithographic artist then in the employ of Pendleton's Lithography. It was used for the frontispiece in the fifth edition of Nathaniel Morton's *New England's Memorial* published in Boston in 1826 by John Davis. In this prephotographic period, the map was redrawn by hand, and thus it has variations and inaccuracies not on the original map.

John left the firm in 1828. He went to New York City for a brief period and then relocated to Philadelphia. While in New York, though, John apparently did some lithographic work. In his book *Manhattan Maps*, Daniel C. Haskell described the map of *The City of New York drawn from actual surveys as furnished by the several city surveyors 1834. Pendleton, lithographer No. 192 B.Way.*[15] This map was reported to be in the collections of the New-York Historical Society. Harry T. Peters, however, gave John's New York address as 137 Broadway. In Philadelphia John became a partner in the lithographic firm of Pendleton, Kearny & Childs in 1829.[16]

Carrying on with Pendleton's Lithography in Boston, William relied upon a staff of competent artists, copyists, and lithographic printers. While proficient in engraving and business management, he had little skill or interest in lithography, which John had taken care of. Among the printers William employed was the Frenchman M. Dubois, who may have been invited to come to Boston by John when he was purchasing lithographic supplies and equipment in Paris in 1825. In 1828 two Pendleton employees, Thomas Edwards and Moses Swett, founded the Senefelder Lithographic Company with the engraving firm Annin & Smith. The Senefelder firm was absorbed by Pendleton's Lithography in 1831.

It is possible that the Pendleton firm prepared other maps for inclusion in books after its reproduction of the Foster map, but there is no record of them. The earliest separate Pendleton maps on record appear to have been printed in 1828 to accompany various reports submitted to Congress. There was no official federal map publishing agency at that time, and the printing of maps to illustrate various congressional reports was contracted to private engravers and lithographers. Pendleton's Lithography, one of these private contractors, printed the maps that accompanied those reports of the commissioners under the Treaty of Ghent pertaining to the northern and northwestern boundary of the United States.[17] The eight Pendleton maps cover the U.S.-Canadian boundary in Lake Huron and Lake Erie (Fig. 18–1). All the maps were drawn on stone by James Eddy.

David Stauffer and Fielding Mantle report that Eddy was born on May 29, 1806, and was the son of Benjamin Eddy, a shipwright, and Sarah James Eddy.[18] He seems to have been among the group of young artists who gravitated to Pendleton's Lithography in the late 1820s, and he may have learned the technique of drawing on stone there. Eddy also prepared a *Map of the Country Embracing the Several Routes Examined with a View to a National Road from Zanesville to Florence* for Pendleton's Lithography in about 1828. It also accompanied a report to Congress. Eddy's name appears as the lithographic draftsman on two other Pendleton maps. The first is a *Map of Lynn and Saugus* (Massachusetts), which was "surveyed and drawn by Alonzo Lewis, Author of History of Lynn." Published in 1829, it includes the names of numerous landowners and was the first of a number of Massachusetts town plans printed by Pendleton's Lithography. The other map was the *Map of the Military Bounty Lands in Illinois, Compiled from official Surveys by Zophar Case, Clerk in the Auditor's Office, (Illinois)* (Fig. 18–2). There is no date on the map, but a statistical table in the lower left corner of the sheet includes the 1830 population figures for Illinois. This map was also probably prepared to illustrate a congressional report.

Pendleton's most ambitious program of cartographic printing was prompted by the legislature of Massachusetts in response to appeals from various scientific bodies and individuals for an authoritative map of the state. As reported in Chapter 6, the legislature passed a resolution requiring Boston and the several Massachusetts towns to make accurate maps of their territories. The resulting large state map was Simeon Borden's 1844 *Topographical Map of Massachusetts*. It was engraved on eight copper plates by George G. Smith of Boston. Smith was for a number of years a partner in the engraving firm of Annin & Smith. He probably learned engraving from Abel Bowen, who employed him around 1815. His association with Annin began about 1820. Smith was early interested in lithography and went to Paris for instruction and materials, probably in 1828. Upon his return to Boston, Annin & Smith set up a subsidiary lithographic firm with Edwards and Swett under the name Senefelder Lithographic Company. It operated for three years before being taken over by Pendleton's Lithography. The addition of the Senefelder equipment increased Pendleton's printing facilities to four lithographic presses and four copper-plate presses.

Although the town maps proved unsatisfactory to Borden as compilation material for the state map of Massachusetts, they did include detailed local data about property owners, roads, physical features, forests, and villages of interest to town occupants. Accordingly, between 1830 and 1836 a number of Massachusetts town plans were published. A few were printed from engraved copper plates, but the majority, numbering more than forty, were reproduced by lithography. Most of these lithographic plans were prepared by Pendleton's Lithography. Some were also done by the Senefelder Lithographic Company and "Annin, Smith & Cos. Lithg."

We do not know who initiated the idea to publish the town plans. Pendleton's and the other lithographic

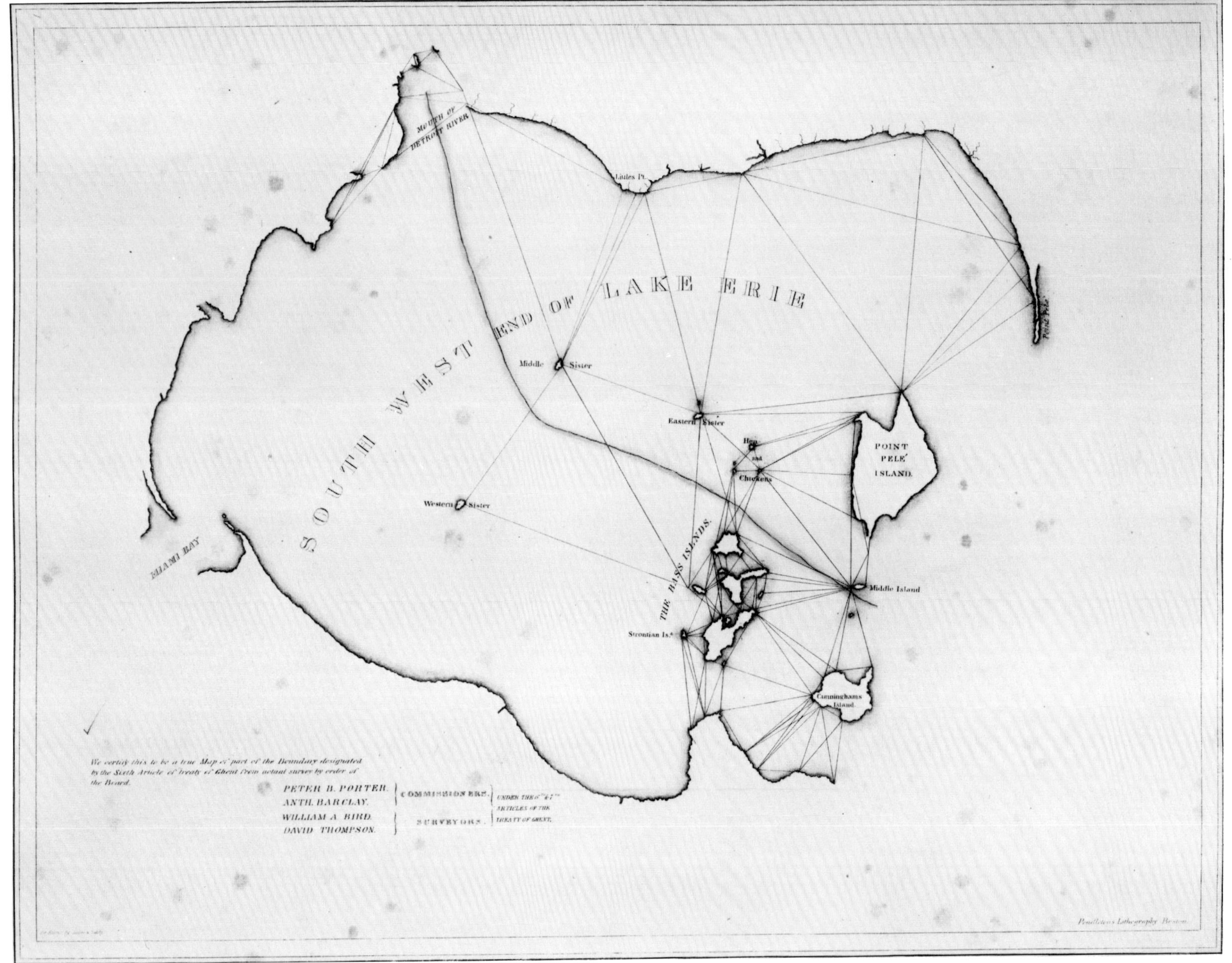

Fig. 18–1. Map of Lake Erie, one of the eight lithographed and printed by Pendleton's Lithography to accompany the report of the commissioners under the Treaty of Ghent. The report and maps were published in 1828.

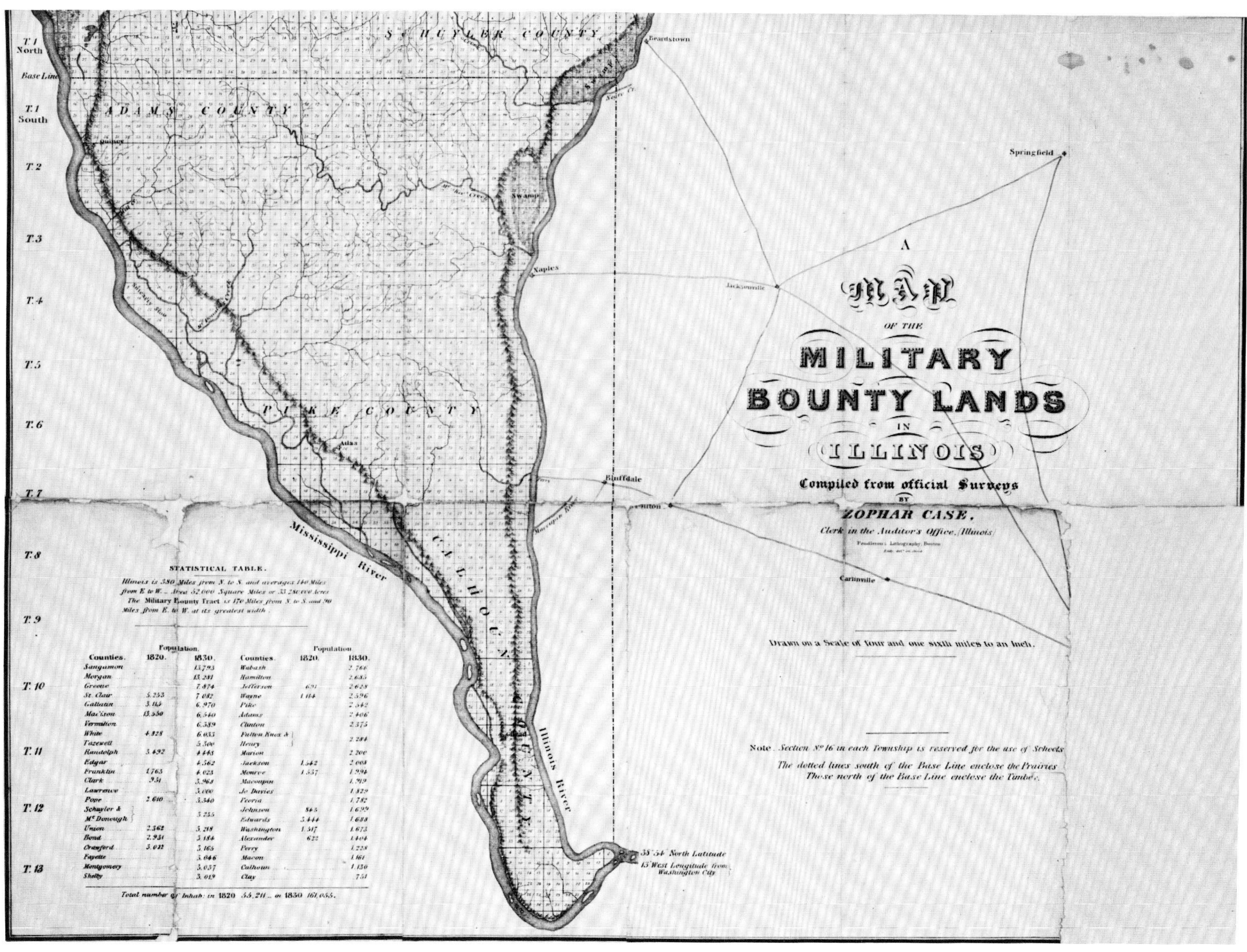

Fig. 18–2. The lower half of Zopher Case's *Map of the Military Bounty Lands in Illinois*. It was lithographed and printed by Pendleton's in about 1830.

Fig. 18–3. Alexander Wadsworth's map of Mount Auburn, Massachusetts. It was lithographed and printed by Pendleton's in 1831.

firms printed the plans on order. This is confirmed by a bill preserved in the collections of the American Antiquarian Society in Worcester, Massachusetts. The bill was made out to a Mr. Stephen Metcalf by W. S. Pendleton and dated February 1, 1833. It covers "lithographing a map of Bellingham, 11 copies, $44.00, 25 extra copies, $2.50, coloring 125 copies, $6.25."[19] This indicates that Metcalf could sell the printed town plans for as little as one dollar and still make a modest profit.

A number of the original manuscript town plans are preserved in the Massachusetts State Archives. Comparing the printed maps with their corresponding manuscript maps reveals some significant differences. In addition, of course, to being in a more finished form with neatly designed title cartouches, all the printed maps have the names of property owners. None of the manuscript maps include such names. A common feature on both the manuscript and printed plans is the symbol for wooded areas; the wooded areas, though, are rarely the same in outline. Despite these changes, all but one or two of the printed maps credit the surveyor who prepared the manuscript plan as the author.

The publisher is named on several of the maps: Moses Pettingell on the plan of Newbury, Newburyport, and West Newbury; Lionel Shattuck on the plan of the town of Concord; C. Harris on the map of Worcester; and Metcalf on the plan of Bellingham. The Newbury plan was registered for copyright by Pettingell and is the only Massachusetts town plan that carries a copyright statement. It is quite probable that local individuals served as the publishers of the town plans. They may have engaged the original surveyor to traverse the town to add the names of landowners, or this information may have been secured from the town clerk. It is very likely that the plans done by Pendleton's Lithography were redrawn by one of the firm's artists. The plans seem to have been drawn with lithographic ink on transfer paper and then transferred to stone. That Pendleton's Lithography had the capability of effecting such transfers is evident from the information in a letter quoted by

Peters in his *America on Stone*. Written by Fred K. F. Hassam on January 1, 1895, to a Mr. Denio, the letter describes the opening dinner of Boston's Tremont House Hotel on October 16, 1829. Hassam noted that "everything was done to make it a positive step forward in the art of Hotel Keeping. . . . When the time came to have the first dinner Pendleton suggested something rare and new in the way of transfer upon stone and printed therefrom. Pendleton wrote the bill of fare with transfer ink upon transfer paper. Bischou transferred it upon the prepared lithographic stone and printed about 200 that were used at the dinner."[20]

The earliest Massachusetts town plan printed by Pendleton was the map of Lynn and Saugus done by Eddy in 1829. This plan, though, was not based on the surveys for the Massachusetts state map. The years 1830 and 1831 were Pendleton's most productive ones for printing the Massachusetts town plans. In 1830 it printed plans of the towns of Andover, Bellingham, Concord, Falmouth Neck, Hingham, Lexington, Newbury, Plymouth, and Stow. The town plans of Bradford, Dorchester, Milton, Groton, Halifax, Holliston, Lancaster, Mount Auburn (Fig. 18–3), Scituate, and Sharon were done in 1831. In 1831 Senefelder Lithographic printed plans of Gloucester, Stoughton, and West Bridgewater. A plan of the town of Newton published in the same year has the imprint "Annin, Smith, & Cos. Lithography." Annin, Smith & Company also printed the town plans of Bridgewater and Ipswich in 1832.

Other plans printed by Pendleton's Lithography include Framingham, Halifax, and Wareham in 1832, Amherst and Worcester (Fig. 18–4) in 1833, and Taunton in 1836. In 1832 the firm also printed a *Geological Map of Massachusetts*, which was prepared by Edward Hitchcock and "Executed under the direction of the Government of the State" (Fig. 18–5). It is uncolored, with the geological formations and minerals identified by twenty-seven different symbols. The geological map was prepared by Hitchcock to illustrate his book *Geology of Massachusetts*, the first edition of which was published

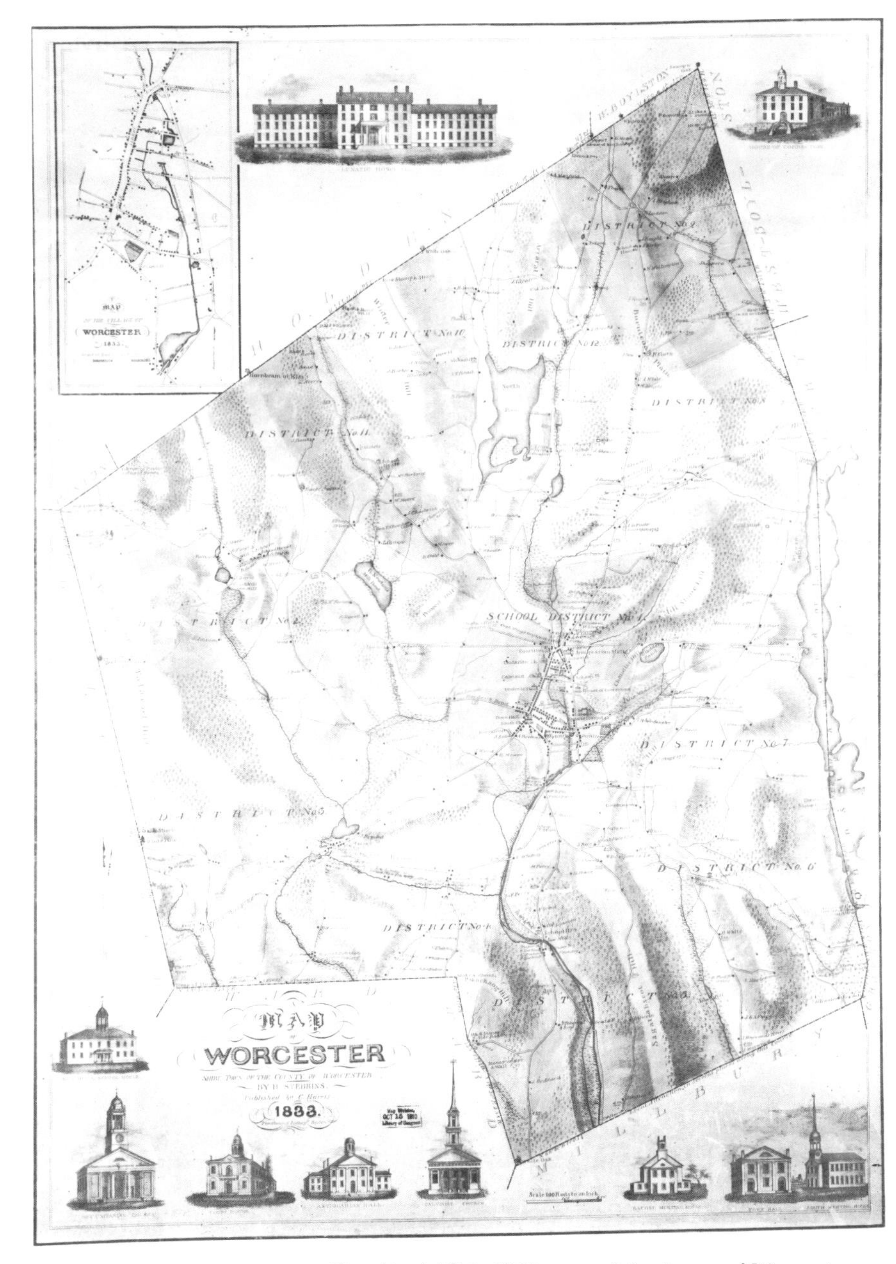

Fig. 18–4. This 1833 map of the town of Worcester, Massachusetts, was printed by Pendleton's. It is distinctive for its depiction of generalized relief and vegetation on the map and of churches and public buildings in the margins.

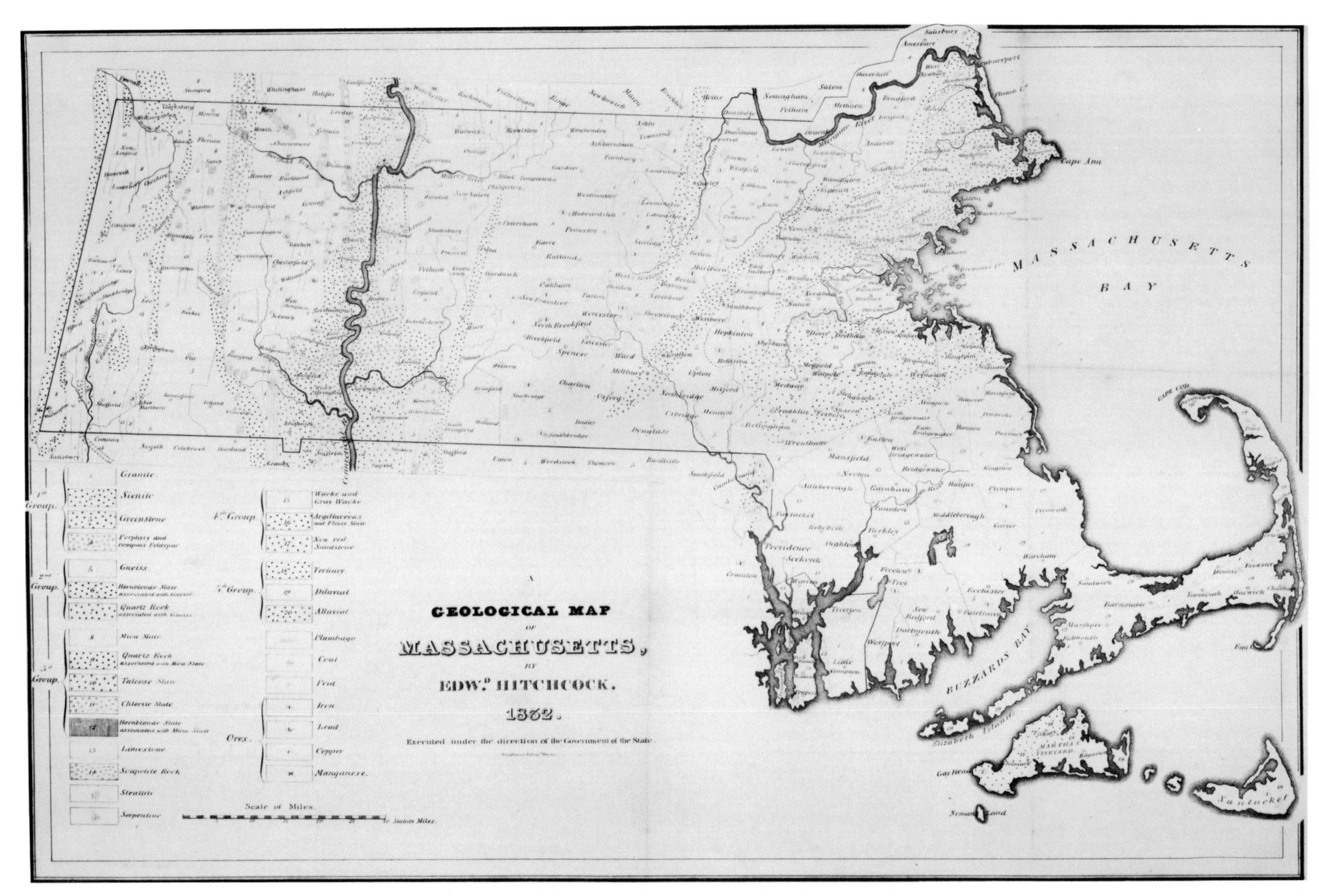

Fig. 18–5. Edward Hitchcock's *Geological Map of Massachusetts*, 1832.

in 1833. A revised edition is dated 1835. A number of views included in both editions have the credit "Mrs. Hitchcock Del." The views, like the map, were printed by Pendleton's Lithography. In the published volumes, the map and views were manually colored.

Although Pendleton's principal cartographic contribution was the series of Massachusetts town plans, the firm also printed other maps, several of which have already been described (Fig. 18–6). Railroad surveying and construction were beginning to have an impact on map printing and publishing by the mid–1830s. Pendleton's first and perhaps only contribution to railroading was its *Plan and Geological Section of a Rail-Road Route from Old Ferry Wharf, Chelsea to Beverly Surveyed under the Direction of Hon. Thos. H. Perkins and Others, By D. Jay Browne Engineer, 1836* (Fig. 18–7). Pendleton also printed a *Map of the Indian tribes of North America about 1600 A.D. along the Atlantic; and about 1800 A.D. westwardly*, which is used as the fold-out frontispiece of volume 2 of the *Transactions of the American Antiquarian Society* for 1836.

In this year William Pendleton sold his lithography business to Thomas Moore, his bookkeeper. It is probable that Moore had managed the lithographic business for the previous four or five years. David Tatham reported that

> the absorption of the Senefelder firm in 1831 marked the culmination of William Pendleton's association with lithography. In that year President Jackson vetoed the rechartering of the Bank of the United States with the result (among others) that in 1833 'pet' banks throughout the country issued new paper currency, each bank using its own designs. The business generated for copper plate engravers was considerable and to take advantage of the opportunity, William Pendleton formed the New England Bank Note Company. Though it was quartered in the same location as the lithographic shop, the bank note firm doubtless claimed a good portion of Pendleton's attention, and it is fair to guess that by 1834 the lithographic business had come to be increasingly under the direction of its chief artist, Robert Cooke, and its bookkeeper, Thomas Moore, who . . . acquired the firm in July 1836.[21]

Moore operated the business until 1840 and in general continued most of the policies and practices established by Pendleton. Among the cartographic products bearing the "Moore's Lithography, Boston" imprint are a *Plan of Lots on Mount-Bowdoin in Dorchester—Laid out by Cornelius Coolidge & Surveyed by Thomas M. Mosely, Sept. 1836* and a *Plan of Land & Water Lots of the Charlestown Wharf Company*, ca. 1838. In addition to printing the nine charts of the coast of Maine (see Chapter 15), Moore's Lithography did several maps for the *Report and Resolves in Relation to the Northeastern Boundary*, which was published by the Massachusetts senate.

Based on the holdings of the Library of Congress Geography and Map Division and the Winsor Memorial Map Room of Harvard University, the Pendleton, Senefelder, and Moore lithographic establishments printed between forty and fifty maps. Assuming that these collections lack some of the products of these firms and that only a few maps printed by them as book illustrations have been identified, we may infer that their total cartographic production between 1826 and 1840 was in the neighborhood of seventy or seventy-five maps. Although most maps published in the United States during these years were reproduced from engraved copper plates, these lithographic firms made a noteworthy impact on American map publishing.

There is one other significant but little known Pendleton cartographic contribution deserving mention. In her *Catalogue of Early Globes*, Ena Yonge described a terrestrial globe made by "Pendleton, Boston," which she tentatively ascribed to William Kimbrough Pendleton with the questionable date of 1840.[22] Yonge's date and assumed author seem to be in error, although it has not been possible to examine the particular item she described. There is, however, a globe in a private Ameri-

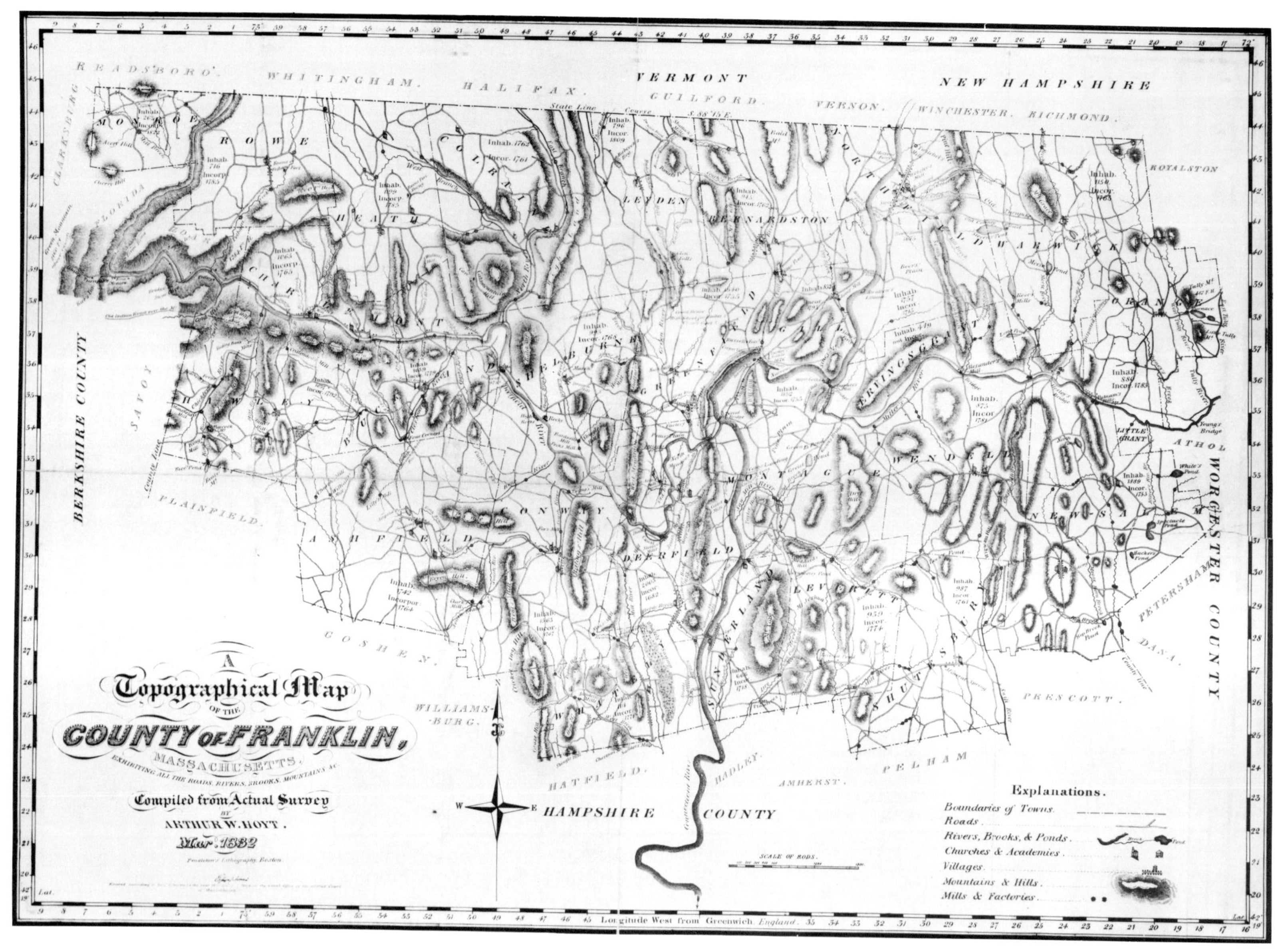

Fig. 18–6. This map of Franklin County, Massachusetts, was published by Pendleton's Lithography and shows generalized topography by means of hachures and crude "caterpillar" hills.

can collection that appears to be identical with or similar to the one described in the Yonge catalog. This globe is also 16 cm. in diameter and with its mounting stands 32.5 cm. high. The cartouche reads, "Pendleton's Lithy. Boston." The same collector has "another globe—same size, same stand but no cartouche, date or author. There are changes in place names, lettering, and delineations but it is hard to tell if it is earlier or later. It is so close to the Pendleton globe it seems as if it would have had to have been made by the same firm."[23] Although the globe with the Pendleton imprint is not dated, it was probably made about 1834 (Fig. 18–8).

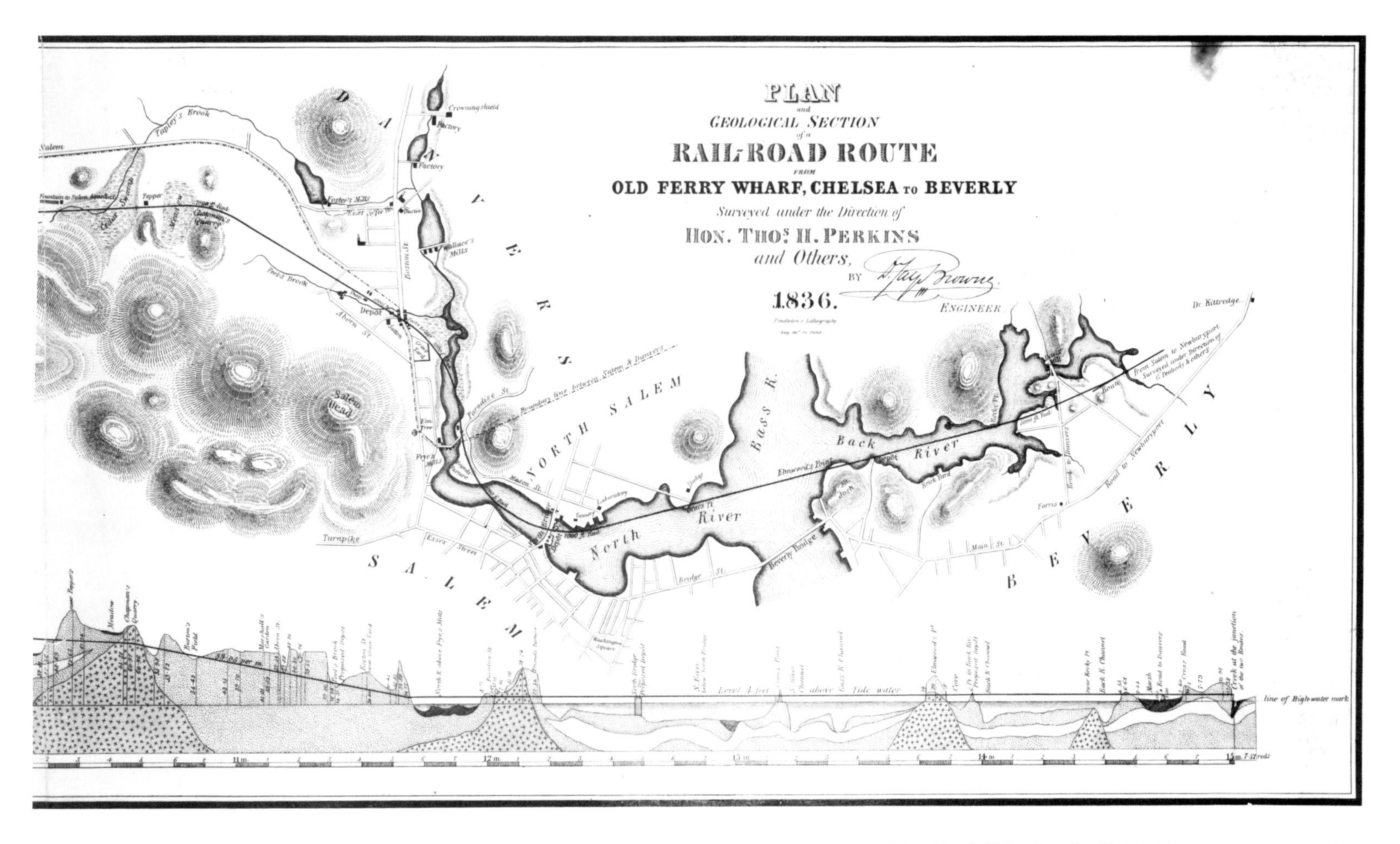

Fig. 18–7. This plan by D. Jay Browne appears to be the only railroad map lithographed in the Pendleton shop. Crude hachuring is used to represent hills.

No information has been uncovered relating to the Pendleton venture into globe manufacture. Our only clue comes from a 1943 article by Alexander Vietor. He stated in *Antiques* magazine that

> William B. Annin worked contemporaneously with Wilson [the first American globe manufacturer]. . . . Annin was most likely a pupil of Abel Bowen, as he worked for that engraver in 1813 and some years afterward. . . . Whether [Annin and Smith] actually sold globes is not perfectly clear. . . . In 1826 Annin took out a patent to make artificial globes. The details of the patent unfortunately remain obscure, due to the destruction of early patent records. Nevertheless it must have been in connection with the method of engraving or printing the plates for the paper gores which, after being printed, were pasted to the ball that was to be a globe. The bookseller, Josiah Loring, was the channel through which Annin globes were sold, under Loring's name.[24]

Fig. 18–8. Two globes in a private collection, one of which bears a Pendleton's Lithography imprint. The similarity of the two globes suggests that the gores of the second one were also lithographed in the Pendleton shop.

There were close relations between William Annin and William Pendleton, and as previously reported Pendleton's Lithography absorbed Annin & Smith's Senefelder Lithographic Company in 1831. We might infer, therefore, that Pendleton was aware of Annin's globe patent and he or one of his associates may have experimented with printing globe gores by lithography. Judging from the small number of Pendleton globes that have survived, it is probable that very few were produced.

Although maps constituted only one part of their varied printing schedules, Pendleton's Lithography and later Moore's Lithography were America's principal proponents of lithographic cartography from 1825 to 1840. There were nonetheless a number of other lithographic printing companies founded in the United States in these years. All did general job printing, which included maps. Boston, New York, and Philadelphia were the principal lithographic centers, with a number of lithographic shops as adjuncts to engraving firms. In a number of instances, skilled lithographers were imported from France or Germany to staff these shops. Regrettably, information about most of these lithographic establishments is meager. In his classic work *America on Stone*, Peters observed "that even if the work of no one of these other establishments is comparable in quality to that of

Currier & Ives, their work as a whole is of far greater interest and importance. It is a vast and absorbing jungle, which can be explored by collectors and students for many years. To my mind it is the greatest existing area of Americana yet unexplored."[25]

Several early lithographic firms were established by former employees of Pendleton's Lithography. Among these was John Pendleton who formed the lithographic firm of Pendleton, Kearny & Childs. Francis Kearny, it will be remembered, had previously been engaged in engraving and was in partnership with Benjamin Tanner and John Vallance. Cephas Childs was born in Bucks County, Pennsylvania, on September 8, 1793. In 1812 he was apprenticed to the Philadelphia engraver Gideon Fairman. He served in the War of 1812, and after further training he began engraving in 1818. Pendleton seems to have interested Childs and Kearny in lithography. During its brief existence, Pendleton, Kearny & Childs engaged in making portraits. Characteristically, Pendleton left the partnership in 1830 and returned to New York City.

In 1831 Childs went to Europe to study lithography, and while there he met P. S. Duval, a French lithographer. Upon Childs's return to Philadelphia he formed a partnership with the painter Henry Inman. They shortly thereafter invited Duval to come to Philadelphia to superintend their lithographic operations. Childs also employed George Lehman, a landscape painter who was born in Lancaster County, Pennsylvania. Childs and Lehman formed a partnership in 1835. It was of short duration for later in 1835 Lehman and Duval established a company that Duval, independently and in various other associations, continued to run until 1879. The Duval firm did various types of lithographic printing, including the reproduction of maps. Duval was among the earliest lithographers to engage in chromolithography and to adapt the rotary steam press to lithographic printing.

An early cartographic product of Lehman & Duval is the *Plan of the City of St. Louis*, which was originally surveyed by City Surveyor Rene Paul in 1823 and revised and corrected in June 1835 (Fig. 18–9). The later version was printed by Lehman & Duval. Also issued under their imprint was the *Map of Part of the Wisconsin Territory* (Fig. 18–10) and Stephen Taylor's 1838 *Map of the Wisconsin Land Districts*. On some copies of this map Lehman's name is omitted. It is also absent on Duval's printing of Henry K. Strong's 1838 *Map of the Swatara Coal Region* (of Pennsylvania). An 1848 map of the United States printed for the U.S. General Land Office also has the Duval imprint. It also appears on W. F. Roberts's *Map of the Anthracite Regions of Pennsylvania*, which was published in March 1849. These examples indicate that lithography was being increasingly employed to print thematic or special subject maps.

In 1828 Pendleton's Lithography took on a fifteen-year-old apprentice named Nathaniel Currier. A native of Roxbury, Massachusetts, Currier remained with Pendleton's until 1833 when he relocated to Philadelphia to study with M. E. D. Brown, a master lithographer. In the following year Currier moved to New York City, where he planned to establish a partnership with John Pendleton. The latter, however, had made other arrangements, and Currier associated himself with a Mr. Stodart. This association proved to be unsatisfactory, and in 1835, at the age of twenty-two, Currier established his own business at 1 Wall Street. As was true of other lithographers, Currier engaged in job printing during his first five or six years. Probably because of his experience at Pendleton's Lithography, maps were among his earliest lithographs. They included the 1835 *Map of Cleveland and Its Environs* (Fig. 18–11) and the 1837 map of the *City of Detroit Michigan from Late and Accurate Surveys* published by Morse & Brother. In 1836 Currier printed, probably for the territorial government, maps of several portions of Wisconsin Territory and the northern part of Illinois. He also printed a map of Port Kalamazoo, Michigan, for K. Lane around 1837. One of his major jobs in 1838 was the reproduction of some sixty page-size district maps for James F. Smith's *The*

Cherokee Land Lottery, Containing a Numerical List of the Names of the Fortunate Drawers in Said Lottery. Currier prepared other lithograph maps for book illustrations in his early years, among them a *Map of the Northern Part of the State of Maine*.

Currier's phenomenal success in 1840 with his print of the fire on the steamboat *Lexington* induced him to focus primarily on historic and popular prints. Few maps were, therefore, printed at his shop after this date. Between 1870 and 1892, however, Currier & Ives published panoramic maps of some of the largest U.S. cities.

Several other New York City lithographic shops

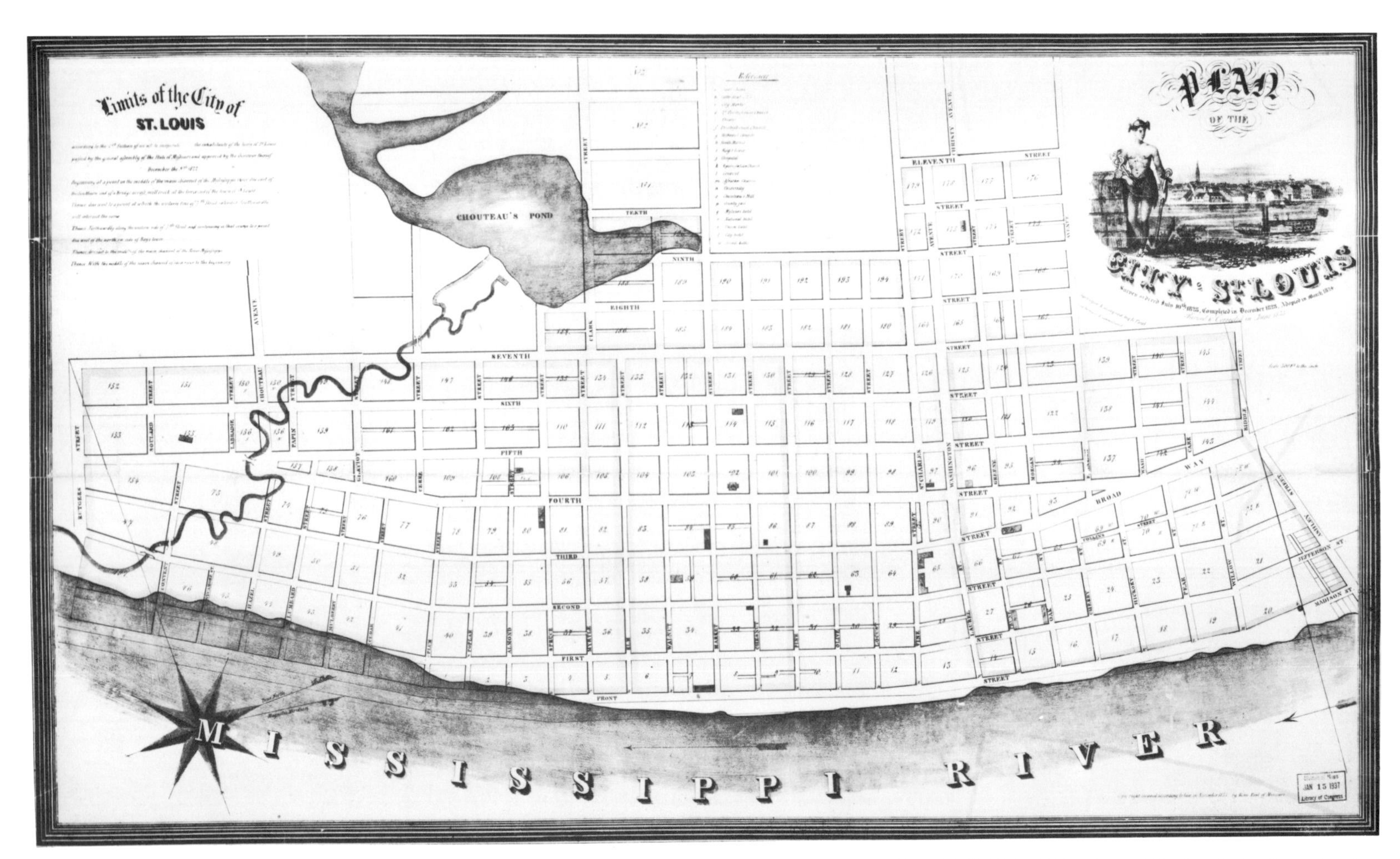

Fig. 18–9. This 1835 plan of St. Louis by Rene Paul is one of the earliest maps lithographed by Lehman & Duval.

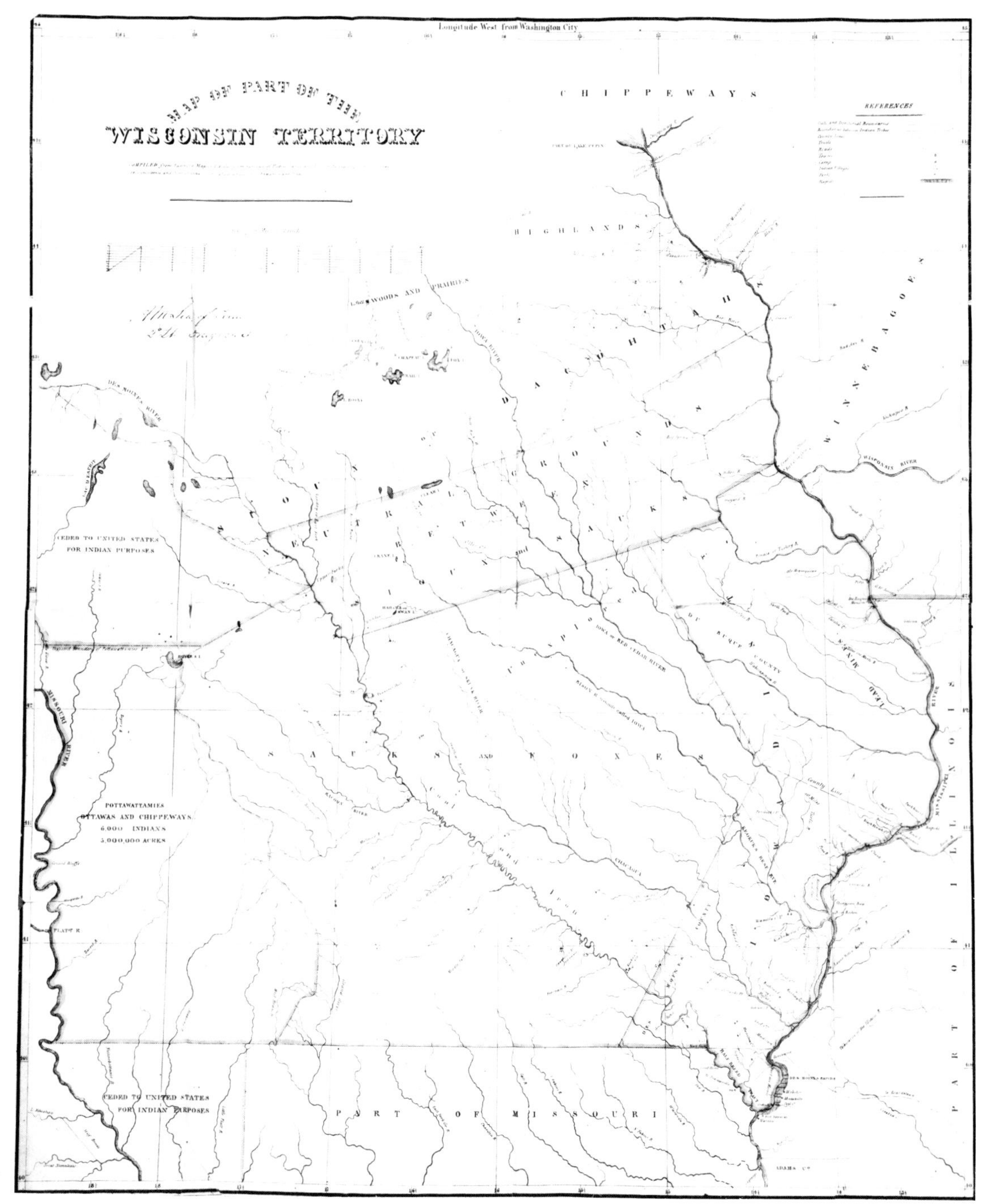

Fig. 18–10. This map, ca. 1836, was probably lithographed by Lehman & Duval to illustrate a congressional report.

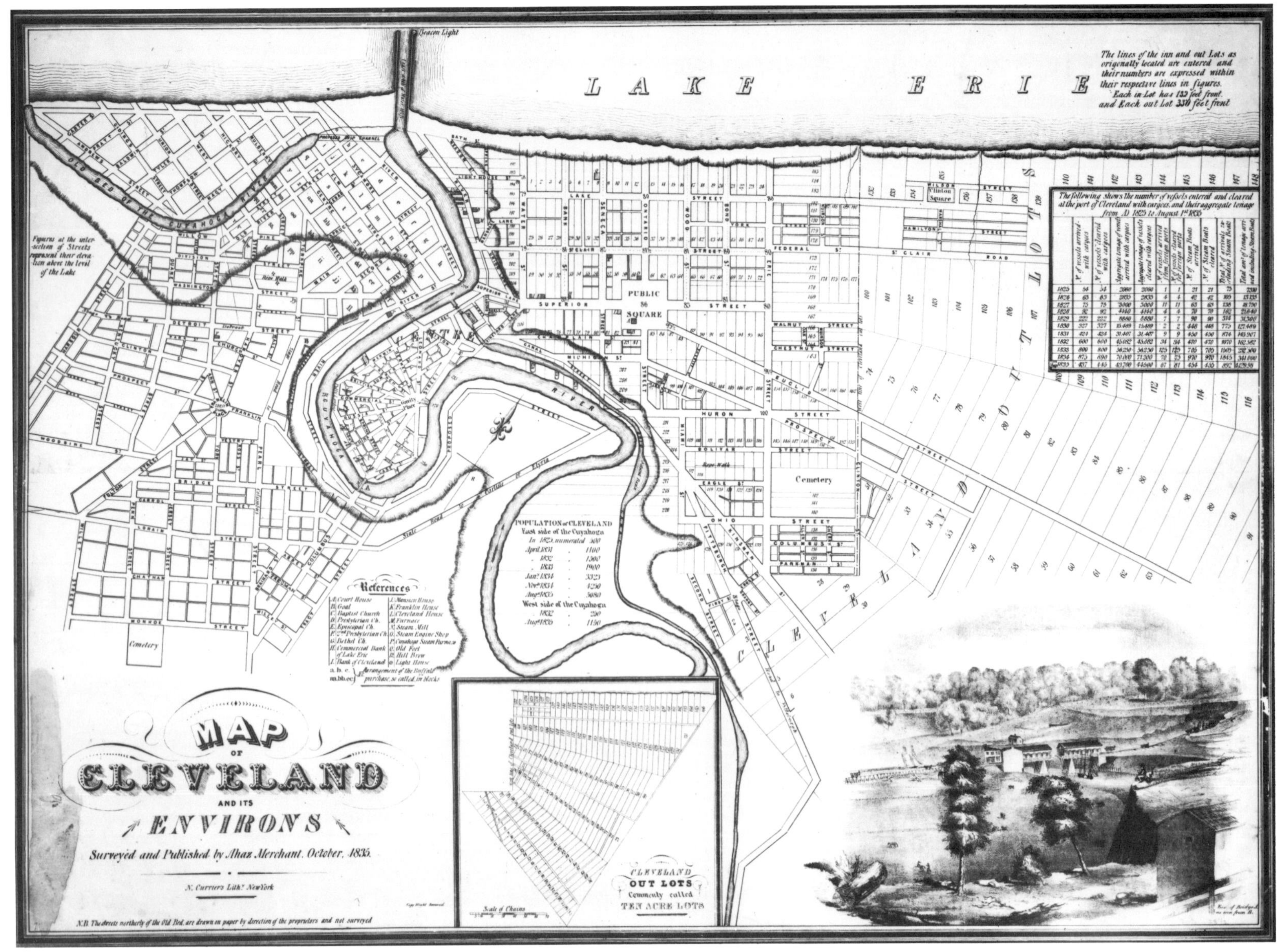

Fig. 18–11. Ahaz Merchant's 1835 map of Cleveland was one of the early products of Nathaniel Currier's lithography shop.

printed maps between 1827 and 1850. One of the earliest maps is the *Map of the Military Bounty Lands of the State of Illinois Shewing the True Boundaries of each County, as fixed by the Legislature in 1825*. It was drawn by H. Ball for the North American Land Agency at 107 Water Street in New York. It was "Drawn on Stone & Printed by M. Williams 65 Canal Street, N. York May 1827." Michael Williams lithographed from several locations in New York City between 1826 and 1834. No other maps printed by him have been identified.

Peter Miller & Company was in business in New York from 1834 to 1869. Harry Peters states that "the house seems to have made maps and plans only. There are for instance: a map of Brooklyn, 1835; . . . a New Haven real estate map, 1835; . . . a folding plan in a book projecting a stone dock by the New York Floating Dock Company, 1845; and a map of lands at De Lancey's Neck."[26] An undated *Map of Lake George* by "Miller's Lith. 102 Broadway, N.Y." was probably printed around 1845.

Prosper Desobry had lithographic shops at various New York City locations between 1824 and 1844. The following Desobry maps have been identified: *Map of the Hudson River, from New York to Albany, with Historical and Descriptive Notes*, which was deposited for copyright on July 13, 1833; *Map of Brooklyn, Kings County, Long Island, from an Entire New Survey by Alexr. Martin 1834* (Fig. 18–12); *Map of Lots in the 5th Ward of the City of Albany . . . , Surveyed Feby. 1834 by George W. Carpenter*; and *Plat of Madison the Capital of Wisconsin, 1836*, which was surveyed by John V. Suydam.

The Mesier lithography shop prepared prints from several locations in New York City between 1835 and 1851. It was run by family members Peter A., Edward S., and Catherine, the wife of Peter. They produced a large volume of sheet music as well as prints and a few maps. One of their maps, undated but probably printed around 1836, is a plan of *Chicago With Several Additions*

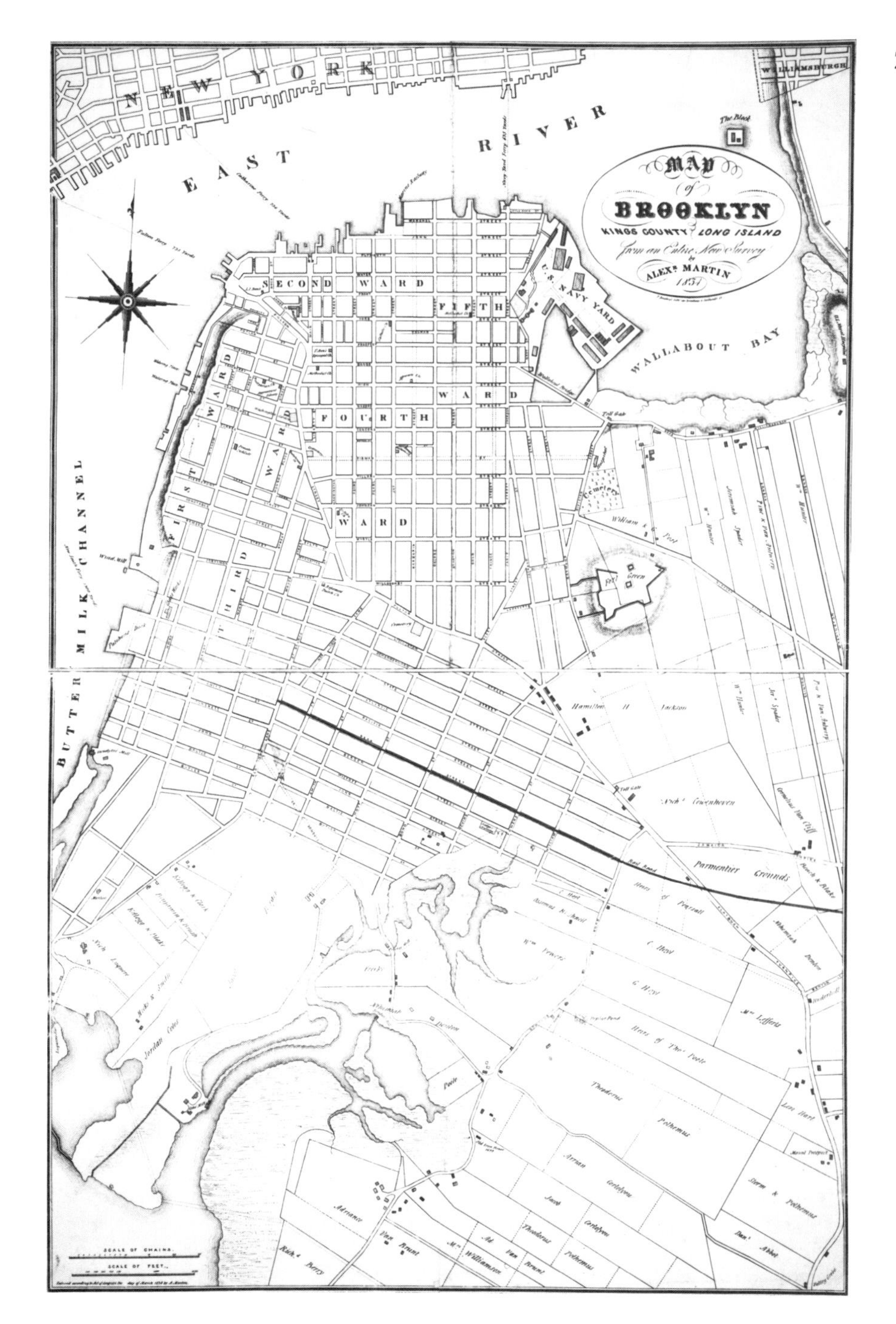

Fig. 18–12. Alexander Martin's 1834 map of Brooklyn that was lithographed by Desobry.

Compiled from the recorded plats in the Clerk's Office, Cook County, Illinois. The Mesiers, Peters disclosed, "were important, early, and their work is scarce and almost always of interest."[27]

William and Francis Endicott operated a lithographic shop for almost forty years at 59 Beekman Street in New York City. Peters stated that "it is hard to summarize the Endicotts. They did everything and did it well . . .[they] worked with and for Currier & Ives, yet in spite of all that much of their work lacks real individuality."[28] An example of Endicott & Company's lithographic cartography is the *Plan of Sacramento City, State of California*, published in 1849.

John T. Bowen operated a lithographic shop at several different locations in New York City from 1834 to 1838. He moved to Philadelphia in 1838. He is primarily remembered for his prints and views, but during his Philadelphia period he printed some maps and diagrams relating to the anthracite trade in Pennsylvania. Prepared by R. C. Taylor, they were probably published around 1839.

Between 1849 and 1900, Julius Bien had an active lithographic business in New York City with a major concentration on cartographic works. Bien was born in Naumburg, near Kassel, Germany, on September 27, 1826. Because of his involvement in the German revolution of 1848, Bien immigrated to America in 1849 and started his business with one hand-operated lithographic press. Peters noted that "soon after his arrival in this country he became interested in improving the quality of maps, and thanks to President Pierce and his administration, he was soon making maps of the new surveys in the west. He produced literally thousands of maps of various parts of the United States, lithographed and engraved, and all for use by state governments and the federal government."[29] During the Civil War Bien equipped a field map printing unit for General William Sherman. Although Bien was primarily a lithographic printer, he did venture into cartographic publishing late in his career. In 1891 he published the *Atlas of the Metropolitan District and Adjacent Counties of New York City* and in 1893 issued the *Atlas of Westchester County, New York*. His *Atlas of the State of New York* was published in 1895, and five years later he produced the *Atlas of the State of Pennsylvania*. A revised edition of this atlas was published in 1901.

The longest-operating lithographic printer in the United States was A. Hoen & Company of Baltimore. The firm was established in 1835 by Edward Weber as E. Weber & Company. A native of Germany, Weber reportedly had learned the lithographic technique from its inventor, Alois Senefelder. Weber's nephews, August and Ernest Hoen, assisted him in the business, and when he died in 1848 they assumed control. In 1853 the firm name was changed to A. Hoen & Company. Relating data provided by August Hoen, Peters noted:

> The firm, from its infancy has devoted much time to scientific illustrations and highly specialized map work. Shortly after 1848 we lithographed the pictorial illustrations of The Fremont Report (Explorations along the 39th Parallel). Later on, we made a series of maps for the Venezuelan Boundary Commission and, in more recent years, made such map work as those covering the Great Lakes Boundary Line. . . .
>
> The making of a series of maps of Latin America on the scale of 1:1,000,000 under the direction of the American Geographical Society has been entrusted to us. It is one of the greatest efforts in geography ever attempted in the new world.[30]

In 1929 Hoen printed the first National Geographic Society map supplement, the beginning of a long and profitable association with the society. By the 1970s, however, the Hoen plant could not print the large number of maps required for the *National Geographic*'s evergrowing circulation. As a result, Hoen's lucrative contract with the society was terminated in 1975. Hoen continued to operate for several years, but in 1981 it was absorbed by the John Lucas Printing Company, also of

Baltimore. Thus ended almost a century and a half of continuous lithographic printing, much of which was of cartographic works.

Notes

1. Alois Senefelder, *A Complete Course of Lithography* (New York, 1968), 27.
2. Michael Twyman, *Lithography 1800–1850, the Techniques of Drawing on Stone in England and France and Their Application in Works of Topography* (London, 1970), 64.
3. Rollo Silver, *The American Printer 1787–1825* (Charlottesville, Va., 1967), 61.
4. Peter Marzio, "American Lithographic Technology before the Civil War," in *Prints in and of America to 1850*, ed. John D. Morse (Charlottesville, Va., 1970), 221.
5. Cadwallader D. Colden, *Memoir, Prepared at the Request of a Committee of the Common Council of the City of New York, and Presented to the Mayor of the City, at the Celebration of the Completion of the New York Canals* (New York, 1825), 354–55.
6. Silver, *American Printer*, 169.
7. Christopher C. Baldwin, *Diary* (Worcester, Mass., 1901), 33.
8. George R. Groce and David H. Wallace, eds., *The New-York Historical Society's Dictionary of Artists in America 1564–1860* (New Haven, Conn., 1957), 497.
9. See, for example, David Tatham, "The Pendleton-Moore Shop, Lithographic Artists in Boston, 1825–1846," *Old Time New England* 52 (1971): 32.
10. Baldwin, *Diary*, 332.
11. Charles Henry Taylor, "Some Notes on Early American Lithography," *Proceedings of the American Antiquarian Society*, n.s., 32, pt. 1 (1923): 77.
12. *Bowen's Boston Newsletter and City Record*, Feb. 4, 1826, 95.
13. George H. Eckhardt, "Early Lithography in Philadelphia," *Antiques* 28 (1935): 250.
14. Mabel M. Swan, "The 'American Kings,'" *Antiques* 19 (1931): 278–86.
15. Daniel C. Haskell, *Manhattan Maps, a Co-operative List* (New York, 1931), 48.
16. Harry T. Peters, *America on Stone* (New York, 1931), 312.
17. *Letter from the Secretary of State Transmitting, pursuant to a Resolution of the House of Representatives, of the nineteenth ultimo, A copy of the Maps and Report of the Commissioners Under the Treaty of Ghent, For Ascertaining the Northern and Northwestern Boundary Between the United States and Great Britain March 18, 1828* (Washington, D.C., 1828).
18. David M. Stauffer, *American Engravers upon Copper and Steel* (New York, 1907) 1:75; Fielding Mantle, *American Engravers upon Copper and Steel, a Supplement* (Philadelphia, 1917), 14.
19. Peters, *America on Stone*, 318.
20. Ibid., 315.
21. Tatham, "Pendleton-Moore Shop," 35–36.
22. American Geographical Society, *A Catalogue of Early Globes Made Prior to 1850 and Conserved in the United States*, ed. Ena L. Yonge (New York, 1968), 54.
23. Personal letter to author from the owner of the two globes, Mar. 25, 1980.
24. Alexander O. Vietor, "Some American Globemakers," *Antiques* (Jan. 1943), 22.
25. Peters, *America on Stone*, 12.
26. Ibid., 285.
27. Ibid., 281.
28. Ibid., 179.
29. Ibid., 94.
30. Ibid., 219.

19. *The S. A. Mitchell and J. H. Colton Map Publishing Companies*

Between 1831 and 1890, general map and atlas publishing in the United States was dominated by the companies founded by S. Augustus Mitchell in Philadelphia and Joseph H. Colton in New York City. Both firms managed to survive the transitions of switching to new reproduction techniques to build themselves into prosperous and successful geographical and cartographical publishing houses.

Social and economic conditions in the country during these years favored enterprises of all kinds, including the publishing of maps, atlases, and other geographical works. By 1831 the young republic had successfully weathered its first half century, including another war with England. It was steadily expanding westward due to the migration from the eastern states and immigration. The internal movement was aided by the construction of turnpikes, canals, and railroads. Between 1810 and 1850, the population of the country increased three fold, and a number of new states were established. Much of the population growth was due to the four million immigrants who entered the country between 1820 and 1860. In the late 1840s Texas was annexed, gold was discovered in California, and Indian wars had removed these native people from their desirable land. The Civil War temporarily curtailed some internal activities, but after peace was reestablished all activities were accelerated.

Public land surveys in the Middle West and the several exploratory surveys beyond the Mississippi provided a steady flow of new geographical and cartographical data that found its way into new and revised editions of maps and atlases. The Industrial Revolution had far-reaching effects on map printing and production. The reproduction of maps from copper-engraved plates gave way to simpler, less costly, and more rapid printing and publishing techniques that were capable of supplying the quantities of cartographic products demanded by an expanding nation. The citizens of this vibrant and dynamic republic had a keen interest in geography, for they faced the challenges of exploring, mapping, and developing an entire continent. This created the fertile market for geographical and cartographical publications that was so effectively cultivated by the Mitchell and Colton firms.

Samuel Augustus Mitchell was born in Bristol, Connecticut, on March 20, 1792. He was the youngest son of William and Mary Alton Mitchell. His father had immigrated to America from Scotland in 1773 and had settled in Bristol. Little is known about Mitchell's schooling, but judging from his later activities, his education must have been better than that enjoyed by the average American youth of his day. Mitchell spent a number of years teaching, probably in Connecticut. He found the available geography textbooks quite inadequate and set himself the task of writing and publishing more satisfactory works. Philadelphia was at the time the leading publishing center in the United States, and apparently for this reason Mitchell moved there in 1829 or 1830.

His first cartographic publication was *A New American Atlas,* which was issued in 1831. This was not an original work but a reprint of Anthony Finley's atlas of the same title that was published in 1826. All fifteen plates in Finley's atlas were drawn by D. H. Vance and engraved by J. H. Young. Little is known about Vance, but Young successfully practiced engraving in Philadelphia from 1817 to 1866. Mitchell's atlas includes the same fifteen plates, but certain changes have been made. Towns and roads have been added, particularly in some of the more recently settled areas in the West and South. Legends or explanation tables are added as new features on some of the maps. The imprint "Philadelphia, Published by A. Finley, 1826" has been replaced on all the plates by "Published by S. Augustus Mitchell, Philadelphia, 1831." Vance's name is deleted on all the maps, but the credit "J. H. Young, Sc." is retained in the lower right corner of each map, although in a different position than on the Finley maps. It is inside the neat line rather than outside, as on the Finley maps. The most obvious alteration on the Mitchell maps, however, is the addition of a decorative border comprised of short horizontal parallel lines. Such decorative borders, which appear on many maps published after 1830, may have been engraved with a geometrical lathe, which was invented by Asa Spencer around 1820. Intended originally to engrave distinctive designs on bank notes, these engraving machines were also used to design decorative borders for maps.

Mitchell first became associated with J. H. Young when working on the *New American Atlas*, and their association proved to be a mutually profitable one. The two men collaborated in producing cartographic publications for several decades. Revised and updated maps from the *New American Atlas* were also published separately on thin bankers paper and folded into pocket-size simulated leather covers. Some of these maps were reduced in size from the plates of the atlas. One series of folded maps introduced in 1834 was identified as the "Tourist Pocket Maps" (Figs. 19–1 and 19–2). Young is listed as the author on some of these, and D. Haines is credited as the engraver on some.

In 1832 Mitchell introduced his *Travellers Guide Through the United States, A Map of the Roads, Distances, Steam Boat & Canal Routes &c. By J. H. Young, Philadelphia. Published by S. Augustus Mitchell.* There are inset maps of the environs of eight cities, and an accompanying sheet containing an index of place names, lists of steamboat and canal routes, a distance chart, and several statistical graphs and tables. The *Travellers Guide* was published in a number of editions for two decades. Later editions included new place names as the frontier shrank. From 1837 on, the *Travellers Guide*'s supplementary data were printed in booklet format rather than on an accompanying sheet.

It is interesting to note that the plate for the *Travellers Guide* map was done by steel engraving. This was a new technique in cartographic reproduction in the United States at that time. Copper-plate engraving was the principal method for reproducing maps, but only a few thousand impressions could be obtained from the soft copper plates. Lithography had not yet made an impact on map printing and publishing, and so engravers and publishers looked for other, more durable, metals like steel on which to engrave and thus obtain more prints of an engraved image.

The Mitchell and Colton companies were the principal utilizers of steel engraving for reproducing maps. It seems pertinent, therefore, to insert here a summary of the invention and development of this technique. While steel engraving proved to be most ideally adapted to bank note engraving, for a brief period it proved to be an important medium for reproducing maps and plates for atlases.

Engraving on steel was not a new idea. In fact, Albrecht Dürer reportedly prepared engravings on steel as early as 1510. The difficulty of engraving in the hard surface of steel and the high breakage rate of tools used for such engraving, however, discouraged its use for most practical purposes. This problem was resolved in the first decade of the nineteenth century by Jacob Perkins.

Perkins was born in Newburyport, Massachusetts, on July 9, 1766. His father was a tailor, and Perkins apparently enjoyed several years of schooling in one of the private schools in the village. In 1778, at the age of twelve, Perkins was apprenticed to goldsmith Edward Davis. The work involved more than making jewelry,

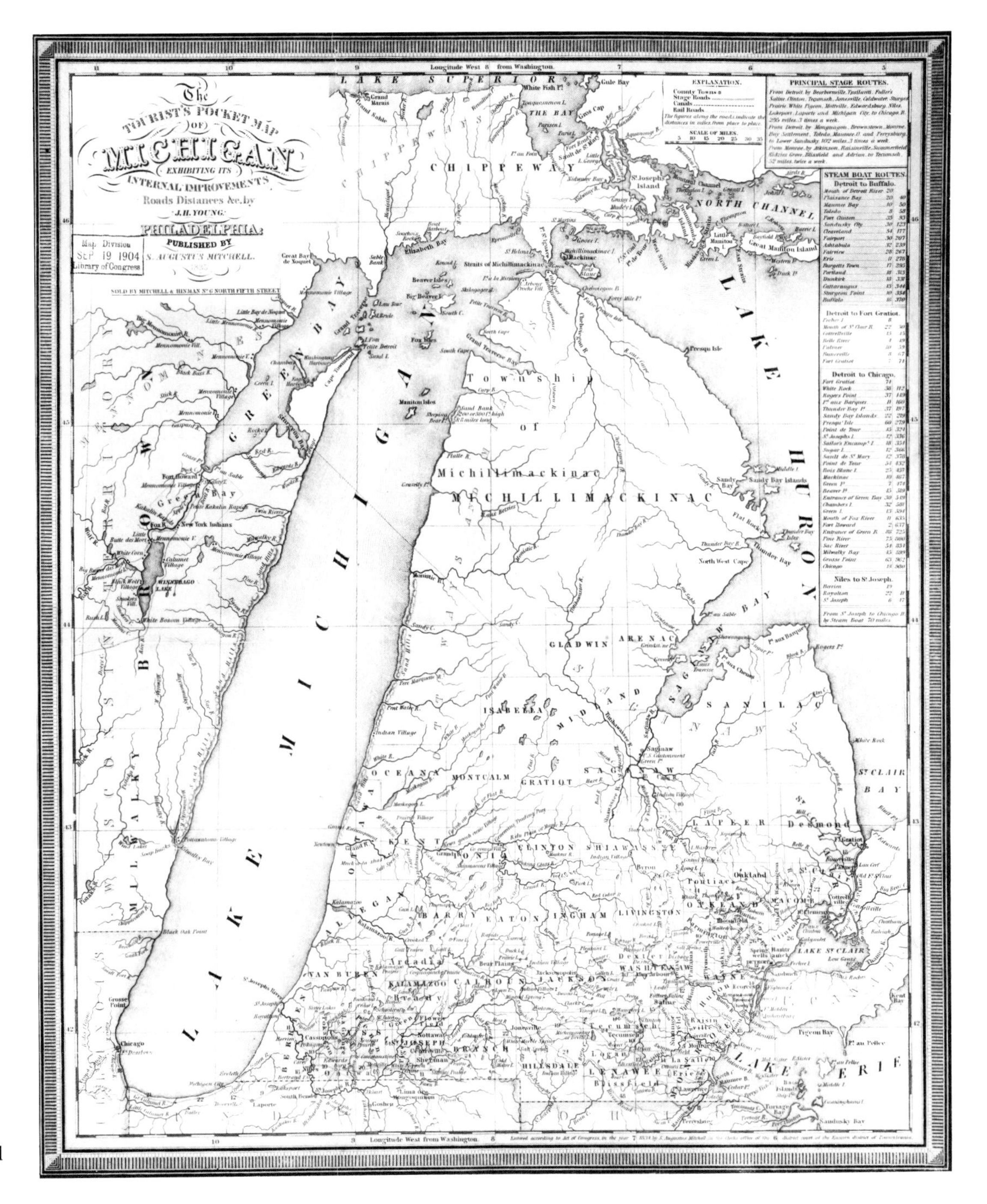

Fig. 19–1. An 1835 Young and Mitchell tourist pocket map of Michigan.

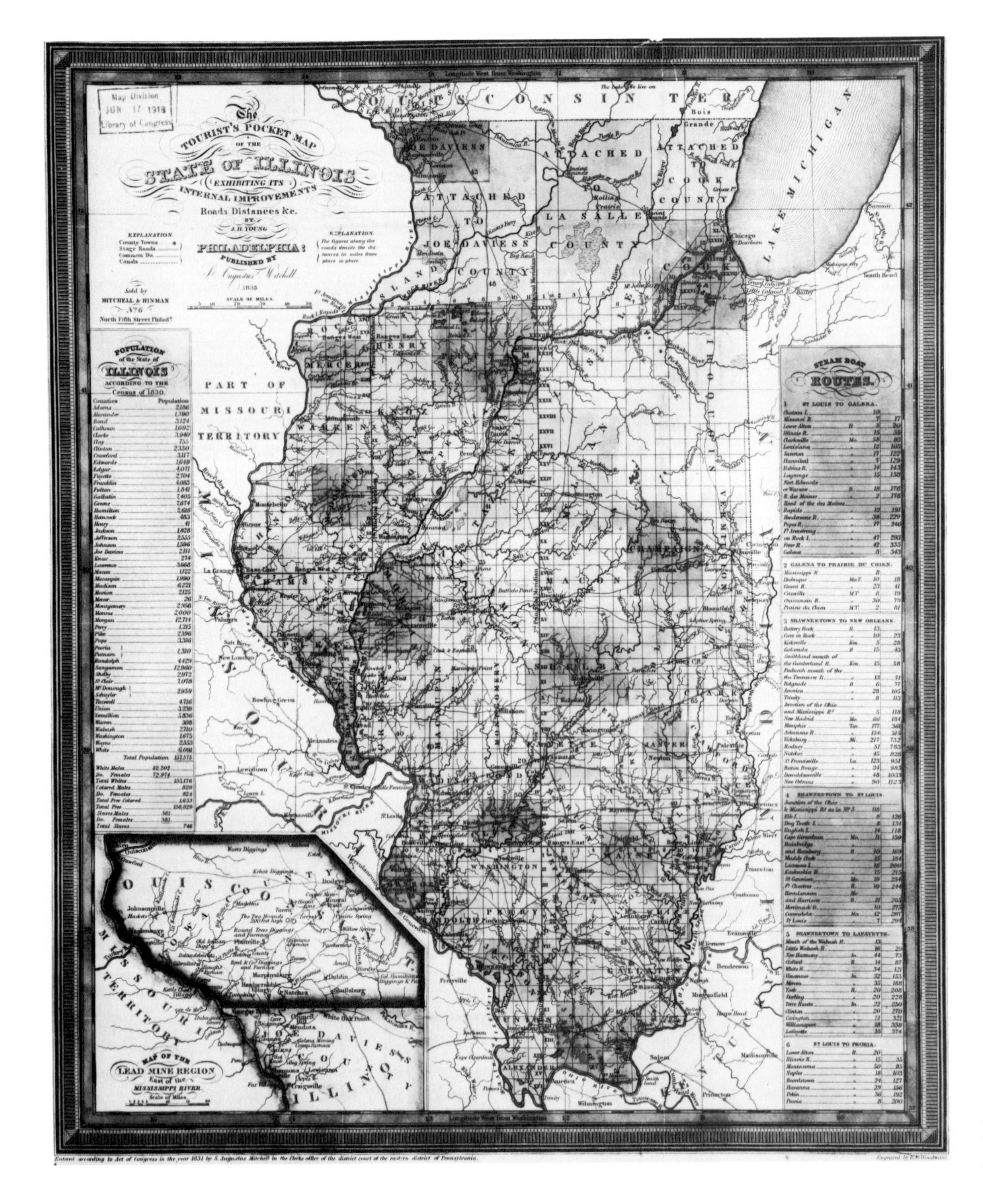

Fig. 19–2. The Young and Mitchell tourist pocket map of Illinois, 1835.

and it stimulated the inventive curiosity of the boy. When Davis died suddenly in 1781, Perkins assumed charge of the shop for Davis's widow. Perkins ably carried out his duties and also turned his attention to several practical inventions. On November 11, 1790, he married Hannah Greenleaf, also of Newburyport. He prepared dies for copper coins for the state of Massachusetts and in 1792 applied for a position in the U.S. Mint in Philadelphia. This effort was unsuccessful, and Perkins again concentrated on inventing. In 1793 he invented a machine for cutting and heading nails, which involved him in a legal suit for several years.

There was, in the first decade of the nineteenth century, a demand for bank notes, and most engravers in the country were occupied at one time or another with this work. Copper plates proved to be unsuited for printing bank notes, though, because of their softness, and because the resulting notes could be easily counterfeited. (A counterfeiter could engrave a duplicate of a bank note in a reasonably short time on copper. Engraving on hard steel took a much longer time, and thus the possible gains were minimized.) Around 1804, Perkins discovered a process for alternately softening (by decarbonization) and hardening (by cementation) steel. When in its decarbonized state it was possible to engrave the essential bank note information on the steel plates. By Perkin's technique, the steel was then rehardened and used to print an unlimited number of impressions. Perkins's helpers employed Asa Spencer's steel lathe and ruling machine to engrave and print distinctive and unique designs on the backs of bank notes.[1]

A report on "The Prevention of Forgery," communicated to the English Society for the Encouragement of Arts, Manufactures and Commerce, in 1819 by Perkins, engraver Gideon Fairman, and English businessman Charles Heath, described in detail Perkins's stereotype steel-plate process. The report also described Spencer's steel lathe and engraving machine. It noted that

> by examining the machine engraving . . . the two styles of work, viz. copper plate and letter-press printing will be seen beautifully combined. This is effected by the process of transferring and retransferring. This kind of engraving is extremely difficult to imitate. This machine, which is denominated the geometrical lathe, was invented in America by Mr. Asa Spencer. Its powers for producing variety are equalled only by the kaleidoscope; but for beautiful patterns it surpasses every thing of the kind. It has one of the peculiarities of the kaleidoscope, viz. that the turning of a screw, like the turning of the kaleidoscope, produces an entirely new pattern, which was never before seen, and perhaps would never be seen again. This pattern, however, may be perpetuated by the transferring process.[2]

In developing, seeking markets for, and exploiting his stereotype steel-plate invention, Perkins had the assistance of a number of friends and associates. He was not himself an engraver, so he relied upon several individuals skilled in this art. One of the first of these was Gideon Fairman, who was an engraver in Albany and then in Newburyport, Massachusetts, where he worked with the engraver William Hooker. It will be recalled that Hooker was a son-in-law of Edmund March Blunt, who published and sold nautical charts, publications, and navigating instruments. Hooker and Fairman both engraved charts for Blunt. Fairman was an early acquaintance of Perkins, and in 1810 the two men collaborated in producing and publishing a series of school copybooks. In this year, Fairman also had a disagreement with Hooker, and they dissolved their partnership. With encouragement from Perkins, Fairman moved to Philadelphia where he joined the established bookplate and bank note engraving firm of Murray & Draper. It is possible that Fairman had the financial backing of Perkins, who wished to establish a business and bank note engraving connection in Philadelphia. More or less continuously until 1821, Fairman was a partner in this firm which was known at different times as Murray, Draper, Fairman & Company and Fairman, Draper, Underwood & Company. Although his name never appeared in the firm's name, it is believed that Perkins was an active, if silent, partner. In fact, Perkins moved to Philadelphia late in 1815, and two years later he established his family there.

Other Perkins associates who were, for various periods of time, affiliated with Murray, Draper, Fairman & Company were Spencer, J. W. Carpenter, an engraver, and Charles Toppan, who was Perkins's nephew by marriage. Toppan, who learned engraving from Fairman, was instrumental in forming the American Bank Note Company in around 1858 and became its first president.

In 1818 the Bank of England offered a prize for a successful method of preventing counterfeiting of bank notes. Perkins and his associates decided to try for the award, and on May 31, 1819, Perkins, Fairman, Carpenter, Toppan, and Spencer sailed for England, landing in Liverpool on June 28 or 29. In London Perkins established a partnership with Fairman and Heath, and together they gave their report on forgery. Although the firm failed to get the Bank of England award, it found a considerable market for the siderographic engraving technique over the next several decades. Perkins settled permanently in England, moving his family there in 1821. Fairman returned to Philadelphia in July 1822 and rejoined Fairman, Draper, Underwood & Company. He died in that city on March 15, 1827. Carpenter and Toppan likewise returned to America. Perkins spent the last three decades of his life in England, where he died July 30, 1839, at the age of eighty-three.

There is no evidence that Perkins's siderographic technique was used for engraving maps on steel in America prior to 1830. The technique, however, was used in England. Shortly after Perkins settled in London, the *Boston Library Gazette* issue of December 1820 reported that "already Perkins, Fairman, and Heath have engaged to Manufacture bank notes on their inimitable plan for several Yorkshire & other banks, and they are also preparing various engravings for popular books, as maps and views for Goldsmith's Geography." Nor is there any suggestion that Murray, Draper & Fairman engraved maps on steel.

In London, around 1829, the Society for the Diffusion of Useful Knowledge began issuing a series of drypoint steel-engraved maps of various parts of the world, including plans of important cities. A number of the steel-engraved city plans are reproduced in Melville C. Branch's 1978 volume *Comparative Urban Design: Rare Engravings, 1830–1843*. Branch notes that "the practice of reproductive engraving on steel arose in the early nineteenth century from the demand for a larger number of prints than could be obtained from a copper plate without wear. Steel is, of course, much tougher and harder to work upon, and mechanical aids, such as ruling machines were resorted to."[3] It is a fair assumption that the steel-engraved maps and plans published by the society were prepared by means of Perkins's siderographic technique.

The *Journal of the Franklin Institute* for January 1829 reprinted, from the *Transactions of the London Society for the Encouragement of Arts, Manufactures, and Commerce*, a report "On Improvements in the Art of Engraving in Steel," by C. Warren. The technique described closely parallels that of Perkin's stereotype steel plates, but Warren also recommended the employment of acid etching as well as inscribing with the burin. His report concluded that

> concerning the great superiority of steel plate over copper plate, for all works that require a considerable number of impressions to be taken, there can exist no doubt: for though the use of the graver, and of the other tools, requires more time on steel than on copper, and though the process of rebiting has not yet been carried to the degree of perfection in the former that it has been in the latter, yet the texture of steel is such, as to admit of more delicate work than copper, and the finest and most elaborate exertions of the art, which on copper would soon wear so as to reduce them to an indistinct smeary tint, appear to undergo scarcely any

> deterioration on steel: even the marks of the burnisher are still distinguishable after several thousand impressions.[4]

Warren also suggested that it was not necessary to recarbonize the steel plates after engraving. It was found that a soft decarbonized steel plate could print several thousand copies without any sensible wear.

Since American engravers comprised a relatively small fraternity, and most of them were located within the limits of a few square blocks in Philadelphia, the use of steel plates for engraving was common knowledge after about 1820. They were intrigued by the prospect of engraving and printing maps by this means. Young was one of these engravers and undoubtedly recognized the advantages of using steel plates for engraving maps. He probably persuaded Mitchell to use the technique, which he did for his *Travellers Guide*, which is 44 by 56 cm. in size and covers the United States east of the ninety-fifth degree meridian. Below the map's title cartouche is the notation "Engraved on Steel by J. H. Young & D. Haines." Little is known about Haines. David Stauffer stated in a one-sentence entry that "D. Haines was an engraver of business-cards, etc., working in Philadelphia about 1820."[5]

There is no evidence that Mitchell was either a cartographer or an engraver. He apparently served as the editor and business manager of his company. Apparently, Young was the company's principal compiler and draftsman as well as chief engraver from 1830 on. In these capacities Young prepared several Mitchell maps of the United States and a series of regional and state maps.

Before the first edition of the *Travellers Guide* was published, Young had started to compile a larger map of the country. Titled *Map of the United States, by J. H. Young*, it was copyrighted by Mitchell on October 10, 1831, and probably published early in 1832. Above the title cartouche there is a steel-engraved illustration of the shield of the United States on a large rock, from which a tree grows. An eagle is perched on the tree's largest branch. To the left of the rock is a harbor view with steam and sailing ships, and to its right is a canal lock. The map, including the decorative border, is 111 by 88 cm. There are a large inset map of North America and six small insets of cities and their environs. Other ancillary information includes charts showing the comparative heights of mountains and lengths of rivers and the lengths of the principal U.S. canals. The map was "Engraved by J. H. Young, D. Haines, and F. Dankworth." Stauffer has no entry for Dankworth, but he seems to have engraved maps for Young and Mitchell between 1831 and 1845. Inasmuch as Young and Haines engraved the *Travellers Guide* map on steel, the same medium was probably used to engrave this map, particularly because the illustration above the cartouche is a steel engraving, as is the decorative border. Like all of Mitchell's U.S. maps, this one extends west only to around the ninety-fifth meridian.

Revised editions of the *Map of the United States* were published to 1844. On some versions townships are separately colored. In all editions they are numbered. Later editions of the map show additions of places in western states and territories. Mitchell appears to have marketed some of his maps through book stores and other retailers, for an 1844 edition of the map has the imprint "Sold by T. & E. H. Ensign, New York."

While this map was still in its early editions, Young compiled *A New Map of the United States*, which was copyrighted by Mitchell in 1833 and offered for sale by the firm Mitchell & Hinman at 6 North Fifth Street in Philadelphia in 1834. (Mitchell formed a temporary partnership with Hinman to facilitate the distribution of the firm's maps.) It is larger than the earlier map, being at the scale of 1:1,500,000 and measuring 132 by 196 cm. with the decorative border. Below the title cartouche is a delicately engraved coastal view which is dominated by an American eagle perched on a seashell boat. The view was "designed by W. Mason." The engravers for this map were Young, Dankworth, E. Yeager, and E. F. Woodward. Of the last two engravers, only Woodward is noticed by Stauffer, who makes no reference to his map engravings. There are a large inset map of North America and ten small insets of city and vicinity maps.

Another version of the *New Map of the United States* was also published in 1834, but with a different title: *Reference and Distance Map of the United States, by J. H. Young*. It has the same view, decorative border, and engravers. Like the *New Map of the United States*, it was sold by Mitchell & Hinman. In 1835 the same map was again published with a different title: *Mitchell's Reference and Distance Map of the United States, by J. H. Young*. The only variation is the addition of Mitchell's name. Under this title, the U.S. map was issued in 1836, 1845, and 1851 editions. A new copyright was taken out in 1845, although the 1851 edition still carries the 1833 copyright registration. Beginning with the 1845 printing, a new decorative border frames the map. The map was accompanied by a four-hundred-page octavo volume with indexes of all the counties, towns, and rivers.

Mitchell published another U.S. map, the *National Map of the American Republic or United States of America*, in 1843. The map proper extends west only to approximately the ninety-fifth degree meridian and is 62 by 85 cm. in size. Framing it are thirty-two inset maps of principal cities and towns. There are small decorative borders around each of the insets, and a large, elaborate border framing the entire map sheet, which measures 98 by 114 cm. The American eagle, astride the official seal of the nation, decorates the four corners. Below the title cartouche there is a table which gives the population of each county.

At the bottom of the cartouche is the credit "Drawn by J. H. Young, Engraved by J. H. Brightly." There is no indication that the engraving was on steel plates, but because this was the medium employed for other Mitchell maps, it is likely that it was also used for this publication. It is interesting to note that Young is credited only with the drawing. Little is known of Brightly. George Groce and David Wallace describe him as a wood engraver who was born in England around 1818 and lived in Philadelphia and New York City between 1841 and 1858.[6] It appears from his association with Mitchell and Young that he had previous experience in engraving maps on steel. The *National Map* was republished in 1846 and 1847 editions and probably others. Some versions are wall maps, as is the one just described, while others were issued folded in covers as pocket maps with an accompanying "Route-Book . . . comprising tables of the principal rail-road, steam-boat and stage routes throughout the United States."

Mitchell and Young also collaborated in compiling and publishing *A Map of the World on Mercator's Projection Exhibiting the Researches of the Principal Modern Travellers & Navigators*. It was sold by Hinman & Dutton at 6 North Fifth Street in Philadelphia. The first edition appears to have been issued in 1837. It carries the credit "Engraved by J. H. Young, Philadelphia, Assisted by F. Dankworth, E. Yeager & J. Knight." There is a decorative border as on other Mitchell-Young cartographic publications, and we may infer that the engraving was again effected on steel plates. The map measures 198 cm. east to west and 137 cm. north to south. A six-hundred-page volume, available as an accompaniment to the map, includes an index of place names. Editions of the map were published as late as 1850.

Although Mitchell's first cartographic publication was the 1831 *New American Atlas*, he seems to have focused his major attention on maps during the succeeding fifteen years. His next published atlas was, like the *New American Atlas*, not an original work. In 1845 Mitchell acquired the copyright for Tanner's *New Universal Atlas* from Carey & Hart, which had previously purchased the copyright from Tanner and had published the atlas in 1843 and 1844 (see Chapter 13). Mitchell informed his customers in an advertisement in the 1846 edition of his *Route Book Adapted to Mitchell's National Map of the American Republic* that "since the first of May, 1845, the Subscriber [i.e., Mitchell] has the entire manufacturing and sale of [Tanner's *New Universal Atlas*]. The plates of this Atlas (costing originally more than Ten Thousand Dollars) have been greatly improved, and the edition now offered is believed to be, according to its extent, correctness, and style of execution, the cheapest work of this kind ever published in the United States."

Mitchell published two editions of the *New Universal Atlas* in 1846. In the first of these the Carey & Hart imprint on the title page has been replaced by "Philadelphia, Published by S. Augustus Mitchell, N.E. corner of Market & 7th Streets, 1846." The Carey & Hart copyright notice remains on the title page as does Tanner's name. The Carey & Hart imprint has, however, been removed from all maps and Mitchell's substituted. Tanner's copyright notice, however, still appears on most of the maps. In Mitchell's second 1846 edition, Tanner's name is omitted on the title page and from the copyright registrations on the separate maps (Fig. 19–3). The copyright claimant on most of the individual maps is "H. N. Burroughs." His name also appears on a chart indicating the lengths of rivers and the heights of mountains (Fig. 19–4). Nothing is known about Burroughs, but he was undoubtedly an employee or associate of Mitchell.

In all the publications of the atlas before Mitchell's 1846 editions, plate marks are clearly evident on the pages of the book. In the two 1846 editions, however, these marks are missing, indicating that copper-plate engraving had been discontinued as the medium of reproduction. Because the maps closely resemble those in earlier editions of the atlas, it is likely that the images were transferred from the copper plates to lithographic stones. The maps are also distinctive in having decorative borders with manually applied green coloring. The lithographic transfer of the atlas maps was probably done in the Philadelphia lithographic shop of Peter S. Duval. His foreman, Frederick Bourquin, a native of Switzerland, was skilled in the transfer technique, and in 1847 his work received commendation from the Franklin Institute.

Editions of the *New Universal Atlas* were issued periodically until 1859. After 1849 the color on the map borders appears to have been printed by chromolithogra-

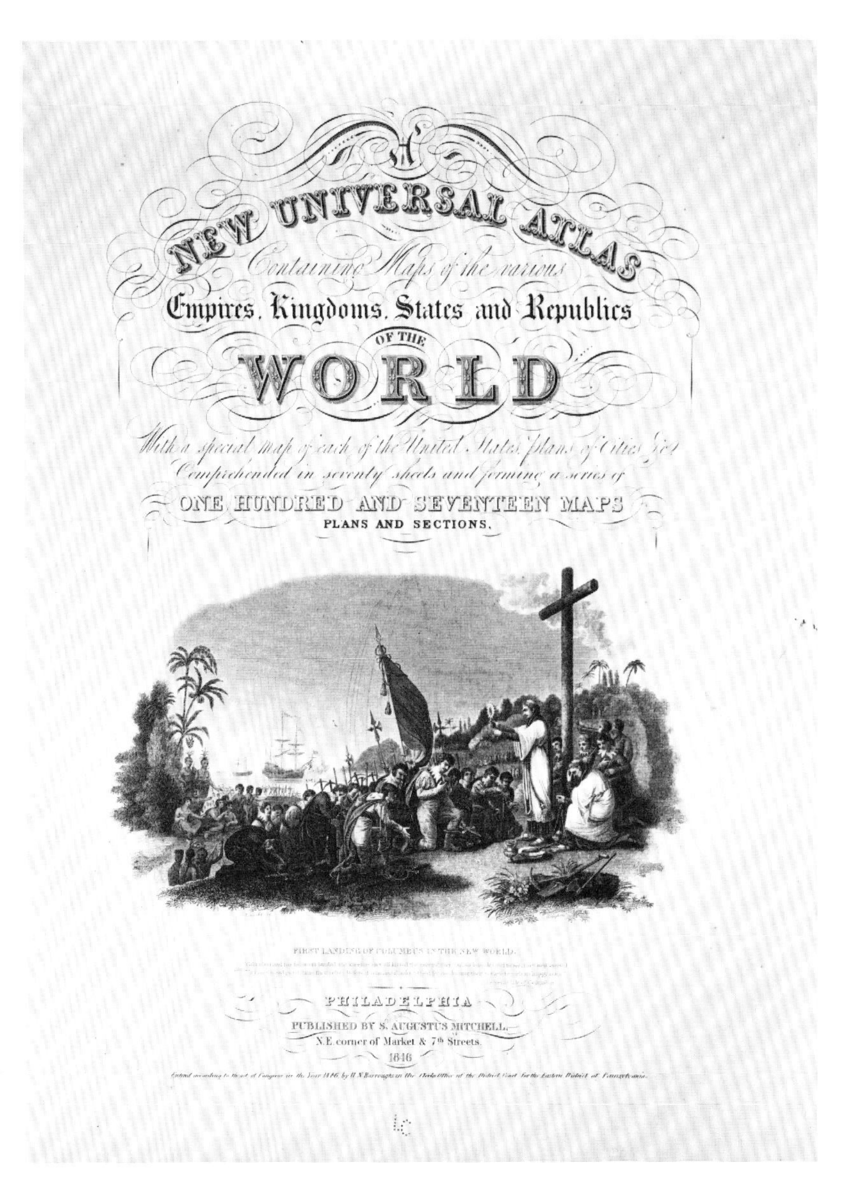

Fig. 19–3. Title page of the second 1846 edition of Mitchell's *New Universal Atlas*.

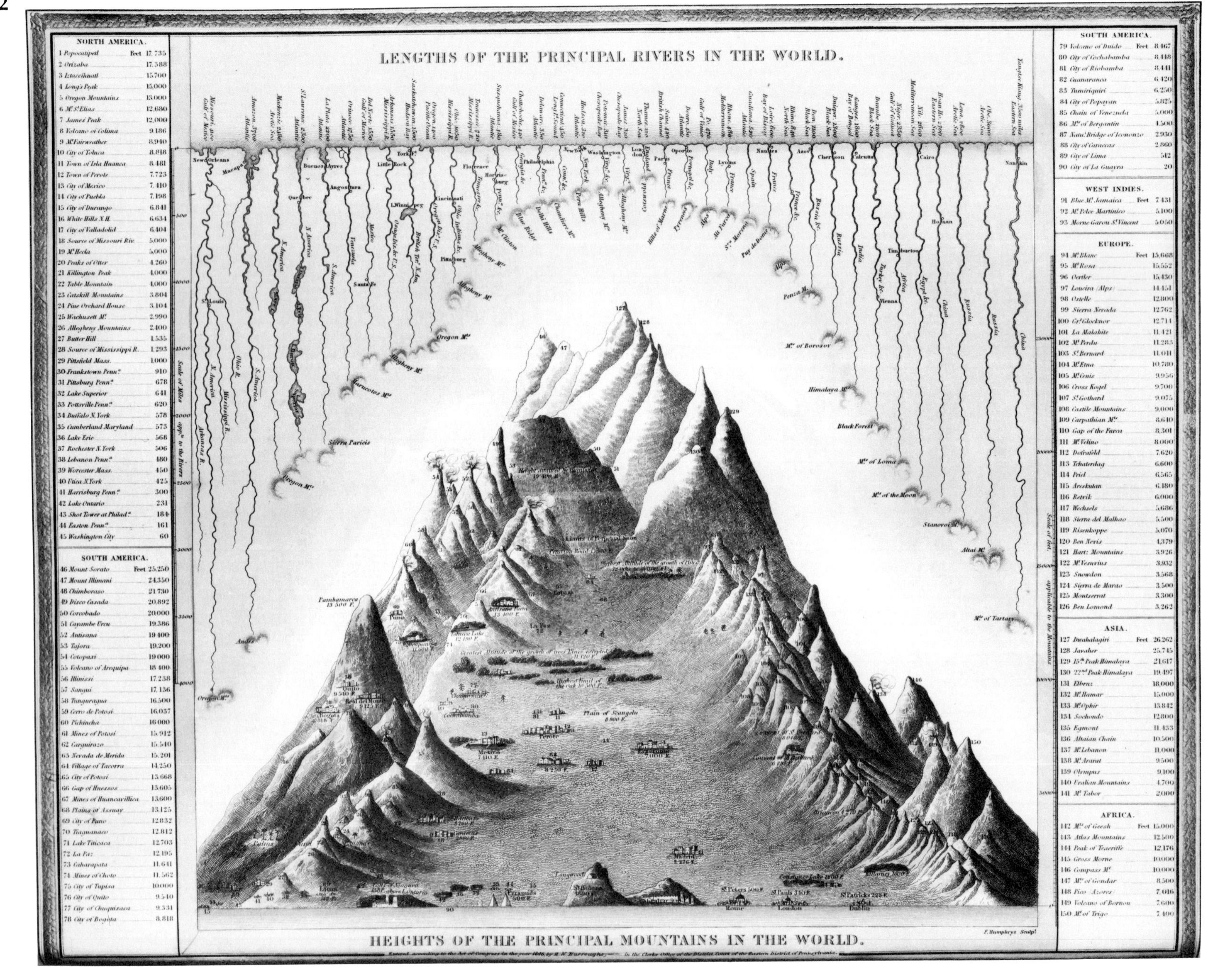
LENGTHS OF THE PRINCIPAL RIVERS IN THE WORLD.
HEIGHTS OF THE PRINCIPAL MOUNTAINS IN THE WORLD.
T. Humphrys Sculp.
Scale of Miles app.n to the Rivers
Scale of feet applicable to the Mountains
NORTH AMERICA.
1 Popocatipetl Feet 17.735
2 Orizaba 17.388
3 Iztaccihuatl 15.700
4 Long's Peak 15.000
5 Oregon Mountains 13.000
6 Mt. St. Elias 12.680
7 James' Peak 12.000
8 Volcano of Colima 9.186
9 Mt. Fairweather 8.940
10 City of Toluca 8.818
11 Town of Isla Huanca 8.481
12 Town of Perote 7.723
13 City of Mexico 7.410
14 City of Puebla 7.198
15 City of Durango 6.841
16 White Hills N.H. 6.634
17 City of Valladolid 6.404
18 Source of Missouri Riv. 5.000
19 Mt. Hecla 5.000
20 Peaks of Otter 4.260
21 Killington Peak 4.000
22 Table Mountain 4.000
23 Catskill Mountains 3.804
24 Pine Orchard House 3.104
25 Wachusett Mt. 2.990
26 Allegheny Mountains 2.400
27 Butter Hill 1.535
28 Source of Mississippi R. 1.293
29 Pittsfield Mass. 1.000
30 Frankstown Penn.a 910
31 Pittsburg Penn.a 678
32 Lake Superior 641
33 Pottsville Penn.a 620
34 Buffalo N. York 578
35 Cumberland Maryland 573
36 Lake Erie 568
37 Rochester N. York 506
38 Lebanon Penn.a 480
39 Worcester Mass. 450
40 Utica N. York 425
41 Harrisburg Penn.a 300
42 Lake Ontario 231
43 Shot Tower at Philad.a 184
44 Easton Penn.a 161
45 Washington City 60
SOUTH AMERICA.
46 Mount Sorato Feet 25.250
47 Mount Illimani 24.350
48 Chimborazo 21.730
49 Disco Casada 20.892
50 Corcobado 20.000
51 Cayambe Urcu 19.386
52 Antisana 19.400
53 Tajora 19.200
54 Cotopaxi 19.000
55 Volcano of Arequipa 18.400
56 Illinissi 17.238
57 Sanqui 17.136
58 Tunguragua 16.500
59 Cerro de Potosi 16.037
60 Pichincha 16.000
61 Mines of Potosi 15.912
62 Carguirazo 15.540
63 Nevada de Merida 15.201
64 Village of Tacorra 14.250
65 City of Potosi 13.668
66 Gap of Huessos 13.605
67 Mines of Huancavillica 13.600
68 Plains of Assuay 13.125
69 City of Puno 12.832
70 Tiaguanaco 12.812
71 Lake Titicaca 12.703
72 La Paz 12.195
73 Caharapata 11.641
74 Mines of Choto 11.562
75 City of Tupisa 10.000
76 City of Quito 9.540
77 City of Chuquisaca 9.331
78 City of Bogota 8.818
SOUTH AMERICA.
79 Volcano of Duido Feet 8.467
80 City of Cochabamba 8.448
81 City of Riobamba 8.441
82 Cuanaranca 6.420
83 Tumiriquiri 6.250
84 City of Popayan 5.825
85 Chain of Venezuela 5.000
86 Mts. of Bergantin 4.500
87 Natl. Bridge of Icononzo 2.930
88 City of Caraccas 2.860
89 City of Lima 542
90 City of La Guayra 20
WEST INDIES.
91 Blue Mt. Jamaica Feet 7.431
92 Mt. Pelee Martinico 5.100
93 Morne Garou St. Vincent 5.050
EUROPE.
94 Mt. Blanc Feet 15.668
95 Mt. Rosa 15.552
96 Oertler 15.430
97 Loucira (Alps) 14.451
98 Ostelle 12.800
99 Sierra Nevada 12.762
100 Gr.d Glocknor 12.714
101 La Malahite 11.421
102 Mt. Perdu 11.283
103 St. Bernard 11.011
104 Mt. Etna 10.780
105 Mt. Cenis 9.956
106 Cross Kogel 9.700
107 St. Gothard 9.075
108 Castile Mountains 9.000
109 Carpathian Mts. 8.640
110 Gap of the Furca 8.301
111 Mt. Velino 8.000
112 Dofrafeld 7.620
113 Tchaterdag 6.600
114 Priel 6.565
115 Areskutan 6.180
116 Retrik 6.000
117 Wechsels 5.686
118 Sierra del Malhao 5.500
119 Risenkoppe 5.070
120 Ben Nevis 4.379
121 Hartz Mountains 3.926
122 Mt. Vesuvius 3.932
123 Snowdon 3.568
124 Sierra de Marao 3.500
125 Montserrat 3.300
126 Ben Lomond 3.262
ASIA.
127 Dwahalagiri Feet 26.262
128 Javaher 25.745
129 15th Peak Himalaya 21.617
130 22nd Peak Himalaya 19.497
131 Elbruz 18.000
132 Mt. Hamar 15.000
133 Mt. Ophir 13.842
134 Sochondo 12.800
135 Egmont 11.433
136 Altaian Chain 10.500
137 Mt. Lebanon 11.000
138 Mt. Ararat 9.500
139 Olympus 9.100
140 Uralian Mountains 4.700
141 Mt. Tabor 2.000
AFRICA.
142 Mts. of Geesh Feet 15.000
143 Atlas Mountains 12.500
144 Peak of Teneriffe 12.176
145 Gross Morne 10.000
146 Compass Mt. 10.000
147 Mts. of Gondar 8.500
148 Pico (Azores) 7.016
149 Volcano of Bornou 7.600
150 Mt. of Trigo 7.400

phy, although the color on the actual maps was still manually applied. Mitchell published the atlas until 1850, when the copyright was purchased by Thomas, Cowperthwait & Company of Philadelphia (Fig. 19–5). This firm enlarged the number of maps in the atlas from its original 117 to the 122 contained in its 1853 edition. Thomas, Cowperthwait & Company also added the new features of a chromolithographed title page and contents page. In 1856 Charles De Silver acquired the atlas copyright and published it in 1857 and 1858. The 1859 edition of the atlas still carries De Silver's copyright notice, but it was published by Cushings & Bailey of 362 Market Street in Baltimore. This Cushings & Bailey edition has 145 maps, including a map by Young of the United States extending westward to the Pacific Ocean. This was the last edition of the *New Universal Atlas* published.

By 1860 Mitchell was sixty-eight years old, and there are indications that he retired from the actual direction of his publishing business, in favor of his son, S. Augustus Mitchell, Jr., in that year. *Mitchell's New General Atlas,* introduced to replace the *New Universal Atlas,* was copyrighted and published in 1860 by Mitchell's son. In a slightly smaller format than the earlier work, the *New General Atlas* includes forty-seven quarto plates, which embrace seventy-six maps and plans. Like the former atlas, the maps in the new volume have decorative borders. They are uncolored, however. Reproduction was by lithography, and the coloring on the maps appears still to have been manually applied. There are generally two or more state maps on each atlas page. Like later editions of the *New Universal Atlas,* there is a list of post offices in the United States and other statistical and descriptive information.

Fig. 19–4. An interesting feature in atlases of the middle and late nineteenth century was the inclusion of charts illustrating the heights of mountains and the lengths of rivers. This chart from Mitchell's *New Universal Atlas* was copyrighted by H. N. Burroughs.

Editions of the *New General Atlas* were issued annually with a progressive increase in the number of map plates to 1887. S. Augustus Mitchell, Jr., continued to be listed as the publisher on the title page until 1879, although the 1870 edition also includes as co-publisher J. W. Willson of New York City. The editions from 1880 to 1887 were published by the Philadelphia firms Bradley & Company or Wm. M. Bradley & Bro. The last edition of the *New General Atlas,* dated 1893, was published under a different title in Philadelphia by the A. R. Keller Company. In addition to 93 map plates, it includes 162 pages of description and statistics. The title of this edition is *Mitchell's Family Atlas of the World, Subscription Edition.*

Other Mitchell atlases with long runs were the *School Atlas,* published from 1839 to 1886, and the *Ancient Atlas, Classical and Sacred,* which dates from 1844 to 1874. Both of these atlases are small-format volumes with eight to twelve pages of maps. Some early editions of these two atlases were published by Thomas, Cowperthwait & Company and then by E. H. Butler & Company of Philadelphia.

Mitchell died in Philadelphia on December 18, 1868. For more than six decades the publishing firm he founded, and which was continued by his son, published a great volume of atlases, maps, geography textbooks, and tourist guides. During its most prosperous years the Mitchell firm was said to have had 250 employees and reportedly sold more than four hundred thousand publications annually.

The other major map publishing firm of this period, J. H. Colton & Company, was probably founded by Joseph Hutchins Colton in New York City in 1831. The earliest publications with Colton imprints that have been found, however, are dated 1833. Colton was born in Longmeadow, Massachusetts, on July 5, 1800 (Fig. 19–6). He apparently had only the most basic education for at the age of twelve, "with his parents' godspeed, he went with friends of the family to Manlius, N.Y., expecting to remain there and lay the foundation for a future home, but extraordinary conditions of hardship

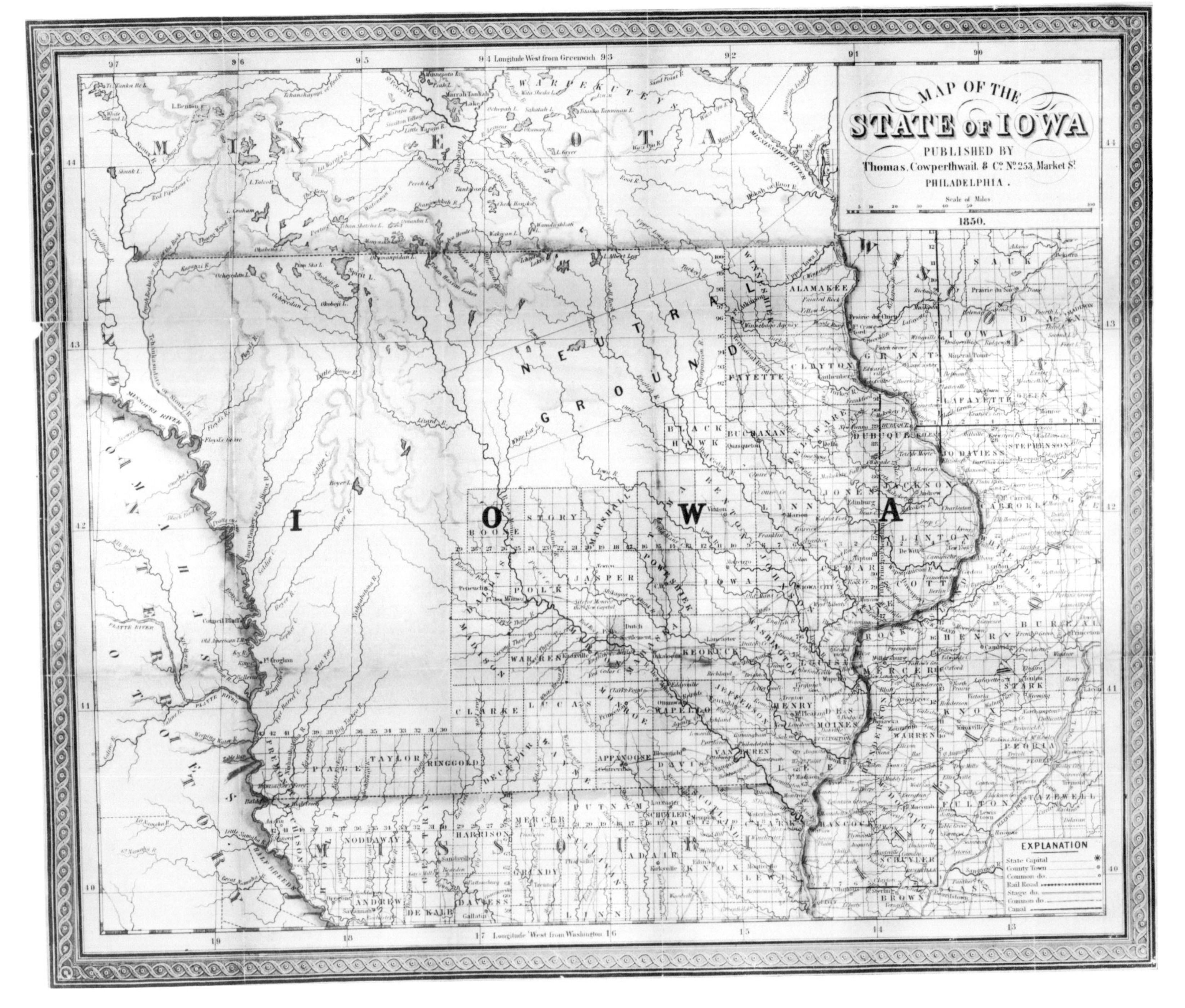
MAP OF THE
STATE OF IOWA
PUBLISHED BY
Thomas, Cowperthwait, & Co. No. 253, Market St.
PHILADELPHIA.
Scale of Miles
1850.
Longitude West from Greenwich
Longitude West from Washington
MINNESOTA
NEUTRAL GROUND
IOWA
MISSOURI
EXPLANATION
State Capital
County Town
Common do.
Rail Road
Stage do.
Common do.
Canal

beyond those incident to all pioneer life led to his return the following year."[7] Three years later, at age sixteen, Colton obtained employment in a general merchandise store in Lenox, Massachusetts, where he remained until the spring of 1829. His next move was to Hartford, Connecticut, where for eighteen months he served as a night clerk in the U.S. Post Office. "In the winter of 1830–31, [Colton moved] to New York City, where from small beginnings he established the business with which his name is identified the world over—the publication of the best grade of geographical publications and the most extensive house in America for many years for the manufacture of maps of every kind, atlases, school geographies, etc."[8]

Like Mitchell, Colton appears to have had little or no education or training in geography or cartography. Their principal contributions to the success of their respective firms, therefore, was in administration, management, and distribution. Also in the pattern of Mitchell, Colton's initial cartographic publications were works prepared by other mapmakers. His first undertaking was the map of New York State, originally published in 1830, by David H. Burr. The map was reengraved for Colton by Samuel Stiles of S. Stiles & Company in New York. Colton then copyrighted and published the map in 1833. It is the earliest Colton imprint to be identified. The address given for the firm was 9 Wall Street, which was also the address of the engraver Stiles. In the following year Colton published Burr's map of the city of New York, which was originally drawn by Burr for the guidebook *New-York As It Is In 1833*, published by J. Disturnell. The map also has an 1833 copyright registration notice and was also engraved by Stiles. Annual editions of the guide, with Burr's map, were published by Disturnell for a decade or more. Both of the Colton-Burr

Fig. 19–5. This 1850 map of Iowa was published by Thomas, Cowperthwait & Company. This firm was in business from about 1838 to 1852 and appears to have published some Mitchell maps and atlases.

Fig. 19–6. Joseph Hutchins Colton, founder of the J. H. Colton map publishing company.

maps have decorative borders that were probably prepared with a ruling machine or an automatic lathe. The maps appear to have been printed from engraved copper plates, for plate marks are clearly visible.

The next Colton cartographic venture occurred in 1835, when he purchased the copyrights to several maps of Michigan and Wisconsin that had been compiled and published between 1825 and 1835 by John Farmer of Detroit. It is possible that Colton learned of Farmer's desire to sell the map copyrights through Stiles. Stiles had earlier been associated with Vistus Balch in Utica, and the Balch & Stiles firm engraved Farmer's 1825 map of Michigan.

In 1836 Colton published the *Topographical Map of the City and County of New York and the Adjacent Country, With Views in the border of the principal Buildings and interesting Scenery of the Island*. It was engraved and printed by S. Stiles & Company. It may have been engraved on steel and has two decorative borders. The inner border is a representation of a surveyor's chain, while the outer one consists of a number of vignettes within entwining vines. The vignettes show views and buildings of the city. Below the map is a view of Nieuw Amsterdam (New York City), in 1639, which originally appeared on Visscher's map of Manhattan, and a view of "Broadway from the Park," which includes the American Museum Building, the Astor House, and St. Peter's Roman Catholic Church.

The map encompasses the entire extent of Manhattan Island as well as adjacent regions. It was almost certainly based on the 1811 commissioner's map of the city of New York drawn by William Bridges and engraved by Peter Maverick. The cartographer or draftsman of the Colton map is not given but is believed to have been Burr. I. N. Phelps Stokes states that "this is one of the most beautiful nineteenth century plans or maps of Manhattan Island, and is full of interesting information. It is perhaps the last example of really artistic mapmaking as applied to Manhattan Island."[9] The map was accompanied by a pamphlet titled *A Summary Historical, Geographical, and Statistical View of the City of New York; together with some notices of Brooklyn, Williamsburgh, &c. in its Environs*. Other editions of the map, with few modifications, were published by Colton in 1837, 1841, and 1851. The map also has a number of features in common with Burr's *Map of the City and County of New York With the Adjacent Country*, which was in Burr's 1829 *Atlas of the State of New York* and was published separately in several editions until 1840. In view of Colton's earlier publication of Burr's maps, it is possible it may be Burr's work. Colton had also published an 1836 map of Ohio by Burr (Fig. 19–7).

J. H. Colton & Company apparently had no principal cartographer and engraver like Young in the Mitchell firm. For the first ten or fifteen years of operation, Colton, therefore, purchased copyrights of maps prepared by other individuals and companies. In 1839 Colton issued the first edition of his *Western Tourist and Emigrant's Guide*, which was originally drawn and copyrighted by J. Calvin Smith and engraved by Stiles, Sherman & Smith. It went through a number of editions to about 1855. The guide includes a map of *Ohio, Michigan, Illinois, Missouri, & Iowa Showing the Township Lines of the United States Surveys*. The guide gives Colton's address as 124 Broadway. In the following year, the firm was located in the Merchant's Exchange; from 1844 to 1854 it was at 86 Cedar Street. In 1855 it moved to 172 William Street, where it remained for some time.

Little is known about George Sherman or Smith, but Colton undoubtedly made contact with them through Stiles. Smith was one of the charter members of the American Geographical and Statistical Society, which was founded in New York City in 1851. In the centennial history of the society, Smith is described as a compiler and publisher of gazetteers. Smith's *New Map of the City of New York With Part of Brooklyn & Williamsburg* was copyrighted by John Disturnell in 1839 and published

Fig. 19–7. This 1836 map of Ohio by David H. Burr was an early Colton publication.

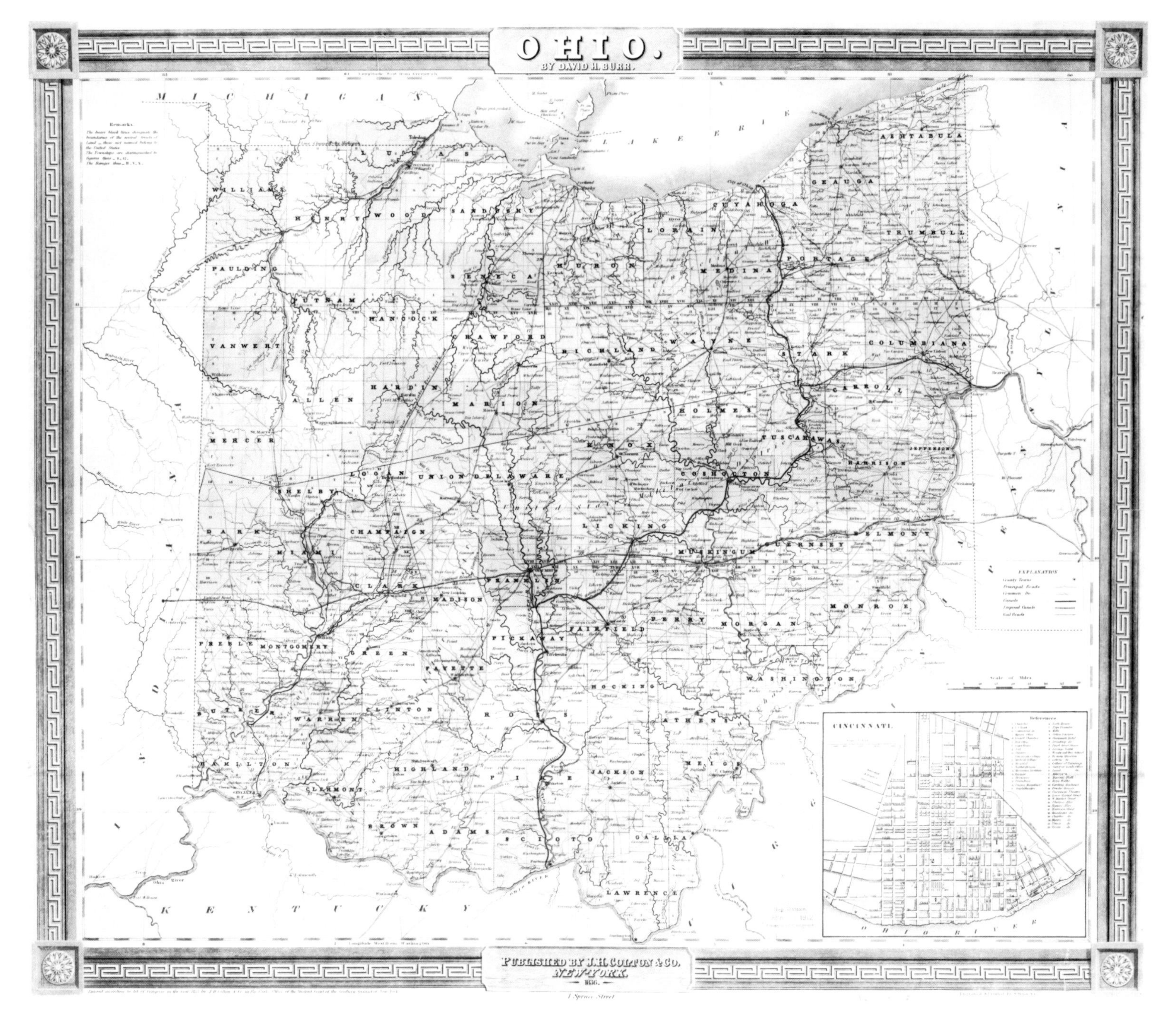
OHIO.
BY DAVID H. BURR.
MICHIGAN
LAKE ERIE
KENTUCKY
CINCINNATI.
OHIO RIVER
LUCAS
WILLIAMS
HENRY
WOOD
SANDUSKY
CUYAHOGA
ASHTABULA
GEAUGA
TRUMBULL
LORAIN
HURON
MEDINA
PORTAGE
PAULDING
PUTNAM
HANCOCK
SENECA
CRAWFORD
VANWERT
RICHLAND
WAYNE
STARK
COLUMBIANA
ALLEN
HARDIN
MARION
CARROLL
HOLMES
TUSCARAWAS
MERCER
KNOX
JEFFERSON
HARRISON
LOGAN
UNION
DELAWARE
COSHOCTON
SHELBY
CHAMPAIGN
LICKING
DARK
MIAMI
MUSKINGUM
BELMONT
CLARK
FRANKLIN
MADISON
MONROE
EXPLANATION
PICKAWAY
FAIRFIELD
PERRY
MORGAN
PREBLE
MONTGOMERY
GREEN
FAYETTE
HOCKING
WASHINGTON
BUTLER
WARREN
CLINTON
ROSS
ATHENS
HAMILTON
HIGHLAND
PIKE
JACKSON
MEIGS
CLERMONT
BROWN
ADAMS
SCIOTO
GALLIA
LAWRENCE
PUBLISHED BY J.H. COLTON & CO.
NEW YORK.

by Tanner & Disturnell in 1846. It was engraved on steel by Stiles, Sherman & Smith. Disturnell, like Smith, was a charter member of the American Geographical and Statistical Society. In fact, the organizational meeting was held in Disturnell's Geographical and Statistical Library at 179 Broadway. John Wright described Disturnell as a good salesman, industrious and aggressive, who was a prolific publisher of guidebooks, maps, handbooks, and distance tables.[10] His junior partner at one time was T. R. Tanner, son of the famous geographer and map publisher.

By 1850 the Colton company was heavily involved in publishing guidebooks and immigrant and railroad maps. Virtually all their maps were framed in decorative borders consisting of intertwining vines, flowers, etc. Encircled within some of the borders are vignette views and buildings. Although the engraver is not given on all the maps, some dated as late as the mid-1850s have the Stiles, Sherman & Smith credit line.

Although most Colton maps are of individual states or groups of states, the company also compiled a number of maps of the United States and the world. An extremely popular publication was *Colton's Map of the United States of America, the British Possessions, Mexico and the West Indies Showing the Country from the Atlantic to the Pacific Ocean*, which was issued in 1846. Revised editions are dated as late as 1884 (Fig. 19–8). Some editions carry the credit "Drawn and Engraved by J. M. Atwood, New York." Whether the engraving was on steel or copper is not indicated. Above the title cartouche, on most editions, is a steel engraving of the American eagle perched on the official shield of the country. A decorative border, consisting of vines, leaves, and grapes frames the map.

In 1849 the firm published *Colton's Illustrated & Embellished Steel Plate Map of the World on Mercator's Projection, . . . Compiled, Drawn & Engraved by D. G. Johnson* (Fig. 19–9). The map was copyrighted in 1848 by Colton. It had been previously published in 1847 as *Johnson's Illustrated & Embellished Steel Plate Map of the World on Mercator's Projection. . . . Engraved & Published by D. Griffing Johnson, 80 Nassau Street, New York*. Again, Colton acquired the copyright of this map rather than individually producing it. It was published in a number of editions to 1868 and has a decorative border with vignette illustrations depicting various countries. Engraved on steel, the map was probably done by the siderographic technique invented by Jacob Perkins.

Colton's Map of the United States of America Including Canada & a large portion of Texas, published in 1853, was a reissue of J. Calvin Smith's *Map of the United States of America*, published by Sherman and Smith in 1843. Both maps are identified as having been printed from steel plates. Colton also published 1854 and 1857 editions of the map.

In 1859 Horace Thayer published, in a small wall-map format, *Colton's New Illustrated Map of the World on Mercator's Projection*, which was engraved by J. M. Atwood and F. H. King. According to Groce and Wallace, Atwood was born in Washington, D.C., around 1818 and was active in New York as an engraver from 1838 to 1852. No data on King has been found. The publisher Thayer was located at 18 Beekman Street in New York. On several Thayer maps published a few years later, Colton's eldest son, G. W. Colton, appears as his junior partner. Groce and Wallace tentatively identify Thayer as a lithographer, and it may be that the younger Colton was learning this technique of map reproduction from him. Editions of *Colton's New Illustrated Map of the World* were published to 1867. Some of the later ones carry the imprint of G. W. and C. B. Colton.

The Colton firm also discovered a ready market for railroad maps and between 1850 and 1887 published and sold thousands of them (Figs. 19–10 and 19–11). "A large segment of commercially produced railroad maps, perhaps as much as 30 per cent, was deposited by the New York City publishing house established in 1831 by Joseph Hutchins Colton."[11] It also issued a number of Civil War maps, most of which were adapted from the firm's standard map series (Fig. 19–12). Several Colton Civil War maps carry the imprint of other publishers such as Lang & Laing, a New York lithographic firm. It is very likely that the Colton maps were transferred to lithographic stones or zinc plates so large runs could be printed.

While many Colton maps were acquired from various individuals or publishers, the company appears to have conceived, planned, and compiled original atlases. Colton's entrance into the atlas field was quite late; its first atlas was not issued until 1855. In that year the first volume of *Colton's Atlas of the World Illustrating Physical*

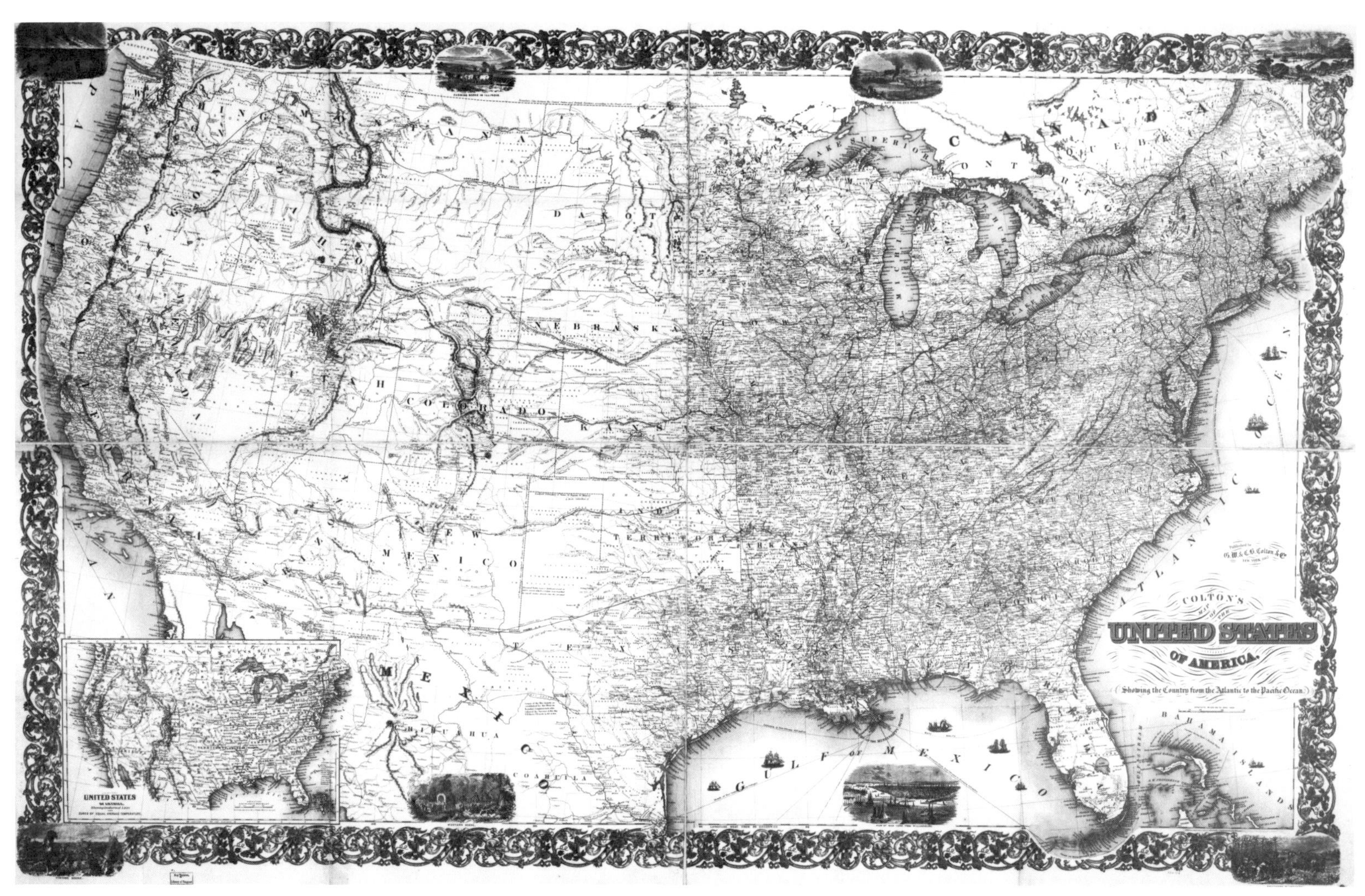

Fig. 19–8. An 1867 edition of Colton's map of the United States.

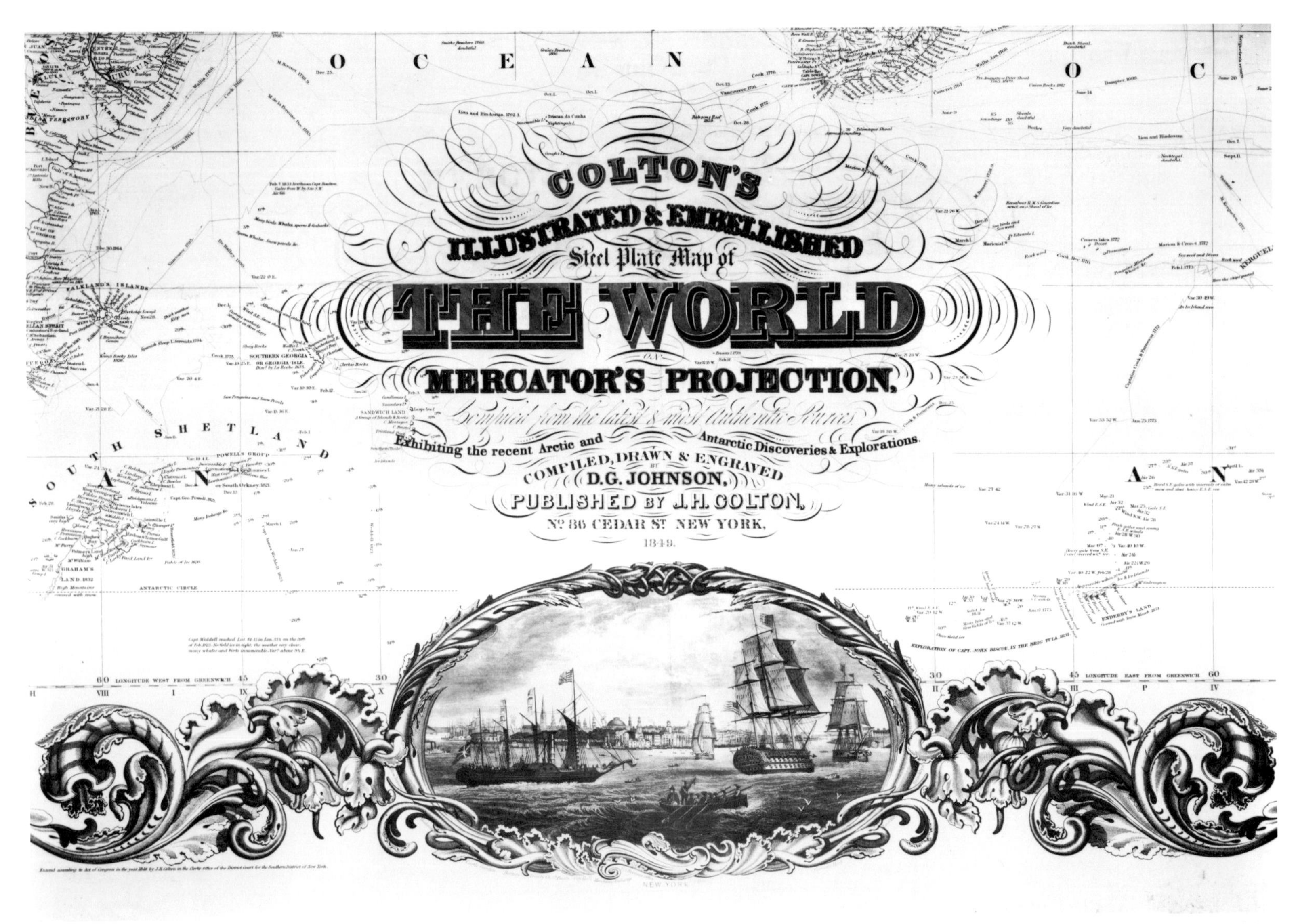

Fig. 19–9. Title cartouche of Colton's 1849 map of the world.

and Railroad Geography, by George W. Colton, Accompanied by Descriptions Geographical, Statistical, and Historical by Richard Swainson Fisher, M.D. was released. A London distributor, Trübner & Company of 12, Paternoster Row, is identified on the title page.

On the atlas's half-title page, there is a steel-engraved illustration of a band of Indians on a cliff viewing, on the plains below, cultivated fields, homesteads, roads, railroads, and other evidence of the advance of white settlers. This volume, which includes North and South

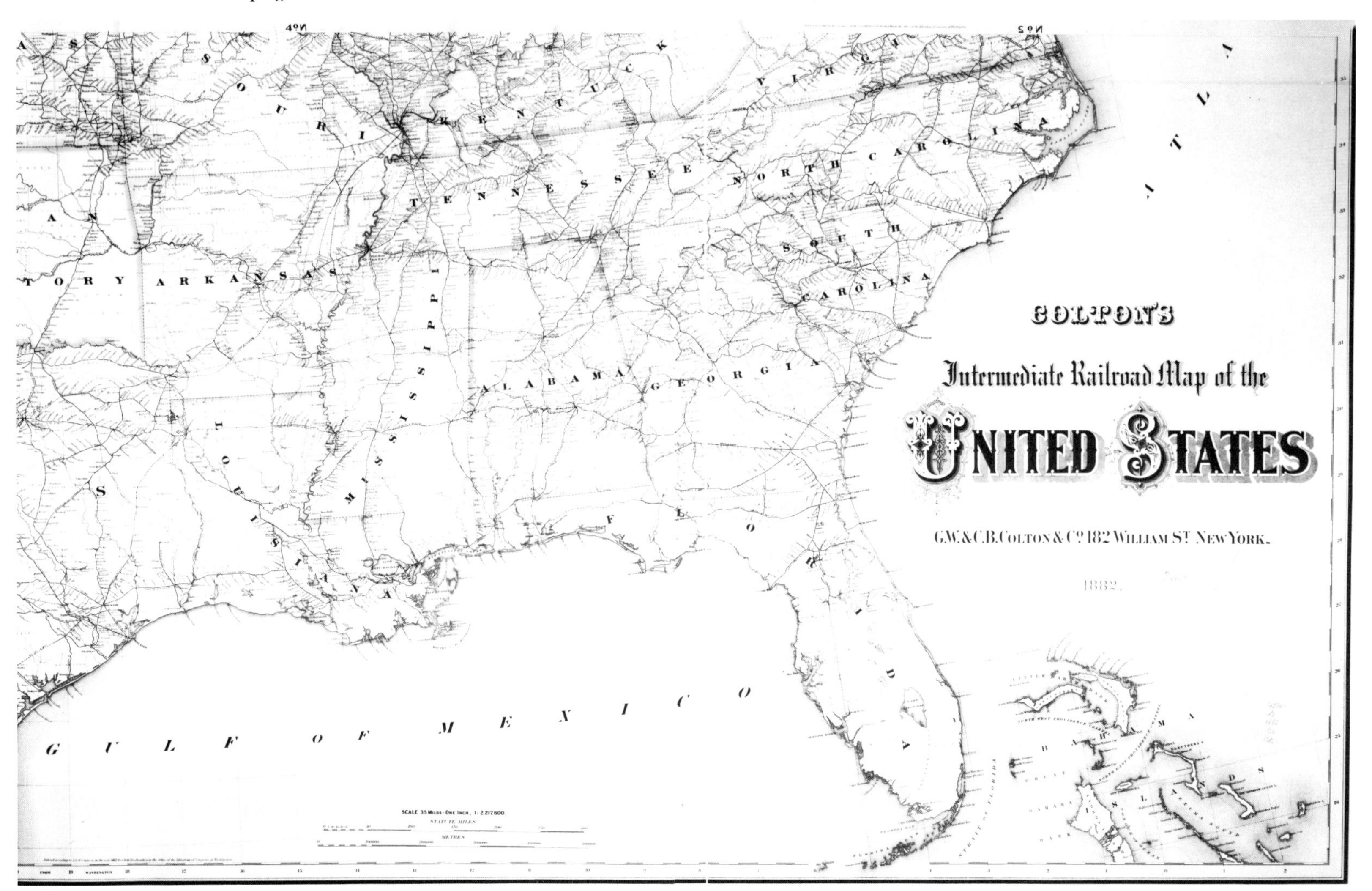

Fig. 19–10. The southeastern portion of Colton's 1882 railroad map of the United States.

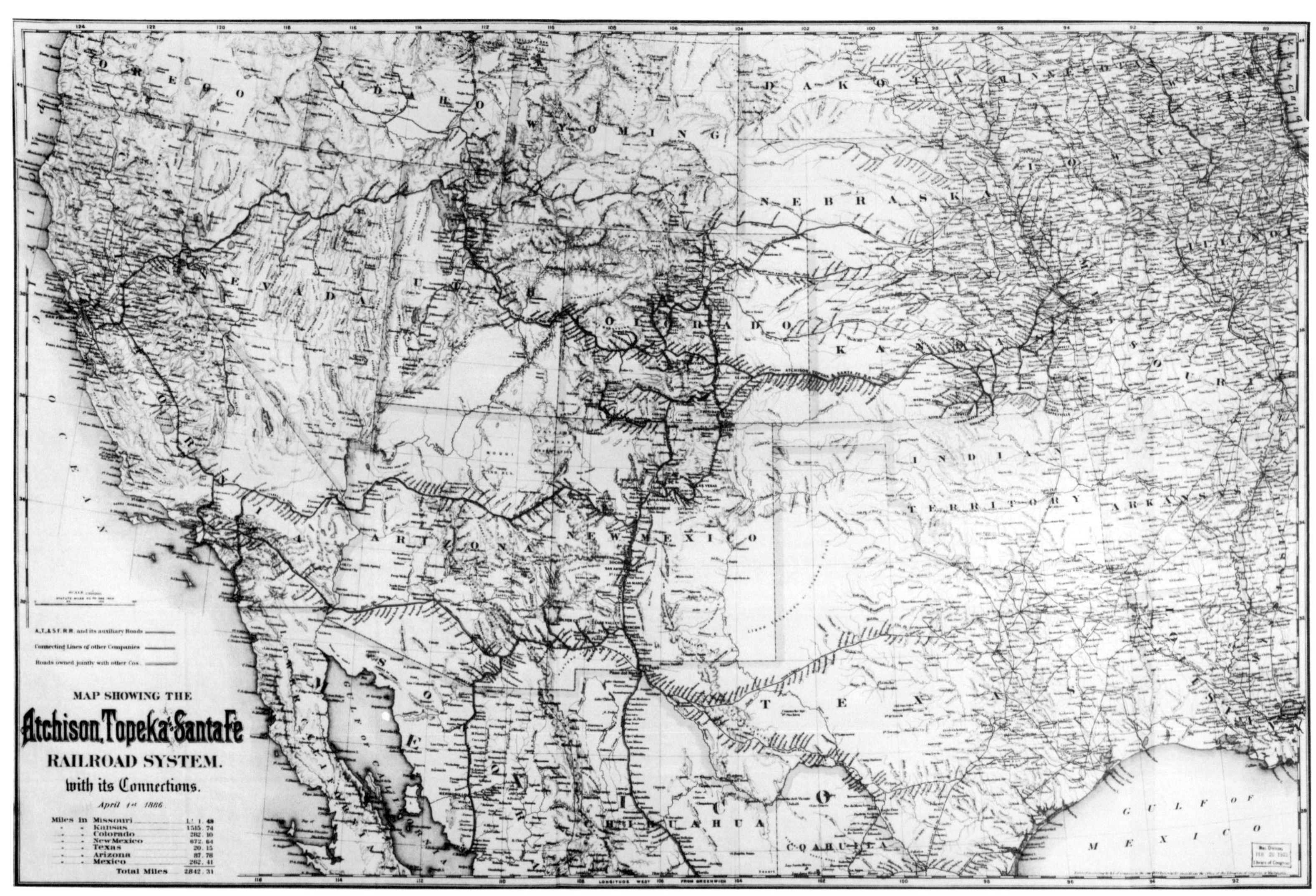

Fig. 19–11. An 1886 map of the Atchison, Topeka & Santa Fe Railroad system. The map was registered for copyright in 1883.

America, lists sixty-three maps in the table of contents. All the maps have decorative borders of an interlacing pattern. They exhibit no plate marks and so the engraved maps were probably transferred to lithographic stones. Color was manually applied. In contrast with some of the atlases issued by Mitchell and other contemporary cartographic publishers, Colton's atlas includes extensive descriptive texts. The essays were prepared by Richard Swainson Fisher, who wrote for many of the Colton gazetteers and historical and geographical books and guides.

Volume two of the *Atlas of the World* was published in

Fig. 19–12. One of Colton's Civil War maps.

1856 and has maps and text for "Europe, Asia, Africa, Oceania, Etc." The half-title page is illustrated with a steel engraving of four women (apparently representing the above-mentioned parts of the world) with surveying instruments and a large globe, which they are examining. Special features of this volume are charts showing the relative heights of mountains and lengths of major rivers in the world. The publisher's advertisement includes the statement that "the artists employed in engraving the maps of the Atlas are the most competent in their profession and the results of their labors compare favorably with those of the most celebrated in their vocation." Notwithstanding the reference to engraving, the maps in this volume show no plate marks. They, too, were probably engraved first and then transferred to lithographic stones or zinc plates for reproduction. A second edition of the two-volume *Atlas of the World*, virtually unchanged, was published in 1856.

The *Atlas of the World* is one of the first publications of the company to carry the name of George W. Colton. He was born in Lenox, Berkshire County, Massachusetts, on September 22, 1827. Although he seemingly had no formal training in geography or cartography, he early demonstrated an interest in his father's business and, in all likelihood, compiled the *Atlas of the World* or directed the compilation of it. He was twenty-eight years old when the first volume of the atlas was published. It is likely that he served henceforth as the Colton firm's principal map compiler, cartographer, and engraver, and within the next decade he and his brother, Charles B. Colton, assumed direction of the company.

The *Atlas of the World* had a short life and was discontinued after two editions. The public may have been deterred from purchasing the large-format two-volume publication because of its fairly high price. At any rate, it was replaced in 1857 by *Colton's General Atlas* in one volume. The title page describes the atlas as "containing One Hundred and Seventy Steel Plate Maps and Plans, on One Hundred Imperial Sheets." This suggests that the *General Atlas* may have been introduced because this type of reproduction had become available. The publisher's advertisement still states that the maps were engraved, but it is not clear whether the maps were engraved on steel or whether they were transferred from copper to steel plates. The latter is more likely. Around 1857 a technique of enfacing copper with steel was invented. It allowed the engraver to work his image on a soft copper plate, upon completion of which the copper plate was coated with a more durable steel surface by means of electroplating.

G. Woolworth Colton is given as the author of the atlas. Colton had obviously stopped using his full first name in favor of the seemingly more distinguished sounding G. Woolworth. Most of the maps in the *General Atlas* duplicate those published earlier in the *Atlas of the World*. In fact, the original 1855 copyright registrations appear on most of them. In order to fit the maps onto a smaller page, however, the maps in early editions of the *General Atlas* do not have decorative borders. The borders were reinstated beginning with the 1860 edition. Like its predecessor, the *General Atlas* includes extensive text and statistical data by Fisher.

The *General Atlas* apparently proved to be a popular publication, and it was reissued in revised editions to 1888. The number of plates were increased to 180 by 1866 and to 212 by 1876. All editions to 1874 include the title-page description, "Containing——Steel Plate Maps and Atlases." This phrase is omitted in editions published after this date, indicating, perhaps, that lithography had been used. The similarity of the maps in post–1874 atlases to those in earlier editions suggest that the original copper or steel engravings may have been transferred to lithographic stones or zinc plates.

We may wonder why Colton atlases were not reproduced by the wax engraving process, which was extensively used by atlas publishers in the United States in the 1870s and 1880s. In his book *The All-American Map*, which relates the story of cerography, David Woodward observed that "a number of possible reasons may be proposed why the atlases of certain American atlas publishers such as Samuel Augustus Mitchell (1792–1868) and George W. Colton (1827–1901) were copper or steel engraved and not wax engraved. . . . The firms of Mitchell and Colton had chosen to widen the market for their products, not by tapping the cheap atlas market in the United States, but by entering the European atlas trade. It was thus essential for them to match the fine quality of the intaglio atlases of France, Britain, and Germany."[12]

There is some evidence that Joseph Colton disposed of his atlas copyrights in 1860. In that year there was published *Johnson's New Illustrated (Steel Plate) Family Atlas, With Descriptions, Geographical, Statistical and Historical*. The credit on the title page reads, "Compiled, Drawn, and Engraved, under the Supervision of J. H. Colton & Alvin Jewitt Johnson." The atlas was "Published by Johnson and Browning, Formerly (Successors to J. H. Colton and Company) Tenth Street (Second Door from Main), Richmond, Virginia, 1860, Copyright by Johnson & Browning." The frontispiece illustration is the one of the Indians on a cliff that was used in earlier Colton atlases. Although a number of maps in the atlas duplicate those used in Colton publications, the decorative borders are of a different design.

Editions of the *Family Atlas* published in 1862 and 1863 give the publisher as Johnson & Ward, "successors to Johnson and Browning, who were (Successors to J. H. Colton and Company)." Although there are descriptive texts in all three early editions of the *Family Atlas*, no author is credited on the title page. Beginning with the 1864 edition, however, the descriptions are ascribed to "Richard Swainson Fisher, M.D. Author of Colton's General Atlas of the World, and Gazetteer of the United States, and Other Statistical Works, and late editor of the Journal of the American Geographical and Statistical Society." Society records indicate that Fisher was elected to its membership in 1855, but numbers of the journal prior to 1864 do not show him as editor. John Wright, however, indicated that Fisher was secretary, member of the council, and librarian of the society in 1859.[13] It is possible that editing the journal was at the time one of the duties of the secretary.

The reproduction technique employed for the Johnson (and very likely the Colton) atlases is indicated at the bottom of the contents page. A note there states, "The Maps are transferred and Printed by D. McLellan & Bros. 26 Spruce Street, New York." Groce and Wallace give this entry for David McLellan: "Lithographer, born in Scotland about 1825, son of James McLellan. He was active in NYC from 1851. From 1859 to 1864 he was associated with his father and thereafter with his brothers."[14] This confirms that the maps for the Colton atlases were engraved on steel plates and then transferred to lithographic stones for reproduction. The title page of the 1864 edition of the *Family Atlas* acknowledges the debt to Colton. The credit line reads, "Maps Compiled, Drawn and Engraved Under the Supervision of J. H. Colton and A. J. Johnson. The new plates, Copyrighted by A. J. Johnson, are made exclusively for Johnson's New Illustrated Family Atlas. Others are the same as used in Colton's General Atlas."

Editions of the *Family Atlas* were published to 1885. The Colton name was dropped from the title page after the 1865 edition. Fisher was credited as the author of the descriptions for several more years. Notwithstanding Colton's sale of the maps to Johnson for use in the *Family Atlas*, Colton's *General Atlas* continued to be published. After 1864, though, the *General Atlas* was published by G. W. and C. B. Colton & Company. Colton was sixty-four years old in 1864 and apparently quit as head of the company in favor of his sons. It is not clear whether he carried on in a less active capacity with the firm. At any rate, he lived almost thirty years after his retirement. He died July 19, 1893, at the Brooklyn home of his son Charles.

In addition to its major atlases described above, the Colton firm published a number of other cartographic volumes that had short runs. In 1854 and 1856 Colton published editions of an *Atlas of America*, which includes only the American maps from the *General Atlas*. Like the

General Atlas, it includes descriptive texts by Fisher. Colton also published several small-format atlases: the *Illustrated Cabinet Atlas of Descriptive Geography*, 1859; *Colton's Condensed Octavo Atlas of the Union*, 1864; and *Colton's Quarto Atlas of the World*, 1865.

Toward the end of its existence, G. W. and C. B. Colton & Company published two single-edition atlases. The *Stand Atlas of Bible Geography*, published in 1891, consists of seven small display maps "designed especially for the use of Families, Secondary Schools and Bible Classes." In 1892 the firm issued a *Complete Ward Atlas of New York City With Key Locating Wards*. The maps in both works appear to have been lithographically reproduced. These seem to be the last cartographical publications of the Colton firm. We have no information as to whether the company expired in 1892 or whether it was taken over by another publisher. At that date, wax engraving had been adopted as a reproduction medium by most of the large American cartographic publishers. Having built their business on engraving and lithography, the Coltons were apparently unwilling to reorganize it. In 1892 George W. Colton was sixty-five years old, and his brother, Charles B. Colton, was sixty. George died in 1901 at the age of seventy-four, and Charles was eighty-four at his death in 1916.

Notes

1. Luther Ringwalt, ed., *American Encyclopaedia of Printing* (Philadelphia, 1876), 345–46.
2. Grenville Bathe and Dorothy Bathe, *Jacob Perkins, His Inventions, His Times, and His Contemporaries* (Philadelphia, 1943), 183.
3. Melville C. Branch, *Comparative Urban Design: Rare Engravings, 1830–1843* (New York, 1978).
4. C. Warren, "On Improvements in the Art of Engraving on Steel," *Journal of the Franklin Institute* 7 (1829): 170.
5. David McNeely Stauffer, *American Engravers upon Copper and Steel* (New York, 1907) 1:111.
6. George C. Groce and David H. Wallace, eds., *The New-York Historical Society's Dictionary of Artists in America 1564–1860* (New Haven, Conn., 1957), 81.
7. George Woolworth Colton, *A Genealogical Record of the Descendants of Quartermaster George Colton: Collected and Arranged from All Available Public and Private Sources* (Philadelphia, 1912), 273.
8. Ibid.
9. I. N. Phelps Stokes, *The Iconography of Manhattan Island 1498–1909* (New York, 1918) 3:687.
10. John K. Wright, *Geography in the Making: The American Geographical Society 1851–1951* (New York, 1952), 16–17.
11. Andrew M. Modelski, comp., *Railroad Maps of the United States: A Selected Bibliography of Original 19th Century Maps in the Geography and Map Division of the Library of Congress* (Washington, D.C., 1975), 4.
12. David Woodward, *The All-American Map, Wax Engraving and Its Influence on Cartography* (Chicago, 1977), 48.
13. Wright, *Geography*, 403.
14. Groce and Wallace, *Dictionary*, 416.

20. *Henry Francis Walling*

Adapting the transfer process to lithographic map reproduction in the United States in the mid-1840s stimulated the compilation and publication of county, town, and city maps. This specialty loomed large in commercial cartography from 1846 to the later decades of the nineteenth century. During this period hundreds of state, county, town, and city maps and atlases were published. A number of individuals participated in this phase of map compiling and publishing, but the contributions of Henry Francis Walling are particularly noteworthy.

Walling was born in Burrillville, Rhode Island, on June 11, 1825. Early in his life the Walling family moved to Providence, where he was educated in the public schools. He prepared for college at Lyons and Friese's Classical School and served as a library assistant in the Providence Athenaeum. Walling also studied mathematics and surveying and, finding these pursuits to his liking, decided against attending college. Instead, he accepted employment in 1844 or 1845 in the office of Samuel Barrett Cushing, a competent and established civil engineer. Cushing made Walling his partner in 1846. The first product of the partnership appears to be a revision of James Stevens's *Topographical Map of the State of Rhode-Island*, which was originally published in 1831. It was printed in 1846 by Isaac H. Cady of Providence. No information has been found on Cady. Judging by the quality of the map, however, he seems to have been a skilled lithographic printer.

With his career well launched, Walling married Maria Fowler Wheeler in 1847. The bride was sixteen years old, the groom twenty-two. In the summer of that year Walling conducted surveys of the town of Northbridge in Worcester County, Massachusetts, from which he drafted a map that was published in 1848. In 1849 Cushing & Walling published *A Map of the City of Providence From Actual Survey* (Fig. 20–1). It was lithographed "at S. & W., 15 Minor Street, Philada." This was the cartographic publishing establishment of Robert P. Smith while he was in a brief, informal partnership with Isaac Jones Wistar. A small view of a portion of Providence decorates the title cartouche on the plan, and on the face of the map, beyond the settled areas, the seal of the city is shown.

Encouraged by the success of the Northbridge and Providence town plans and sure that local officials might be interested in having town or city maps of their jurisdictions, Walling severed his relations with Cushing and established his own engineering and surveying office around 1850. He did, however, maintain contact with his former partner. He conducted surveys in collaboration with the new firm of Cushing & Farnam and drafted a *Map of the Proposed Providence Water Works*, which was published in 1853. Extending his surveys to Massachusetts towns, Walling initially focused his attention on Bristol County located immediately to the east of Providence. His surveying efforts, probably conducted during the summer of 1850, resulted in maps of five towns that were published toward the end of that year. The following year was particularly productive with the publication of ten Massachusetts town plans, including towns in Bristol, Essex, Norfolk, and Worcester counties. With this increase in business, Walling took on assistants and in 1852 established his mapping headquarters at 81 Washington Street in Boston. For printing some of the earlier town plans he still relied upon Philadelphia lithographers and publishers such as Robert P. Smith, Friend & Aub, and A. Köllner. Walling apparently prepared the town plans on contract for town officials. After supplying these officials with a specified

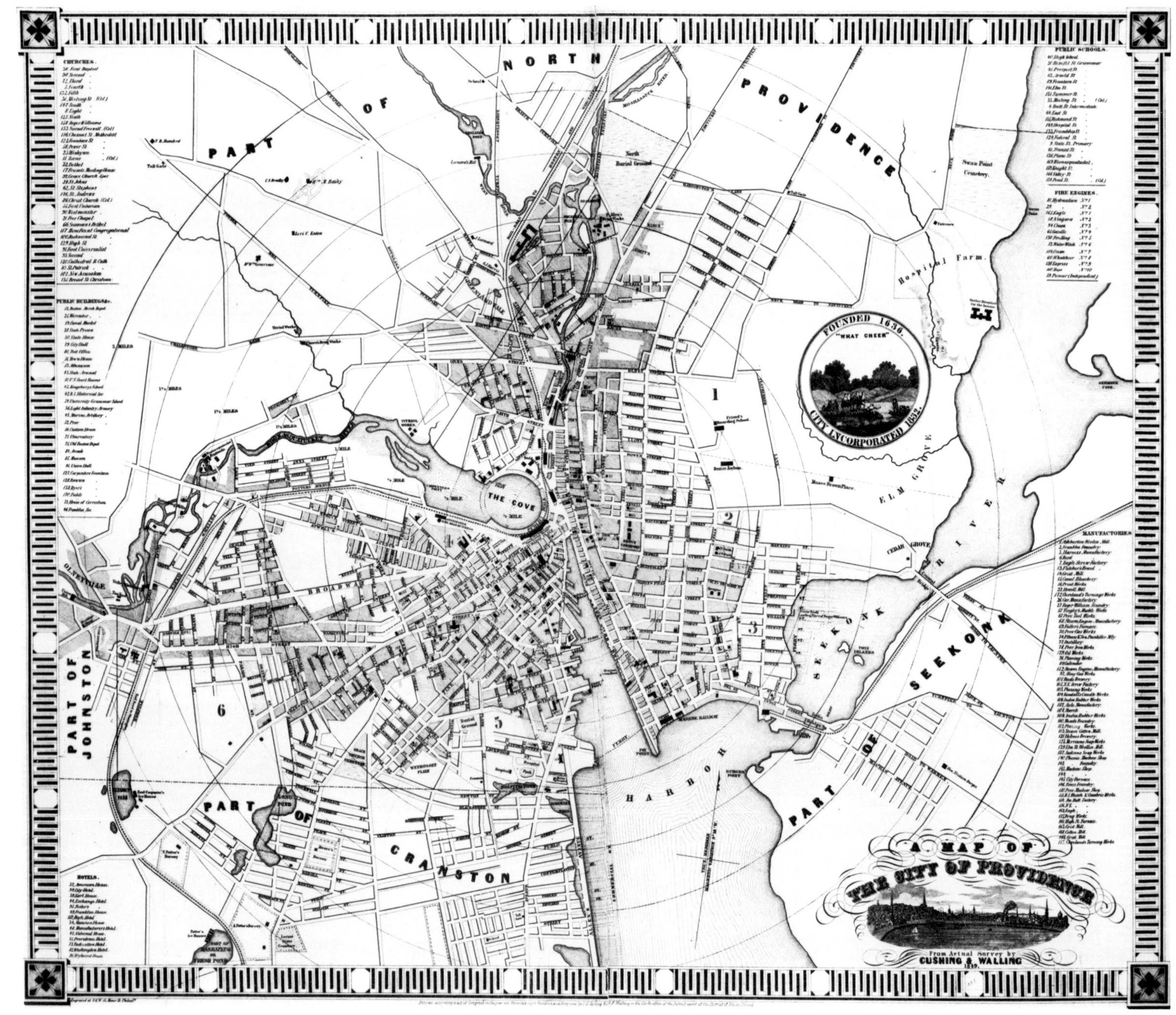

Fig. 20–1. Cushing & Walling's 1849 map of Providence, Rhode Island.

number of the maps, he was free to sell additional copies on his own.

Walling published seven maps in 1853 for towns in Essex, Middlesex, Norfolk, Plymouth, and Worcester counties. On the Concord plan (Fig. 20–2), there is a notation that "the Town Lines are laid down principally from old Surveys. White Pond & Walden Pond from Surveys by H. D. Thoreau, Civ. Engr." Three Massachusetts towns were mapped by Walling in 1854, eight each in 1855 and 1856, and four in 1857. No less than fifty Massachusetts town plans were prepared by Walling and his associates between 1850 and 1857.

From 1855 on, Walling is identified on his maps as the "Superintendent of the State Map." In that year, by direction of the Massachusetts legislature, he was appointed by the governor to update and revise Simeon Borden's map of Massachusetts, which was published in 1842. The Joint Special Committee on the State Map summarized previous official action, noting the inaccuracies in the town plans and county maps that were used in compiling the earlier state map. The committee emphasized that

> the inconveniences and disadvantages arising from these numerous inaccuracies have been very great. It is extremely desirable to remedy them. Mr. Walling, who has been commissioned by the State authorities to superintend the revisal of the State Map, a gentleman who is understood by the Committee to be competent for the task, offers, for the very modest sum of twenty thousand dollars, accurately to renew the local surveys both of the counties and the towns. He is enabled to do it for this price only because he relies upon numerous sales of town maps, and from the fact that he has already finished three counties and twenty-five additional towns at his own expense.
>
> This proposition, however, does not include such a topographical survey as will introduce upon the map contour lines which shall display at once the amount and extent of the elevations, the direction and degree of the slope, as well as the

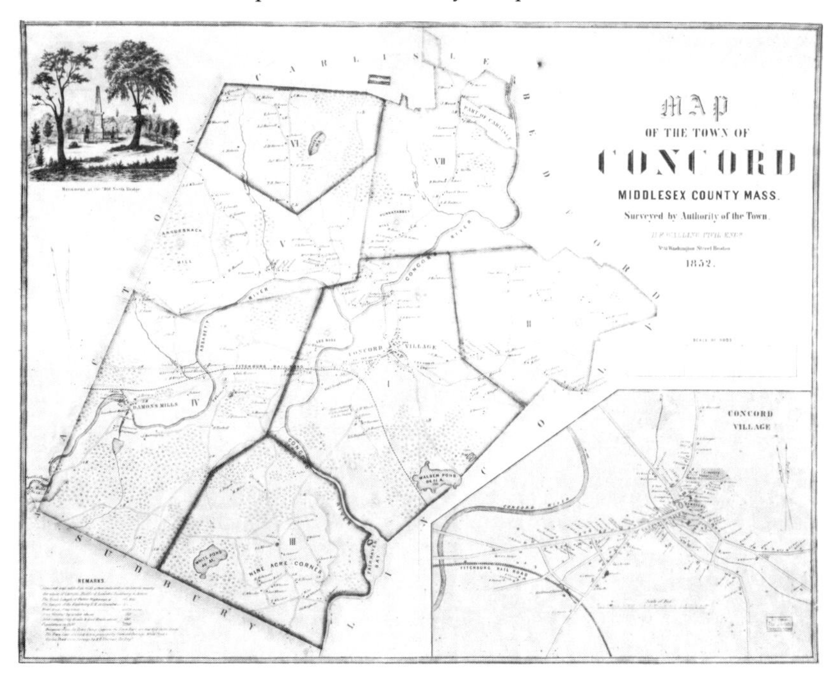

Fig. 20–2. Walling's 1852 map of the town of Concord, Massachusetts, was actually published in 1853. The residence of Ralph Waldo Emerson is shown, near the east margin on the inset map of Concord Village, in the lower right corner of the map sheet.

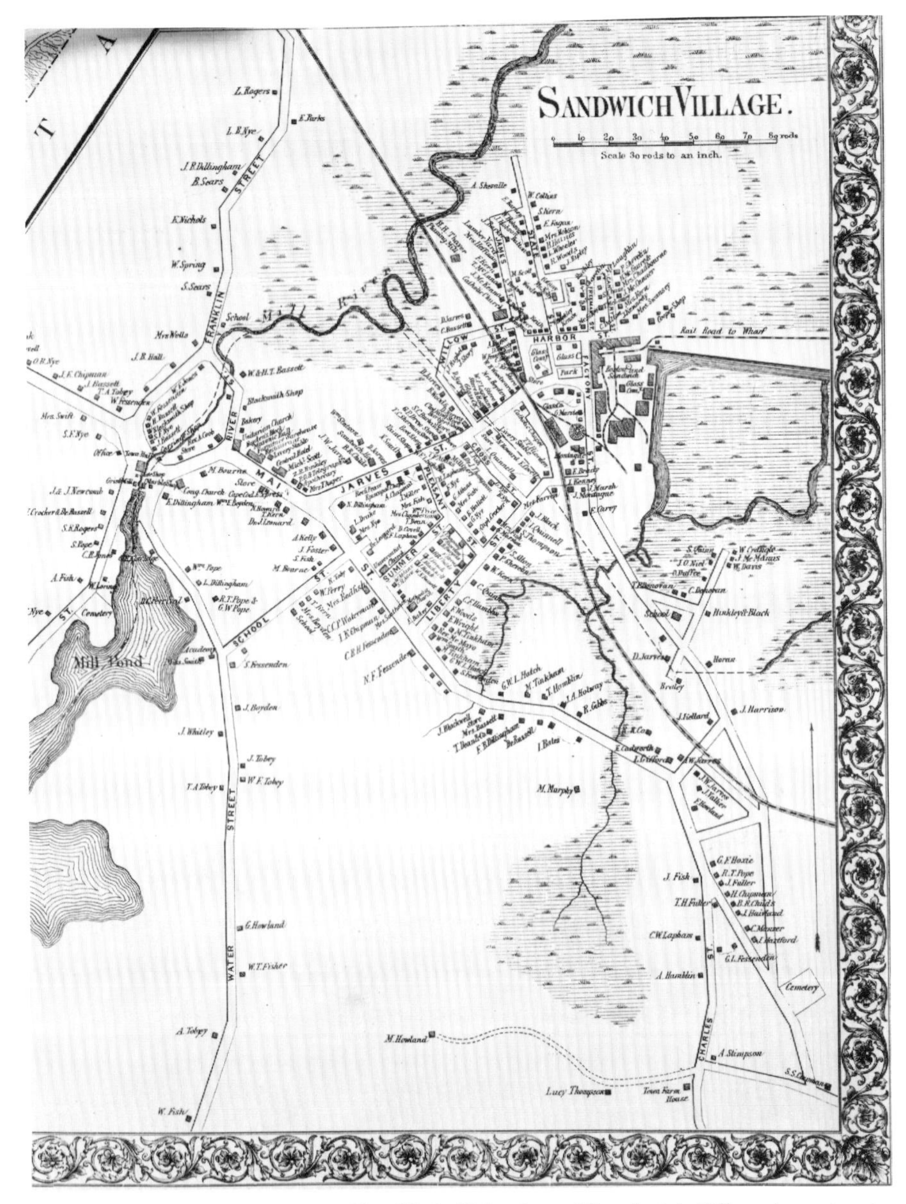

Fig. 20–3. This plan of Sandwich Village is an inset on Walling's 1857 map of Barnstable County, Massachusetts.

> lines of level throughout the Commonwealth. . . . No reliable evidence was presented to determine the probable additional expense; although Mr. Walling was ready to make his surveys for the first year with reference to this addition of the elevations, leaving it for future legislatures to ascertain their value and expense.[1]

Because the committee favored preparing a topographical map of the state rather than revising the old Borden map, it resolved "that it is inexpedient to make a new State Map at the present time, but that the subject be recommended to the favorable consideration of the next legislature."

The subsequent legislature authorized Walling to proceed with revising the map. Years later, in 1871, Walling related that

> in 1860 a thorough revision of the [state] map was made by H. F. WALLING, in accordance with an Act of the Legislature. Under his direction a complete re-survey of the entire state was made in detail, and separate maps upon a large scale were constructed and published of all the counties in the State, the trigonometrical survey being used as a basis. The former errors in the location of roads, streams, town lines, etc. were eliminated and the work brought up to that time by adding new roads, railroads, etc. Although very heavy expenses were incurred in the field work, the preparation of the maps, and in the engraving of the numerous corrections upon the copper-plates, no direct pecuniary assistance was received by Mr. Walling from the State, but the privilege of publishing the map was granted to him, a privilege, however, which had been found by previous publishers to be of little or no value. Owing to derangements produced by the late war, the publication of the map at that time was discontinued, and heavy losses were incurred by Mr. Walling.[2]

As a basis for revising the state map, Walling surveyed and published maps of all fourteen Massachusetts counties between 1855 and 1860 (Figs. 20–3 and 20–4). He also extended his activities to other New England states in this period. He issued seven or more Vermont county maps (Fig. 20–5), five of New Hampshire counties (Fig. 20–6), six of Maine, and two of Connecticut. He also published state maps of Maine, Vermont, and Rhode Island, and city plans of Providence and Boston.

Walling later summarized, in the journal of the Association of Engineering Societies, his experiences in surveying and mapping New England towns and counties. Segments of the paper provide a valuable perspective on town and county mapping.

> From the year 1850 up to the commencement of the civil war in 1861, and for some ten or twelve years after the close of the war, the demand for maps giving more local detail than had been shown upon previous maps became sufficient to induce surveyors to engage in their publication, a portion of the expence being in some cases defrayed by State, county or municipal appropriations. Maps of towns and counties, afterward compiled into State maps, were made in this way, almost entirely from original surveys, in all of the New England States, New York, Pennsylvania, Ohio, and Maryland. Large maps of some of the States west of Ohio were also published from time to time. Since the construction of these Western maps consisted principally in compilation from the United States Land Survey plans, the labor in preparing them was much less than in the Eastern States. . . .
>
> Injustice has been done to a number of local map makers in New England and Middle States, who at their own cost and risk, during the last thirty-five years, have materially added to the previous knowledge of local topography, so far, at least, as it pertains to horizontal relations. . . .
>
> Having previously made surveys and maps of a number of towns and cities in the State, which were generally paid for by municipal appropriations, I undertook in 1854 the construction of a separate map of each county in the State, excepting the small counties of Suffolk, Duke, and Nantucket, which were included on sheets containing larger adjacent counties. . . . The scope of the work I had undertaken included the representation of roads, railroads, streams, ponds, marshes, the sea-coast, with its capes, bays, inlets and islands, and of important buildings, including dwellings, churches, school-houses, mills, manufactures, stores, etc.[3]

Walling could make his many surveys with the aid of a surveyor's compass for determining directions and an odometer for measuring distances. The odometer consisted of a small brass circular box that housed a series of cogwheels which regulated the motion of an index on a dial plate fixed to its exterior. The dial recorded the number of revolutions of a wheel attached to the box. The length of the perimeter of the wheel multiplied by the number of revolutions gave a reasonable indication of the distance traversed. In discussing the instrument, Walling acknowledged that although the odometer was "less accurate than a chain in the hands of a skillful chainman, it has the merit of economy, since one person with a compass and odometer can do the work of a party of three with compass and chain. Thus, the inaccuracies due to the inequalities of the surface are made less on common roads, than . . . would be supposed."[4] The odometer and compass were the principal instruments used by surveyors who prepared the many city, town, county, and state maps and atlases published between 1850 and 1880.

While Walling's earlier plans and maps were lithographed and printed in Philadelphia, beginning in 1855, some Walling maps were reproduced by Ferdinand Mayer & Company, which was located at 96 Fulton Street in New York City. Since Walling had mapped most of the towns and counties in Massachusetts and in other New England states, he was planning to expand his activities to the middle and trans-Appalachian states. Thus, in 1856 Walling relocated his headquarters to 90

Fulton Street in New York City, down the block from the Ferdinand Mayer lithographic company. H. F. Walling's Map Establishment remained at this location until 1858 when it relocated to 358 Pearl Street.

Walling had mapped a few counties outside of New England prior to 1856, among them being Ontario and Wayne counties in New York State. From his New York City headquarters he greatly expanded his activities in

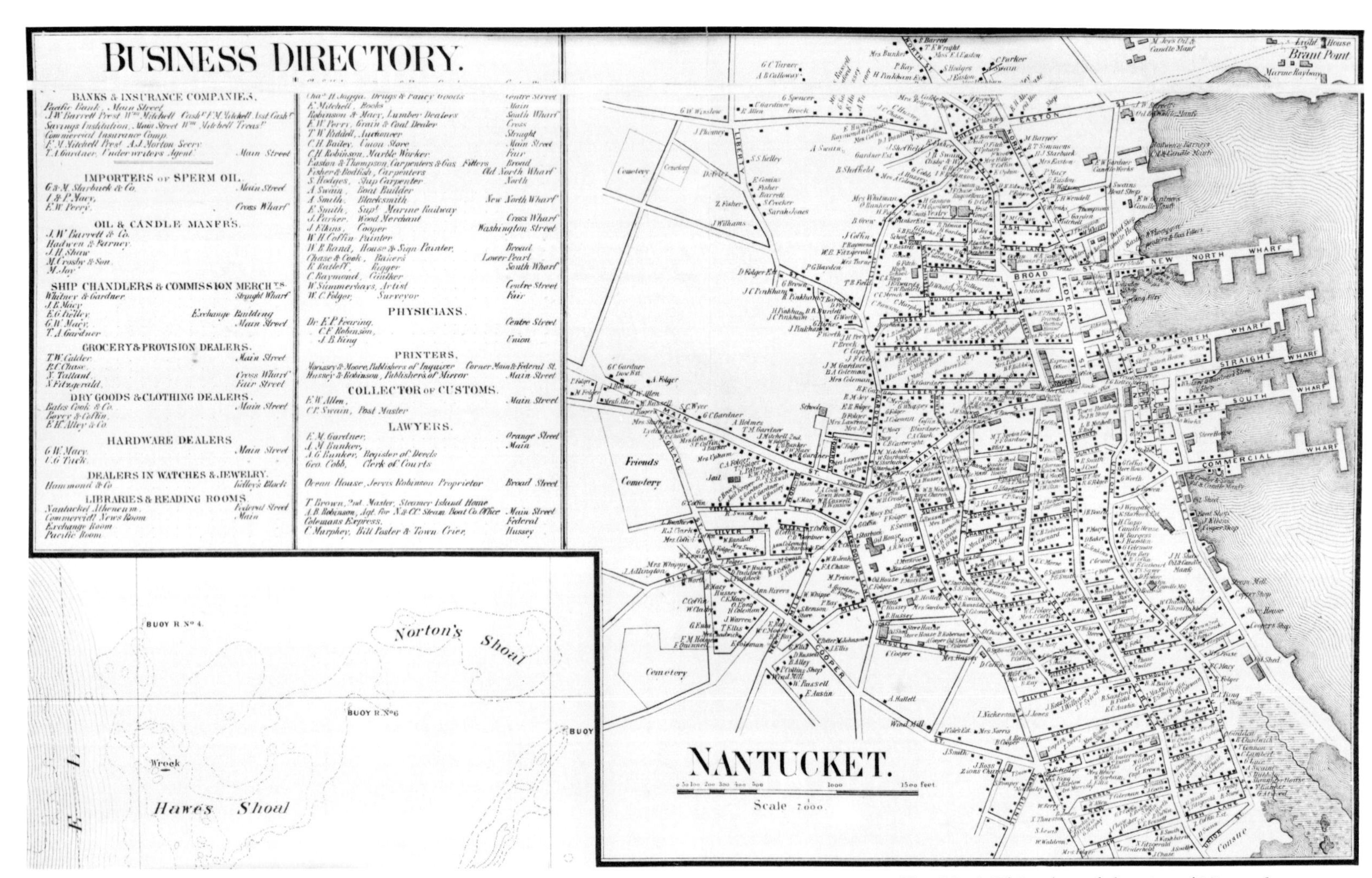

Fig. 20–4. This plan of the city of Nantucket appears as an inset on Walling's 1858 map of Barnstable, Dukes, and Nantucket counties, Massachusetts.

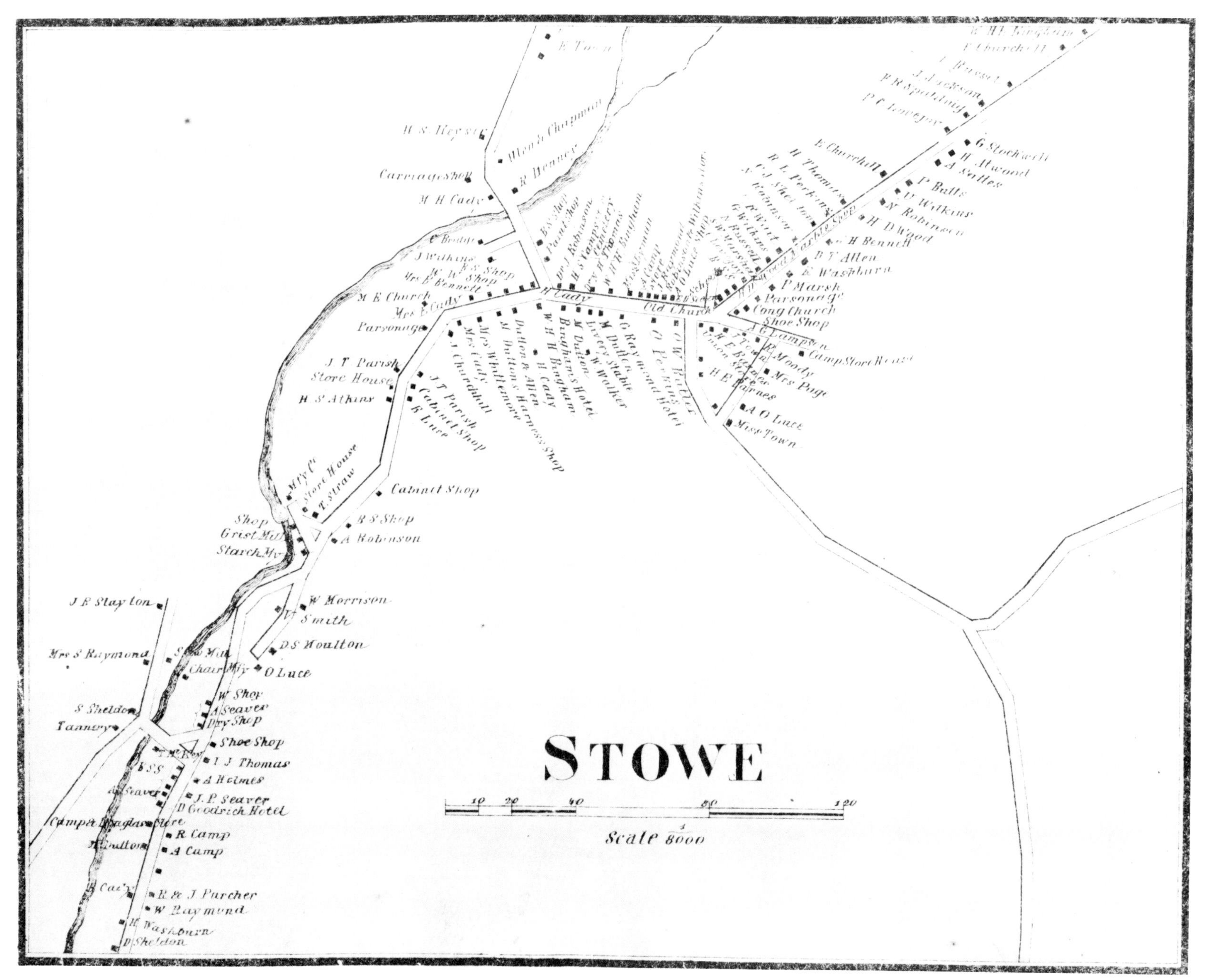

Fig. 20–5. This map of the village of Stowe is an inset on Walling's 1859 map of Lemoille and Essex counties, Vermont.

TOPOGRAPHICAL MAP
OF
CARROLL
COUNTY
NEW HAMPSHIRE
FROM
Actual Surveys under the Direction of
H. F. WALLING
Published by
SMITH AND PEAVEY
Nos. 356 358 & 360 Pearl Street New York
1860
NORTH CONWAY
TAMWORTH VILLAGE
JACKSON
CHATHAM
BARTLETT
CONWAY CENTRE
CONWAY CORNER
SANDWICH CENTRE
ALBANY
CONWAY
SOUTH WOLFBOROUGH
MOULTON MILLS
OSSIPEE CORNER
WATER VILLAGE
MADISON
EATON
MOULTONBOROUGH CORNER
SANDWICH
TAMWORTH
OSSIPEE CENTRE
FREEDOM VILLAGE
FREEDOM
MELVIN
MOULTONBOROUGH
OSSIPEE
EFFINGHAM
TUFTONBOROUGH
LORD'S CORNER
WAKEFIELD CORNER
UNION VILLAGE
WOLFEBOROUGH
SMITH'S BRIDGE VILLAGE
WAKEFIELD
BROOKFIELD
TABLE OF DISTANCES

the middle states, and his surveyors were operating in New Jersey, Pennsylvania, New York, and Ohio by the late 1850s. Before the Civil War a few Walling maps had been published for counties in Indiana, Illinois, and Wisconsin. Maps issued in this period usually state that the surveys were made "under the direction of H. F. Walling." Many have the imprint "engraved, Printed, Colored & Mounted at H. F. Walling's Map Establishment." Some have a second imprint, "Walling & Rice, Publishers," with the same address as the Map Establishment.

The Civil War was a serious deterrent to the profitable operation of Walling's Map Establishment. C. O. Boutelle, one-time director of the U.S. Coast and Geodetic Survey, noted in a memorial to Walling published in 1891 that Walling's establishment had been "brought into successful and profitable operation, when, in 1861–62, the war suspended and nearly ruined him. The class of men in his employ were precisely those most needed in the country's service and most ready to give both service and life to the country when called upon."[5]

Notwithstanding this major setback, Walling retained an office in New York City and continued his work as a civil engineer and mapmaker until 1868. He published some county maps as late as 1864, but they were very likely based on surveys made before 1861. Several of the maps were issued in collaboration with other publishers. He also apparently established a partnership with Ormando W. Gray, with offices in New York City and Danielsonville, Connecticut. This partnership produced atlases of a number of states in the late 1860s and early 1870s. In the mid-1860s Walling also occupied himself in compiling city, steamboat, and railroad guidebooks illustrated with maps that were published in 1867 by Taintor Brothers & Company of New York City.

Although Walling did not have a college education, he was appointed in 1868 to the chair of civil and topographical engineering in Lafayette College's Pardee Scientific School in Easton, Pennsylvania. During his tenure at Lafayette, he continued his cartographic activities, with special attention to compiling state atlases. As this activity became more demanding, Walling resigned his professorship in 1872 and reestablished his cartographic publishing business in Cambridge, Massachusetts. He again specialized in compiling and publishing county maps, city and town plans, and state atlases.

Fig. 20–6. Walling's 1860 map of Carroll County, New Hampshire.

Walling had a strong interest in topographical, or scientific, surveys, but his cartographical publications were almost exclusively of the cadastral type. Thus, he felt constrained to defend the publication of cadastral maps and plans. In a paper published in 1886 he decried

> the wholesale condemnation which certain scientific critics have from time to time bestowed upon commercial maps, as such, not discriminating between those which have been carefully made from original work and other mere compilations from Government explorations, etc. which perpetuate from year to year the original errors of their sources of information, indicates an unjustifiable lack of knowledge on the part of the critic, of the amount of original and really valuable information which some of the commercial maps embody. They have rendered valuable service to railroad and hydraulic engineers, to county and town officers, and even to geologists and other naturalists and scientific investigators. Besides these, many intelligent citizens, realizing the value of the maps for their own more private uses, have, by bestowing their patronage upon the surveyors, enabled them to publish these maps. If competent experts found on examination that these maps were erroneous and unworthy of credit, it would be their duty, in behalf of the public to explore their unworthiness by pointing out the inaccuracies etc. But it is hardly justifiable to condemn them *en masse* merely on account of their self-sustaining origin or because the mechanical execution was

> not upon so expensive a scale as might have been possible under more favorable conditions.[6]

While engaged in private surveying and map publishing, Walling established cordial relations with officials in federal mapping agencies. In April 1875 he asked, in a letter to J. E. Hilgard, the assistant in charge of the U.S. Coast Survey, if he could use the survey's triangulation data as a base for his mapping of the city of Pawtucket, Rhode Island. His request was granted, and later in the year Walling acknowledged the loan of a plane table.[7] Continuing his correspondence with Hilgard, Walling proposed in a letter dated October 16, 1877, that "to stimulate towns and cities to have more careful surveys the state might undertake to pay half the expense of having plane table surveys carried on at a cost which might be less for the towns and cities themselves to incur, according to their requirements, it being understood that in all cases the triangulation would be furnished by the Coast Survey." Walling also broached another subject. "You were kind enough," he wrote, "to intimate some time ago that the department might employ me, if such a survey were initiated in R.I., to make the triangulation. I feel sufficiently interested in it to work pretty hard this fall to bring something of the kind about. May I consider that in the case of Boston or some of the suburban towns engaging in a survey somewhat like my co-ordinate plan, that the triangulation can be furnished in this way."[8]

Apparently this work for the Coast Survey in Rhode Island did not materialize. However, when the sales for his cadastral maps and atlases faltered, Walling accepted employment with the Coast Survey in Washington, D.C., in July 1880 as a map compiler at a salary of $150 a month.[9] The report of the superintendent of the U.S. Coast and Geodetic Survey for the fiscal year ending June 1881 provides the information that "the work assigned in July to Mr. H. F. Walling was the compilation of a map of that part of the Appalachian chain of mountains extending from the southern part of Maryland to the northern part of Georgia and Alabama. . . . The area over which work has been carried during the fiscal year comprises that portion of Maryland lying between Parris Ridge on the east and North Mountain Range on the west, with some of the adjacent portion of Virginia and West Virginia."[10]

Walling remained with the Coast and Geodetic Survey until 1883, when he transferred to the U.S. Geological Survey. In arranging the transfer there had been a question about Walling's title with the Geological Survey. On March 20, 1883, he wrote to Major J. Wesley Powell, director of the survey, about this subject. "The work charms me," he stated, "and . . . I am anxious to engage in it under your direction, but I feel obliged, unless you are disposed to allow me to rank as a 'geographer', to ask for more time for consultation before accepting the offer as it now stands."[11] This matter was referred to Henry Gannett, the chief geographer of the survey, who "satisfactorily arranged" it. When Walling actually entered into duty with the Geological Survey on June 21, 1883, it was as a topographer at an annual salary of two thousand dollars. In this position, Walling wrote a letter to Survey Director Powell, dated January 8, 1884, reporting on a suggestion he had made to Massachusetts's newly elected governor about federal-state cooperation in topographic mapping. Walling also broached this topic in a May 16, 1884, letter to Hilgard, who by this time had become superintendent of the Coast and Geodetic Survey. Writing from Cambridge, Massachusetts, Walling remarked,

> I suppose you have heard through Mr. Whiting, if not otherwise, of the recently proposed state survey of Massachusetts. Just before the inauguration of the present governor it occurred to me that a much better survey than I was making would be likely to result from a cooperation between the state and general governments. Accordingly I sug-

> gested in an informal way, such a plan to the governor elect who was struck by it and he mentioned it in his inaugural message. The subject was referred to a committee who will probably report a bill the special provisions of which are not yet announced. . . . It has always seemed to me that the triangulation for state surveys should be performed by the trained parties of the Coast Survey . . . and in the consideration which I have presented to the committee, this has been made prominent. I had no direct instructions from the Geological Survey to present such a view, but as no objections have been made by Maj. Powell or Mr. Gannett who are aware of what I have done, I conclude that it has not incurred their disapproval. . . .
>
> In general would it not be good policy to establish a triple co-operation of this kind . . . for which the Coast and Geodetic Survey might obtain special local appropriations, incurring no responsibility for the topographical details. In this way and perhaps in no other it may become possible to gradually construct a map of the whole country much better adopted for its immediate wants than those now in existence.[12]

Walling's suggestions were taken seriously, for the Geological Survey's annual report for the year 1883–84 noted that "it was decided to commence a topographical map of New England to be published upon a scale of 1:125,000. In accordance with this decision Professor H. F. Walling was appointed . . . for the purpose of taking charge of that work, and field work was shortly instituted under his direction in western Massachusetts. . . . Professor Walling's intimate acquaintance with the work which had previously been done in the state, rendered him particularly fitted for the difficult task of handling this material."[13] The survey's report for the following year related that "before the close of the fiscal year Professor Walling had been placed in the field in the southwestern part of Massachusetts. He commenced work in the town of Mt. Washington and was engaged until August in extending a minute triangulation in that and the neighboring towns and adjoining portions of New York and Connecticut."[14]

To direct and supervise the cooperative mapping program with the Geological Survey, Massachusetts established the office of the Commissioners of the State Topographical Survey in 1884. In their first report the commissioners related that "party No. 1, in charge of Henry F. Walling, commenced field work on or about August 22d, in the mountain district in the southwesterly corner of the State. . . . The area covered is about 122 square miles." In 1885 the commissioners reported that "in addition to the field parties of the United States Geological Survey, the services of Mr. Henry F. Walling of that Department, have been given to the Commission in connection with the town boundary line determinations." The commissioners' 1886 report included Assistant Geological Survey Director Henry Gannett's statement that "to aid the Commission in its difficulties and at the same time contribute to the supply of points necessary for the topography, the Director of the United States Geological Survey assigned Mr. H. F. Walling, of his department to the service of the Commission for this special triangulation. . . . Under this arrangement Mr. Walling has been in active service since April 1886."[15]

On December 28, 1887, the commissioners reported "the substantial completion of the field-work in connection with the topographic map of Massachusetts" and summarized these accomplishments:

> total area mapped during the year, in Massachusetts and immediately adjacent parts of New Hampshire Rhode Island and Connecticut, all on a scale of 1–30,000, and with contour interval of twenty feet, is estimated at 2,530 square miles, and the cost of the same for salaries, subsistence, field materials and supplies, etc., of all men employed, $19,215. . . . The horizontal scale of the map, as

> published, is 1–62,500, or about one mile to an inch. Relief is expressed by contour lines, at vertical intervals of 20 feet. . . . Determination of boundary lines of towns . . . has been in charge of Mr. H. F. Walling.[16]

The 1889 report of the commissioners noted that "the work connected with the town boundary survey has been much embarrassed during the past year by the loss of Prof. Henry F. Walling, whose sudden death, by heart disease, occurred on the 8th of April last. Mr. Walling has been identified with the State Survey from its inception to the time of his death; in fact it was mainly due to his personal effort that the survey was inaugurated. He was at the time a member of the staff of the Geological Survey, which he finally left for the exclusive service of the Commonwealth. . . . To Walling more than to any one else, is due the appreciation of good maps which is now bearing fruit in the national survey."[17]

For more than forty years Walling vigorously engaged in surveying, mapmaking, and map publishing. A memorial published in the journal of the Association of Engineering Societies in 1889 reported that "between the years 1848 and 1888 [Walling] prepared and published, either alone or with others, maps of no less than 20 States and Provinces, 280 counties, and over 100 cities, towns and special localities. [Walling's works] were commercial maps, and subject to the defects of such. . . . But they were fully up to the requirements of the times when they were issued, and to the conditions and resources of the country represented."[18] Although the greater part of his life was spent in producing commercial maps, Walling is also known for his significant role in advocating scientific and topographical surveys under federal supervision and in promoting the concept of federal-state cooperation in surveying and mapping. Such cooperative surveys still comprise a large proportion of the topographic mapping conducted by the U.S. Geological Survey in its national mapping program.

Notes

1. Massachusetts, House, Joint Special Committee on the State Map, *Documents*, 1855, H. Doc. 294.
2. Henry F. Walling and Ormando W. Gray, *Official Topographic Atlas of Massachusetts* (Boston, 1871), v–vi.
3. Henry F. Walling, "Topographic Surveys of States," *Journal of the Association of Engineering Societies* 5 (1886): 163–66.
4. Ibid., 167.
5. C. O. Boutelle, "Henry Francis Walling [Memorial]," *Bulletin of the Philosophical Society of Washington* 11 (1891): 492–96.
6. Walling, "Topographic Surveys," 165.
7. National Archives, U.S. Coast Survey, Miscellaneous A–Z, Applications for Maps and Reports, vol. 1, 1875.
8. National Archives, U.S. Coast Survey, Miscellaneous Correspondence of Office of Assistant in Charge, vol. 4, 1877.
9. National Archives, U.S. Coast and Geodetic Survey, Personnel Records 1816–1881.
10. U.S. Coast and Geodetic Survey, *Report of the Superintendent . . . of the Work during the Fiscal Year Ending with June, 1881* (Washington, D.C., 1883), 26–27.
11. National Archives, U.S. Geological Survey, Letters Received 1883.
12. National Archives, U.S. Coast and Geodetic Survey, Records of Office of Assistant in Charge, Departments of Treasury, War and Navy, Post Office, Interior, and Agriculture, vol. 3, 1884.
13. U.S. Geological Survey, *Annual Report 1883–84* (Washington, D.C., 1884), 3–4.
14. Ibid., *Annual Report 1884–85* (Washington, D.C., 1885), 9.
15. Massachusetts, Commissioners of the State Topographical Survey, *Reports for the Years 1884–1895* (Boston, 1895), 9, 12, 22.
16. Ibid., *Report for the Year 1887* (Boston, 1888), 16.
17. Ibid., *Report for the Year 1888* (Boston, 1889), 31.
18. C. W. Folsom, F. O. Whitney, and E. L. Brown, "Henry Francis Walling," *Journal of the Association of American Engineering Societies* 8 (1889): 28.

21. *Robert Pearsall Smith*

Another prolific publisher of local maps in the fifteen years preceding the Civil War was Robert Pearsall Smith (Fig. 21–1). Unlike Walling, Smith had no professional training or experience in surveying and mapping. His cartographical contributions were primarily concerned with the technical processes of map reproduction and printing and with his role as a liaison between surveyors and printers. Smith's map publishing career was relatively brief. It began in 1846 and ended in 1864. Of some five hundred county maps and city plans published during this period, Smith was associated with more than one third of them. Notwithstanding his considerable output, Smith has received meager attention in the study of mid-nineteenth-century commercial map publishing. This is in part because his cartographic business was an early interest which was overshadowed by later episodes in his interesting and eventful life. Mainly, however, Smith's accomplishments were eclipsed by the dramatic and unconventional careers of the members of his immediate family.

Smith gained a respectable living from his cartographic ventures and was a prominent figure in commercial map publishing during the eventful years when lithography supplanted copper-plate engraving as the principal medium for reproducing and printing maps. He also played a significant role with his father, John Jay Smith, in introducing a variation of the lithographic process to the United States.

In April 1845 Smith, a senior at Haverford School (now Haverford College) near Philadelphia (Fig. 21–2), was "ordered by the best medical advice of Philadelphia . . . to take a sea voyage."[1] College records suggest that his ailment was "inflammation of the eyes," a euphemism sometimes used to explain scholastic difficulties.[2] His father, heavily immersed in editorial and publishing projects in addition to his responsibilities as the head of the Library Company of Philadelphia, welcomed the prospect of a trip to Europe and a temporary escape from his routine chores. We know much about their European trip from John Jay Smith's journal and his letters to his eldest son, Lloyd Pearsall Smith, who printed them in *Smith's Weekly Volume*, the literary journal he edited and published. The journal and letters were then published in 1846 in a two-volume book entitled *A Summer's Jaunt across the Water*.

By virtue of his editing and publishing contacts and his position as the head of one of the most distinguished American libraries of the period, John Jay Smith was welcomed, with his son, by printers, publishers, scholars, and librarians in Britain and on the continent. Their interest and curiosity concerning all facets of bookmaking and publishing moved them to attend a lecture by Michael Faraday at the Royal Institution in London on the subject of anastatic printing. The potential applications of this new reproduction process greatly interested and intrigued the father and son. John Jay Smith recorded in his journal that "my son, as well as myself, passed as much time as we could spare in acquiring a knowledge of the manipulations necessary to teach the art. It is one of great [potential] importance in the United States, as by it copies of wood-cuts, and even of copper and steel engravings, are readily taken; it has not yet arrived at its destined perfection, nor has the steam-press been entirely adapted to the work of printing from the zinc plates rapidly; the English patentees have no doubt of ultimate success, and, in the mean time are employing several presses at work."[3]

Faraday's presentation on anastatic printing was well publicized in technical journals. There was a keen interest in all applications of the principles of lithographic

Fig. 21–1. Robert Pearsall Smith, ca. 1870.

printing, which had been invented by Senefelder more than four decades earlier but had only recently been introduced into Britain. Anastatic printing was one of many adaptations and modifications of the lithographic technique that were developed and perfected by European printers and technicians during the half-century after its discovery. Because each new discovery was jealously guarded, ideas were often conceived or developed by different individuals. Thus, Carl Halbmeier has noted that the "anastatic method devised in 1840 is little more than a renewed effort sponsored [earlier] by Senefelder [who] called this method autography."[4]

The Smiths followed up the Faraday lecture with a careful study of articles and reports on anastatic printing which appeared in the January, February, and June 1845 issues of the *Art-Union, Monthly Journal of the Fine Arts and the Arts Decorative Ornamental* and in the February 8 and March 1, 1845, issues of *Chambers's Edinburgh Journal*. John Jay Smith excerpted several of these articles in his journal account that was published in *Smith's Weekly Volume* on November 5, 1845.

Although there had been similar adaptations of the lithographic process some years earlier, anastatic printing, as revealed to John Jay and Robert Pearsall Smith, was invented by Charles Frederick Christopher Baldamus of Berlin. Four years later, Joseph Woods took out a patent in England on Baldamus's process. Basically, the anastatic process involved transferring a page of letterpress or an illustration to a lithographic stone or a zinc plate for reproduction. Initially, the technique was hailed as the means for multiplying illustrations that had been originally reproduced from copper or steel plates. It was soon learned, however, that original illustrations drafted on paper under the proper conditions could also be duplicated by the anastatic process.[5]

In his lecture Faraday demonstrated that letterpress laid upon white paper and rubbed at the back left the letters imprinted in reverse. He explained that in a similar manner the print or illustration could be transferred to a zinc plate, from which multiple copies could be printed. As reported in the *Art-Union*, a dilute nitric acid "applied to the back of the letterpress, passed through the paper, but not through the printer's ink." The excess acid was absorbed with blotting paper and

> the acidified sheet was then placed upon the zinc plate, and passed once under a small hand-press, when on the removal of the paper the printing was found transferred in reverse to the plate, which now presented a dull appearance, the polish having been destroyed by the acid, which so readily attacks zinc . . . the letters thus transferred were left consequently very slightly in relief—indeed, so slightly that the effect was imperceptible. The plate was then rubbed with gum in solution, which . . . strengthened the [impression]. The next proceeding was the application of ink by rubbing in the same manner; the result of which was that this ink attached itself to the film [of ink] already deposited on the zinc by the pressure of the roller. The plate is then washed over with phosphatic acid, which has an especial effect upon the whole. . . . The printing surface was then ready for the press; it was inked by a common leather roller in the ordinary way, and with as much rapidity; and impressions were produced within the time proposed for the whole process—twenty minutes. It is to [be] observed that the first impressions are not the best; but the perfection of the invention soon becomes obvious, and justly merits the epithet—anastatic, or reproductive.[6]

The *Art-Union* account also noted that "the printing from which the plate had been prepared was of recent date; but the age of the typography presented no obstacle to the success of the operation, since to transfer letterpress printed say a hundred years ago, perhaps, the principal difference in the treatment might be to subject it for a longer time to the action of the nitric acid."

In its report, *Chambers's Edinburgh Journal* predicted that

> in another department of relief printing, there is no question that the anastatic process will cause a complete revolution, and that very speedily; namely, in illustrative and ornamental printing. Wood-engraving will be entirely superseded, for no intermediate process will now be necessary between the draughtsman and the printing of the design. . . . Now . . . all the draughtsman will have to do will be to make his drawing on paper, and *that*, line for line, will be transferred to the zinc, and produce, when printed, exactly the same effect as his original draught. A pen is recommended for the purpose. . . . The requisite ink is a preparation made for the purpose, and may be mixed to any degree of thickness in pure distilled water, and should be used fresh and slightly warm when the effect is to be given. . . . A drawing thus produced can be readily transferred to the zinc in the manner above described for typography.[7]

By the time they left London, the Smiths were well instructed in the principles and techniques of anastatic printing. John Jay Smith had also made preliminary arrangements to secure the American license and patent for the process. Official copies of the license in English

Fig. 21–2. An inset view of Haverford School from the map of Delaware County, Pennsylvania, published by Robert P. Smith in 1848.

and German are preserved in the Library Company of Philadelphia's John Jay Smith papers. Transacted in Berlin on January 20, 1846, the license was signed by Ernest Werner Siemens, Charles Frederick Christopher Baldamus, and Charles William Siemens. The license specified that

> as joint proprietors of an invention made by us, relative to preparation and restoration of printed surfaces, and machinery and tools required for the use of same (anastatic printing) [we constitute] John Jay Smith, Esquire, of Philadelphia, our general and special attorney, in our name to do whatever may be required to introduce, dispose of, and render productive our aforesaid invention throughout the whole extent of the United States; to act as our representative in all transactions whatsoever with private individuals as well as public bodies; to include every kind of contract, to give for us binding declarations, and especially to apply for a patent for the invention, in such case to give security, and in short to do, without exception, for our interest whatever we might be authorized or required to do, were we personally present, including the right to substitute others under him. Whatever our said attorney in virtue of this power shall, or may do we promise to approve.[8]

There is no mention of Smith's obligations to the proprietors, financial or otherwise, or any reference to the remuneration he might receive for services performed. Anastatic printing, however, proved to be the medium through which the Smiths, and particularly Robert, entered the field of commercial map publishing.

John Jay Smith observed in his journal that in England the anastatic press had an "abundance of work to do for the railway companies, whose plans of roads, bridges, elevations, termini, etc. are reproduced with a magic rapidity; . . . it is only necessary that any plan should be drawn in lithographic ink, brought to the anastatic office, and in a few moments, facsimiles are handed to the artist. Of course the art has a thousand applications. . . . So with a map; . . . he has copies struck off, and attains his object."[9] Smith added a footnote to this account before *A Summer's Jaunt* went to press: "Since the foregoing was written, the author brought with him to Philadelphia the necessary apparatus and information to pursue the business sufficiently to offer patents for the different States for sale, and a variety of work has been executed at the anastatic office in Philadelphia, to the great satisfaction of artists, architects, surveyors, and draughtsmen."[10]

John Jay Smith apparently exerted a strong influence upon his sons and their careers. It was certainly his idea to establish an anastatic printing shop in the United States, but the actual operation of it was entrusted to Robert. The office was opened, probably in February or March 1846, at a location not far from the library, which then was on Fifth Street facing Independence Square. The John Jay Smith papers contain a broadside which announced,

> Robert P. Smith agent for the Patentees of the new process of Anastatic Printing has his presses at the Sunday School Union Building No. 144 Chestnut St. third story where he is prepared to execute the orders with which he may be favored. To Architects, Artists, Draughtsmen and Conveyancers this art recommends itself by the perfection, facility and cheapess [*sic*] with which designs and drawings of buildings; plans, maps, sketches and writings are copied and multiplied from a single original on common paper when drawn in ink supplied at the office.
>
> Persons interested are invited to call at the Anastatic Printing Office and inspect specimens of transfers of sketches and writing, type, plans, and pictures.

The broadside is in neat, cursive script and was undoubtedly reproduced by the anastatic process.

The press in the Anastatic Printing Office was probably small and of the flatbed type, similar to the standard lithographic press. Reproductions were made from zinc plates, and most of the jobs performed were on relatively small sheets and with small runs. John Jay Smith's inventive mind was, however, focusing on wider utilization of the process. An interest in cemeteries and their ornamentation led to the publication in New York in 1846 of his *Designs for Monuments and Mural Tablets*. The twenty-six plates in the volume were reproduced on the anastatic press.

Robert Pearsall Smith's first venture into mapmaking in 1846 utilized this new reproduction process. His father and his brother, Lloyd, also participated in the project. At this time, Lloyd was operating a legal bookstore. He observed that lawyers had frequent occasion to consult, in real estate negotiations, the earliest published maps of Philadelphia and the settled parts of Pennsylvania. As original copies of these maps were in short supply, Lloyd decided to publish, in conjunction with his father and brother, a facsimile reproduction of Thomas Holme's *Map of the Province of Pennsilvania*, which was originally surveyed in 1681 under the direction of William Penn.

John Jay Smith refers to this map project in *A Summer's Jaunt across the Water*. In discussing anastatic printing, he noted that

> an interesting application was made at the suggestion of a number of gentlemen, the result of which is now for sale. The oldest map of the province of Pennsylvania, begun in 1681 and completed under Penn, known as Holmes's [*sic*] Map, had been so extremely scarce as to make its production in courts, or reference to it by lawyers and scriveners, extremely difficult. Under these circumstances an experienced draughtsman carefully traced and copied his tracings in lithographic ink. It was printed off in sections, coloured, and mounted, and, being a perfect fac simile of the original, met with a rapid sale. To make the matter a little curious, as the first anastatic map ever issued, but two hundred copies were printed and the plates were destroyed. It was as large as the usual maps of the United States, hung upon rollers, and is a curiosity and a rarity.[11]

Of the original two hundred copies of the Holme anastatic reproduction, there are only a few extant survivors (Fig. 21–3). It should be noted that this was not a true facsimile, for in copying the original printed map, the draftsman made a number of errors and modifications. The title, for example, was executed in plain block letters in contrast with the more ornate lettering on the original engraved map. Further, the lengthy and de-

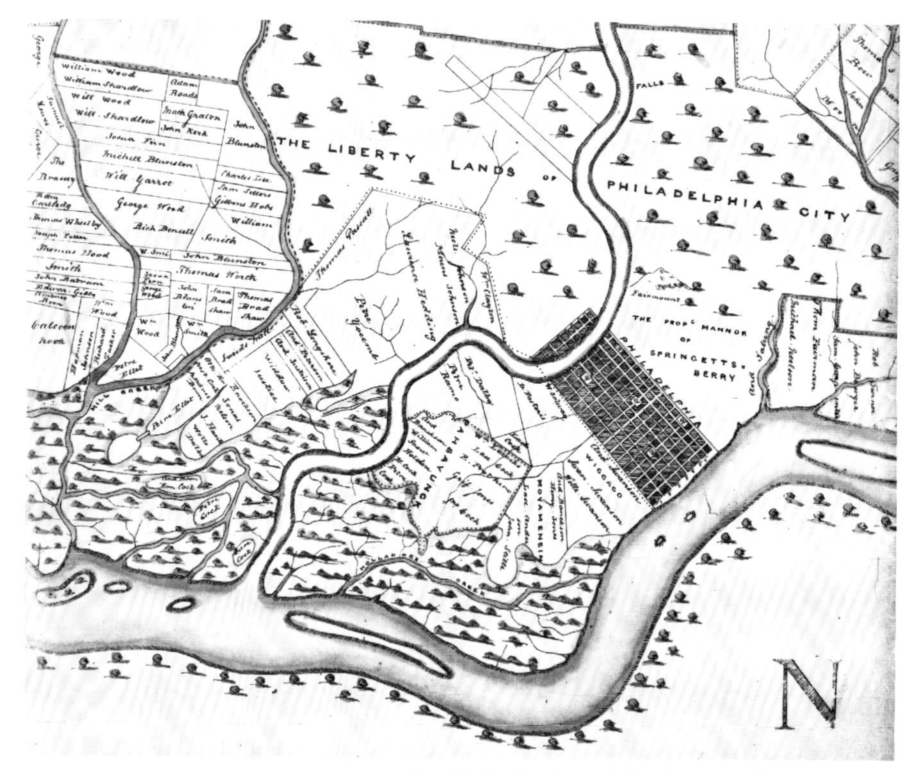

Fig. 21–3. Portion of the Smiths's 1846 anastatic reproduction of Thomas Holme's 1681 map of Pennsylvania.

tailed "General Description of the Province of Pennsylvania in America" in the bottom margin of the original is omitted on the facsimile. There are a number of other omissions as well as spelling modifications and misspellings. Below the Penn coat of arms and the dedication to William Penn, the facsimile edition imprint data are set within a box: "Facsimile of Holmes-Map of the Province of Pennsylvania with the names of the original purchasers from William Penn begun in 1681. Reproduced from the original in the Philadelphia Library by the ANASTATIC PROCESS. 200 copies only are printed. Published at the Anastatic Printing Office 144 Chestnut Street and by Lloyd P. Smith Law Bookseller, Philadelphia 1846." This facsimile, which measures 84 by 141 cm., may have been printed in as many as eight sections on the small anastatic press. Because the copies I have examined are mounted on cloth, it is not possible to identify the original sheet lines.

In the same year the Smiths published an anastatic facsimile of John Reed's *Map of the City and Liberties of Philadelphia,* which was originally published in 1774. Like the Holme map, the Reed map was not a direct reproduction, but was redrafted for transfer to the anastatic press's zinc plate. Consequently, it, too, includes a number of inaccuracies because of copyist errors. The Reed reproduction was also restricted to two hundred copies.

The success of the two anastatic map facsimiles encouraged the Smiths to undertake other cartographic projects. John Jay Smith's *Recollections* gives information on these. "Having in my employ at the library, as a sub, a clever civil engineer from England J. C. Sidney," he wrote, "I kept him at work in the morning in the upper library rooms, making maps of the city, of ten miles around it, etc. I reproduced . . . a fine map of Philadelphia, and many county maps of merit, all of which were successful ventures. I had no office-rent to pay, very few expenses, and Sidney worked with tolerable accuracy."[12] It is interesting to note that John Jay Smith takes credit for a number of the map publications, but many of them carry the imprint or copyright registration notice of Robert Pearsall Smith. In the early phases of Smith's career, it is possible that he was dependent upon his father's financial support.

The map Smith specifically referred to was the map of *Ten Miles Around the City of Philadelphia*, which he classed "the most successful of our maps."[13] It is laid out in a circle 51 cm. in diameter and is printed on a 56-by-56-cm. sheet. A brief title at the top reads *Sidney's Map of Ten Miles round.* In the upper left corner is the more complete title: *Map of the Circuit of Ten Miles Around the City of Philadelphia With the Names of Villages, Roads, Mills, Property Owners, Taverns, &c. From Original Surveys by J. C. Sidney, C.E.* (Fig. 21–4). The map also notes, "Delaware Co. by Dr. Ash. Robert P. Smith Publisher 144, Chestnut Street Philadelphia 1847." It was "Engraved by N. Friend." Although the word engraved was used, the map was in fact lithographed. In the early years of lithography, it was not uncommon to refer to the process as engraving, or lithographic engraving. The illustrations in the lower left and right corners are, respectively, of Girard College and Laurel Hill Cemetery. Both reflect John Jay Smith's interests. He was a supporter of the college and one of the founders of the cemetery. Both illustrations are credited to A. Kollner, who like Friend was one of Philadelphia's early lithographic artists. It is interesting to note that the map was printed at Peter Duval's establishment. The Smiths had begun, as early as 1847, to shift from the anastatic printing process to other lithographic reproduction techniques.

Anastatic printing never fully achieved the success that was foreseen by Michael Faraday, John Jay Smith, and others. In 1862 it was reported that

> the anastatic process of lithography, which caused considerable sensation when first invented, was

Fig. 21–4. J. C. Sidney's circular 1847 map of Philadelphia appears to be the earliest Smith lithographic publication.

Sidney's Map of Ten Miles round.
MAP
OF THE CIRCUIT OF TEN MILES AROUND
THE CITY OF
PHILADELPHIA
With the names of Villages, Roads, Mills,
Property Owners, Taverns, &c.
From Original Surveys by
J. C. SIDNEY, C.E.
Delaware Co. by Dr. Ash.
ROBERT P. SMITH
Publisher
144. Chestnut Street
PHILADELPHIA
1847.
DELAWARE
RIVER
PETTYS ISLAND
RICHMOND
CAMDEN
LAUREL HILL CEMETERY

> expected to prove a most valuable extension of the art, since it was supposed to combine the advantages of lithographic drawing or transfer, with ordinary letter-press printing. It was found that drawings made by lithographic ink on common drawing-paper, or by lithographic chalk on granulated paper, as well as old impressions of copperplates, lithographs, woodcuts and letter-press printing, could be easily transferred to zinc, and it was confidently expected that it would be practicable to print from such transfers by means of cylinder machines. In practice, however, the plan was not found to answer so as to be remunerative, and for the present it appears to have fallen in abeyance.[14]

The first anastatic printing presses were the flatbed type, but it was early realized that the zinc plates could be curved and used in rotary presses. In 1847 William Smith Williams reported that "without going into a minute description of the Anastatic printing machine . . . it will suffice to say that the zinc plate is rolled round the cylinder of the press, and the plate is washed, or moistened and inked during its revolution by means of cylindrical rollers. The inking roller is the kind ordinarily used in lithographic printing. The washing-roller is composed of sponges covered with linen, and has a double movement, longitudinal as well as rotary, thus producing the wavy motion of washing with the hand."[15]

The original objective of the anastatic process was to reproduce illustrations and text that had originally been printed from copper plates or movable type. The Smiths seem to have been among the first in the United States to adapt the technique to the reproduction of illustrations and maps from original drawings on paper. As the adaptations and improvements of lithography multiplied, the transfer process and the use of zinc plates, which were basic to the anastatic technique, were incorporated into the standard lithographic processes. By 1850 the rotary steam press was in operation, glazed paper was introduced, and chromolithography had been perfected. Within the next decade the principles of photography were applied to lithography, with further benefits to the art. The more lithography was refined, the more firmly it became established in the United States and particularly in Philadelphia. The city boasted no fewer than sixteen lithographic firms by 1856.[16]

All these developments contributed to the speedy reproduction of maps, and cartographic publishing boomed. City, town, and county maps were produced in great quantities, and Robert Pearsall Smith was a major part of this growth. In Smith's early career, his father apparently made the major decisions, probably because Smith was only nineteen years old when they opened the Anastatic Printing Office. The elder Smith made the circular map of Philadelphia the success it was. In his *Recollections*, John Jay Smith noted that after the paper editions of *Ten Miles Around the City of Philadelphia* had been profitable, his

> suggestive thoughts came in aid of a good sale and profit. The printer made copies [of the map] on silk pocket-handkerchiefs, which for a considerable time sold as fast as they could be printed. When the sale of these drooped, I procured linen, and then muslin, by the box, and run them off on two or more presses, until almost every retail dry-goods shop presented a display in the window of these *mouchoirs*. They were literally sold by the thousands. Unfortunately, no printer could supply fast colors, and after everybody had the map, the sale became less and less, and the printings were stopped.

There are, regrettably, no extant examples of this map novelty. Printed with impermanent ink, the maps faded or were washed out. Smith goes on to say that "of the books, maps, etc., which were thus published, I have unfortunately rarely kept copies; a *last* copy or two is apt to be given away or sold by the owner. . . . In all this, which may appear like turmoil, I was happy in generally having good agents, and very few knew who it was that issued books and maps in such variety and numbers. It was a necessary duty to provide an education for my children."[17]

Although John Jay Smith probably retained a financial interest in his son's publishing business, Robert gradually became more independent of his father. In 1850 the elder Smith moved to a new home in Germantown, and in the following year he retired from his position as librarian of the Library Company to spend the remainder of his life working on writing and editorial projects. His vacating his post of librarian, though, is further evidence of his strong influence over his sons. "My reason for resigning this honorable hereditary position," he wrote, "was the promotion of the welfare of my eldest son, Lloyd, who, like myself having little pleasure in business, agreed to leave his situation in a mercantile house to be my assistant on the morning attendance, which fell too heavily upon me with only a bungling assistant."[18]

Unlike his father and brother, however, Smith had little interest in such intellectual pursuits as writing and editing. Rather, he experienced his greatest success in activities where his expansive personality could be most effectively utilized. Thus, he was best as a salesman and promoter of maps. Smith's personal attributes were remembered some years later by his son Logan Pearsall Smith. "My father," he wrote, "was a man of fine presence, and of a sanguine, enthusiastic temperament, too impulsive to manage his own affairs by himself; however . . . his gifts of imagination were made to contribute to the firm's prosperity. He was, above all, a magnificent salesman; . . . he exercised upon [customers] the gifts of persuasion and blandishment, almost by hypnotization, which were destined later, in European and more exalted spheres, to produce some startling results."[19] In referring to the firm, Logan erroneously associated his father's experience as a salesman with the Whitall-Tatam Glass Company (a company he later became a partner of), rather than with map publishing and promotion.

As head of the Anastatic Printing Office, Smith became acquainted with many of the printers and publishers in these exciting and eventful years when lithography was becoming established in Philadelphia. Among his associates were several Europeans who had brought to the United States their skill and experience in lithographic techniques. One of them was Peter Duval. When Smith joined the printing and publishing fraternity, Duval was the head of his own firm and was regarded as one of the foremost lithographic printers in Philadelphia. By the late 1840s he was lithographing in color, and by 1850 his printing plant was utilizing zincography and the rotary steam press.

For many new developments Duval was indebted to his shop foreman, Frederick Bourquin. There is some evidence that Smith and Bourquin were associated in some ventures. Bourquin was, undoubtedly, much interested in the reproduction technique introduced by the Smiths in their Anastatic Printing Office and apparently sought to improve on the process. His 1847 award from the Franklin Institute for some steel and copper engravings he had reproduced by the transfer process were possibly an adaptation of this step in anastatic printing. The report of the committee on exhibitions noted that "these [engravings] are highly meritorious: not for any new invention or discovery; but for the great neatness, beauty, and accuracy of execution."[20] Two years later Bourquin and Duval began using zincography, possibly also an adaptation from the anastatic process.

By the end of 1847, Smith decided there was no great future in the anastatic process and terminated the Anastatic Printing Office. The tedium of operating the office, moreover, did not appeal to him. In the year or two that it was in existence, however, Smith had become interested in map printing and publishing. In 1848, therefore, he established (quite likely with the financial support of his father) a map publishing company located at 15 Minor Street, within two short blocks of the Library Company. He was soon engaged in publishing city and county maps. Smith apparently contracted the engraving and printing, with most of the jobs going initially to the Duval firm. He served as a middleman between surveyors and printers, for his name appears most frequently on maps in the role of copyright registrant. In the Library of Congress collections there are a number of maps on which Smith's name does not appear as publisher, but as the registrant.

Smith's shop was apparently opened late in the year, for the only Smith map dated 1848 is the *Map of Delaware County, Pennsylvania, by Joshua W. Ash, M.D.* The lithographer was Gustavus Kramm. Pictured in a marginal illustration is Smith's alma mater, Haverford School, which is located within Delaware County. Such personal associations are found on several of the early Smith maps.

By the middle of the nineteenth century Philadelphia had attained some prominence as a center of the insurance industry, and Smith was interested in expanding his business. In 1848 he proposed to the Philadelphia Contributorship, a leading fire insurance company, to prepare a series of insurance maps. These were to exhibit "in each and every Square of the City of Philadelphia and the improved parts of the Liberties, every Lot with its number on it (without reference to accurate measurements)."[21] It is doubtful whether this series was ever published.

The Smith maps published in 1849 are identified as "Published by Smith and Wistar." This partnership was never officially consummated. Smith offered it as an inducement to Isaac Jones Wistar so that he would continue performing the business and shopkeeping functions, while Smith negotiated with surveyors for the rights to their maps and with lithographers and printers to reproduce them. Wistar was a Haverford classmate of Smith, and, like Smith, was a member of the Society of Friends. Wistar had also left Haverford before graduating. He spent the next several years in various unfruitful but adventurous pursuits. After an unhappy attempt at farming, Wistar joined Smith. He recorded in his autobiography:

> I proceeded to engage myself, in 1848, to a Philadelphia map publisher who undertook to teach me how to keep his accounts, for which I was to receive the compensation of three dollars a week. As I did not prove an inept scholar, this was soon increased to four and then to five dollars, upon which I managed to pay board and live till November, when my cousin, Dr. Mifflin Wistar, who with his wife was about to travel in the South for health, invited me to accompany them, and though my master [i.e., Smith] offered a partnership interest, and did actually print my name on some of his maps, the temptation [to travel] was much too strong to be resisted.[22]

Wistar subsequently worked his way to the Pacific coast, where he followed various pursuits, including those of Indian fighter and lawyer. He returned to the East to participate in the Civil War, in which he was wounded on several occasions and attained the rank of brigadier general.

Several maps were apparently in press when Wistar left Smith in November 1848 and were published in 1849 with the Smith & Wistar imprint. One of these was a *Map of Burlington County, New Jersey*, which was made "mostly from original surveys by J. W. Otley and R. Whiteford." John Jay Smith was born in Burlington

Fig. 21–5. Portion of Sidney's large 1849 map of Philadelphia. Near the center of this section of the map is the Library Company of Philadelphia, of which John Jay Smith was librarian. To the northwest of the library is Minor Street, site of Robert Smith's map publishing establishment.

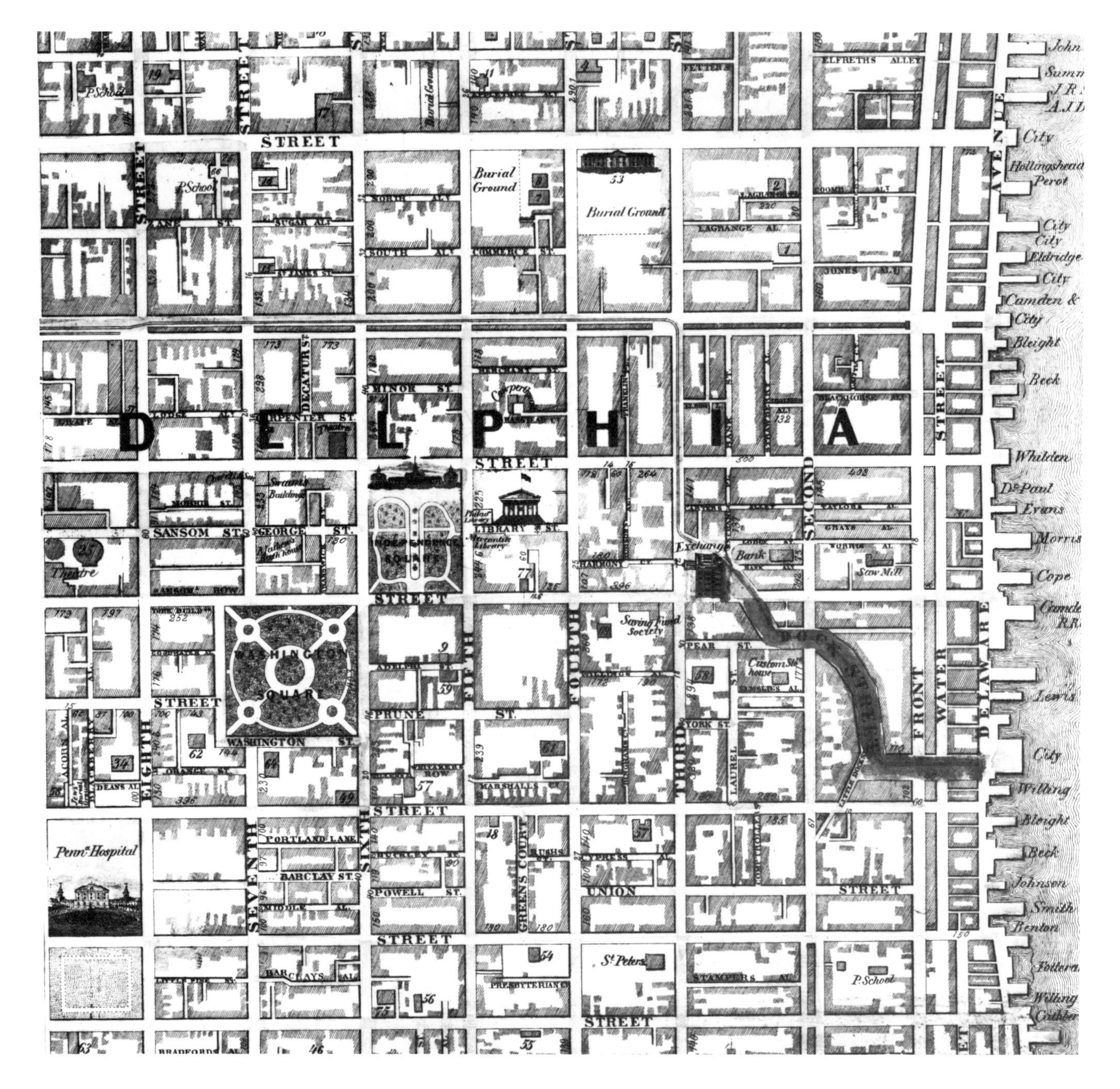
D E L P H I A
STREET
Burial Ground
Burial Ground
INDEPENDENCE SQUARE
WASHINGTON SQUARE
Penna. Hospital
Exchange
Saving Fund Society
Custom House
SECOND
THIRD
FOURTH
FIFTH
SIXTH
SEVENTH
EIGHTH
FRONT
WATER
DELAWARE
AVENUE
UNION
St. Peters
PRESBYTERIAN
P. School
Theatre
Saw Mill
Bank

County, and the Smiths had many relatives and friends there.

Another 1849 Smith & Wistar release was the *Map of the City of Philadelphia Together With All the Surrounding Districts Including Camden, New Jersey From Official Records, Plans of the District Surveyors and Original Surveys by J. C. Sidney* (Fig. 21–5). This is a magnificent and detailed representation of Philadelphia in the mid-nineteenth century. The plan is 173 by 173 cm. in size and is enclosed within an ornate border. The scale is 1:5,500. Illustrations of a number of important buildings are inserted at their appropriate locations on the map. Some eighty additional landmarks are referenced by number to a marginal list. The plan, which was printed at Duval's establishment, appears to have been reproduced by chromolithography. District boundaries have, however, been hand colored. In the lower left corner of the plan is the dedicatory note, "To the Inhabitants of the Region Here Delineated This Map is Respectfully Dedicated by their Obedient Servant, J. C. Sidney, Civil Engineer & Surveyor." Sidney, the former "sub" at the Library Company, displays a craftsmanship and artistry on this plan far superior to that on the Holme and Reed facsimile maps. In the intervening years, Sidney had greatly perfected his technique.

Maps of East and West Goshen in Chester County, Pennsylvania, Montgomery County, Pennsylvania, New Castle County, Delaware, and Salem and Gloucester counties, New Jersey (Fig. 21–6), also carry 1849 Smith & Wistar imprints. All these places are within thirty miles of Philadelphia, indicating that Smith's map publishing was still primarily of local significance. One exception was Walling's and Cushing's *Map of the City of Providence, Rhode Island*, which notes that the map was "engraved at S & W, 15 Minor St., Philadelphia." As Smith & Wistar did not have their own lithographic facilities at this time, the lithographic work was probably contracted to one of the regular lithographic establishments. Among the surveyors whose names appear on the Smith & Wistar maps are James Keily, William E. Morris, Samuel M. Rea, and Alexander C. Stansbie.

The Smith & Wistar credit does not appear on maps published after 1850. Some carry the imprint "Published by R. P. Smith No. 15 Minor St. Philadelphia," while others have only Smith's notice of copyright registration. Except for one or two Walling maps of New England localities (such as the map of Dighton, Massachusetts), the 1850 Smith publications are within a radius of two hundred miles from Philadelphia. They include the city and county of Baltimore, Maryland, Bucks County, Pennsylvania, Dutchess County, New York, Richmond County on Staten Island, New York, and Essex County, New Jersey. Several include the names of local publishers, an unquestionable aid in marketing the maps. Duval probably printed most of the maps, although his firm name appears on only a few. Friend is credited as the engraver on several maps.

The maps by this date were more finished in appearance, reflecting the improvements made in lithographic printing. Illustrations of scenic sights and public or private buildings decorate the margins of most of them. Many were issued in both colored and uncolored editions. The colored editions were often backed with cloth and mounted on rods, with varnish applied to the face of the map. Smith still relied principally upon the surveyors who had prepared his earlier maps, among them Sidney, Morris, and Keily. In several instances they collaborated with local surveyors in compiling maps. By this time Sidney was surveying and mapmaking full time.

An interesting 1851 Smith production is the *Map of the Vicinity of Philadelphia from Actual Surveys* (Fig. 21–7). It appears to be a revision of Sidney's 1847 map of ten miles around Philadelphia. In contrast with the earlier map, this one is rectangular rather than circular. It gives no surveyor, lithographer, printer, or publisher credits. On the verso of one Library of Congress copy is the manuscript notation "Philadelphia R. P. Smith Nov.

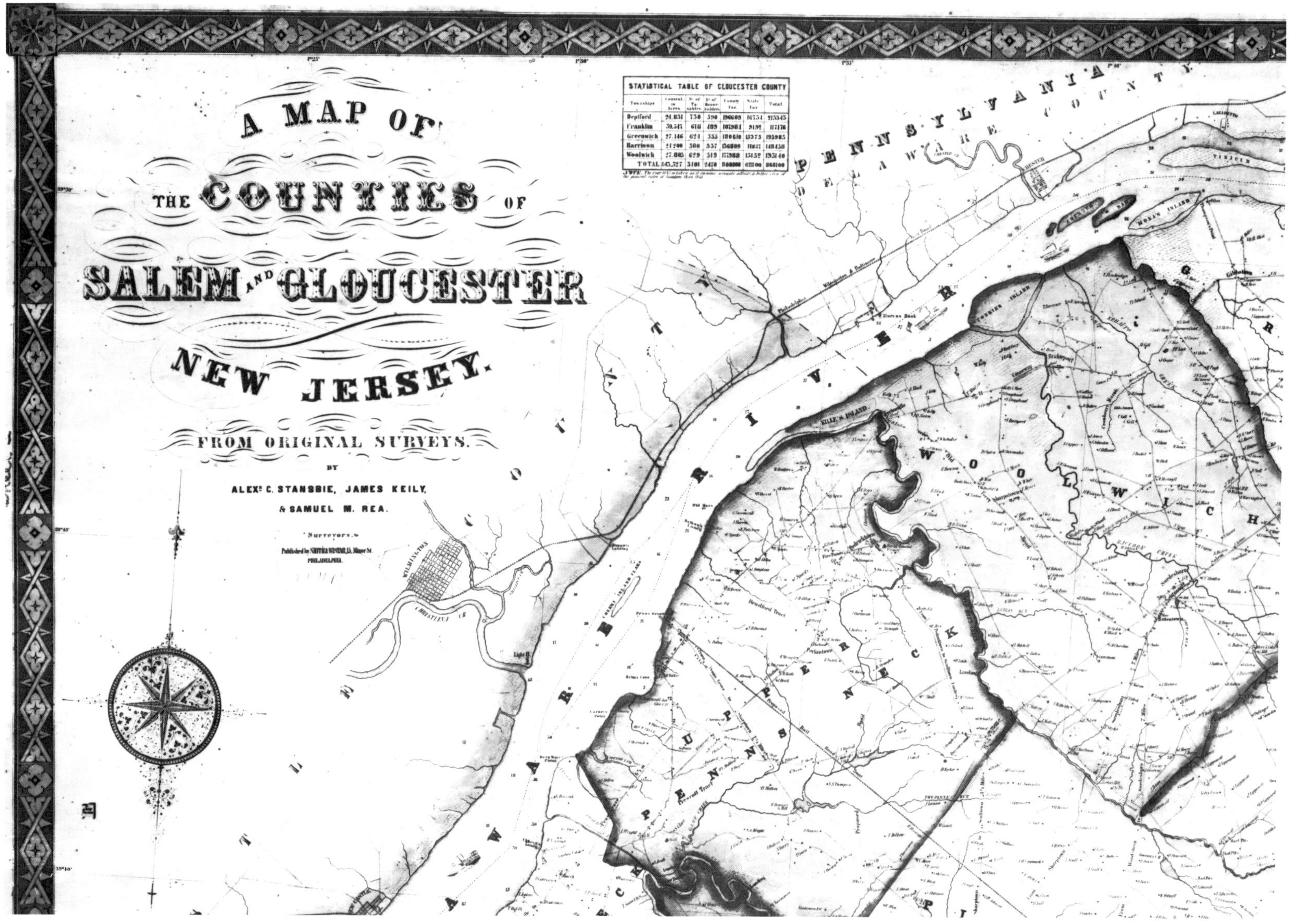

Fig. 21–6. Segment of Smith & Wistar's 1849 map of Salem and Gloucester counties, New Jersey.

'51." Maps of West Chester County, New York, and Monmouth, New Jersey, are the only other 1851 maps that can be positively assigned to Smith. There are also few Smith maps dated 1852 or 1853. Among those deposited for copyright in 1853 are maps of Dinwiddie, Henrico, and Loudoun counties, Virginia (Fig. 21–8). Keily surveyed for the maps of Dinwiddie and Henrico counties. For all three of the Virginia county maps Smith had co-publishers.

There are several possible explanations for Smith's decreased output of maps in these years. In 1852 Walling established his own map publishing company and no longer needed Smith as an agent for his printing and publishing connections. Also, on June 25, 1851, Smith

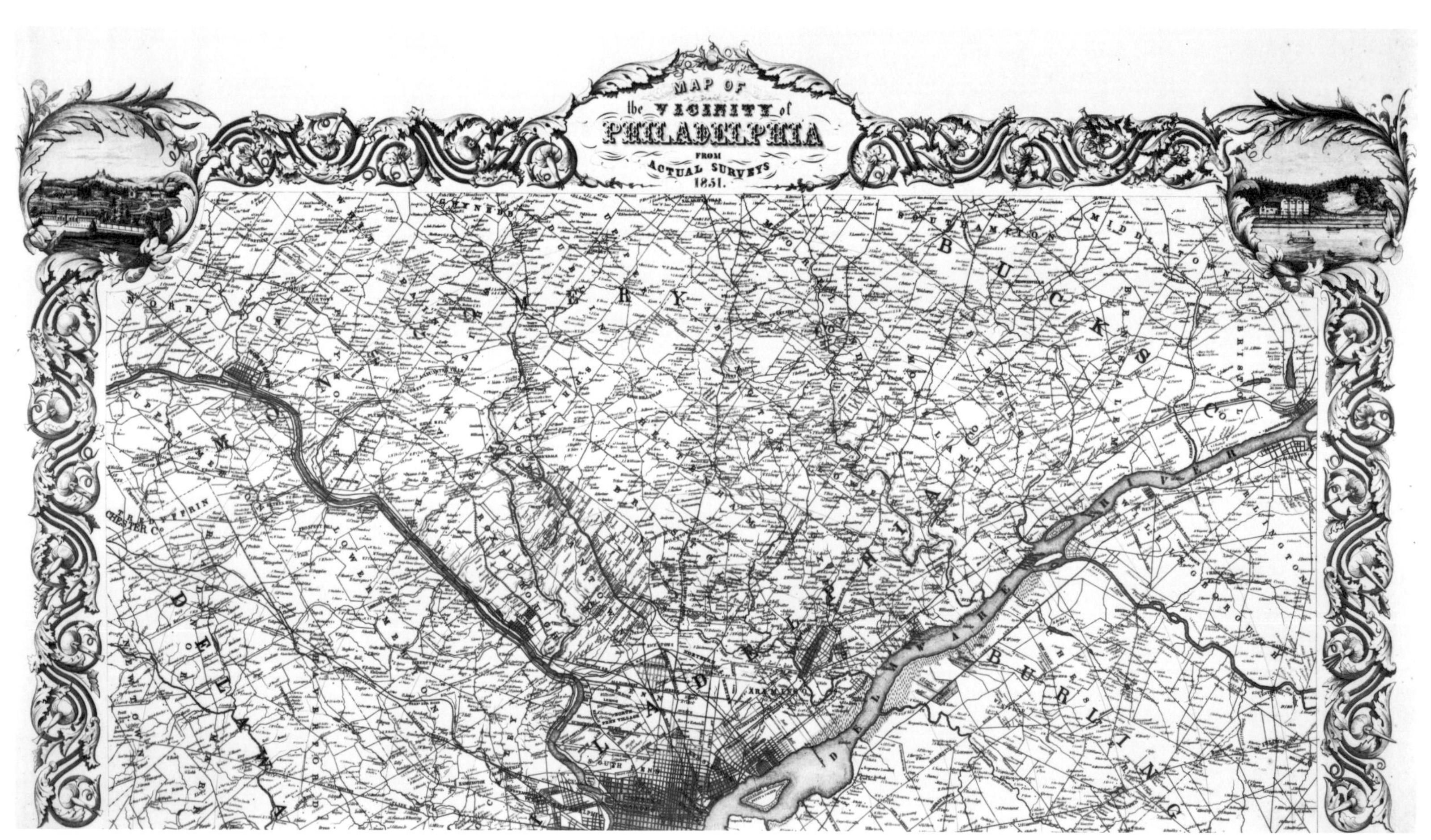

Fig. 21–7. The north half of Smith's large 1851 map of Philadelphia and its vicinity.

married Hannah Whitall in the Philadelphia Quaker Meeting House. During these years, Smith was also involved in a book publishing venture with his father that limited the amount of time he could devote to map production. The project was the American edition of François Andre Michaux's *North American Sylva, or Description of the Forest Trees of the United States, Canada and Nova Scotia*. Smith published it in three volumes dated 1852, 1853, and 1855. Even more contributory to his reduced output, though, was his involvement in the ambitious project to supply state and county maps to each school district in New York State and to compile a gazetteer of the state. This consumed a considerable amount of his time and energy between 1851 and 1853 and required numerous contacts with legislators and private citizens. This project is described in detail in the following chapter.

Notes

1. John Jay Smith, *A Summer's Jaunt across the Water* (Philadelphia, 1846), 16.
2. Haverford College Alumni Association, *Matriculates of Haverford College, 1833–1900* (Philadelphia, 1900), 31.
3. Smith, *A Summer's Jaunt*, 131.
4. Carl Halbmeier, *The History of Lithography* (New York, 1926), 114.
5. Walter W. Ristow, "The Anastatic Process in Map Reproduction," *Cartographic Journal* (June 1972), 37–42.
6. "Anastatic Printing," *Art-Union, Monthly Journal of the Fine Arts and the Arts Decorative Ornamental* 7 (June 1845): 152.
7. "Discoveries in Printing," *Chambers's Edinburgh Journal*, n.s., 3 (Mar. 1, 1845): 138.
8. John Jay Smith Manuscripts, vol. 14: 61, Historical Society of Pennsylvania, Philadelphia.
9. Smith, *A Summer's Jaunt*, 130–31.

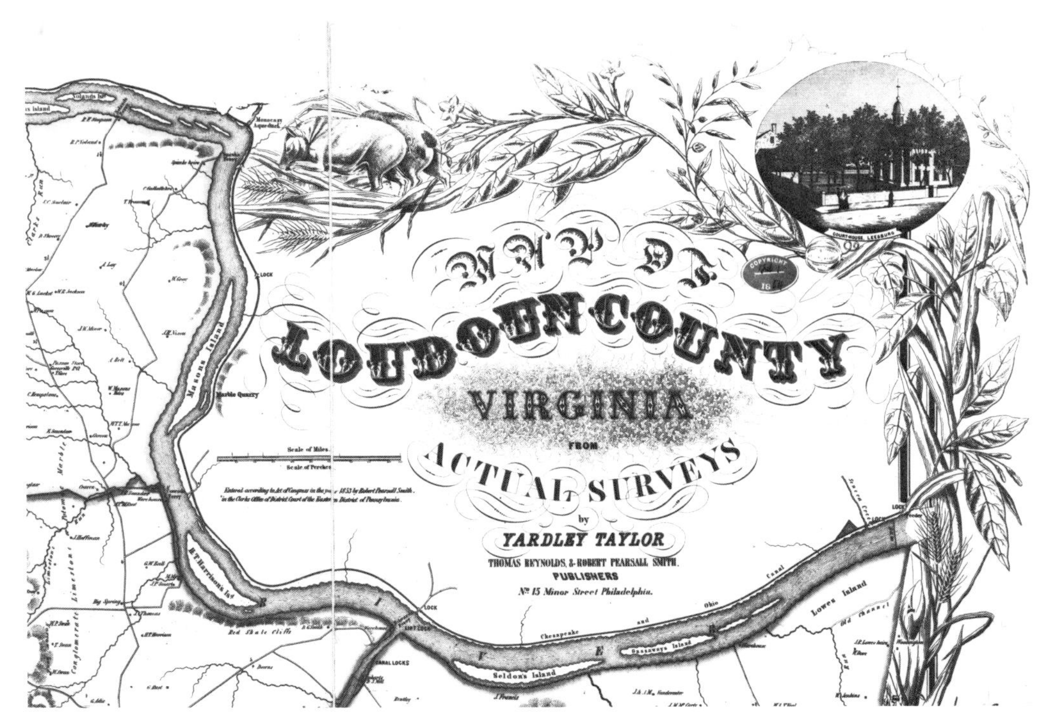

Fig. 21–8. Title cartouche of Smith's map of Loudoun County, Virginia. Yardley Taylor was a local surveyor.

10. Ibid., 131–32.
11. Ibid.
12. John Jay Smith, *Recollections*, ed. Elizabeth Pearsall Smith (Philadelphia, 1892), 224.
13. Ibid.
14. Society for Promoting Christian Knowledge, *The History of Printing* (London, 1862), 214.
15. William Smith Williams, "On Lithography," *Transactions of the Society of the Arts* (London), pt. 2 (May 1849): 226–50.
16. Nicholas Wainwright, *Philadelphia in the Romantic Age of Lithography* (Philadelphia, 1958), 73.
17. Smith, *Recollections*, 224–25.
18. Ibid., 95.
19. Logan Pearsall Smith, *Unforgotten Years* (London, 1938), 29.
20. Franklin Institute, Committee on Exhibitions, Report No. 1217, "Transferred Engravings," *Journal of the Franklin Institute* 54 (1847): 379.
21. Philadelphia Contributorship minutes of Nov. 1, 1848, as quoted in Philadelphia Contributorship, *229th Annual Report 1981* (Philadelphia, 1982), 24–25.
22. Isaac Jones Wistar, *Autobiography* (Philadelphia, 1937), 38–39.

22. *The French-Smith Map and Gazetteer of New York State*

Various events in the 1850s, such as a growing interest in accurate state maps, induced Robert Pearsall Smith to widen his horizons. By the middle of the nineteenth century, scientists and the more technically advanced surveyors were concerned with the need for geodetic and triangulation surveys in the United States. With few exceptions state, county, and town surveys previous to this period were based upon crude compass and chain, or odometer, surveys. The youthful American Association for the Advancement of Science (AAAS) concerned itself with remedying this problem. At the association's sixth annual meeting, which convened in Albany, New York, in August 1851, the membership approved a resolution presented by Lieutenant Edward B. Hunt "that the President of the Association be requested to appoint a committee of seven members, to prepare a memorial in the name of the Association, to be addressed to the Governor and Legislature of New-York, urging the speedy commencement of a geographical survey of that State and presenting a matured project of that kind of survey deemed most desirable, with a careful estimate of its cost."[1]

Hunt was a native of Portage, New York, and graduated from the U.S. Military Academy in 1841. As second-ranking member of his class, he had received a commission in the prestigious Army Corps of Engineers. After this commission, he returned to West Point in 1846 to spend the next three years as an assistant professor in the Department of Engineering. From 1851 to 1858 Hunt was assigned to the U.S. Coast and Geodetic Survey in Washington. While with the survey, Hunt became familiar with scientific surveying, mapping, and charting and with the latest technical developments in cartography. He was especially impressed with the triangulation and topographic surveys that had been completed for a number of European countries during the first half of the nineteenth century. It became his firm belief that such surveys should be carried out in the United States, and this conviction moved him to present his resolution at the AAAS meeting.

In his proposal Hunt emphasized that "geography is no longer content with the vague information of first reconnaissance, but demands the aid of accurate methods and the most perfect instruments." He felt the time was ripe for New York "to undertake an accurate geographical survey of her whole territory." To ensure that there was no question concerning the type of scientific survey he had in mind, Hunt specified that "a base be measured in Western New-York, and made the starting line for a system of primary, secondary and tertiary triangulation, extending towards Pennsylvania and New-England. A connection will be obtained in the Hudson valley with the Coast Survey triangulation. . . . In point of accuracy and style, the work should not fall essentially below that of the Coast Survey, and might perhaps well be assimilated to the operations for a single section of the coast." He concluded that "circumstances seem particularly to point to [New York] as the State most needing such a survey, and best able to undertake it."[2]

Not surprisingly, the president of the AAAS appointed Hunt to the seven-member resolution committee. His committee associates were William M. Gillespie, O. M. Mitchell, Charles Huckley, Elias Loomis, Samuel Ruggles, and Alexander D. Bache. The memorial prepared by the committee was submitted to New York's governor, Washington Hunt (Edward Hunt's older brother), on February 17, 1852. The proposal stressed that

> a trigonometrical survey of the State of New-York is of unusual interest, from the great extent of its territory and the varied character of its surface. This extent is such that with the present sparseness of population a correct map cannot be expected for many years to come unless through State enterprise, presenting a strong reason for immediate action by the Legislature. The diversified character of this region not only affords the finest possible field for a topographical survey, but suggests the idea that the vast unemployed resources of the State can never be fully known and developed until a perfect map of the surface has been given to the world.[3]

The memorial further emphasized to the governor and legislature that an accurate topographical survey would benefit education in the state. Precise maps, used in the "common and higher schools," would "raise the standard of geographical instruction and give a strong impulse to the science of physical geography, which is about to become an important branch of elementary instruction."[4] Education at this time was a major topic of concern because the educational system was being reorganized and expanded. The federal supervision of schools was transferred from the secretary of state to the superintendent of public instruction of the newly created Department of Public Instruction. Financial aid to the states for education was also greatly increased.

Governor Hunt transmitted the memorial to the state legislature on February 18, 1852. In an accompanying memorandum, the governor stressed that "an early commencement and judicious prosecution of the proposed undertaking would redound to the honor of the State, and promote its highest interests."[5] Despite Hunt's strong support, the proposal received no action by the New York State legislature. At the seventh annual meeting of the AAAS held in Cleveland in 1853, the resolution committee reiterated the proposal's history and reported that "at the recent regular session of the Legislature of New-York Governor Seymour [Hunt's successor] recommended in his message a similar survey; but neither at this session, nor at the extra one held since, has any action, it is believed, been taken on the matter."[6] The committee, having carried out its mission, albeit with no success, was respectfully discharged.

There is no clear indication why the legislature failed to consider the proposal. The times were turbulent, and the legislators perhaps believed that other matters had higher priorities in their deliberations. A proposal to enlarge the Erie Canal was before them, for instance. In 1851 Governor Hunt was able to push through an appropriation of nine million dollars to improve the canal, but the act was declared unconstitutional. It was not until 1854 that a state constitutional amendment made canal improvements possible. As a one-term governor, Hunt was limited in opportunities to have the memorial acted upon favorably. His successor, Horatio Seymour, also supported the map proposal, but the odds against it were too great. The high cost of conducting scientific triangulation surveys was one deterrent. Moreover, the number of American engineers and surveyors who were technically competent to carry out such surveys was, at the time, extremely limited. Perhaps the major reason for the legislature's neglect, however, was the prevalent belief in mid-nineteenth-century America that mapmaking and map publishing could be adequately handled by private individuals and commercial publishers.

With no official support forthcoming, various concerned individuals and private societies did begin to promote and prepare county and town maps. One of the first such endeavors was a survey of Washington County undertaken by Asa Fitch, a practicing physician. It was authorized by the New York State Agricultural Society in 1847 for a report published in 1849 and 1850 in the society's *Transactions*.[7] The second part of the report, published in 1850, includes a fold-out map (36 by 13 cm.) of Washington County. Fitch had used the map of the county in David Burr's 1829 *Atlas of the State of New York* as a base but made "corrections and additions without number in the minor details." The map was engraved and printed by John E. Gavit. Fitch's survey and map were one-time efforts, and he made no further contribution to the geography or cartography of the state.

This was not true of John Delafield, a longtime proponent of mapmaking endeavors, who compiled the 1850 article "A General View and Agricultural Survey of the County of Seneca."[8] His survey included a fold-out map of the county by William T. Gibson that carried names of landowners. A larger and more detailed edition of the map was published separately in 1852 (Fig. 22–1). Delafield was president of the New York State Agricultural Society in 1851 and president of the Seneca County Agricultural Society from 1846 to 1853. He was an enthusiastic and effective promoter of agriculture in New York State. Born in New York City in 1786, three years after his parents emigrated from England, Delafield graduated from Columbia College in 1805, spent several years in shipping, and then engaged in banking in London. Because of his American citizenship, he was technically a prisoner during the War of 1812, but he remained in London until 1819, when he suffered financial reverses. Washington Irving wrote about Delafield's and his wife's misfortune in his story "The Wife," which is in the *Sketchbook of Geoffrey Crayon, Gentleman*.

In 1820 the Delafields returned to New York, where Delafield again found employment in banking and ultimately became president of a bank. When the panic of 1837 financially ruined him once more, he moved to Seneca County and took up farming. Within a short period Delafield's farm was one of the models of the state, and he became a leader in state and county agricultural societies. He was a knowledgeable and respected citizen who was immeasurably helpful in promoting plans for producing a map of New York State.

Several surveyors had prepared maps of New York counties prior to Robert Pearsall Smith's interest in that state. Among these was Smith's early associate James C. Sidney, who published maps of Richmond and Dutchess counties in 1850 and of Orange County in 1851 (Fig. 22–2). Similarly, P. J. Browne published a map of Monroe County, New York, in 1852 (Fig. 22–3) as well as one of Cortland County in 1855.

By 1852 Smith had become heavily involved in publishing county and city maps, including several counties in southern New York State. The AAAS proposal and its failure with the New York legislature did not escape his attention. He saw in the legislature's inaction an opportunity for private enterprise: a state map could be prepared from the county surveys already being made by a number of local surveyors. With this objective in mind, Smith made contact with Delafield and with some of the more active county mappers. Smith may have met Delafield through his father, John Jay Smith, who edited a professional journal specializing in agriculture. He may also have been referred to Delafield by some members of the New York State legislature he had contacted about his proposed plans. At any rate, before the end of 1852 Smith and Delafield, in affiliation with the New York City printing firm of Collins, Bowne & Company, had drafted a proposal "to encourage a survey of the State [of New York] by supplying the common schools with county and State maps." The proposal was transmitted to the legislature for consideration at its seventy-sixth session. It requested an act requiring each school district in the state to purchase a copy of a map of the county in which it was situated, as well as the state map that was to be compiled from the county surveys and published by Smith and his associates.

Before the legislature's session convened, the two sponsors had solicited support for their proposal from influential local administrative and political leaders, agricultural societies, citizens' groups, boards of education, school superintendents, teachers, and other prominent individuals. Smith had also by the beginning of 1853 negotiated agreements with a number of local map

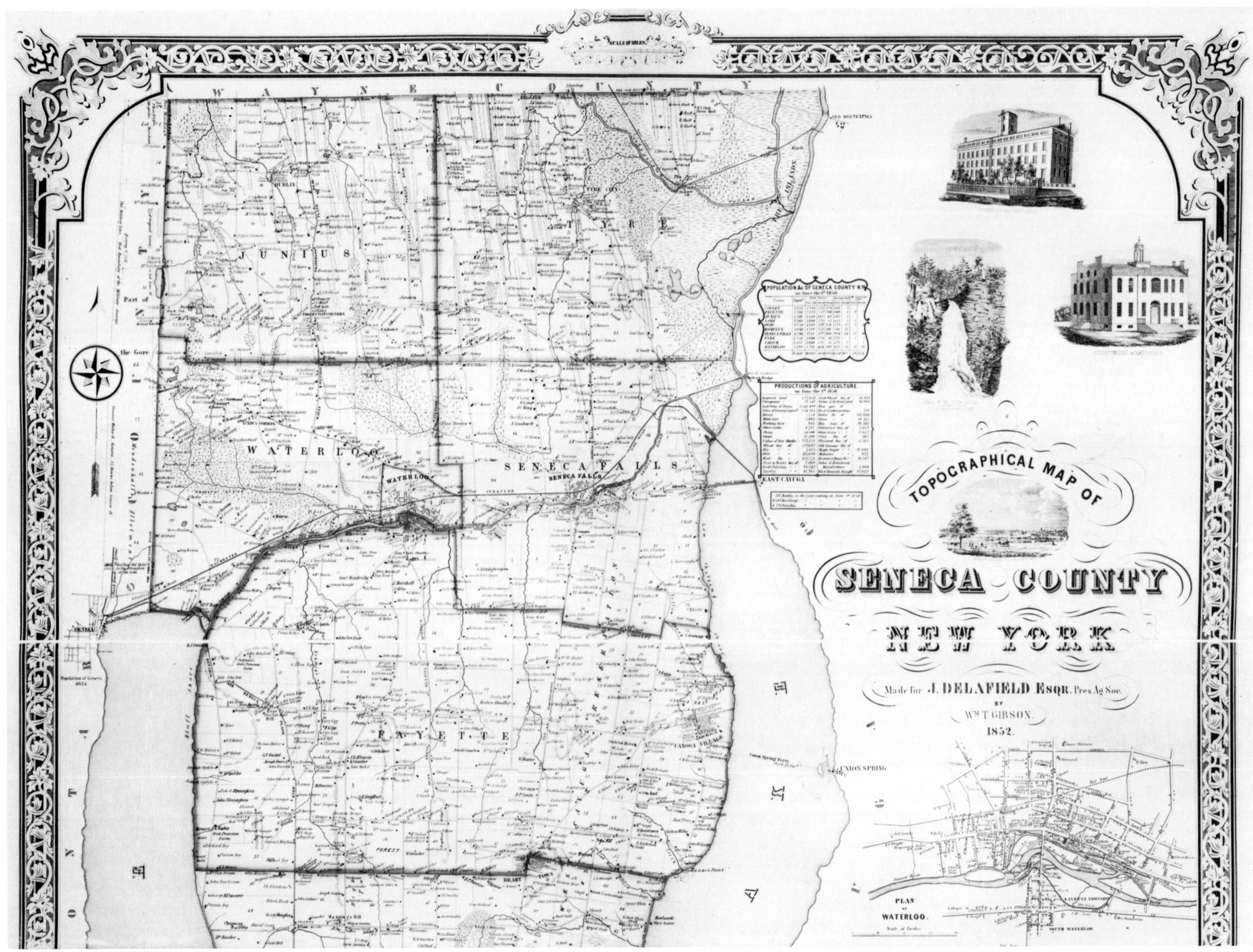

Fig. 22–1. The northern half of William T. Gibson's 1852 map of Seneca County, New York.

men. His plan of operation is evident from a letter dated January 4, 1853, and addressed to Oliver P. Tillson and Peter Henry Brink, two county surveyors. From Albany, where he was temporarily staying, Smith wrote:

> As arranged in our interview today I hereby agree to furnish an engraving [i.e., lithographic reproduction] of your new map of Ulster County New York equal in style to my map of Orange County without charge or cost to you. In repayment you give me the full use of the said map of Ulster County for use and reduction for a State map of New York. You also give me the privilege of taking as many impressions as may be necessary for supplying schools by a contract with the State of New York—which use for a State map and copies for schools are to be without charge to me, the engraving furnished being full compensation therefor. The engraving is to include four views, reductions of plans of villages, for the margins & any other matter you may wish to insert. The copyright is to be taken out and owned by me you having the privilege of making any town maps from these surveys. I am privileged to give away a few copies but am not to sell any copies except for schools.[9]

Smith provided the local surveyors with the appropriate transfer ink to use in drawing county maps. The manuscript maps could then be easily transferred from paper to stone or zinc lithographic plates for printing. Most of the lithographers and printers patronized by Smith were in Philadelphia, among them N. Friend, Peter S. Duval, Frederick Bourquin, and Gustavus Kramm.

The Smith-Delafield proposal was officially received by the New York State Assembly on January 13, 1853,

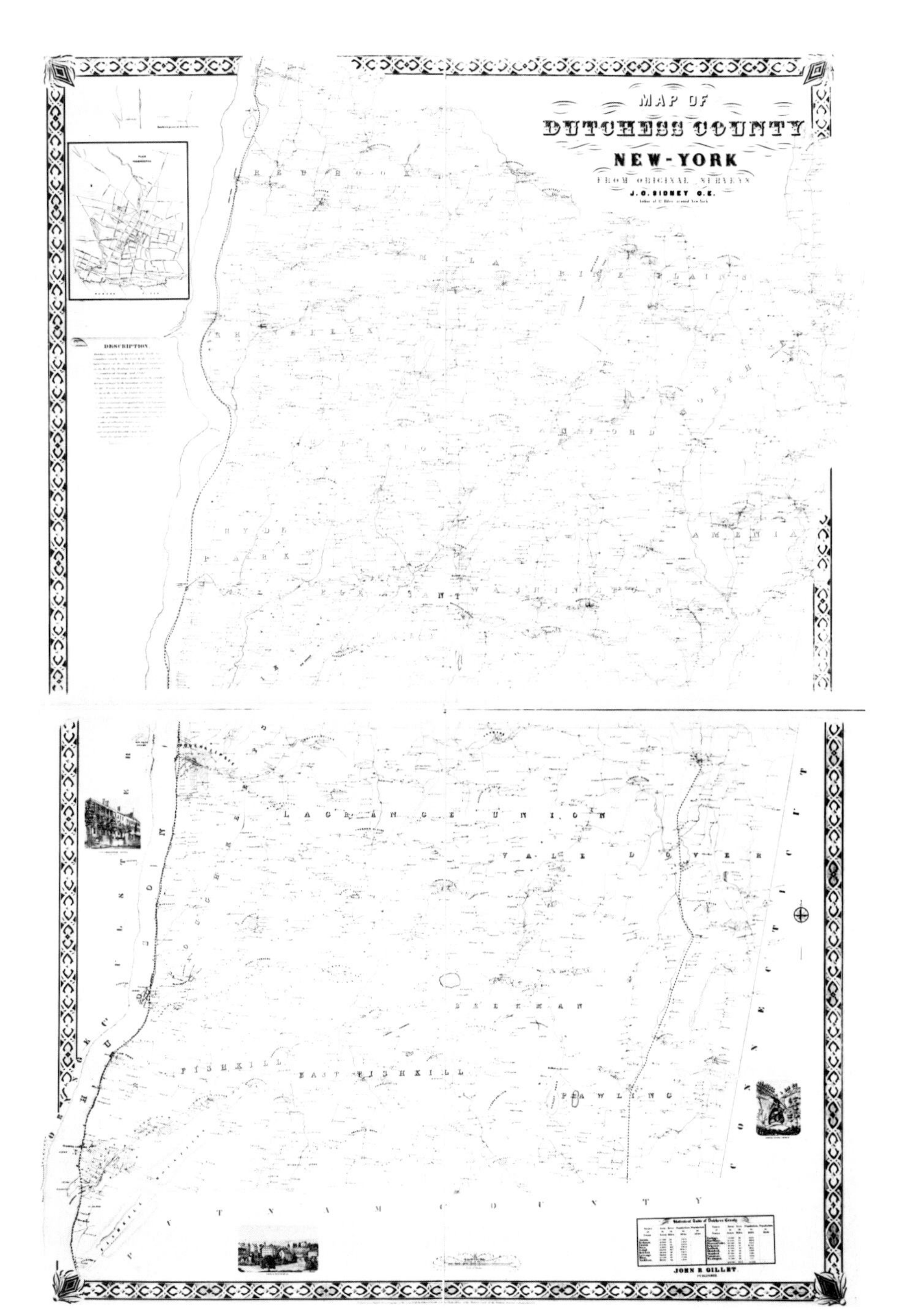

Fig. 22–2. This 1850 map of Dutchess County was one of the earliest of a New York county printed by lithographic procedures. Although Sidney, who prepared the map, was associated with Robert Smith, the latter did not publish the map.

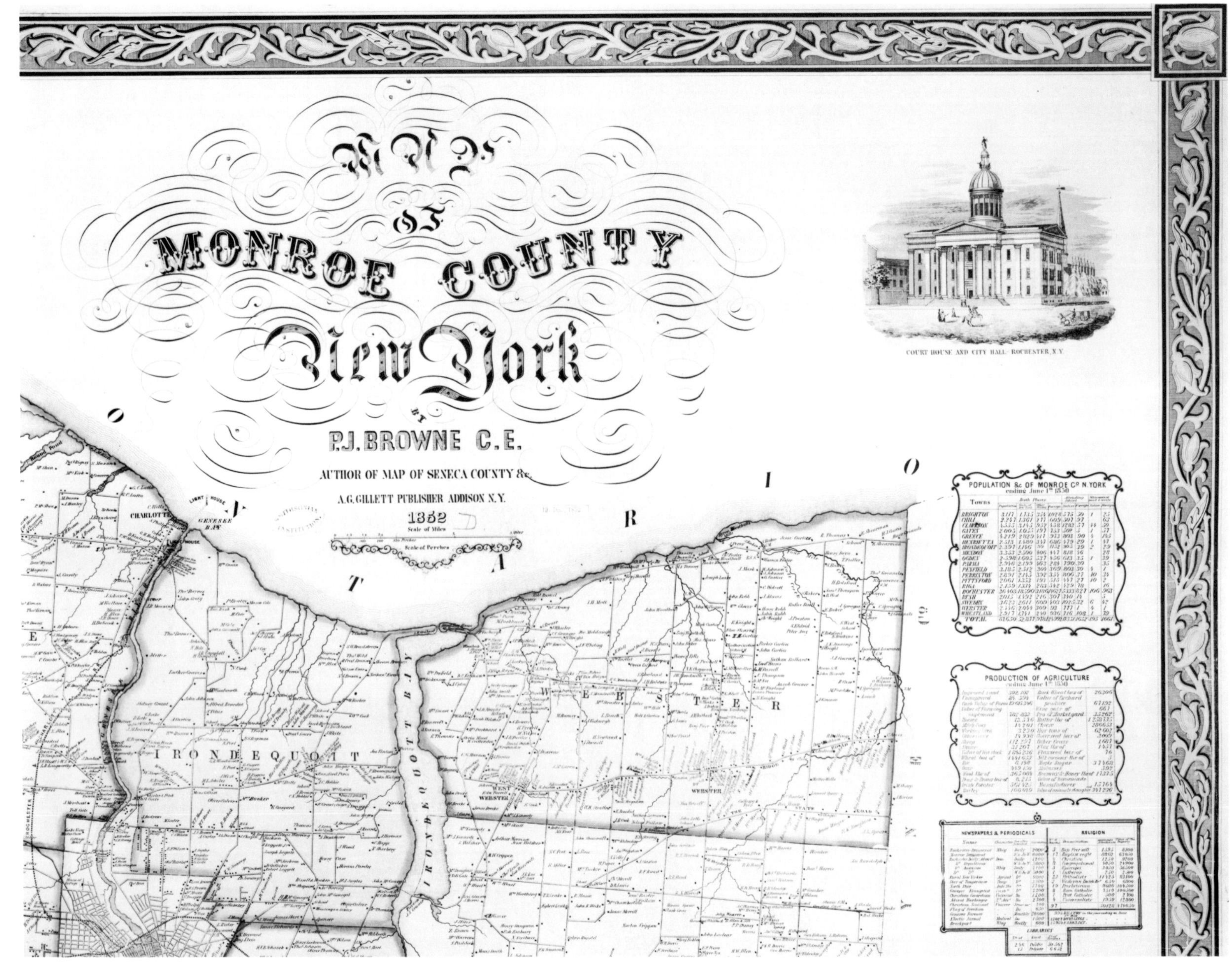

Fig. 22–3. Cartouche and portion of P. J. Browne's 1852 map of Monroe County, New York, which was published prior to Smith's interest in the mapping of New York State.

when Sperling G. Hadley "presented the petition of John Delafield and others, praying for a law relative to the surveying of the several counties of the State, &c., which was read and referred to the committee on colleges, academies and common schools."[10] Between this time and March 31, 1853, more than twenty-five supporting petitions were submitted to the assembly. The petitions came from county supervisors, sheriffs, town officials, superintendents of schools, teachers, citizens, boards of education, common councils of towns and cities, and agricultural societies. At least twelve counties and two towns and cities were represented in the petitions. Their similarity in wording suggests that the petitions were solicited, probably by Smith, Delafield, or their agents. Three examples follow.

> Mr. Hastings presented the petition of supervisors and others of Monroe Co., praying for a law in favor of the survey of the counties and the State of New-York, which was read and referred to the committee on agriculture.
>
> Mr. Temple presented the memorial of Madison County agricultural society, praying for the proposed distribution of county and State maps in district schools, which was read and referred to the committee on colleges, academies and common schools.
>
> Mr. Streeter presented the petition of 88 names of Wayne Co., praying for aid to complete a survey of the topographical features of the several counties of the State, and the distribution of the same in every school district in the State, which was read and referred to the committee on agriculture.[11]

The use of the word "topographical" in the last petition is of interest and raises the question of whether the petitioners referred to the type of survey proposed by Edward Hunt and the AAAS committee in 1851. This is probably not the case, though, for in the mid-nineteenth century "topography" was often used synonymously with "description" when referring to land areas.

As indicated in these petitions, the Smith-Delafield proposal was referred to the assembly's Committees on Colleges, Academies, and Common Schools and on Agriculture. It was principally considered by the former committee, although the reports of both were published in the assembly's *Documents*. The report of the agriculture committee is valuable bcause it quotes much of the Smith-Delafield petition. It notes that

> the memorial of John Delafield, Collins Bowne & Co., and Robert P. Smith, respectfully sheweth, that an examination and survey of the county of Seneca, in the State of New-York, was made in the year 1850, at the request and under the direction of the New-York State Agricultural Society. . . . In the progress of this survey it was discovered that no boundary lines of any county in the State were delineated and recorded in any offices of the State . . . that, with the exception of the city of New-York, no astronomical observations had been made to establish the longitude of any place within this State, that the boundaries of counties described in the Revised Statutes are indefinite, devoid of mathematical indications, and do not afford any data for a proper delineation; that the longitudes of places mentioned in some of the public documents of this State are evidently incorrect. . . . [The] memorialists further show, that they have been engaged in the survey of several counties in this State, and have accomplished the survey of an area equal to about one-third of the whole State; they have in their employ a large corps of civil engineers, draughtsmen and engravers, with every appliance needful for an early completion of a survey of the whole area of the State. . . . Your memorialists are prepared and willing to establish an observatory; to cause careful and repeated stellar observations to be made for the determination of longitude of places within the State of New-York; and to engrave not only a perfect series of county maps, embracing the

> mathematical and physical geography of each county, but also a map of the whole State, from surveys, which shall delineate all the public roads, as well as other topographical and useful information.[12]

The agriculture committee report also states that the petitioners asked for "such encouragement as may be afforded by the State . . . in such form only as shall at once give to the people a full, valuable and sufficient consideration." In return for furnishing each school district with a map of its own county and a map of the entire state (all maps being subject to the approval of resident surveyors or others who could attest to their accuracy), the petitioners asked to be reimbursed by the state for the "expense incurred in the survey, preparation and completion of the above-named maps; the cost of which in the aggregate, shall not exceed the sum of four dollars for each county map delivered, and five dollars for each State map delivered." The report concluded with the statement that the committee was "disposed to concur in the general views of the petitioners, and do recommend that copies of the maps of counties be placed in the several school houses, as also a map of the State; and that the patronage of the State be extended to the memorialists as prayed for."[13]

The report of the Committee on Colleges, Academies, and Common Schools, after relating some of the problems that result from the lack of accurate surveys, observed that "no one, we think, can doubt the importance and necessity of an intimate knowledge of the topography of our State, and, as a consequence, the necessity for accurate delineations of every county and town. . . . As an object of educational interest, the proposed survey of the State . . . [is] to exhibit, also, the formations and arrangements of its surface, constituting its geological features. . . . It is intended to add, also, agricultural statistics." The committee noted in its report that "we regard it [the project] as a measure carried on by *private capital*, and all experience tends to prove that private enterprise possesses economical advantage over works executed at the public expense. . . . With a conviction that judicious economy, with a full remuneration value to the State, will be secured by extending the patronage prayed for, your committee respectfully reports the bill, presented herewith."[14]

Despite these favorable reports, there appears to have been no further action on the proposal until March 14, 1853, when Sylvanus S. Smith, chairman of the Committee on Colleges, Academies, and Common Schools, moved "that Assembly bill No. 116, being the bill to encourage the survey of the State by supplying the common schools with county and State maps . . . be taken from committee of the whole [including the entire assembly] and referred to a select committee to report complete."[15] Smith, Ashbell Patterson, chairman of the agriculture committee, and Assemblyman William Taylor were named to the assembly's select committee. Then, for reasons unknown, the Smith-Delafield proposal was caught up in parliamentary maneuvering. On the following day, "on motion by S. S. Smith, and by unanimous consent, [it was] resolved that the select committee to which was referred Assembly bill No. 116, in relation to State and county maps, . . . with power to report same complete, be discharged from . . . further consideration . . . and be recommitted to the committee of the whole, retaining [its] place in the general orders."[16]

More than two weeks passed before there was further consideration of the map bill. In the house, on March 31, "Mr. William Cary moved a suspension of the rules, and that the bill entitled 'An act to encourage the survey of the State by supplying the common schools with county and State maps,' be taken from the general orders and referred to a select committee, with power to report complete. Mr. Speaker put the question . . . to the said motion of Mr. Cary, and it was determined in the negative, a majority not voting in favor thereof, under joint rule."[17] The bill's supporters were not yet ready to abandon it, and on April 5 Patterson moved that the rules be suspended and the act referred to a select committee. This motion was also defeated. On April 12, Sylvanus Smith "from the [assembly's] select committee to which was referred the senate bill entitled, 'an act to furnish the State and county maps as proposed by Collins, Bowne & Co., and R. P. Smith' . . . reported that the committee had gone through said bill, made the amendments thereto, and saw no reason why the same, as amended, should not be passed into a law. [With unanimous consent it was] *Ordered* that said bill be read a third time."[18]

There was still strong opposition to the bill. Its proponents tried various parliamentary procedures to bring it to a vote in the assembly, but two months passed before the third reading on the bill was scheduled. On June 9, when it was finally being read, Assemblyman Case moved to recommit it to the committee of the whole, and, after some debate, this was done. The legislative session was drawing to a close, and the bill's opponents used such delaying tactics to forestall a vote before adjournment.

Following Case's motion, Patterson, one of the map bill's staunchest supporters, "moved to make said bill a special order for Monday next immediately after the reading of the journal."[19] When the bill was brought up on the appointed day, however, a quorum was not present. On the motion of Taylor the house recessed, and when it reconvened, the necessary number of members was present, and the house resolved itself into a committee of the whole. There was a day-long debate on the bill. By the end of the day no agreement had been reached, and Assemblyman Clapp, one of the bill's supporters, moved that the committee of the whole sit again to debate it. This motion failed to pass, and in a last desperate attempt to keep the bill alive, Taylor moved to refer the bill to the Committee on Agriculture. No vote was taken on this motion before the legislature adjourned for the day. There was no further consideration of the Smith-Delafield bill, and it expired with the summer adjournment of the assembly on July 21.

Since the official legislative journals provide no information on the stance of the bill's opposition, we can only conjecture on its reasons. One may have been that the state map Smith and his associates planned to publish was somewhat below the standard proposed in the AAAS memorial. Some sources have suggested that the bill failed because Delafield's sudden death on October 22, 1853, deprived the bill of its cosponsor and one of its most ardent supporters. While it is true that as president of the New York State Agricultural Society, Delafield was able to marshal statewide support for the bill and to solicit many of the petitions submitted to the legislature, he died more than four months after the assembly let the bill expire. The cost of providing a copy of the state map and a copy of the pertinent county map to each school district in the state may also have given pause to some legislators. There were in 1859 sixty counties in New York State and more than 11,600 school districts. The total cost to the state of purchasing the essential number of maps might well have been in excess of one hundred thousand dollars.

It is possible, too, that the Smith-Delafield bill was caught up in sectional politics. Its principal support apparently came from upstate legislators and from educators and agricultural societies in the central and northern counties. Assembly representatives from the southern, urban part of the state appear to have been its principal opponents. Their opposition may have been encouraged by New York City map lithographers and publishers, who were very likely concerned that such a lucrative contract was being awarded to a Philadelphia promoter and publisher.

Defeat of the bill and the death of Delafield notwithstanding, Smith did not abandon the project to compile a map of New York State. He had already completed negotiations with a number of local surveyors to prepare county maps, and other surveyors were placed under contract during the next several years. In 1855 he engaged John Homer French to superintend the compilation and editing of the state map and gazetteer, a comprehensive geographical dictionary with brief descriptions and historical summaries for New York counties, cities, towns, and villages (Fig. 22–4).

French was born in Batavia, Genesee County, New York, on July 27, 1824, and was educated in local public schools and at Cary Collegiate Institute in Genesee County and Clarence Academy in Erie County. He began teaching in district schools at the age of seventeen,

Fig. 22–4. John Homer French, ca. 1865.

and by the time he was twenty-one he had started revising Daniel Adams's *Arithmetic*, one of the standard textbooks of the period. During the next seven or eight years French held positions as teacher or principal in various public and private schools. Some biographical accounts state that he conducted surveys and made maps of several New York counties before working on the state map and gazetteer with Smith. There is no evidence of this, however, and his name does not appear on town maps until 1856 or on county maps until 1858. The county maps were apparently resurveys, prepared under his direction, of counties for which earlier surveys were unacceptable for compiling the state map.

French began work on the map and gazetteer in the summer of 1855 at Smith's newly established headquarters in Syracuse. His key staff members were Franklin B. Hough, James Johonnot, and Francis Mahler. Hough and Johonnot are listed in the *Gazetteer of the State of New York* as foremen of the statistical department, and Mahler served as foreman of the drafting department. Hough was born July 20, 1822, in Martinsburg, Lewis County, New York. He graduated in 1843 from Union College, afterward teaching briefly in an Ohio academy. He attended Cleveland Medical College and from 1848 to 1852 practiced medicine in Somerville, St. Lawrence County, New York. His interests in mineralogy, geology, natural resources, and history led him to undertake the activities that brought him to the attention of Smith and French. In 1851 he presented a lecture on the early history of St. Lawrence County, which was published in 1853 with a history of Franklin County. In 1854 he published a history of Jefferson County and in 1856 one of Lewis County. Hough was in 1855 appointed superintendent of the New York State census. Along with Johonnot he was probably responsible for most of the descriptive and statistical data in the gazetteer. Johonnot, born in 1823, was also an experienced teacher, geographer, and author of textbooks. It is likely that French, with his experience in education, was responsible for engaging both of these men for the project.

Mahler, who is referred to in the gazetteer as a lieutenant, was most likely born in Germany, where he was educated as a topographical engineer. He brought to the map and gazetteer project a decade of experience in mapping and drafting. He supervised a staff of three, including French's brother, Frank, in compiling and drafting the map of New York State and some of the county maps. Smith probably engaged him through some of his contacts in Philadelphia. The years from 1856 to 1858 were the most active ones in compiling the map and gazetteer, and Mahler and his small staff drew upon the county maps in constructing the large state map. At times this presented problems. French reported in the preface to the gazetteer that

> at the commencement of the re-survey, maps of fifty-one counties had been completed and published; six counties had been surveyed, but the maps were not yet published; and three counties were still unsurveyed. Of the fifty-one published maps, twelve were found to be so deficient in matters essential to the State map according to the plan fixed upon, as to render new surveys of these counties necessary. Surveyors were sent into the remaining counties, with copies of the published or manuscript maps in hand, with instructions to visit every town, to correct every error that should be found on the maps, to make additions of new roads, note changes in boundaries, and, in short, to return the maps properly revised and corrected for use in the preparation of the State Map. In many instances new surveys of parts of towns, town lines, roads, and streams were found necessary, and also countless changes in the location of boundary lines, roads, streams, and bodies of water, and in the representation of the topographical features of the country. The surveyors were instructed to obtain copies of manuscript and other local maps, as far as practicable, as these were generally found to contain metes and bounds, and, being plotted to large scales, were of value in laying down boundary lines. Draftsmen were also sent to the several private Land Offices in the State, and to Albany, and all maps of any value in the offices of the State Engineer and Surveyor and the Secretary of State were copied, to be laid under contribution in preparation of the State Map. The Superintendent of the United States Coast Survey and the Secretary of the Interior also furnished copies of all maps in their Departments pertaining to the State.

The difficulties outlined by French resulted largely because each county map was based on independent compass and odometer surveys, with no coordinated geodetic controls. The maps were also laid out on a variety of scales and were made by more than fifty individuals of varying backgrounds, training, experience, and standards. It is no surprise that some were unacceptable as compilation sources for the state map. It is possible to identify the counties that were surveyed or resurveyed under French's direction by their publication dates of 1857, 1858, or 1859. Counties requiring new surveys were the sparsely settled ones in the Adirondack Mountains. Resurveys were necessary for the counties that were inadequately mapped before 1853.

Most of the New York county maps were lithographed and printed in Philadelphia. A number of publishers' names appear on the imprints. In some instances publishers listed addresses in the county covered by their map. For a majority of the maps copyrighted by Smith, however, the publisher's address is given, before 1857, as either 15 Minor Street, 17 Minor Street, or 19 Minor Street, Philadelphia; after 1857 the publisher's address is given as either 517 Minor Street, 519 Minor Street, or 521 Minor Street. (In 1857 the system of numbering buildings was changed in Philadelphia, thus accounting for the different addresses.) Smith's various addresses were used by a number of different publishers. This was due to the fact that Smith acted as their agent and supplied them with this work.

Minor Street at that time ran east and west for one

block between Fifth and Sixth streets and was one short block south of, and parallel to, Market Street. Smith's offices were just a few doors away from Duval's lithographic printing shop at Fifth and Minor streets. This region housed a number of lithographers, printers, and publishers. On April 30, 1856, a fire broke out in the Jessup & Moore paper warehouse on Ninth Street, three blocks north of Minor Street. Before it was extinguished, the fire destroyed or damaged forty buildings in this area. Duval's shop suffered serious damage, although Smith's escaped it. Following the fire, Duval's foreman Frederick Bourquin established his own firm with the support of Smith. Smith turned over to Bourquin much of the lithographic business the New York project generated. Bourquin's name did not always appear on the maps, and Smith's was shown only as the copyright registrant. At this time, Smith seems to have been a silent partner of Frederick Bourquin & Company. By 1858, however, they are listed as co-partners in William H. Boyd's *Philadelphia City Business Directory*.

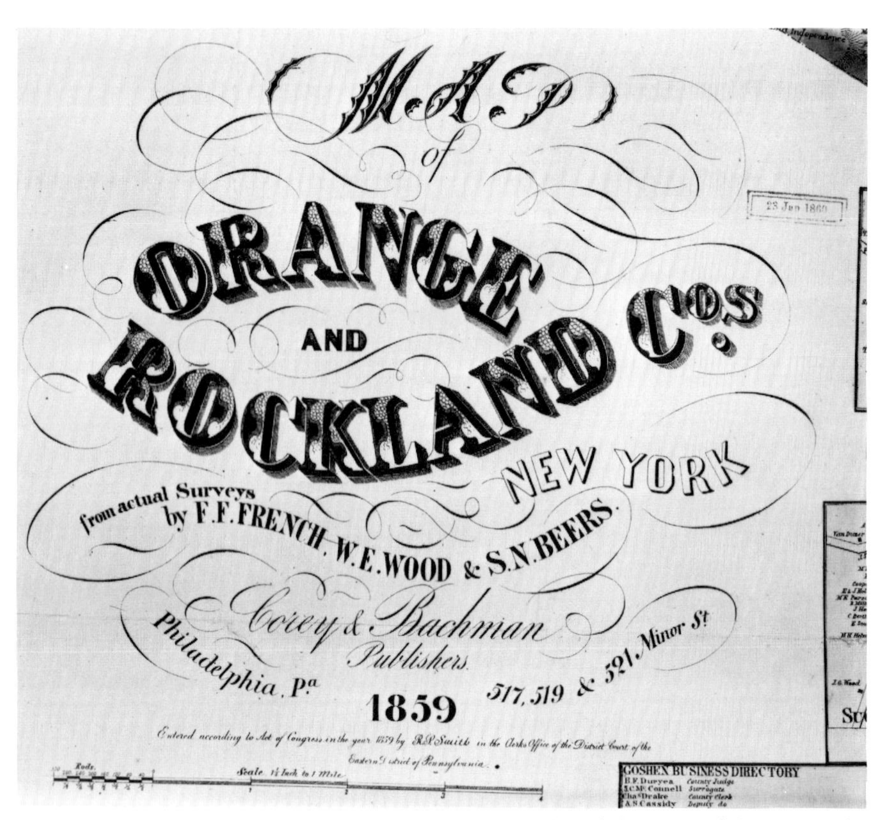

Fig. 22–5. Title cartouche of the map of Orange and Rockland counties, one of the last published for the New York State map project. The surveyor, F. F. French, was the brother of J. H. French.

Although Smith had negotiated contracts with local surveyors for maps of most New York counties, a small number of local surveyors apparently preferred to copyright and publish their own maps. After 1854, however, virtually all the county maps were published under the direction of French and Smith. Their maps were, in general, at larger scales and of higher quality than the ones independently produced. By the time the last maps were produced, they had become quite standardized in appearance, although there was no uniform scale or size. Scales varied from 1:63,360 to 1:40,000 (Fig. 22–5). The maps were all of wall-hanging dimensions, with the largest measuring 168 by 160 cm. They were generally mounted on cotton muslin, with rods tacked to the tops and bottoms.

The maps usually have decorative borders and ornate, elaborate title cartouches. Local administrative units are shown, often with their borders delineated in color. Names of landowners, roads, towns, villages, and such physical features as rivers, lakes, swamps, and prominent land forms (indicated by crude hachures) appear on the maps. Although a number include the word "topographic" in their titles, they are more properly classified as cadastral or land-ownership maps. Marginal space is filled with insets of towns and villages, statistical tables, and illustrations of historic sites, scenic features, public buildings, churches, and private dwellings. The maps could be ordered highly colored, with border tints only, or entirely without color. If requested, the face of the map would be varnished as a preservative.

While his competent staff moved ahead with the map and gazetteer compilation in Syracuse, Smith engaged in various promotional and political maneuvers. In 1856 there was again hope for possible state support from the legislature in its seventy-ninth session. Smith promptly followed this up, and at the January 28, 1856, meeting of the assembly, it was proposed "that each member of the committee on the erection and division of towns and counties be furnished with a map of the State for their use."[20] No action could be taken on this proposal, however, because there was as yet no up-to-date state map available. During the legislature's eightieth session in 1857, there was considerable discussion about a state map. One of the first proposals, introduced by Assemblyman Varnum on February 2, resolved

> that the Secretary of State be authorised and required to purchase, at a price not exceeding twenty-five dollars each, so many copies of the "New-York Atlas," formerly issued under the superintendence and direction of Simeon DeWitt (revised edition), in which the latest surveys and divisions of the State shall be delineated, with all the railroads and canals, so as to meet the approbation of the State Engineer and Surveyor, together with statistics of population according to the census of 1855, and the several acts of the Legislature of 1856 and 1857 affecting the same, as shall suffice to make distribution to the Governor and Lieutenant Governor of the State, and to the State libraries of the several States and Territories of the United States, to the Smithsonian institute, library of Congress and library of the Executive mansion at Washington, one each; also fifty copies of the same to be delivered to the Regents of the University for distribution as they may direct.[21]

Varnum's resolution was tabled following debate. This was a fitting action, for Varnum was not well informed on the state of mapmaking in New York. Simeon De Witt, New York State's surveyor general from 1784 to 1834, had published a map of the state but did not compile an atlas. Varnum was apparently referring to the *Atlas of the State of New York*, which was compiled by David H. Burr and published in 1829. A revised edition had been issued in 1838.

Several days later, Varnum's resolution was referred to the joint committee on the library, whose chairman was Delegate Abraham G. Thompson, a staunch supporter of Smith's proposal. More knowledgeable than Varnum about New York's map situation, Thompson introduced a substitute motion requesting that the state engineer and surveyor report on whether a New York statute, requiring that a state map indicating town and county boundaries fixed or altered by the legislature be retained in his office, had been complied with. If not, the state engineer was to see if information for such a map existed; whether an accurate state map could be obtained under the powers allowed in the existing laws; and determine what kind of legislation would be necessary to permit him to prepare an accurate atlas of all New York counties, with statistical information, and a revised map of the state (which he had already recommended doing in his annual report). Furthermore, the engineer should estimate the cost and make any other suggestions to achieve these ends.[22] This resolution was approved by the house, and on February 18, 1857, the assembly sent a request to Silas Seymour, the state engineer and surveyor, for the information.[23]

It is possible that Thompson, in substituting his resolution for Varnum's, hoped to call to the legislature's attention how lacking the state was in up-to-date maps and thus ensure a more favorable reception for Smith's proposal. On February 26, 1857, Seymour replied that there was no such map as described in the assembly's directive and that "the only approximation towards it is an old state map and atlas, by David H. Burr . . . published . . . in the year 1839." He further observed that "there is no authenticated record or map of the bounds of all the towns or counties . . . on file in this office from which an accurate map can be compiled. And the State Engineer has no knowledge of the existence of such information, except that some enterprising map publishers have recently published maps of the counties of the State, and it is believed that materials for the publication of maps of nearly all the counties have been, and are now being prepared by these publishers." He continued:

> It has occurred to the State Engineer that the information already obtained by these gentlemen may be made available to the county clerks, and to the State for a consideration which the State can consistently grant by securing to these parties the use of the copy right of the atlas and maps of the State for a certain number of years. . . .
>
> The State Engineer has no means of judging of the accuracy of the surveys and other information obtained by these parties, but there will be no difficulty in deciding, after a proper investigation, as to whether they are sufficiently accurate for the purposes contemplated.[24]

Seymour also outlined provisions for "An Act to provide for the completion of an accurate map and atlas of the State," which would require the supervisors of all towns and wards of the state to prepare accurate surveys of their jurisdictions and forward them to their respective county clerks. The county clerks were charged with compiling, from the local surveys, "a true and authentic map of said county, which county map shall be made in all respects to conform with directions furnished . . . by the State Engineer and Surveyor."[25] The county maps would then be forwarded to the state engineer and surveyor for use in compiling a state map.

These procedures closely paralleled those followed by Smith and French in their state map and gazetteer project. Moreover, one of the final paragraphs in Seymour's proposal provided that

> if any parties . . . shall have made the necessary surveys for the township and county maps with accuracy as to meet the requirements of this act, and shall furnish the results of such surveys to the supervisors and county clerks throughout the State without charge; the State Engineer and Surveyor may contract with such parties for the said publication and the enjoyment of the said copy rights on such equitable terms as will give said parties a fair compensation for their surveys and secure to the State a sufficient number of copies of the said atlas and map for use in all public offices and for general distribution and exchange.[26]

Smith, fully aware of the legislature's interest in maps of the counties and the state, wrote to Seymour, informing him that

> I have actual and accurate surveys of nearly all the counties composing the State, and can in a few weeks complete the whole number. They have been made at a cost of about five thousand dollars, and embrace an amount of topographical information that is invaluable to the citizens. I now propose to reduce these county maps to an uniform scale, and make a more complete map and Atlas of New-York, than is possessed by any State in America.
>
> To do this as thoroughly as should be done for New-York, accurate observations of latitude and longitude, and topography, are needed on a scale of expense far beyond the means of private enterprise. Could the Engineer's Department be authorized to have these executed by competent parties the work would be of practical and permanent benefit, and the State would possess at a cost of $20,000 a far more valuable and complete map than was obtained by the much smaller State of Massachusetts at a cost of between one and two hundred thousand dollars.[27]

Smith supported his appeal with letters from New York Senator William H. Seward, A. Dallas Bache, superintendent of the U.S. Coast Survey, J. Peter Lesley, one of the country's foremost engineers, and Lorin Blodgett, a recognized meteorologist. Seward observed that "it would be a valuable addition to the map, if the latitude and longitude of various points in the State exactly ascertained could be noticed on it. Can you suggest any mode by which this could be done through the aid of the observers engaged in your bureau?" Bache wrote that he believed that "the longitude of the most important county seats upon or near the railroads, can be determined by telegraph and be referred to the Dudley Observatory at Albany," which was just about to open. Bache also stated that "the latitude of all the county towns of the State, and the approximate longitude, could be determined for about fifteen thousand dollars, and this would fix accurately, geographically, so many (59) points in the State map to serve as points of standard reference for the rest of the map." He promised that if the secretary of the treasury agreed, he would organize the system himself, without remuneration.[28]

Lesley, with engineering considerations in mind, wrote in support of adding geologic and topographic data to the state map, noting that the strata of the Paleozoic regions found in southern or central New York could be "adequately expressed only by the topography or relief structure." Smith's county maps and surveys provided a basis, and "now, the topographer with one or two assistants in a single season can lay down upon these [county maps], and afterwards upon a reduced scale upon the State map, accurately all the features of the surface—at the same time the out crops of the formations will go in according to the directions given to them by the contour lines of the surface."[29]

Finally, the meteorologist Blodgett reminded Seymour that the statistics about climate collected at colleges and academies for more than thirty years made it possible to clearly define the climate of every part of the state. "The mode of doing this by isothermal, and rain or hyetal charts," he explained, "has recently been perfected under the auspices of the Surgeon General's office of the War department at Washington, and not only the scientific public and the State, but the whole country would be greatly interested in seeing this system applied to the State of New-York, giving a geography of the climates as well as of the surface." The cost of engraving special maps could be saved by including climatic information on general maps of New York, he suggested. Indeed, for the proposed map, "a sum of two thousand five hundred dollars would furnish all the drawings . . . in the most complete form, with the proper statistics and text in explanation."[30]

Impressed by such support for the map, the state engineer wrote a letter to Smith and French on February 23, 1857, stating, "I understand that you have in your possession accurate surveys of fifty-seven counties in this State, and that the surveys of the remaining three counties are now in course of preparation, and also that you have it in contemplation to publish a State map and atlas which shall be based on these surveys." Seymour called their attention to the legislature's inquiry concerning the lack of authentic maps of the several counties and requested Smith and French to "make a proposition stating the terms upon which you will furnish to the clerk of each county and city in this State, such information as will enable him to furnish to the office of the State Engineer and Surveyor an authenticated map of his respective city or county . . . as may be required by the clerk in order to enable him to furnish to the State Engineer such a map of the county or city as may be required by law."[31]

Smith and French replied one day later. They offered to give the clerks of the twenty-six counties where they had lot and reservation lines from their own surveys all the information necessary for them to make accurate maps of the towns in their counties. This they would do free of charge. For the counties where they did not have lot and reservation lines, they offered to do topographical surveys and to furnish the information to the county clerks for twenty-five dollars a town. Another option Smith and French proposed was to draft maps of all the towns in the state and to furnish them to the appropriate county clerks at twenty-five dollars a town. On the publication of these maps they would give free copies of the state atlas to every state official in Albany and to each county clerk. In addition, they would give fifty copies to the state for exchange purposes. In return for supplying such information and material gratis to county and state officers, Smith and French requested

> the exclusive right to publish said county and State maps as the same may be authenticated by the State Engineer under authority of the Legislature for the full term of the copy right of the said map and atlas. Also, [they requested] for the map and atlas the results of the astronomical observations proposed to be made for the State of New-York, by the superintendent of the United States coast survey. Also, any geological and Meteorological statistics in possession of the State previous to the issue of the atlas and map in the several editions.[32]

It is interesting that at this time, Smith and French had envisioned publishing an atlas of the state as well as a map.

On February 28, 1857, the state engineer's report, with supplementary documentation, was presented to the legislature and laid on the table. A week later, on March 6, Assemblyman Abraham G. Thompson introduced a bill entitled "An act to provide for the compilation of an accurate map and atlas of this State," which was referred to the joint committee of the library. On March 9, the committee "reported in favor of the passage of the [bill], with an amendment; which was agreed to and said bill committed to the committee of the whole." Thompson moved, on March 11, "that the Clerk be requested to have the bill reported by the joint library committee, entitled 'An act to provide for the compilation of a correct map and atlas of the State' printed and placed on the files of the members without delay."[33]

Thompson followed up, on March 23, with a motion "that the bill no 366, entitled 'An act to provide for the compilation of an accurate map and atlas of the State,' reported by the joint library committee, be considered in the third committee of the whole under the general orders." On April 4, the chairman of the Committee on Public Printing, in a related action, reported in favor of a resolution "that 250 extra copies of the report of the State Engineer and Surveyor, with the accompanying bill, in answer to a resolution of the Assembly relative to a map of the State, be printed for the use of the Engineer and Surveyor." The assembly concurred. Three days later, the house resolved itself into a committee of the whole to consider several bills, among them the map bill. The assembly *Journal* reports no action on the bill on that day or during the remainder of the eightieth session. There is, moreover, no reference to the proposed bill in the journals documenting the legislature's eighty-first session.[34]

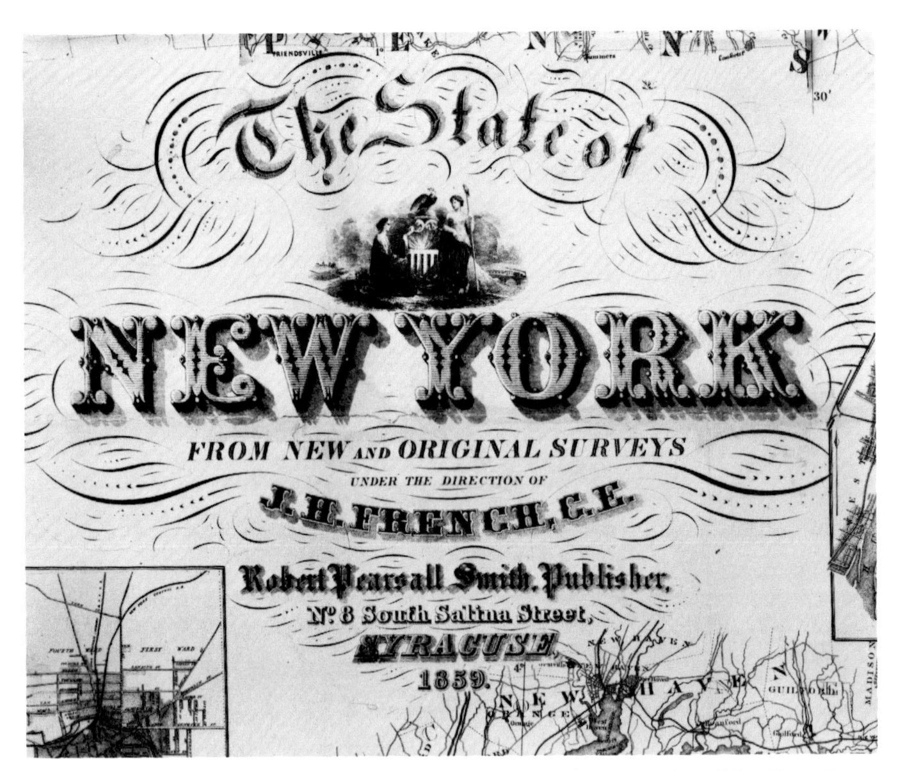

Fig. 22–6. Elaborate title cartouche of the French-Smith map of New York State.

Despite this extensive legislative maneuvering over several years, Smith and French received no financial aid from the state. They continued their work on the map and gazetteer, however. Between 1857 and 1859 French and his staff mapped the counties they had not previously surveyed and resurveyed those for which there were inadequate and inaccurate maps. More than twenty county maps, most of them prepared under French's direction, were published during these years. Smith was probably occupied in making arrangements for printing the map and gazetteer, soliciting subscriptions, and negotiating financing. All this activity culminated with the publication of the map and the gazetteer in 1859.

The map is titled *The State of New York from New and Original Surveys under the Direction of J. H. French, C.E.* and the publisher's imprint reads "Robert Pearsall Smith, Publisher, No. 8 South Salina Street, Syracuse, 1859" (Fig. 22–6). The map was "Entered according to Act of Congress in the year 1859 by Robert Pearsall Smith in the Clerk's Office of the District Court of the Northern District of New York." It is at the scale of 1:300,000 and is of a large wall-map size (168 by 183 cm.). Decorative entwined grapevines border the map (Fig. 22–7). Inset maps show the geology and meteorology of the state: Long Island, which occupies space immediately below the southeastern part of the conterminous portion of the state; a plan and panoramic map of New York City; and plans of the cities of Buffalo, Rochester, Syracuse, Oswego, Hudson, Schenectady, Poughkeepsie, Albany, Auburn City, Troy, and Utica. There is also a chart which indicates the times in various cities in the state when it is noon in Albany. The map includes twelve scenic and urban views and pictures of public buildings, significant institutions, and selected industries (Fig. 22–8). These illustrations, reproduced from engraved steel plates, also illustrate the gazetteer. Four large lithographic stones, each measuring 97 by 86 cm., were required to print the state map.

The Library of Congress has copies of the 1859 and 1860 editions of the map. It is possible that other editions were published to 1864, when it is believed Smith terminated his map publishing activities, but there is no record of this. An 1865 edition of the map was published by H. H. Lloyd & Company of 21 John Street in New York City. The imprint remains the same, except for the substitution of Lloyd's name for Smith's. Smith's original copyright registration notice of 1859 is unchanged. There are a few variations, such as the transposition of the inset maps of Auburn City and Albany. The printer is not indicated on any of the Smith or Lloyd copies. The maps were probably, at least initially, reproduced by Frederick Bourquin. Most copies of the map were colored, apparently manually. County and borough boundaries are accentuated with wide colored bands, and townships are differentiated with varying tints. The coloring for a particular township may differ from map to map. It seems strange that the printer resorted to manual coloring, for by this time chromolithography had been utilized by some Philadelphia printers for eight or ten years.

The lengthy gazetteer title is *Gazetteer of the State of New York Embracing a Comprehensive View of the Geography, Geology, and General History of the State, and a Complete History and Description of Every County, City, Town, Village, and Locality. With Full Tables of Statistics. By J. H. French, Ll.D., Member of the American Association for the Advancement of Science; Corresponding Member of the New-York Historical Society; of the Albany Institute, etc. Illustrated by Original Steel Engravings, and Accompanied by a New Map of the State from Actual Surveys. R. P. Smith, Publisher, 8 Sth Salina Street, Syracuse.* The gazetteer carries the same 1859 copyright registration notice as the map. Other credits include "Stereotyped by L. Johnson & Co., Collins, Printer, Perry, Binder." The printer Collins was a partner in the New York firm of Collins, Bowne & Company, one of the original sponsors, with Smith and Delafield, of the New York State map project.

The gazetteer has 752 pages, the last 12 of which contain advertising, most of it for New York educational

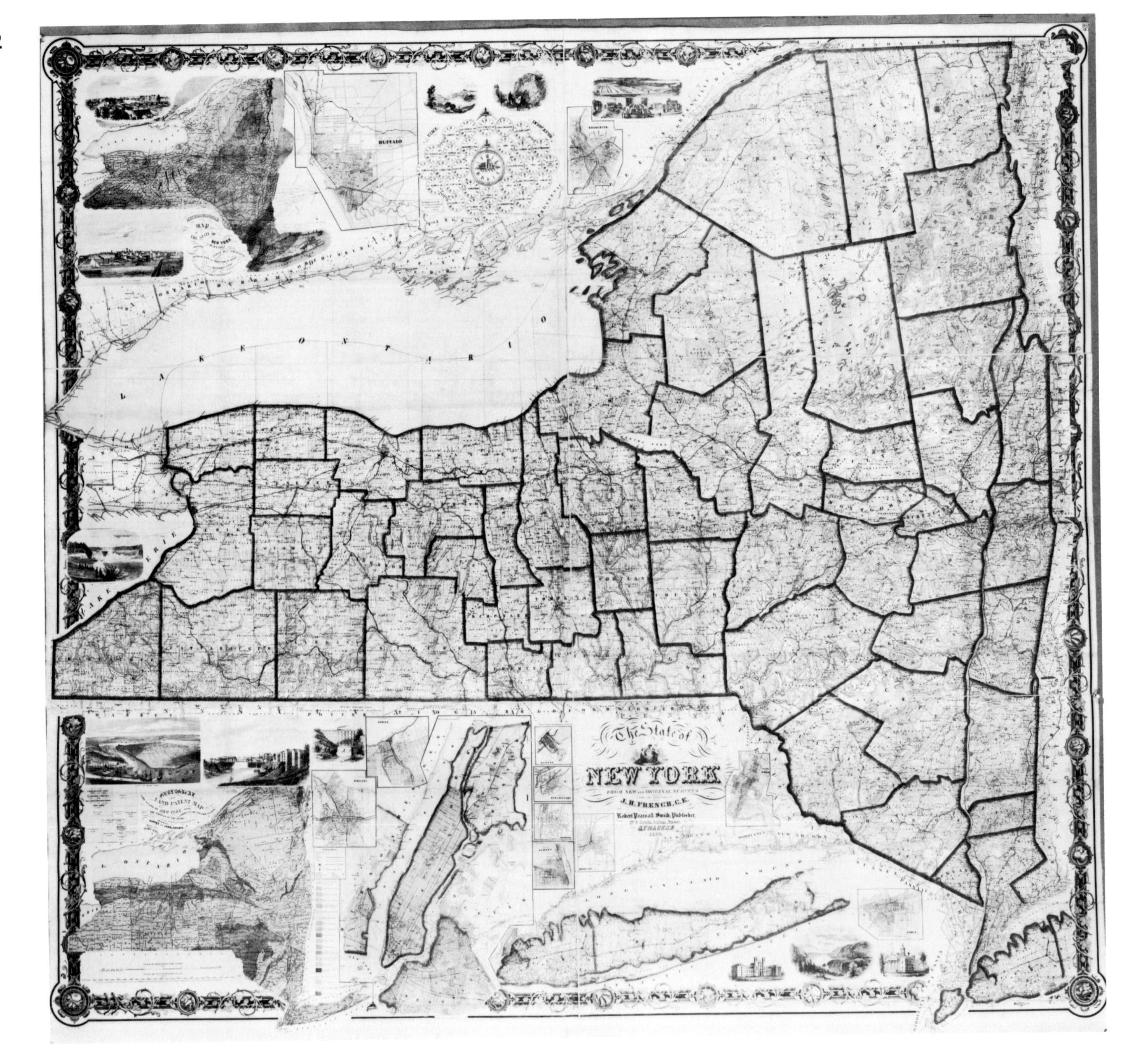
The State of
NEW YORK
FROM NEW AND ORIGINAL SURVEYS
J. H. FRENCH, C.E.
Robert Pearsall Smith, Publisher,
SYRACUSE
LAKE ONTARIO
LAKE ERIE
BUFFALO
ROCHESTER
LAND PATENT MAP
L. ONTARIO

institutions. Approximately 20 percent of the volume is devoted to the state as a whole, with detailed sections on boundaries, geology, commerce, agriculture, government, railroads, corporations, schools, and so on. Descriptive summaries for all of the counties fill the remainder of the pages. The utility of the volume is enhanced by indexes of subjects and geographical names.

A most welcome feature of the gazetteer is a one-page "List of Persons Employed upon the Construction of the State Map and the Preparation of the Gazetteer." More than sixty persons are named "who have been employed for considerable lengths of time." In addition to the superintendent (French) and foremen, the names of draftsmen, surveyors and statisticians, lithographers, steel and wood engravers, and artists are listed. Most numerous are the surveyors, many of whom worked on the county maps. We will look more closely at some of them in a later chapter.

In the preface French gives useful information about the organization and planning of the map and gazetteer and relates how Smith's and his objectives were successfully realized. He expresses appreciation to his associates and others who contributed to the map and gazetteer projects, but he focuses particular attention on Smith. He asserted that

> the intelligent citizens of the State of New York cannot fail to appreciate the liberality of the publisher in the great expenditures he has made in bringing out these works. The cost of the original surveys for the county maps was about $48,000, and the expenditures on the works from the commencement of the re-survey to the date of publication have reached about $46,000 more, making a total investment of $94,000. The whole time spent in surveys, collection of materials, writing, engraving, proofreading, &c., has been equal to the time of one person 125 years. It is believed that no similar enterprise of equal extent, and involving

Fig. 22–7. The French-Smith map of New York State.

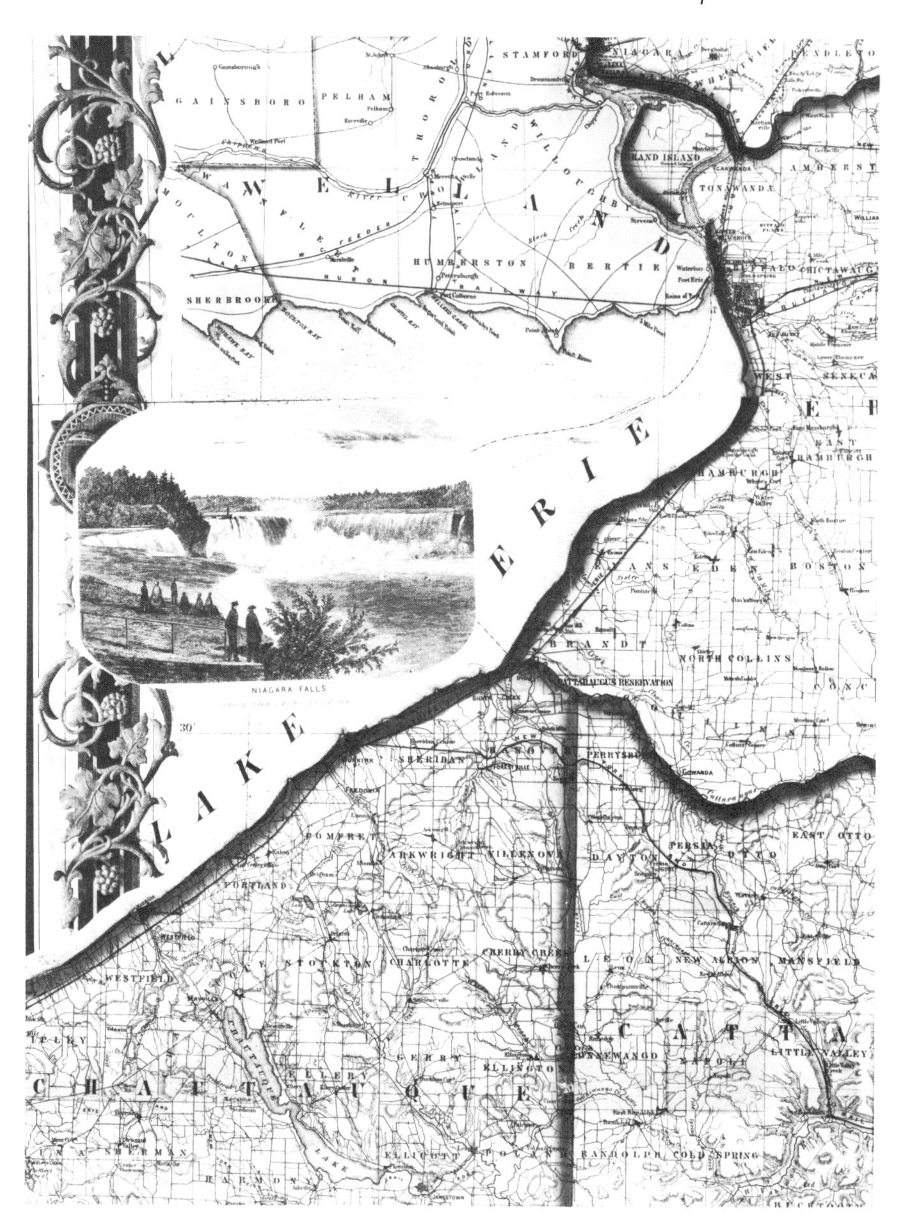

Fig. 22–8. Detail of the western section of the French-Smith New York map showing a portion of Lake Erie and a view of Niagara Falls.

Fig. 22–9. Title page of a 1860 edition of the French-Smith gazetteer.

> the outlay of so large a capital, has ever been undertaken at private expense in this or any other country. Time, talents, and money have alike been devoted to the production of a Map and Gazetteer that it is hoped will be found every way worthy of the Empire State.

By 1860 the gazetteer had gone through eight editions, with minor revisions, including title changes (Fig. 22–9). Two editions were issued in 1871, and a revised edition, edited by Franklin B. Hough, was issued in 1872.

Because the map and gazetteer were not officially supported, no state agencies, departments, schools, or libraries received copies of it. This prompted a series of petitions to the legislature requesting funds to purchase them. On February 11, 1861, state senator Ephraim Goss "gave notice that he would at an early day ask leave to introduce a bill to provide for furnishing the district school libraries of this State with State Map and Gazetteer." Two days later, when Goss did introduce the bill, it was referred to the committee on literature.[35] During the next month, eight petitions "of school commissioners and boards of education" from various counties were submitted. There is no record in the journals of the house or senate that an act was ever passed authorizing the funds to purchase the map and gazetteer. On January 22, 1862, however, the senate resolved "that the Clerk be directed to procure a map of the State of New York for the use of the Senate."[36]

Although a fair number of copies of the New York State map and gazetteer were sold, the returns very likely did not offset the heavy investment of Smith and his associates. Shortly after they were published the United States was plunged into the bitter and costly Civil War, and both individuals and the state had higher priority projects for available funds. Furthermore, many surveyors joined the Union forces, and paper, ink, and lithographic presses were required for the production of military maps.

From the perspective of more than a century we may critically examine the French-Smith map and gazetteer. A contemporary reviewer believed that "the new Gazetteer and Map of the State of New York are far better than anything that has preceded them."[37] We can certainly agree with this evaluation, for few state maps or gazetteers published in the pre-Civil War period can compare with the French-Smith publications. The state map did not, however, meet the standards proposed by Edward B. Hunt and his associates in the AAAS. This type of map, based on scientific triangulation surveys, would have been too expensive an undertaking without state financial support.

Because the county maps from which it was compiled were prepared, for most part, on uniform standards, the French-Smith state map was of somewhat higher quality than the maps of other states published prior to the Civil War. In addition to receiving an acceptable state map, the citizens of New York also acquired a creditable series of maps for all counties. With their detailed information on landowners and land holdings, the county maps constitute a most valuable cartographic record of local settlement and history of a century and a quarter ago. For this data the citizens of the state are indebted to the vision, enterprise, and dedication of Smith, French, and their associates.

After the completion of his large, time-consuming project, Smith continued to promote mapmaking and surveying. In 1864, at the request of J. Peter Lesley, then secretary of the American Philosophical Society, he prepared a communication respecting the need for published maps of U.S. counties. The article was printed in the society's *Proceedings*. In it, Smith called attention to the lack of good maps available to military commanders during the Civil War. "For the long campaigns in Virginia," he noted, "the utmost annoyance has been experienced for want of maps. . . . It is to be hoped that on the return of peace, this greatly desired contribution to science, the extension of the Pennsylvania Appalachian topography southward, will be made, and with an advantage not enjoyed by those who did the work in Pennsylvania, namely, with the well-constructed county maps, done with an odometer, like those of the Northern States."

Smith supplemented his article with a map that identified counties for which maps had been published (Fig. 22–10). Lesley noted that "Mr. Smith has kindly colored for me a map of the United States to show the parts covered by these odometer surveys. They are 300 in number. They have formed the basis of the recently published and very correct State Maps of New York, Pennsylvania, and New Jersey." (Smith's original hand-colored, annotated map, which identifies counties mapped before 1864, is preserved in the Library of Congress). Continuing his discussion, Smith further emphasized that "for eighteen years, this slow discussion of the boundaries, streams, roads, and houses of the surface of the United States, has been carried on by Smith and others, with a continually improving organization, and increasing rapidly, until about two-thirds, of the well-settled north has been delineated. The field work seems rude to the physicist, engaged in discussing the figure of the earth, and to the chief of a survey of an arc of the meridian. But the results are perfectly satisfactory to the naturalist, the county surveyor, the soldier, and the geologist." (And so they were until the U.S. Geological Survey began publishing, some two decades later, its topographic maps based on original triangulation surveys).

Deploring the scant attention that had been accorded to county maps, Smith wrote that "it would be natural to expect to find complete sets of the county maps of each State in the archives of its capital. Strange to say, none such is known to exist except at Albany. Stranger still, no set of these maps, no record of all this labor done, is to be seen at the Capitol of the Nation, neither in the Library of Congress, nor in the Bureau of the Interior, nor at the Bureau of the Coast Survey. A few of

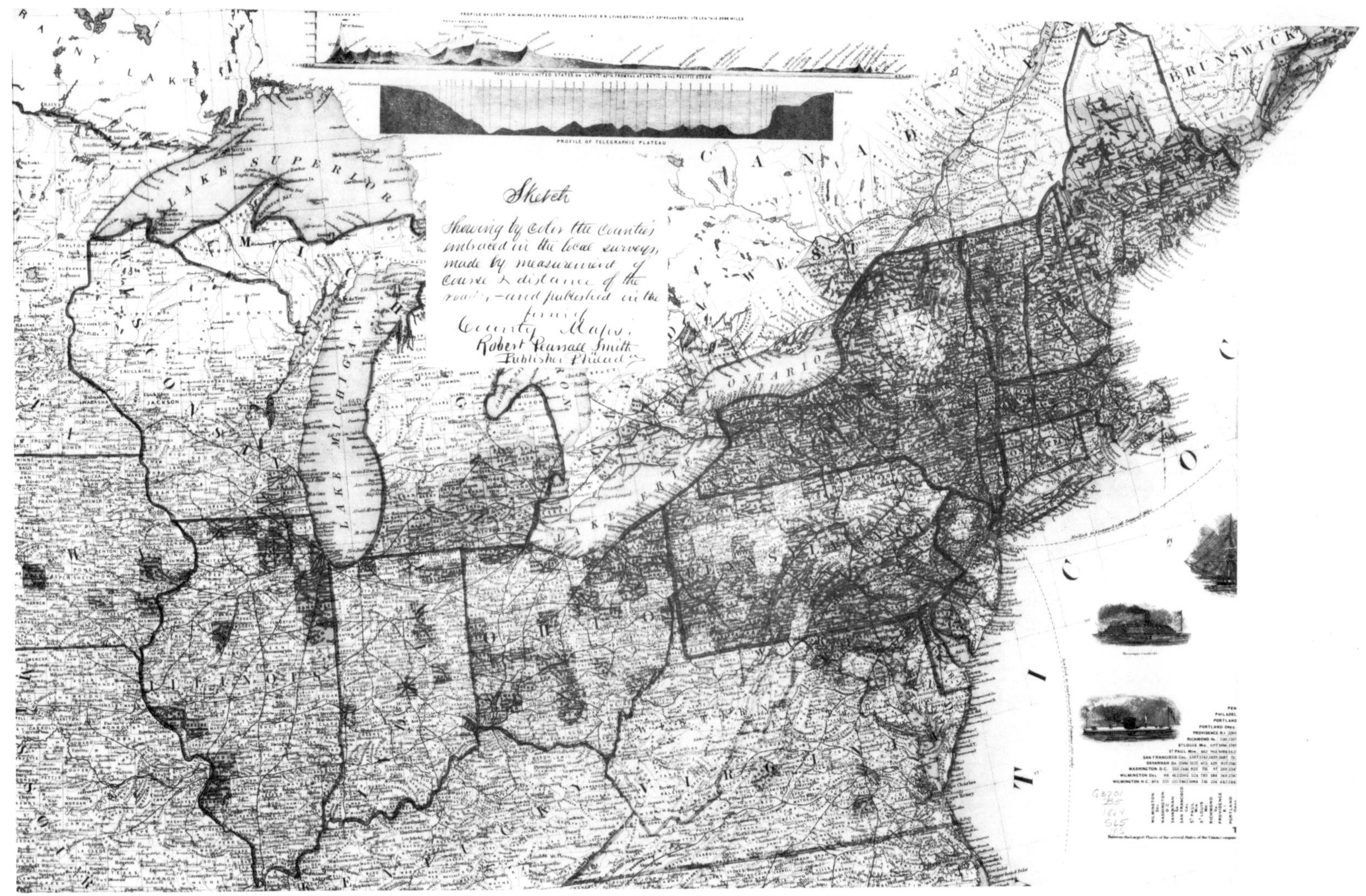

Fig. 22–10. The map Smith prepared to accompany his article in the *American Philosophical Society Proceedings*.

them, the number amounting perhaps to one-twentieth of the whole, are on file in the Engineer Department of the United States."[38]

Perhaps due to his financial losses on the map and gazetteer or to the disruptions caused by the Civil War, Smith terminated his map interests in 1865. H. H. Lloyd acquired the copyright for the New York State map and issued several editions of it in its original size and format and in a reduced-scale version. Other Smith map publications were apparently taken over by Bourquin, who continued to be heavily involved in map and atlas printing and publishing until his death in 1894.

John Mickle Whitall, Smith's father-in-law, helped him following the termination of his business. Whitall placed Smith in charge of the Whitall-Tatam Glass Company's Millville, New Jersey, plant. Smith and his wife, Hannah, lived in Millville until 1871. In that year the couple experienced a religious conversion that induced them to leave Millville and the glass factory to become lay evangelists. Smith was particularly impressed with William E. Boardman's Higher Life Movement, and in 1873 he joined Boardman in the British Isles to preach the new gospel. His family followed him in 1874. According to one historian, Boardman's movement had in the Smiths "its most capable propagators and it was through them that it attained its widest extension and its most lasting influence."[39]

These were exciting times for the Smiths, particularly for Robert, who basked in the warmth of popular esteem and adulation. The culmination of his evangelical work came in 1875 with a moving eight-day conference at Brighton, over which Smith presided with magnificent skill and authority. In its closing days, however, rumors circulated among the conferees imputing questionable and nonspiritual relations between Smith and a female disciple. Though the charges were never documented, the evangelist's honor and integrity were sullied, and the Smiths, discredited and disheartened, returned to America and the glass company.

Smith never fully recovered from the experience and spent the remainder of his life on the fringes of the social and intellectual eddies which swirled about his celebrated wife and children. His wife was more resilient and established a reputation as a well-read author, lecturer, and feminist.

In 1888 the Smiths returned to England, where their children had previously migrated. They lived there for the remainder of their lives. The younger Smiths circulated freely with young British intellectuals and liberals; Hannah continued her writing and lecturing. Their daughter Alys became the first wife of Bertrand Russell. Her older sister, Mary, married Benjamin Conn Costeloe, an Irish scholar and liberal, and, after his death, she wed the distinguished expatriate American art critic Bernard Berenson. The Smiths' son, Logan Pearsall, achieved fame as a writer and literary critic.

Notes

1. "Resolutions and Acts of the Association," in *Proceedings of the American Association for the Advancement of Science* (Washington, D.C., 1852), 402.
2. E. B. Hunt, "Proposal for a Trigonometrical Survey of New-York," *Proceedings of the American Association for the Advancement of Science* (Washington, D.C., 1852), 383–84.
3. New York, *Senate Documents*, 75th sess., 1852, vol. 1, no. 41, 4.
4. Ibid., 6.
5. Ibid., 1–2.
6. A. D. Bache, "Report of the Committee Appointed to Memorialize the Legislature of New-York in Regard to a Trigonometrical Survey of that State," in *Proceedings of the American Association for the Advancement of Science* (Cambridge, Mass., 1856), 280.
7. Asa Fitch, "Survey of Washington County, New York," *Transactions of the New York State Agricultural Society* 8 (Albany, 1849), and 9 (Albany, 1850).
8. John Delafield, "A General View and Agricultural Survey of the County of Seneca," in New York, *Assembly Document* 150, 74th sess., 1851, and in *Proceedings of the County Agricultural Societies* 10 (1850): 356 ff.

9. Oliver J. Tillson Papers, Cornell University Library, Collection of Regional History and University Archives, 1441, 1458.
10. New York, *Assembly Journal*, 76th sess., 1853, vol. 1, 82.
11. Ibid., 122, 150, 175.
12. New York, *Assembly Journal*, 76th sess., 1853, vol. 2, *Documents*, no. 33, "Report of the Committee on Agriculture, in Relation to the Topographical Survey of the State," 1–2.
13. Ibid., 2–4.
14. New York, *Assembly Journal*, 76th sess., 1853, vol. 2, *Documents*, no. 32, "Report of the Committee on Colleges, Academies, and Common Schools, on the Topographical Survey," 2–4.
15. New York, *Assembly Journal*, 76th sess., 1853, vol. 1, 502.
16. Ibid., 504.
17. Ibid., 787–88.
18. Ibid., 1033.
19. Ibid., 1263.
20. Ibid., 79th sess., 1856, 172.
21. Ibid., 80th sess., 1857, 269.
22. Ibid., 412.
23. New York, *Assembly Documents*, 80th sess., 1857, no. 114, 1.
24. Ibid., 2–3.
25. Ibid., 4.
26. Ibid., 6–7.
27. Ibid., 7–8.
28. Ibid., 8–10.
29. Ibid., 10–11.
30. Ibid., 12.
31. Ibid., 13.
32. Ibid., 14–15.
33. New York, *Assembly Journal*, 80th sess., 1857, 587, 620, 649.
34. Ibid., 819, 1126, 1149.
35. New York, *Senate Journal*, 84th sess., 1861, 164, 172.
36. Ibid., 85th sess., 1862, 73.
37. *Journal of the American Geographical and Statistical Society*, no. 2 (1860): 135.
38. R. Pearsall Smith, "United States County Maps," *American Philosophical Society Proceedings* 9 (Mar. 1864): 350–52.
39. Benjamin B. Warfield, *Perfectionism* (New York, 1931) 2:494.

23. *Jay Gould as Surveyor and Mapper*

The French-Smith map project served as a vehicle for training many local surveyors. Some individuals who had their first mapping experience on the project continued in this profession following publication of the state map and gazetteer. For others it was an interesting, but temporary, interlude and an entrée to more ambitious and lucrative endeavors. In this latter category was Jay Gould, who was one of the wealthiest and most despised men in the United States when he died on December 3, 1892. In an era of free enterprise and rugged individualism, the undersized and frequently ailing Gould proved to be a most cunning and crafty manipulator.

In an age famed for its lusty and acquisitive robber barons, very few of Gould's contemporaries approached his "genius for trickery and thimble-rigging, his boldness in corruption and subornation, his talent for strategic betrayal, his mastery over stock and bond rigging, his daring in looting a company and defrauding its stockholders."[1] His ruthless amassing of wealth was the antithesis of the traditional all-American boy in the inspirational stories of Horatio Alger. Gould's boyhood and youth were, however, very much in the pattern of Alger's simon-pure heroes, and his initial successes were achieved in surveying and mapping.

Jason Gould, as he was christened, was born in West Settlement, Delaware County, New York, on May 27, 1836, the sixth child and first son of John Burr Gould and Mary Moore Gould. His mother died when Gould was six years old, and his care and training were assumed by older sisters and two successive stepmothers. Until he was fourteen years old, Gould attended Beechwood Academy in the nearby village of Roxbury. His father was a founder and patron of the school. One of Gould's classmates at Beechwood was John Burroughs, the distinguished naturalist and author. The Burroughs' farm, like the Goulds', was located on the lower slopes of the Catskill Mountains, some four miles north of Roxbury. It was not as productive as the Gould dairy farm, and Burroughs recalled, some years later, that "the Goulds were very prosperous, and naturally stiff-necked, and they lived in a little better style than the other farmers."[2] Gould had little taste for farm life and work. Although his father hoped that he would help run the farm after completing his studies at the academy, Gould continued his education at Hobart Academy, some fifteen miles northwest of Roxbury. His father permitted him to go with the provision that he support himself.

In the autumn of 1850, Gould obtained a job as a clerk in a store near Hobart and contracted to work for his room and board. Gould's scholastic interests were primarily in mathematics and surveying, and he soon acquired all the knowledge in these subjects that his instructors could offer. To supplement this rudimentary training, he often arose at three o'clock in the morning to study engineering and surveying books. He would then begin his duties in the store at six o'clock. When his father sold the dairy farm and bought a residence and hardware store in Roxbury in 1851, Gould returned home to keep the books and help manage the store. In his spare time, he continued his surveying studies. Gould remained in Roxbury until the spring of 1852, when he was offered a job as a surveyor's assistant.

His new employer, John J. Snyder, agreed to pay Gould, then sixteen years old, twenty dollars a month and his keep for conducting surveys of Ulster County, which borders Delaware County on the east. Snyder, however, was heavily in debt and unable to pay his young assistant. Gould accordingly joined in a partner-

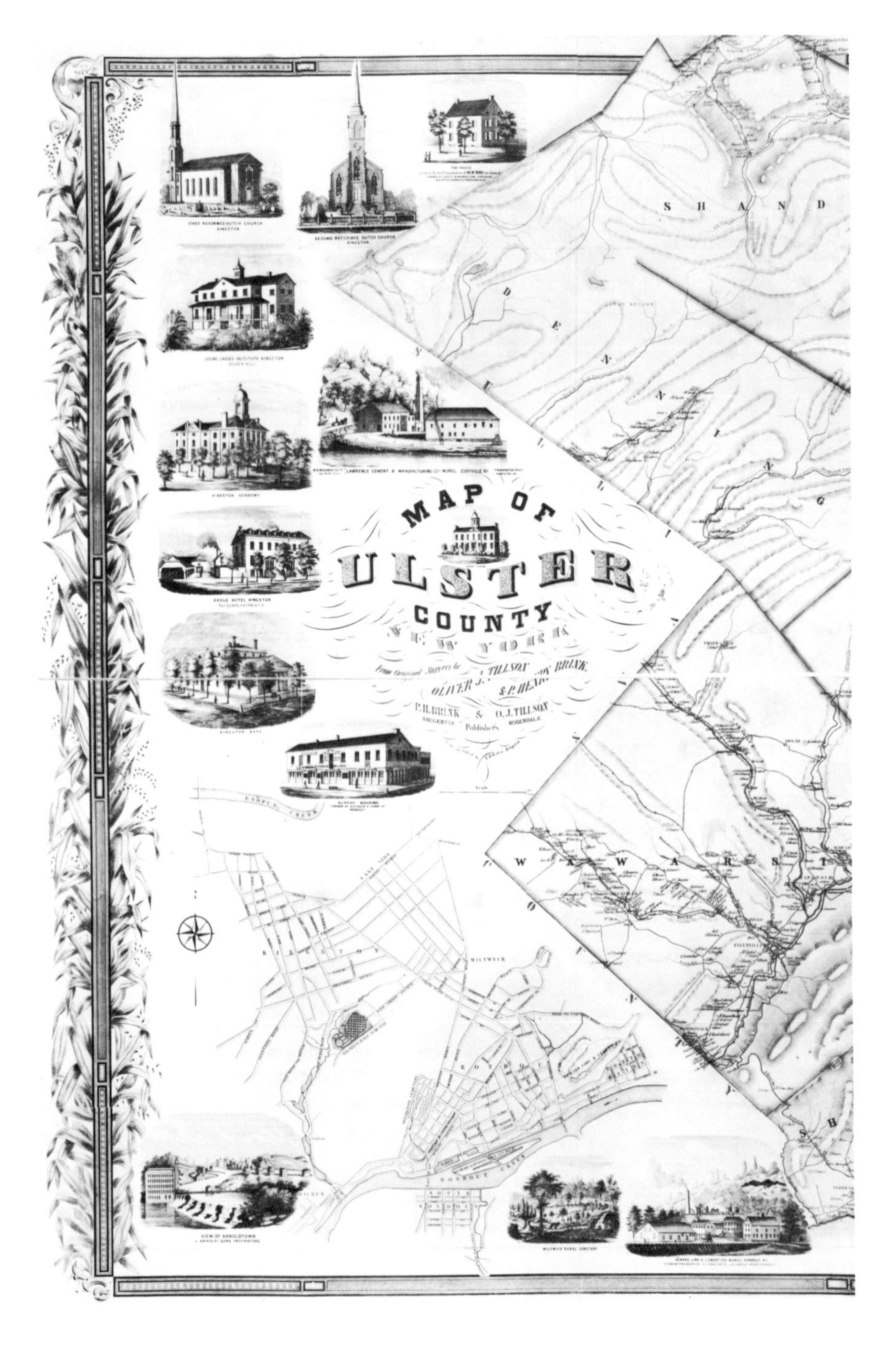

ship with two other Snyder assistants, Oliver J. Tillson and Peter Henry Brink, to complete the Ulster County map. Their headquarters were set up in the home of Tillson's parents, some ten miles south of Kingston, the county seat. In a letter to Gould's daughter, Helen Gould Miller, dated August 3, 1894, Tillson described the procedures of the three young surveyors. "Our manner of work," he wrote, "was to start out in the forepart of the week, each taking a separate road to survey and return to my father's home the last of the week. We used an instrument called an Odometer, (similar to a wheelbarrow), which measured distances, and a Compass to take angles and directions. In about two months the surveys were completed after which we made the map, at my father's house."

After this, Tillson related, Gould "became discontented and wished to sell his interest. He did not wish to incur the expense of his share of the engraving and publishing charges. . . . We therefore auctioned off between us the various books and instruments, etc., and I personally bought his interest in the map and took his receipt, dated Dec. 27, 1852."[3] Tillson also recalled that he bought Gould's interest in the map for forty dollars. Other accounts state that Gould received as much as five hundred dollars. The actual payment received by Gould was probably around two hundred dollars, the amount Tillson later paid Brink for his interest in the map.

The *Map of Ulster County New York from Original Surveys by Oliver J. Tillson & P. Henry Brink* was published in 1853 (Fig. 23–1). Tillson and Brink are also listed as the publishers. Gould's name does not appear anywhere on the map, which measures approximately 102 by 115 cm. Townships are outlined, and the names of all landowners are given. The map includes roads, canals, and rivers, as well as generalized topography. There are inset maps of the villages of Saugerties, Rondout, and Kingston and fifteen border illustrations of churches, ceme-

Fig. 23–1. Title cartouche and segment of the 1853 map of Ulster County.

teries, industrial plants, banks, academies, and public buildings.

The lithography and printing were contracted to Robert Pearsall Smith, who served as their liaison. The cost of printing the Ulster County map was two hundred dollars, plus sixty dollars for the views and nine dollars for proof alterations. A second, revised, edition of the map was published in 1854. Unlike the first edition, it carries a copyright registration in Tillson's name. In October 1853 four hundred copies of the map were delivered to Tillson, who apparently sold them directly to users or distributed them to dealers in the several cities and towns in the county. The retail price of the map was five dollars.

When Smith was preparing to compile his New York State map and was negotiating with various local surveyors to prepare county maps, Tillson and Brink were among the first surveyors with whom he spoke. Tillson and Brink apparently did not sign a contract with Smith, for Smith is not the copyright registrant of the Ulster map. Smith did, however, attract Gould to his project. Gould had moved to Albany after dissolving his partnership with Tillson and Brink and in 1852 had enrolled in the Albany Academy. While there Gould became acquainted with Smith, Delafield, and their plan. Helen Gould, Gould's niece, wrote that "while surveying Albany County, Uncle Jay had become interested in a legislative bill sponsored chiefly by a Mr. John Delafield, the passage of which would make possible the completion of a topographical survey of New York State. . . . He felt that the passage of the bill in question would be a genuine service to the cause of education, as well as to the future of New York State. And he hoped, too, that, should it pass, he might himself be the one to get the work, the commission to do the surveying, which he felt able to do as well, at least, as anyone else." Gould had hoped to make enough money from these anticipated surveying assignments to cover his expenses at Yale University. The failure of the Smith-Delafield proposal was a bitter blow to him, and he "required time to think things through. Soon, however, his decision was definitely made. He would turn his back on disappointment, and . . . finish the job at hand."[4]

The job at hand was the map of Albany County, and during the summer of 1853, Gould and his cousin I. B. Moore carried out surveys (Fig. 23–2). Gould drew the map and probably spent the latter months of 1853 at this task. He also apparently solicited advance subscriptions for the map in Albany as well as in some of the larger villages and towns. He seems to have promoted the idea of including marginal sketches of the homes of affluent residents on a fee basis. The *Map of Albany County New York* has thirty-one illustrations, eighteen of which are of private residences. There is no copyright registration notice on the first edition of the map. A second edition, which has few changes, was registered for copyright in 1854 by Smith. This confirms that Smith and Gould were business associates, but it seems unlikely that Gould, as his niece suggested, was a major collaborator with Smith and Delafield.

Gould also had several commissions, including one from the Cohoes Manufacturing Company to prepare a map of the village of Cohoes, which was north of Albany. Gould conducted this survey during the summer of 1853. The map, which is undated, was probably published in 1854. It is titled *Map of Cohoes New York* (Fig. 23–3). It was "Published by Gould and More" and "Drawn by J. Gould." The map was probably completed before Gould's contact with Smith, for it was lithographed and printed by Sarony & Major of New York City. There is a panoramic view of Cohoes, a sketch of Great Cohoes Falls, and a Cohoes business directory in the margins. Gould received six hundred dollars for his work on the map.

In 1853 Gould also surveyed part of the route for the Albany to Shakers plank road. While engaged in the survey Gould reported to a former surveying instructor, James Oliver, that "I succeeded in everything without any trouble until I came to making the estimates of cost, embankments, excavations, and culverts, especially the

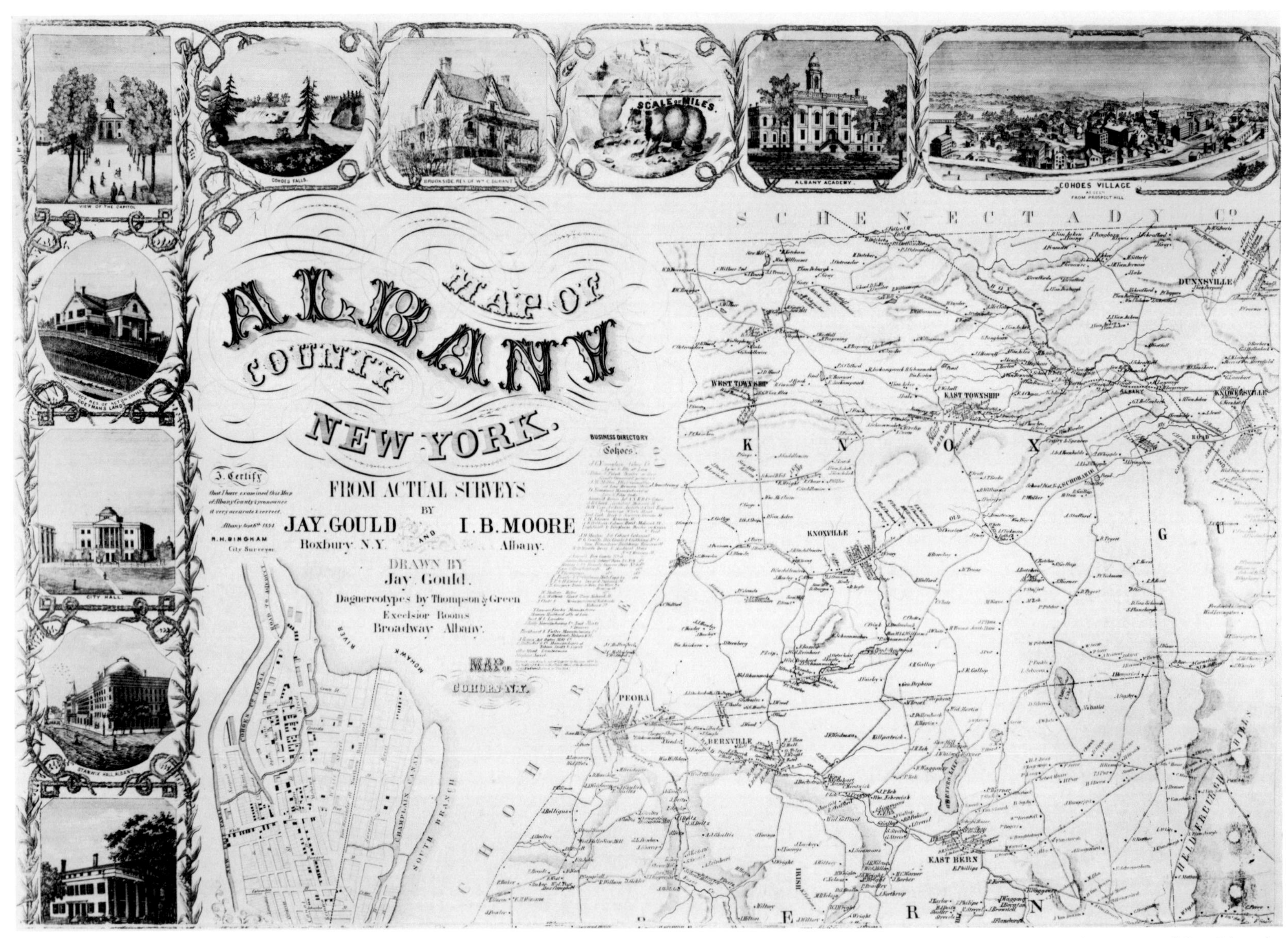

Fig. 23–2. Title cartouche and portion of Gould and Moore's map of Albany County.

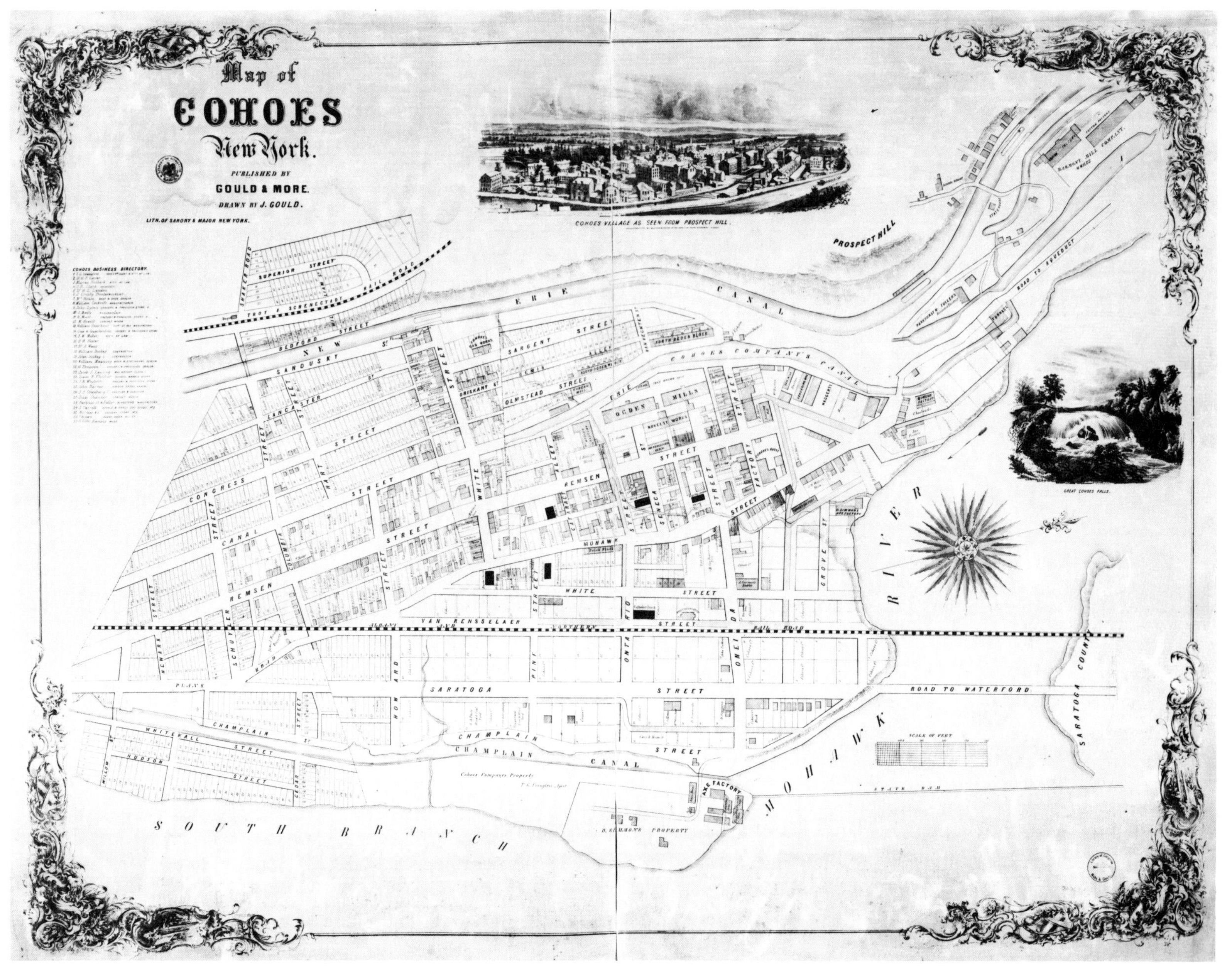

Fig. 23–3. Gould was also assisted by his cousin Moore (misspelled More) on surveys for this map of Cohoes, New York.

embankments and excavations, when I came nearly being floored for a second time. But just then Gillespie [the author of a surveying textbook] stepped in with his *Roads and Railroads* where I found the proper information."[5] This job was completed in March 1853, and, with no others forthcoming, Gould decided to prepare a map of his home county, Delaware.

He returned to Roxbury and set up a small office above his father's hardware store. To insure a good reception for his proposed map, Gould prepared a prospectus which he mailed on April 30, 1854, to residents throughout the county. He engaged several assistants to do the actual surveying for, as he once wrote to a friend, "to tell the plain truth, my education in surveying is made up of the leisure time I had in the tinshop the winter before I went to Ulster, and all I have gained since has been of a practical character, so you see I do not possess a thorough knowledge of the important principles that are involved in this pursuit."[6] One of his assistants was John Champlin, the husband of Ellen Moore, one of Gould's cousins. Champlin surveyed for a short time, and, after completing work on the Delaware County map, he moved to Grand Rapids, Michigan, to read law in the office of an older brother. He subsequently became a professor of law at the University of Michigan and eventually became chief justice of the Michigan State Supreme Court.

In promoting his map, Gould also sought the aid of local newspaper editors. The technique he employed was a portent of his subsequent financial maneuvering. On his seventeenth birthday, Gould made a small cash contribution to a Delaware County newspaper, the *Bloomsville Mirror*, which was edited and published by his good friend Simon B. Champion. Several months after his act of generosity, Gould wrote to Champion that "thus far in this State, without exception, these [county] maps have met with unbounded success, having been liberally and, some of them, munificently countenanced by the Supervisors and sometimes by individuals. In Delaware County the Supervisors ought to encourage it by having a map for each of the School Districts. I want you to give me an editorial to this effect."[7]

This letter suggests that Gould was also advancing the proposal of Delafield and Smith. It is very probable that Gould undertook the task of surveying and mapping Delaware County as part of Smith's and Delafield's overall plan for compiling a map of New York State from county maps. That Gould was an active collaborator on the project is confirmed by the inclusion of his name in the gazetteer's list of the people who worked on it. Gould's name appears in the list of surveyors and statisticians. Tillson and Brink are not included.

While engaged in soliciting subscriptions and directing his assistants working on the Delaware map, Gould also gathered information and anecdotes about the early history and prominent residents of Delaware County. He planned to include this information in a book on the county. In the summer and autumn of 1854 he conducted surveys for a portion of a proposed railroad between Newburgh and Syracuse, New York. Several Gould biographers also state that he conducted surveys for maps of Greene and Sullivan counties, both of which, like Ulster County, adjoin Delaware County on the east. Maps of these two counties were copyrighted by Smith and were published in 1856. Neither, however, has any reference to Gould in its title or credits. It is possible that Gould planned the surveys of Greene and Sullivan counties and engaged assistants to carry out the fieldwork. Because of his several illnesses and his diminishing interest in surveying, he apparently sold his interests in these maps before they were published.

In 1855 the work on the Delaware County map and on the book manuscript about the history of Delaware County was completed. Rights to the map were sold to Smith, who registered the map for copyright. Gould reportedly received one thousand dollars for the map, in addition to the reimbursement for the wages of his assistants and other expenses. The map appeared in 1856 under the title *Map of Delaware Co. New York from actual*

Fig. 23–4. Gould's map of Delaware County.

MAP
OF
DELAWARE CO.
NEW YORK
From actual survey
BY JAY GOULD
Published by COLLINS G. KEENEY No. 17 & 19 Minor St.
PHILADELPHIA
1856
MEREDITH SQUARE
DOWNSVILLE
CROTON
SIDNEY PLAINS
FRANKLIN
ANDES
LANSINGVILLE
HAMDEN
HARPERSFIELD CENTRE
MASONVILLE
HANCOCK
STAMFORD
WALTON
DELHI
BOVINA CENTRE
ROXBURY
GRIFFINS CORNERS
BLOOMVILLE
MARGARETTVILLE
J. CHURCHILL DAGUERREAN ARTIST
STATISTICS OF DELAWARE CO.
BUSINESS DIRECTORY OF DELHI
TABLE OF DISTANCES
DELAWARE LITERARY INSTITUTE FRANKLIN DELAWARE CO. NEW YORK
PENNSYLVANIA
SULLIVAN COUNTY

survey by Jay Gould (Fig. 23–4) and was "Published by Collins G. Keeney No. 17 & 19 Minor St. Philadelphia 1856." Although Keeney was a Roxbury printer, the publisher's address is actually Smith's. The map measures 142 by 142 cm., includes the names of all land and property owners, shows generalized relief, roads, and administrative boundaries, and has inset maps of a number of towns and villages. On the map of Roxbury the shop and residence of Gould's father can be located. There are twenty marginal illustrations, among them sketches of the residences of affluent citizens. The map is said to have sold well, and Smith and his associates used it in compiling the map of New York State.

The manuscript for the book was delivered early in 1856 to the print shop of Robb, Pile & McElroy on Lodge Street in Philadelphia. On April 30 of that year there was a disastrous fire in Philadelphia which destroyed or seriously damaged a number of printing establishments, including that of Robb, Pile & McElroy. In Chapter Ten of his book Gould observed that "the reader . . . has already been made aware of the almost total destruction of this work by fire. And this misfortune appears nowhere more evident than in the present chapter. Several of the first pages of the manuscript, containing much valuable and important information are thus irremediably lost."[8] The volume was published later that year under the title *History of Delaware County and Border Wars of New York, Containing a Sketch of the Early Settlements in the County and a History of the Late Anti-Rent Difficulties in Delaware with Other Historical and Miscellaneous Matter, Never Before Published. By Jay Gould. Roxbury: Keeny & Gould, Publishers, 1856.* Gould is said never to have referred in later years to this, his sole literary work. There was, moreover, no copy of the book in the fairly extensive personal library he had assembled.

In 1856 Gould tired of mapping and publishing and sold his rights to the book as he had with his several maps. Altogether, he reportedly received five thousand dollars for his cartographic works and the *History of Delaware County*. With this capital accumulation, Gould established a partnership with Zadok Pratt, a former congressman and a wealthy industrialist, to operate a tannery in eastern Pennsylvania. Pratt invested one hundred and twenty thousand dollars in the project while Gould contributed five thousand dollars and his services, talents, and managerial abilities. The tannery was immensely successful, and Gould, without the knowledge of his absentee partner, speculated heavily with the firm's funds. When Pratt became aware of the situation he proposed dissolving the partnership, offering to sell out for half his investment or to buy Gould's interest.

To Pratt's surprise, Gould chose to acquire his interest. Gould obtained the necessary money from Charles Leupp, a New York City leather dealer, who became Gould's second unfortunate partner. Leupp, too, was outwitted and swindled by Gould, and, despondent and ruined financially, he committed suicide. Gould retained the tannery for several years, resisting efforts by Leupp's brother-in-law to gain control. In 1860 he closed the plant, moved to New York, and began in earnest his financial maneuverings. The remainder of his career bears little resemblance to the bright years of his youth, when he practiced the respected professions of surveying and mapping.

Notes

1. Richard O'Connor, *Gould's Millions* (Garden City, N.Y., 1962), 13.
2. Ibid., 17–18.
3. Oliver J. Tillson Papers, Cornell University Library, Collection of Regional History and University Archives, 1458.
4. Alice Northrup Snow, *The Story of Helen Gould* (New York, 1943), 81.
5. Ibid., 82.
6. Robert Irving Warshaw, *Jay Gould, the Story of a Fortune* (New York, 1928), 34.
7. Ibid.
8. Jay Gould, *History of Delaware County and Border Wars of New York* (Roxbury, N.Y., 1856), 226.

24. *More about County Maps and Mappers*

Henry F. Walling and Robert Pearsall Smith unquestionably made the greatest contributions to town and county mapping in the fifth, sixth, and seventh decades of the nineteenth century. A number of other individuals, however, played significant roles in this phase of commercial map publishing in the United States. While some of them operated independently, many owed part of their success to their associations with Walling and Smith. Unfortunately, for many of these surveyors and mappers the only personal data available is printed on the maps they produced.

James C. Sidney, John Jay Smith's former clerk, began publishing county maps in 1850 with the maps of Baltimore County, Maryland, Essex County, New Jersey, and Dutchess and Richmond counties, New York. In the following year, in collaboration with James Neff, he published maps of Westchester County, New York, and Allegheny County, Pennslyvania. He also independently produced a map of New York's Orange and Onondaga counties, which he surveyed with Neff and Lawrence Fagan and published in 1852 and 1855 editions.

Apart from the information that Sidney was born in England, little is know about his background. He must have had some training in surveying and drafting, for in 1850 he published a four-volume work entitled *American Cottage and Village Architecture*. A fifth volume appeared in 1855.[1] He seems to have given up surveying and mapping around 1852 in favor of architecture and landscape planning. In 1850 he also published the *Description of Plan for the Improvement of Fairmont Park* (in Philadelphia). Sidney focused his attention on school architecture in the late 1860s and prepared plans for a score or more buildings. In partnership with Frederick Merry, Sidney also designed a number of homes in Philadelphia. He is believed to have died in 1881.

Samuel M. Rea and J. W. Otley were two producers of county maps who each had an early association with Smith. Rea's first cartographic contribution appears to be a map of New Castle County, Delaware, which was published in 1849 by Smith & Wistar. Smith & Wistar also published in the same year a map of Salem and Gloucester counties, New Jersey, which was a joint production of Rea, James Keily, and Alexander C. Stansbie. Smith had apparently enlisted Rea in the New York State map project. Maps of Chemung and Chautauqua counties, New York, surveyed by Rea and A. V. Trimble, were published in 1853 and 1854, respectively. Rea also teamed up with Otley to produce an 1854 map of Genesee County and an 1858 map of Livingston County. Otley's earliest county map was of Burlington County, New Jersey, which was surveyed in association with R. Whiteford and published by Smith & Wistar in 1849. In partnership with Keily, Otley continued to focus his attention on New Jersey, for the next two years. Their map of Mercer County was published in 1849, and in the following year their maps of Middlesex and Somerset counties were issued. Otley then transferred his interest to New York State where, in association with F. W. Keenan, he surveyed Columbia County, the map of which was published in 1851 by J. E. Gillette of Philadelphia. This seems to have been Keenan's only venture into county mapping.

J. Chace, Jr., another county mapper, conducted surveys of more than twenty counties in seven states between 1854 and 1860. Although little is known about Chace, the imprint on several of his early maps identifies him as "J. Chace, Jr., Civ. Eng. Troy, N.Y." Much of his work, however, was done in New England where he seems to have mapped counties that were not covered by Walling. He did no maps for counties in Massachusetts or Rhode Island and only one, Fairfield, in Connecticut. The Fairfield map, done in collaboration with William J. Barker, was issued in 1856 and 1858. Barker's address is given on the map as North Hector, New York. This is the only Chace-Barker association known. Between 1856 and 1860, Barker published maps of seven or eight Pennsylvania counties that were prepared by various surveyors. Barker's imprints on some of these maps also give a Philadelphia address.

During this period, Chace, working independently and in collaboration with other surveyors, prepared maps of four counties in New Hampshire, two in Vermont, and five in Maine. These maps were issued by various publishers in Philadelphia and Boston. On several of them Chace is listed in the publisher's as well as in the surveyor's imprint. Chace also produced maps of Erie County, Pennsylvania, in 1855, Darke County, Ohio, in 1857, Fulton County, New York, in 1856, and Suffolk and Warren counties, New York, in 1858. All of the New York maps were registered for copyright by Smith, indicating that Chace participated in the French-Smith map project. His name, however, does not appear in the list of surveyors published in the gazetteer. There is no record of any cartographic contribution by Chace after 1860. Nothing else is known about him.

A. E. Rogerson's name does appear in the New York State gazetteer, and like Chace he contributed three county maps to the French-Smith project. With E. J. Murphy, he conducted surveys of Oneida County for a map published in 1852. Rogerson independently produced maps of Rensselaer and St. Lawrence counties, which were published, respectively, in 1854 and 1858. The two men collaborated again on a map of Greene County, Ohio, which was published in 1855 by A. D. Byles of Philadelphia. There is no record of further cartographic work by Rogerson or Murphy.

Concurrent with Walling's early town surveys in Massachusetts, a number of Connecticut town plans were published, the majority of them by Richard Clark of Philadelphia. They included nine or ten Connecticut towns surveyed by E. M. Woodford that were published between 1851 and 1853 (Fig. 24–1). Woodford also compiled a map of Hartford County, Connecticut, which H. & C. T. Smith of Philadelphia published in 1855. In addition, Woodford tried his hand at publishing and issued an 1855 town map of Canaan, New Hampshire, which was surveyed by William C. Eaton (Fig. 24–2). Clark also published, in 1852, 1853, and 1854, a number of Connecticut town plans based on surveys by Lawrence Fagan (Figs. 24–3 and 24–4).

Fagan, whose name appears on the list of "Surveyors and Statisticians" in the gazetteer, had conducted surveys for a map of Onondaga County, New York, as early as 1852. Collaborators on this map (of which there is also an 1855 edition) were Sidney and Neff. In the following year Fagan produced a map of Tompkins County, New York, after which he seems to have shifted his activities for the next two years to Connecticut towns. He returned to New York State in 1855, apparently as a member of the French-Smith team of surveyors, and produced maps of Chenango (1855) and Schenectady (1856) counties. He returned to New England to prepare a map of Cheshire County, New Hampshire, in 1858. Two years later his map of Berks County, Pennsylvania, was published. Groce and Wallace identify Fagan as a native of Ireland who in 1860 was listed as a Philadelphia artist, thirty-five years of age, and owning realty valued at three thousand dollars. Judging from the dates of his published maps, Fagan probably immigrated to the United States in around 1850, when he was twenty-five years old.[2]

Franklin Gifford, John F. Geil, and Samuel Geil are three other individuals whose names appear on the gazetteer list of surveyors. The inclusion of John Geil raises questions, for his name does not appear on any New York county map. He surveyed maps of Lorain and Me-

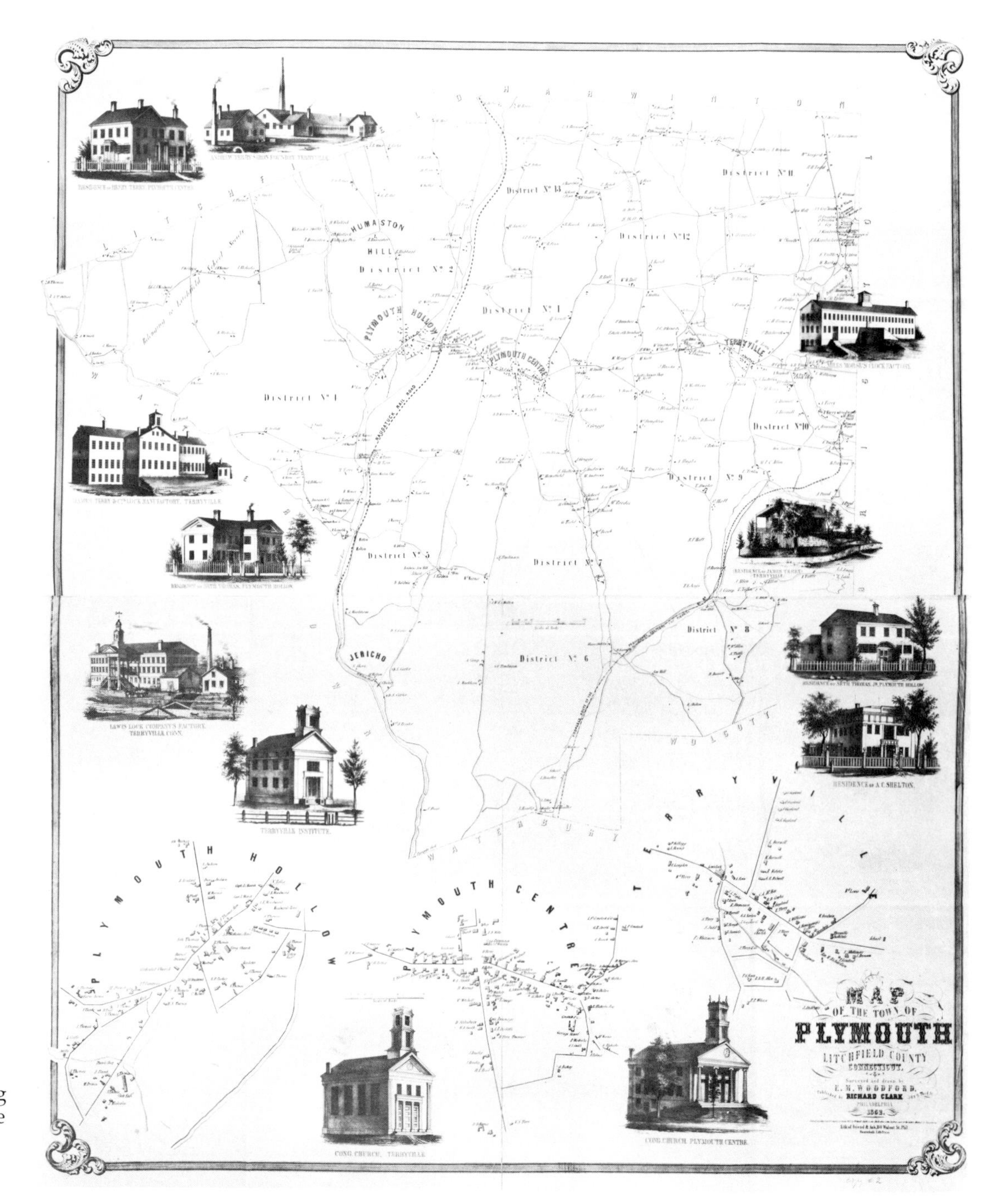

Fig. 24–1. Walling's town mapping did not include Connecticut. Woodford and his associates apparently established surveying priorities in that state. This 1852 map of the town of Plymouth in Litchfield County, Connecticut, was prepared by Woodford and published by Clark.

dina counties, Ohio, that were published in 1857 by Matthews & Taintor of Philadelphia. His relationship to Samuel Geil is not known, but they may have been brothers. Samuel Geil and Gifford collaborated on surveying Niagara County, New York, with the resulting map, dated 1852, bearing the imprint "Franklin Gifford, Publisher Wilson, N.Y." The map was registered for copyright by Smith. An inset "Plan of the Village of Wil-

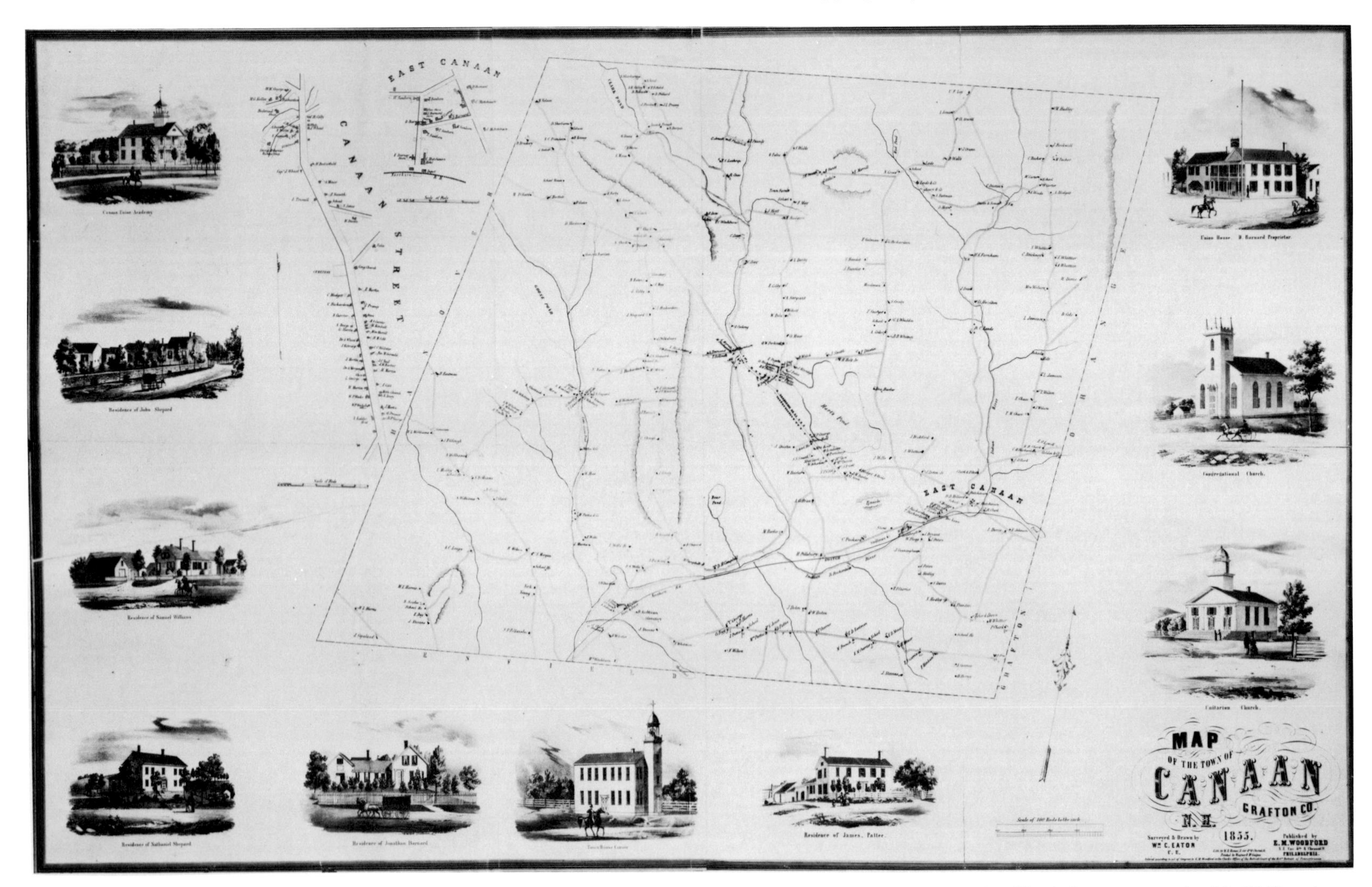

Fig. 24–2. Woodford also competed with Walling in New Hampshire. This map of Canaan, New Hampshire, was an 1855 publication of Woodford.

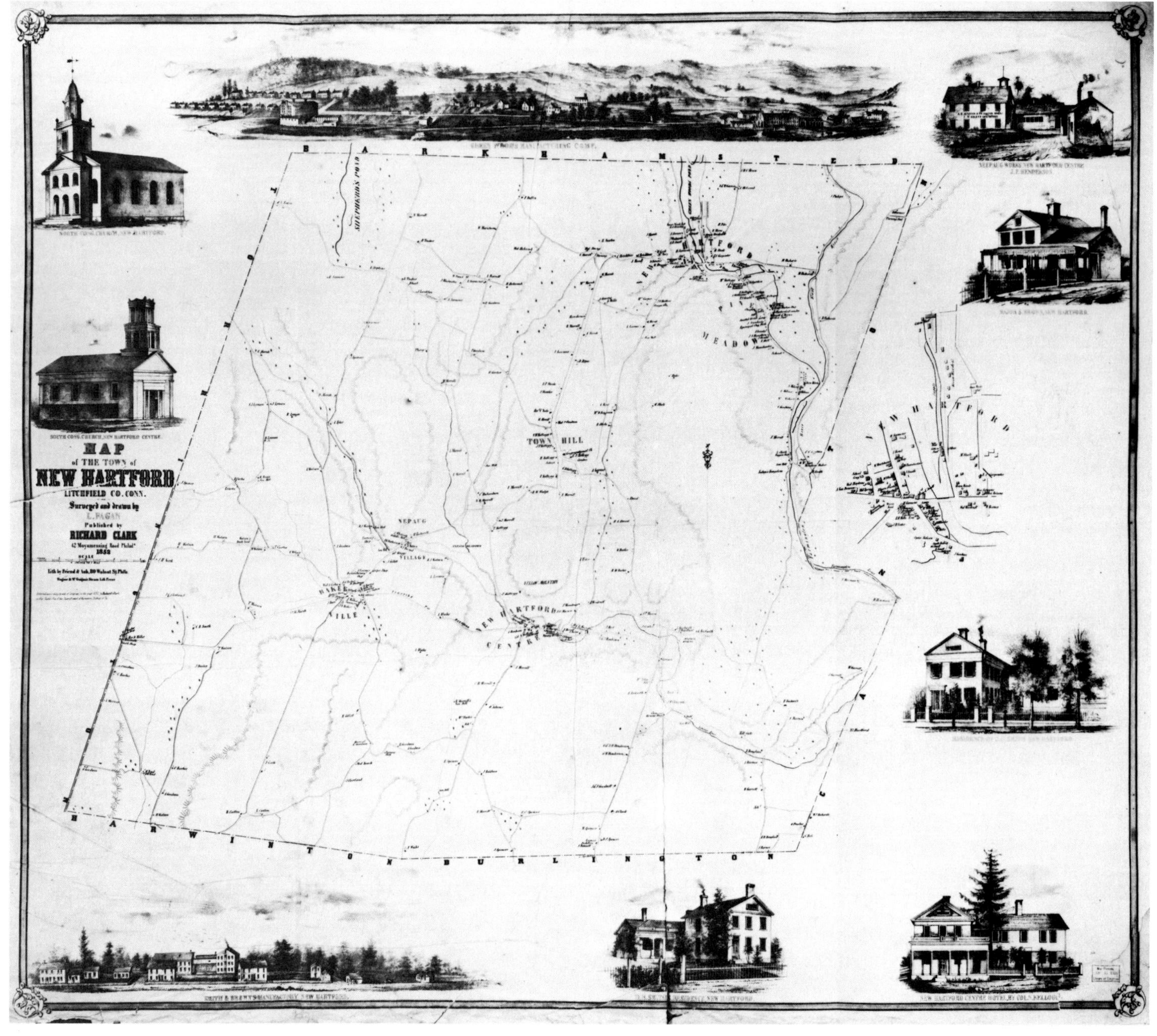

Fig. 24–3. Fagan's map of New Hartford, Connecticut, which was published by Clark in 1852.

son, Drawn by Franklin Gifford, Surveyor" locates Gifford's residence in the village. Samuel Geil is credited as compiler of the inset "Plan of the Village of Lewiston, Niagara Co. N. York." He also surveyed and drafted the inset plan of the village of Niagara Falls.

Also published in 1852 was a *Map of Orleans County, New York From Actual Survey by Lightfoot & Geil Civil Engineers Philada*. The publisher was Lloyd Van Derveer, of 15 Minor Street, Philadelphia, and the map was registered for copyright by Smith. Little is known about the surveyor Jesse Lightfoot. His first cartographic effort was a map of Monmouth County, New Jersey, which was published in 1851. He also collaborated with Samuel Geil to produce a map of Morris County, New Jersey, that J. B. Shields published in 1853.

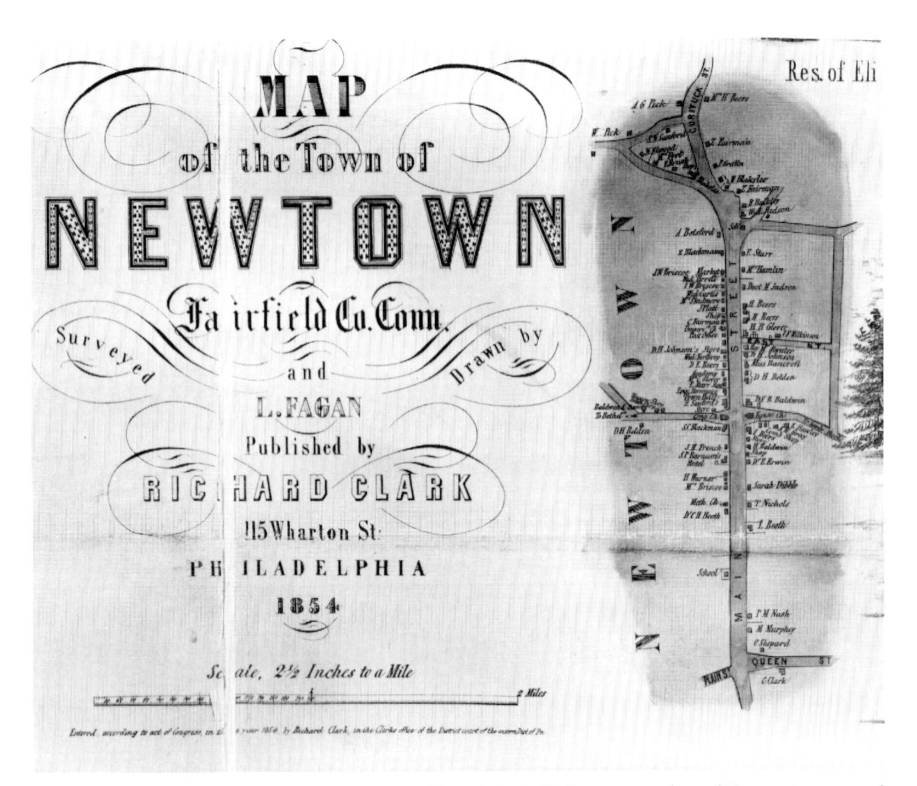

Fig. 24–4. Title cartouche of Fagan's map of Newtown, Connecticut.

Samuel Geil then joined forces with Gifford and S. K. Godshalk to survey Cayuga County, New York, for a map published by Geil in Philadelphia in 1853. Geil independently prepared a map of Erie County, New York, of which there are 1854 and 1855 editions. B. J. Hunter collaborated with Geil on maps of Montgomery and Oswego counties, New York, published, respectively, in 1853 and 1854. Geil's map of Tioga County was published in 1855, and his maps of New York's Greene, Saratoga, and Cattaraugus counties were issued in 1856. Godshalk assisted Geil on the Cattaraugus map. Geil moved into new territory in 1857 to produce with several associates maps of twenty Michigan counties. He also published maps of Elkhart and La Porte counties, Indiana, in 1861 and 1862. As a publisher, he is listed variously as S. Geil, Geil & Jones, and Geil, Harley & Sivard. There is no record of Geil publications after 1864.

Also influential in local mapping during the last half of the nineteenth century were various members of the Beers family. At least eight members of this clan engaged in map and atlas surveying, production, and publishing between 1855 and 1915. The Beerses originated in or near the town of Newtown in Fairfield County, Connecticut, and were descendants of Anthony Beers, who emigrated from England in around 1630. John Homer French almost certainly introduced them to surveying and mapping. It will be recalled that prior to joining Smith in 1855 French taught in various schools and academies in New York State and New England. In 1852 he was engaged by the trustees of the Newtown Academy in Newtown, Connecticut, to serve as headmaster and teacher of mathematics (Fig. 24–5). At the time, French was described "as an excellent disciplinarian, a superior teacher, and excelled as a mathematician."[3]

During his second year at the academy, French encouraged his students to publish a small newspaper called the *Academician*. Because of financial inadequacies the journal had a run of only one year, but its pages recorded the names of faculty and students during the 1853–54 school year. Among the listed students are several with whom we are concerned in this chapter, including Silas N. Beers (Fig. 24–6), Silas's brother, Daniel G. Beers (Fig. 24–7), and John Hobart Beers. Several members of the Beers family not listed among the students also chose mapmaking as a profession. Other Newtown Academy students such as D. Jackson Lake and Augustus Warner were likewise drawn to cartographic careers.

French left the academy in the summer of 1855 to become the general superintendent of the New York map project then being organized by Smith. We do not know how French became known to Smith, but by the time Smith contacted him, he had already taught for a number of years and had co-written a popular series of arithmetic books. From his mapmaking headquarters in Syracuse, French reviewed and inventoried the maps of New York counties that had previously been published. Some of these maps were not up to French's standards, and he set up a program to remap certain counties and to make original surveys for counties not yet mapped. To assist him in this project he enlisted several of his former Newtown Academy students.

The first of these was Silas N. Beers, who was born in Newtown on September 3, 1837, and was the son of Charles Henry Beers and Mary Elizabeth Glover Beers.[4] Beers probably joined French in the summer of 1856, when he was nineteen years old. His first assignment was to survey Schuyler County, located approximately sixty miles southwest of Syracuse. The Schuyler County map, at the scale of approximately 1:40,000, was published in 1857 "by J. H. French, 8, South Salina St., Syracuse, N.Y." It was lithographed and printed in Philadelphia. The map, which does not carry the Smith copyright registration, has marginal illustrations of prominent buildings and homes. This practice of including illustrations apparently set the pattern for subsequent county maps prepared under French's direction. The cartouche acknowledges that the map was surveyed and drawn by Beers.

Probably in the summer of 1857 French assembled a crew of three young men to assist him in surveying Oneida County, located about twenty-five miles to the east of Syracuse. In addition to Silas Beers, the group included Beers's cousin Frederick W. Beers and D. Jackson Lake. Frederick W. Beers, the son of James Botsford Beers, was born August 17, 1839, and resided with his family in Brooklyn. His family retained close ties with the Newtown Beerses, and there are indications that Frederick, during his teen years, spent his summers on the farm of his uncle, Horace Beers, in Newtown. It was probably during such a visit that Frederick learned about the French-Smith map project and joined Silas in this venture.

Lake was of an old New England family which had settled in Newtown early in the 1700s. Although his birth date is not known, he was most likely born in 1839 or 1840. The cartouche of their published map reads *Gil-*

Fig. 24–5. French's home and the Newtown Academy are among the illustrations on Fagan's *Map of the Town of Newtown*.

lette's Map of Oneida Co. New York From actual Surveys under the direction of J. H. French By S. N. Beers, D. J. Lake, and F. W. Beers, 1858. The address of the publisher is given as 517, 519, and 521 Minor Street, Philadelphia. This was, of course, the address of Smith, who registered the map for copyright. It is possible that French used the Oneida County map project as a training ground for his three young surveyors, and he probably spent a day or two in the field with them each fortnight.

Silas Beers also prepared maps of Columbia, Ontario, Orange, Rockland, Rensselaer, and Yates counties for French and Smith. Frederick Beers and Lake assisted him with the surveys for the Columbia County map, and Lake collaborated with him on the Rensselaer map. His maps of Chenango and Cortland counties and Yates County were published in 1863 and 1865, long after the French-Smith New York State map and gazetteer were issued. It is possible that the surveys had been completed and the maps drafted but that publication had been delayed because of the Civil War. Silas Beers was aided in surveying for the Chenango-Cortland map by Frederick Beers, Augustus Warner, Charles S. Warner, and Beach Nichols. The Warners and Nichols were also former students of French. The map was published by A. Pomeroy of 517, 519, and 521 Minor Street in Philadelphia. Pomeroy also registered the map for copyright. Worley & Bracher is listed as the lithographer, and Frederick Bourquin is listed as the printer. The Yates County map was surveyed by Silas Beers, Daniel G. Beers, A. B. Prindle, and H. A. Hawley. Again, Worley & Bracher was the lithographer and Bourquin the printer. The publisher and copyright registrant, however, was Stone & Stewart of 600 Chestnut Street, Philadelphia.

After completing their surveys of New York counties, Frederick and Silas Beers collaborated on preparing maps of several counties in Pennsylvania and New Jersey. Their maps of Monmouth County, New Jersey, and Washington County, Pennsylvania, were published in 1861; their maps of Allegheny and Dauphin counties, Pennsylvania, and Cumberland County, New Jersey, were issued in the following year. Lamson B. Lake (D. Jackson Lake's cousin and also a resident of the Newtown area) and Charles Warner assisted them with the Cumberland County survey, and Henry F. Walling collaborated on the Monmouth map. For the Allegheny map (published by the Philadelphia firm Smith, Gallup & Hewitt) the Beerses were aided by Prindle, Charles Warner, and another cousin, James M. Beers. James Beers, the son of John G. Beers, was born on August 12, 1841. This map is his only work listed in Richard W. Stephenson's *Land Ownership Maps*.[5] There is no record of any other cartographic works by him, except that he is listed as an assistant on an atlas of the oil region of Pennsylvania done by Frederick in 1865 (see Chapter 24). In 1864 James Beers married Emily Julia Beach of New Haven, Connecticut, where the couple established their home. There he was a machinist and manufacturer of wagon wheels. He died in New Haven on January 18, 1918.

There is a break in Beers cartographic activity after 1862, and we may assume that the young men served in the Union army. Frederick and James Beers are the only ones for whom we have a military record. Frederick served in the Third Regiment of the Pennsylvania Heavy Artillery under Colonel Joseph Roberts and was discharged on February 17, 1863, with the rank of first lieutenant. It seems likely that he enlisted in Pennsylvania after completing his surveys of some of the state's counties. James enlisted in the Twenty-third Regiment of the Connecticut Volunteers on September 11, 1862, and was mustered out on August 31, 1863. Following his military service Frederick resumed county mapping, primarily in Pennsylvania, New Jersey, and Michigan. He produced maps individually and in collaboration with various associates. Among his collaborators were A. D. Ellis and G. C. Soule, with whom he surveyed a map of Venango County, Pennsylvania in 1865 (Fig. 24–8).

Charles S. Warner and Augustus Warner were the sons of Charles C. and Ann Clark Warner. Augustus

Fig. 24–6. Silas N. Beers.

was born August 10, 1839, and Charles was probably born a year or two later. Augustus, after attending the Newtown Academy, went to the Connecticut Normal Institute, from which he graduated in 1858. He taught for several years, but poor health prompted him to choose the out-of-door profession of surveying and mapping. He was joined by Charles, and, in 1865, another brother, L. C. Warner, was added. They surveyed and mapped seventeen or eighteen counties in Indiana and a half dozen in Ohio from 1862 to 1867. Also assisting them on several of the surveys were S. B. Hayes, D. Jackson Lake, A. Y. Peck, William H. Fraser, H. H. Simmons, and H. H. Hardisty. Peck may have been from Newtown. For a number of the maps prepared by the three Warners, Charles is also listed as the publisher, often in association with C. A. O. McClellan, E. B. Gerber, S. L. Yandes, T. B. Tucker, J. Willard, or A. Shoemaker. Willard's, Shoemaker's, and Warner's locations are usually given as Newtown, while the others are either Philadelphia or Waterloo or Ligonier, Indiana.

D. Jackson Lake, an assistant to the Beerses on the Oneida and Rensselaer, New York, maps, was included on the list of surveyors and statisticians in the gazetteer. Before he became involved with the New York project and while he was working on it, he assisted in surveying several counties in Pennsylvania. He surveyed Westmoreland County, in the southwestern part of the state, with N. S. Ames, and the resultant map was published in 1857 by William J. Barker of North Hector, New York. Barker was also the copyright registrant. Lake then mapped Bradford County, in the northeastern part of Pennsylvania bordering New York State, with Ames and

Fig. 24–7. Daniel G. Beers.

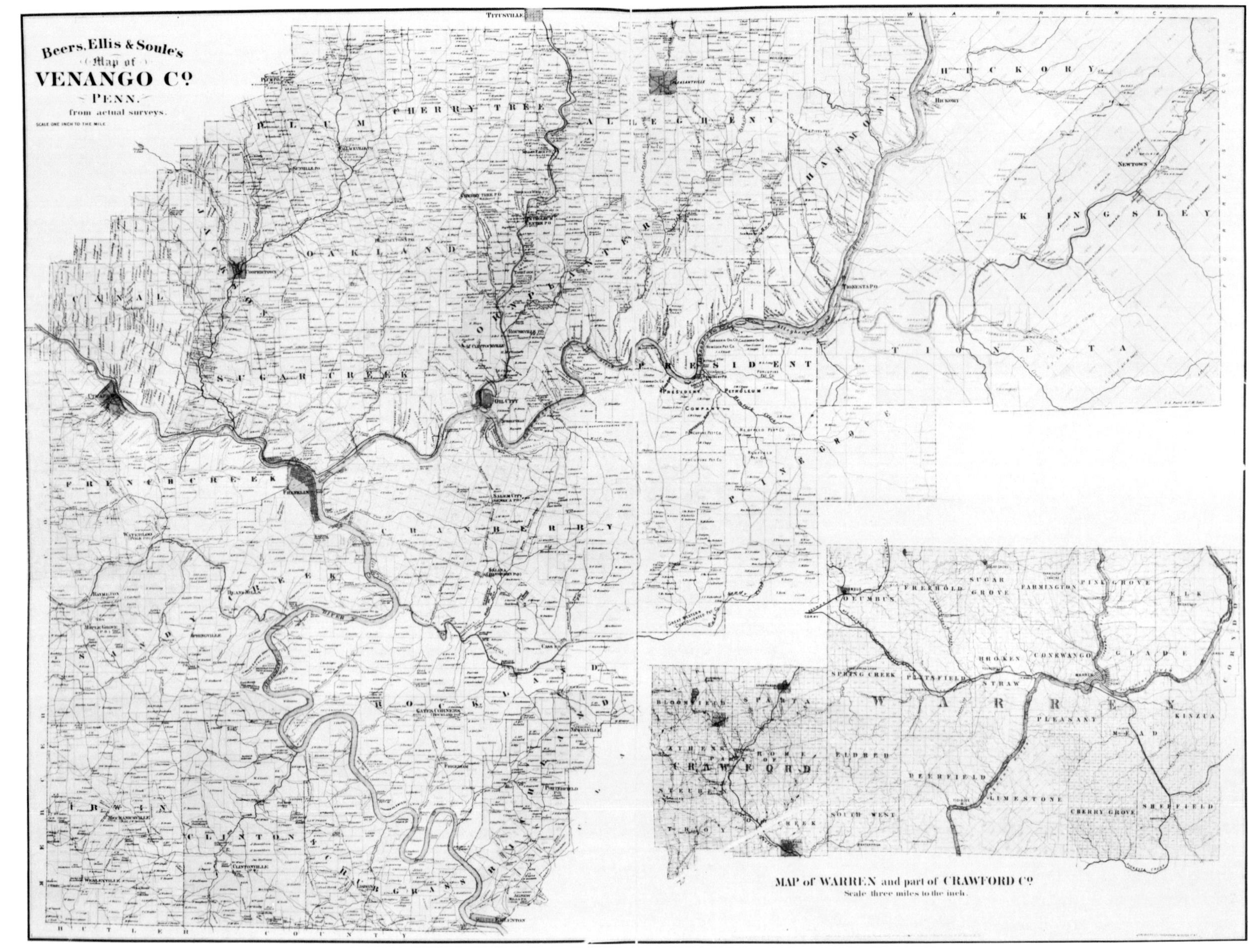

Fig. 24–8. Beers, Ellis & Soule collaborated on surveying this map of Venango County, Pennsylvania, which was published by F. W. Beers & Company in 1865.

D. H. Davison. This map was published and copyrighted by Barker in 1858. In 1860 Lake published *Shearer's Map of York County, Pennsylvania, From Actual Surveys by D. J. Lake* with W. O. Shearer. Their address was 517, 519, and 521 Minor Street, Philadelphia. Shearer and Lake also copyrighted the map. This appears to have been Lake's first experience in cartographic publishing. It is not clear how he became involved in surveying these three widely separated counties in Pennsylvania particularly because the surveying of the two New York counties was interspersed between the mapping of them.

Lake, with Lamson Lake and Silas, Frederick, and Daniel Beers, also compiled a *Map of the Vicinity of Philadelphia, From actual surveys by D. J. Lake and S. N. Beers, Assisted by F. W. Beers, L. B. Lake, and D. G. Beers. John E. Gillette & C. K. Stone Publishers 517, 519, 521 Minor Street Philadelphia*. It was registered for copyright by Gillette and printed by Bourquin. The map, which is an impressive 175 by 195 cm., gives names of property owners and identifies all townships. There are insets of Philadelphia, Wilmington, Delaware, and a number of smaller towns and villages. The map has a decorative border. Eight different editions of the map were issued, each featuring a separate group of insets focused upon a specific city, county, or suburban area.

There is no record of any Lake map publications in 1862, but he reappears in 1863 with a map of Hancock County, Ohio, which he surveyed with Charles Warner. During the next seven years, in collaboration with Augustus Warner, Charles Warner, J. Silliman Higgins, or R. Thornton Higgins, Lake surveyed some ten counties in Ohio and several in Indiana.

J. S. Higgins and R. T. Higgins were natives of Stepney, a neighboring village of Newtown. They were, apparently, too young to participate in the New York project. J. Silliman Higgins's first entrance on the mapping scene was in 1865, when a map of Boone and Clinton counties, Indiana, which he surveyed with Augustus Warner, was published in Philadelphia by Cowles & Titus. In 1866 maps of Tippecanoe County and Shelby and Johnson counties, Indiana, were published. Higgins and Warner, with R. M. Sherman, G. P. Sanford, and R. H. Harrison, surveyed them both. Sanford, Harrison, and Higgins also collaborated on the 1867 map of Warren County, Ohio. R. Thornton Higgins joined his brother and D. Jackson Lake in 1868 to produce maps of Belmont and Butler counties, Ohio, and in 1869 they did maps of Montgomery and Hardin and Marion counties, Ohio. A number of early Ohio county maps were also prepared by local surveyors. Among these surveyors was George C. Eaton of Delaware, Ohio, whose map of Muskingum County was published by W. P. Bennett of Zanesville in 1852 (Fig. 24–9).

Another surveyor, who worked independently from the New York State map project, was Simon J. Martenet. He served for some years as the city surveyor of Baltimore. According to Edward B. Mathews, "during the financial crises of 1857, when his business had somewhat fallen off in the general depression of that period, to fill up his time he commenced the survey of Cecil County [Maryland] with the design of making a map of the same which should be only one of many representing the counties of the state."[6] Before the Civil War broke out, he completed maps of Cecil, Howard, Kent, Anne Arundel, and Prince Georges counties. Maps of Allegany, Garrett, Carroll, Montgomery (Fig. 24–10), and Harford counties were ready after the war. Maps of some of the counties he did not survey had been previously prepared by others, such as J. C. Sidney's map of Baltimore (1850), Bond's map of Frederick (1858), Dilworth's map of Talbot (1858), and Taggart & Downin's map of Washington (1859).

Martenet planned to compile a Maryland state map from the county maps, and in 1865 he received a grant from the legislature for that purpose. He published the map in the same year. Mathews notes, however, that "the financial aid of greatest value to the publication of this map was obtained about ten years later, when another bill was passed to supply copies of his map to each

Fig. 24–9. George C. Eaton's 1852 map of Muskingum County, Ohio.

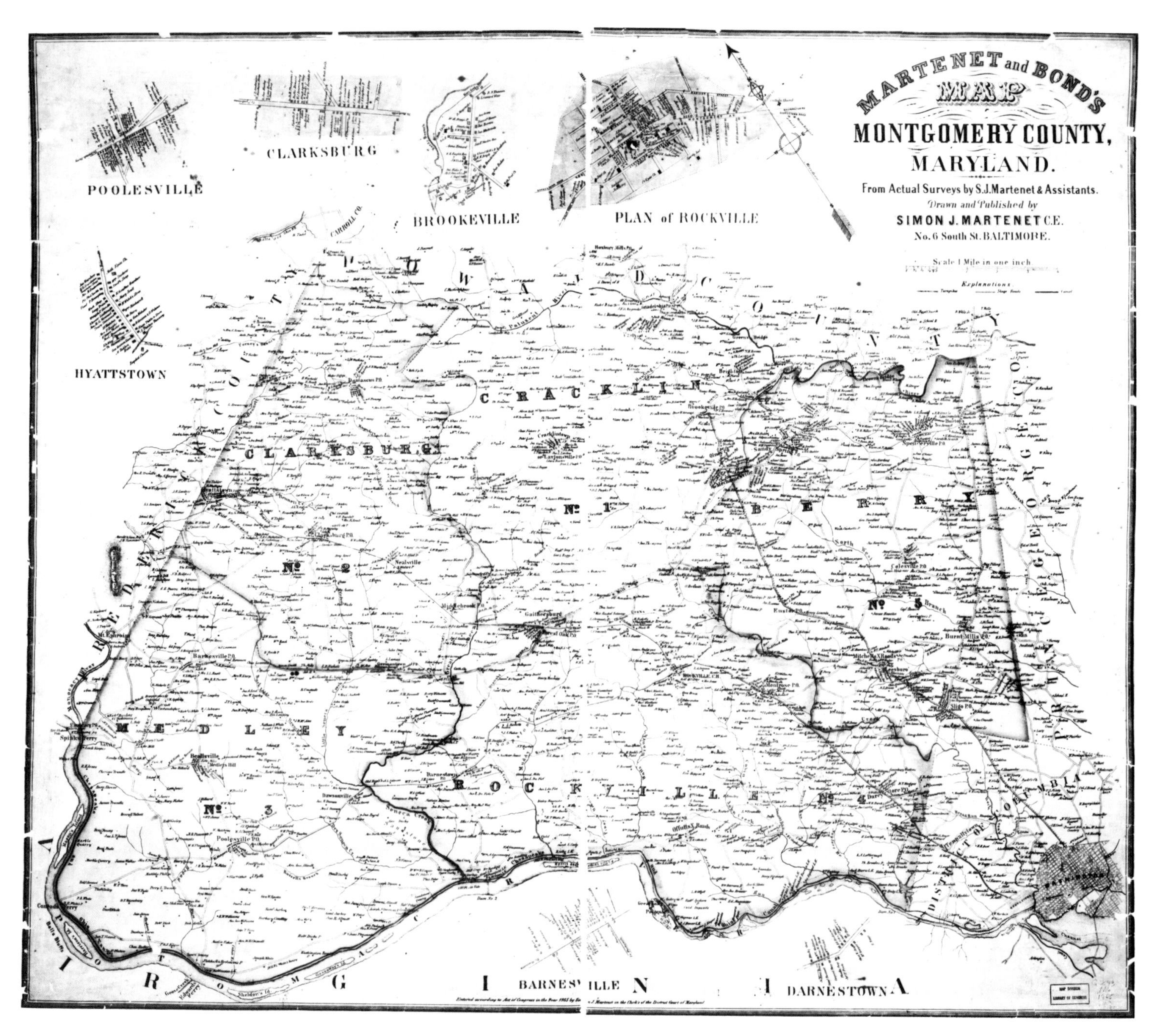

Fig. 24–10. Simon J. Martenet's 1865 map of Montgomery County, Maryland.

of the counties as official maps of the state, and the superintendent of the schools to furnish only the Martenet map to county schools making requisitions for a state map. This bill immediately introduced a considerable demand for sheets, which were sold at the rate of $7 apiece."[7] Editions of the map issued in 1885, 1886, and 1898 were published by J. L. Smith of Philadelphia.

Griffith M. Hopkins also surveyed and compiled many county maps. We know little about his personal life. The date of his birth is unknown, but his younger brother, Henry W. Hopkins, was born in 1838, so Griffith was probably born around 1836. His first cartographic contribution appears to be *A Topographical Map of Lincoln Co. Maine from actual Surveys by G. M. Hopkins. Publishers Lee & Marsh 17, 19, & 21 Minor St. Philadelphia 1857*. The copyright registration of this map was in Robert Pearsall Smith's name. In 1858 Hopkins's *Map of Cuyahoga County Ohio From Actual Surveys & County Records, under the Supervision of G. M. Hopkins, Jr. C. E.* was published by S. H. Matthews of 517, 519, and 521 Minor Street in Philadelphia. In a box near the bottom of the map is the imprint "R. Pearsall Smith, Wholesale Map Manufactory Nos. 517, 519, and 521 Minor St. & No. 2 Chestnut Street, Philadelphia." This imprint is one of only a few instances in which the name Wholesale Map Manufactory is used.

Hopkins's maps of Adams, Northumberland, and Susquehanna counties, Pennsylvania, were also published in 1858. Although there is a different publisher's name on each, the publisher's address (517, 519 and 521 Minor Street) is the same on all three. Hopkins's sole 1859 publication is *Clark's Map of Litchfield County, Connecticut From Actual Survey by G. M. Hopkins*. This is the same Richard Clark who published several Connecticut town plans prepared by E. M. Woodford and Lawrence Fagan between 1851 and 1854. The Litchfield County map is one of the few prepared by Hopkins with which Smith was not associated. One of Hopkins's most productive years was 1860, with the publication of his maps of Columbia, Montour, Mercer, Northampton, and Wayne counties in Pennsylvania, and Sussex and Warren counties in New Jersey. His maps of Bergen and Passaic counties, New Jersey, and Armstrong County, Pennsylvania, were published in 1861, and a map of the three counties of Perry, Juniata, and Mifflin, Pennsylvania, is dated 1863. His Fayette County, Pennsylvania, map, probably delayed until the Civil War ended, was issued in 1865. Hopkins collaborated with his brother, Henry, in surveying for the 1860 map of Warren County, New Jersey, and the 1861 map of Armstrong County, Pennsylvania. The brothers later formed their own cartographic firm. Historian Jefferson Moak stated that Griffith and Henry Hopkins founded their own publishing house in Philadelphia in 1865, although the earliest identified imprint of their company is 1870.[8] In the company's early years, the brothers were involved in county atlas production, but gradually they focused on city plans and atlases (see Chapter 16).

Prior to the Civil War, county map production and publishing, as we have seen, was primarily in the hands of eastern individuals. By the late 1850s their representatives had crossed the Appalachians to survey and map counties in Ohio and Michigan.

Beyond the Appalachians, M. H. Thompson of Geneva, Illinois, was prominent in county map production. He published several maps of counties in Illinois and Iowa. His earliest recorded work is the *Map of Knox County Illinois compiled, Drawn & Published from County Records, Actual Surveys & Personal Examination by: M. H. Thompson C.E. & Map Publisher 1861* (Fig. 24–11). Advertisements of local businessmen are set into the map's decorative border, and there are a number of illustrations of residences, public buildings, and commercial and industrial establishments.

The imprint on Thompson's 1862 map of McHenry County, Illinois, explains that it was done "from . . . actual surveys and personal examination by M. H. Thompson & Brother." This map was lithographed by Charles Shober of Chicago. A note on Thompson's 1865 map of Clinton County, Iowa, indicates that it was "pub-

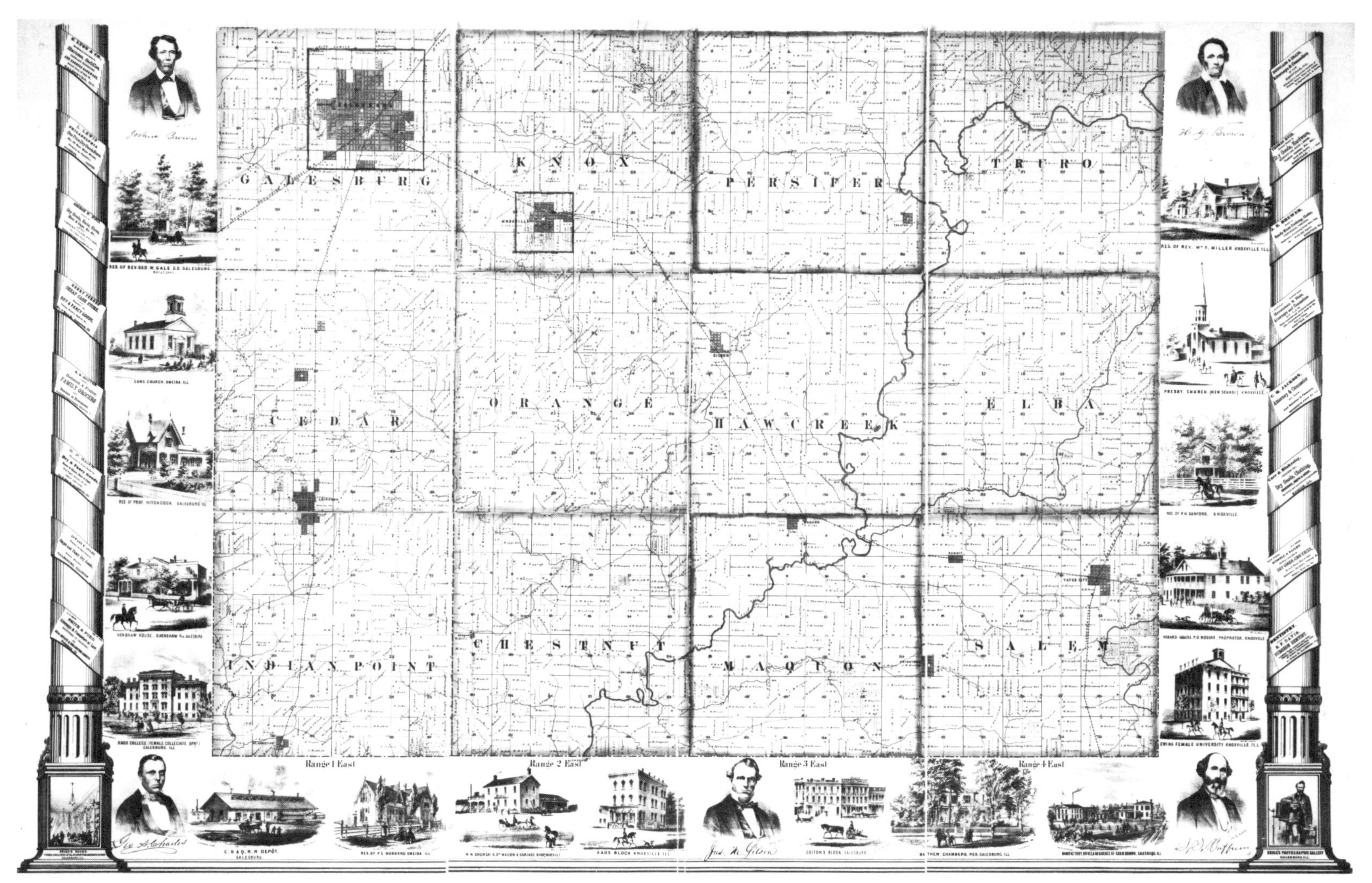

Fig. 24–11. The southern half of M. H. Thompson's 1861 *Map of Knox County Illinois*. Thompson was one of the few midwestern surveyors who engaged in county map surveying and publishing prior to the Civil War.

lished From Surveys & Actual Examination of Title Records Made by C. H. Thompson 1865." It also identifies Thompson's headquarters as being located in Dundee, Illinois. As county surveyors and mappers in the Midwest could draw upon the maps prepared by the U.S. General Land Survey, the Thompsons did so. The cartouche on the Clinton County map recognizes their indebtedness by noting that it was "Surveyed Drawn and Published from Government and actual Surveys made expressly for this Map."

For some of the men described in this chapter surveying was a temporary pursuit that they regarded as a healthful and profitable summer occupation. Others, as we shall see in the following chapter, were sufficiently interested in surveying, mapping, and publishing to establish their careers in cartography.

Notes

1. Much of the information about Sidney was derived from my personal correspondence with Jefferson M. Moak, a research historian with the Philadelphia Historical Commission.
2. George C. Groce and David H. Wallace, eds., *The New-York Historical Society's Dictionary of Artists in America 1564–1860* (New Haven, Conn., 1957), 218.
3. Jane Eliza Johnson, *Newtown's History and Historian, Ezra Levan Johnson* (Newtown, Conn., 1917), 228.
4. Much information on the various members of the Beers family is found in *The Beers Genealogy, The Descendants of Anthony Beers of Fairfield, Connecticut through His Son John*, comp. Mary Louise Regan (Palatine, Ill., 1974), and in *Commemorative Biographical Record of Fairfield County, Connecticut* (Chicago, 1899).
5. County maps prepared by the surveyors and mappers discussed in this chapter are indexed in *Land Ownership Maps, A Checklist of Nineteenth Century United States County Maps in the Library of Congress*, comp. Richard W. Stephenson (Washington, D.C., 1967).
6. Edward Bennett Mathews, *The Maps and Map-Makers of Maryland* (Baltimore, 1898), 444.
7. Ibid.
8. Jefferson M. Moak, "The G. M. Hopkins Company," *Mapline*, no. 10, June 1978.

25. *The County Atlas*

The popularity of county land ownership maps, of which some 350 were published between 1851 and 1860, reflected the administrative and political importance of the county and town in the lives of Americans. By 1860 most of the prosperous counties in the New England and Middle Atlantic states had been mapped while only insignificant coverage had been given to Ohio, Indiana, Illinois, Iowa, Michigan, and Wisconsin. After the Civil War, which greatly curtailed the cartographic industry, mapmakers began to concentrate on these states. There were advantages in mapping the midwestern states. The U.S. General Land Office had by the 1850s completed official surveys of most of the land between the Appalachians and the Mississippi. County map publishers had only to copy these surveys and supplement them with information derived from horse and carriage and odometer surveys. In contrast, publishers of county maps for the mid-Atlantic and New England states had to conduct complete compass, chain, and odometer surveys, for these states were settled prior to the establishment of the General Land Office.

Middlewestern farmers were also relatively prosperous. They profited from high prices during and following the Civil War, and their well-being was further enhanced by the introduction of various types of farm machinery in the postwar decades. Between 1850 and 1870 there was a great movement of settlers into the midwestern states. Some came from the East, while others were part of the European immigrant influx during these years. Thus, there was a large reservoir of labor from which to draw. The Industrial Revolution, stimulated in part by wartime demands, also helped to create a receptive environment for map publishers and salesmen. Map publishers, too, were encouraged by such technical improvements as the development of the steam rotary printing press, the substitution of zinc printing plates for heavy and fragile lithographic stones, the perfection of chromolithography, and the invention of cheap paper.

The idea of assembling county maps into a bound format seems to have occurred to several individuals at about the same time. All the above conditions contributed to making the time propitious for such a format. In addition, rereleasing previously published maps in atlases created a new, profitable market. There was no limit to the number of pages an atlas might contain, so it was possible to include much more descriptive, statistical, historical, biographical, and illustrative material. The inclusion of biographical sketches, views, and portraits provided the mapmakers with the opportunity to play upon the vanities of their customers, which resulted in profitable returns. Atlases were also more practical and durable. The wall map, the most common format in the pre-Civil War years, was difficult to handle and examine and was subject to rapid wear and attrition. Some county maps, as we have seen, measured as much as 175 by 175 cm.

With all these factors favoring sectioning county maps and binding them within protective covers, it is not strange that the atlas format had a wide appeal. The earliest county atlas on record is the *Map of Berks County Pennsylvania from Actual Surveys by L. Fagan*, which was published in Philadelphia in 1861 by Henry F. Bridgens. This pioneer volume is a segmented and bound version of Lawrence Fagan's wall map of Berks County that was published in 1860. An 1862 edition of the atlas, in a more orthodox atlas format, is titled *Township Map of Berks County, Pennsylvania*. It contains individual maps of forty-one townships, plans of seven towns and cities, a map of the county, a list of post offices, and statistical tables. There are no illustrations.

Fagan and Bridgens both prepared county maps prior

to their Berks County atlas collaboration. In addition to his work on the French-Smith New York State project, Fagan also mapped counties in Connecticut and New Hampshire. Bridgens's earliest cartographic products were maps of Cumberland (1858) and Lebanon (1860) counties in Pennsylvania. Two states of the Cumberland map were published in 1858. The earlier one has no borders and no printer or lithographer imprints. The second state has a decorative border, lists names of property owners on the inset plan of the city of Carlisle and includes the imprints, in the lower left and right margins, respectively, "Friend & Aub Lith. 330 (old 80) Walnut St., Phila." and "Printed by Wagner & McGuigan 38 Franklin Place Phila." The latter imprint also includes Bridgens's copyright registration notice. In addition to the Berks County atlas, Bridgens published the *Atlas of Lancaster Co. Penna.* in 1864 (Fig. 25–1) and the *Atlas of Chester County, Pennsylvania* in 1873 and 1874 editions. The Chester County atlas was done in collaboration with "A. R. Witmer and Others."

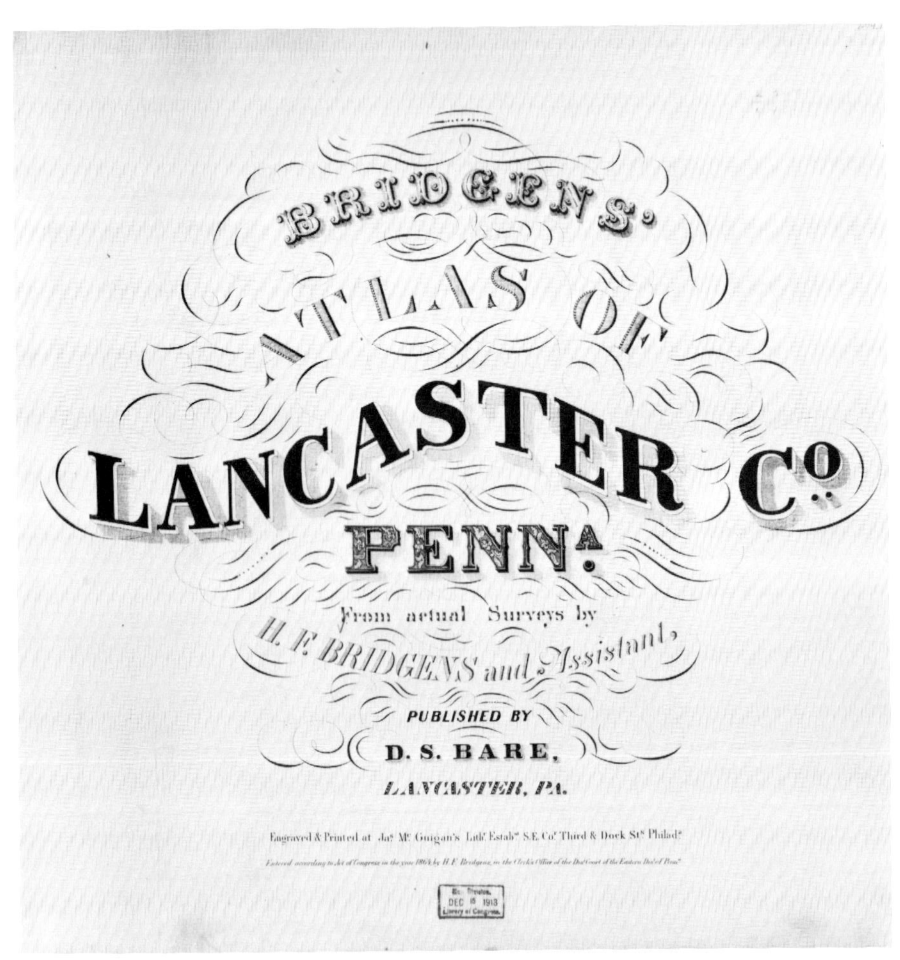

Fig. 25–1. Title page of Bridgens's atlas of Lancaster County, Pennsylvania, one of the earliest American county atlases.

Most county atlases for the New England and Middle Atlantic states were issued by established publishers located, with few exceptions, in Philadelphia or New York City. The atlases appear to have been offered for sale by the publishers and their distributors, who unlike their midwestern counterparts, did not sell their atlases by subscription. It was not until the mid-1870s, when some trans-Appalachian atlas producers extended their activities to the East, that atlases of eastern counties were sold by advance subscription.

A large percentage of the eastern county atlases were published by members of the Beers family. Silas Beers, Daniel Beers, and their assistants collaborated on a survey of Jefferson County, New York, the results of which were published in an atlas by C. K. Stone of Philadelphia in 1864. It includes individual township and town maps, a table of distances, and a title page. There is no separate map of the county and only one illustration. During the next three years the Beerses were involved in the preparation of atlases for nine counties in New York and four in Pennsylvania. Those dated 1865 and 1866 were published in Philadelphia by Stone & Stewart. Surveys for most of them were credited to Silas and Daniel Beers and assistants. Those atlases published in 1865 and later include a map of the pertinent county, and some have several illustrations, usually placed near the front of the volume. Silas Beers's final cartographic work appears to have been on the atlas of Westmoreland County, Pennsylvania, which was published in Philadelphia in 1867 by A. Pomeroy. In that year Silas Beers returned to Newtown, Connecticut, where he engaged in farming and practiced architecture. The new Trinity Church, erected in Newtown in 1870, was one of his major architectural achievements. Beers and his wife, Sarah Nichols Beers, had one daughter, Susan Lynn, who was born April 18, 1865. Beers died in 1873; his widow died in 1933.

Daniel Beers continued in the atlas business, directing surveys and compiling atlases of Franklin, Union, and Schuyler counties, Pennsylvania, in 1868 and of Allegany and Cattaraugus counties, New York, in 1869. The imprints on the Pennsylvania atlases cite the publisher as Pomeroy & Beers, and D. G. Beers & Company of New York City is identified as the publisher on the New York atlases. Little is known about Pomeroy. His name appears on county maps as an associate publisher as early as 1858 and on county atlases until about 1876. The lithography on some Pomeroy atlases was done by Worley & Bracher, with the printing done by Frederick Bourquin. Both firms were located at 32 Chestnut Street in Philadelphia.

Beers established his own publishing company, Daniel G. Beers & Company, at 95 Maiden Lane in New York City in 1868. From this address he published his atlases of Cattaraugus and Allegany counties. In 1870 or 1871 Beers moved the company to 31 South Sixth Street, Philadelphia. From those headquarters the firm published an atlas of Essex County, Massachusetts (1872), as well as atlases of the New York counties of Steuben (1873), Columbia (1873), Oneida (1874), Wayne (1874), Lewis (1875), Madison (1875), and Franklin (1876). Between 1876 and 1878, Daniel G. Beers & Company also published about fifteen wall maps of counties in Kentucky and Tennessee. Although his publishing business was located successively in New York City and Philadelphia, Beers reestablished his residence in Newtown around 1870. Biographer Mary L. Regan notes that he was an inventive man and held many patents, most of which probably related to carriage tops, which he manufactured in his later years. One of his patents, dated 1877, was entitled "Improvements in Clothes Wringers." Regan discloses that "he was very much a participant in the affairs of Newtown and was elected a Selectman in 1908. At the time of his death he had been a Trustee of the Newtown Savings Bank for 35 years and Vice President for 18. He was president of the library association and the Men's Club. For Trinity Church he was, in turn, Clerk, Vestryman, Junior Warden and Senior Warden."[1] Beers and his second wife, Anabella S. Fitch, had four daughters. He died in Newtown on February 12, 1913. In 1905 Beers published the *Bi-Centennial Map of the Town of Newtown, Connecticut*, with "Plans & Supervision by D. G. Beers." The map is framed with photographs of homes, churches, and factories.

Silas and Daniel Beers's cousin Frederick apparently returned to western Pennsylvania after the Civil War to resume his surveying. In 1865 his map of Crawford County was published in Philadelphia by A. Pomeroy & S. W. Treat. In the same year the *Atlas of the Oil Region of Pennsylvania, from actual surveys under the direction of F. W. Beers, C.E., Assisted by Beach Nichols, J. M. Beers, A. Leavenworth, C. S. Peck, C. A. Curtis & Geo. Stewart* was issued. It was published by Beers and A. D. Ellis and G. G. Soule. This appears to be the earliest cartographic publication of Beers, Ellis, & Soule of 43 John Street, New York City.

Following completion of the oil region volume, Beers turned his attention to county atlases. His earliest publication in this format was the *Atlas of Erie County, Pennsylvania*, which was published in 1865 by Beers, Ellis & Soule. It is possible that surveys for this work were conducted concurrently with those done for the oil region atlas. In 1865 Beers and his assistants also prepared surveys in central Ohio, north and east of Columbus, which resulted in atlases of Delaware, Licking, and Muskingum counties. All were published by Beers, Ellis

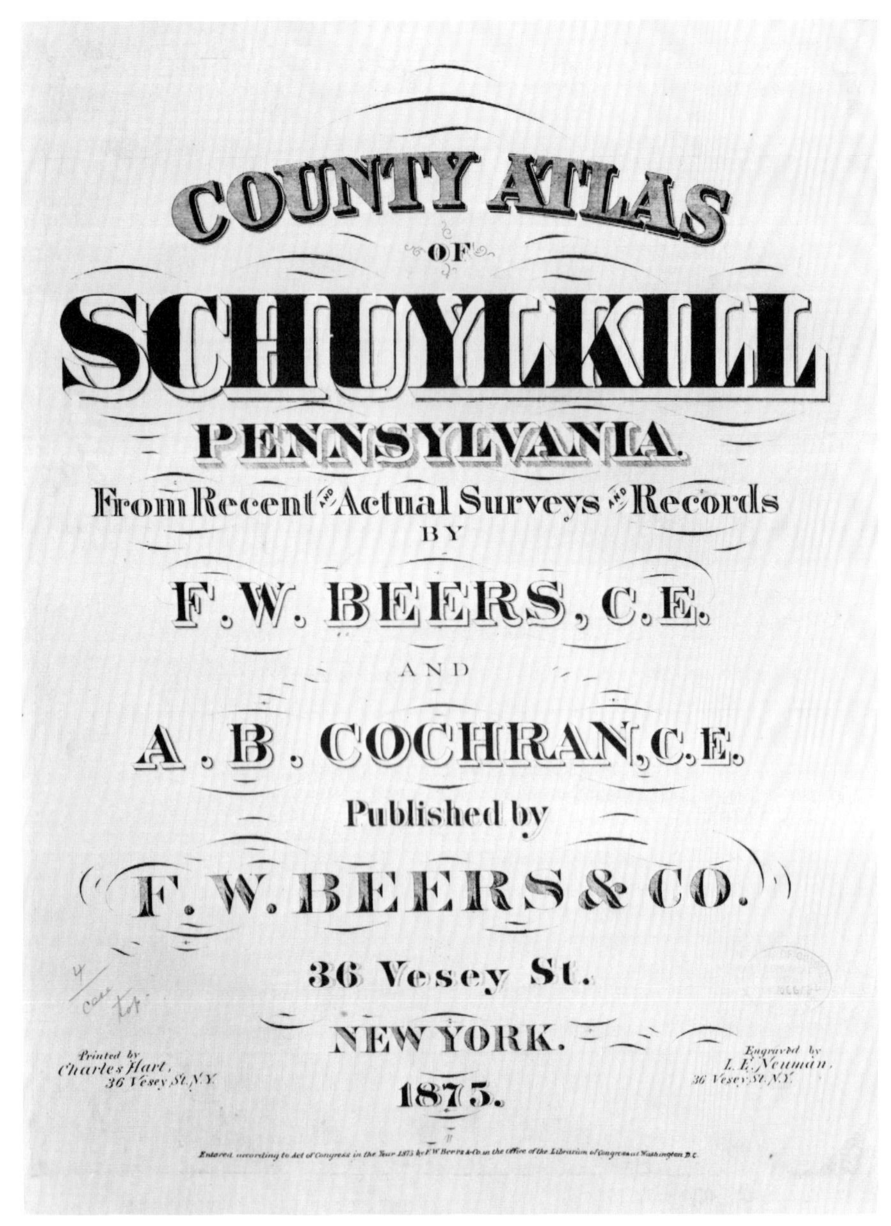

COUNTY ATLAS

OF

SCHUYLKILL

PENNSYLVANIA.

From Recent and Actual Surveys and Records

BY

F. W. BEERS, C. E.

AND

A. B. COCHRAN, C. E.

Published by

F. W. BEERS & CO.

36 Vesey St.

NEW YORK.

1875.

Printed by Charles Hart, 36 Vesey St. N.Y.

Engraved by L. E. Neuman, 36 Vesey St. N.Y.

Entered according to Act of Congress in the Year 1875 by F. W. Beers & Co. in the Office of the Librarian of Congress at Washington D.C.

Fig. 25–2. Title page of Beers and Cochran's *County Atlas of Schuylkill Pennsylvania.*

& Soule in 1866; they were lithographed and printed in Philadelphia.

Beers, Ellis & Soule's output in 1867 was limited to the *Atlas of Greene County, New York* and the *Atlas of New York City and Vicinity*. The latter work was also issued in an 1868 edition. The publisher's imprints on these atlases give the firm's address as 95 Maiden Lane, New York City, which was the address of Daniel G. Beers & Company. Since Daniel Beers's company did not relocate to Philadelphia until 1870 or 1871, Frederick and Daniel Beers may have shared publishing facilities. By 1869 Beers, Ellis & Soule published some fifteen or twenty atlases, principally for counties in New York, Pennsylvania, Connecticut, and Vermont. The partnership appears to have dissolved in 1869 or early 1870, for atlases published in 1870 carry the imprint of F. W. Beers & Company of 93 and 95 Maiden Lane. It is possible that Beers took over operation of D. G. Beers & Company from his cousin. The new company's publications in 1870 include atlases of Hampden and Worcester counties, Massachusetts, and Stark County, Ohio. F. W. Beers & Company continued at its Maiden Lane location through 1872, publishing atlases of counties in Vermont, Maine, Massachusetts, Pennsylvania, Ohio, New York, and Michigan. Frederick Beers appears to have established temporary partnerships with certain local surveyors to publish some county atlases. One such individual was A. B. Cochran, with whom Beers published the *County Atlas of Schuylkill Pennsylvania* in 1875 (Figs. 25–2 and 25–3). Little is known about Cochran, except that he was a civil engineer and probably a resident of Schuylkill County.

In 1873 the company moved to 36 Vesey Street, New York City. Atlases of counties in the New England and Middle Atlantic states dominated the output of the company, but the company did map counties in Michigan and one or two in Kansas. In 1875 or 1876 F. W. Beers & Company appears to have undergone reorganization. While some county atlases published between 1875 and 1881 still carry the imprint F. W. Beers & Company, 36

Fig. 25–3. View of Brookside Farm from Beers and Cochran's *County Atlas of Schuylkill Pennsylvania*.

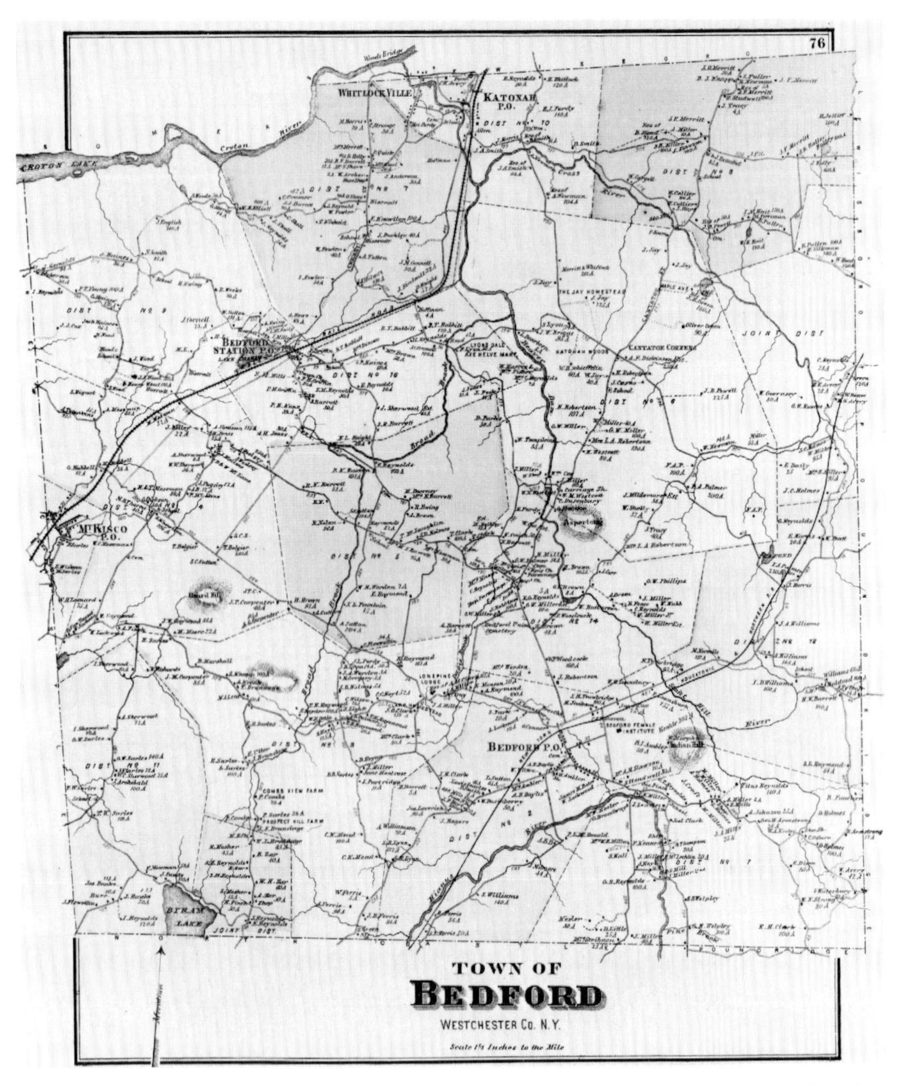

Fig. 25–4. Map of Bedford from the *County Atlas of Westchester, New York* published in 1872 by J. B. Beers & Company.

Vesey Street, others published at the same address have such varied imprints as J. B. Beers & Company (the firm of Beers's father), Walker & Jewett, and R. T. White & Company. An F. W. Beers atlas of Orange County, New York, was also published in 1875 by Andreas, Baskin & Boor of the Lakeside Building in Chicago.

It is difficult to follow the varied career of Frederick Beers. He apparently joined his father, James Botsford Beers, in founding the J. B. Beers & Company publishing house in 1870, while maintaining many other business ventures. His father was born in Newtown on October 16, 1811. Shortly after his marriage to Huldah Clark in 1835, he moved to Berlin, Maryland, where his three sons, John Clark, Albert, and Frederick, were born. It is not known how James Botsford Beers was employed while in Maryland. In around 1845 he moved his family to Brooklyn because he had business interests in New York City. Regan states that "the firm of Beers and Clark on John Street, specialized in gold pens. The company name was changed to J. B. Beers and Sons and later, in the 1870s, to J. B. Beers and Company, on Vesey Street. Their interest was then in maps and publishing. Sons John Clark and Frederick W. joined in his business at various times, and his grandson James Lemuel [son of John Clark] eventually became a member of the firm."[2]

Atlases under the J. B. Beers & Company imprint were published of Westchester County, New York, in 1872 (Fig. 25–4), Middlesex County, Massachusetts, in 1875, the city of Newton, Massachusetts, in 1886, and Staten Island, New York, in 1887. *A Catalogue of Maps & Atlases For Sale By J. B. Beers & Co., New York*, published in around 1878, lists more than 150 county and city atlases, some 60 maps, and 15 county histories. Works of other publishers were included as well as those issued by the Beers company. In addition, Frederick Beers continued to compile and publish atlases under his own imprint. He published atlases for more than 80 counties in some 10 states and produced several city atlases. In 1891 his *Atlas of the Hudson River Valley from New York City to Troy* was published by Watson & Company of New York City.

Elaborating on Frederick's other business ventures, Regan noted that "he appears to have been always associated with J. B. Beers and Co. of New York City. . . . In the 1860s he also had an interest in a lamp company called Beers, Judson and Beers, with his brother-in-law, Henry L. Beers. In addition to these he was employed for 35 years in the office of the Commissioner of Records of Brooklyn, N. Y., where he was the chief of the map division for many years. . . . He published, among others, the *Farm Line Atlas of the County of Kings and the Atlas of Long Island*. While he was basically a resident of Brooklyn, N.Y., he lived in Danbury, Conn., in 1870, in New Jersey in 1872, and in Naples, Ontario Co., N.Y. in 1880."[3] Beers died in Bridgeport, Connecticut, on September 8, 1933, at the age of ninety-four. He retired from publishing when he was ninety years old.

Another Beers, John Hobart Beers, a distant cousin of Frederick, Silas, and Daniel Beers, was born on July 13, 1840. He was the son of Hermon and Phebe Sherman Beers. Regan records that John "began his career by working two years for the Bridgeport City Bank, two for a Newtown merchant, and two for a merchant in Baltimore, Md. From 1863 to 1870 he worked as a bookkeeper in Camden, N.J. and Philadelphia, Pa., with the exception of 1865, when he was in the map business in Pennsylvania and Ohio. In 1870 he moved to Chicago, Ill. and became a publisher, starting as a member of the firm of Warner, Higgins and Beers, with Augustus Warner, J. S. Higgins of Stepney, Conn. and his brother, William Hermon Beers."[4] Following the Chicago fire of 1871, John Beers and Warner reorganized the firm. Renamed Warner & Beers and located on Clark Street, it published maps and atlases until 1879, when the name was changed to J. H. Beers & Company. Warner & Beers relocated to the Lakeside Building in Chicago in 1873.

Notwithstanding Regan's claim that Beers was in the map business in 1865, we have not been able to identify any Ohio or Pennsylvania county maps with which he was associated. Under the Warner, Higgins & Beers imprint, atlases of five Illinois counties were published between 1870 and 1873. The reorganized firm, Warner & Beers, issued about seventeen atlases of Illinois counties between 1872 and 1876, as well as one county atlas each for Indiana and Iowa (Figs. 25–5, 25–6, and 25–7). The most productive year for the firm was 1875, with ten atlas publications. As J. H. Beers & Company the firm published fifteen atlases between 1877 and 1884, two for Illinois counties and the remainder for counties in Indiana. The company also published county histories and biographies, probably in an effort to grasp a new market since atlases had been published of most of the prosperous midwestern counties by the mid-1880s. One such county history issued by J. H. Beers & Company was the *Commemorative Biographical Record of Fairfield County Connecticut*, which was published in Chicago in 1899. It is one of the most authoritative historical and biographical records of the Beers family.

Between 1864 and 1885 the several members of the Beers family, singly or in association with each other and other surveyors and publishers, produced more than 125 atlases, most of them for counties in the New England and Middle Atlantic states. This number accounts for two-thirds of the county atlases produced for these states during this period. More than 80 of these atlases were published by Frederick Beers, individually or in collaboration with associates and partners. During these years, Beers family members also surveyed or published more than 50 county land ownership maps.

Augustus Warner, who had helped survey a number of counties in Pennsylvania, New Jersey, Ohio, and Indiana between 1862 and 1867, invented in 1868 a combination county atlas that included maps of the world, the United States as a whole and its major divisions, the pertinent state, the featured county, and detailed maps of the townships and towns in that county. The general world and U.S. maps, most of which were prepared by H. H. Lloyd & Company of New York City, were used repeatedly in different county atlases. Lithographing

Fig. 25–5. Surveyors and a view artist at work are pictured on the title page of Warner & Beers's 1874 *Illustrated Historical Atlas of St. Clair Co. Illinois.*

and printing for most of Warner's early county atlases was done by Frederick Bourquin in Philadelphia. After 1872 he used Chicago lithographers and printers. Some of his combination atlases have illustrations, others do not. Warner continued in the atlas business until 1877 and was, for a time, the most successful publisher of county atlases in the United States.

In 1878 Warner visited China and Japan and spent considerable time in the latter country. When he returned to Chicago he opened a Japanese curio store with J. A. Spooner, but it operated only briefly. Warner also engaged in the manufacture of barbed wire and owned the *Chicago Daily American* briefly. He remained single until after he retired from the atlas business. On January 15, 1880, he married Rissa J. Beers, the youngest daughter of Cyrenius Beers, who lived in Chicago but was also a native of Newtown. Cyrenius Beers had left Newtown at the age of twenty-two and had settled in Chicago in 1835. Regan notes that "he was a dominant figure in the early days of Chicago and highly respected among members of the Beers family."[5]

Another county mapper and county atlas publisher was George E. Warner, who entered the cartographic field in 1869. He was probably born in around 1845 and was probably not a member of Augustus Warner's family. The first maps with which he was associated are of Wapello, Jasper, Jefferson, and Mahaska counties, Iowa, all of which he surveyed. The Wapello County map is dated 1870, and the other three were issued in 1871. All four were published by Warner and his partner, H. H. Harrison, under the name Harrison & Warner. The firm was variously located in Ottumwa, Marshalltown, Fairfield, and Oskaloosa, Iowa. The four maps were lithographed by Worley & Bracher and printed by Bourquin. Warner and Harrison also published atlases of Iowa and Wisconsin counties before they terminated their partnership in 1875.

In the following year, Warner established an association with Charles M. Foote. Foote, who was born in 1849, appears not to have been engaged in map or atlas publishing prior to his partnership with Warner. Their first publication was a map of Sedalia County, Missouri. Over the next decade Warner & Foote produced maps of counties in Wisconsin, Minnesota, and Iowa. They focused on counties that had been omitted in the pre-Civil War era of mapping and which were at that time still sparsely populated. Their *Atlas of Grant County, Wisconsin*, published in 1877, appears to be the earliest Warner & Foote publication. The following year they published atlases, or as they titled them, plat books, of several Minnesota counties. Over the next seven or eight years they produced a dozen or more atlases for counties in Iowa and Minnesota. During these years they also published a number of maps of counties in these states. Their headquarters were located in Minneapolis, but their lithography, printing, binding, and mounting were done by Philadelphia firms.

In contrast with the atlases published by Augustus Warner and his associates, Warner & Foote county atlases have no illustrations. An explanation for this practice is found in the introduction of the *Testimonial Book* published by Warner & Foote in 1880. "Imposters," the publishers remarked, "are to be found in every business and profession, the more honorable callings being often the most infested. Quacks have made so-called maps a medium for introducing views of imaginary 'improvements' and cuts of impossible 'blooded stock,' but they cannot detract from the value of a true map any more than the counterfeiter depreciates the government coin." Continuing, they observed that "engraving, printing and coloring are the handiwork of skilled artizans [*sic*], but before they can begin their labor, the cartographer, or map maker, must collect and arrange the materials; upon his facilities and judgment depends the whole value of the completed map; his errors, as well as his excellencies, will be perpetuated in every copy; no personal consideration or political bias should find expression, there should be but one object and aim, acquiring, to impart, correct geographical knowledge."[6]

Most of the *Testimonial Book* consists of laudatory let-

Fig. 25–6. Two farmsteads as pictured in Warner & Beers's *Illustrated Historical Atlas of St. Clair Co. Illinois*. These views show considerable rural activity as well as a passenger train rolling across the flat Illinois farm land.

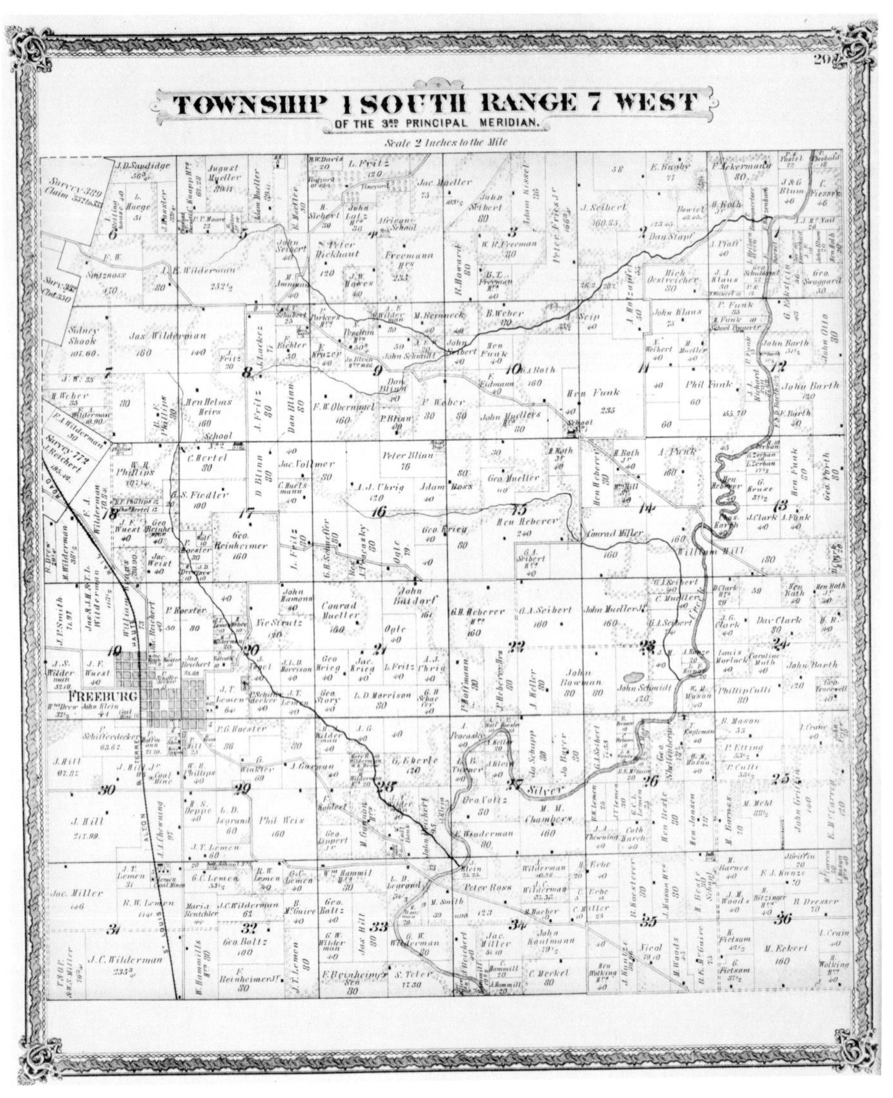

Fig. 25–7. The rectangular survey pattern of townships in the Middle West is neatly illustrated on this township plat from the *Illustrated Historical Atlas of St. Clair Co. Illinois.*

ters and statements about Warner & Foote maps. The *Sedalia* [*Missouri*] *Daily Bazoo* of December 26, 1876, for example, noted this about the map of Pettis County, Missouri, one of the first to be published by Warner & Foote:

> Some eight or nine months ago Geo. E. Warner and C. M. Foote, well known county map publishers, commenced the compilation of a farm map of Pettis County. The result of their labors is now before us, and we can only say that they have produced for us a splendid work, fully demonstrating that the recommendations which they brought from other counties was well deserved. The gentlemen have in their employ a large corps of assistants, experienced in their several branches of collecting information, drafting, engraving, printing and coloring. Everything is systematically done, and their completed maps are not only a credit to themselves, but a standing advertisement of the enterprise and business of the county where published.[7]

The *Red Wing* [*Minnesota*] *Argus* issue of April 9, 1879, reported that "Messrs. Warner & Foote publish maps only of counties where they can obtain enough orders in advance, to pay the cost of compiling the work, making the engraving &c. No maps are kept on sale or printed, except for those ordering of the agent who corrects the plats in advance of publication, and everyone ordering can depend upon getting just the map, in just the manner described in the prospectus. Commissions and residents of counties where maps are needed, should examine those put up by Warner & Foote before making arrangements with any other publishers."[8]

Warner & Foote continued in business until 1886, after which date Warner's name no longer appears in the company's name. It is not known whether he retired or died. Foote carried on the business in association with Edwin C. Hood and John W. Henion at times and independently at other times until 1899. His son Ernest B.

Foote took over the business and published county atlases between 1900 and 1903. After the departure of Warner, the company focused on publishing county atlases rather than atlases and maps. Wisconsin, Minnesota, Michigan, and Iowa were the principal states serviced.

D. Jackson Lake, who had prepared and published a number of maps of Pennsylvania, Ohio, and Indiana counties between 1862 and 1870, entered the field of county atlas publishing in 1870 and became one of the most prolific producers of such works. In his first six years in the business, he published atlases of twenty-five counties in Ohio, ten in Michigan, two in Indiana, and one each in Maryland and Pennsylvania. With another Newtown native and graduate of Newtown Academy, Bruce N. Griffin (sometimes spelled Griffing), Lake published an atlas of Brown County, Ohio, in 1876. Later that year the two formed a firm with another partner. The firm, Lake, Griffing & Stevenson, published atlases of seven Maryland counties, as well as for one Ohio and one Kentucky county.

The firm was reorganized in 1879 under the name D. J. Lake & Company. Griffin left the firm in that year and returned to Newtown, where he joined his brother, George, and brother-in-law, Marshall H. Sears, in operating the family button factory. In 1889 Griffin became a partner in the business. Lake meanwhile shifted his company's focus farther west. His output during the next five years included twenty Kentucky county atlases, ten atlases of Indiana counties, three of Illinois counties, and one for an Ohio county. In less than fifteen years, Lake and his associates published some seventy-five county atlases.

Most Lake atlases dated before 1876 were lithographed and printed by C. O. Titus or Titus, Simmons & Titus, located at several Philadelphia addresses. After 1876, most of Lake's county atlases were lithographed by Worley & Bracher and printed by Frederick Bourquin. Both firms were located at 320 Chestnut Street in Philadelphia. County atlases prepared by Lake and his associates before 1875 had few or no illustrations. After this date they began to follow the general practice of midwestern publishers and generously filled their atlases with lithographic illustrations of public buildings, churches, scenic and historic places, industrial establishments, banks, exteriors and interiors of businesses, farmsteads and residences, and portraits of landowners, their wives, children, prize horses, cattle, and hogs.

As the most easterly of midwestern states, and with most of its counties well populated and prosperous in the post-Civil War years, Ohio received major attention from county atlas publishers. By 1890 more than ninety atlases had been published of Ohio counties. A number of counties had atlases compiled by two different publishers; several counties were covered by three atlases published by three different publishers. The most productive publisher after Lake and his associates was Louis H. Everts who, independently and in collaboration with others, is credited with about twenty Ohio county atlases. Although their areas of interest occasionally overlapped, Lake generally mapped counties in the northeastern, eastern, and southern parts of Ohio, while Everts focused on northern, western, and central counties. Some eight or ten Ohio counties were also covered by atlases bearing the imprint of J. A. Caldwell. These counties were principally in the north central and central parts of the state.

Writing in 1911, C. E. Sherman described the activities of county atlas publishers in Ohio:

> During this period [from 1870 to 1885] a large majority of the counties of the State were mapped from original surveys, in part, and in part from compilation. The method was to take the outline of the public land surveys from the Land Office returns, and add the location of the wagon roads and other artificial features, such as buildings, cemeteries, orchards, from a wheel traverse of the roads.
>
> In many instances these traverses were made on foot, the surveyor carrying his outfit on a light

> wheelbarrow equipped with a large wheel. He counted the revolutions of this to get distances, and took bearings with a compass. Private land lines were added from the county records. The results from these and other sources were fudged until they agreed, and published to a scale of usually two miles per inch, on sheets approximating 15 × 18 inches in size, a township to each sheet. The townships for each county were bound together in an atlas, together with more or less descriptive matter, prefaced with a state and county map. Notwithstanding their crude method of making, these atlases are very valuable for many purposes still. They show public and private land lines, buildings, bridges, streams, ponds, orchards, cemeteries, roads, canals, railroads, towns, cities and political districts as they existed at the time of survey. They also occasionally show a limited amount of natural features such as woods, marshes and relief, and usually give the township range and section numbers in addition to property owners' names.[9]

Beach Nichols, another Newtown native and surveyor of county maps, also entered the county atlas publishing business. After surveying counties in 1862, he enlisted in Company C of the Twenty-third Regiment of the Connecticut Volunteers. He was honorably discharged on August 31, 1863, and around 1865 he married Adelia Fairchild, also of Newtown. Nothing is known of Nichols's activities during the next several years, but in 1867 or 1868 he again turned his attention to surveying and mapping, focusing on atlases. In 1868, in partnership with J. Jay Stranahan, Nichols published atlases of Herkimer County, New York, and Montgomery and Fulton counties, New York. Surveys for both atlases were made by and under the direction of Nichols. The publisher's address was 95 Maiden Lane in New York City. Neither atlas has illustrations.

Nichols's publications in 1873 included atlases of Blair and Huntington counties, Pennsylvania, and Lycoming County, Pennsylvania. Both were published by A. Pomeroy in Philadelphia, with lithography by Worley & Bracher and printing by Bourquin. Atlases of Ontario and Schuyler counties, New York, based on surveys by Nichols and his assistants, were 1874 publications of Pomeroy, Whitman & Company. Worley & Bracher was the lithographer and Bourquin the printer. The same publisher issued Nichols's atlas of Chenango County, New York, in 1875. It has six pages of illustrations. Atlases of Armstrong and York counties, Pennsylvania, were 1876 Pomeroy publications, again with the surveys done under Nichols's direction. The Armstrong County atlas includes five pages of illustrations. The last recorded Nichols-Pomeroy atlas is of Perry, Juniata, and Mifflin counties, Pennsylvania, which was published in 1877. Like most Nichols-Pomeroy atlases, it is not illustrated. Nichols appears to have returned to Newtown in 1877, but did not give up surveying entirely. In her book *Newtown's History and Historian, Ezra Levan Johnson*, Jane E. Johnson includes the following report on a town meeting:

> Whereas at a special meeting of the Town of Newtown held Saturday, January 5, 1878, at one o'clock P.M., it was voted that the school district lines be defined by the selectmen, and the Town clerk make copy of the same in a book kept for that purpose. Now, therefore, we the selectmen of the Town of Newtown for the time being have performed said duty with the assistance of Beach Nichols as surveyor and do hereby define and fix the lines of the following districts in the words and figures here-in-after set down.[10]

There were marked differences between eastern and midwestern county atlas surveyors and publishers. Many of the eastern surveyors were trained civil engineers, and, after gaining experience in preparing county maps, they began to compile and publish atlases. A general characteristic of the midwestern atlas men was their lack of permanent headquarters, their tendency to change partners frequently, and the almost universal practice of soliciting subscriptions for atlases, and for the illustrations and biographical sketches included in them, prior to publication. Some fifty or sixty persons engaged in county atlas production in the midwestern states between 1870 and 1885. The majority published fewer than four atlases each, while less than a dozen individuals were responsible for more than four hundred midwestern county atlases or approximately 80 percent of the total atlas output for these states.

One of these prolific midwestern county atlas publishers was Louis H. Everts. He was born on April 14, 1836, in the town of Otto, in Cattaraugus County, New York. In 1851 his family moved to Illinois, where his father, Samuel Everts, purchased land in Kane County. When Everts was twenty-one years of age, he entered the mercantile house of Potter Palmer in Chicago. The outbreak of the Civil War interrupted his career, and late in 1861 he was back in Kane County, where he assisted in raising a regiment of soldiers. He was rewarded for this service with the commission of lieutenant in Company D of the Fifty-second Illinois Infantry. Before the end of 1861, Everts had been promoted to captain and assistant adjutant general in the Second Division of the Sixteenth Army Corps, which was commanded by Brigadier General Thomas W. Sweeny. Everts served with distinction and was with General William T. Sherman on the historic march from Atlanta to the sea.

He was mustered out of the army with the rank of major in 1865. Shortly thereafter Everts formed a partnership in Geneva, Illinois, with Captain Thomas H. Thompson, an army associate, with the objective of publishing county maps. Prior to the war, Thompson, in association with his brother, M. H. Thompson, had published maps of Knox County (1861), and McHenry County (1862), both in Illinois. Among the early publications of Thompson & Everts is an 1867 map of Jackson County, Illinois, and maps of Carroll, Delaware, and Linn counties, Illinois, that were published in 1869. In 1870 Thompson & Everts began publishing atlases, releasing in that year what they called "combination atlas maps" of La Salle County, Illinois, and Henry and Johns counties in Iowa, from their headquarters on the second floor of the Masonic Building in Geneva. The firm then published atlases of De Kalb and Kane counties, Illinois, in 1871 and 1872, respectively. In the 1870 atlases lithographic illustrations were placed at the rear of the volumes. After 1870 the illustrations are interleaved with township maps throughout the atlases. The lithographic printing for Thompson & Everts atlases was done by N. Friend of 332 Walnut Street, Philadelphia. The atlases were sold by subscription by traveling salesmen.

Thompson and Everts seem to have dissolved their partnership in late 1872 or early 1873, for in the latter year the *Combination Atlas Map of Will County Illinois* was published under the imprint of Thompson Brothers & Burr (Fig. 25–8). In the following year, this firm published an atlas of Du Page County, Illinois. Everts, meanwhile, associated himself with O. L. Baskin and David Stewart and with them published atlases of Rock and Walworth counties, Wisconsin, in 1873. In the next year an atlas of Sandusky County, Ohio, was issued under the imprint of Everts, Stewart & Company and combination atlas maps of Lenawee and Washtenaw counties, Michigan, were published by Everts & Stewart. As noted earlier, Everts also independently published atlases of about twenty Ohio counties in 1874 and 1875. In 1876 Everts published atlases of counties in New York, Pennsylvania, New Jersey, and Michigan under the firm names of Everts, Ensign & Everts, Ensign, Everts & Everts, and Everts & Stewart.

All of the combination atlas maps issued by Everts and his associates are modeled on the same pattern. They generally include maps of the United States, the pertinent state, and the featured county, a history of the county that is several pages in length, plats of individual townships, and lithographic illustrations. Illustrations of farmsteads and town residences are particularly numerous, and virtually all aspects of rural and urban life are portrayed with pictures of factories, commercial es-

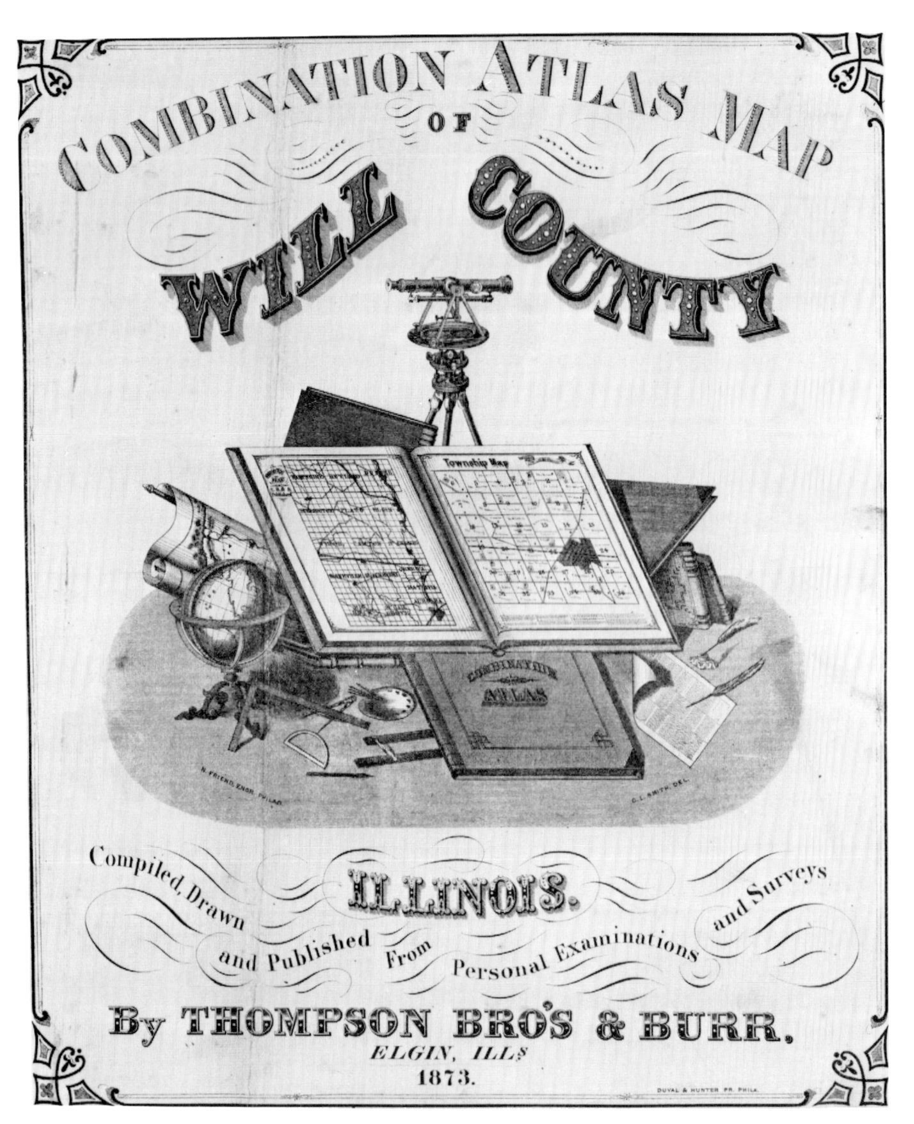

Fig. 25–8. Title page of the 1873 *Combination Atlas of Will County Illinois.*

tablishments, interior and exterior views of retail stores, churches, and public buildings. The views show various types of carriages and wagons, as well as horses, cattle, sheep, and hogs in neatly fenced fields. Some later volumes have portraits of prominent businessmen and farmers and, in some instances, of their wives and children. Biographical sketches are also a feature of some atlases. Proprietors and individuals paid prescribed fees for having their homes or businesses pictured in the atlases.

In 1876 Everts also began publishing county and city histories and biographies and achieved considerable success during the next decade. In 1884 L. H. Everts & Company published Thomas Scharf's and Thompson Westcott's *History of Philadelphia 1609–1884,* which includes the following biographical sketch:

> Maj. Louis H. Everts, the publisher of this history of the great city of Philadelphia, has qualified himself by long experience and large enterprise for the preparation and issue of local historical works, in which line of business he is not excelled by any competitor in the United States. He had but recently left the military service of his country, when, in 1866, he established a publication house in the West, and as his operations extended, a removal to a more central point of business facilities was found indispensable, and he therefore transferred his headquarters to Philadelphia in 1872. Since then he has been a very busy and successful man. In all he has issued over two hundred local histories embracing cities, counties, towns, etc. in New York, Pennsylvania, Michigan, New Jersey, Massachusetts, Connecticut, Maine, Ohio, Maryland, Indiana, Tennessee, Missouri, New Hampshire, and California. In these many volumes are preserved, in an enduring and concise form, valuable records of many American cities or other political divisions, from the day when they came into existence to that of the issue of the books.[11]

When Thompson and Everts began to solicit buyers for their maps in 1867, they engaged as one of their canvassers another former army associate, Alfred T. Andreas. Andreas, after a sluggish start, became one of their most successful map and atlas subscription salesmen. He soon realized, though, that more money could be made as an atlas publisher than as a salesman. With his brother-in-law, John M. Lyter, therefore, he established the firm of Andreas, Lyter & Company in Davenport, Iowa, in 1869 or 1870. From these headquarters they published, between 1870 and 1873, eleven or twelve atlases of Illinois counties (Fig. 25–9).

In 1873 Andreas moved his operations to Chicago, where he took offices in the recently completed Lakeside Building, which was located at the corner of Adams and Clark streets. Lyter remained in Davenport, and the firm name became A. T. Andreas. The Lakeside Building housed lithographic, printing, and binding facilities, permitting all phases of atlas manufacturing to be accomplished under one roof. The lithographer Charles Shober also moved his firm, the Chicago Lithographing Company, into the building. It will be remembered that J. H. Beers & Company had offices in the Lakeside Building, too. From this location, Andreas published fifteen or more atlases of counties in Illinois, Iowa, Indiana, and Ohio between 1873 and 1875. Several of the later atlases were issued under the imprint of Andreas & Baskin. In 1873 Andreas also embarked on the very ambitious project of preparing an atlas of the state of Minnesota, which will be dealt with in Chapter 26.

J. Silliman Higgins also published several atlases. He had worked on maps of Indiana and Ohio counties between 1865 and 1869, several in association with Jackson Lake and his brother, R. Thornton Higgins, before joining with Augustus Warner and J. H. Beers in producing county atlases in Chicago in 1870. In 1874 he became a partner of Howard R. Belden, and as Higgins, Belden & Company they published ten to twelve illustrated historical atlases of counties in Indiana, Michigan, and Wisconsin. The atlases are heavily illustrated with sketches of farmsteads, public buildings, churches, and some portraits. They also include short biographies of well-to-do farmers and townsmen. Several atlases fea-

Fig. 25–9. Illustrations of a prosperous farmer's properties from the 1871 *Atlas Map of Fulton County, Illinois* published by Andreas, Lyter & Company.

ture biographical sketches and views of the residences of particularly eminent landowners. These landowners paid handsome fees to be so immortalized. There is some suggestion that in around 1877 Higgins, Belden & Company sold out to, or were absorbed by, J. H. Beers & Company. Since Higgins, Belden & Company was also located in the Lakeside Building, the Beers Company simply took over their offices. Higgins, Belden & Company's 1875 *Illustrated Historical Atlas of Henry County, Indiana* was the inspiration for Ross Lockridge, Jr.'s well-known 1947 novel *Raintree County*.[12]

Another energetic publisher of midwestern county atlases was Wesley Raymond Brink. Between 1873 and 1877, under the imprints W. R. Brink & Company, Brink, McCormick & Company, and Brink, McDonough & Company, he published illustrated historical atlases of about two dozen counties in southwestern Illinois and in scattered parts of Missouri. Brink also published histories of more than fifteen Illinois counties during this period and up to 1883.

Born in Frenchtown, New Jersey, on February 23, 1851, Brink was the tenth child of Thomas Lequear and Elizabeth Thatcher Brink.[13] He received his early education in New Jersey, attended Princeton University, and around 1871 moved to Cordova, in Rock Island County, Illinois, where an older brother, George Washington Brink, was practicing law. Wesley had studied law and was reportedly admitted to the Illinois bar. Before setting up a law practice, though, he was employed on a temporary basis by an atlas publisher. Brink became so impressed with the opportunities in this field that he gave up plans of being a lawyer and established an atlas publishing company in Edwardsville, Illinois. Before long, he was heading a firm with some thirty employees.

The first Brink publication was the *Illustrated Encyclopedia and Atlas Map of Madison County, Illinois*, which was published in 1873 by Brink, McCormick & Company. The firm's location is identified as being in the Democrat Building in St. Louis, Missouri. There are indications that McCormick was associated with the *St. Louis Democrat* newspaper, at whose printing plant the atlas was published. Edwardsville, the county seat of Madison County, Illinois, is located across the Mississippi River from St. Louis. The Madison County atlas includes a number of illustrations, portraits, and biographical sketches. All Brink atlases have these features. In 1874 the firm produced atlases of five additional Illinois counties, and its output the following year included five more Illinois county atlases and two of Missouri counties (Fig. 25–10). In 1876 and 1877 it produced at least seven more atlases of Missouri counties. After 1877 Brink appears to have given up atlases in favor of publishing county histories, in which he had great success. Although no county histories dated after 1884 carry a Brink imprint, he is said to have continued to compile them for other publishers until his death in 1902.

An article in the *Edwardsville Intelligencer* of October 27, 1875, gives some insight into the soliciting techniques used by atlas publishers and their agents.

> Most of our readers in this county will remember Mr. Robert Frazier as the humorous gentleman who solicited biographical sketches, obituary notices, pictures, farm scenery, etc. for Brink McCormick & Cos. Atlas (of Madison County, Illinois). If the biographies were cast in the same mold; and if the pigs' tails were of the same length and all turned in the same direction; and if the same croquet party appeared simultaneously in all the front door yards of the county residences; if all the shrubbery and trees are as alike as two peas; and if the same little dog plays with the same little ball in all parts of the county at the same time; and if the portraits resemble creoles and mulattoes, and if Old Joe cannot be distinguished from Thos. Kennedy; and if the subscribers cannot find their own land;—if, we say, all these things occur, it is certainly no fault of Mr. Frazier's. His duty was to take contracts at the highest possible rate and retain his commission, and by a careful diagnosis of that smile of his we doubt not but he received his

full per cent. The fault of the work lays with the blacksmith. He didn't hammer it out right. The metal was either too hot or too cold and he didn't have the right kind of coal."[14]

John P. Edwards, publishing independently and under the Edwards Brothers imprint, was an atlas publisher who focused his attention on counties in Missouri and Kansas. From 1875 to 1884 Edwards Brothers did a fairly profitable business, publishing atlases of some forty-five Missouri and Kansas counties (Fig. 25–11). The firm's first effort, entitled *An Illustrated Historical Atlas of Boone County, Missouri*, set the pattern for their subsequent publications. It includes a history of the county, statistical data, an advertising directory, a list of patrons of the atlas, and a few biographical summaries, portraits, and views. In the back of the volume there are maps of the state, the United States, and the world. The number of illustrations in Edwards Brothers atlases are far fewer than in Brink publications. In fact, some Edwards Brothers's county atlases of the early 1880s are without illustrations. It is possible that town and rural residents in the states covered by the firm were less willing, or able, to pay the fees to have their portraits or views of their farmsteads included in the atlases. Edwards Brothers continued in business until 1885. Several of the later atlases of Kansas counties carry the imprint "John P. Edwards, Quincy, Illinois." Earlier atlases were credited to the "Edwards Brothers of Missouri," but the firm's general office address was given as 209 S. Fifth Street in Philadelphia. This was, very likely, the location of the lithographic printing offices.

Lyman G. Bennett of Chicago and A. C. Smith published a *Map of Ramsey County, Minnesota* (actually an atlas) in 1867 that was lithographed and printed by Charles Shober & Company, proprietors of the Chicago Lithographing Company. In the same year, they published a *Map of Winona County, Minnesota* (also an atlas), which was also lithographed by the Shober company. This volume includes an ornately lettered title page,

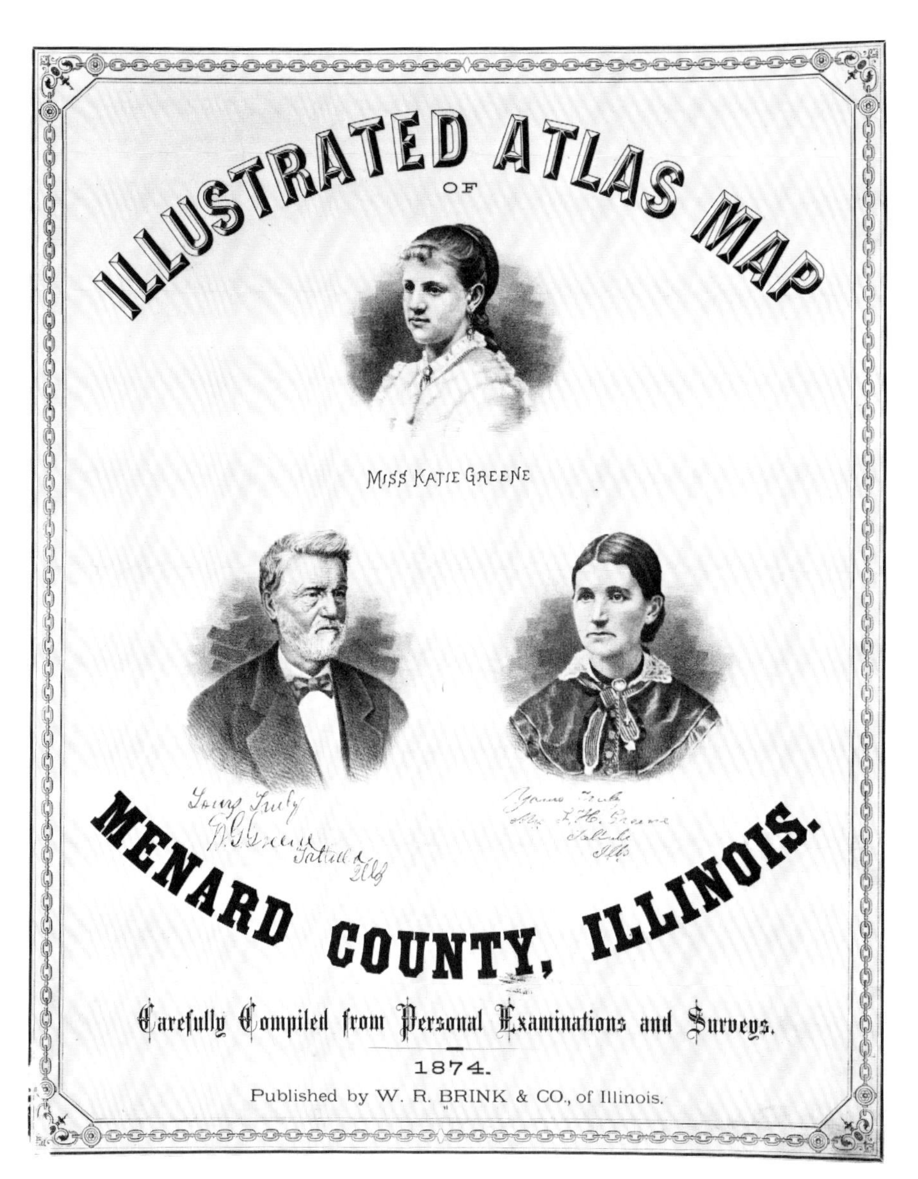

Fig. 25–10. Title page of Brink's *Illustrated Atlas Map of Menard County, Illinois*, 1874. Prominent citizens paid premium prices to have their portraits reproduced on the title page.

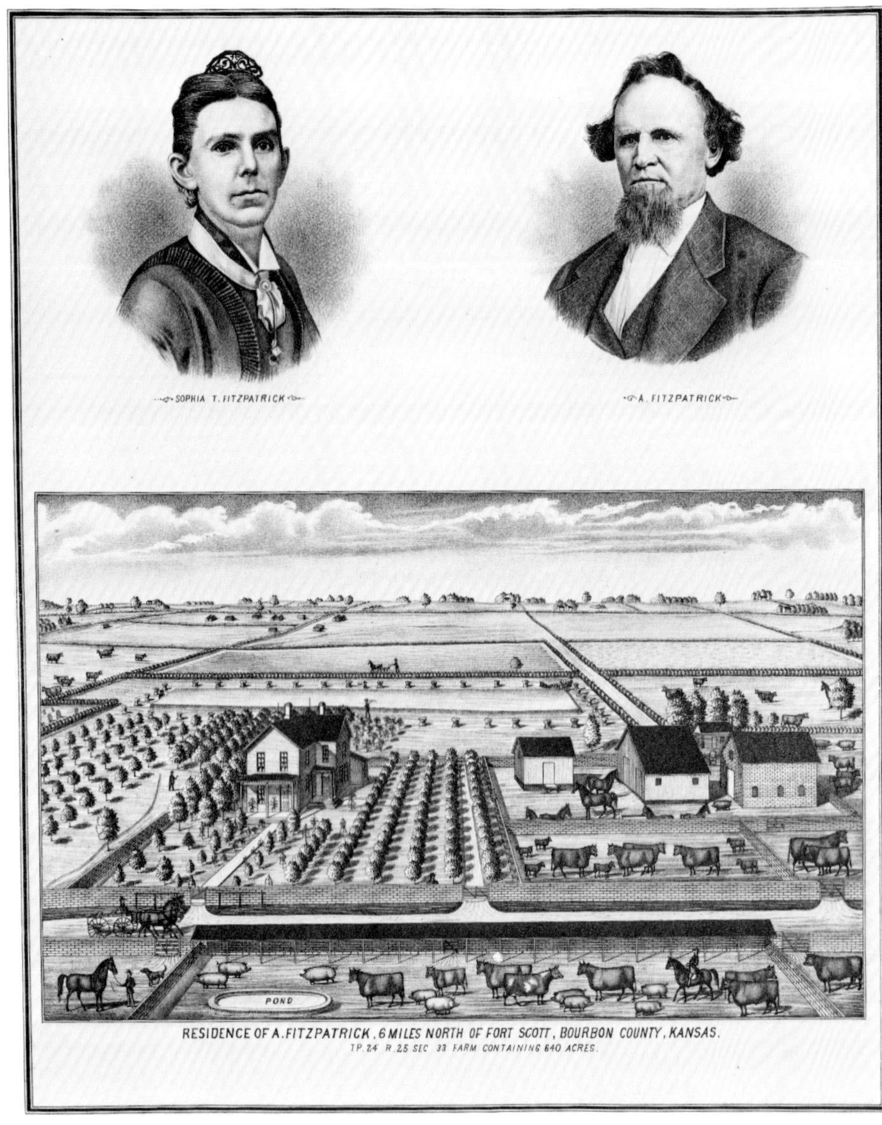

Fig. 25–11. These portraits of the Fitzpatricks and their farm are from the Edwards Brothers's 1878 *Atlas of Bourbon County, Kansas*. The Edwards Brothers firm was a latecomer to the field of county atlas publishing.

nineteen page-size township maps, fifteen lithographic illustrations of public and institutional buildings, scenic sites, prominent residences, and panoramas of villages. The frontispiece is a large fold-out panoramic color map of the city of Winona. Bennett and Smith's compilation of these two early county atlases is of especial interest because at the time they were published Minnesota was still in its first decade of statehood. They appear to be the only atlases compiled by Bennett and Smith. Mai Treude states that "both publications are unassuming volumes lacking the familiar directories of land owners. However, they do contain lithographs of residences, businesses, and various institutions."[15]

The production span for the illustrated lithographic county atlas extended from about 1862 to 1886. During this period, atlases were published for more than seven hundred counties. The county atlas had its most productive development in the states north of the Ohio River and in the upper Mississippi River valley. These regions accounted for two-thirds of the total atlas output, with the remaining third covering counties in the New England and Middle Atlantic states. More than one hundred atlases were published of Ohio counties, seventy for Illinois, sixty-five for Pennsylvania, and sixty-one for New York. Other states with fifty or more were Indiana and Missouri. More than two-thirds of the total production of atlases was accomplished in the 1870s. Almost one hundred county atlases were published in 1875, the peak year.

As noted earlier, atlases of the western counties were sold by subscription by canvassers who received a percentage of their sales. Subscription selling has a long history dating back as far as the fifteenth century. This method of marketing books, maps, and periodicals was particularly common in the post-Civil War decades. During these years, as F. E. Compton has noted, "the American people had money in their pockets and an insatiable desire for at least a veneer of 'culture.' Even those who never expected to open a book wanted a home library as a hallmark of social standing."[16] After the war, demobilized soldiers found they could make a living by canvassing for various popular publications, among them the eminently successful memoirs of Ulysses S. Grant. Soliciting county and state maps and atlases was one of several lucrative branches of subscription selling. The general isolation and loneliness of rural life induced farmers and their wives to welcome warmly the itinerant peddler. Thus, a May 1878 entry in the diary of a young Iowa farm girl reports that "a map peddler came along and wanted to stay all night. We kept him. He showed us his map this morning and father concluded to take one. He is a very pleasant and intelligent gentleman and his name is Allison."[17]

In promoting sales, atlas publishers and canvassers judiciously cultivated local newspaper editors by taking ads in their journals and offering them complimentary copies of completed volumes. In exchange, the journalist would write a favorable story about the projected atlas. For example, under the heading "Our New Atlas," the May 7, 1873, *Burlington, Iowa Hawkeye* disclosed that "Capt. W. H. Saunders will soon commence an experimental canvass for the new atlas map of Des Moines County. Should he find sufficient encouragement, the maps will be ready for publication and delivery in about one year. This map, once made, will be of incalculable benefit to the people of the county. It will furnish information interesting to everybody. It would be well for our people to take an interest in the project to help the work along and secure a copy of the map." Most county atlas canvassers were quite successful, and there were few farm owners or town residents in the Middle West who did not subscribe to one or more atlases. It is estimated that eighteen hundred to four thousand copies of an atlas were printed for each county for an average per county of around twenty-five hundred copies. The total number of volumes sold between 1865 and 1890 was approximately two million. Prices varied from six to twelve dollars per copy.

By the mid-1870s county atlases for the midwestern states had become quite standardized. They generally included a map of the pertinent state, a map of the county, and page-size maps or plans of individual townships with enlarged plans of towns and cities. Township plats were usually copied from official U.S. General Land Office surveys and records that were on file in the county seat. The records included the names of landowners and the extent of their land holdings in acres, section lines, roads, woodlots, rivers, and streams. Town and city plans located churches, schools, colleges, doctors' residences and offices, retail establishments, factories, warehouses, and railroads and railroad yards. A map of the world, of the United States, and climatological and geological maps were also included in some county atlases.

Other features of most western-type atlases were a history of the county, biographical sketches of prominent citizens, a business directory by town, and a list of patrons or subscribers. What adds distinction and color to many of these county atlases are the numerous lithographic views and portraits. They were, of course, inserted on a fee basis. Many a farmer and townsman succumbed to the high pressure salesmanship of the canvassers and to personal vanity and signed contracts to have pictures of himself, his wife, his farmstead and, perhaps, his prize livestock printed in the county atlas. Portraits were usually sketched from photographs supplied by the individuals. Views of residences and farms were drawn by sight, usually in oblique perspective.

The view artist also prepared illustrations of hotels, churches, schoolhouses, public buildings, mills, cooper shops, cemeteries, foundry and machine shops, interiors of jewelry, clothing, and shoe stores, wagons and carriages of every description, quarries, railroad trains, many types of fences and hedges, barns, houses of various architectural styles, agricultural machinery, windmills, breweries and distilleries, livery stables, undertaking establishments, blacksmith shops, parks, and recreation centers. In fact there was little pertaining to the life of American rural and town residents in the Victorian age that was not pictorialized in the county atlas.

Although we know the names of about one hundred surveyors and map and atlas publishers, there is little information about the artists who prepared the views and portraits. In some atlases the lithographic artists are named on the title page. The initials or full names of others are found in the lower corners of views in some atlases published by Brink, Everts, Andreas, Warner & Beers, and a few others. A. B. Greene, F. J. Howell, J. R. Buckingham, and J. W. Smith are among the illustrators identified in Brink atlases. Brink's younger brother, Edgar, worked briefly as an illustrator. While still in his early twenties, he suffered a sunstroke while sketching on a hot summer day. Some brain damage resulted and, after several years of partial incapacitation, he took his life.[18]

A number of Everts's atlases were illustrated by C. H. Radcliffe, Clarence L. Smith, E. A. Sumner, A. G. Snell, H. A. Mills, B. F. Goist, and G. W. Salisbury, among others. Artists identified on Warner & Beers atlases are H. C. Furness and L. C. Corwine. Others whose illustrations appear in one or more county atlases are H. G. Howland, Charles Gasche, William Engel, T. Mathews, J. H. Headington, Edward Blue, Joe K. French, and I. D. Forgy. While there are some variations in the quality of the drawings, they are characteristically quaint and lacking in artistic excellence. All, however, include a great amount of detail, although the illustrators at times used artistic license in presenting houses, barns, and fences in excellent repair and fields and garden plots neatly manicured. John C. Parish described his reaction to viewing one of the illustrated atlases of the 1870s.

> And how we fed our eyes upon the pictures with which these pages of maps were interlarded. Here the artist and lithographer had nobly portrayed Iowa. We found the residences of the leading citizens of our town—and of other towns. There were pictures without end of farm residences. . . . Everywhere trim wooden fences enclosed those gabled houses of half a century ago, and almost everywhere the lightning rod salesman had made his visit.
>
> The historical data of prime importance was that which the atlas makers presented with no idea of recording history—the detailed maps of the counties in 1875, and the pictures of the homes and business houses and public institutions of a day that is gone. To be sure we must make allowance for certain distortions due to state and community pride. For example, in the pictures of Iowa farms there were pigs, large and round, who did not wallow or lie asleep in the mud, but stalked about in stately or dignified fashion or gazed reflectively at the gigantic cows, who, disdaining the grass, stood at attention in the foreground. The horses were of the prancing variety with upraised hoof and overflowing mane and tail. They drew brand new wagons up the road, or buggies in which rode be-parasolled and curiously dressed ladies.
>
> I used to wonder why cattle and horses and hogs were always drawn with their fat profiles toward the front of the picture—as if a strong wind had blown straight across the page lining them up like weather vanes. Now I know that the glorified live stock was an expression of Iowa ideals in 1875—and that fact in itself is of historic importance.[19]

Although some individuals were dissatisfied with their portraits, county atlas subscriptions sold well, and fees in payment for portraits and views inflated the publishers' incomes. Such publishers as Andreas, Everts, Warner & Beers, Brink, and Edwards Brothers had a number of productive and prosperous years.

As evidence of the profitable returns, see the adjacent table for Bates Harrington's estimates of the sales of Andreas's *Atlas of Peoria County, Illinois*, published in 1878. Andreas's production costs (including commissions, artists' fees, printing costs, etc.) amounted to $15,663, and with sales totaling $33,218, he was left with a profit of $17,555 (less some loss in uncollected subscriptions).[20] Mrs. Florence Brink Alarco, daughter of W. R. Brink,

2140 subscribers at $9 each	$19,260
6 full-page views at $145 each	$870
77 half-page views at $76 each	$5608
41 eight-inch views at $60 each	$2460
50 six-inch views at $36 each	$1800
12 four-inch views at $28 each	$336
2 portraits at $250 each	$500
8 portraits at $100 each	$800
63 columns of biographies (2.5¢ per word, 8 words per line)	$1134
3 pages of business notices	$450
total	$33,218

estimated that in 1873 her father cleared forty-three thousand dollars on his atlas publications.[21] Between 1869 and 1877, it is believed that publishers made three million dollars on Illinois county atlases alone, and several publishers are reported to have each grossed over a million dollars a year between 1870 and 1880.[22]

The title pages of county atlases were printed in large type, and many included illustrations or portraits. A montage of atlases, globes, and surveying and drafting instruments was a popular motif for such title page illustrations. Another frequently used illustration depicts several surveyors at work with an artist nearby sketching the surrounding landscape and a railroad train passing in the background. Atlases were bound between hard, cloth-covered boards with the spine lettering stamped in gold.

There was a rapid decline in atlas publishing in the 1880s, and by 1890 the illustrated lithographic county atlas had virtually disappeared from the market. Among the reasons for this phenomenon is that by 1880 most of the prosperous counties, particularly in the Middle West, had been supplied with one or more atlases. The panic of 1873 also contributed to the drop-off in production, although there was a lag of several years before its full effects were felt. New techniques of reproduction and printing also contributed to the phasing out of the lithographically reproduced atlas. Most significant perhaps was the invention and perfection of the halftone process of reproducing photographs around 1890. As early as 1894 county atlases were published with halftone reproductions of portraits and photographs of public buildings, industrial and commercial establishments, and farmsteads. Another technical innovation of importance was the adaptation of chromolithography to map printing. Maps in county atlases published in the 1860s and 1870s were usually manually colored, generally with the aid of stencils. By the late 1880s chromolithography was quite advanced and was employed in most county atlases. Color printing required a hard, glossy-finished paper, the use of which characterizes post-1890 atlases.

The wax engraving technique for printing maps appears to have had only an indirect effect on county atlas production. In the 1870s it became common practice to include in county atlases maps of the world, the United States, and the state in which the county was located. Such maps were usually supplied by large commercial cartographic publishers and were printed by the wax engraving process. It is possible that this process may have been used for township maps published in post-1890 county atlases.[23] However, Frederick Bourquin and other Philadephia lithographic printers continued to print some county atlases well into the 1890s.

These technical developments resulted in a new breed of county atlas publisher. Two of the most prolific producers of county atlases between 1890 and 1920 were George A. Ogle & Company of Chicago and the Northwest Publishing Company of Philadelphia and Minneapolis. Ogle published more than five hundred atlases, primarily of counties in the upper Mississippi valley and in the plains states. Ogle atlases dated prior to 1895 have no illustrations. After this date they have many portraits and photographic views, as well as advertising displays illustrated with photographs.

The Northwest Publishing Company, which was located in Philadelphia from 1892 to 1899 and in Minneapolis from 1900 to 1910, published some 120 plat books of counties in Illinois, Iowa, Kansas, Minnesota, Missouri, Nebraska, Ohio, and Wisconsin. The plat books include a map of the pertinent county, individual plats of the townships, and maps of the world, the United States, and the state in which the county is located. The maps appear to have been lithographically reproduced and manually colored. The plat books include no illustrations.

Despite their limitations and inaccuracies, nineteenth-century county atlases nonetheless preserve a detailed cartographical, biographical, and pictorial record of a large segment of rural America in the Victorian age. As Compton remarked, "these histories, [atlases] and collected biographies are today important source works for historical research. . . . The subscription publishers who fathered them may have had no other aim than to turn an honest (or almost honest) penney, but they did a work of lasting value."[24] John Maass echoes Compton in his book *The Gingerbread Age*. "The nineteenth century state and county atlases," he wrote, "contained large-scale township maps. The publishers would also immortalize substantial citizens—for a fee, they included their biography, portrait and pictures of their home or place of business. The artists often seemed to have trouble with perspective, but they made up for it by the lovingly drawn details and the lively sketches of people, horses, cattle, dogs, wagons, buggies and railroad trains. These lithographs are both a documentary record of nineteenth century life and delightful specimens of American folk art."[25]

Notes

1. Mary Louise Regan, comp., *The Beers Genealogy, The Descendants of Anthony Beers of Fairfield, Connecticut, through His Son John* (Palatine, Ill., 1974) 2:155–56.
2. Ibid., 128.
3. Ibid., 154.
4. Ibid., 81–82.
5. Ibid., 111.
6. Warner & Foote, *Testimonial Book* (Minneapolis, Minn., 1880), 4.
7. Ibid., 35–36.
8. Ibid., 19.
9. C. E. Sherman, *Progress Report of the Ohio Cooperative Survey* (Springfield, Ohio, 1911), 13–14.
10. Jane Eliza Johnson, *Newtown's History and Historian, Ezra Levan Johnson* (Newtown, Conn., 1917), 98.
11. J. Thomas Scharf and Thompson Westcott, *History of Philadelphia 1609–1884* (Philadelphia, 1884) 3:2332.
12. In a footnote to his article "Cadastral Survey and County Atlases of the United States," in the June 1972 issue of the *Journal of the British Cartographic Society*, Norman J. W. Thrower wrote that "in graciously granting permission to use the quotation from *Raintree County*, written by her late husband, Vernice Lockridge Noyes wrote the author of this article: 'You may be interested to know that the atlas which Mr. Lockridge was describing was actually entitled *An Illustrated Historical Atlas of Henry County, Indiana*, published by Higgins Belden & Co., Chicago, in 1875. This was in the library of his grandfather, John Wesley Shockley, whose name appears on page 44.'"
13. For more detailed information on Brink see Betty A. Spahn and Raymond J. Spahn, "Wesley Raymond Brink, History Huckster," *Illinois State Historical Society Journal* 58 (Summer 1965): 117–38.
14. Ibid., 124.
15. Mai Treude, *Windows to the Past, a Bibliography of Minnesota County Atlases* (Minneapolis, Minn., 1980), 10.
16. F. E. Compton, *Subscription Books* (New York, 1934), 36.
17. Merrill E. Jarchow, "Social Life of an Iowa Farm Family, 1873–1912," *Iowa Journal of History* 50 (Apr. 1952), 123–54.
18. B. and R. J. Spahn, "Wesley Raymond Brink," 130–31.
19. John C. Parish, "An Old Atlas," *The Palimpsest* 1 (Aug. 1920): 61–63.
20. Bates Harrington, *How 'Tis Done, A Thorough Ventilation of the Numerous Schemes Conducted by Wandering Canvassers together with the Various Advertising Dodges for the Swindling of the Public* (Chicago, 1879), 69.
21. B. and R. J. Spahn, "Wesley Raymond Brink," 131.

22. Gerald Carson, "Get the Prospect Seated . . . and Keep Talking," *American Heritage* 9 (Aug. 1958): 80.
23. The technique of wax engraving and its development are comprehensively presented in David Woodward's *The All-American Map* (Chicago, 1977).
24. Compton, *Subscription Books*, 17–18.
25. John Maass, *The Gingerbread Age* (New York, 1957).

26. *The New State of State Atlases*

There was a transitory interest in state atlases in the latter half of the 1820s. Between 1825 and 1829, bound collections of maps for the states of South Carolina, New York, and Maine were respectively published by Robert Mills, David H. Burr, and Moses Greenleaf. Three decades passed before compilers and publishers again ventured into this branch of commercial cartography. As might be expected, the most successful publishers of state atlases had previously been involved in preparing county maps or atlases. Between 1866 and 1890, atlases were published for more than twenty states. For some states such volumes were issued by several different publishers in two or three editions.

Simon J. Martenet used his surveys for his maps and atlases of Maryland counties and his 1865 Maryland state map as compilation sources for his *Map of Maryland, Atlas Edition,* which he registered for copyright in 1866. The volume includes a Maryland state map, a brief geographical description of the state, a plan of Baltimore, and twenty-one county maps. Each of the county maps is followed by a one-page description of the county. The only illustration in the atlas is a view of the Maryland Military and Naval Academy located in Oxford.

Seven years later, in collaboration with Henry F. Walling and Ormando W. Gray, Martenet prepared a *New Topographical Atlas of the State of Maryland and the District of Columbia . . . Together With Maps of the United States and Territories.* It was published in Baltimore in 1873 by Stedman, Brown & Lyon. It is larger and more comprehensive than Martenet's 1866 atlas and includes lengthy geological, zoological, and climatological descriptions and maps. There is also an alphabetical list of cities, districts, villages, post offices, etc., a population table by county of Maryland and the other states, and a "Classified Directory of the Principal Business Men of the City of Baltimore." Cartographic contents include plans of Baltimore, Cumberland, Annapolis, Frederick City, Hagerstown, Georgetown, Washington, D.C., and Alexandria, and twenty-three county maps.

Daniel G. Beers also compiled and published several state atlases. In 1868 he published with partner A. Pomeroy of Philadelphia the *Atlas of the State of Delaware from actual surveys by and under the direction of D. G. Beers.* The atlas was lithographed by Worley & Bracher and printed by Frederick Bourquin. The street address given for Pomeroy and Beers, 32 Chestnut Street, was the same as those listed for Bourquin and Worley & Bracher. This volume includes a map of Delaware and more than forty plates with plans and maps of towns and counties. There is also a table of distances between cities, towns, and villages in Delaware and a page of population and agricultural statistics. Little is known about Pomeroy, who, independently or in association with others, engaged in map and atlas publishing in Philadelphia between 1858 and 1876. Because the address on his imprints is often the same as Bourquin's, he may have had some association with him.

Beers's *Atlas of the State of Rhode Island and Providence Plantations, From actual surveys and official records* was compiled and published by D. G. Beers & Company in 1870. J. H. Goodhue and H. B. Parsell are listed as co-compilers on the title page. Worley & Bracher prepared the lithographs, and the volume was printed by Bourquin. The address of D. G. Beers & Company is now given as 320 Chestnut Street in Philadelphia, the same address as the lithographer and printer. In format, contents, and appearance, the atlas of Rhode Island is similar to that of Delaware. Neither volume includes illustrations. These are the only two state atlases compiled and published by Beers.

Beers's cousin, Frederick W. Beers, also tried his hand at state atlas compilation. His *State Atlas of New Jersey Based on State Geological Survey and From Additional Surveys By and Under the Direction of F. W. Beers* was published in 1872 by Beers, Comstock & Cline of 36 Vesey Street in New York City. The Beers, Comstock & Cline imprint appears on several of his atlases published in 1872 and 1873. The Vesey Street address is the same one used for the publishers employed by Beers for his other publications. The maps for the New Jersey atlas were lithographed by Louis E. Neuman and printed by Charles Hunt, both of 36 Vesey Street.

The atlas contains maps of the United States, New Jersey, a geological map of the state, maps of individual counties, and a number of town, village, and city plans. Also included are a distance graph, a section on the state's geography, tables of the area and composition of each county, a historical sketch of New Jersey, and geological, agricultural, and climatic summaries. There is also a list of New Jersey post offices, county population tables, and information from the 1870 general census of the United States. Following each county map there is a page or two of "Business Notices." These listings probably served as inducements in selling the atlas. Like Daniel Beers's two state atlases, the New Jersey volume has no illustrations.

The most active publisher of state atlases between 1868 and 1885 was Henry F. Walling, who was one of the earliest producers of town and county maps. Walling, unlike a number of other county map publishers, did not venture into county atlas production. He had, however, assembled a large personal collection of county maps, including his own publications as well as those of other surveyors and mappers. Some of the maps were cut in sections and assembled in bound volumes or portfolios for easy consultation and better preservation. This collection of county maps was the principal source of data, drawn upon by Walling and his staff in compiling his state atlases. By about 1870 it was one of the most comprehensive collections of local American maps.

Ohio had early invited Walling's interest, and before the Civil War surveyors under his direction had mapped five or six Ohio counties. The state was also well covered by other surveyors, and by 1864 maps had been published for some sixty counties in the state. Utilizing these maps as well as his personal surveys, Walling directed the compilation of a map of Ohio, which was published in New York City by A. R. Z. and L. G. Dawson in 1861. The credit line "Drawn by Thos. W. Barker & Melville Clemens" is printed in the lower right corner of the map. There are twenty-one marginal inset maps of towns and cities, a distance graph, and climatological and geological maps of Ohio. An 1866 edition of the map was published by Walling under his own imprint. His address was listed as 95 Liberty Street in New York City.

This 1866 map supplied the major source data for the *Atlas of the State of Ohio From Surveys Under the Direction of H. F. Walling*, which was published by Henry S. Stebbins of 229 Broadway, New York City. Although the date on the title page is 1868, the copyright registration note reads "Entered according to Act of Congress, in the year 1867, by Henry F. Walling." The atlas includes thirteen pages of description and statistics and sectional maps of portions of the state that each embrace from six to ten counties. As on the state map, there are plans of a number of towns and cities. Supplementing the Ohio maps are maps of the United States as a whole and of its major subdivisions. The atlas contains no illustrations.

A second 1868 edition of Walling's atlas includes the supplementary title page information following Walling's name: "To which is added An Atlas of the United States, Published for Henry S. Stebbins By H. H. Lloyd & Co. 21 John Street, New York." The sectional maps of Ohio, with some additional data, are the same as in the first 1868 edition. In the Stebbins edition, though, there are more supplementary pages. Also, all the maps in this edition have the imprint "Published for H. S. Stebbins, by H. H. Lloyd & Company, New York." We have little information about Stebbins. H. H. Lloyd & Company engaged in map publishing at several addresses in New York City from around 1860 to 1878. In 1976 the Bookmark, of Knightstown, Indiana, published a facsimile of Walling's Ohio atlas, omitting the general map supplement and including a preface and an introductory essay on "Ohio Land Grants and Surveys."

Walling formed a brief partnership with Robert Allen Campbell, a civil engineer, which produced *Campbell's New Atlas of the State of Illinois With Descriptions Historical, Scientific and Statistical*. It was published in 1870 in Chicago by Campbell (of 131 South Clark Street) and in Philadelphia by S. Augustus Mitchell. The atlas includes maps of the world, North America, the United States, Illinois, and groups of Illinois counties; a plan of Chicago; and climatological and geological maps of the state. There is a place name index and essays on geology, agriculture, and education. There are no illustrations.

Apart from the information revealed in his publications, we know little about Campbell. In 1873 he produced a *New Atlas of Missouri*, which was published at his offices in St. Louis, Missouri and by Mitchell in Philadelphia. In format, content, and arrangement it resembles the Campbell-Walling atlas of Illinois. A larger proportion of the Missouri atlas is, however, devoted to descriptive matter. An essay on entomology includes several illustrations.

Then, with Ormando W. Gray, Walling compiled the 1871 *Official Topographical Atlas of Massachusetts From Astronomical, Trigonometrical, and Various Local Surveys*, which was published by Stedman, Brown & Lyon. In his preface Walling summarizes the history of surveying and mapping in Massachusetts and lists the official U.S. Coast Survey charts and the published city, town, and county maps which he drew upon in compiling the Massachusetts map. Walling had, of course, prepared all of the county maps of the state as well as many of the state's town and city plans. Walling also noted that "the atlas form has been adopted in this edition [of the Massachusetts map], in accordance with the expressed wishes of many who prefer this form on account of its convenience for reference, etc., and the maps of each county will be found by itself. Preceding these maps are some general descriptive articles which have been kindly prepared by gentlemen eminently familiar with the subjects upon which they have written. Several of these are accompanied by appropriate general maps of the entire State." These descriptive essays were written by Albert H. Hoyt (on Massachusetts state history), Edward Appleton (on railways), C. H. Hitchcock (on geology), and Lorin Blodget (on climate). In addition to the county maps and town plans, the atlas includes general maps of the United States, New England, and Massachusetts. There are no illustrations.

Not much is known about Gray. His earliest recorded atlas, of Windham and Toland counties, Connecticut, was published in 1869. The title page indicates that it was "From Actual Surveys by O. W. Gray, C. E., Danielsonville, Conn." The directory of Danielsonville on the map of that borough lists Gray as a civil engineer and topographical surveyor. Gray's firm, O. W. Gray & Son, published atlases of Dutchess and Essex counties, New York, in 1876.

Walling and Gray's *New Topographical Atlas of the State of Pennsylvania* was published by Stedman, Brown & Lyon in 1872. The title page gives Walling's address as Lafayette College, Easton, Pennsylvania, and Gray's as 55 North Sixth Street, Philadelphia. In format and contents this atlas is closely patterned after the one of Massachusetts. There are separate botanical, climatological, geological, mining, and railroad maps of the state, with essays on these subjects. Maps of contiguous counties and a plan of Philadelphia comprise the major cartographical portion of the atlas. An alphabetical list of cities, boroughs, townships, post offices, and railroad and telegraph stations is a useful feature of the volume. On the verso of the title page, beneath the copyright registration notice, is the imprint "J. Fagan & Son, Electrotypers, Philad'a." This suggests that this atlas, and perhaps the Massachusetts atlas, was reproduced by the wax engraving process rather than by lithography. A facsimile edition of Walling and Gray's Pennsylvania atlas was published in 1977 by the Bookmark in Knightstown, Indiana.

Walling and Gray also collaborated in 1872 on the *New Topographical Atlas of the State of Ohio, With Descriptions Historical, Scientific, and Statistical, Together With Maps of the United States and Territories*. Again the publisher was Stedman, Brown & Lyon. It is interesting that the publisher's location is given as Cincinnati rather than Philadelphia. The atlas is similar to their Pennsylvania and Massachusetts atlases. The copyright registration is by "H. F. Walling, and O. W. Gray and H. H. Lloyd & Co." In addition to the series of maps in county groupings, there are plans of Cincinnati, Cleveland, Columbus, Dayton, Springfield, and Toledo. The U.S. maps were very likely prepared by H. H. Lloyd & Company, which published general maps and atlases. Wax engraving and electrotyping were very possibly the reproduction techniques employed in printing the atlas. Also included in the atlas are essays on agriculture, canals, climatology, education, geology, public lands, railroads, topography, turnpikes, and zoology. Historical, railroad, and agricultural maps of Ohio are new to this edition. The sectional maps of Ohio are similar to those in the 1868 edition. Like the latter, that of 1872 has a supplementary section of maps of the United States as a whole and of its principal subdivisions. Another feature of the 1872 edition is a "Classified Business Directory of the patrons of the Atlas," arranged under the cities of Cincinnati, Columbus, and Toledo. This edition likewise has no illustrations.

This atlas appears to be the last on which Walling and Gray collaborated, although they did join Martenet in compiling the 1873 atlas of Maryland. Gray also worked independently, and in 1873 Stedman, Brown & Lyon published *Gray's Atlas of the United States, with general Maps of the World*. Revised editions of this work were published annually to about 1879. O. W. Gray & Son issued in 1875 *The National Atlas, Containing Elaborate Topographical Maps of the United States and the Dominion of Canada*. Revised editions were issued to 1889. Beginning in 1880, special editions were issued with emphasis on specific states.

Walling independently compiled and edited the *Atlas of the State of Michigan*, which was published in 1873 by R. M. & S. T. Tackabury in Detroit. The textual portion is similar to that in the Walling-Gray state atlases. It contains essays on topography and hydrography, history, education, forest and mineral wealth, railroads, geology, and climate, each illustrated with a pertinent map of the state. The Michigan atlas, unlike earlier Walling-Gray works, has individual maps of most counties, with no more than two counties on any one map. There are also plans of Detroit and Grand Rapids, general maps of the United States and Europe, and maps of the congressional, judicial, and senatorial districts of Michigan. There are a list of cities, towns, and villages in the state and a table showing the population of the United States and its territories. Included also, as an inducement to purchase the atlas, is a list of the business addresses of a number of merchants and professional people by city and county.

In the preface of the atlas Walling explained the value of the maps for immigrants, for promoting public improvements, and for general and educational uses. He also acknowledged the U.S. General Land Office surveys that he used as a basis for his maps and the "earlier state and county maps." Walling informed the atlas's readers that

> most of the engraving has been done upon stone by Mr. Louis E. Neuman, of New York; the lithographic printing by Messrs. Forbes & Co. of Boston, H. Seibert & Brothers, of New York, Calvert Lithographing Co., of Detroit, and Julius Bien, of

> New York, who photo-lithographed the counties of the Upper Peninsula from drawings by H. S. Packard; the coloring by Miss Helen D. Findlay, of New York, and the letter press printing by the Claremont Manufacturing Co., of Claremont, N. H., who also made a portion of the paper. We are indebted to G. W. & C. B. Colton & Co., of New York, for the included maps of the United States and Europe; to Eugene Robinson, Esq., for the map of Detroit; and to the Calvert Lithographing Co., for the District maps. The binding was done by H. D. Houghton & Co., of the Riverside Press, Cambridge, Mass.

To promote subscriptions for the atlas, the publishers conducted a complete canvass of the state.

Facsimile editions of Walling's atlas of Michigan were published by Treasure Land of Mt. Clemens, Michigan, in 1972 and by the Bookmark, Knightstown, Indiana, in 1977. Both editions are uncolored and reproduce only the county maps and the general map of Michigan. The Treasure Land reproduction also omits much of the textual matter.

Following publication of the Michigan atlas, Walling turned his attention to Canada. His 1875 *Atlas of the Dominion of Canada* was published in Montreal by George N. Tackabury. The title page includes a photograph of the Parliament buildings in Ottawa. The textual material and maps are similar to Walling's state atlases. In the preface Walling indicates that work on the atlas was commenced in the autumn of 1871.

Walling also joined with Tackabury to compile and publish the *Atlas of the State of Wisconsin* in 1876. It closely follows the pattern set by the atlas of Michigan. Among the local maps listed as source material in the preface are Walling's maps of Milwaukee, Waukesha, and Kenosha counties, which were published, respectively, in 1858, 1859, and 1861. There are geology, railroad, climatic, and voting district maps of the state, all with descriptive essays. As in the Michigan atlas, most of the counties are shown separately. There are also plans of ten cities. The population of the United States and its territories is given in a table, and there is a list of Wisconsin cities, towns, and villages. Like the Michigan atlas, this volume has a business directory arranged by city and county.

Walling's last state atlas, the *Atlas of the State of New Hampshire Including Statistics and Descriptions*, was published in 1877 by Comstock & Cline of New York. Charles H. Hitchcock, a professor of geology at Dartmouth College (who wrote an essay on geology for an earlier Walling atlas), compiled and edited much of the descriptive matter for this atlas. Warren Upham contributed essays on river systems and railroads, J. H. Huntington wrote one on climatology, and William F. Flint considered the distribution of trees. Contour lines, drawn by Upham under the direction of Hitchcock, are shown on all the county maps. Like all of Walling's atlases, this one does not include illustrations.

Following the completion of his state atlases and to procure badly needed funds, Walling sold his large private map collection to the U.S. Coast and Geodetic Survey in 1884. The Walling Collection remained in the library of the survey until October 1900 when it was transferred to the Library of Congress. At that time it included thirty-two county atlases, forty-three county maps sectioned and pasted in covers, and six county wall maps. These maps and atlases are presently in the custody of the library's Geography and Map Division.

Augustus Warner and J. Silliman Higgins also published state atlases. Their firm evolved successively from Warner & Higgins to Warner, Higgins & Beers to Warner & Beers and finally to J. H. Beers & Company. In 1869 Warner & Higgins published an *Atlas of the State of Illinois to which is added an atlas of the United States, maps of the hemispheres &c., &c., &c.* It includes political, geological, climatological, and railroad maps of the state, a plan of Chicago, and sectional maps embracing from two to five counties each. The state maps are supplemented by maps of the United States and of its major subdivisions. These maps were derived from H. H. Lloyd & Company's atlas of the United States. Early editions of the state atlas have no illustrations. The lithography was done by Worley & Bracher and the printing by Bourquin. Associates listed on the title page include Joseph H. Cox, W. C. Anderson, Howard R. Belden (who later became Higgins's partner), William H. Beers (John Hobart Beers's brother), T. J. L. Remington, John I. Moore, John W. Clarke, and John W. Lincoln.

Editions of the Illinois state atlas were published in 1870, 1871, 1872, and 1876. The atlas was gradually expanded to include historical essays, statistics, and illustrations. The imprint on the title page of the 1876 edition of the *Atlas of the State of Illinois* is "Union Atlas Co. Warner & Beers, Proprietors Publishers, Lakeside Building Cor. of Clark & Adams Sts. Chicago." In the preface of this edition the publishers wrote, "in presenting our *Illustrated Atlas of Illinois* to the public, we desire to thank our Subscribers throughout the State for their liberal patronage and generous encouragement. Although the extraordinary financial depression of the past year has reduced our subscription list to some extent, yet it is fully as large as we could have reasonably expected in times like the present."

Fig. 26–1. The Lakeside Building in Chicago.

This final edition of the atlas includes general maps of the world, the United States, England, France, Germany, and Scandinavia as well as sixteen thematic maps of the country showing the distribution of various agricultural products, population densities and origins, and the distribution of selected diseases. There are also extensive statistical sections based on the 1870 census, biographical sketches of prominent citizens, a list of patrons of the atlas, and a business directory arranged by county.

The Lakeside Building, which housed the Warner & Beers publishing company, was constructed after the Chicago Fire of 1871 as a center for map and atlas lithographing, printing, and publishing (Fig. 26–1). An illustration of the building is featured in this atlas, with the following caption:

> We present on this page an engraving of The Lakeside building, an elegant stone structure, located on Clark and Adams streets, Chicago, opposite the new Custom House. Its imposing appearance and chaste style of architecture have attracted no little attention, being notable even among the many wonderful business palaces of beautiful New Chicago. But this building is distinctive in the purpose to which it is devoted, as well as in outward appearance. From top to bottom it is almost exclusively occupied by the various departments of publishing. It did not require

a prophet to foresee that Chicago was destined to be the great western center of literature as well as of commerce, and The Lakeside Company determined to provide the amplest facilities for the publication of atlases, maps, gazetteers, books, magazines, newspapers, etc., etc., to supply the ever increasing demands of the people. The result was the erection of a massive six-story and basement stone building 125 × 100 feet, filled with every appliance required in printing, binding, lithographing, mapmaking and coloring, engraving, etc., etc.

The first floor is occupied for stores and by Geo. Sherwood & Co., school book publishers. The second floor is set apart for offices, among which are those of Donnelley, Loyd & Co., book and directory publishers, also of Baskin, Forster & Co., and H. Belden & Co., map and atlas publishers. The third floor (reached by steam elevator) contains the principal office of Warner & Beers, proprietors of the Union Atlas Company; also their engraving, lithographing and map-coloring rooms. Here, also are the offices of Higgins Bros. & Co., atlas publishers; A. C. Fisher & Co., directory publishers; A. Maas & Co., wood engravers; and The National Live Stock Journal. The fourth floor is occupied by Chas. Shober & Co., proprietors of The Chicago Lithograph Company. The fifth floor contains A. J. Cox & Co's. book binding establishment. The sixth floor is occupied by the manufacturing department of The Lakeside Publishing and Printing Company, book and job printers. In the basement can be found A. H. Reeve's gold-beating works.

The advantages of such concentration of all branches of publishing under one roof are obvious, and whatever is undertaken by any one of the firms in this building has the hearty cooperation of all the others in their several departments, and their publications, in consequence, are executed with an economy, promptness and beauty otherwise unattainable.

D. C. Edwards of Chicago republished Warner & Beers's atlas of Illinois, with few, if any, modifications in 1879. In 1972 Mayhill Publications of Knightstown, Indiana, published in a facsimile edition the county maps, the plans of Chicago and other cities, and a sampling of the illustrations from the 1876 atlas. The preface notes that "in the original book, most counties were printed with other counties. We have separated these counties and printed them in alphabetical order."

The most active among the midwestern atlas publishers was Alfred T. Andreas (Fig. 26–2). After publishing more than twenty-five county atlases, he expanded his activities to include the publication of the *Illustrated Historical Atlas of the State of Minnesota* in 1874 (Fig. 26–3). This work introduced a new atlas format and an innovative distribution policy for such publications.

Fig. 26–2. Alfred T. Andreas.

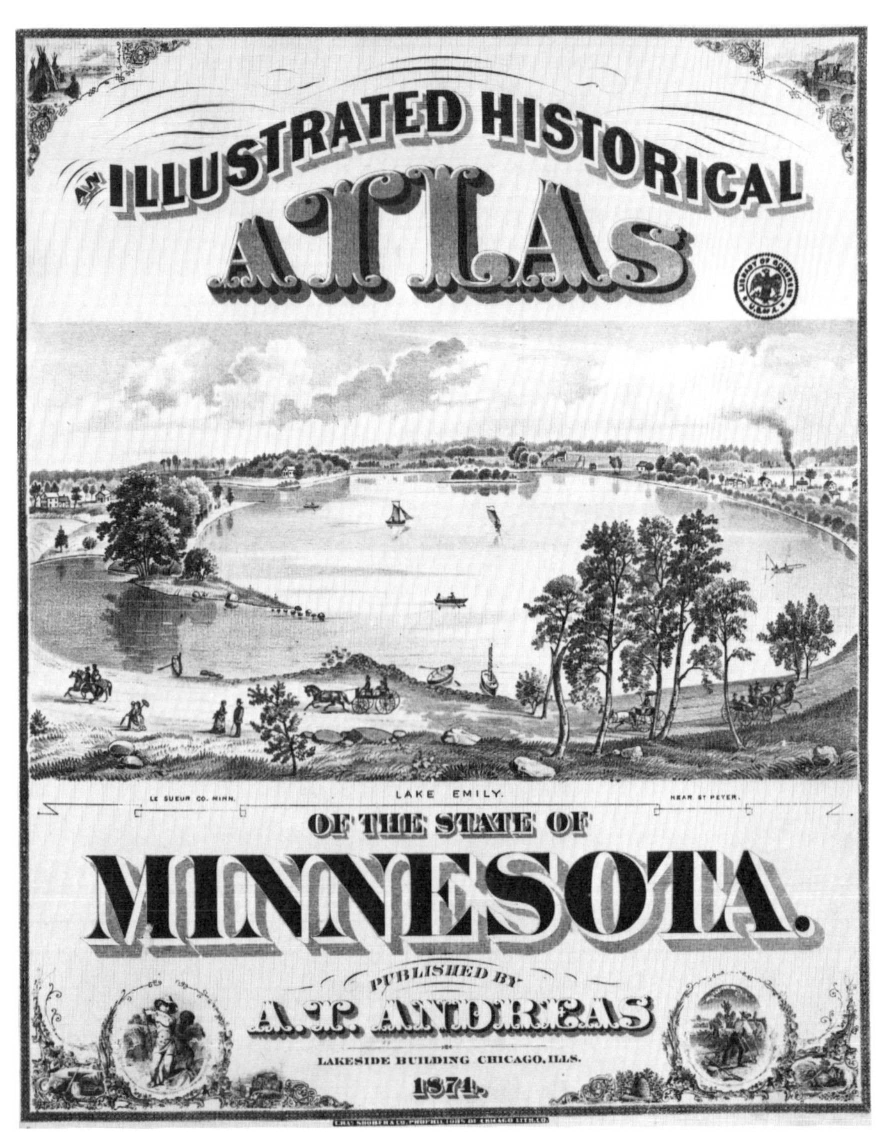

Fig. 26–3. Title page of Andreas's atlas of Minnesota.

State atlas publishing was an extension and refinement of the highly successful business of county atlas publishing, which in turn evolved from the county map industry that flourished before and after the Civil War. Bates Harrington credits Andreas with originating both the illustrated county atlas and the illustrated state atlas.[1] While this is a bit overstated, it is unquestionably true that both cartographic formats attained their maximum development under his guidance.

Andreas had served in an Illinois army unit during the Civil War, and, after being mustered out in 1865, he married Sophia Lyter of Davenport, Iowa. Following several unsuccessful business ventures and his work with Thomas H. Thompson and Louis H. Everts, he formed his own publishing company. He mainly produced county atlases, and his success in this field moved him to employ similar procedures to compile and publish state atlases. In 1873 he moved into the Lakeside Building and shortly thereafter began planning the production of the Minnesota atlas. It was a daring and challenging choice, for the state had entered the union only fifteen years earlier. Compilation data were meager; maps and atlases had been published for only two or three Minnesota counties. This also had a positive angle, however, for subscription agents would be canvassing in virgin territory.

Andreas organized and directed an elaborate collecting, surveying, compiling, and drafting operation. Surveyors were sent into the field to traverse the roads and to draw on their maps section and property lines, streams, railroads, towns, villages, churches, schoolhouses, quarries, and other features of the landscape. For areas previously surveyed by the U.S. General Land Office, the official plats were used as bases. Concurrent with these mapping operations, the state was thoroughly worked by skilled and aggressive canvassers who solicited subscriptions for the atlas and contracts for biographies, portraits, and sketches. By subscribing to the atlas for fifteen dollars an individual was guaranteed a listing among its patrons and his name was inscribed on the map at the location of his farm, residential, or commercial property. Agents received a commission of three dollars on each atlas sale to a town resident and an additional fifty cents for a sale to a rural resident. A canvasser's weekly earnings ranged from twenty-five to sixty dollars.[2]

At the peak of operations more than one hundred Andreas employees were engaged in various activities throughout the state. This large work force was under the competent direction of Thomas H. Thompson, who had several years earlier introduced Andreas to the mapping business. The scope and magnitude of the project were detailed by Andreas on one of the preliminary pages of the volume under the heading "What It Takes to Make a State Atlas." The lithography, printing, typesetting, coloring, and binding were done by firms "located in LAKESIDE BUILDING, which was erected especially for the publishing business, and particularly for the publishing of Atlases. . . . The coloring is the work of Warner & Beers and will speak for itself as to the careful manner in which they have accomplished their part of the work. If one person had done all the coloring it would have taken him forty-five years."[3] Some of the maps appear to have been printed by chromolithography, while others were clearly hand colored, very probably with the use of stencils. Among the maps done by chromolithography are a number of thematic and statistical maps of Minnesota and the United States. The publisher's statement further notes that seventy tons of paper and seventeen tons of cardboard (for the covers) were required for the atlas. Names and addresses are listed in the volume for more than one hundred persons who were "engaged in our office and on the field work on the Minnesota State Atlas."

The recession of 1873 made normal financing difficult, and Andreas had to seek assistance from Benjamin F. Allen, a wealthy banker who had shortly before moved to Chicago from Des Moines, Iowa. When Allen's financial structure in Illinois collapsed, Andreas reportedly lost $130,000. He was forced to reorganize with several of his principal creditors, and his firm became the Andreas Atlas Company.[4] Despite these financial problems, the *Illustrated Historical Atlas of the State of Minnesota* was published on schedule and was distributed to subscribers beginning in December 1874. The wheat crop was poor that year, and, pressed for cash, many farmers reneged on their contracts or offered notes in payment. Some ten thousand copies of the atlas were, however, ultimately delivered to residents of the state.[5]

In the volume's preface Andreas expressed "the sense of relief we feel from the responsibilities which have weighed upon us during the preparation of a work so unique and voluminous." He acknowledged "the liberality of the citizens of Minnesota, and the generous manner in which they have supported our undertaking. . . . This generous support has not been without its effect on us, in making us the more earnest and determined in our efforts, that our patrons should not be disappointed, but should have delivered to them within the specified time, an Atlas in all respects worthy of their liberal patronage."

For fifteen dollars the subscriber received a volume measuring 44 by 36 cm. and containing just under four hundred pages. It included double-page maps of the state, the United States, and the world; five pages of statistical maps of Minnesota and the United States; and seventy pages of county maps and plans of towns and cities (Fig. 26–4). The volume also contained more than one hundred pages of lithographic sketches, portraits, views, and landscapes. Geographical and historical essays, statistical tables, and biographical data comprised another one hundred pages. Of particular importance to the financial success of the atlas was the extensive list of "Patrons of the Minnesota State Atlas" that filled some thirty pages. In 1979 Unigraphic, Inc., of Evansville, Indiana, published a slightly reduced facsimile edition of the atlas.

Andreas anticipated that the Minnesota volume would be the first in a series of illustrated state historical atlases. As soon as his surveyors and canvassers had

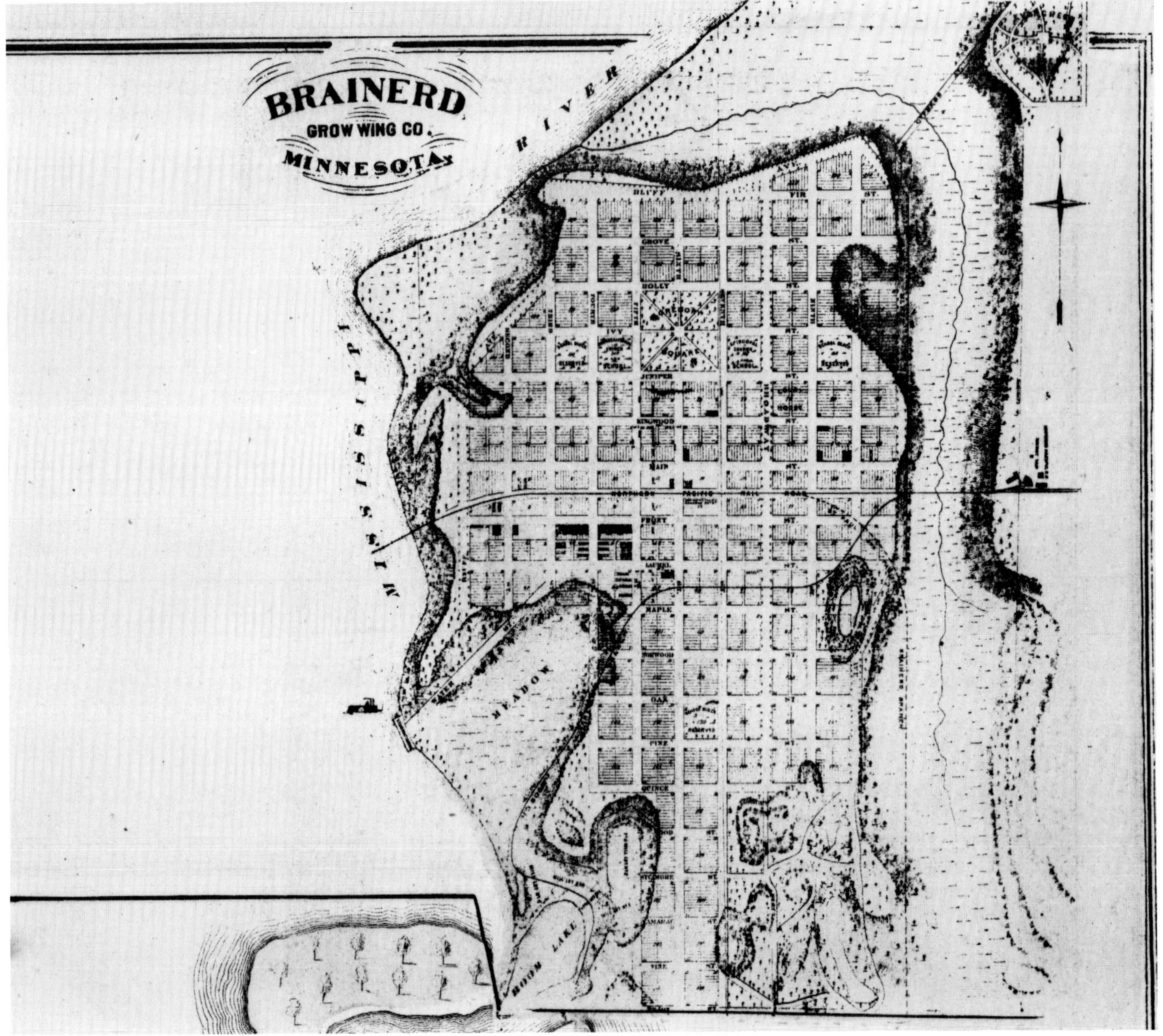

Fig. 26–4. Town plan of Brainerd, Crow Wing County, in Andreas's *Illustrated Historical Atlas of the State of Minnesota*.

completed their work in Minnesota, therefore, he shifted them to Iowa to undertake the preparation of an atlas of that state. Procedures were similar to those employed for the Minnesota atlas. Notwithstanding the serious financial depression and the fact that illustrated atlases had been published for a number of Iowa counties, more than 22,000 copies of the 1875 *Illustrated Historical Atlas of Iowa* were sold at fifteen dollars apiece. Receipts from pictures and biographies amounted to seventy thousand dollars. Production costs exceeded three hundred thousand dollars, however, and cancellations greatly reduced the publisher's profit.[6]

Prior to publication, Andreas received favorable publicity in a number of Iowa newspapers. Typical was his personal appeal printed in the October 15, 1874, issue of the *Anamosa Eureka*:

> TO THE PUBLIC—I propose publishing an Atlas of the State of Iowa, containing a map of every County in the State, showing Sections, Timber, Prairie, Roads, and Railroads, Streams, School Houses, Churches, etc., etc. Fine maps, 14 × 28 inches of the State of Iowa, United States and Territories, and of both Hemispheres. Plans, with histories of the Cities, Towns, Villages, and Counties of the State, Biographies of a large number of early settlers and prominent men in the State. Also a condensed political history of the State, giving votes, etc. Six maps of Iowa, so colored as to show the Geological and Climatological condition of the State, with also the Congressional, Senatorial and Representative Districts. Sixteen maps of the United States, colored in five fine shades, to show the amount of Wheat, Corn, Hay, Cotton and Tobacco raised in proportion to acres cultivated. Also to show deaths by consumption and other diseases, in proportion to the deaths by all diseases, and to show density of population, and proportion of colored, and various foreign nationalities in the United States.
>
> An immense amount of very useful statistical information, covering about 50 square feet of closely printed matter, in every Atlas. To the patron of the work is published his name, residence, business, nativity, post-office address, and when he came to the State, besides locating name and residence on his land. The whole will be illustrated by fine lithographic views of hundreds of public buildings and private residences in both town and county and portraits of prominent men.
>
> A large force of experienced men will commence immediately an experimental canvass; and if sufficient encouragement is received, I hope to complete the work sometime during 1875. Yours truly.
>
> A. T. Andreas, Publisher
> Chicago, Ill.

Even C. C. Carpenter, Iowa's governor, was moved to make this endorsement of the atlas in the November 19, 1874, issue of the *Belle Plaine Union* and in other newspapers in the state:

> To Whom it may Concern:—I have examined the proof sheets of the Minnesota Illustrated Historical Atlas, by Capt. A. T. Andreas, of Chicago, and I regard it as a work of superior merit, and it seems to me any citizen of Minnesota could hardly do without it. I understand from E. T. Phelps that Capt. Andreas is now taking the preliminary steps preparatory to publishing a similar Atlas of Iowa, with such improvements as past experience in this business naturally suggests to his mind.
>
> Having known Mr. Andreas intimately and well for many years, I have no hesitation in assuring all who may read this that he will do all he proposes, and that the public will find his work fully equal to the promises of his circular and advertisements.

Such testimonials are reprinted in William J. Peterson's historical introduction to the facsimile edition of the Iowa atlas that was published in 1970 by the State Historical Society of Iowa.

As another promotional feature, the *Illustrated Historical Atlas of Iowa* was published in nine different editions, one for each congressional district of the state. Each of these editions contains, in addition to the standard maps, text, and illustrations, from twenty to more than one hundred views of the pertinent district. The inconsistency in the number of illustrations is explained in the publisher's preface.

> Citizens in some portions of the State may complain that they are not equally represented with other portions in some departments of the work. This is explained by inequality of patronage. The publishers would have been glad, had it been practical, to have made every part of the State alike in the representation in all parts of the Atlas; but owing to previous publications of County Atlases, etc., in some localities, a feeling existed prejudicial to our work, and it was found impossible to induce Agents to work in those localities with any assured prospect of success. We feel confident that, if the citizens of those portions of the State could have understood the character and value of the Atlas we have actually produced, they would have cheerfully given us the encouragement necessary to have represented them as fully in the View, Biography and Portrait Departments, as other portions are represented.
>
> It should be borne in mind that the enormous expense of getting up these departments would not be incurred in those sections of the State which yielded us barely subscribers enough for canvassing the territory. But in the Map and Historical Departments, which did not depend upon the cooperation of patrons, the work will be found equally complete for all parts of the State.

Andreas also stated that "the *Historical Atlas of Iowa* is the largest publication ever issued in one edition, absorbing One Hundred and Twenty-two Tons of supersized and calendered paper, and requiring the services of Three Hundred Men for an average of one year and a quarter." The atlas was lithographed and printed by Charles Shober & Company. With a few exceptions, all the maps seem to have been printed by chromolithography.

Because of unfavorable economic conditions the Iowa atlas was also financially unsuccessful, and Andreas once again was forced to reorganize his atlas company, this time under the name Baskin, Forster & Company. Because most of his creditors were also residents of the Lakeside Building, it is possible that they acquired interests in the atlas company in lieu of cash payments. Shortly after copies of the Iowa atlas were distributed, Baskin, Forster & Company, very likely under the guidance of Andreas, launched a project to compile an *Illustrated Historical Atlas of the State of Indiana*, which was published in 1876 (Figs. 26–5, 26–6, and 26–7).

This atlas is similar in size and format to those of Minnesota and Iowa, although it has fewer illustrations. Apparently, separate editions were not issued for each congressional district, but views, biographies, and portraits were allotted most generously to those areas that had yielded the most subscriptions and portrait and view contracts. In the preface the publisher states that "the work has been in progress for more than a year past, and although it has been a period of comparative financial depression, reducing our subscription list considerably below what we might have expected under more favorable circumstances, yet we have spared neither pains nor expense to fulfill our engagement with our patrons and make the work as complete as possible."

Indiana had already been quite thoroughly covered by publishers of county maps and atlases during the previous two decades. This fact, and the financial recession, limited subscriptions to twelve thousand. The atlas was, therefore, a financial disaster for Baskin, Forster & Company and Andreas. Andreas never fully recovered. He engaged in several enterprises during the next seven or eight years, including the compilation and publication of historical and biographical books. In 1881 he published the *History of Milwaukee* and the *History of Northern Wisconsin*. His *History of Nebraska* was issued in

1882, the *History of Kansas* in 1883, and the *History of Cook County, Illinois* in 1884. The best known of his historical works is the *History of Chicago*, which was published in three volumes from 1884 to 1886. It is still recognized as the best historical record of Chicago in the nineteenth century.

Andreas's last cartographical publication was the 1884 *Historical Atlas of Dakota*. The publisher's imprint on the title page is that of R. R. Donnelley & Sons of the Lakeside Press in Chicago. The atlas includes county maps and historical, biographical, and descriptive sections, but no illustrations. In the preface Andreas again acknowledged that "the labor and expense in preparing this atlas and history of the Territory of Dakota have been much greater than the publisher anticipated." Following this final setback, Andreas limited himself to publishing city directories. He remained in Chicago until 1897, when he moved to Boston. Two years later he relocated to New Rochelle, New York, where he died on February 10, 1900.

During his varied publishing career, Louis H. Everts also issued a few state atlases. After attaining considerable success in publishing county atlases between 1874 and 1876, he moved his business from Chicago to Philadelphia, where he was very successful in publishing county histories and biographies. Everts turned his attention to state atlases in 1885, when, under the firm name of Everts & Kirk, he published *The Official State Atlas of Nebraska*. He noted in the preface that

> the Government Surveys have been followed in the compilation of each county, except where changes were found necessary, and reliable data for such alterations were obtainable. . . . [The atlas] embraces a general map of the State, showing the relative position of each County . . . the location of Railroads, Cities and Villages, the principal Streams and much topography, not heretofore obtainable. This is followed by separate maps of the different counties of the State drawn on a much larger scale, on which the details are more thor-

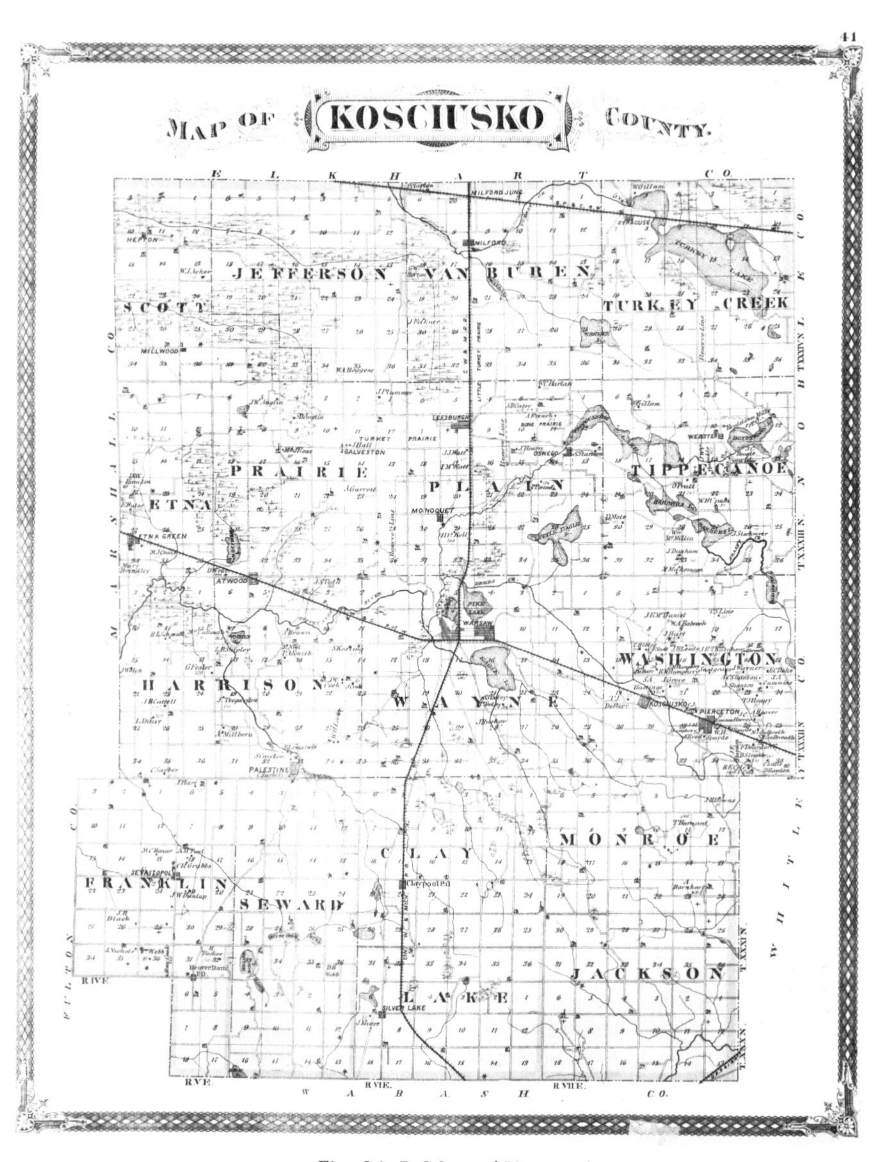

Fig. 26–5. Map of Kosciusko County from Baskin, Forster & Company's 1876 *Illustrated Historical Atlas of the State of Indiana.*

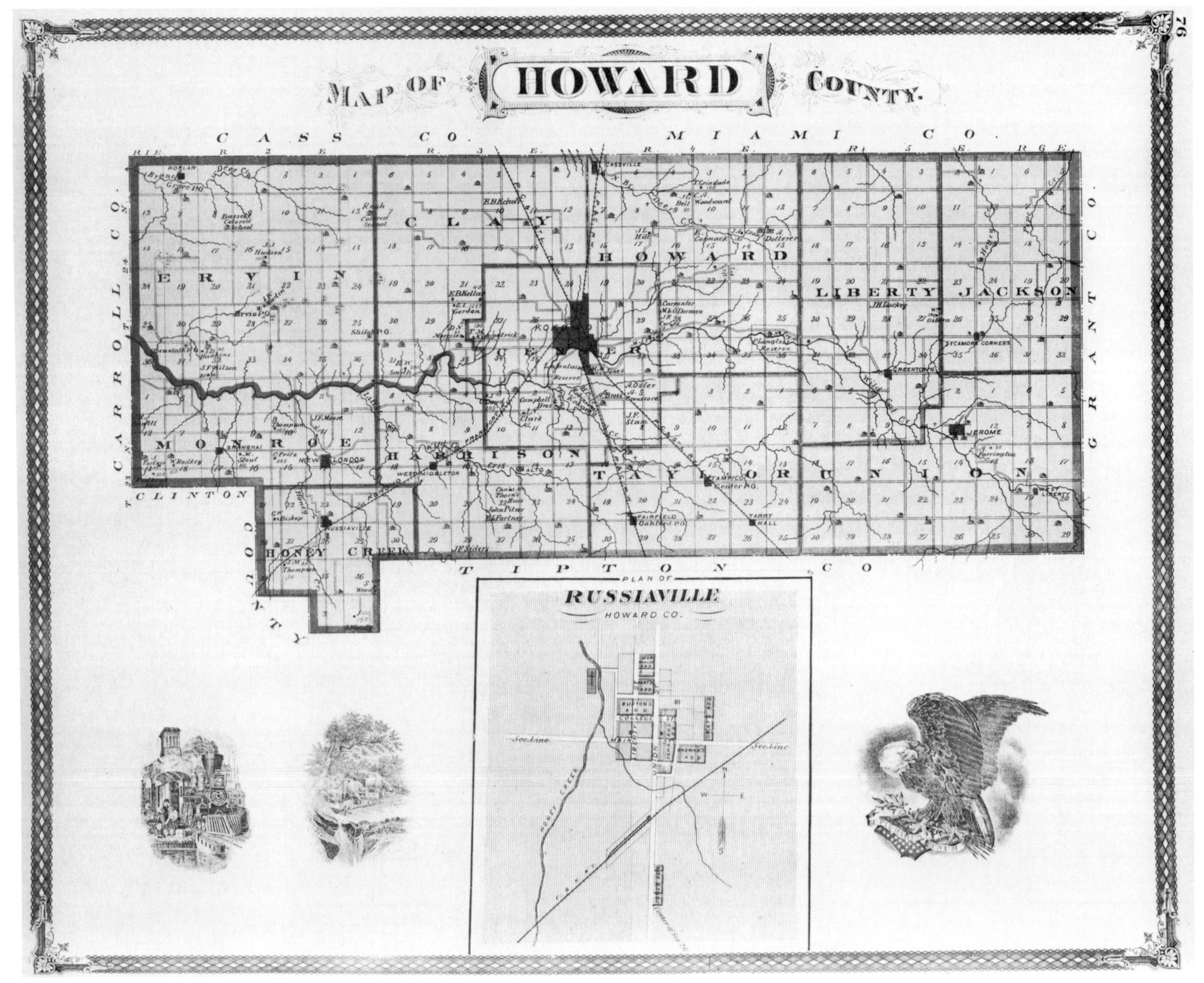

Fig. 26–6. Map of Howard County from the *Illustrated Historical Atlas of the State of Indiana*.

oughly given; there are also plans of the leading Cities and Villages throughout the State, which have been carefully prepared from the records of each.

All the maps were reproduced by chromolithography. There are twenty-six pages of illustrations, principally of stock farms and private residences. Among the former is Louiland Farms, the property of Louis H. Everts.

Two years later, L. H. Everts & Company published *The Official State Atlas of Kansas* (Fig. 26–8). Although slightly larger than the Nebraska atlas, it is very similar in appearance and contents. Like the Nebraska volume, compilers of the Kansas atlas relied heavily upon the official government surveys. The increased size is principally due to the greater number of illustrations, for which there is a classified list. The illustrations, which are based on artists' sketches, may have been reproduced by photolithography.

Asher & Adams, a New York City publisher of general maps and atlases, produced three different versions of a New York State atlas. The first, published in 1869, is titled *Asher & Adams' New Topographical Map of the State of New York Made from Official Records and Actual Surveys*. It was reproduced by Sage, Sons & Company, a lithographing, printing, and manufacturing firm located in Buffalo, New York. The atlas, which is based on a wall map of the same title, includes general geological and climatic maps of the state, a plan of New York City, and twelve sectional maps, each embracing from one to twelve counties. It also contains a "Classified Business Directory of New York City," which was previously printed on the face of the wall map.

Of a slightly larger format is the second version, titled *Asher & Adams' New Topographical Atlas and Gazetteer of New York*. It was published in 1870 by Asher & Adams of 335 Broadway, New York, and by Asher, Adams & Higgins of Indianapolis, Indiana. It is not known whether the Higgins in the Indianapolis firm was J. Silliman Higgins. A note on the title page of this atlas reads "Engraved on Copper Plate, and Lithographed

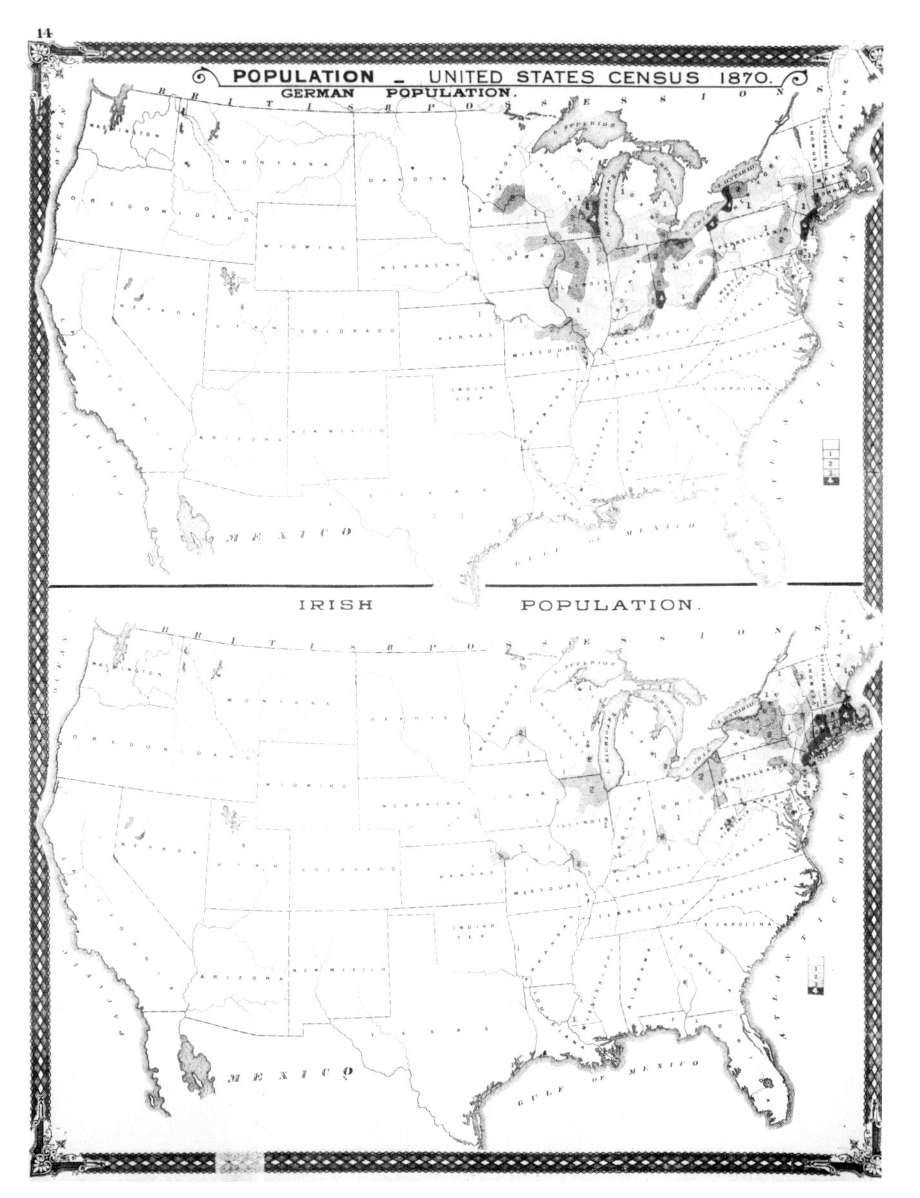

Fig. 26–7. The *Illustrated Historical Atlas of the State of Indiana* includes a number of thematic maps, among them these showing the population distribution of ethnic groups.

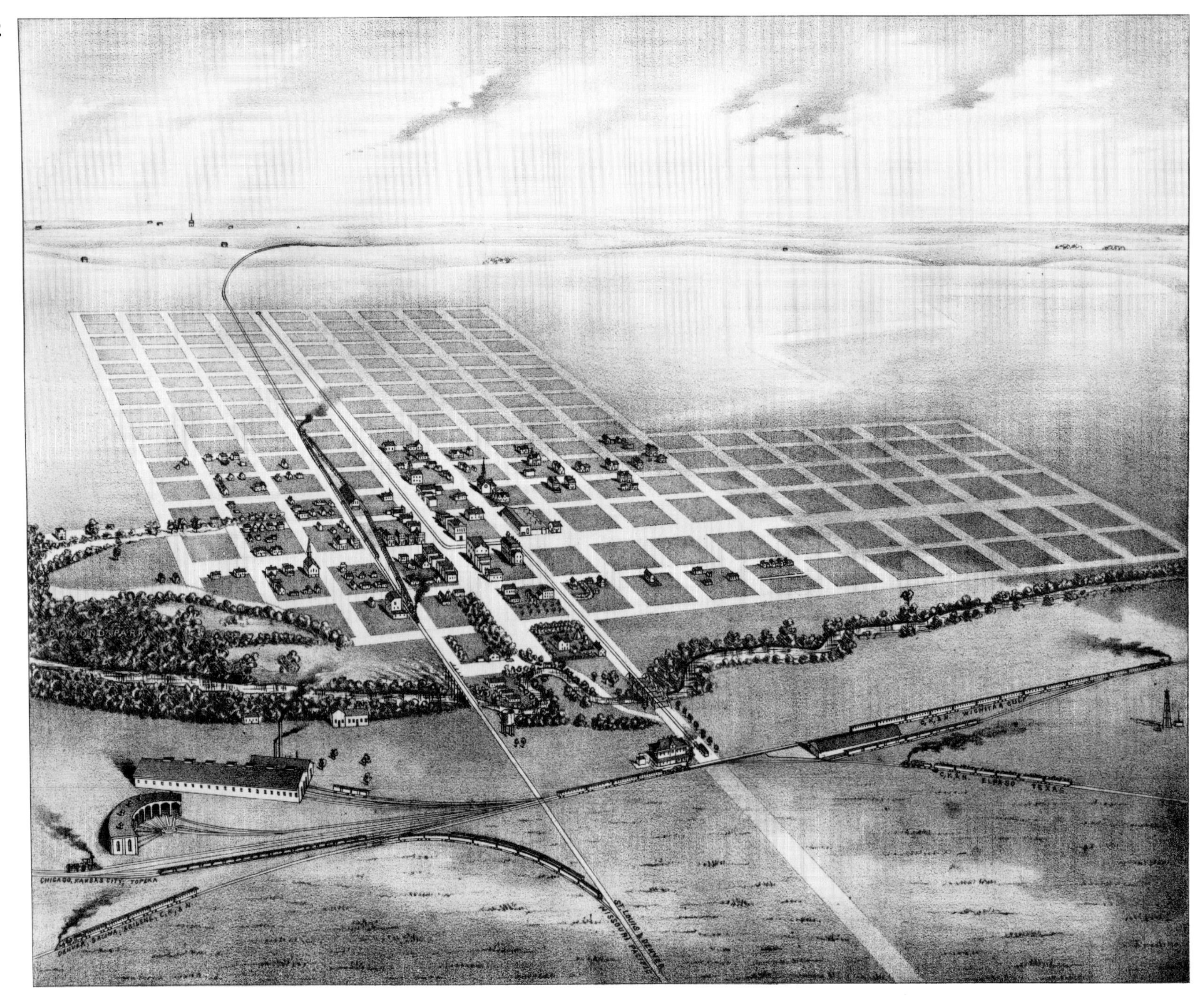
DIAMOND PARK
C.K.&N. WICHITA & GULF
C.K.&N. EL PASO TEXAS
CHICAGO, KANSAS CITY, TOPEKA
DENVER, SALINA, ABILENE, C.K.&N.
ST. LOUIS & DENVER
MISSOURI PACIFIC

from Original Drawings." This may indicate that the maps were printed from electroplates, the images on which had been derived from wax engravings. A map of the United States is the only cartographic addition to the 1870 edition, but there are new supplementary texts and a statistical section titled "Gazetteer of New York With Introductory Sketch of its Topography, Geology and History." The third edition was published in 1871. The only change from the previous year's edition appears to be the addition of a map of Europe, the elimination of some of the historical material from the textual section, and the revision of some statistical data to reflect the 1870 census.

In 1870 the firm Higgins & Ryan published in Indianapolis the *New Topographical Atlas and Gazetteer of Indiana, Comprising a Topographical View of the Several Counties of the State Together with a Railroad Map of Ohio, Indiana, and Illinois*. I have not examined a copy of this atlas, but it is very probably similar to a volume with the same title that was published in 1870 and 1871 by Asher, Adams & Higgins of 335 Broadway in New York. The electrotyping was done at the Franklin Type Foundry in Cincinnati, Ohio, and the atlas was printed by I. W. Field of 410 Walnut Street in Philadelphia. This atlas resembles the New York atlases in contents and appearance. It contains a map of the United States, a railroad map of seven midwestern states, and county maps presented in groups of three to eight counties per map. Gazetteer and statistical sections are also included.

It was earlier noted that Henry F. Walling and George N. Tackabury published the *Atlas of the State of Wisconsin* in 1876. At about the same time that Walling and his men were conducting surveys in Wisconsin, Snyder, Van Vechten & Company of Milwaukee was also occupied in compiling an atlas of the state. The Walling-Tackabury volume was published first, and when the Snyder-Van Vechten volume was issued in 1878, its sales were limited. The atlas was titled the *Historical Atlas of Wisconsin Embracing Complete State and County Maps City & Village Plats, Together With Separate State and County Histories Also Special Articles on the Geology, Education, Agriculture, and Other Important Interests of the State*. In contrast with the Walling atlas, the Snyder-Van Vechten volume contains a number of lithographic illustrations. The historical and descriptive departments are also quite comprehensive. The preface explains that "the maps of counties are based on the national surveys, those of cities and villages on official records. In order to present correctly the various features given in the separate county maps, additions were made by the aid of engineers' plans and maps of the different railroad companies; by county and town records, and by personal investigations of competent surveyors and draughtsmen, sent to nearly every county for this purpose."

Fewer than five thousand copies of the Snyder-Van Vechten atlas were sold. As Harrington reported, "the outcome was a very heavy loss. Mr. Snyder was a very fine gentleman, and went into the State of Wisconsin to make them a very superior atlas; but he was met coldly from the beginning, on account of the dissatisfaction with county atlases and maps. He worked unceasingly for three years on that atlas, and until his health gave entirely out. When the delivery commenced, he was taken down from exhaustion, and died."[7]

The map plates for the Snyder-Van Vechten atlas were then acquired by H. R. Page & Company of Chicago, which published the *Illustrated Historical Atlas of Wisconsin* in 1881. In addition to the general and county maps and the city plans found in the 1878 atlas, the 1881 volume includes maps of thirteen states besides Wisconsin. The history and textual sections are far more extensive than those in the Snyder-Van Vechten work, and the illustrations are different.

Griffith M. Hopkins, who was involved in publishing county and city atlases, also contributed in a minor way to state atlas production. In 1873 he published the *Combined Atlas of the State of New Jersey* in two versions. The

Fig. 26–8. Map of Herrington from L. H. Everts & Company's 1887 *Official State Atlas of Kansas*.

first incorporated maps and plans of Hudson County, and the second included maps and plans of the city of Newark. Hopkins's address is given on the title pages of both works as 320 Walnut Street, Philadelphia. Both volumes contain maps of the state, nine sectional maps of varying groups of counties, and a number of detailed maps and plans of the city of Newark and the urban regions of Hudson County.

Notes

1. Bates Harrington, *How 'Tis Done; A Thorough Ventilation of the Numerous Schemes Conducted by Wandering Canvassers together with the Various Advertising Dodges for the Swindling of the Public* (Chicago, 1879), 55.
2. Ibid., 81.
3. Alfred Andreas, *An Illustrated Historical Atlas of the State of Minnesota* (Chicago, 1874), 7.
4. L. F. Andrews, *Pioneers of Polk County, Iowa* (Des Moines, 1908), 55.
5. Harrington, *How 'Tis Done*, 82.
6. Ibid., 82.
7. Ibid., 86.

27. *Mapping the Trans-Mississippi West*

Surveys of portions of the vast territory lying between the Mississippi River and the Pacific coast had been made by Spanish, French, English, and Russian explorers and traders from the middle of the sixteenth century to the end of the eighteenth. Many of the maps resulting from these surveys, however, remained in manuscript format, and general maps of North America as late as the beginning of the nineteenth century showed little geographic detail for the trans-Mississippi West.

It was little wonder, therefore, that many Americans regarded the fifteen million dollars the United States paid to France to acquire the Louisiana Territory as an unduly high price for obtaining a geographic pig in a poke. As Carl Wheat has pointed out, at the time the sale was consummated, "neither seller nor buyer possessed any real understanding either of the extent of [the] territory . . . or of its extraordinary content and unimagined complexity."[1] President Thomas Jefferson was not one to tolerate for long such a lack of information about the newly purchased land that more than doubled the size of the Republic. Thus, the ink was scarcely dry on the official transfer documents when he commissioned two young Americans, Meriwether Lewis and William Clark, to explore and map the unknown lands of the Louisiana Purchase.

In authorizing the Lewis and Clark Expedition, "Jefferson set in motion a succession of geographical explorations and surveys that were largely responsible for securing to the United States the country to the Pacific coast."[2] During the six or seven decades after Lewis and Clark's historic expedition, the federal government supported a dozen or more exploring and surveying expeditions in the trans-Mississippi territory. Private and commercial printers and publishers played an important role in presenting to the public the cartographic results of these official surveys.

Following their return, Lewis and Clark were rewarded with government appointments; the former became governor of the Louisiana Territory and the latter superintendent of Indian affairs for the territory. Because of their official duties and the death of Lewis in October 1809, the official report of their expedition was delayed. In 1810 the records and diaries were turned over to Nicholas Biddle, a Philadelphia writer and lawyer, who prepared the narrative report. The final editing and polishing was performed by Paul Allen, a journalist. The two-volume work was published in 1814 by Bradford & Inskeep of Philadelphia. Biddle's name does not appear on the title page.

Published with the book was Clark's *Map of Lewis and Clark's Track Across the Western Portion of North America from the Mississippi to the Pacific Ocean By Order of the Executive of the United States in 1804, 5 & 6*. Clark had forwarded the manuscript map to Biddle in December 1810. In addition to information Clark derived from the expedition, the map included data he had acquired between 1806 and 1810 from various explorers, traders, and trappers who had visited him in his capacity as superintendent of Indian affairs for the Louisiana Territory. Clark's manuscript draft was neatly redrawn for printing by Samuel Lewis, the skilled Philadelphia cartographer. It was engraved by Samuel Harrison.[3] "In Clark's published map," John Allen observed, "may be seen the general accuracy and character of the new images of the Northwest and the contrast between those views and the ones held prior to the successful completion of the bulk of the expedition's objectives."[4]

Though the American public anxiously awaited publication of the Lewis and Clark report and map, it had to wait until 1814, when, as Wheat reminded us, "the public [was] at last given full information on those epoch making discoveries. Meanwhile the commercial cartographers were left to stew in the juice of their thoroughly outdated geography."[5] The mapmakers were quick to make use of the new information.

One of the first commercially published maps to include Lewis and Clark's data was Samuel Lewis's map of *Missouri Territory formerly Louisiana*, which was incorporated in the 1814 edition of Mathew Carey's *General Atlas*. The map extends from the Mississippi River to the Pacific Ocean and features, as Wheat noted, "highly up-to-date Lewis and Clark geography."[6] It was reprinted without change in subsequent editions of the *General Atlas*.

Because Samuel Lewis drafted the map for the official report of the Lewis and Clark Expedition, he obviously drew upon Clark's manuscript map in compiling his own map. He did not, however, appropriate such data for his *New and Correct Map of the United States of North America* that was published in Philadelphia in 1816 by Emmor Kimber. This large wall map extends only to beyond the western limits of the state of Louisiana. Lewis's *A New and Correct Map of the United States, including great portions of Missouri Territory*, published in 1819 by Henry Charles of Philadelphia, does, however, include Lewis and Clark data to the 110th meridian, the western limit of the map.

The cartographic publication that best publicized for the American people the data derived from the Lewis and Clark Expedition and Zebulon Pike's exploration of the southwest in 1806 and 1807 was John Melish's 1816 *Map of the United States with the contiguous British & Spanish Possessions*. In the 1815 prospectus announcing his pending publication, Melish affirmed that the map would be compiled from "the best and latest State maps . . . materials in various Travels through the country, particularly Pike's and Lewis' and Clark's, and various materials in the public offices at Washington."[7] Because Melish printed only one hundred copies of his U.S. map at a time to allow him to incorporate new and corrected geographic data, there are twenty-four variant states of this map dated between 1816 and 1823. These variants reflect the acquisition of new geographical information about the trans-Mississippi West.

William Darby published in 1818 a small *Map of the United States Including Louisiana*, which only extends west to the 115th meridian. The map was engraved by James D. Stout and printed in Brooklyn by Harrison & Bashworth. It incorporates Lewis and Clark data in the Missouri valley, but, in the western Texas portion of the map is imprinted this statement: "This part of Provincias Internas is but imperfectly known; no scientific traveller having ever explored, the wide range from St. Antonie de Behar to Red River."

In 1822 Henry S. Tanner published his large map of North America, which was in the following year incorporated in his *New American Atlas*, sectioned on four plates. In the "Geographical Memoir," an extremely useful feature of the atlas, Tanner summarized the source material he used in compiling the North America map. For the U.S. portion, this material included Lewis and Clark data as well as the information gained by Major Stephen H. Long's explorations in 1819–20. Long's party "explored the front range of the Rockies, made known what was later called Long's Peak, looked upon Pike's 'Highest Peak' . . . and came home by way of the Canadian [River], now at last found to be a branch of the Arkansas, rather than the Red." Tanner stated, "to the labours [of Long] I am indebted for the means of correcting some of the innumerable errors which have, until now, maintained their ground in all our most esteemed maps of the countries explored by Major Long and his party."[8] A slightly revised edition of this map was published in 1839. In 1829 Tanner had published the large map, the *United States of America*, which only extended slightly beyond the headwaters of the Missouri River. He probably excluded the western extremities of the country because his North American map had covered them.

Much of the cartographical data resulting from the Lewis and Clark, Pike, and Long expeditions was presented to the public on a series of commercial maps published between 1816 and 1840, but not always accurately. Herman R. Friis pointed out that "most of the maps of the western United States in the 1820's and 1830's that were available to and were viewed by the public included much misinformation that was based on fancy rather than on fact. . . . This lack of fundamental accuracy is perhaps best illustrated by House Geographer David H. Burr's official 'Map of the United States of North America with Parts of the Adjacent Countries,' compiled and printed for the House of Representatives of the United States in July, 1839."[9] Wheat differed from Friis's evaluation of Burr's map. It, he noted, "is a map of great importance, for it carries Jedediah Smith through all of his adventuring and pictures the West through Smith's eyes. It should have been better known in its day, and its comparative lack of influence is hard to explain. . . . The . . . Map of the United States is . . . the most complete that had yet appeared."[10]

Beginning with the Lewis and Clark Expedition, army engineers made major contributions to the exploring and mapping of the American West. The formal creation of the Corps of Topographical Engineers within the War Department by a congressional act of July 5, 1838, "served to coordinate and expand official geographical exploration and mapping by the Federal Government, especially in the West."[11] Thus, official military surveys dominated exploration, surveying, and mapping in the trans-Mississippi West during the 1840s and 1850s. These activities have been intimately detailed in official reports as well as in a number of scholarly publications.[12]

As the federal government had only limited facilities for reproducing maps prior to the Civil War, it relied on commercial concerns to engrave and print maps for official congressional and governmental reports. A number of the official maps of the western surveys were lithographed during the 1840s and 1850s. Among the lithographers who worked on these publications were Julius Bien and Ackerman & Sarony of New York City, P. S. Duval, T. S. Wagner, and Chillas of Philadelphia, C. B. Graham of Washington, D.C., and Edward Weber of Baltimore. Bien and Weber were especially active in lithographing maps and other graphics for these reports. One of Weber's notable works in this area was John Charles Frémont and W. H. Emory's *Military Reconnaissance from Fort Leavenworth, in Missouri to San Diego, in California*, which was published in 1848. Although Bien did varied lithographic work, Harry Peters stated that he "will always be remembered chiefly as the first great scientific cartographer in the United States. Soon after he arrived in this country he became interested in improving the quality of maps, and thanks to President Pierce and his administration, he was soon making maps of the new surveys in the west. He produced literally thousands of maps of various parts of the United States, lithographed and engraved, and all of high quality, for use by state governments and the federal government."[13] Bien also lithographed maps for the reports of the Pacific Railroad surveys, among others. "Probably Bien's outstanding work prior to the Civil War," stated Richard Bartlett, "was Lieutenant C. K. Warren's wonderful map of the territories between the Mississippi and the Pacific."[14]

Interest in the West grew as more information about it became available through the reports and maps of the several Frémont expeditions. It was further stimulated by the successful settlement of Oregon, San Francisco, and Salt Lake City, the Mexican War of 1846–48 (which added more territory to the rapidly expanding United States), and the discovery of gold in California. All these events placed pressure on the federal government to step up in its survey programs and to examine possibilities for constructing roads and railroads from the Mississippi River to the Pacific coast. Friis wrote that

> the U.S. government, especially through the Topographical Bureau of the War Department, responded quickly and efficiently to this clamor for geographical information by directing a large number and a wide variety of expeditions whose primary objective was topographical surveying and landscape description. Leading civilian scientists and topographical engineers and their assistants combined their talents for observation, analysis, and interpretation to produce a surprisingly large treasure trove of objective, on-the-spot information about the New West that significantly modified the then current concept of an inhospitable, resourceless, sterile environment.[15]

For some of these expeditions there are published reports and maps. For others, the data remain in manuscript form, in the files of the National Archives. Friis disclosed that

> perhaps the most important product of these extensive field surveys and office activities directed to the publication of the scientific results of the geographical explorations was the compilation of an accurate landform 'Map of the Territory of the United States from the Mississippi to the Pacific Ocean' on a scale of 1:3,000,000 or about fifty miles to an inch. This was Lieutenant Warren's immediate and primary responsibility. On March 1, 1858, Warren transmitted to his superior, Captain A. A. Humphreys, a letter and a report of memoir to accompany the foregoing map. In his letter he notes: 'In compiling the map . . . my instructions were to carefully read every report and examine every map of survey, reconnaissance, and travel which could be obtained, to ascertain their several values, and to embody the authentic information in the map. This duty is now to the best of my ability completed.' . . . In his brief, straightforward manner Warren transmitted to his superiors a memoir and a map that must rank as one of the most significant contributions to the history of American cartography.[16]

Wheat confirmed "that the information on which this map was based was sufficient to afford a reasonable over-all picture of the American West, that its compiler made effective use of that information, and that subsequent efforts in the way of maps may properly be deemed merely the filling in of detail."[17]

Commercial map publishers were keen to meet the demand for maps of the West. But, as Wheat observed, "by far the majority of commercial maps of the West during the early eighteen-forties were of dubious geographic legitimacy. The cartographers should not be blamed, however, since they still had little to go on except Lewis and Clark, Pike, Long, and Humboldt."[18] Most of the commercial maps of the United States published before 1846 by such well-known publishers as Tanner, Phelps & Ensign, and S. A. Mitchell did not extend to the Pacific coast. Some of the maps, however, included inset maps of North America, which presented the basic information about the West available prior to publication. Among such are J. Calvin Smith's large 1843 *Map of the United States of America, Canada and Texas . . . Compiled from surveys of the United States Land Office*. It was printed from engraved steel plates. S. A. Mitchell's 1845 *Reference and Distance Map of the United States* only extends west to a little beyond the limits of Missouri, but has the large inset "General Map of the United States with the Contiguous British and Mexican Possessions" in the lower right corner of the map sheet. There is, similarly, an inset "Map of North America Including All the Recent Discoveries" on J. H. Young's 1844 *Map of the United States*. In 1845 Phelps & Ensign published J. L. Woodbridge's *Map of the United States With The Recent Counties, Cities, Villages and Internal Improvements in the Western States* on which there is an inset "Map of North America including the West Indies." The

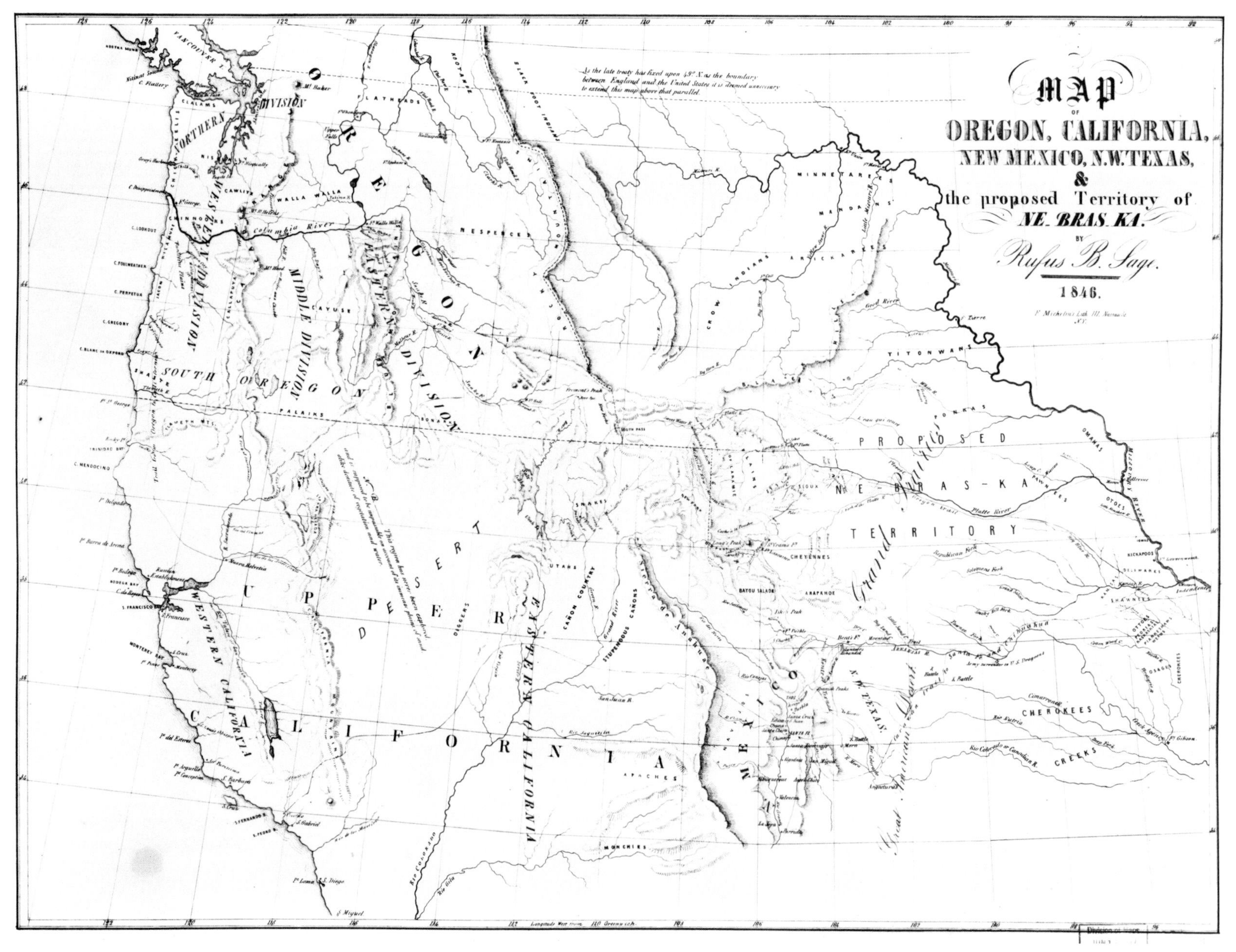

Fig. 27–1. This map was prepared by Rufus B. Sage to accompany his book, *Scenes in the Rocky Mountains*, published in Philadelphia in 1846.

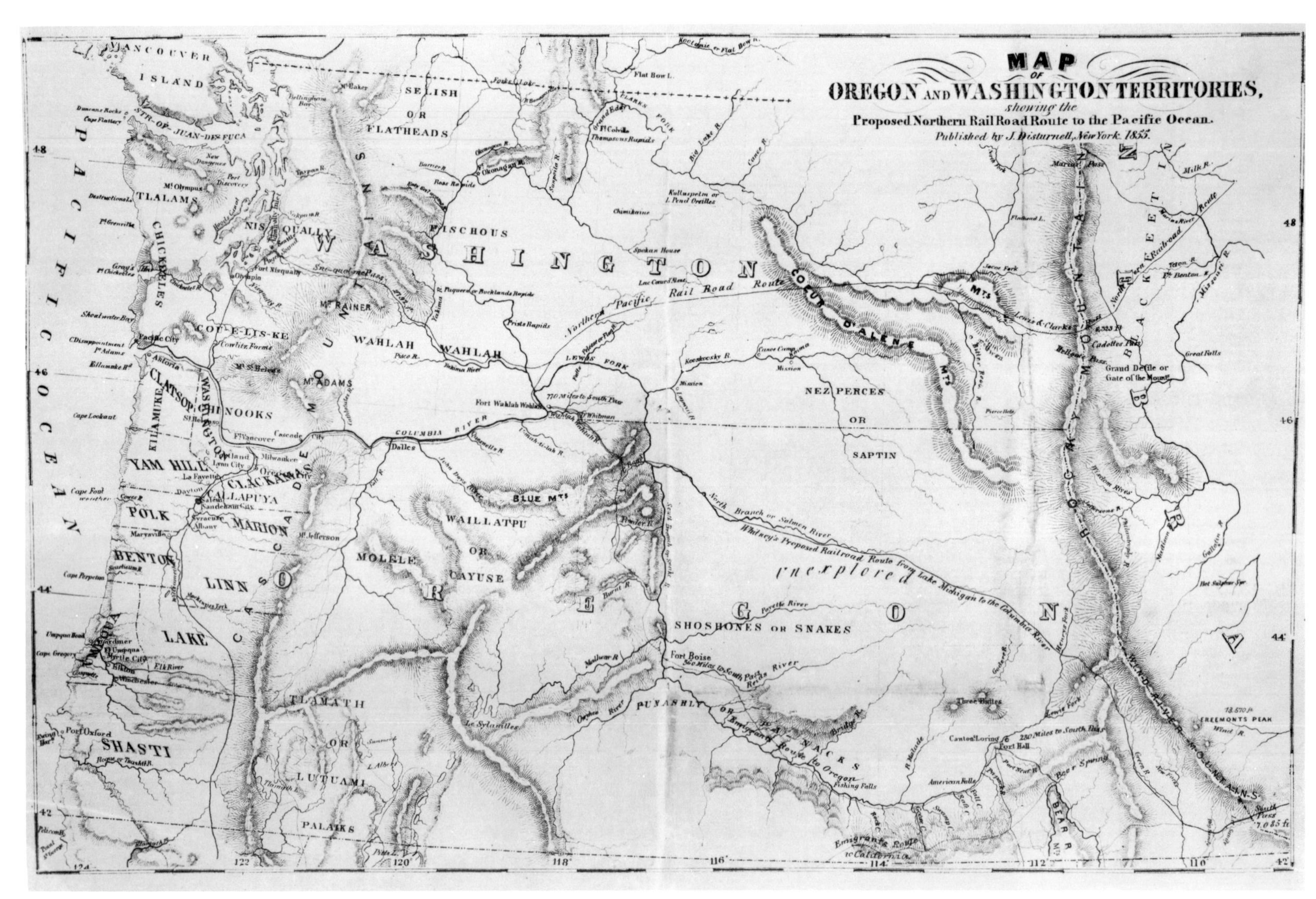

Fig. 27–2. John Disturnell issued this map of Oregon and Washington in 1855.

cartographic data for the trans-Mississippi West is quite incomplete on this inset.

Strangely, S. A. Mitchell's *New Map of the Western United States*, published in 1845, only extends west beyond the lower course of the Missouri River, although the same publisher's 1846 *Reference and Distance Map of the United States* has the inset "New Map of Texas, Oregon and California." The reference and distance map shows the routes to Oregon and Santa Fe and incorporates data from Emory's map, *Texas and the Country Adjacent*, published in 1844. Phelps's 1847 *Ornamental Map of the United States and Mexico* is rather crude cartographically, but it does show the proposed route of the great Oregon railroad, Frémont's route, and several other routes and trails.

A number of individuals, some of them writers, visited the western lands during the first half of the nineteenth century. Many of the books that described these visits included maps, such as one of Oregon, California, New Mexico, northwest Texas, and Nebraska that illustrated Rufus B. Sage's *Scenes in the Rocky Mountains*, which was published in Philadelphia in 1846 (Fig. 27–1).

The demand for maps of the West by Americans increased with the outbreak of the Mexican War in 1846. This was reflected in the inclusion of Mexico on U.S. maps and in the publication of separate maps of Mexico with adjoining states of the Union. Particularly significant, because it was used in negotiating the peace treaty of February 2, 1848, that brought the Mexican War to a close, was John Disturnell's *Map of the United States of Mexico*, the first edition of which was published in 1847.[19] Disturnell also published a number of regional maps, among them the map of *Oregon and Washington Territories* in 1855 (Fig. 27–2). Also of interest because it was published by a western lithographer, Julius Hutawa of St. Louis, is the *Map of Mexico, New Mexico, California*

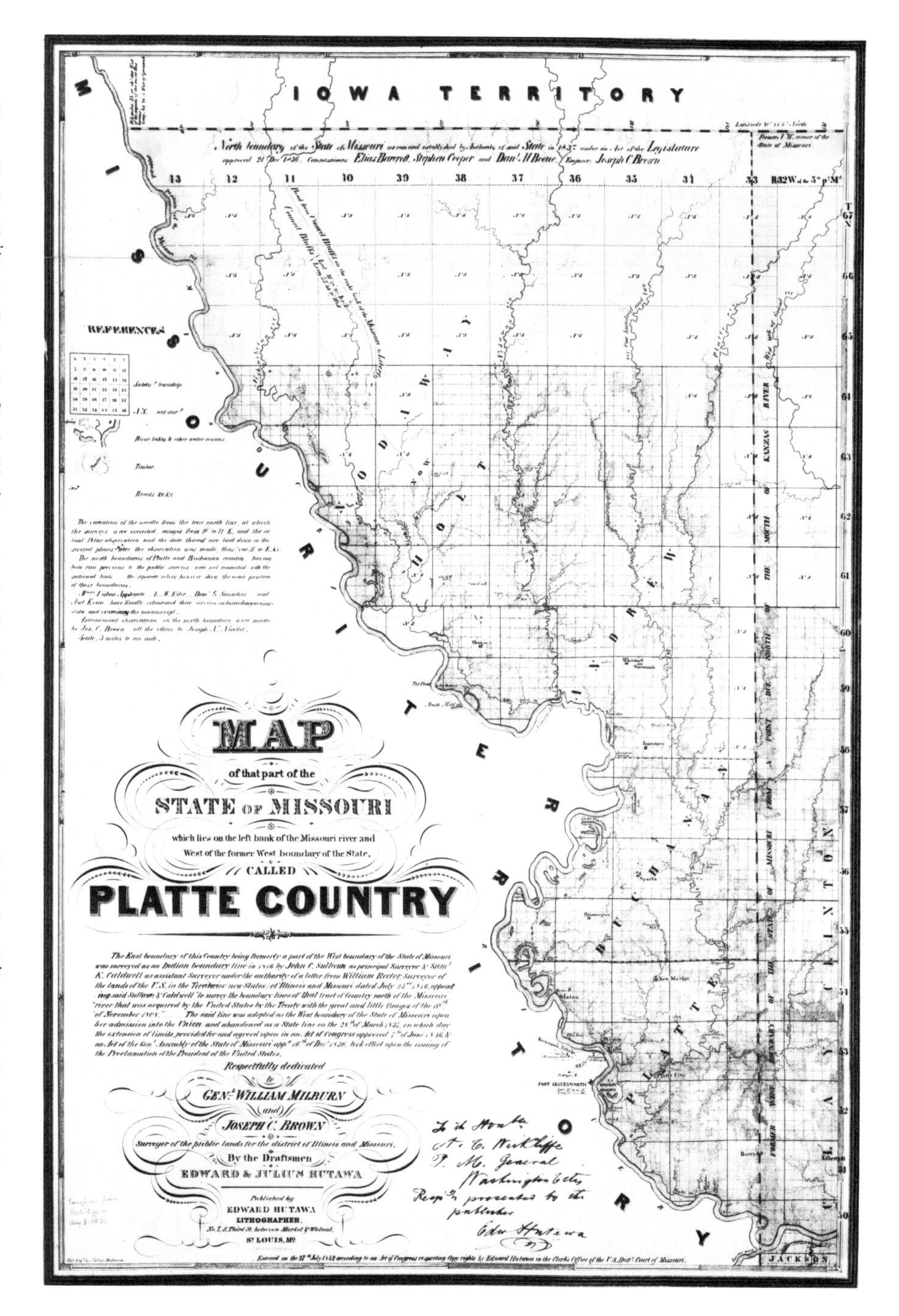

Fig. 27–3. Edward Hutawa published this map of the Platte Country in Missouri in 1842. He compiled it with his brother, Julius Hutawa.

& Oregon, which was issued as a supplement to the October 1, 1847, issue of the *Missouri Republican*. The map includes much information from Frémont's surveys, as well as the routes of various other explorers. Several years earlier, in 1842, Hutawa had drafted a map of the Platte country in Missouri with his brother, Edward Hutawa, which also was published by Edward (Fig. 27–3).

Published in the same year was Colton's large *Map of the United States of America, the British Provinces, Mexico and the West Indies*, which was drawn and engraved by Sherman & Smith of New York. The map, which has a decorative floral border, was probably printed from engraved steel plates. Comparable to Colton's map in size and content is the *New Map of the United States and Mexico*, which was published in 1847 by Monk & Sherer of Cincinnati. It was drawn and engraved by the Cincinnati firm Doolittle & Munson. The engraving was probably on steel.

The events of 1848, the ending of the Mexican War that brought California and New Mexico into the United States, the California gold rush, and the proclaiming of Oregon as a territory, all contributed to the increasing numbers of maps published on the West. Of importance to commercial map publishers as a compilation source was Charles Preuss's *Map of Oregon and Upper California from the Surveys of John Charles Frémont and Other Authorities*. It accompanied Frémont's *Geographical Memoir on Upper California*, which was published as a Senate Executive Document in 1848. The map synthesizes data from other sources in addition to Frémont's discoveries. The Preuss-Frémont map, noted Wheat, "gave to the American West its first reasonable published cartographic reflection."[20]

The discovery of gold in California set off a mass migration from the eastern states and resulted in pressing demands for guides, maps, and descriptive publications about the West. An indication of the cartographical response to these needs is evident from Wheat's classic volume *The Maps of the California Gold Region 1848–1857*.[21] Wheat described 323 pertinent maps, including official maps from Frémont's expeditions and from Senate and House documents, commercial maps and guides published by Sherman & Smith, John Disturnell, J. H. Colton, Ensign & Thayer, S. Augustus Mitchell, J. Calvin Smith, Henry S. Tanner, and lesser known publishers, and maps published in Great Britain, France, Germany, the Netherlands, and other countries. Maps of the gold fields that appeared in contemporary books and newspapers are also described.

Among the earliest of the commercial gold rush maps is a small lithographic reproduction entitled "Map of California and the Routes to Reach It" that illustrated Henry I. Simpson's *The Emigrant's Guide to the Gold Mines*, which was published in New York in 1848. Commercial publishers also revised and updated previously published maps to take advantage of the gold field interest. One such map was Colton's 1849 *Map of the United States, the British Provinces, Mexico &c Showing the Routes of the U.S. Mail Steam Packets to California and a Plan of the Gold Region*. This was only one of a number of Colton gold field maps.

John Disturnell, a prolific publisher of guidebooks and maps, issued the 1849 *Map of California, New Mexico and Adjacent Countries Showing the Gold Regions &c.* with his *Emigrant's Guide to New Mexico, California & Oregon*. Some of Disturnell's maps and guides were issued in Spanish editions.

Some lesser known publishers also sought to profit from the boom in gold field maps. Of particular interest in this group is the *Correct Map of the Bay of San Francisco and the Gold Region From Actual Survey June 20th 1849 for J. J. Jarves*. It was published by James Munroe & Company of Boston and lithographed by J. H. Bufford & Company. J. J. Jarves had gone to the Hawaiian Islands in 1838 and two years later became editor of a Honolulu newspaper. In 1848 he was commissioned by the Hawaiian government to negotiate treaties with the United States. It was apparently on this visit that he made his map. Subsequently, Jarves settled in Italy where he was the U.S. consul in Florence from 1880 to 1882.

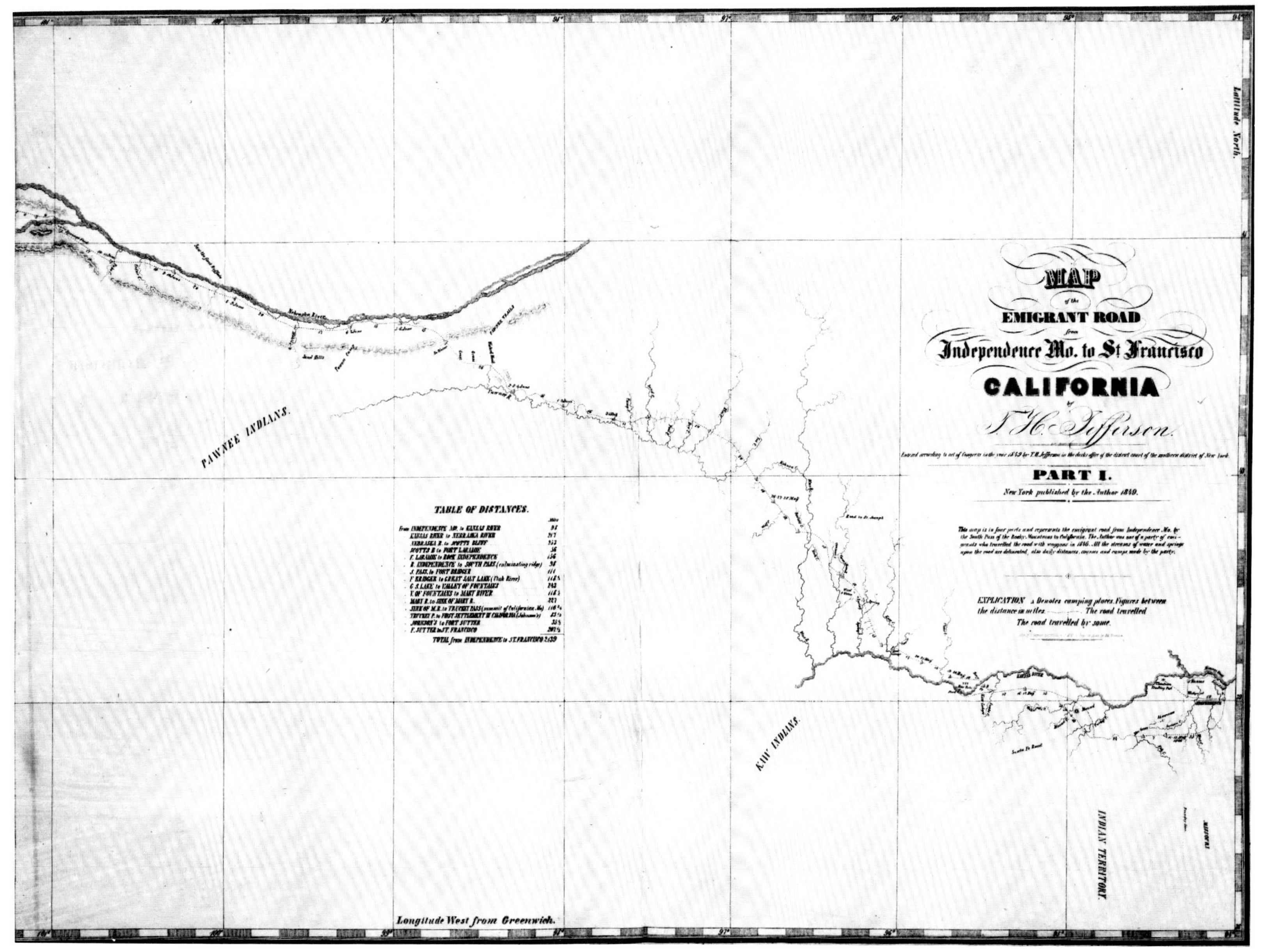

Fig. 27–4. The first sheet of Jefferson's 1849 four-sheet *Map of the Emigrant Road*.

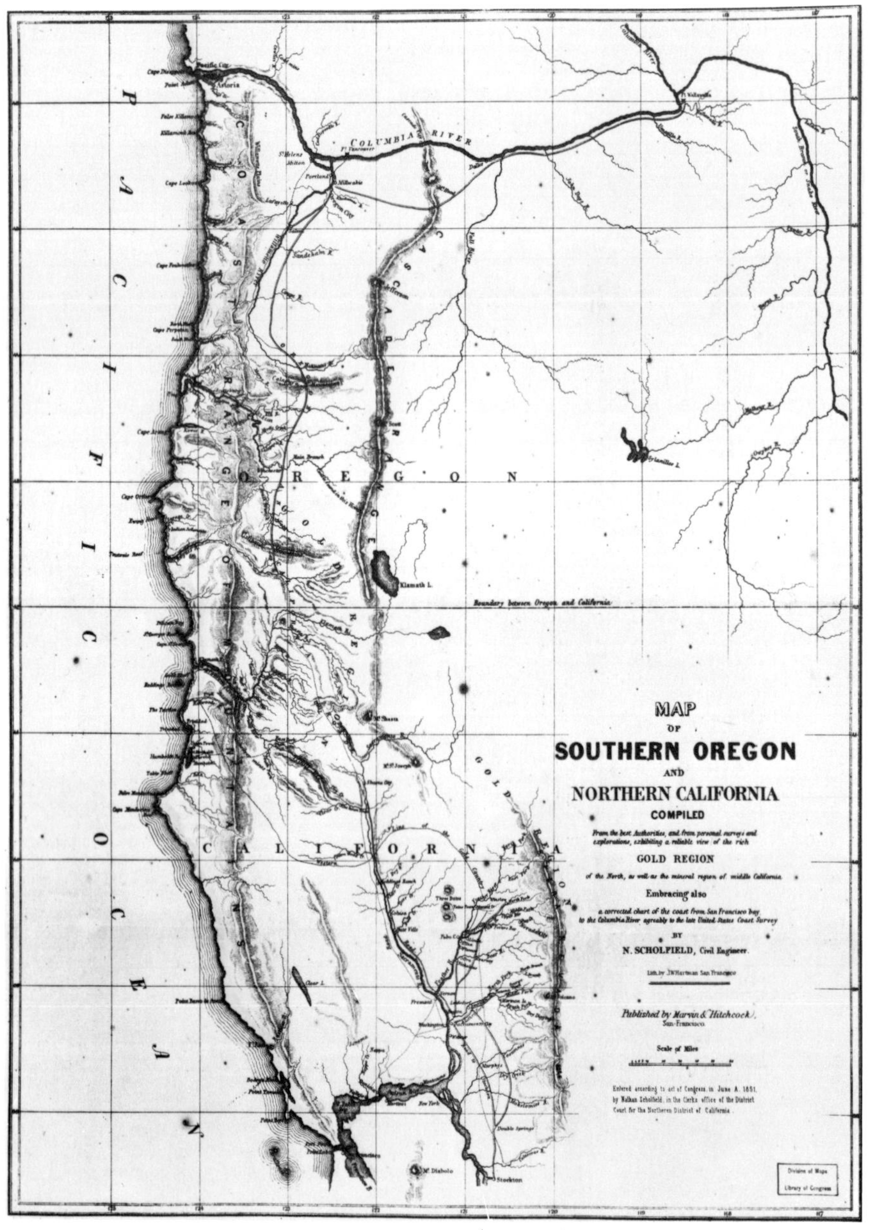

Fig. 27–5. Nathan Scholfield's 1851 map of southern Oregon and northern California.

T. H. Jefferson published his own four-sheet *Map of the Emigrant Road from Independence Mo. to St. Francisco California* in 1849 as an accompaniment to his guidebook with the same title (Fig. 27–4). In the guide Jefferson identifies himself as "one of a party of emigrants who travelled the road with wagons in 1846." Little is known about Jefferson apart from what is revealed by the map and guide, which are extremely rare. Happily, a facsimile edition of the map and guide, with introduction and notes by George R. Stewart, was published in 1945 by the California Historical Society.

A number of gold field maps and guides were also published in the West. Among these is the *Map of the Route to California, Compiled from Accurate Observations and Surveys by Government*. It was engraved by Joseph E. Ware and published in St. Louis by J. Halsall in 1849. It is largely derivative from Frémont's. Ware also compiled the 1849 *Emigrant's Guide to California*, which was also published in St. Louis. In as early as 1850, San Francisco began to become a cartographic publishing center. Cooke & Le Count published its *Map of the Gold Region of California* there in 1850. The map was also lithographed in the city by Pierce & Pollard. The map is extremely rare. Also issued in San Francisco was the 1851 *Map of the State of California . . . Also a Complete Delineation of the Gold Region*, which was lithographed and published by B. F. Butler. A portion of this map was reprinted the same year under the title *Map of the Gold Region*.

Nathan Scholfield compiled the *Map of Southern Oregon and Northern California . . . exhibiting a reliable view of the rich Gold Region*, which was lithographed by J. W. Hartman of San Francisco and published in that city in 1851 by Marvin & Hitchcock (Fig. 27–5). Also with a San Francisco imprint is J. B. Tassin's 1851 *Newly Constructed and Improved Map of the State of California. . . . With a Cor-*

Fig. 27–6. Two early San Francisco residents, Thomas Tennent and Alexander Zakreski, produced this 1853 map of Oregon and California.

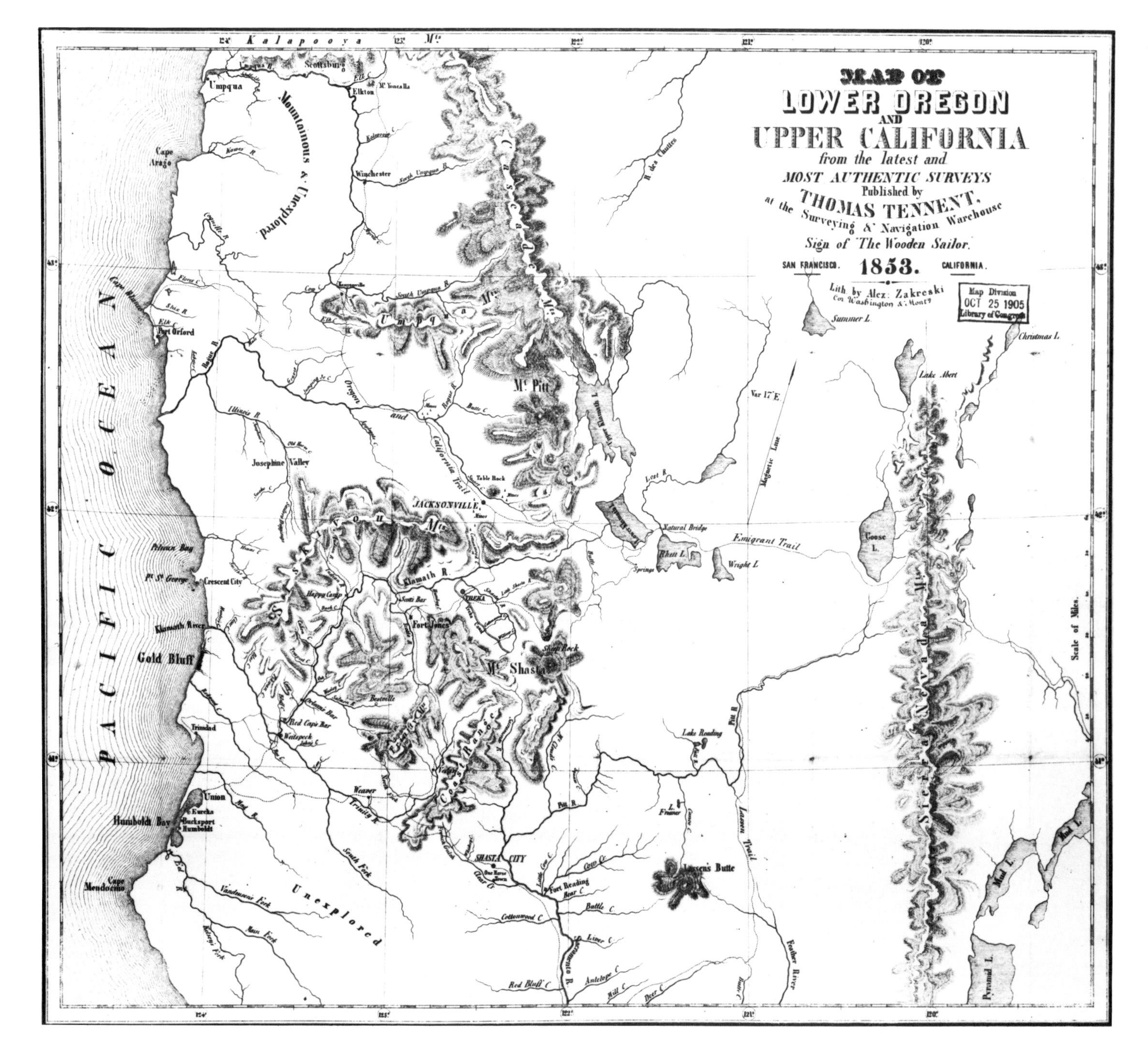

MAP OF
LOWER OREGON
AND
UPPER CALIFORNIA
from the latest and
MOST AUTHENTIC SURVEYS
Published by
THOMAS TENNENT,
at the Surveying & Navigation Warehouse
Sign of The Wooden Sailor.
SAN FRANCISCO. 1853. CALIFORNIA.
Lith by Alex: Zakreski
Cor. Washington & Mont.y
Map Division
OCT 25 1905
Library of Congress
Kalapooya Mts
PACIFIC OCEAN
Umpqua
Scottsburg
Elkton
Mt. Yoncalla
Cape Arago
Mountainous & Unexplored
Winchester
Port Orford
Mt. Pitt
Josephine Valley
Table Rock
JACKSONVILLE
Oregon and California Trail
Natural Bridge
Emigrant Trail
Rhett L.
Wright L.
Goose L.
Lake Abert
Christmas L.
Summer L.
Var. 17° E.
Magnetic Line
Pelican Bay
Pt. St. George
Crescent City
Happy Camp
Klamath R.
Scotts Bar
Fort Jones
Klamath River
Gold Bluff
Mt. Shasta
Shasta Butte
Trinidad
Red Cap's Bar
Weitspeck
Lake Reading
Pit R.
Union
Humboldt Bay
Eureka
Bucksport
Humboldt
Cape Mendocino
Weaver
Trinity R.
Coast Range
Sierra Nevada Mts.
SHASTA CITY
One Horse Town
Fort Reading
Lassen's Butte
Lassen Trail
Unexplored
South Fork
Cottonwood C.
Battle C.
Cow C.
Liver C.
Antelope C.
Red Bluff C.
Mill C.
Deer C.
Sacramento R.
Feather River
Pyramid L.
Mud L.
Scale of Miles.

rected and Improved Delineation of the Gold Region. It was lithographed by Pollard & Peregoy and published by Cooke & Le Count. The 1853 *Map of Lower Oregon and Upper California* was published by Thomas Tennent and lithographed by Alexander Zakreski (Fig. 27–6). Tennent, who was born in Philadelphia in 1822, was trained as a manufacturer of mathematical instruments. He immigrated to California in 1849, and, after an unsuccessful attempt at gold mining, he opened an office and sales center for marine and surveying instruments and publications. We know little concerning the early life of Zakreski, who operated a lithographic shop in San Francisco between 1852 and 1857. In 1853 he also published *The Only Correct and Fully Complete Map of San Francisco,* which was "Drawn & Lith. by Alex: Zakreski, Cor. Washington & Montgomery at the Topographical Office." In 1858 Zakreski was employed as a draftsman in the U.S. Surveyor General's Office in San Francisco.

One of the most prolific and successful of San Francisco's early lithographic firms was Britton & Rey. Joseph Britton was born in Yorkshire, England, in 1825 and came to America with his parents when he was ten years old. He lived in New York City until he was twenty-four years old, and apparently he learned lithography there. Prints by him are as early as 1847. In 1849 Britton joined the thousands who migrated to the gold fields. Unsuccessful in the mines, he went to San Francisco in 1852 and, in partnership with C. J. Pollard, established a lithographic printing shop. This association was short-lived, however, and later in the same year he joined with Jacques J. Rey in a partnership that flourished for almost three decades. Rey was born in Alsace, France, in 1820 and early studied art and lithography. He, too, was attracted by the lure of the gold fields and in around 1850 went to California via the Panama route. Little is known about his activities before he formed his partnership with Britton. Three years after joining Britton, he married Britton's sister. Peters noted that "Joseph Britton remained a bachelor all his life. He lived always with the Rey family and . . . throughout the long association of those two men, in business and in family life, they were the most amiable friends. In their lithographic business, Rey has been spoken of as the artist of the firm, and Britton as the business man, although he, too, could sketch."[22]

In addition to doing a large number of prints, Britton & Rey lithographed and published maps of San Francisco, the gold fields, and California. In 1853 they published the *Map of the State of California,* which was "drawn and compiled from recent surveys by J. B. Trask." Wheat stated that "this is one of the best, as well as one of the rarest maps of the period. Its maker, John Boardman Trask (b. Roxbury, Mass., 1824, d. San Francisco, 1879), came to California in 1850, joined the Mexican Boundary Survey and became first State Geologist of California. On May 16, 1853, he was one of the eight who founded the California Academy of Sciences. While preparing his 'Report of the Sierra Nevada or California Range' . . . , he published his map of California and also that of the mineral districts."[23] Trask's 1853 *Topographical Map of the Mineral Districts of California* was also lithographed and published by Britton & Rey.

In 1855 Britton & Rey lithographed and published the *Map of the Mining Region of California.* Then the firm published the 1857 *Map of the State of California Compiled from the U.S. Land and Coast Surveys . . . , By George H. Goddard.* Of this Goddard map Wheat remarked that "nothing cartographically comparable to this map had hitherto appeared, except possibly the Baker maps of 1855 and 1856 and the Britton and Rey map of 1855, all of which embraced much smaller areas, and disclosed less artistic draughtsmanship."[24] Goddard, who was born in Bristol, England, in 1817, was trained as an architect and surveyor. He immigrated to California in 1850, settling in Sacramento where he was employed as a government surveyor. Goddard moved to San Francisco in 1862, where he remained until his death in 1906 at the age of eighty-nine. He produced a number of sketches, paintings, and maps. Among his cartographic works are two panoramic maps of San Francisco that were published by Britton & Rey in 1868. Britton & Rey also published a number of plans of San Francisco and other California cities, including some attractive panoramic maps and bird's-eye views. Although most of their maps were of California or its parts, Britton & Rey published in 1865 a map of the Boise Basin, in Idaho (Fig. 27–7).

Another significant California mapmaker of the mid-nineteenth century was George Holbrook Baker, who was born in Boston on March 9, 1827. He worked with an artist in New York City for several years preparing sketches and maps. Tiring of this, he enrolled in the National Academy of Design and was a student there when word of the discovery of gold in California reached New York. He joined the westward movement, and on May 28, 1849, he arrived in San Francisco. After a brief and unsuccessful attempt at mining, he engaged in mercantile pursuits in San Francisco. In 1852 he moved to Sacramento where he continued these activities. Two years later, in partnership with a man named Barber, he went into the business of making woodcut illustrations. In around 1856 he was editing and publishing several newspapers. A few years later, Baker set up a lithographic printing shop in Sacramento. Following the great Sacramento flood of 1862, he relocated to San Francisco where he again engaged in lithographing, continuing in that business until his death in 1906. A cartographic product of his brief association with Barber is the *Map of the Mining Region of California*, which was drawn and compiled by Baker and published in 1855 by Barber & Baker. There is also an 1856 edition of the map. The lithography for both was done by R. W. Fishbourne of San Francisco. Fishbourne, independently and in association with others, practiced lithography in San Francisco between 1851 and 1858.

Another western cartographer, Benjamin F. Butler, worked as a lithographer in New York City between 1838 and 1849. He relocated his business to San Francisco in about 1850 and is credited by some as having been the first man to establish a lithographing plant there. In 1851 he published the small *Map of the District, Night of May 3d, 1851*. Also in that year Butler published the *Map of the State of California*, which completely delineates the gold region. The firm also issued in 1851 the *Official Map of the City of San Francisco Full and Complete to the Present Date* by William M. Eddy, the city surveyor. It was lithographed by B.F. Butler's Lithography. There is another edition of this map that was lithographed by Britton & Rey. The *Map of the Northern Portion of San Francisco County* by Clement Humphreys, county surveyor, was an 1852 Butler publication.

Although not from a local press, the *Map of the City of San Francisco Compiled from Records & Surveys by R. P. Bridgens, C. E.* should be mentioned. It was a large wall map "respectfully Dedicated to the Citizens by the Publisher M. Bixby." Published in 1854, it was lithographed by Friend & Aub of 80 Walnut Street, Philadelphia, and printed by "Wagner & McGuigan's Lith. & Steam Press, Phila." Bordering the map are illustrations of about twenty public and commercial buildings. We have no information about R. P. Bridgens and do not know whether he was related to the H. F. Bridgens who was involved with county maps and atlases in the late 1850s and early 1860s.

The most significant development of the 1850s in the trans-Mississippi West was the series of detailed surveys made between 1853 and 1857 to determine routes for possible rail lines. The "scientific results of these surveys," wrote Friis, "represent a major contribution to the exploration and mapping of the West. . . . By the end of the 1850s the Mississippi Valley and its western fringe, the Great Plains, had been surprisingly well mapped and explored by reconnaissance methods."[25] As noted earlier, the information from these railroad surveys and others was utilized by C. K. Warren in compiling his authoritative 1857 *Map of the Territory of the United States from the Mississippi to the Pacific Ocean*. This map and those included in the reports of the railroad surveys served as basic compilation data for commercial map publishers for the next decade or so.

Most active among these publishers was J. H. Colton

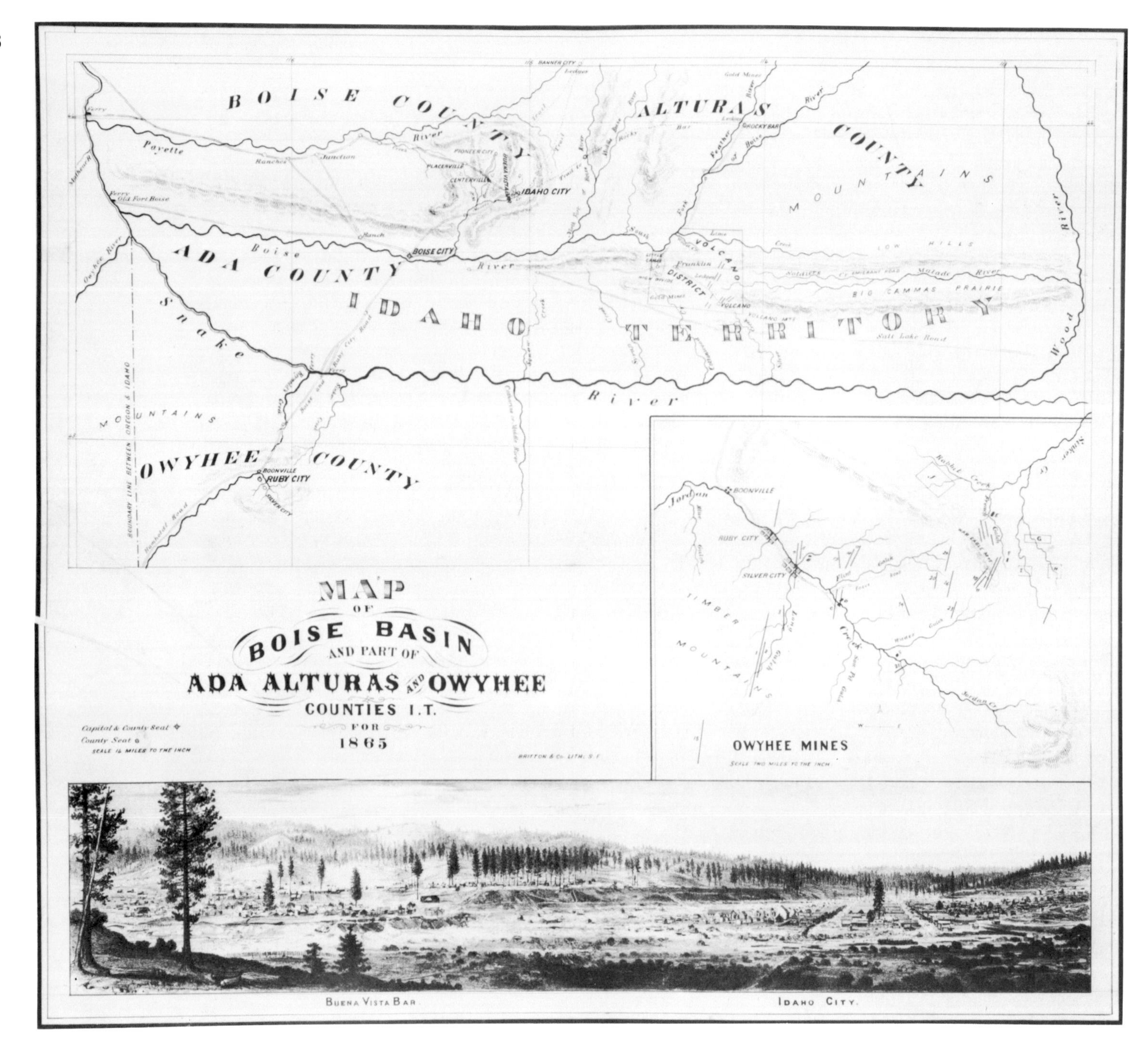
BOISE COUNTY
ALTURAS COUNTY
ADA COUNTY
IDAHO TERRITORY
OWYHEE COUNTY
IDAHO CITY
BOISE CITY
RUBY CITY
BOONVILLE
SILVER CITY
PLACERVILLE
CENTERVILLE
PIONEER CITY
ROCKY BAR
BANNER CITY
Old Fort Boise
Payette River
Boise River
Snake River
Owyhee River
Wood River
VOLCANO DISTRICT
BIG CAMMAS PRAIRIE
Salt Lake Road
MOUNTAINS
LOW HILLS
BOUNDARY LINE BETWEEN OREGON & IDAHO
MAP OF BOISE BASIN AND PART OF ADA ALTURAS AND OWYHEE COUNTIES I.T. FOR 1865
Capital & County Seat
County Seat
SCALE 16 MILES TO THE INCH
BRITTON & Co. LITH. S.F.
OWYHEE MINES
SCALE TWO MILES TO THE INCH
Jordan Creek
TIMBER MOUNTAINS
BUENA VISTA BAR
IDAHO CITY

& Company, which issued *The United States of America* (1855), which shows the explorations for the Pacific railroads, and *Colton's Map of the United States of America, the British Provinces, Mexico and the West Indies* (1859), which was published by Thayer & Colton in three different sizes. Also worthy of mention are D. Griffing Johnson's 1857 *New Illustrated & Embellished County Map of the Republics of North America*, which was published in New York and Washington by D. G. & A. J. Johnson and is 190 by 78 cm. in size; *Mitchell's New National Map, Exhibiting the United States With the North American British Provinces*, published in Philadelphia in 1856; and the 1859 *Map of the United States West of the Mississippi Showing the Routes to Pike's Peak Overland Mail Route to California and Pacific Rail Road Surveys* compiled by D. McGowan and George H. Hildt and lithographed by Gast & Brothers of St. Louis, Missouri.

An early commercial portrayal of Texas following its accession by the United States is *J. De Cordova's Map of the State of Texas, Compiled from the Records of the General Land Office of the State by Robert Creuzbaur, Houston, 1840*. It was engraved by J. M. Atwood of New York. John M. Atwood was born in Washington, D. C., in 1818 and was active as an engraver in New York City from 1838 to 1852. De Cordova's map carries the signed endorsements of four Texas senators and representatives and the commissioner of the Texas General Land Office. Robert Creuzbaur was official draftsman for the land office. The map was, however, copyrighted by De Cordova and published as an accompaniment to his book *The State of Texas; Her Capabilities and Her Resources*, which was published in Galveston. Below the map's title cartouche is De Cordova's statement that "without my signature all copies of this map have been fraudulently obtained." In the lower right corner of the map sheet there is an inset map that shows Texas and the bordering areas of Mexico and the United States. Editions of De Cordova's map were also published in 1850, 1851, 1853, 1856, and 1857. The 1856 and 1857 editions were published by J. H. Colton & Company and carry the note "Revised and Corrected by Charles W. Pressler."

Fig. 27–7. Britton & Rey published this map of the Boise Basin of Idaho in 1865.

The Civil War had little cartographical impact on the trans-Mississippi West because no battles were fought there. The war did, however, put an end to the military surveys that had been so actively carried on during the 1840s and 1850s. The establishment of the territories of Colorado, Nevada, and Dakota early in 1861 did create a need for new maps, which was filled in part by the established S. A. Mitchell and J. H. Colton companies. Western lithographers and publishers were becoming more numerous and active in the early 1860s and also responded to the need for maps of these areas.

One particularly prolific cartographic publisher was H. H. Bancroft & Company of San Francisco. Its founder, Hubert Howe Bancroft, was born in Granville, Ohio, on May 5, 1832. From 1848 to 1852 he was employed in his brother-in-law's bookstore in Buffalo, New York. In 1852 he went to California and established his own bookstore and publishing house. He published the *Map of the State of California* in 1858. It was copyrighted in Philadelphia by H. Cowperthwait & Company. In 1862 Bancroft published the *Map of the Washoe Silver Region of Nevada Territory* in San Francisco and under his own imprint (Fig. 27–8). It is based on a manuscript map compiled and drawn by William H. Knight. In the following year he published *Bancroft's Map of the Pacific States*, which was also compiled by Knight, and *Bancroft's Map of the Colorado Mines*. Wheat disclosed that the *Map of the Pacific States* "proliferated in 1864 into a whole family of notable maps, automatically establishing the rising H. H. Bancroft house as the greatest map publisher of the Pacific Coast."[26] Bancroft's map of the Colorado mines illustrated his *Guide to the Colorado Mines*.

Bancroft also published the *Map of Oregon, Washington, Idaho, and British Columbia* and the *Map of California, Nevada, Utah and Arizona*, both of which were compiled by Knight and issued in 1864. The same compiler pre-

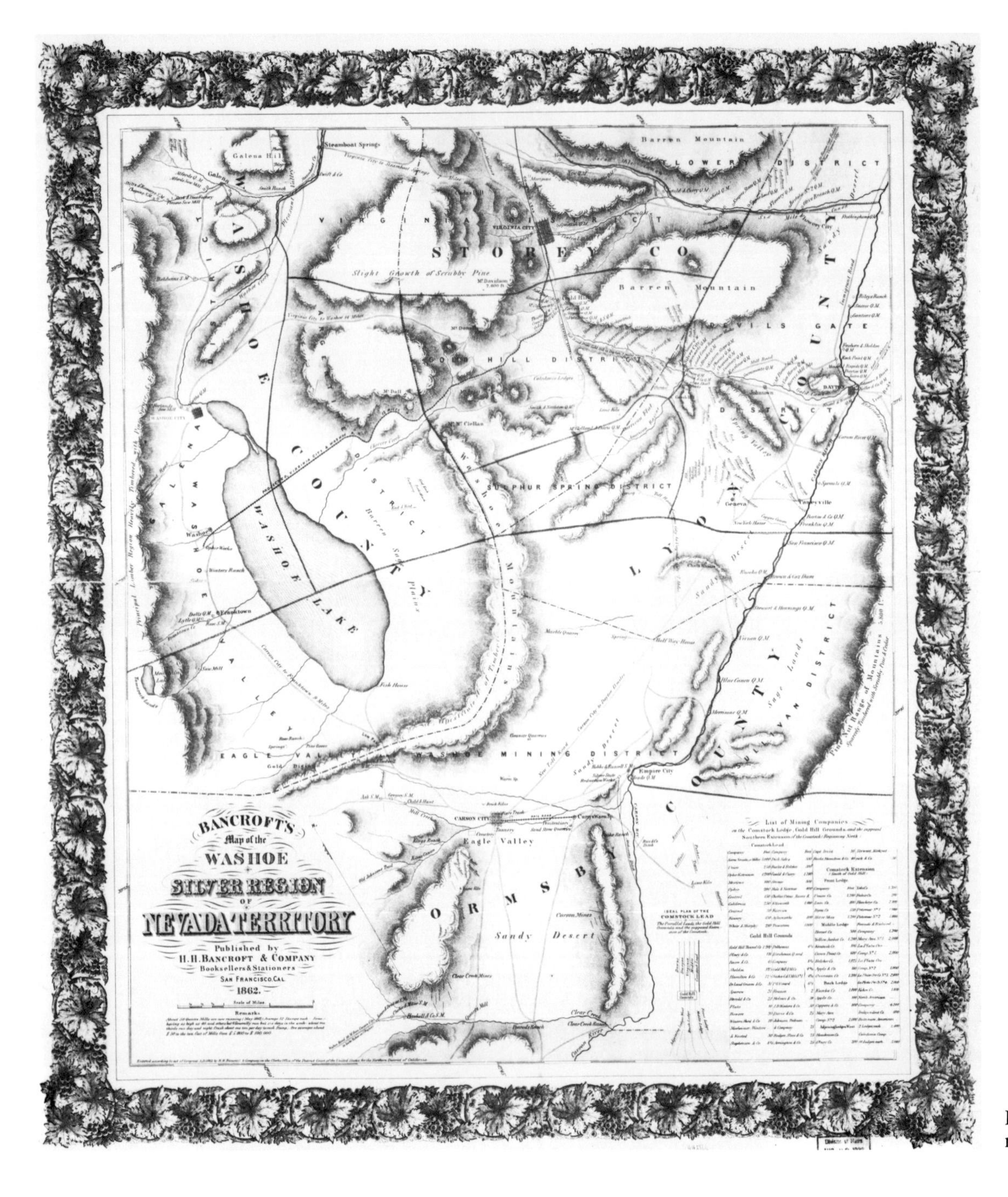

Fig. 27–8. H. H. Bancroft & Company's map of the Washoe region in Nevada.

pared Bancroft's *Map of the Rocky Mountain States and the Pacific Coast*, which is dated 1866. Bancroft published a new edition of the *Map of the Pacific States* 1866 and another in 1868. New editions of the Oregon and California maps were also released in 1868. The Oregon map was reissued in 1872, the same year in which Bancroft published the *Map of California, Showing the Principal Railroads Completed and in Course of Construction*. The lithographer is not indicated on any of the Bancroft maps. Bancroft continued publishing local and general maps to about 1890. In addition to his bookstore and publishing interests, he was a historian and an avid collector of material relating to California and the West. His collection of more than sixty thousand volumes was presented to the University of California, Berkeley, in 1905.

The A. L. Bancroft & Company, which appears to have had no relationship with H. H. Bancroft & Company, lithographed and published maps in San Francisco during the 1820s. Among these were street maps of San Francisco, maps of the bay counties, and bird's-eye views of six or eight cities and towns in California and Washington.

Warren Holt published maps in San Francisco between 1862 and 1875. The earliest Holt cartographic publication appears to be the 1862 *New Map of the State of California*, which seems to be derived from an 1861 map by Leander Ransom done for the California Academy of Natural Sciences. In 1863 Holt published *De Groot's Map of Nevada Territory Exhibiting a Portion of Southern Oregon & Eastern California*. (Henry De Groot's data were also used for Bancroft's 1863 *Map of the Colorado Mines*.) Another 1863 Holt publication was *A New Map of the State of California and Nevada Territory*, which was compiled by Ransom and A. J. Doolittle.

Holt's *Map of the Owen's River Mining Country Compiled and Drawn from the Most Reliable Information by Arthur W. Keddie* was published in 1864. The lithographer is not indicated on most Holt maps. A notable exception is the attractive 1866 *Map of the State of Nevada*, which has the credit "Grafton T. Brown, Lithographer, 543 Clay St. San Francisco." Brown, who operated a lithographic shop from about 1872 to 1877 was, according to Peters, a negro craftsman. In 1866 Holt also published a new edition of Ransom and Doolittle's map of California and Nevada. This map was published again in a revised edition in 1868. In 1869 Holt published the revised and greatly enlarged *Map of the States of California and Nevada*, which was compiled by Charles D. Gibbes, Julius H. Von Schmidt, and Arthur W. Keddie. It was lithographed by S. B. Linton of Philadelphia. Reduced approximately one-fourth in size, the map was published in 1873 and 1875 editions.

Los Angeles lagged behind San Francisco as a center for map publishers. William Robinson observed that "map-making . . . seems almost to have been an unknown art among Los Angeles citizens prior to 1849." Because of this situation, California's Governor Bennet Riley "recommended that Army lieutenant Edward O. C. Ord be hired by the city for the [city] surveying job."[27] Ord completed his survey of Los Angeles in 1849, which also covered the heart of the Pueblo lands. The map was never printed and exists today only in several hand-drawn copies. The second major survey of Los Angeles was completed in 1858 by Henry Hancock and his deputy surveyor, George Hansen. This map was also not printed and exists today only in manuscript form. In fact, it was not until the 1870s or 1880s that printed maps of Los Angeles were generally available. Robinson noted that popular in the 1880s were

> two . . . maps . . . excellent in coverage and usefulness. One, undated but published probably in 1875, by men active in Los Angeles in the 1870s, is a composite map showing Ord's Survey in the center, encircled by Hancock's and Hansen's Donation Lots. It was published by Bancroft & Thayer, a Los Angeles real estate firm made up of C. A. Bancroft and John S. Thayer, and W. H. J. Brooks, a title searcher. The lithographer was A. L. Bancroft & Co., San Francisco. The other, issued in November of 1887, was the work of V. J. Rowan and Theo. G.

> Koeberle, expert surveyors and map-makers. It is in black and white, and minute in detail.[28]

Although San Francisco was the principal center for cartographic lithography and publishing in the West, St. Louis had also achieved some early prominence in these fields. In as early as 1838, the *Plan of the City of St. Louis,* was drafted and constructed by Edward and Julius Hutawa in St. Louis. The map was dedicated to General William Milburn and was published by the lithographic printing establishment of Charles Friederich & Company of St. Louis. I have not examined a copy of this edition, but the 1842 edition was reproduced in 1853 by N. L. Wayman. Wayman also reproduced an 1851 edition of the map, which was "Published by Julius Hutawa, Lithographer, North Second St. No. 45, St. Louis, Mo." On an undated map of St. Louis that was probably published around 1865, the Julius Hutawa Lithography Company is noted as having been established in 1835. This map also carries the inscription, "Since I constructed and published the first map of St. Louis in 1838, this the Ninth Edition wich [*sic*] I respectfully lay before the Public." Peters ascribed to Julius Hutawa the 1846 publication of the *Map of the City of St. Louis* by compilers J. N. Nicollet and John C. Frémont, although he added that he had only seen maps that Hutawa had compiled himself.[29]

In 1844 Edward Hutawa copyrighted the large *Sectional Map Of The State of Missouri Compiled From The United States Surveys and other sources by the Publisher*. The map was "Engd. on Stone by Julius Hutawa." Wheat also lists an 1847 *Map of Mexico, New Mexico, California & Oregon Compiled . . . from the latest authorities . . . by Juls. Hutawa, Lithr. Second St. No. 45 St. Louis, Mo. 2(nd) edition*. Two copies of a map with the same title and also marked as a second edition, but dated 1863, are in the collections of the Library of Congress. The 1847 map was issued as a supplement to the *Missouri Republican* for the October 1, 1847, issue. Wheat noted that "if all the commercial maps had been so generally satisfactory as this Hutawa map of 1847, the public could have had little complaint."[30] Wheat also credited Julius Hutawa with having drawn and lithographed the 1849 *Map and Profile Sections Showing Railroads of the United States,* which accompanied the report by J. Loughborough on the "Project for a Pacific Railway," given before a St. Louis convention on October 15, 1849. Hutawa also lithographed I. Williamowicz's *Map of the Cairo and Fulton Railroad*, which was published around 1853. Hutawa also compiled for a January 1854 isue of the *Missouri Republican* the *Map of the United States Showing the Principal Steamboat Routes and Projected Railroads Connecting With St. Louis.*

Another early St. Louis lithographic firm was Gast Brothers, which included Augustus, John, and L. Gast. One of the firm's cartographic products is the map of *Part of the Mineral Region State of Missouri Constructed Under the direction of H. King, M. D., Geologist, by Wm. Kossak, Surv. & Civil Engr., St. Louis, Mo. 12 May 1853.*

Active in compiling and drafting maps of Utah and Salt Lake City between 1870 and 1900 was Bernard Arnold Martin Froiseth. Wheat, after gleaning information from Froiseth's obituary in the *Salt Lake Tribune* on November 5, 1922, noted that Froiseth was born in Trondheim, Norway, in 1839. He immigrated to Minnesota as a boy, and was trained as a civil engineer at the University of Montreal. He apparently was employed by the Department of the Interior before the Civil War and during the conflict was commissioned a colonel in the northern army and detailed for special duty. He moved to Salt Lake City in 1869. Froiseth married Jennie Anderson in Brooklyn in 1871, and he brought her to Salt Lake City, where the couple made their home.[31] Froiseth was employed by the surveyor general of Utah and is credited with having made the first map of the territory of Utah and the first map of the state of Utah (which entered the Union in 1896). His *New Sectional & Mineral Map of Utah* went through a number of editions from 1871 to 1898. All except the 1898 edition were lithographed by A. L. Bancroft & Company in San Francisco.

Wheat, describing the 1898 edition, stated that "executed on a large scale, 8 miles to an inch, the map is a pioneer masterpiece, wonderful in its detail. Froiseth had made full use of the best government maps."[32] Froiseth had prepared this last edition of his map by commission of the Utah state legislature.

Froiseth's earliest cartographic work, copyrighted in 1869 and 1870, was the *Maps of Utah Territory, Great Salt Lake Valley and Salt Lake City*, all three of which are on one sheet, along with a portrait of Brigham Young and Young's certification. It was printed by the American Photolithographic Company of New York City. In 1871 Froiseth prepared the *New Mining Map of Utah*, which was also lithographed by the American Photolithographic Company. In 1885 Froiseth published a plan of Salt Lake City, and in 1898 a new edition of the *Sectional & Mineral Map of Utah* appeared. It was published by W. B. Walkup & Company of San Francisco. Froiseth left government service shortly after 1871 and was involved in real estate and map publishing for the remainder of his career. He died on November 4, 1922, at the age of eighty-three.

Of particular interest is the *Map Showing the Extent of Surveys in the Territory of Utah 1856*, which was "Examined and approved the 30th Sept. 1856 Sur. Genl. David H. Burr of Utah." This map was lithographed and printed in Philadelphia by P. S. Duval & Son. Burr, who had prepared the 1829 map and atlas of New York State, was surveyor general of the Utah Territory from 1855 to 1857. His Utah map was published as part of Senate Executive Document 5 of the third session of the 34th Congress.

The trans-Mississippi West was still comparatively unsettled when county map and atlas publishing flourished in the midwestern and northeastern states during the late 1860s and the 1870s. Despite this, Thomas H. Thompson, who had prepared maps and atlases of Illinois and Iowa counties, migrated to California around 1876. There he published atlases of Sonoma County (1877), Fresno County (1891), and Tulare County (1892). In partnership with West, he issued atlases of Santa Clara County (1876), Alameda County (1878), and Solano County (1878). All are illustrated with lithographic views and sketches similar to those in the atlases of midwestern counties (Fig. 27–9). The four California county atlases published in the 1870s were lithographed and printed by Thomas Hunter of Philadelphia. The printer is not given in the two atlases published in the 1890s, and the illustrations in these volumes are printed in a sepia tone. In all of the atlases, relief is shown by crude hachures on some of the local maps.

The production of panoramic maps and bird's-eye views of American cities had been well developed by the 1860s in the midwestern states. The western publishers, however, soon began publishing them, too. As John Reps noted, "when the rapid settlement of the American West began in the middle of the nineteenth century . . . most [of the mapmakers] began to include western town and city views among their publications but they soon faced competition from newly-organized western publishing firms, notably those established in San Francisco shortly after the Gold Rush."[33] The two San Francisco lithographic firms that were most active in printing panoramic maps were Britton & Rey and A. L. Bancroft & Company. Britton & Rey printed views of a number of California cities, as well as of a few urban centers in Nevada, Idaho, and other western states between 1856 and 1901. Some of Britton & Rey's earliest panoramic maps were prepared by Charles Kudrel and Emil Dresel. The firm also printed such views prepared by George H. Baker, George Goddard, and, in later years, by Augustus Koch. Bird's-eye views were A. L. Bancroft & Company's principal output in the late 1870s. All of the views were drawn by Eli S. Glover. Most of the Bancroft lithographs depicted California cities, but some were of cities in Washington, Oregon, and Montana.

Following the Civil War, the federal government again turned its attention westward. It sponsored four geographical and geological surveys between 1867 and

Fig. 27–9. Illustrations from Thompson and West's *Atlas of Alameda County*, 1878.

1879. The War Department directed two of the surveys: the U.S. Geological Exploration of the Fortieth Parallel, headed by civilian Clarence King, and the U.S. Geographical Surveys West of the One Hundredth Meridian, headed by Lieutenant George Wheeler. The Department of the Interior sponsored the other two: the U.S. Geological and Geographical Survey of the Territories, directed by Ferdinand V. Hayden, and the U.S. Geographical and Geological Survey of the Rocky Mountain Region, headed by John Wesley Powell. As Bartlett has written, "these four surveys are often grouped together, and referred to as 'The Great Surveys.' They were great in the sense of the vast territories they examined, in their breadth-embracing topography, geology and the natural sciences—and in the span of years in which they operated. Their contributions to the knowledge of the American West were enormous."[34] The reports of the Great Surveys comprised the compilation material for commercial map publishers for a number of years.

When the Great Surveys had ceased to function in 1879, many still felt it necessary to coordinate further governmental explorations and surveys. As a result, the U.S. Geological Survey was established within the Department of the Interior, and within a few years the survey assumed major responsibility for preparing the basic topographic maps of the country. In 1876 the U.S. General Land Office also began publishing maps of those states that contained public lands. Henceforth, commercial map publishers would have as basic source material the detailed, accurate maps and charts prepared by official federal agencies for compiling their own maps and atlases.

Notes

1. Carl I. Wheat, *Mapping the Transmississippi West*, Vol. 2, *From Lewis and Clark to Frémont 1804–1845* (San Francisco, 1958), 1.
2. Herman R. Friis, "Highlights in the First Hundred Years of Surveying and Mapping and Geographical Exploration of the United States by the Federal Government 1775–1880," *Surveying and Mapping* 18 (1958): 186–206.
3. William I. Harrison, *William Harrison, Sr. and Sons, Engravers, A Check List of Their Works* (South Yarmouth, Mass., 1978).
4. John Logan Allen, *Passage through the Garden, Lewis and Clark and the Image of the American Northwest* (Urbana, Ill., 1975), 382.
5. Wheat, *Transmississippi West* 2:12.
6. Ibid., 13.
7. John Melish, *Prospectus of a Six Sheet Map of the United States and Contiguous British and Spanish Possessions* (Philadelphia, 1815).
8. Wheat, *Transmississippi West* 2:77.
9. Herman R. Friis, "The Image of the American West at Mid-Century (1840–60): A Product of Scientific Geographical Exploration by the United States Government," in *The Frontier Re-examined*, ed. John Francis McDermott (Urbana, Ill., 1967), 50.
10. Wheat, *Transmississippi West* 2:167.
11. Herman R. Friis, *Geographical Exploration and Topographic Mapping by the United States Government, 1777–1902, as Reflected in Official Records* (Washington, 1952), 3.
12. For a comprehensive study see William H. Goetzmann, *Army Exploration in the American West 1803–1863* (New Haven, Conn., 1959).
13. Harry T. Peters, *America on Stone: The Other Printmakers to the American People* (New York, 1931), 94.
14. Richard A. Bartlett, *Great Surveys of the American West* (Norman, Okla., 1962), 101.
15. Friis, "Image of the American West," 49.
16. Ibid., 58–59.
17. Carl I. Wheat, *Mapping the American West 1540–1857* (Worcester, Mass., 1954), 165.
18. Ibid., 99.
19. See Lawrence Martin, "John Disturnell's Map of the United Mexican States," in *A La Carte*, comp. Walter W. Ristow (Washington, D.C., 1972), 204–21.
20. Wheat, *Mapping the American West*, 116.
21. Carl I. Wheat, *The Maps of the California Gold Region 1848–1857* (San Francisco, 1942).
22. Harry T. Peters, *California on Stone* (Garden City, N.Y., 1935), 63.

23. Wheat, *Maps of the California Gold Region*, 112.
24. Ibid., xliii.
25. Friis, *Geographical Exploration and Topographic Mapping*, 5. For more information on the official maps resulting from these surveys see Wheat, *Transmississippi West*, volume 4.
26. Wheat, *Transmississippi West* 5:104.
27. William W. Robinson, *Maps of Los Angeles from Ord's Survey of 1849 to the End of the Boom of the Eighties* (Los Angeles, 1966), 2, 6.
28. Ibid., 18.
29. Peters, *America on Stone*, 228.
30. Wheat, *Transmississippi West* 3:46.
31. Ibid., 5:278.
32. Ibid., 281.
33. John W. Reps, *Cities on Stone* (Fort Worth, Tex., 1976), 10.
34. Bartlett, *Great Surveys*, xiv.

28. *Rand McNally & Company*

The name Rand McNally is today synonymous with maps, and Rand McNally & Company is one of the foremost cartographic publishing houses in the world. Although the company achieved its greatest success and growth in the twentieth century, its foundations were firmly laid prior to 1900. As we have seen in previous chapters, commercial map publishing during much of the nineteenth century had its principal centers in Philadelphia and New York City. During the last several decades of that century, cartographic printing and publishing branched out to the Middle West in Cincinnati, St. Louis, Milwaukee, and, particularly, Chicago.

The westward migration of the industry was related to the post-Civil War agricultural prosperity in the Upper Mississippi and Ohio valley states, heavy immigration into these regions by northern Europeans, the popularity of county and state maps and atlases, and the importance of Chicago as a center for the burgeoning railroad network. In 1856, however, Chicago was still a young city of some 85,000 inhabitants when William H. Rand opened a small print shop on Lake Street (Fig. 28–1). Rand, who had learned the printing trade in Boston, had joined the gold rush to California in 1849. Failing to achieve success in the gold fields, he located in Los Angeles, where he worked as a newspaper reporter and printer. He visited New England briefly in 1853 to get married, after which he returned to Los Angeles and lived there for two more years. Why he left the Pacific coast is not known, but he moved to Chicago in 1855. It was in this city that he achieved his greatest success.

In his shop, Rand had initially engaged in printing and preparing directories and guidebooks, including an 1859 guide to the gold districts of Kansas and Nebraska. In 1858, Rand employed Andrew McNally, a young Irish immigrant printer (Fig. 28–2). Ten years later the two became partners under the firm name Rand McNally & Company. Several years earlier they had begun to print tickets for the railroads. This entrée into the rapidly expanding railroad industry was to have a great influence on the company. Soon Rand McNally was printing annual reports for several railway companies as well as timetables, coupons, and tickets. In 1869 Rand McNally advertisements boasted a printing capacity of one hundred thousand tickets a day.

The firm expanded into publishing in July 1871 with

Fig. 28–1. William H. Rand. Courtesy of Chicago Historical Society (#IChi 12254).

the monthly *Western Railway Guide, the Travelers' Hand Book to All Railway and Steamboat Lines of North America*. For the August issue the title was shortened to *Rand McNally & Company's Railway Guide*. The company had published only two more monthly issues when the Chicago fire of October 1871 destroyed its shop and most of its printing equipment. Within several days, however, Rand McNally was back in business in rented quarters, and by May 1872 the *Railway Guide* was again being published. An advertisement in the guide's October issue of that year announced that Rand McNally was capable of engraving and printing maps.

In referring to engraving, Rand McNally did not mean the traditional method of engraving on copper or steel plates, but the technique of wax engraving and electrotyping that the company called relief line engraving. The inventor of wax engraving, or cerography, Sidney E. Morse, first announced his discovery in the June 29, 1839, issue of the *New-York Observer* (of which Morse was the editor). Morse and his engraver, Henry Munson, had experimented with cerography as early as 1834, but the process was not patented until 1848, by which time wax engraving was well established in England and the United States.[1] David Woodward described the process in his work *The All-American Map*:

Fig. 28–2. Andrew McNally. Courtesy of Rand McNally & Company.

> Wax engraving was a simple and ingenious process by which a metal printing plate in relief could be produced from an engraved mold. The engraver applied a thin layer of wax to a smooth plate, usually made of copper. He then engraved lines and symbols in the wax through to the metal beneath, using special tools of varying thickness and shape. The lettering could either be hand engraved or produced by pressing metal type through the wax. Fine line tints were often engraved with ruling machines.
>
> After the engraving, the spaces between the lines were built up with wax to give depth to the subsequent printing plate. This plate was then cast from the wax by the process of electrotyping, a technique analogous to electroplating. Graphite was dusted on the mold to render the surface electrically conductive. The mold was then placed in an electrotyping tank where a copper or nickel shell was electrochemically deposited. When thick enough, the shell was removed from the mold, backed up with type metal, mounted on wood, and used as a printing plate. The result was a relief-printing plate that could be printed on letter press printing machines, a property that accounted in large measure for the versatility of the process in the nineteenth and early twentieth centuries.[2]

Woodward doubted that Morse used electrotyping as part of the cerographic process prior to 1840. He believed rather that Morse's plates were prepared by "casting a fusible metal against [the wax impression] to give a relief printing surface directly." Woodward concluded that "the contribution of Sidney Morse was important in cartography in that it unquestionably demonstrated the use of wax as a molding medium to produce a relief-printed surface of sufficient quality for maps; hence the term cerography. There is also no question about his taking advantage of electrotyping when it was developed enough for him to do so, and at least as early as 1848, the date of his patent."[3]

Morse was so extremely sensitive of his invention that except for his own cerographic publications the technique was little used by other American publishers for several decades. Cerography also was in direct competition with lithography, which, by the middle of the nineteenth century, had largely supplanted copper-plate engraving for map reproduction. Morse showed little interest in cerography after 1850. In around 1855, however, wax engraving was revived, or rediscovered, by the firm of Jewett & Chandler in Buffalo, New York. The company prepared a few maps by the process, but the use of the technique in cartographic publishing was limited before 1870.

After 1870 and for the next six or seven decades, wax engraving became one of the most important reproduction techniques in the publication of commercial maps and atlases in the United States. The distinguished American cartographer and geographer J. Paul Goode noted that wax engraving

> has had the largest influence on map making in this country. In this process a polished copper plate is given a thin coat of hardened beeswax and sensitized with silver chloride. The drawing is photographed upon the wax, and the wax under the lines cut away by a graver. Then the wax is warmed, to soften it, and the names, set in ordinary type, are pressed through the wax to the copper plate. Then an electrotype is made from this surface, backed with type metal and mounted for use in an ordinary printing press. All our better wall maps and school geography and atlas maps are done by this process. It has done more than any other one thing in America to put the map to work among millions of people.[4]

Rand McNally & Company was one of the principal beneficiaries of wax engraving as a technique for reproducing commercial maps. With its already well established relations with the transportation companies, Rand McNally adapted the process to the production of maps and cartographically illustrated timetables and guides. As Andrew McNally III has recalled, "the huge growth in railroad travel had created a tremendous demand for maps. There were many other map makers in America at the time, but none employed the modern methods of making map engravings *in wax*, which was an idea adopted by Rand McNally to accelerate and facilitate correction work. The introduction of this single technique was responsible for the Company's instantaneous success in map making. It made it possible for us to draft and correct maps for public sale and for railroad promotional use at a fraction of former costs."[5] It is interesting to note that wax engraving was introduced to Rand McNally by Charles H. Waite, who had previously been employed by Jewett & Chandler.

Woodward cited several reasons why wax engraving was particularly adapted to railroad maps.

> In the first place, since each station maintained by the railroad in question had to be shown on the map, with a large number of small names necessary to identify them, the technique's ability to employ small letterpress was especially significant. The railroad map consisted essentially of lines and points, without the need for halftone. Large printing runs were required, often at short notice. The blocks were often used in a variety of forms to be printed with type, as in timetables in book form or as large wall posters, and consequently had to be produced by a relief process. Finally, the ease with which a large number of corrections could be made on a wax engraving was clearly an asset for this kind of work.[6]

Rand McNally's initial announcement of its ability to engrave maps in the December 1872 *Railway Guide* was followed by a progress report a few months later. "With much pride and a very great deal of satisfaction," the April 1873 *Railway Guide* called its users' "attention to the constantly increasing number of Map illustrations appearing in the *Guide* from month to month. The maps of the lines of the Chicago & Northwestern, St. Louis & Iron Mountain, and Cairo & Fulton railroads in the present number, are refined to as beautiful specimens of the Engravers' Art, and serve well to illustrate the perfection attained in this department by RAND MCNALLY & CO."[7]

Editions of the *Railway Guide* prior to 1872 had carried maps of the United States prepared by the Gaylord Watson Company of New York and Chicago that were also reproduced by the wax engraving process. Andrew McNally III related in 1956 that

> by the mid-1870s, Rand, McNally & Co. was not only producing the largest and most up-to-date railroad maps of individual lines; it had also entered the venturesome field of general map publishing—a field in which it was to achieve top national ranking within ten years, and to hold that position throughout the decades that followed. Two map publications appeared in 1875: one stimulated by the great flow of travelers, emigrants, visitors and new residents into and through the city of Chicago; the other a response to the wave of national interest created by the discovery of gold in the Black Hills of Dakota. These were the first of the firm's famous *Pocket Maps; Rand McNally & Co's. Guide Map of Chicago*, 38 pages and map; and *Rand McNally & Co's. New Map and Guide to the Black Hills*, 32 pages of text and folded map.[8]

Prior to the publication of the pocket maps, however, Rand McNally issued the *Railroad Map of the United States and Canada*, which was published in the May 1873 *Railway Guide*. The editor noted that "we are enabled in this number to present . . . our new Railway Map of the United States and Canada. In our judgment it possesses more practical value, as a General Railway Map, than any yet published. Its convenient size, considering the territory represented, its illustration of the various railroad lines shown in the Guide, and with all the perfect clearness with which it was printed, accords to it the first place among the several Railway Guide Maps yet issued." The uncolored map included all of the country from the eastern seaboard to the western borders of Montana, Wyoming, Colorado, and New Mexico. Separate copies of the map were offered for twenty-five cents.

The February 1874 *Railway Guide* featured a new colored railroad map of the United States and Canada. The editors believed that

> upon critical examination, it will be found the most reliable map and the best representation of the completed lines of this country now published. . . . Every section of the new lines of Canada, and the 3,833 miles of road completed in the United States in 1873, is carefully located, and our entire railroad system, now embracing nearly 73,000 miles of roads in operation, can readily be traced.
>
> This map, in size 18 × 36 inches, was drawn and engraved . . . at an expense of nearly three thousand dollars. No similar publication ever displayed such enterprise, and no such map was ever produced by any publisher, and sold at less than three or four times the price of this publication.

Separate copies of the map printed on fine cardboard were priced at twenty-five cents each.

The company's map business came of age in 1876 with publication of *Rand McNally & Co's New Railroad and County Map of the United States and Canada, Compiled from the Latest Government Surveys, and Drawn to an Accurate Scale* (Fig. 28–3). The large colored map, which extends from the Atlantic to the Pacific, is at the scale of

1:2,027,520 and measures 134 by 244 cm. It was reproduced by the wax engraving process and, in the lower left corner, carries the credit, "engraved under the direction of C. H. Waite." At the bottom of the title cartouche is printed the statement "Drawn, Engraved and Printed in Colors (under Letters Patent), by RAND MCNALLY & CO., Chicago. The entire map is printed from electrotype plates, sections of which can be used for Special Railroad Maps, Publishers' Premium Maps, Maps to accompany reports, pamphlets, etc., etc., and for various

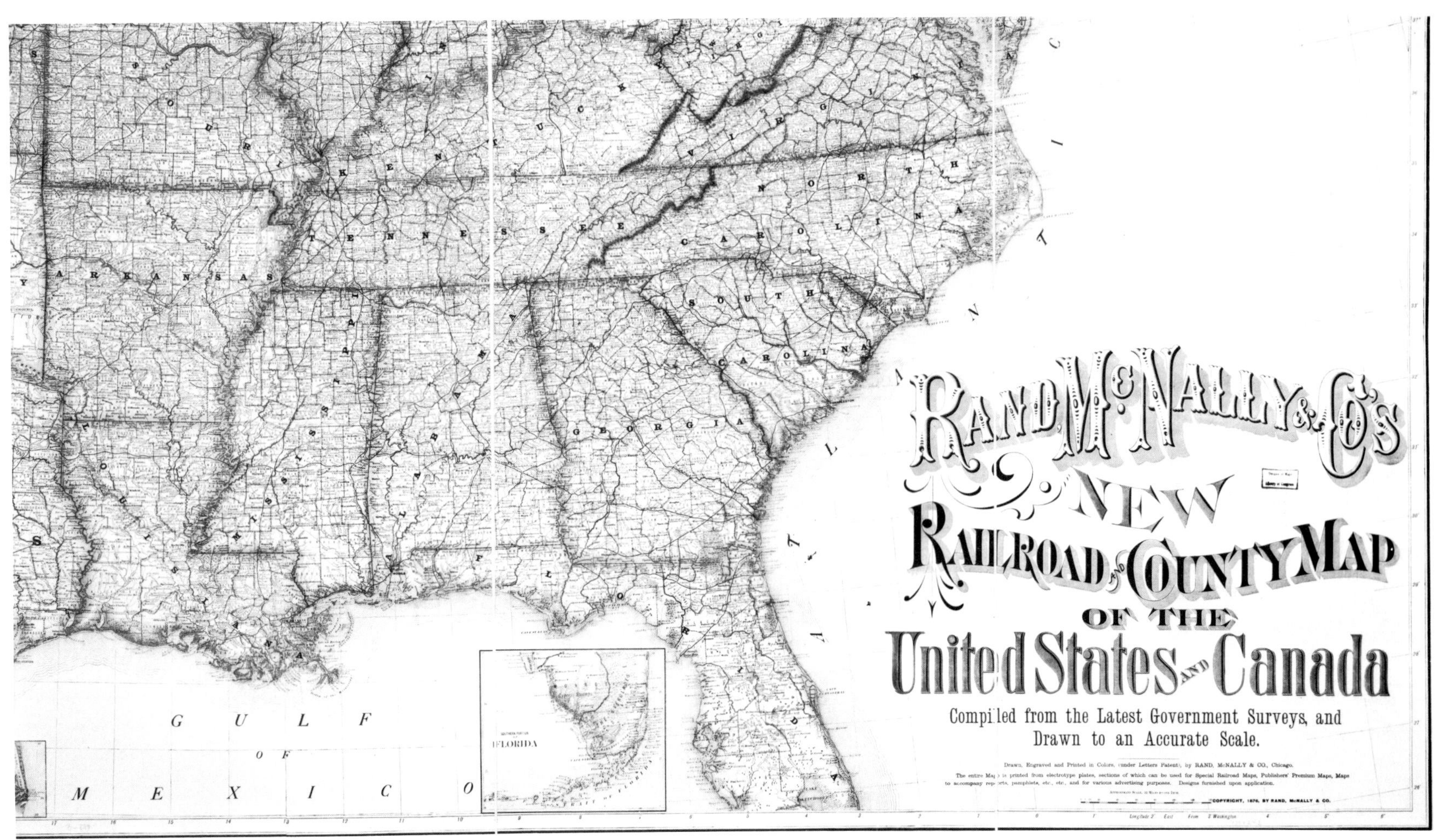

Fig. 28–3. Title cartouche and segment of Rand McNally & Company's 1876 new railroad and county map.

advertising purposes. Designs furnished upon application." The map sold for fifteen dollars. Copyrighted and published in the same year, and bearing the same title, is a map which shows the country only as far west as the 116th meridian. It sold for five dollars.

The large U.S. map was designed to be used as a basic map from which various segments could be extracted to form individual maps to fit the needs of specific customers. Andrew McNally III remarked that it was "the foundation and cartographic reservoir for the commercial map business of Rand McNally. For sectioned into States and Territories, it became in atlas form Rand McNally & Co's. *Business Atlas of the Great Mississippi Valley and Pacific Slope*, and in pocket map form Rand McNally & Co's. *Series of Indexed Pocket Maps*, consisting by 1878 of some forty different titles covering the States and Territories of the United States and the Provinces of Canada."[9]

The large size, comprehensive detail, and cartographic excellence of Rand McNally's *New Railroad and County Map of the United States and Canada* invites the question of who was responsible for its planning, design, compilation, and cartographic execution. Charles H. Waite directed the engraving of the map. He had come to Rand McNally in 1872 and for some years thereafter headed the company's map engraving department. In addition to his proficiency in wax engraving, Waite also invented a technique for coloring maps mechanically. In December 1874 he transferred to Rand McNally's board of directors his interest in the map coloring invention. Waite was also associated with the company through marriage. His sister had apparently joined him in Chicago shortly after he joined Rand McNally & Company, and she subsequently married James McNally, the brother of the co-founder of the firm. Waite served with Rand McNally until his death in May 1903 at the age of sixty-eight. At the time of his death, he was a superintendent of the company and had worked for it for thirty-one years.

In 1869 Rand McNally had engaged Robert A. Bower to establish its map department, and for almost forty years he served as the department's director. Bower, who was born in Brown County, Ohio, in 1841, early moved to Illinois, where he attended the State Normal University in Bloomington. He taught school for several years before joining Rand McNally. He spent the remainder of his career with the company, advancing to the office of vice-president before retiring in 1908. Bower had married Charlotte E. Cuyler in 1876. Among his achievements he is credited with having originated the map indexing system.[10] During its early years, Rand McNally did not give credit to compilers, cartographers, or draftsmen on its maps, but it is doubtless likely that Bower played a major role in the preparation of the U.S. railroad maps and other cartographic works published during his tenure with the company.

A small promotional booklet issued by Rand McNally in around 1879 and reissued in a facsimile edition in 1946 gives information, with illustrations, about the company's early cartographic production. Under the heading "Map Drawing and Designing Room" is the statement that "probably more *original* map projections have been made in our map drawing room than have ever been produced in the United States. It is not generally known that our large railroad and county map, which is 58 × 100 inches, is the second *original* projection of a United States map ever made. Our United States and Canada Atlas is made from the same projection." The booklet also notes that "it is generally admitted that our map engraving is not excelled in the world for clearness of outline and beauty of execution. We have frequent orders from Europe for map work, and the liberal patronage received from all sections of the United States, attests in a gratifying manner the superiority of our work in this line. Our new wall map of the United States and our Business Atlas, is the largest work ever attempted by our process of engraving. It required the services of ten compilers and engravers nearly two years, and cost about $20,000."

While the 1876 *New Railroad and County Map of the*

United States and Canada was without question an original Rand McNally compilation, it did draw upon previously published maps of the country and upon government surveys. It was noted that early issues of the *Railway Guide* were illustrated with maps prepared by the Gaylord Watson Company. That firm produced maps and atlases between 1871 and 1885, initially in New York City but also in Chicago after 1883. Between 1871 and 1874 Gaylord Watson published several editions of a *New County and Railroad Map of the United States and of the Dominion of Canada*, which extended west only to the 103d meridian. It also published a map of the western states and territories in 1871. The similarity in appearance, content, and in map title of Gaylord Watson maps to the 1876 Rand McNally maps suggests the possible existence of a cooperative relationship between them. The Gaylord Watson maps were also reproduced by the wax engraving process, some of them by Fisk & Russell of New York City. Like a number of other firms, Fisk & Russell was an offshoot of the Jewett & Chandler company. Several Gaylord Watson maps have ornately decorated title cartouches. In 1875 Gaylord Watson published two different versions of a *New Commercial County and Railroad Atlas* of the United States and Canada. The firm also published the *New Indexed Family Atlas of the United States, with Maps of the World* (1883) and the *New and Complete Illustrated Atlas of the World* (1885). Both of these atlases have New York and Chicago imprints. The Chicago imprints are respectively listed under the firm names of Tenney & Weaver and R. A. Tenney.

As reported earlier, Rand McNally & Company's 1876 *Business Atlas* was one of the first derivatives of the *New Railroad and County Map of the United States and Canada* (Fig. 28–4). The publication of this atlas in Chicago, as Dale Morgan has observed,

> marked a major westward shift of the cartographical center of gravity in the United States, for nearly all previous atlases had issued from Philadelphia or New York. This in turn reflected a west-

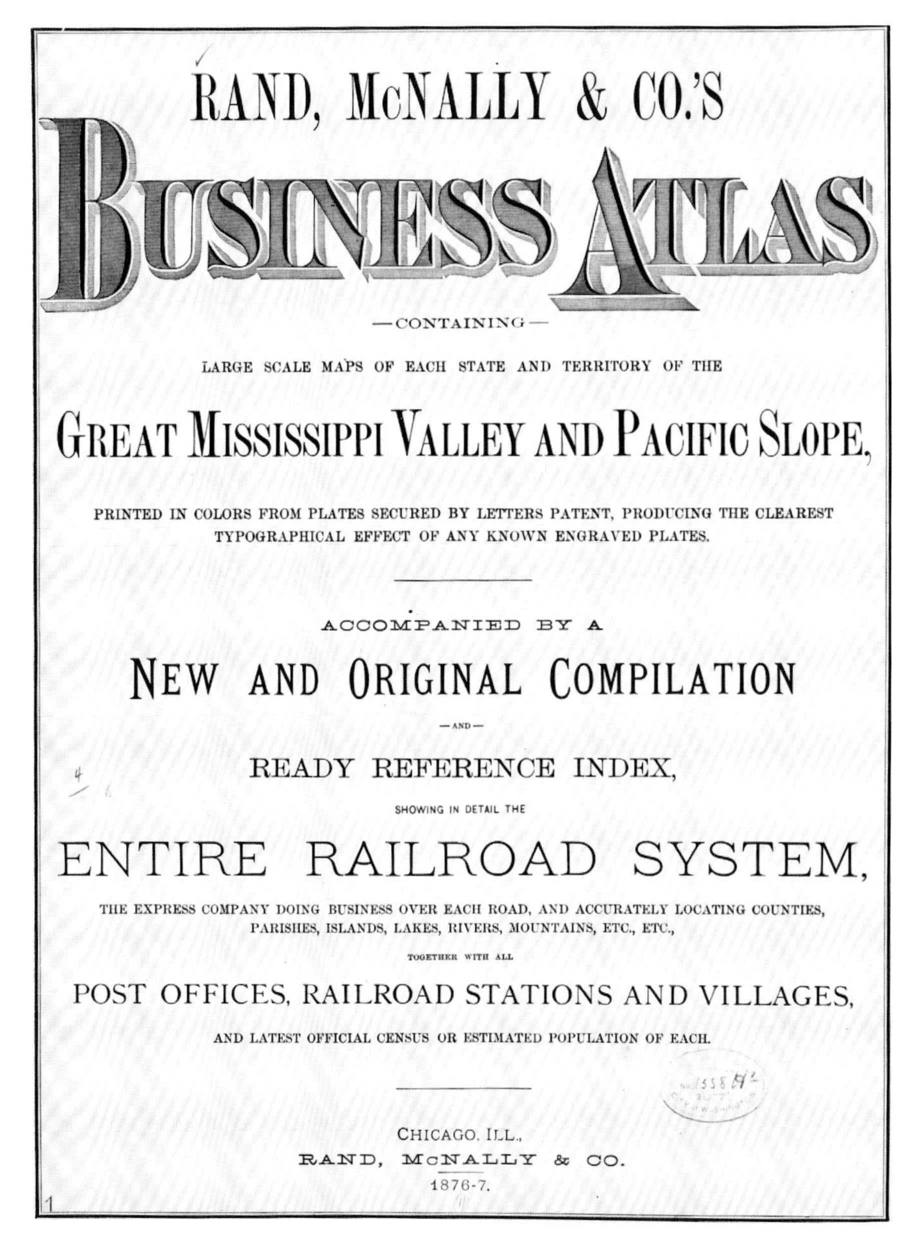

RAND, McNALLY & CO.'S

BUSINESS ATLAS

—CONTAINING—

LARGE SCALE MAPS OF EACH STATE AND TERRITORY OF THE

GREAT MISSISSIPPI VALLEY AND PACIFIC SLOPE,

PRINTED IN COLORS FROM PLATES SECURED BY LETTERS PATENT, PRODUCING THE CLEAREST TYPOGRAPHICAL EFFECT OF ANY KNOWN ENGRAVED PLATES.

ACCOMPANIED BY A

NEW AND ORIGINAL COMPILATION

—AND—

READY REFERENCE INDEX,

SHOWING IN DETAIL THE

ENTIRE RAILROAD SYSTEM,

THE EXPRESS COMPANY DOING BUSINESS OVER EACH ROAD, AND ACCURATELY LOCATING COUNTIES, PARISHES, ISLANDS, LAKES, RIVERS, MOUNTAINS, ETC., ETC.,

TOGETHER WITH ALL

POST OFFICES, RAILROAD STATIONS AND VILLAGES,

AND LATEST OFFICIAL CENSUS OR ESTIMATED POPULATION OF EACH.

CHICAGO, ILL.,
RAND, McNALLY & CO.
1876-7.

Fig. 28–4. Title page of the *Business Atlas*, the first atlas published by Rand McNally & Company.

> ward shift of the center of gravity of American business, occasioned by the accelerating growth of the Midwestern and Western States and Territories. Still it seems a happy turn of fortune that Rand McNally invested the company's future in a first atlas which frankly embraced the West. Totally disregarding the Atlantic seaboard, confident of the vitality and economic vigor of the old and new West, Rand McNally offered the American business world a group of maps depicting the Western States and Territories with a wealth of detail such as had characterized no previous atlas.[11]

The cost of the *Business Atlas* was twelve dollars. The atlas was also released in an abridged version in 1876. This volume was titled the *Business Atlas . . . of the Great Mississippi Valley and Pacific Slope*. Facsimile reproductions of the maps and indexes from this atlas are included in *Rand McNally's Pioneer Atlas of the American West*, which was published in 1956, Rand McNally's centennial year. The *Business Atlas*, in both its large and abridged versions, was published annually under that title to 1910. It was then published as the *Commercial Atlas of America* to 1936. Since 1937 it has appeared in annual editions as *Rand McNally's Commercial Atlas and Marketing Guide*. Its 115th edition was published in 1984.

The *New Railroad and County Map of the United States and Canada* was also the principal source for a series of indexed maps of individual states and territories, such as the 1876 *Railroad Map of Utah* (Fig. 28–5). These were published in a folded, pocket-size format with accompanying booklets in either hard or soft cover. The booklets contained an index of towns and cities, population statistics arranged by county, and information about the railroads. These indexed pocket maps sold for fifty cents; hard cover editions were an additional ten cents. By 1878 about forty indexed pockets maps had been published (Fig. 28–6). In some editions of the maps the index was printed on the map sheet rather than in an accompanying booklet. Enlarged editions of the state and territory maps were available mounted on cloth with rollers at top and bottom for seventy-five cents each. The indexed maps were periodically revised and updated.

Rand McNally's map production was not limited to these cartographic works. In 1877 the firm published the large *New Map of the Black Hills*, which it claimed to be "the only Map of the Black Hills ever compiled from actual survey." The surveyor of the map was Major George Henckel, deputy U.S. surveyor. On the right and left margins of the map sheet there are illustrations of Indians, wildlife, and western scenes. An edition of the map without illustrations was also published. The company also issued city plans and guides, the first of which was titled the *Map Showing the Boulevards and Park System and Twelve Miles of Lake Frontage of the City of Chicago*. It was published with an accompanying booklet in 1880. Revised editions of the plan were issued in subsequent years. The company continued to compile and publish maps for various railroad companies as well as state maps showing the rail networks (Fig. 28–7). In 1880 Rand McNally began publishing globes and school wall maps.

The company then published in 1880 the *New Indexed Atlas of the Northwest*, which includes a map of the United States and individual maps of eleven midwestern and plains states and the Canadian province Manitoba. This volume was the precursor of the large 852-page *Indexed Atlas of the World, Containing Large Scale Maps of Every Country and Civil Division Upon the Face of the Globe, Together With Historical, Statistical and Descriptive Matter Relative to Each*, which Rand McNally published in 1886. The state and territory maps resemble those in the *Business Atlas*, and like the *Business Atlas* maps were based on the company's *New Railroad and County Map of the United States and Canada*. The maps of foreign countries, however, appear to have been newly compiled, drafted, and engraved for the *Indexed Atlas of the World*. With the map of each country there is a descriptive article on its geography, economics, and social conditions. The introduction of the atlas places great

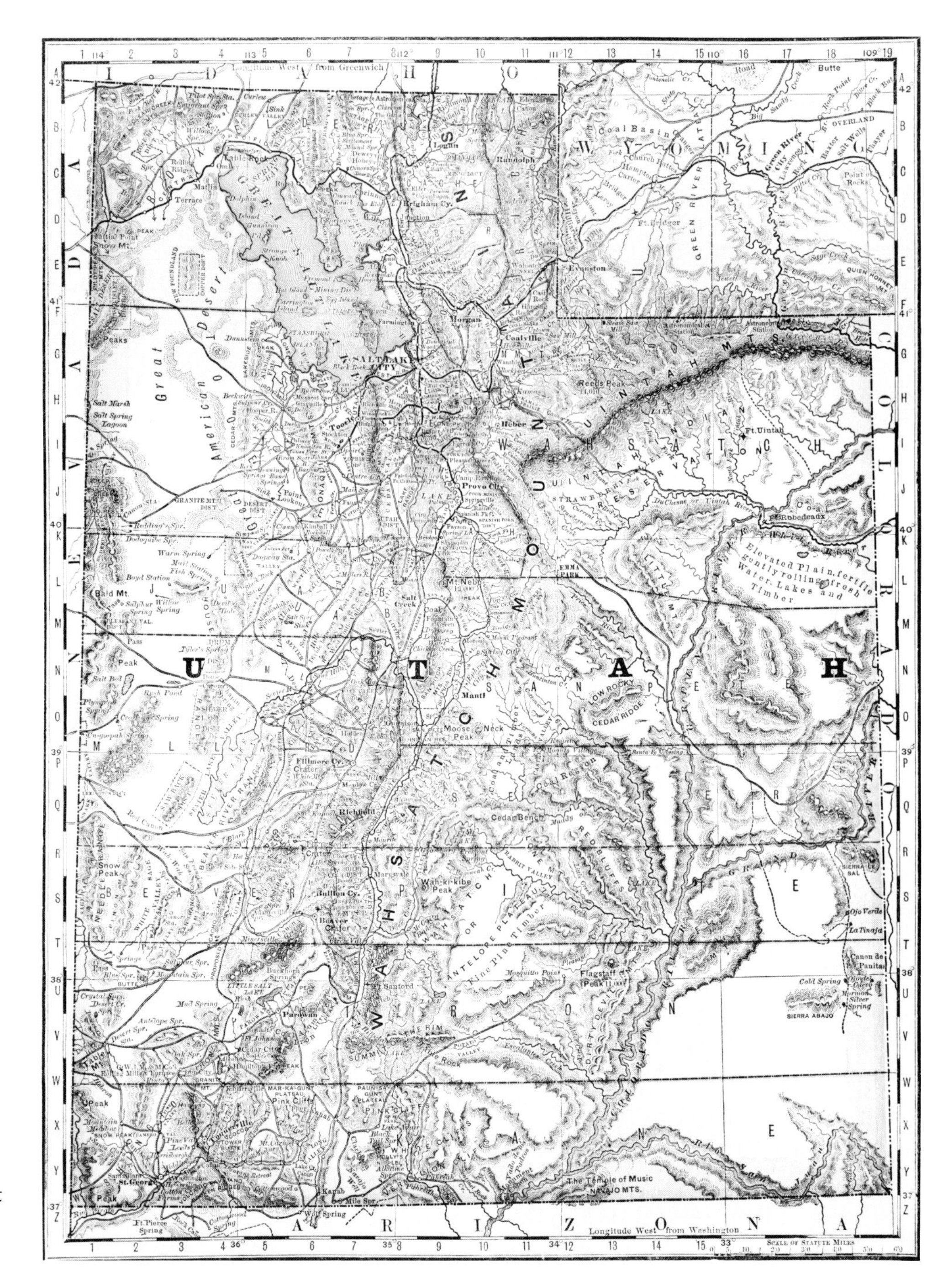

Fig. 28–5. Railroad maps were among the earliest cartographic publications of Rand McNally & Company, such as this 1876 *Railroad Map of Utah*.

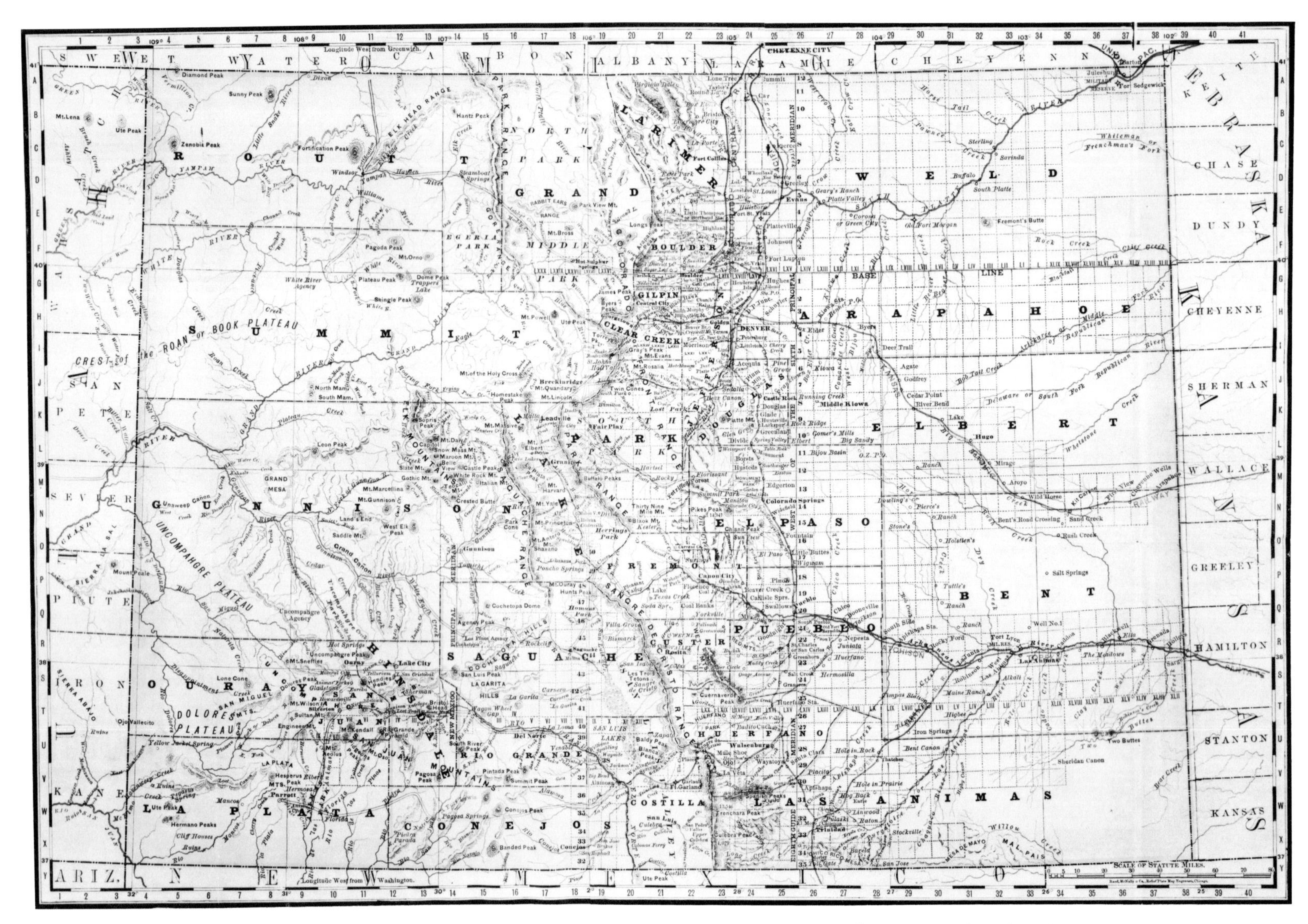

Fig. 28–6. Colorado railroads are shown on this 1879 Rand McNally map.

emphasis on the indexing system employed in it and stresses "the great use which has been made of colored diagrams."

The *Indexed Atlas of the World* was popular, and a number of one-volume editions were published to 1892 (Fig. 28–8). Later editions were expanded to include as many as 904 pages. From 1894 to 1908, the atlas was published in two- and four-volume editions, in enlarged folio formats. To distribute the *Indexed Atlas of the World* and other publications, Rand McNally established, in 1886, the Continental Publishing Company. The imprint of this firm appears on some Rand McNally atlases and maps until 1904.

Rand McNally also supplied most of the maps for Godfrey Jaeger & Company's 1881 *Historical Hand-Atlas Illustrated, Containing Large Scale Copper Plate Maps of Each State and Territory of the United States and the Provinces of Canada . . . and History of Ottawa County, Ohio*. Notwithstanding the title's reference to copper-plate maps, those published in the atlas are Rand McNally's standard wax engraved reproductions derived again from the *New Railroad and County Map of the United States and Canada*.

The expanding map company soon realized that its standard map series could be packaged in various atlas formats to meet the requirements and economic capacities of different markets. Thus, in 1885 the company issued a *Pocket Atlas of the World*, with maps presented at reduced scales on pages measuring 15 by 8 cm. Facing each map plate is a page of pertinent text. The *Pocket Atlas* appeared periodically in revised editions until 1924. Some were issued under other imprints, such as the 1888 *New York Tribune's Pocket Atlas of the World*.

Rand McNally's *General Atlas of the World* was first published in 1887 and had a format of 31 by 23 cm. It contained 129 pages of maps followed by 86 pages of text. The *General Atlas* apparently was not published again until a similar volume, the *New General Atlas of the World*, appeared in 1905 and 1921.

HOW THE PUBLIC DOMAIN HAS BEEN SQUANDERED

Map showing the **139,403,026** acres of the people's land—equal to

871,268 FARMS OF 160 ACRES EACH

Worth at $2 an acre, $278,806,052,

GIVEN BY

Republican Congresses to Railroad Corporations

This is more land than is contained in New York, New Jersey, Pennsylvania, Ohio, and Indiana.

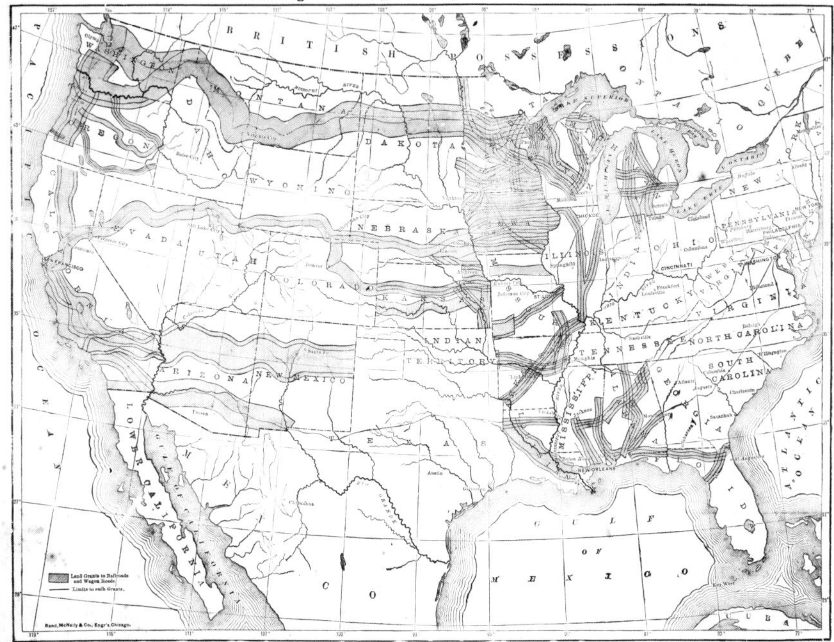

We believe that the public lands ought, as far as possible, to be kept as homesteads for actual settlers; that all unearned lands heretofore improvidently granted to railroad corporations by the action of the Republican party should be restored to the public domain; and that no more grants of land shall be made to corporations, or be allowed to fall into the ownership of alien absentees.

DEMOCRATIC PLATFORM, 1884.

Division of Maps MAR 31 1939 Library of Congress

Fig. 28–7. A Rand McNally map on a 1884 Democratic campaign poster.

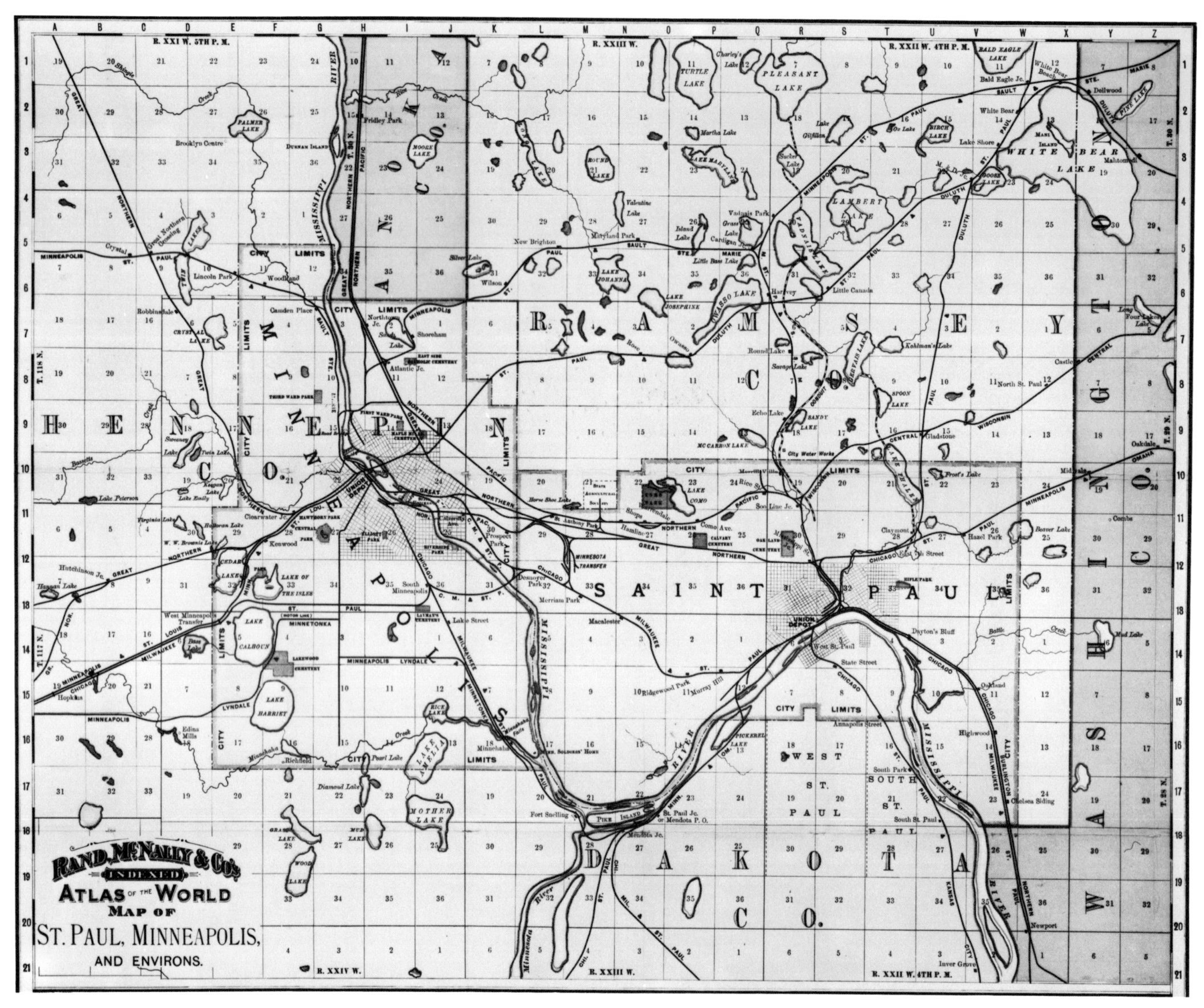

Fig. 28–8. Map of St. Paul and Minneapolis from the 1891 edition of Rand McNally's *Indexed Atlas of the World*.

Slightly larger than the *General Atlas* was the *Rand McNally Standard Atlas of the World* issued in 1888 and 1890. It features a number of black-and-white illustrations as well as colored maps and text. Similar in format and content is the Rand McNally *New Family Atlas of the World*, which was issued in 1888 with the imprint of the People's Publishing Company, Lakeside Building, Chicago. Somewhat larger in size than the *Pocket Atlas* was the 1889 *Model Atlas of the World and Cyclopedia of Useful Information*. In addition to a series of small, uncolored maps, the atlas includes illustrations of various national flags, portraits of U.S. presidents, and an assortment of miscellaneous information on first aid, exercise, and how to carve different cuts of meat, among others. No other editions of the *Model Atlas* have been identified. With a page size of 20 by 14 cm., Rand McNally's 1892 *New Handy Atlas of the World* is larger than the *Model Atlas* but contains less encyclopedic information.

Other world atlas titles issued by Rand McNally before the close of the nineteenth century were *Universal Atlas of the World* (1893), *Pictorial Atlas of the World* (1896), *Twentieth Century Atlas of the World* (1896), *Popular Atlas of the World* (1896), *Household Atlas of the World* (1898), and *Unrivaled Atlas of The World* (1899).

Rand McNally's cartographers and salesmen were also cognizant of the market for domestic maps, and in the last two decades of the century the company published several atlases of the United States. Among these was the 1884 *New Dollar Atlas of the United States and Dominion of Canada*, with state maps and text. Slightly larger in format and containing more statistical information was *The New Handy Family Atlas of the United States and Canada*, which was published in 1885 under the imprint of F. B. Dickerson & Company of Detroit and Cincinnati. The *New Pocket Atlas of the United States and . . . Dominion of Canada* was published in 1892 and 1893. The *Family Atlas of the United States* (30 by 23 cm.) was compiled by Rand McNally for the World's Fair Educational Association in 1892 "and sold only to the Participants in the map-drawing contest."

The Spanish-American War and the Boer War at the close of the nineteenth century inspired several war atlases. In large folio format, the *Rand McNally Standard War Atlas With Marginal Index* was published with sixteen maps in 1898. An edition of the same atlas carrying the imprint of the *Washington Post* has the same date. The *Rand McNally War Atlas*, in a somewhat smaller format, survives in several editions. The imprints on these editions cite a variety of companies, including the Long Island Title Guarantee Company and the New Jersey Foundry & Machine Company.

By the time the twentieth century began, Rand McNally & Company had achieved first rank among commercial cartographic publishers in the United States. In 1886 the company had added lithography to its reproduction facilities and had opened a branch office in New York City. There were also significant changes in the company's management around the turn of the century. William Rand retired in 1899, and Andrew McNally died on May 7, 1904. By 1900 Rand McNally & Company had outgrown its dependence upon the railroads. Although many of its maps of the country and individual states still showed railroad lines, new modes of transportation, such as the bicycle, automobile, and airplane, began to command cartographic attention. Decades later, Rand McNally became an active participant in producing maps and charts for cyclists, motorists, and aviators.

Notes

1. For a comprehensive review of the invention and development of cerography and wax engraving see David Woodward's *The All-American Map, Wax Engraving and Its Influence on Cartography* (Chicago, 1977).
2. Ibid., 1.
3. Ibid., 21–23.
4. J. Paul Goode, "The Map as a Record of Progress in Geography," *Annals of the Association of American Geographers* 17 (1927): 12.

5. Andrew McNally III, *The World of Rand McNally* (New York, 1956), 10.
6. Woodward, *All-American Map*, 32–33.
7. Rand McNally & Company, *Railway Guide* (Apr. 1873): xv.
8. Andrew McNally III, *Rand McNally's Pioneer Atlas of the American West* (Chicago, 1956), 6.
9. Ibid.
10. Albert N. Marquis, *The Book of Chicagoans* (Chicago, 1911).
11. *Rand McNally's Pioneer Atlas of the American West*, 7.

Index

Walter W. Ristow received the Ph.D. degree in geography from Clark University in 1937. He has served as the chief of the Map Division of the New York Public Library and for more than thirty years held administrative positions in the Library of Congress's Geography and Map Division. Retiring as its chief in 1978, he is presently a consultant in the fields of the history of cartography, map collecting, and map librarianship. Dr. Ristow is the author of the *Guide to the History of Cartography* and *The Emergence of Maps in Libraries* and editor of *A La Carte: Selected Papers on Maps and Atlases* and the facsimile edition of *A Survey of the Roads of the United States of America, 1789* by Christopher Colles. He has also written numerous articles. Dr. Ristow was honored in 1979 with the festschrift *The Map Librarian in the Modern World: Essays in Honour of Walter W. Ristow*.

The manuscript was prepared for publication by Anne M. G. Adamus. The book was designed by Edgar Frank. The typeface for the text and display is Palatino, based on a design by Herman Zapf about 1950. The text is printed on 70-lb. Sterling Matte paper and bound in Holliston Mills' Roxite linen finish cloth over 100 pt. binder's boards.

Manufactured in the United States of America.

DATE DUE

ÉCHÉANCE